石油管材及装备材料服役行为与结构安全国家重点实验室科研成果汇编

（2016 年）

中国石油集团石油管工程技术研究院
石油管材及装备材料服役行为与结构安全国家重点实验室
中国石油集团石油管工程重点实验室
陕西省石油管材及装备材料服役行为与结构安全重点实验室
编

石油工業出版社

内 容 提 要

本书汇编了中国石油集团石油管工程技术研究院、石油管材及装备材料服役行为与结构安全国家重点实验室、中国石油集团石油管工程重点实验室和陕西省石油管材及装备材料服役行为与结构安全重点实验室2016年在国际国内刊物和学术会议上发表的论文，反映了近几年石油管工程的科研成果及进展。内容涉及输送管与完整性评价、油井管与管柱、腐蚀与防护等方面。

本书可供从事石油管工程的技术人员和石油院校相关专业师生参考。

图书在版编目（CIP）数据

石油管材及装备材料服役行为与结构安全国家重点实验室科研成果汇编. 2016年 / 中国石油集团石油管工程技术研究院等编. — 北京 ：石油工业出版社，2017. 12

ISBN 978-7-5183-2221-3

Ⅰ. ①石… Ⅱ. ①中… Ⅲ. ①石油管道-管道工程-文集 Ⅳ. ①TE973-53

中国版本图书馆 CIP 数据核字（2017）第 267095 号

出版发行：石油工业出版社

（北京安定门外安华里 2 区 1 号 100011）

网 址：www. petropub. com

编辑部：（010）64523583 图书营销中心：（010）64523633

经 销：全国新华书店

印 刷：北京中石油彩色印刷有限责任公司

2017 年 12 月第 1 版 2017 年 12 月第 1 次印刷

787×1092 毫米 开本：1/16 印张：42. 25

字数：1080 千字

定价：200. 00 元

（如出现印装质量问题，我社图书营销中心负责调换）

《石油管材及装备材料服役行为与结构安全国家重点实验室科研成果汇编（2016年）》编委会

前　言

石油管材及装备材料服役行为与结构安全国家重点实验室成立于2015年9月，与中国石油集团石油管工程重点实验室和陕西省石油管材及装备材料服役行为与结构安全重点实验室两个省部级重点实验室并行运行，是我国在石油管材及装备服役安全研究领域的科技创新基地、人才培养基地和学术交流基地。

实验室依托中国石油集团石油管工程技术研究院，设置输送管与管线安全评价、油井管与管柱失效预防、石油管材及装备腐蚀与防护、先进材料及应用技术四个研究方向，围绕我国油气工业发展战略，特别是大口径高压输气管道建设和复杂工况油气田勘探开发的技术需求，以管材及装备服役过程中的断裂、变形、泄漏、腐蚀、磨损、老化等突出失效行为为对象，深入开展石油管材及装备材料服役领域的基础和应用基础研究，积累服役性能数据，创新研究试验方法，突破国外的技术封锁和壁垒，形成自主创新的技术体系和知识体系，为我国油气重大工程的选材、安全评估与寿命预测提供科学技术支撑。

2016年，实验室全体科技人员通过潜心研究，取得了丰硕的科研成果。获省部级科技奖励11项，其中特等奖1项，二等奖6项；发表论文96篇，出版著作和研究文集4部；授权发明专利42件；完成国家标准征求意见稿5项，发布行业标准7项，制修订行业标准10项；新立项课题15项，其中国家课题5项，中国石油天然气集团公司纵向课题7项。研究成果在我国油气田勘探开发和重大工程中获得广泛应用，解决了一系列重大工程关键技术和瓶颈问题。输送管与管线研究领域，研究形成X90显微组织分析鉴定图谱，并且在X90管材成分、组织、性能、工艺相关性研究的基础上，提出X90管材技术条件；形成了ϕ1422mm X80管材低温韧性关键技术指标和技术条件；油气输送钢管全尺寸气体爆破试验技术逐步完善，建立了爆破试验规范，并圆满完成了两次试验，填补了国际上X90爆破试验数据的空白，为输送管断裂控制技术研究和发展奠定了基础。在油井管与管柱失效预防研究领域，开展了材料强度温度效应、交变载荷环境下套管材料应变疲劳特性研究，获得材料服役行为特征规律；针对四川盆地深层油气勘探工程风险探井和目标开发试验区块工况，完成三高气井管柱材质评价优选方案设计，并开展气井套管柱井口抬升及油管柱密封失效机理研究，查明了失效影响因素，提出了三高气井油管密封及腐蚀评价方法；建立了管柱气密封螺纹接头优选方法和气密封螺纹接头分级选用技术，完善了储气库注采管柱设计理论方法，并建立了低含水率下管柱腐蚀选材评估方法。在石油管材及装备腐蚀与防护研究领域，确定了加氢换热器典型腐蚀失效形式为铵盐结晶导致换热器管束堵塞、腐蚀穿孔，明确加氢换热器的腐蚀规律，形成了出换热器管束表面渗铝技术及阻垢缓蚀剂两种有效的防腐措施；开发了咪唑啉曼尼希碱复合缓蚀剂，成功解决了塔里木油田含硫外输管线的局部腐蚀穿孔问题；全尺寸管柱腐蚀/应力腐蚀试验方法得到进一步完善，研究表明该方法下的腐蚀机理和腐蚀过程演变发展规律有别于小试片高温高压模拟研究方法，并且更加接近实际管柱在井下的尺寸、结构、拉伸/内压载荷。先进材料及应用技术研究领域，通过复合管成型工艺的开发与优化，制备出内径150mm、工作压力20MPa、抗拉强度20tf非金属管样管，在完成拉伸、压缩、外压挤毁等关键参数室内评价试验的基础上，国内首次完成了海洋柔性管的动态疲劳实验和海试工作；通过有限元分析与宏观力学相结合的方法，形成了较为完善的复合材料增强管线钢管结构设计技术，开发了制备工艺技术，并分别以 ϕ508mm 和 ϕ1219mm 的管线钢管为基础，制成了复合材料增强管线钢管样管；以 X80 钢级，ϕ1219mm 制成的复合材料增强管线钢管，其承压能力超过了 X100 钢级水平，成本下降约5%。

本书收集了重点实验室于2016年在国内外重要刊物和学术会议上发表的论文，并介绍了2016年授权专利及省部级以上获奖成果，这些资料从一个侧面反映了实验室近期所取得的研究成果。可以为从事油气管道工程、油气井工程、石油工程材料、安全工程等方面的工程技术人员、研究人员和管理人员提供参考。

由于编者水平有限，经验不足，加之时间仓促，错误和不妥之处在所难免，敬请广大读者批评指正。

编　者

2017年4月

目　录

第一篇　论文篇

一、输送管与完整性评价

二、油井管与管柱

三、腐蚀与防腐

四、非金属与复合材料

五、其他

第二篇　成果篇

第一篇　论文篇

一、输送管与完整性评价

我国高钢级管线钢和钢管应用基础研究进展及展望

冯耀荣　霍春勇　吉玲康　李鹤林

（石油管材及装备材料服役行为与结构安全国家重点实验室；
中国石油集团石油管工程技术研究院）

摘　要： 回顾了我国油气输送管道工程与高钢级管线钢管研发应用历程和成效，提出了我国油气管道工程和高钢级管线钢管的发展方向。我国油气管道工程发展和高钢级管线钢管研发应用具有研发周期短、应用速度快、实施效果好的显著特点，经过近20年的努力，在油气输送管特别是大口径高压输送管线的管型、钢级、材质、尺寸等参数和高钢级管材研发应用方面达到国际领先水平。研发形成了X70/X80钢级材质选用及针状铁素体管线钢组织分析鉴别与评定技术、管型选用及螺旋埋弧焊管残余应力控制技术、高压输气管道断裂控制技术、地震断裂带等基于应变设计地区管材变形控制技术、高钢级厚壁管线钢及钢管试验检测评价技术、高钢级管线钢焊接热影响局部脆化区的脆化机理、断裂规律与防止技术，自主研发了X70/X80系列热轧板卷和大口径厚壁螺旋埋弧焊管制造技术、宽厚板和大口径厚壁直缝埋弧焊管制造技术、感应加热弯管和管件成分设计及制造工艺技术，并将Ⅰ类地区设计系数提高到0.8，将X80管材口径扩大至1422mm，初步完成X90管线钢和钢管的研制。为更好地满足大口径高压输气管道工程发展需求，必须进一步深化应用基础研究，联合进行产品研发与技术攻关，形成大口径高钢级输气管道输量与其他参数（压力、钢级、设计系数、管径、壁厚等）的优化技术，建立或完善管道应变设计、可靠性设计、高压输气管道断裂控制等理论和方法，研究解决X90~X120高钢级管线钢和钢管屈强比、应变时效、低温韧性、屈服强度测试等关键技术难题，掌握高钢级管材成分/组织/性能/工艺相关性，联合冶金和制管企业形成批量生产能力，研究形成X90~X120管材现场焊接技术，进一步提高环焊缝质量性能水平。实现高钢级管线钢和钢管生产和应用技术的全面突破。

关键词： 高钢级管线钢；高钢级管线钢管；管线；应用基础研究；研究进展；发展展望

1　我国管道工程与高钢级管线钢和钢管研发应用成效

石油、天然气占全球一次能源的57%，我国陆上70%石油和99%天然气依靠管道输送，油

作者简介：冯耀荣，男，出生于1960年，毕业于西安交通大学材料科学与工程专业，博士，教授级高工。一直从事石油管材及装备的应用基础研究和重大工程技术支持工作。现任中国石油集团石油管工程技术研究院总工程师，石油管材及装备材料服役行为与结构安全国家重点实验室主任。地址：陕西省西安市锦业二路89号；邮编：710077；电话：029-81887699/13700220576；邮箱：Fengyr@ cnpc. com. cn。

气管道是国民经济的“生命线”。根据国家能源规划，中国石油集团全面布局油气管道和管网建设，特别是“十一五”以来开始建设与西北中亚、东北俄罗斯、西南缅甸相连的三大陆上油气通道，统筹国内外资源与市场，基本上形成了连通海外、覆盖全国的油气骨干管网。

为了满足长距离大输量高压力天然气管道输送的需要，中国石油集团整体部署，科技管理部门组织开展了若干重大科技攻关，开展了高钢级管线钢管研发和应用关键技术研究，使我国在较短的时间内输气管线的钢级由 X52、X60、X65 发展到 X70 和 X80，在过去的螺旋埋弧焊管的基础上发展了直缝埋弧焊管，产品质量、性能达到国际先进水平，研发成功系列 X70 和 X80 大口径、厚壁螺旋埋弧焊管和直缝埋弧焊管及弯管和管件，形成规模化生产制造能力，使输气管线压力从 6.3MPa 逐步提升至 8.4MPa、10MPa 和 12MPa，单管输送能力达到 $300\times10^8m^3/a$。在管线钢和钢管的研发和应用方面，我们用不到 20 年的时间走过了发达国家将近 40 年的发展道路（图 1）[1-6]。

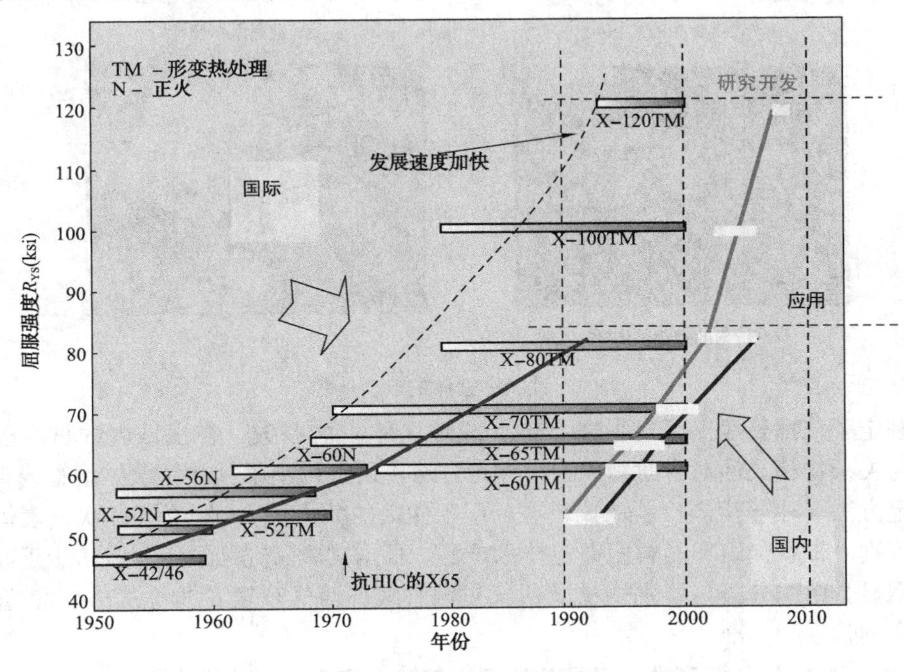

图 1　管线钢和钢管研究应用进展国内外对比

我国管道工程发展和高钢级管线钢及钢管研发应用具有研发周期短、应用速度快、实施效果好的显著特点，特别是建成了全世界瞩目的西气东输管线和全球规模最大的西气东输二线和三线，材料和设备基本实现国产化，我国的 X70、X80 钢管制造技术及应用规模达到国际领先水平。材料及重大装备的国产化带动了产业升级，推动了民族工业发展。我国高压大口径输气管道使全国 5 亿人民受益，为我国能源安全和生态文明建设做出了重大贡献。

2　我国高钢级管线钢和钢管应用基础研究的主要进展

2.1　X70/X80 钢级材质选用及针状铁素体管线钢组织分析鉴别与评定技术[7-11]

系统对比实验研究了不同组织状态 X70/X80 管线钢的性能特点，研究揭示了管线钢成分、组织、性能、工艺之间的相关性，结果表明针状铁素体型管线钢具有韧性好、抗氢致开裂（HIC）性能好、形变强化抗力高和包辛格效应敏感性低等特点，确定了在重要大口径高压输气管线上采用针状铁素体型管线钢的技术路线。在对国内外文献进行系统调研的基础上，采用光学金相和电子金相等手段系统研究揭示了针状铁素体管线钢的组织特征（图2），提出了便于工程研究和检验的简化的组织鉴别和评定方法（即将针状铁素体管线钢组织简化为铁素体（多边和准多边形铁素体）和贝氏体（针状铁素体AF、粒状贝氏体），研发了可提高组织中各组成相的反差的特殊腐蚀显示技术。

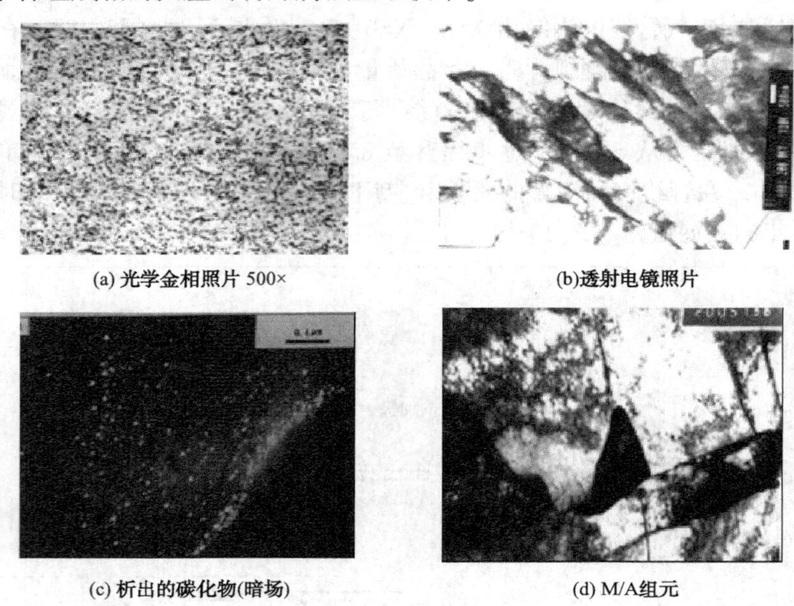

(a) 光学金相照片 500×　(b)透射电镜照片

(c) 析出的碳化物(暗场)　(d) M/A组元

图2　针状铁素体的典型形貌

研究制定了依据管线钢硬组织带（M/A或珠光体）的条数、在视域内的贯穿程度、连续性以及与夹杂物相关性对带状组织进行评定方法。研究提出了针状铁素体型管线钢铁素体晶粒度评定方法（用于生产检验的比较法、用于仲裁的截点法）。解决了针状铁素体管线钢组织分析鉴别、带状组织和晶粒度评定技术难题。根据实验研究结果编制、出版了高强度微合金管线钢显微组织分析与鉴别图谱及西气东输管线、西气东输二线等X70/X80管道工程用管线钢和钢管系列标准。

2.2　X70/X80钢级管型选用及螺旋埋弧焊管残余应力控制技术[8,12]

对国产大口径SSAW焊管、JCOE直缝焊管和进口UOE焊管进行了系统对比评价和试验研究，结果表明国产大口径SSAW焊管和JCOE直缝焊管质量水平可以达到进口UOE焊管水平，螺旋埋弧焊管存在残余应力大且不稳定及焊缝表面质量差的缺点；研究提出了提高国产SSAW焊管质量的措施，制定了高压大口径输气管道国产SSAW焊管技术要求；研究提出了大口径高压输气管线管型选用指导意见，提出在一般情况下（非酸性环境下），只要螺旋焊管规格尺寸满足要求，螺旋焊管和直缝焊管可以等同采用，主张螺旋焊管和直缝焊管联合使用。通过采取一定的工艺措施和材料补充要求，可进一步扩大国产螺旋焊管使用范围。

系统研究了螺旋焊管和直缝焊管残余应力的大小和分布特点及成型工艺、水压试验对残余应力的影响规律；在成型工艺调整合适的情况下，SSAW焊管的残余应力水平可以达到或接近直缝焊管的水平；研究揭示了环切试验中切口位置对残余应力的影响，建立了切口张开量与残余应力之间的对应关系（图3），提出了用切口张开量预测和控制焊管残余应力的方法及判据并在焊管生产中得到应用。

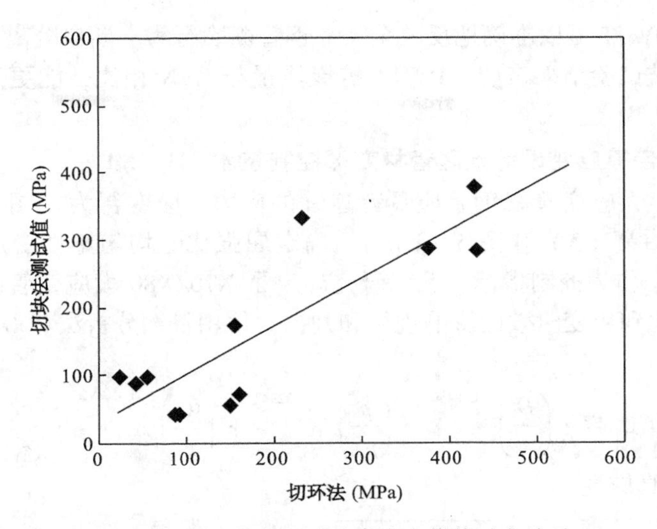

图 3　切块法与切环法残余应力值的对应关系

2.3　X70/X80 高压输气管道断裂控制技术[2,5,6,13]

采用理论分析、数值计算、实验研究等方法系统研究了高压输气管道的动态断裂与止裂行为，在对国际上相关成果分析研究及对管线钢管止裂韧性测试和分析研究的基础上，提出了高钢级输气管线断裂控制的新参量 M 及止裂判据：$M_c = C_V^{1/2}/y > 0.0229$ 及其修正公式 $M = A + FDt/3$（A 为分布参数）。结合国际上相关成果、我们的研究结果及西气东输管道及国内管材生产的实际情况，制定了西气东输管线的断裂控制方案。

针对西气东输二线管道的具体特点，在对影响管道止裂韧性的主要因素（管道几何参数、输送压力、天然气组分、沿线压力和温度分布等）进行分析研究的基础上，基于国际上全尺寸钢管气体爆破试验数据库、国内 X80 管材性能数据库、Battelle 双曲线模型以及 GasDecom 软件对西气东输二线的止裂韧性要求进行了系统研究，综合考虑输气管道安全可靠性与经济可行性，提出了西气东输二线管道管材止裂韧性要求。

自主研发了适用于富气组分的天然气减压波分析程序 DecomWave，分析研究了西气东输二线气源的气质组分以及在冬夏季不同的操作工况对减压波特性的影响，提出了西气东输二线安全运行参数控制要求。

在实验室研究的基础上，开展了模拟西气东输二线实际工况的 X80 管道的实物气体爆破试验（图 4），验证了西气东输二线管道止裂韧性指标的合理性及管道的安全性。实验结果也表明，螺旋埋弧焊管的止裂能力优于直缝埋弧焊管。

(a) 爆破试验场景

(b) 爆破口形貌图

图 4　管线爆破试验场景及爆破口形貌

研究还发现：DWTT 可以准确地反映全尺寸钢管断裂行为，断口分离与钢管止裂能力成反比关系。两次爆破试验结果表明，DWTT 扩展能量与 CVN 相比，能更好地表征钢管的止裂能力。

2.4 地震断裂带等应变设计地区管材变形控制技术[5,6]

经过系统研究，大应变管线钢管应具有连续的应力—应变行为［图 5（a）］，组织结构为铁素体+贝氏体+M（A）［图 5（b）］，确定屈强比、均匀塑性变形延伸率和应力比 $R_{t2.0}/R_{t1.0}$ 和 $R_{t5.0}/R_{t1.0}$ 作为控制指标。系统研究确定了 X70/X80 大应变管线钢和钢管的关键技术指标、形变硬化和应变时效的影响规律和判据。使用量纲分析法获得钢管的临界屈曲应变的预测公式：

$$\varepsilon_b^{crit} = 0.070547 \cdot \left(\frac{D}{t}\right)^{-0.84508} \cdot \left(\frac{p}{p_y}\right)^{-0.0223329} \cdot \left(\frac{\sigma_y}{E}\right)^{-0.138764} \cdot \left(\frac{e_{5.0}}{e_{1.0}}\right)^{6.15051}$$

式中 e——材料的真应变。

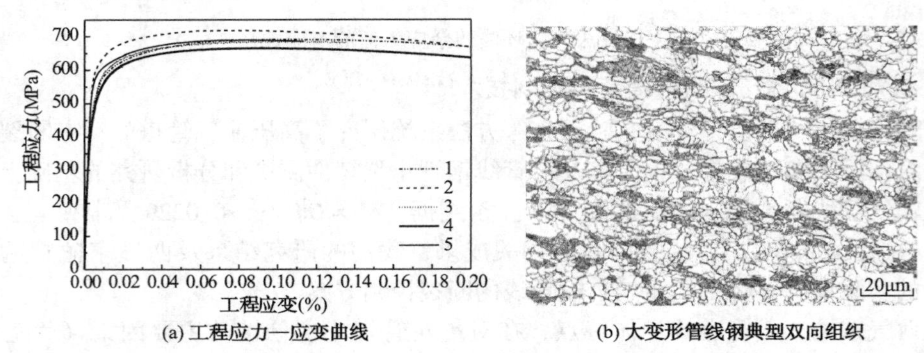

(a) 工程应力—应变曲线　　(b) 大变形管线钢典型双向组织

图 5　X70 抗大变形管线钢管的组织和性能

发明了能够全面准确反映载荷状态、材料性能、钢管规格、几何特性等因素的屈曲应变预测模型和方法，经过千余次数值模拟分析研究，确定为了使管体的屈曲应变（以 2D 平均压缩应变表达）达到 1.5%的水平，管体材料的流变应力比 $R_{t2.0}/R_{t1.0}$ 和 $R_{t5.0}/R_{t1.0}$ 分别需要超过 1.04 和 1.08（图6）。

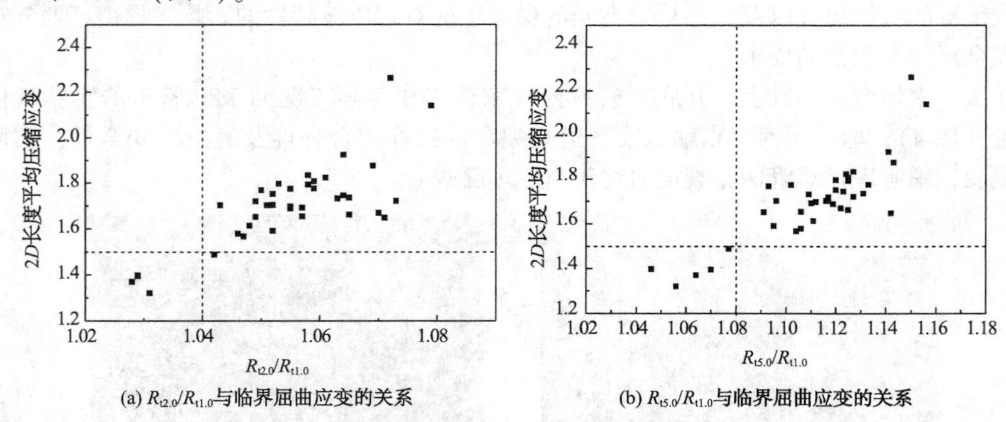

(a) $R_{t2.0}/R_{t1.0}$ 与临界屈曲应变的关系　　(b) $R_{t5.0}/R_{t1.0}$ 与临界屈曲应变的关系

图 6　应力比与弯曲载荷下临界屈曲压缩侧 2D 长度平均压缩应变的关系

自主开发了能够确定大应变钢管内压+弯曲极限承载能力的全尺寸试验评价系统，并进行了实物性能试验评价。经过大应变钢管的工业性试验，各项技术指标合理可行。研究制定了西气东输二线、中缅管线基于应变设计地区使用直缝埋弧焊管技术条件，在大应变钢管试验评价、检验验收和管道工程中批量应用。

2.5　高钢级厚壁管线钢及钢管试验检测评价技术[5,6,11,14,15]

2.5.1　异常断口分析评判技术

针对 X70/X80 壁厚管线钢和钢管落锤撕裂试验（DWTT）中出现的异常脆性断口问题，对 DWTT 试验断口特别是异常断口进行了系统的分析研究，综合分析研究了影响异常断口产生的原因及其与载荷性质和材料品质之间的联系，研究提出了 DWTT 试验断口和异常断口的分类方法和异常断口的评判方法与标准（图 7）。上述断口分类和评定方法纳入了西气东输、陕京二线和 X80 管线钢管工程应用标准中，并被川气东送管线、X80 工程应用段采用，同时纳入石油天然气行业标准和国家标准草案中。解决了高钢级管线钢和管线钢管应用中的技术难题。

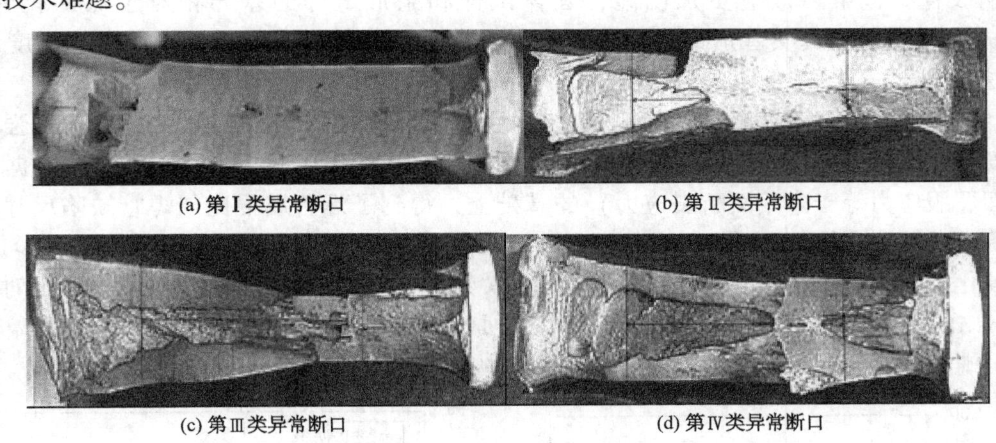

图 7　DWTT 异常断口分类

2.5.2　断口分离分类及评判技术

经过对 X80 管线钢的断口分离现象进行系统研究，发现夹杂物、带状偏析、微观织构等是引起断口分离的主要原因；研究制定了断口分离分级方法，采用该方法对西气东输二线试制的 X80 管材断口分离情况进行了分析，对断口分离严重的钢厂提出改进建议，以及减轻或控制断口分离的措施。同时，对出现三角区（Arrowhead Marking）的 DWTT 试验断口特征进行了分析研究，确定了评判方法（图 8 和表 1）。这些研究成果已经作为补充评定方法应用于西气东输二线管材的断口评定中，研究成果还为减轻各制造厂家的断口分离现象提供了技术依据。

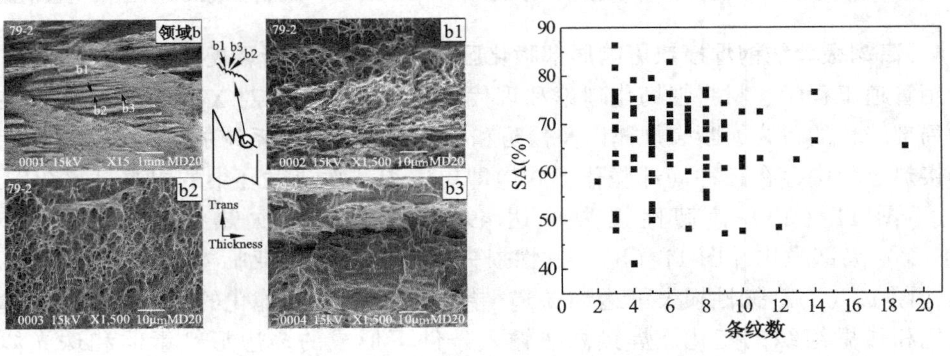

图 8 三角区剪切面积的统计分析

表1　三角区剪切面积统计分析结果

项目	三角区剪切面积统计分析结果（%）		
	最小值	最大值	平均值
条纹数	2	19	6.5
SA（%）	40.9	86.3	65.8
三角区数量	95		

2.5.3　拉伸试验方法及屈强比

（1）通过对百余套 X80 板卷/螺旋钢管、钢板/直缝钢管纵向和横向不同形式拉伸试样的试验研究，得到板/管之间的矩形和圆棒试样屈服强度对比关系，结果表明，X80 钢级管材包申格效应更加明显，圆棒试样和条形试样的差异相对于低钢级更大，圆棒试样更接近于真实管道的情况，提出了采用圆棒试样进行 X80 横向拉伸强度测试的方法。

（2）根据管材的受力情况，研究了临界缺陷长度和屈强比之间的关系，并对屈强比与管道安全性进行了分析研究，结果见图 9 和图 10。在基于先屈服后断裂（YBB）判据（即全截面屈服应力等于净截面断裂应力）的基础上，得到了焊管屈强比与焊管缺陷尺寸和宽板试验试样几何之间的关系，并分析了钢管常规力学性能、抗拉伸应力应变行为、均匀延伸率、硬化指数、钢管承压能力与屈强比之间的关系。最终根据理论计算及试验分析，对不同壁厚的 X80 钢管屈强比进行了科学规定。研究成果已经纳入西气东输二线 X80/X70 板卷、钢板、螺旋钢管、直缝钢管等相关技术条件。

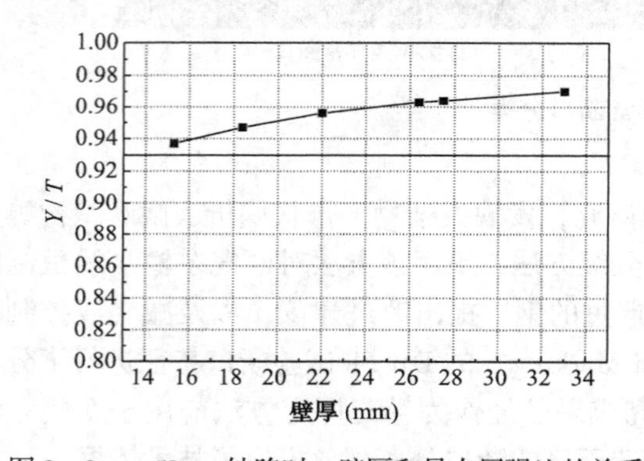

图 9　3mm×50mm 缺陷时，壁厚和最大屈强比的关系

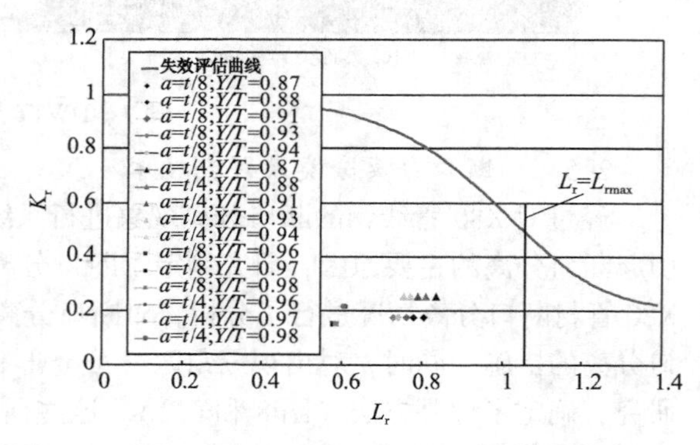

图 10　屈强比对 X80 管道安全性的影响

2.6　高钢级管线钢焊接热影响局部脆化区的脆化机理、断裂规律与防止技术[14]

运用管道工程学、材料强韧化理论和现代物理测试技术，对 X70/X80 管线钢在焊接热过程中局部脆化的行为、断裂规律以及预防和控制措施进行了系统的研究。结果表明：管线钢在焊接热过程中存在三种局部脆化形式，即单道焊一次热循环中的粗晶区（CGHAZ）局部脆化［图 11（a）］，韧性损失可达 49%；多道焊二次热循环中的临界粗晶区（IRCGHAZ）局部脆化［图 11（b），韧性损失可达 69%］和亚临界粗晶区（SCGHAZ）局部脆化［图 11（c），韧性损失可达 61%］。导致 CGHAZ 局部脆化的原因是焊接热过程中的晶粒粗化和显微组织的变化。焊接高热输入条件下形成的多边形铁素体和珠光体，致使 CGHAZ 韧性恶化；中等焊接热输入促使针状铁素体生成，使 CGHAZ 韧性损伤程度降低。

导致 IRCGHAZ 局部脆化的原因是焊接热过程中形成的粗大、富碳的 M-A 组元和表现出来的组织遗传现象。导致管线钢 SCGHAZ 局部脆化的原因是焊接热过程中，基体内碳化物的析出粗化和残余奥氏体的热失稳分解。

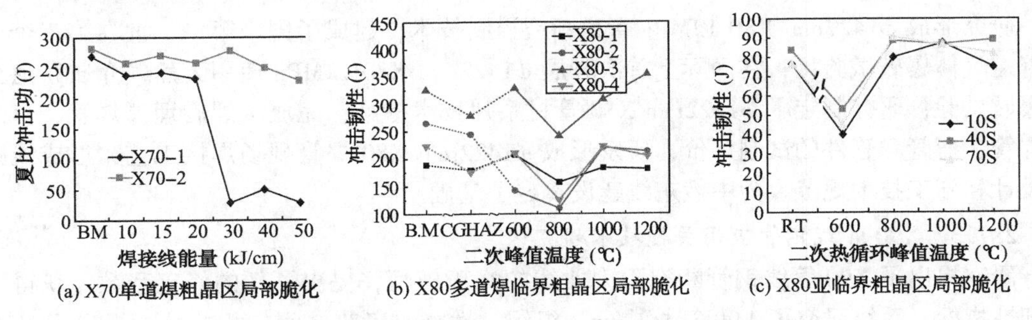

图 11　管线钢在焊接热过程中的三种局部脆化形式

管线钢局部脆化区中裂纹的形核具有三种方式，即夹杂物形核、贝氏体铁素体板条与 M-A 组元界面处形核以及 M-A 组元内部形核。原奥氏体晶界、贝氏体铁素体板条束界可改变裂纹的扩展方向，从而降低裂纹的扩展速度。管线钢的显微组织形态对裂纹扩展有不同的作用。针状铁素体使裂纹扩展速度降低；块状铁素体对裂纹的阻止作用较小；M-A 组元对裂纹的扩展没有阻止作用。基于实验观察和有限元数值分析的结果，提出了管线钢焊接局部脆化区的断裂模型。

在有关管线钢焊接局部脆化机理和断裂规律研究的基础上，提出了预防和控制焊接局部脆化的工艺途径和措施。

2.7　第三代管线钢和钢管应用关键技术研究

我国天然气长输管道工程的发展趋势是进一步增大输量并尽可能地降低建设成本。最近，我国相继与中亚国家和俄罗斯签订新的油气供应协议。今后需要建设输量为（450~600）$\times10^8m^3/a$ 的天然气管道。第二代（X70 和 X80）天然气管道工程技术已经不能满足超大输量的需求。需要发展新一代（第三代）管道工程技术（钢级 X90/X100，管径 1219~1422mm，压力不低于 12MPa，输量（300~600）$\times10^8m^3/a$，设计系数 0.8）。“十二五”期间，中国石油集团设立第三代管线钢和钢管应用研究重大科技专项，取得了重要的理论和技术进展。

2.7.1　输气管道提高强度设计系数工业性应用研究[16-19]

分析研究了提高强度设计系数对管道临界缺陷尺寸、刺穿抗力、应力腐蚀开裂敏感性、失效概率及运行风险的影响，表明我国输气管道在一类地区采用 0.8 设计系数是可行的（图 12）。系统研究确定了 0.8 设计系数管材关键性能指标和质量控制要求，制定了西气东输三线 0.8 设计系数管道用 X80 螺旋缝埋弧焊管技术条件，在西三线示范工程中成功敷设 261km，节约管材 12.6×10^4t，节约采购成本约 1 亿元。

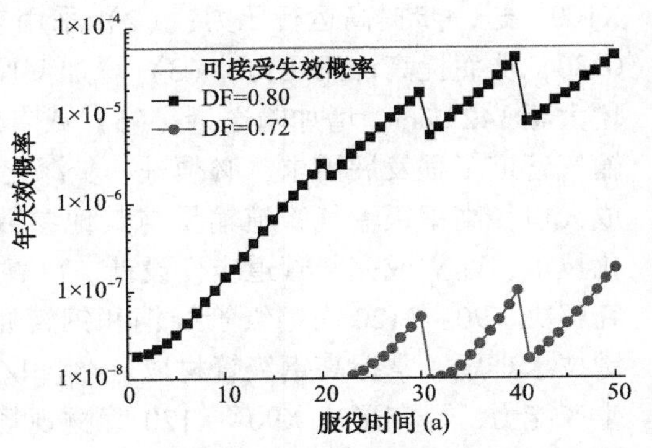

图 12　两种设计系数 X80 管道年失效概率与服役时间的关系

系统分析研究了 X80 厚壁三通的断裂抗力和极限承载能力，在确保三通极限承载能力不小于

3.5倍管道设计压力的情况下，管件壁厚减薄20%~30%，减小了厚壁管件设计的过度保守性，降低了制造难度，在西气东输三线管道工程中得到应用。

2.7.2 φ1422mm X80管线钢管应用研究[20]

研究形成φ1422mm X80 12MPa管道断裂控制技术，建成了国内第一、世界第三个全尺寸管道气体爆破试验场，在国际上首次开展φ1422mm X80 12MPa使用天然气介质管道全尺寸爆破试验；研究制定了φ1422mm X80管材系列技术标准，完成了螺旋埋弧焊管、直缝埋弧焊管、弯管和管件的试制评价；研发形成φ1422mm X80焊管现场焊接工艺，形成线路工程设计和施工技术规范。为中俄东线建设奠定了基础。

2.7.3 X90管线钢管应用关键技术研究[21-23]

研究提出了X90管线钢管断裂阻力曲线的修正模型，提出了新的修正系数，获得了止裂韧性指标。系统研究了X90管材成分、组织、性能、工艺之间的相关性，提出了X90管材的关键技术指标和检测评价方法及配套系列标准。基本完成了X90焊管和管件的研发及试验评价。为X90管线钢和钢管的工程应用奠定了基础。

3 我国管道工程和高钢级管线钢及钢管发展展望

随着世界经济的不断发展，能源消费将持续增加。相对应的，尽管世界经济发展速度放缓，但全球仍处于油气管道建设的持续增长期，油气输送管线钢和钢管的需求量呈现波动式增加。我国天然气资源大约60%集中在西部地区，主要是塔里木和长庆天然气、新疆煤制气。进口天然气包括中亚天然气、俄罗斯西伯利亚和东伯利亚天然气。上述天然气资源大部分需要通过管道实现“西气东输”和“北气南运”。预测我国油气长输管道建设每年需要高性能钢管（100~300）$\times 10^4$t。

我国天然气长输管道工程的发展趋势是进一步增大输量并尽可能地降低建设成本。近年来，我国相继与中亚国家和俄罗斯签订新的油气供应协议。今后需要建设输量为（450~600）$\times 10^8 m^3/a$的天然气管道。第二代（X70和X80）天然气管道工程技术已经不能满足超大输量的需求。需要发展新一代（第三代）管道工程技术（钢级X90/X100，管径1219~1422mm，压力不低于12MPa，输量（300~600）$\times 10^8 m^3/a$，设计系数0.8）。第三代管线钢和钢管的研究已取得重要阶段性成果。

实现超大输量天然气输送的主要途径有：（1）进一步提高管材钢级，如采用X90或X100，进一步提高运行压力；（2）采用较高的设计系数，如由目前的0.72增加到0.77或0.80，从而提高许用压力；（3）增加壁厚，以增大承压能力；（4）增大管径，如可将管径增大至1422mm，增加输送量；（5）或者以上途径的组合。今后，为更好地满足大口径高压输气管道工程发展需求，必须进一步深化应用基础研究，联合进行产品研发与技术攻关，形成大口径高钢级输气管道输量与其他参数（压力、钢级、设计系数、管径、壁厚等）的优化技术，建立或完善管道应变设计、可靠性设计、高压输气管道断裂控制等理论和方法，研究解决X90~X120高钢级管线钢和钢管屈强比、应变时效、低温韧性、屈服强度测试等关键技术难题，掌握高钢级管材成分/组织/性能/工艺相关性，联合冶金和制管企业形成批量生产能力，研究形成X90~X120管材现场焊接技术，进一步提高环焊缝质量性能水平。实现高钢级管线钢和钢管技术的全面突破。

目前，我国超大输量天然气管道建设和高钢级管材的研发应用进入了新的发展阶段，管径、压力、设计系数、钢级的进一步增大或提高，面临的风险将进一步增大，这就需要对具体管线的设计参数进行更加科学的论证，开展更为深入的应用基础研究，综合考虑管道长期服役的安全可靠性和经济性、管材制造能力和质量水平、现场施工技术和能力等因素，特别是要将管道全寿命周期的安全可靠性摆在首位，使管道安全可靠性和经济性得到高度兼顾和优化，确保我国长距离大输量高压输气管道工业又好又快发展。

参 考 文 献

[1] 冯耀荣，陈浩，张劲军，等．中国石油油气管道技术发展展望［J］．油气储运，2008，27（3）：1-8.

[2] 冯耀荣，霍春勇，吉玲康，等．我国管线钢和钢管研究应用新进展及发展展望［J］．石油管工程，2013，19（6）：1-5.

[3] 霍春勇．高钢级管材的发展［C］．石油设备材料国产化会议，北京，2013.3.

[4] 冯耀荣，庄传晶．X80级管线钢管工程应用的几个问题［J］．焊管，2006，29（1）：1-5.

[5] 西气东输二线X80管材技术条件及关键技术指标研究［R］．西安：中国石油集团石油管工程技术研究院．

[6] 西气东输二线管道断裂与变形控制关键技术研究［R］．西安：中国石油集团石油管工程技术研究院．

[7] 冯耀荣，霍春勇，马秋荣，等．西气东输管道及钢管相关技术研究［C］．西气东输管道及钢管应用基础研究论文集．西安：陕西科技出版社，2004：18-32.

[8] 冯耀荣，陈浩，张劲军，等．油气输送管道工程技术进展［M］．北京：石油工业出版社，2006：29-68.

[9] 李鹤林，郭生武，冯耀荣，等．高强度微合金管线钢显微组织分析与鉴别图谱［M］．北京：石油工业出版社，2001.

[10] 冯耀荣，高惠临，霍春勇，等．管线钢显微组织的分析与鉴别［M］．西安：陕西科技出版社，2008.

[11] 冯耀荣，李鹤林．管道钢及管道钢管的研究进展与发展方向（上）［J］．石油规划设计，2005，16（5）：1-7.

[12] 冯耀荣，李鹤林．管道钢及管道钢管的研究进展与发展方向（下）［J］．石油规划设计，2005，17（1）：11-16.

[13] 黄维和．油气管道输送技术［M］．北京：石油工业出版社，2012：12-43.

[14] 冯耀荣，等．影响油气管道安全性的材料因素［M］．北京：石油工业出版社，2007.

[15] 冯耀荣，李洋，吉玲康，等．屈强比对X80焊管力学性能和安全使用的影响［C］//冯耀荣．X80管线钢和钢管研究与应用文集．西安：陕西科技出版社，2011：60-67.

[16] 吴宏，张对红，罗金恒，等．输气管道一级地区采用0.8设计系数的可行性［J］．油气储运，3013，32（8）：799-804.

[17] 赵新伟，罗金恒，张广利，等．0.8设计系数下天然气管道用焊管关键性能指标［J］．油气储运，2013，32（4）：355-359.

[18] 吴宏，刘迎来，郭志梅．基于验证试验法的X80钢级大口径三通设计［J］．油气储运，2013，32（5）：513-516.

[19] 刘迎来，吴宏，井懿平，等．高强度油气输送管道三通试验研究［J］．焊管，2014，37（3）：28-33.

[20] 李丽锋，罗金恒，赵新伟，等．ϕ1422 mm X80管道的风险水平［J］．油气储运，2016，35（4）：25-29.

[21] 史立强，牛辉，杨军，等．大口径JCOE工艺生产X90管线钢组织与性能的研究［J］．热加工工艺，2015，44（3）：226-229.

[22] 刘刚伟，毕宗岳，牛辉，等．X90 高强度螺旋埋弧焊管组织性能研究［J］．焊管，2015，38（10）：9-13.
[23] 王红伟，吉玲康，张晓勇，等．批量试制 X90 管线钢管及板材强度特性研究［J］．石油管材与仪器，2015，1（6）：44-51.

X80 钢级 ϕ1422mm 大口径管道断裂控制技术

霍春勇[1,2]　李　鹤[1,2]　张伟卫[1,2]　杨　坤[1,2]　池　强[1,2]　马秋荣[1,2]

（1. 中国石油集团石油管工程技术研究院；2. 石油管材及装备材料服役行为与结构安全国家重点实验室）

摘　要： 在中俄东线天然气管道工程中采用高钢级（X80 钢级）、大口径（外径为 1422mm）的管道进行高压输送（压力为 12MPa）可以有效增加天然气输送量，满足我国能源战略的需要。然而随着钢级、输送压力、管径及设计系数的不断提高，管道的延性断裂成为断裂的主要方式，止裂控制便成为研究的重点。为此，对止裂韧性计算的主要方法 Battelle 双曲线（BTC）方法以及断裂阻力曲线和减压波曲线进行了深入研究，明确了 BTC 方法的原理及其适用范围，同时分析了 BTC 方法应用于高强度、高韧性管线钢时所存在的问题，给出了目前国际上针对 BTC 计算结果常用的修正方法。回顾了俄罗斯 Bovanenkovo-Ukhta 外径为 1422mm X80 管道断裂的控制方案，针对中俄东线管道的设计参数，对 BTC 计算结果进行了修正，进而制订出符合中俄东线管道安全要求的止裂韧性值（245J）。

关键词： 中俄东线天然气管道工程外径 1422mmX80 钢级断裂控制；止裂韧性；BTC 方法；减压波

随着天然气需求量的与日俱增和管线钢技术的不断进步，采用 X80 及以上级别高强度钢管进行高压、大输量、长距离输送天然气已经成为世界天然气管道输送技术发展的主流趋势。然而随着钢级、输送压力、管径及设计系数的不断提高，管道的延性断裂成为了断裂的主要方式，已经严重威胁管道安全并成为制约高钢级焊管广泛应用的瓶颈问题[1]，止裂控制也成为了研究的重点。

通常采用 Battelle 双曲线方法（BTC）对输气管道的止裂韧性进行预测。然而当管道钢级达到 X80 以上，止裂韧性达到 100J 以上时，BTC 方法预测的准确性便会急剧下降。因此，对于高压（10MPa 及以上）、大口径（外径为 1219mm 及以上）、富气 X80 钢管道，就需要对 BTC 结果进行修正来确定止裂韧性。为此笔者就 BTC 的方法原理、存在的问题及中俄东线 ϕ1422mm 的 X80 钢管道止裂韧性确定方法进行深入的分析和阐述。

1　止裂韧性预测技术

1.1　BTC 方法

API 5L 和 ISO 3183 规定了计算钢管延性断裂止裂韧性的 4 种方法，如表 1 所示。可见

作者简介：霍春勇，男，1966 年生，教授级高级工程师，博士；主要从事长输管道断裂控制方面的研究工作。地址：陕西省西安市锦业二路 89 号。电话：（029）81887999。ORCID：0000-0002-0028-3958。E-mail：huochunyong@ cnpc. com. cn。

当压力达到 12 MPa，钢级达到 X80 时只有 BTC 方法适用，但是当 BTC 的计算值超过 100J 时，则需要对止裂韧性计算结果进行修正。

表 1　钢管延性断裂止裂韧性计算方法表

方法	适用范围
EPRG	$p \leq 8$ MPa，$D \leq 1430$ mm，$t \leq 25.4$ mm，钢级≤X80，贫气
Battelle 简化公式	$p \leq 7$ MPa，$40 < D/t < 115$，钢级≤X80，贫气 预测 $C_v > 100$J 需修正
BTC	$p \leq 12$MPa，$40 < D/t < 115$，钢级≤X80，贫气/富气 预测 $C_v > 100$J 需修正
AISI	$D \leq 1219$ mm，$t \leq 18.3$ mm，钢级≤X70，贫气 预测 $C_v > 100$ J 需修正

注：表中 p 表示压力；D 表示钢管外径；t 表示钢管壁厚：C_v表示夏比冲击功

BTC 方法的原理是通过比较材料阻力曲线（J 曲线）和气体减压波曲线来确定止裂韧性。当这两条曲线相切，代表在某一压力下裂纹扩展速率与气体减压波速率相同，达到止裂的临界条件，与此条件相对应的韧性（C_v，夏比冲击功）即为 BTC 方法确定的止裂韧性。

1.1.1　材料阻力曲线

BTC 方法中用来计算材料阻力曲线的基本模型如下所示。

$$v_c = c \times \frac{\sigma_{flow}}{\sqrt{R}} \times \left(\frac{p_d}{p_a} - 1\right)^m \tag{1}$$

$$p_a = \frac{t}{r}\sigma_{arrest} = \frac{t}{r}\frac{2\sigma_{flow}}{\pi M_T}\arccos\left[\exp\left(\frac{-10^3 \times \pi \times E \times R}{8\sigma_{flow}^2 \times C_{eff}}\right)\right] \tag{2}$$

式中　v_c——裂纹扩展速度，m/s；

R——断裂阻力，J/mm^2；

σ_{flow}——流变应力，MPa；

p_d——裂纹尖端动态压力，MPa；

p_a——止裂压力，MPa；

c、m——回填常数；

t——钢管壁厚，mm；

r——钢管半径，mm；

σ_{arrest}——止裂应力，MPa；

M_T——膨胀因子，无量纲；

E——钢管弹性模量，MPa；

C_{eff}——有效裂纹长度，mm。

其中 R、σ_{flow}、c、m 按表 2 进行计算和取值。

表 2　BTC 方法中对于不同参数的定义表

参数	R	σ_{flow}	C_{eff}	M_T	m	c
定义/取值	C_v/A_c	σ_{YS}+68.95MPa	$3\sqrt{Dt/2}$	3.33	1/6	有土壤回填情况下取 0.2750； 无回填情况下取 0.3795； 海底管道取 0.2350

注：表中 A_c 表示冲击试样韧带面积；σ_{YS}表示钢管的屈服强度

BTC 方法在 20 世纪 70 年代由美国著名的研究机构巴特尔纪念研究所（BattelleMenorial Institute，以下简称为 Battelle）建立，成功地预测了在不同工况条件下 X70 及以下级别管线钢延性断裂止裂所需 C_v。然而对于现代高强度（X80 及以上）、高韧性（100J 以上）的管线钢，BTC 方法预测的准确性则随着钢级和止裂韧性的升高而下降，对此国内外学者进行了大量的研究，主要观点如下：

（1）BTC 方法发展之初，主要用于 X70 及以下级别管线钢的止裂韧性预测，其流变应力定义为屈服强度加上 68.95MPa。随着管道钢级的提高，材料的屈强比也随之上升，采用传统方法计算的流变应力将接近或者超过材料的抗拉强度。因此对于 X80 及以上级别管线钢，采用屈服强度和抗拉强度的平均值或引入加工硬化指数来重新定义流变应力将更为合理。同时，大量的研究成果也表明不同的流变应力定义会优化计算结果，但不是 BTC 方法不准确的根本原因。

（2）BTC 方法中的土壤回填常数 m 和 c 是 Battelle 根据早期全尺寸气爆试验结果回归计算得到，如图 1 所示。回归计算结果如表 2 所示，在任何情况下 m 都为 1/6，而当回填时 c 为 0.2750，无回填时 c 为 0.3795。早期的气爆实验主要采用 X52 和 X65 管线钢管，其韧性都低于 100J，回填时采用黏土和沙土，回填高度为 0.762m。而高钢级（X70、X80、X100 及 X120）、高韧性（大于 100J）管线钢问世之后进行的全尺寸爆破试验中，一方面钢级和韧性不同于 BTC 方法建立时的研究对象（低强度、低韧性管线钢管）；另一方面回填深度和回填土类型也有所不同，例如回填深度增加到 1.2m。试验钢级和回填参数的不同会导致不同的回填常数，例如日本高强度管线钢管委员会通过进行 7 次直径为 1219mm，壁厚为 18.3mm 的 X70 管线钢管全尺寸气爆试验而确定的 m 和 c 分别为 0.67 和 0.393（回填情况）[2]。回填常数既反映了回填土对于裂纹扩展的约束作用，也反映了不同性能钢管对于约束的不同响应。在无约束（例如不回填）和低约束（例如低的回填深度）的情况下，裂纹将扩展得更快。因此 Battelle 早期计算得到的 m 和 c 已不再适用于高钢级管线钢管以及新的回填工艺。

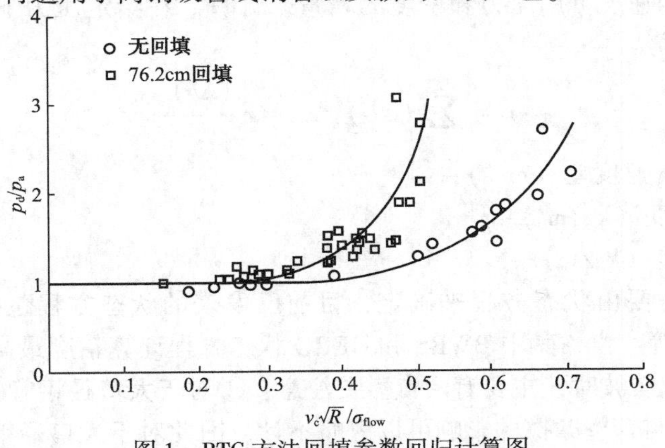

图 1　BTC 方法回填参数回归计算图

（3）BTC 方法中材料断裂阻力 R 是一个非常重要的参数，它反映了材料对于裂纹扩展的阻力。C_v本来不是断裂力学中的断裂韧性参数，但是 Battelle 在早期的研究中，发现低钢级、低韧性管线钢（100J 以下）单位面积 C_v与平面应力下的应变能释放率 G_c呈 1∶1 的线性关系[3]。因此，用 C_v替代 G_c将断裂阻力 R 表征为 C_v/A_c。然而众多的研究结果表明，随着现代管线钢韧性的增加，当超过 100J 后单位面积 C_v与平面应力下的应变能释放率 G_c不再表现为 1∶1 的线性关系[3-4]，如图 2 所示。因此如何正确地表征 BTC 公式中的断裂阻力 R 成为了当前研究的热点和难点问题。

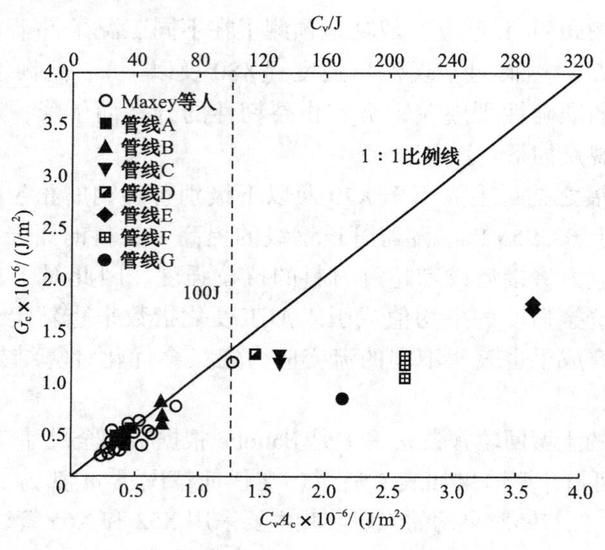

图 2　G_c与 C_v/A_c关系图[3]

1.1.2　减压波曲线

减压波的预测模型基于一维等熵流动，因此与管径无关，只与气质组分、温度和压力有关。减压波速度定义为：

$$W=C-U \tag{3}$$

式中　W——减压波波速（m/s）；

C——局部声速（m/s）；

U——介质流出的平均速度（m/s）。

声速和局部平均速度可用压力和密度的增量来计算。在给定压力下的平均速度 U 是各个变化量的总和：

$$u=\sum_{\rho_1}^{\rho_2}(\Delta U)_s\ \Delta u=c\frac{(\Delta\rho)_s}{\rho} \tag{4}$$

式中　Δu——介质的速度差（m/s）；

ρ——介质密度（kg/m³）；

$\Delta\rho$——密度差（kg/m³）。

密度和声速都需要由状态方程来确定，目前已发表的状态方程包括 AGA-8、BWRS、SRK、PR 和 GERG 等[5-10]，其中 BWRS 和 GERG 状态方程计算精度最高，常用于天然气减压波计算。在计算减压波时，主要有两点需要注意：①对于大口径钢管（如外径为 1422mm 的钢管），外流气体与管壁摩擦的影响可以忽略不计，因此对于大口径钢管无需考虑钢管内表面的粗糙度；②含有重烷烃的富气在减压过程中会出现由气相到液相的相变，造成气液两相共存的现象，从而产生减压波平台。减压波平台的出现使裂纹尖端的压力长时间保持在高位而无法降低，增加了止裂的难度。

1.2 基于BTC模型的主要修正方法

当钢管止裂韧性超过100J时，需要对BTC方法进行修正，主要的修正方法如下。

（1）Leis修正，如式（5）和式（6）所示，其中式（5）适用于X70管线钢，式（6）适用于X80管线钢。

$$\mathrm{Leis1CVN_{arrest}} = \mathrm{CVN_{BTCM}} + 0.002\mathrm{CVN_{BTCM}^{2.04}} - 21.18 \tag{5}$$

$$\mathrm{Leis-2CVN_{arrest}} = 0.985\mathrm{CVN_{BTCM}} + 0.003\mathrm{CVN_{BTCM}^{2}} \tag{6}$$

式中 $\mathrm{CVN_{BTCM}}$ 为BTC方法计算的CVN能量；$\mathrm{CVN_{arrest}}$ 为经修正后的止裂韧性值。

（2）Eiber修正，如式（7）所示，适用于X80管线钢。

$$\mathrm{CVN_{arrest}} = \mathrm{CVN_{BTCM}} + 0.003\mathrm{CVN_{BTCM}^{2.04}} - 21.18 \tag{7}$$

（3）Wilkowski修正，如式（8）所示，适用于X80管线钢。

$$\mathrm{CVN_{arrest}} = 0.056\times(0.1018\times\mathrm{CVN_{BTCM}} + 10.29)2.597 - 16.8 \tag{8}$$

（4）线性修正，如式（9）所示，K 为常数：

$$\mathrm{CVN_{arrest}} = K\mathrm{CVN_{BTCM}} \tag{9}$$

2 ϕ1422 mm的X80钢管道断裂控制方案

2.1 俄罗斯ϕ1422mm的X80钢管道断裂控制方案

为了建设总长度达1106 km的Bovanenkovo-Ukhta长输管线，从2008年3月至2009年1月，俄罗斯共进行了10次X80钢管道全尺寸气爆实验[11,12]。实验钢管包括壁厚为23.0mm、27.7mm和33.4 mm的ϕ1420mm的X80钢直缝埋弧焊管，实验中在起裂管两侧各排列3根等CVN能量测试钢管，根据EPRG标准要求在3根钢管内止裂。在第4次实验中，裂纹穿过所有3根测试钢管而无法止裂。这些钢管的断口特征如图3所示，表现为韧脆混合断口而不具备典型的45°剪切断裂特征。这些钢管的DWTT形貌如图4所示，可见具有明显的断口分离特征。

图3 俄罗斯扩展管的爆破试验断口特征

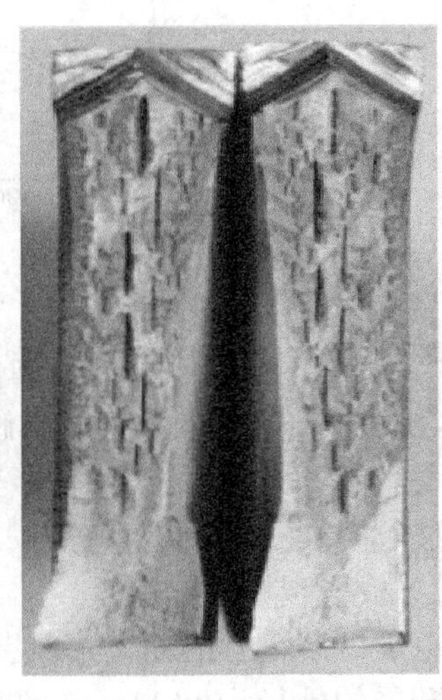

图4 俄罗斯扩展管的DWTT断口形貌

Bovanenkovo-Ukhta 管道的设计温度为-20℃，BTC 修正计算后的止裂韧性为 200J，在此温度下进行的 CVN 试验表明裂纹扩展管和止裂管的 CVN 能量区别不大，而在-40℃进行试验则会使裂纹扩展管 CVN 试样的断口分离现象明显增强，并表现为 CVN 能量的急剧下降。而止裂管即便在-40℃进行试验，其 CVN 能量下降也很小。因此，为了能有效鉴别具有断口分离特征的扩展管，在 Bovanenkovo-Ukhta 管道的技术指标中，将 CVN 试验温度规定为-40℃。

2.2　中俄东线外径为 1422mm 的 X80 钢管道断裂控制方案

中俄东线天然气管道输送的天然气组成为：C_1的摩尔分数为 91.41%，C_2的摩尔分数为 4.93%，C_3的摩尔分数为 0.96%，C_4的摩尔分数为 0.41%，C_5的摩尔分数为 0.24%，N_2的摩尔分数为 1.63%，CO_2的摩尔分数为 0.06%，He 的摩尔分数为 0.29%，H_2的摩尔分数为 0.07%。一类地区设计参数为：管道钢级为 X80，管径为 1422 mm，压力为 12MPa，管道壁厚为 21.4mm，设计系数为 0.72。

中俄东线的设计压力为 12MPa，但是只有在压气站的出气口处压力才会达到 12MPa，而在下一个压气站的进气口处压力会显著降低。中俄东线最低冻土层温度为-1.5℃，在正常输送的情况下，压气站出气口的温度会显著高于-1.5℃，而下一个压气站进气口处的温度也会在-1.5℃以上。只有在最恶劣的情况下（管道埋于冻土层内及长时间停输），管道内的气体温度才会降至地温，但此时管道内压力也会下降（约 1MPa）。综合考虑，在进行止裂韧性计算时选取 12MPa 和 0℃作为计算参数。

图 5 为中俄东线 BTC 计算结果，计算中采用 BWRS 状态方程进行减压波计算。可见中俄东线气质组分存在明显的减压波平台，止裂韧性 CVN 能量计算值为 167.97J，由于 BTC 计算值超过 100J。因此必须进行修正。

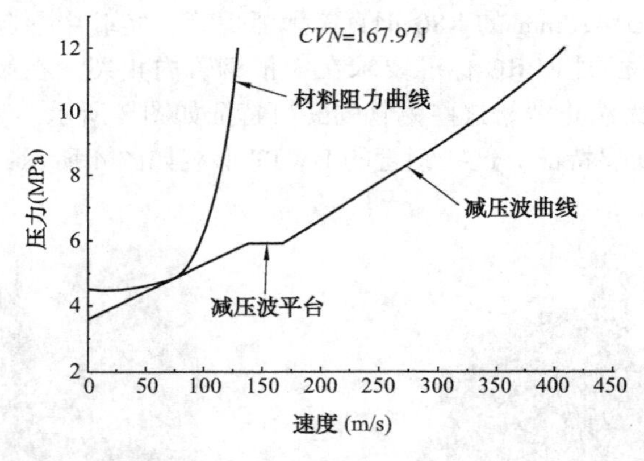

图 5　中俄东线 BTC 计算结果图

经不同方法修正后得到的止裂韧性结果为：BTC 预测值为 167.97 J；1.46 倍修正值为 245 J；Leis-2 修正值为 250 J；Eiber 修正值为 251 J；Wilkowski 修正值为 286 J。其中 Leis-2、Eiber 和 1.46 倍修正的结果基本一致。考虑到 1.46 倍修正可以较好地将全尺寸爆破试验数据库中的裂纹扩展点和止裂点分开，如图 6 所示[13]。进而最终将止裂韧性指标确定为 245 J，此指标为单根止裂韧性指标。

图 7 为单炉试制钢管的断口形貌，可见试样不存在严重的断口分离。最近在国内全尺寸爆破试验场开展的外径为 1422 mm 的 X80 钢管道全尺寸爆破试验同样表明，国内生产的外径为 1422 mm 的 X80 钢管爆破断口形貌为 45°剪切断口，可以依靠自身韧性进行止裂。

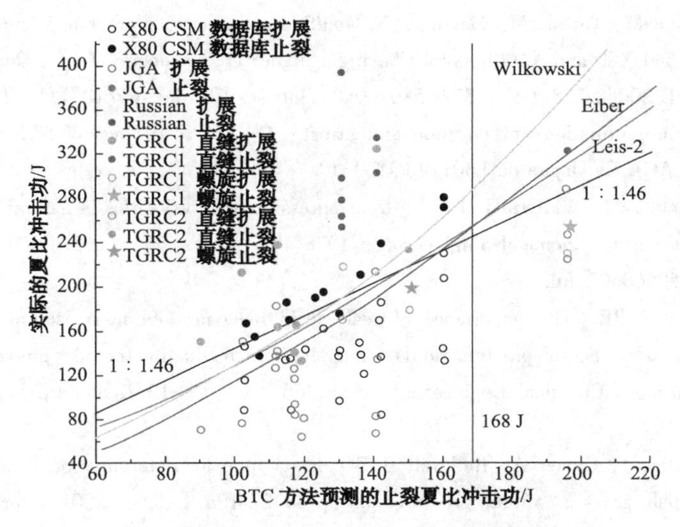

图 6　X80 钢管道全尺寸气体爆破试验数据库图

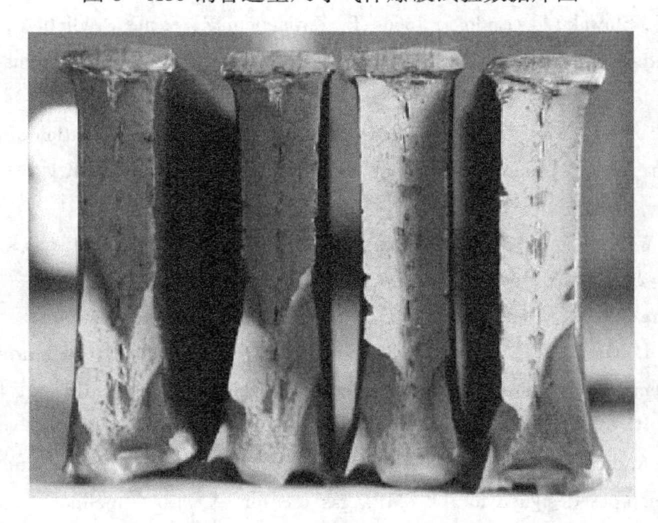

图 7　中俄东线单炉试制钢管 DWTT 断口形貌图

3　结论及建议

（1）Battelle 建立的 BTC 方法不能直接应用于现代高强度高韧性管线钢的止裂韧性计算，其模型的改进有待进一步深入研究解决。

（2）采用 BTC 结合 1.46 倍修正的方法确定中俄东线外径为 1422 mm、X80 钢管道的止裂韧性指标为 245J。

参　考　文　献

［1］Mannucci G，Demofonti G. Control of ductile facture propagation in X80 gas linepipe［C］//Pipeline Technology Conference，12-14October，2009，Ostend，Belgium.

［2］Makino H，Takeuchi I，Higuchi R. Fracture propagation and arrest in high-pressure gas transmission pipeline by ultra high strength line pipes［C］//2008 7th International Pipeline Conference，Calgary，2008. DOI：http：//dx. doi. org/10. 1115/IP C2008-64078.

[3] Kawaguchi S, Ohata M, Toyoda M, Hagiwara N. Modified equation to predict leak/rupture criteria for axially through-wall notched X80 and X100lipepipes having a higher charpy energy [J]. Journal ofPressure Vessel Technology, ASME 2006, 128 (4): 572-580. DOI: http://dx.doi.org/10.1115/1.2349570
[4] Maxoy WA. Fracture initiation, propogation and arrest [C] //Proceedings of 5th Syrrposium on Line Pipe Research. 1974, AGA Catalogue no L30174, J1-J31.
[5] Groves T K, Bishnoi P R, WallvridGe J M E. Decompression wave velocities in natural gas in pipelines [J]. The Canadian Journal of Chemical Engineering, 1978, 56 (6): 664 - 668. DOI: http://dx.doi.org/10.1002/cjce.5450560602/full
[6] Picard D J, Bishnoi PR. The importance of real - fluid behavior and non - isentropic effects in modeling decompression characteristics of pipeline fluids for application in ductile fracture propagation analysis [J]. The Canadian Journal of Chemical Engineering, 1988, 66 (1): 3-12. DOI: http://dx.doi.org/10.1002/cjce.5450660101.
[7] Botros K K, Geerligs J, Glover A, Rothwell B. Expansion tube for determination of the decompression wave speed for dense/rish gases at initial pressures of up to 22MPa [C] //2001 International Gas Research Conference, Amsterdam, Netherlands, 2001.
[8] Makino H, Kubo T, Shiwaku T, Endo S, Inoue T, Kawaguchi Y, et al. Prediction for crack propagation and arrest of shear fracture in ultra high pressure natural gas pipelines [J]. ISIJ International, 2001, 41 (4): 381-388.
[9] Eiber R J, Bubenik A T, Maxey W A. Gasdecom computer code for the calculation of gas decompression speed that is included in fracture control technology for natural gas pipelines [R]. Houston: American Gas Association Catalog, 1993.
[10] Kunz O, Wagner W. The gerg-2008 wide-range equation of state for natural gases and other mixtures: An expansion of gerg-2004 [J]. Journal of Chemical &Engineering Data, 2012, 57 (11): 3032-3091. DOI: http://dx.doi.org/10.1021/je300655b
[11] Pyshmintsev I Y, Lobanova T P, Arabey A B, Sozonov P M, Struin A O. Crack arrestability and mechanical properties of 1420mm X80 grade pipes designed for 11.8MPa operation pressure [C] //Pipeline Technology Conference, 12-14October, 2009, Ostend, Belgium.
[12] Pyshmintsev I Y, Arabey A B, Gervasyev A M, Boryakova A N. Effects of microstructure and texture on shear fracture in X80 linepipes designed for 11.8MPa gas pressure [C] //Pipeline Technology Conference, 12-14October, 2009, Ostend, Belgium.
[13] 李鹤，王海涛，黄呈帅，等．高钢级管线焊管全尺寸气体爆破试验研究［J］．压力容器，2013，30（8）：21-26.

ϕ1422mm X80 钢管断裂韧性指标研究*

赵新伟[1]　池　强[1]　张伟卫[1]　杨峰平[1]　许春江[2]

（1. 中国石油集团石油管工程技术研究院，石油管材及装备材料服役行为与结构安全国家重点实验室；2. 中国石油西部管道公司）

摘　要：研究提出了 ϕ1422mm X80 焊管满足管道断裂控制要求的母材、焊缝及热影响区韧性指标。对于设计压力为 12MPa 的 ϕ1422mm X80 天然气管道，为防止钢管启裂，要求焊缝及热影响区夏比冲击韧性最小平均值为 80J；满足管道止裂要求，要求钢管母材夏比冲击韧性最小平均值为 245J。对开发试制的 ϕ1422 mm X80 焊管冲击韧性测试评价和统计分析结果表明，试制焊管整体上韧性水平较高，可以满足韧性指标要求。

关键词：ϕ1422mm X80 焊管；天然气管道；启裂韧性；止裂韧性

随着我国国民经济的快速发展，以及为降低环境污染对清洁能源的需求大幅度增加，天然气在一次能源的占比在逐年提高。自 2000 年以来，我国天然气消费量年均增长 16%，目前已达每年 $1300\times10^8 m^3$，而且消费量还在以“加速度”递增，预计 2020 年我国天然气消费需求将达 $3600\times10^8 m^3$。我国大约 60%的天然气资源集中在西部地区，主要是塔里木天然气、新疆煤制气。进口天然气主要包括中亚天然气、俄罗斯西伯利亚天然气等。上述天然气资源均需要通过管道实现“西气东输”。高压大输量长距离输送已成为我国天然气管道输送的发展趋势。天然气管道设计和建设中，在不影响管道安全可靠性的前提下，如何最大限度地降低管道建设成本和提高管道输送效率，一直是管道建设投资者和管道运营企业长期关注的问题。

目前我国对天然气管线的工程需求突出表现在两个方面：大输量和低成本。在西气东输二线中采用的 X80 与 X70 相比，节约了钢材 10%，降低了成本。在西气东输二线中采用 ϕ1219mm、12MPa、X80，但其经济输量范围为（250~300）$\times10^8 m^3/a$，最经济输量为 $280\times10^8 m^3/a$，最大输气量只能达到 $330\times10^8 m^3/a$，不能满足中俄天然气管线、西气东输四线、五线等超大输量（超过 $400\times10^8 m^3/a$）的需求。通过技术经济综合分析，目前可采取三种主要技术方案提高管输效率和降低管道成本，一是管道设计系数和规格不变，采用 X90/X100 超高强度管线钢管；二是设计系数和钢级不变，管径增加到 1422mm；三是管道规格和钢级不变，设计系数由 0.72 提高到 0.8。为满足中俄天然气管线、西气东输四线等超大输量管线建设需求，按照上述技术方案，从“十二五”开始，中国石油天然气集团公司设立重大科技专项，组织开展了“第三代大输量天然气管道工程关键技术”的研究攻关，在以上三

基金项目：中国石油天然气集团公司重大科技专项（2012E-28）。

作者简介：赵新伟，男，1969 年出生，博士，教授级高工，院长助理。长期从事油气输送管和管道完整性技术研究工作。E-mail：zhaoxinwei001@cnpc.com.cn　电话：029-81887566，13609265830。

个技术方案的研究攻关上都取得了重要突破。其中，通过 ϕ1422mm X80 管线钢管应用关键技术攻关，研究提出了 ϕ1422mm X80 管道断裂控制方案，制定了 ϕ1422mm X80 板材、管材技术条件，成功开发了 ϕ1422mm X80 焊管（包括 HSAW 和 LSAW）以及配套的弯管和管件，为大输量天然气管线工程建设做好了技术储备。

近几年，在西气东输一线、二线等高压输气管线建设的推动下，国内在管线钢的冶金技术、制管技术、管道施工技术以及质量控制水平有了长足进步[1,2]，尤其是通过西气东输二线工程建设，在 X80 大口径管道建设配套技术研究和工程实践上积累了较丰富的经验，高钢级管线管的实物质量和产品标准已经达到国际先进水平，另外，管道完整性管理技术快速发展[3,4]，管道安全管理水平显著提高，我国大输量输气管道采用 X80 ϕ1422mm 钢管已具备了良好的基础和条件。中国石油即将开工建设的中俄天然气管线已确定采用 X80 ϕ1422mm 设计方案。断裂控制是 ϕ1422mm X80 管线钢管需要解决的应用关键技术问题之一。钢管断裂韧性是 X80 ϕ1422mm 管道断裂控制研究的核心内容，也是保障管道本质安全的关键性能指标。

本文针对 API X80，直径为 1422 mm，设计压力为 12 MPa 的天然气管道，研究并提出了钢管焊接接头和母材的断裂韧性指标。研究结果已纳入中俄东线天然气管道工程用外径 1422mm X80 螺旋埋弧焊管和直缝埋弧焊管技术条件。

1　ϕ1422mm X80 焊管焊接接头断裂韧性和启裂韧性指标

大量的管道失效案例统计分析结果表明，焊接钢管启裂一般都发生在钢管焊缝或热影响区，所以用焊接接头断裂韧性指标作为焊管启裂韧性指标。偏于安全起见，国际上的作法（如 ASME 和 EPRG）是假设钢管焊缝或热影响区存在深度为 $t/4$（t 为钢管名义壁厚）的表面裂纹缺陷，采用断裂力学分析方法获得裂纹不发生扩展的临界断裂韧性值，将此临界断裂韧性值作为焊管的启裂韧性指标。

计算分析采用英国 CEGB 最早提出的以弹塑性断裂力学为基础的失效评估图（FAD）技术，即双判据方法，如图 1 所示。基于失效评估图的双判据法综合考虑了钢铁结构的断裂失效和塑性变形失效两种失效模式，确定的临界断裂韧性与单纯断裂判据相比更为可靠，在结构设计中，既可防止断裂失效，又可防止过量塑性变形失效，从而有效地降低结构的失效概率，是国内外在结构断裂评定规范中普遍采用的一种方法。采用 SY/T 6477—2014《含缺陷油气输送管道剩余强度评价方法》计算分析，FAD 图中的失效评估曲线 FAC 方程如下：

$$K_r = (1 - 0.14(L_r^p)^2)(0.3 + 0.7\exp[-0.65(L_r^p)^6]) \tag{1}$$

$$K_r = K_I/K_C$$

$$L_r = \sigma_{ref}/\sigma_y$$

式中　K_I——应力强度因子；
　　　K_C——材料断裂韧性；
　　　σ_y——材料屈服强度；
　　　σ_{ref}——参考应力，与裂纹尺寸和主应力有关。

当载荷比 L_r 满足公式（2）给出的条件时，按照公式（1）在设定表面裂纹深度 a 为 $t/4$ 后，进行不同裂纹长度 $2c$ 下的断裂韧性敏感性分析，将所定义的无限裂纹长度 $2c$ 对应的断裂韧性 K_C 值作为断裂韧性要求值，即启裂韧性指标。

$$L_r \leqslant (\sigma_y + \sigma_u)/2\sigma_y \tag{2}$$

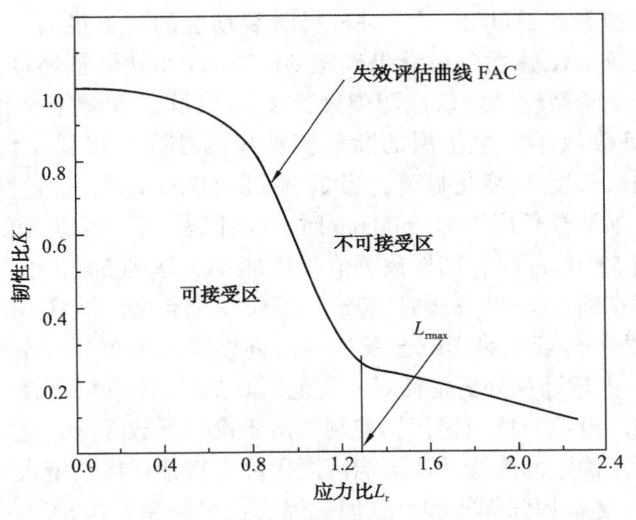

图 1　失效评估技术的示意图

由于断裂韧性 K_C 测试费用高且周期长，为便于工程应用，可以利用研究获得的 K_C 和 CVN 经验关系式，将断裂韧性 K_C 指标转化为夏比冲击功 CVN 指标。本文采用了 API RP 579[5] 推荐的以下经验公式：

$$K_C = 8.47 \times (\mathrm{CVN})^{0.63} \tag{3}$$

以一级地区管道为例，计算钢管焊接接头所需启裂韧性。钢管外径 1422 mm，输送压力 12MPa，钢级为 X80，按照一级地区 0.72 设计系数，钢管设计壁厚为 21.4mm。屈服强度取标准规定的管体屈服强度最小规定值 555MPa，抗拉强度取标准最小规定值 625MPa。

裂纹深度取 $t/4=5.35$mm，考虑轴向半椭圆外表面裂纹、轴向半椭圆内表面裂纹、轴向外表面长裂纹、轴向内表面长裂纹四种裂纹类型，进行不同裂纹长度 $2c$ 下的断裂韧性敏感性分析。其中，对于前两种表面裂纹 $a/c \geqslant 0.031$，对于后两种长表面裂纹 $a/c< 0.031$，定义为长裂纹。四种表面裂纹分别有相应的应力强度因子 K_I 和参考应力 σ_{ref} 的计算公式。作为例子，下面给出了轴向内表面长裂纹的应力强度因子 K_I 和参考应力 σ_{ref} 的计算公式。

$$K_I = \frac{pR_o^2}{R_o^2 - R_i^2}\left[2G_o - 2G_1\left(\frac{a}{R_i}\right) + 3G_2\left(\frac{a}{R_i}\right)^2 - 4G_3\left(\frac{a}{R_i}\right)^3 + 5G_4\left(\frac{a}{R_i}\right)^4\right]\sqrt{\pi a} \tag{3}$$

式中　p——内压；

R_o 和 R_i——钢管的内半径和外半径；

G_0、G_1、G_2、G_3、G_4——与管径、壁厚、裂纹几何尺寸有关的系数。

$$\sigma_{ref} = \frac{gp_b + [(gp_b)^2 + 9(M_s p_m)^2(1-\alpha)^2]^{0.5}}{3(1-\alpha)^2} \tag{4}$$

$$p_m = \frac{pR_i}{t} \tag{5}$$

$$p_b = \frac{pR_0^2}{R_0^2 - R_i^2}\left[\frac{t}{R_i} - \frac{3}{2}\left(\frac{t}{R_i}\right)^2 + \frac{9}{5}\left(\frac{t}{R_i}\right)^3\right] \tag{6}$$

式中　a、g 和 M_s——与裂纹尺寸相关的系数。

通过不同裂纹长度 $2c$ 下的断裂韧性敏感性分析，发现在轴向半椭圆外表面裂纹、轴向半椭圆内表面裂纹、轴向外表面长裂纹、轴向内表面长裂纹四种类型裂纹中，轴向内表面长裂纹最为苛刻，同样的裂纹长度 $2c$ 下，对材料断裂韧性的要求最高。因此，为保守起见，基于轴向内表面长裂纹的敏感性分析结果确定 ϕ1422mm X80 焊管焊接接头断裂韧性指标。图 2 是对于轴向内表面长裂纹启裂韧性随裂纹长度 $2c$ 变化的预测结果。可以看出，随着裂纹长度 $2c$ 增加，保证裂纹不发生扩展的断裂韧性预测越高，在裂纹长度 $2c$ 小于 1500mm 时，韧性需求值随裂纹长度 $2c$ 变化显著，当 $2c$ 超过 1500mm 时，韧性需求值随裂纹长度增加不再有明显提高。如裂纹长度为为 1500mm 时，保证裂纹不发生扩展的韧性预测值为 51J，当裂纹长达度增加到 5350mm 时，韧性预测值仅增加 4J，达到 54J。根据工程实际，出厂钢管焊接接头不可能存在超过 1500mm 的长裂纹。偏于安全起见，钢管焊缝和热影响区启裂韧性取 60J。产品标准中，焊缝及热影响区夏比冲击韧性要求单个试样最小值 60J，最小平均值为 60J/0. 75＝80J。上述计算分析是针对一类地区 0. 72 设计系数钢管，对二、三、四类地区钢管焊缝及热影响区也统一按上述指标控制，由于设计系数降低，这样处理更加保守，安全裕度更大。以上研究确定的焊缝及热影响区夏比冲击韧性指标已纳入中俄东线天然气管道工程用 ϕ1422mm X80 螺旋埋弧焊管和直缝埋弧焊管技术条件（Q/SY GJX 147）。

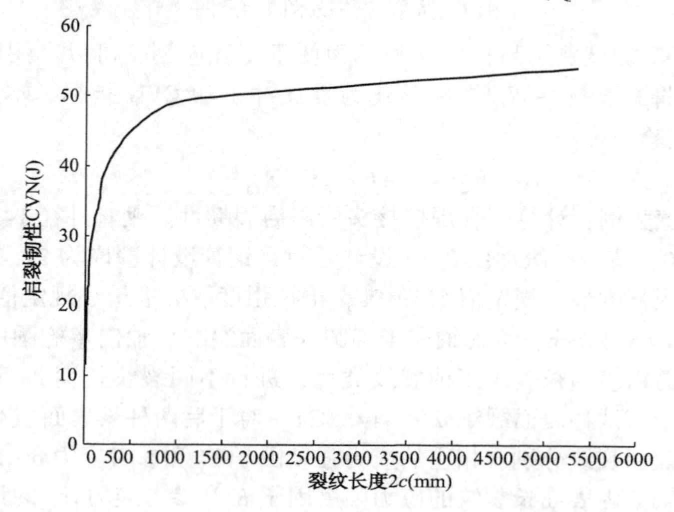

图 2　不同裂纹长度下焊缝启裂韧性预测结果

2　钢管母材止裂韧性指标

为防止天然气管道开裂后发生延性裂纹的长程扩展，管道材料必须有足够的韧性以保证天然气管道一旦开裂能在一定长度范围内止裂。随着钢级、输送压力、管径及设计系数的不断提高，管道的延性断裂止裂问题也更加突出，是高钢级管线钢管应用的瓶颈技术问题。下文针对中俄天然气管线，采用 Battelle 双曲线法（BTC）方法，并引入修正系数，计算了一类地区 ϕ1422mm X80 管道止裂的韧性需求，提出了焊管母材的止裂韧性指标。

2.1　止裂韧性计算方法

API5L[9] 推荐了 4 种钢管延性断裂止裂韧性的计算方法，包括 EPRG（欧洲管线研究组织）准则、Battelle 简化公式、Battelle 双曲线（BTC）方法和 AISI 方法，4 种计算方法的适用范围见表 1 所列。对于设计压力 12MPa 的 ϕ1422mm X80 的天然气管道，只能采用 BTC 模型并引入修正系数的方法来计算止裂韧性。

表 1　API5L 推荐的天然气管道止裂韧性计算方法

序号	方法	适用范围
1	EPRG	$p\leqslant 8$MPa；$D\leqslant 1430$；$t\leqslant 25.4$；$Grade\leqslant$X80；贫气
2	Battelle 简化公式	$p\leqslant 7$MPa；$40<D/t<115$；Grade≤X80；贫气预测 CVN>100J 需修正
3	BTC	$p\leqslant 12$MPa；$40<D/t<115$；$Grade\leqslant$X80；贫气/富气预测 CVN>100J 需修正
4	AISI	$D\leqslant 1219$；$t\leqslant 18.3$；$Grade\leqslant$X70；贫气预测 CVN>100J 需修正

Battelle 双曲线方法基本原理是通过计算和比较材料阻力曲线（J 曲线）和气体减压波曲线来确定止裂韧性，如图 3 所示。当这两条曲线相切，代表在某一压力下裂纹扩展速率与气体减压波速率相同，达到止裂的临界条件，与此条件相对应的韧性（夏比冲击功）即为 Battelle 双曲线法确定的止裂韧性。如果材料阻力曲线和减压波曲线没有交点，则在任何条件下减压波速率都大于裂纹扩展速率，裂纹尖端的压力将一直下降至零，随着驱动力的逐渐下降（裂纹尖端压力），裂纹将最终停止扩展。

BTC 模型中用来计算材料阻力曲线的基本公式如下：

$$v_c = c \times \frac{\sigma_{flow}}{\sqrt{R}} \times \left(\frac{p_d}{p_a} - 1\right)^m \tag{7}$$

$$p_a = \frac{4}{M_T \times \pi} \times \frac{t}{D} \times \sigma_{flow} \times \cos^{-1}\exp\left(\frac{-10^3 \times \pi \times E \times R}{8 \times C_{eff} \times \sigma_{flow}^2}\right) \tag{8}$$

式中　v_c——裂纹扩展速度；

R——裂纹扩展阻力，取 C_V/A_c；

C_V——夏比冲击功；

A_c——冲击试样韧带面积；

σ_{flow}——材料流变应力，取 σ_y+68.95MPa；

p_d——裂纹尖端动态压力；

p_a——止裂压力；

c 和 m——回填常数；

t——钢管壁厚；

D——钢管外径；

σ_{arrest}——止裂应力；

M_T——膨胀因子；

E——钢管弹性模量；

C_{eff}——有效裂纹长度，取 $3\sqrt{Dt/2}$ 。

减压波与天然气组分、温度和压力有关。减压波速度定义为：

$$W = C_S - U \tag{9}$$

式中　W——减压波波速；

C_S——局部声速；

U——介质流出的平均速度。

声速和介质流出的平均速度可用压力和密度的增量来计算，密度和声速都需要由状态方程来确定。本文采用 GASDECOM 软件来预测气体减压波曲线。

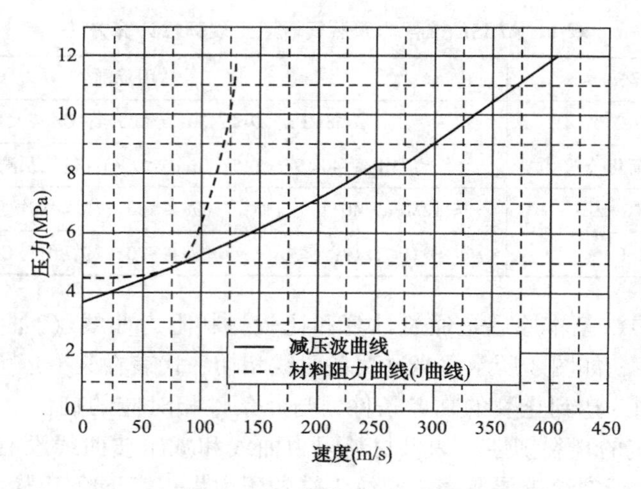

图3 BTC方法原理示意图

BTC方法成功预测了在不同工况条件下X65及以下级别管线延性断裂止裂所需夏比冲击功（CVN），然而当钢级达到X65以上，夏比冲击功达到94J以上时，BTC方法预测的准确性随着钢级的提高而下降。对于高压（10MPa及以上）、大口径（ϕ1219mm及以上）以及富气的X80管道，需要对BTC预测结果进行修正来确定止裂韧性，修正系数取值是基于全尺寸气体爆破试验数据库的分析。通过国际上已发布的全尺寸爆破试验数据库和理论预测值的对比分析，考虑到1.46倍修正可以较好的将全尺寸爆破试验数据库中的裂纹扩展点和止裂点分开，最终采用系数为1.46的线性修正。

2.2 止裂韧性计算结果

止裂韧性计算考虑的天然气组分和管道设计参数见表2和表3，温度取0℃。图4为BTC计算结果。由图4可见，中俄东线气质组分存在明显的减压波平台，CVN止裂韧性计算值为167.97J。由于BTC止裂韧性计算值超过100J，必须进行修正。经1.46倍的线性修正，将止裂韧性指标确定为245J。需要指出的是，计算得到的245J是单根钢管的止裂概率为100%的韧性要求值。参考美国DOT 49 CFR Part 192的规定，裂纹能在5~8根止裂，对应的止裂概率达到95%和99%即可，按照这一原则，止裂韧性指标确定为245J应该是偏于保守和安全的。最终在Q/SY GJX 147—2015《中俄东线天然气管道工程用ϕ1422mm X80螺旋埋弧焊管技术条件》中，母材夏比冲击功CVN平均最小值定为245J，单个试样最小值定为185J。上述止裂韧性指标经过全尺寸气体爆破试验验证（2015年12月，新疆哈密）是可靠的。

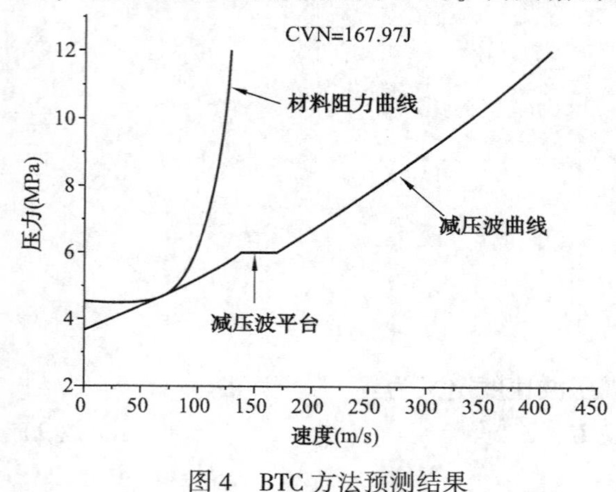

图4 BTC方法预测结果

表2 止裂韧性预测考虑的天然气组分

组分	C_1	C_2	C_3	C_4	C_5	N_2	CO_2	He	H_2
摩尔分数（%）	91.41	4.93	0.96	0.41	0.24	1.63	0.06	0.29	0.07

表 3　设计参数

钢级	管径（mm）	压力（MPa）	壁厚（mm）	设计系数
X80	1422	12	21.4	0.72

3　试制 ϕ1422mm X80 焊管冲击韧性测试分析结果

结合中国石油天然气集团公司重大科技专项“第三代高压大输量油气管道建设关键技术研究”攻关，由西部管道公司和管研院牵头组织，国内宝钢、首钢、鞍钢、邯钢、太钢、湘钢、沙钢等钢铁企业以及宝鸡钢管公司、渤海装备公司等多家制管企业联合开展了 21.4mm/25.7mm/30.8mm 系列壁厚的 ϕ1422mm X80 板卷、钢板和焊管的开发与试制。从单炉试制产品的性能测试和综合评价结果来看，满足 ϕ1422mm X80 产品技术条件要求，为下一步批量试制以及中俄天然气管线工程建设奠定了技术基础。以下是 ϕ1422mm X80 焊管管体和焊缝及热影响区韧性测试评价结果。

3.1　ϕ1422mm X80 焊管管体冲击韧性

图 5 为不同厂家试制的 21.4mm 螺旋埋弧焊管（HSAW）的管体冲击韧性试验统计结果，图中红色直线指示的 CVN 值为标准规定的单个试样最小值 185J。由图 5 可见，螺旋焊管管体的夏比冲击功分布在 240～480J 之间，大部分集中在 320～360J。从不同厂家试制的直缝埋弧焊管 21.4mm、25.7mm 和 30.8mm 直缝埋弧焊管（LSAW）的管体冲击韧性测试统计结果来看，直缝焊管管体的冲击韧性整体较高，最低值 240J，最高值达到 520J，以 25.7mm 后的直缝焊管为例（图 6），夏比冲击功大部分在 440～520J 之间，但与螺旋焊管相比，管体冲击韧性波动较大，这是因为各钢铁企业在钢板化学成分和轧制工艺控制上存在一定差异。试制的 ϕ1422mm X80 焊管管体冲击韧性达到技术条件中规定的止裂韧性指标要求。

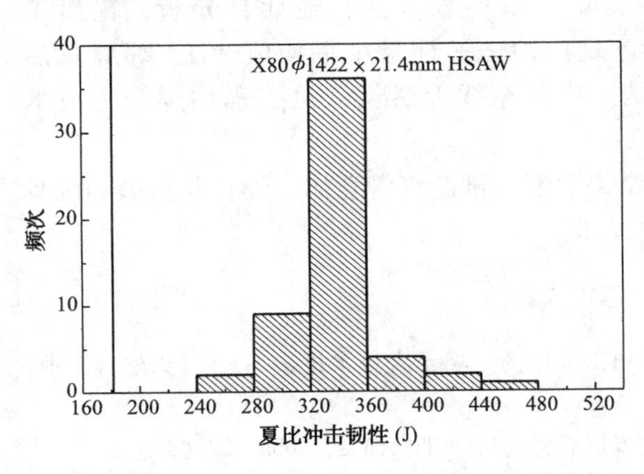

图 5　厚度为 21.4mm 的 HSAW 焊管管体 CVN 值分布

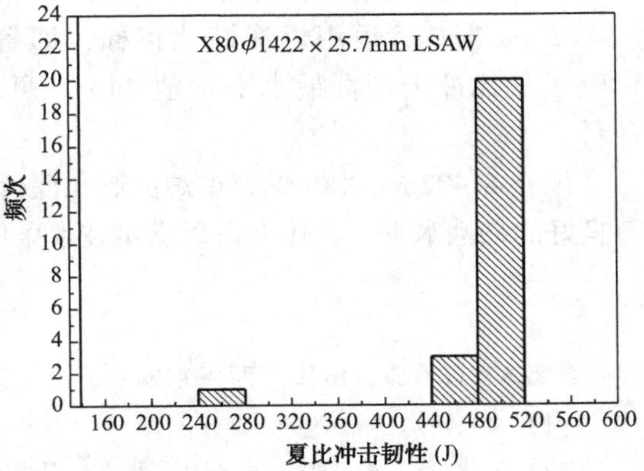

图 6　厚度为 25.7mm 的 LSAW 焊管管体 CVN 值分布

3.2　ϕ1422mm X80 焊管焊缝及热影响区冲击韧性

各厂家试制的 ϕ1422mm X80 焊管焊缝及热影响区夏比冲击韧性测试和统计结果表明，厚度为 21.4mm 的 HSAW 焊管热影响区以及厚度为 21.4mm、25.7mm 和 30.8mm 的 LSAW 焊管热影响区冲击韧性全部满足技术条件要求，HSAW 焊缝热影响区夏比冲击功在 160～360J 之间，LSAW 焊缝热影响区夏比冲击功在 160～480J 之间，作为例子，图 7 给出了厚度为21.4mm的HSAW焊缝热影响区CVN测试统计结果（图中竖线指示的CVN值为标准规定的单个试样最小值60J）；厚度为21.4mm和30.8mm的LSAW焊管焊缝冲击韧性测试结果全部满足技术条件要求，夏比冲击功在80~260J之间，个别厂家试制的厚度为21.4mm的HSAW焊管和厚度为25.7mm的LSAW焊管焊缝冲击韧性单个试样最小值低于60J（作为例子，图8给出了厚度为25.7mm的LSAW焊缝CVN测试统计结果），需要进一步优化焊接工艺，避免出现粗大的焊缝组织而造成焊缝韧性降低。整体上，除个别制管企业需要进一步优化焊接工艺提高焊缝韧性水平外，绝大多数制管企业试制的ϕ1422mm X80焊管焊缝及热影响区具有较高的冲击韧性，达到技术条件规定的指标要求。

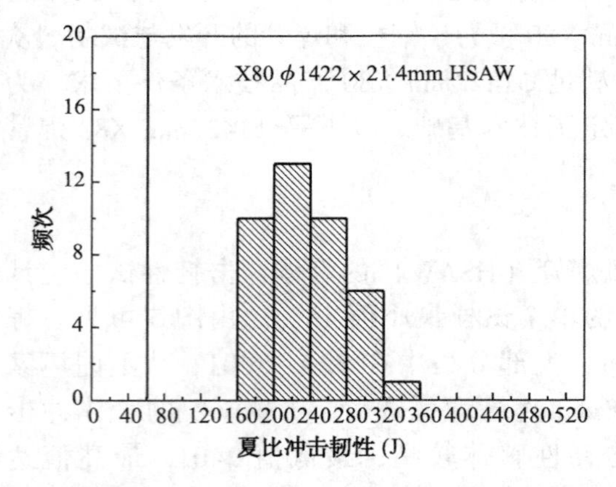

图7 厚度为21.4mm的HSAW焊缝热影响区CVN值分布

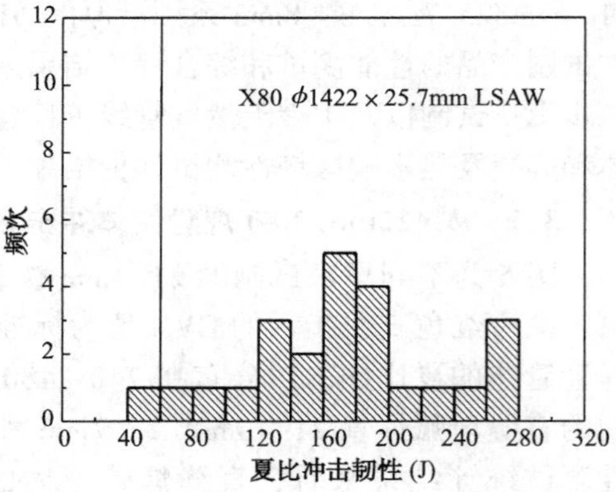

图8 厚度为25.7mm的LSAW焊缝CVN值分布

4 结论

（1）采用失效评估图（FAD）技术和Battelle双曲线方法，经计算分析，提出了ϕ1422mm X80管道用焊管韧性指标，即管体夏比冲击韧性最小平均值80J，焊缝及热影响区夏比冲击韧性最小平均值245J，并纳入中俄东线天然气管道工程用焊管的技术条件。

（2）ϕ1422mm X80焊管单炉试制和评价结果表明，试制钢管管体、焊缝及热影响区具有良好的韧性水平，整体上可以满足技术条件要求。

参考文献

[1] 李鹤林，吉玲康，田伟．西气东输一、二线管道工程中的几项重大技术进步［J］．天然气工业，2010，30（4）：84~90.

[2] 冯耀荣，陈浩，张劲军，等．中国油气管道技术发展展望［J］．油气储运，2008，27（3）：1~8.

[3] 赵新伟，李鹤林，罗金恒，等．油气管道完整性管理技术及其进展［J］．中国安全科学学报，2006，16（1）：129~135.

[4] 董绍华，韩忠晨，刘刚．管道系统完整性评估技术进展及应用对策［J］．油气储运，2014，33（2）：121~128.

[5] The American Society of Mechanical Engineers，American Petroleum Institute. API RP 579-1/ASME FFS-1 Fitness-For -Service［S］. Washington D. C：ASME and API，2007

[6] 中国石油天然气股份有限公司管道建设项目经理部．中俄东线天然气管道工程用外径1422mm X80螺

旋埋弧焊管技术条件：Q/SY GJX 147 [S]. 北京：石油工业出版社，2015.
[7] 中国石油天然气股份有限公司管道建设项目经理部. 中俄东线天然气管道工程用外径 1422mm X80 直缝埋弧焊管技术条件：Q/SY GJX 149 [S]. 北京：石油工业出版社，2015.
[8] American Petroleum Institute. API SPECIFICATION5L Specification for Line Pipe [S]. Washington D. C：API，2012.

X100 管线钢管的技术要求及研究开发

马秋荣

（石油管材及装备材料服役行为与结构安全国家重点实验室；
中国石油集团石油管工程技术研究院）

在“十二五”期间，国内管道还将保持快速增长的趋势，预计每年建设的管道长度在7000km以上。因此，进一步提高国内天然气管道的输送效率成为天然气管道工业发展的方向。

继建设西气东输一线、二线之后，西气东输三线即将开工。未来将建设多条西气东输管线，规划研究的一个重要内容是如何进一步提高单管输气能力。由于西气东输二线采用的X80钢级、ϕ1219mm，12MPa工作压力的方案只能达到$300\times10^8m^3/a$的输气能力，要将输气能力进一步提高到（450~500）$\times10^8m^3/a$，需要进一步提高输送压力和管径。采用X100钢级管线钢管是实现提升输气能力的选择之一，但X100钢管实现工程应用还有很多工作要做。

1 国际上X100钢管研究开发及应用现状

1.1 X100钢管研究开发概况

从1990年起，国际上就开始X100管线钢管开发及工程应用的研究。英国BP公司与钢铁、制管企业合作，开发了X100管线钢管，并进行了冶金、理化性能评价，可焊性评估以及钢管现场弯曲试验。为确定X100管材对钢管长程开裂的止裂能力，还进行了多次全尺寸爆破实验。欧洲钢管公司也已生产出钢板厚度为20mm的X100管线钢，用来制造口径914mm的钢管，目前还试制了25.4 mm厚的钢板。JFE也对X100的研究应用做了大量的工作。从1995年开始JFE就着手X100钢管的试制工作，到2006年为止，他们已经生产试制了3300t X100管线钢管，做了包括力学性能、焊接、涂覆、爆破、止裂器应用等在内的大量研究。

1.2 X100管线钢的止裂性能

CSM和Advantica等已进行了多次X100级钢管全尺寸爆破试验。通过这些昂贵的试验，CSM逐步建立了X100级钢管全尺寸爆破试验的数据库并取得一定的成果。

1.3 X100试验段应用

2000年前后，钢管开发水平得到了很大的提高，在全尺寸气体爆破实验之外，开始小规模建设试验段和试运行。这些试验段的建设旨在展示X100的经济效益、技术可靠性以及建设超高压输送管道的可行性。

国外试验段/示范段建设的目的主要是获取X100的制造、试验和建设（焊接和现场冷弯）经验，评估极地寒冷环境施工技术（焊接和冷弯），研究钢管拉伸和压缩应变行为、基于应变的设计方法，进一步增大规模，证明X100钢管的适用性，检验X100钢管抗第三方损伤、疲劳、应力腐蚀等方面的服役性能。

X100试验段大多是在现有管线上铺设，一般不超过5km长，虽然已经成功建设了多条X100试验段。但是迄今为止所有的管道试验段都是在低应力系数水平下运行，没有在真正X100的设计应力工况下运行。建设试验段的目的集中在管道的设计、X100管线钢管组织生产、施工技术的考核和改进上。对钢管强度、韧性和可靠性的考核主要依靠实验室试验和试验场试验。

2 X100钢管技术要求

ANSI/API SPECOFOCATON 5L 44版/ISO3183：2007（Modified）标准规定了X100需要满足的基本理化性能（表1和表2）。

表1 ANSI/API SPECOFOCATON 5L 44版/ISO3183：2007
标准规定的产品化学成分要求

成分	C	Si	Mn	P	S
质量分数（%）	≤0.1	≤0.55	≤2.1	≤0.02	≤0.01

表2 ANSI/API SPECOFOCATON 5L 44版/ISO3183：2007
标准规定的拉伸性能要求

屈服强度（MPa）		抗拉强度（MPa）		屈强比最大值
min	max	min	max	
690	840	760	990	0.97

ANSI/API SPECIFICATION 5L 44版/ISO 3183：2007（Modified）附录G规定了适用于CVN冲击试验的PSL2钢管和订购用于输气管线管体抗延性断裂扩展PSL2钢管的补充条款，同时也为确定钢管延性断裂止裂CVN冲击功值提供了指南。

表3是附录G中提供的5种方法和其应用范围的说明，包括方法1：Battelle简化公式；方法2：Battelle双曲线法；方法3：AISI法；方法4：EPRG（欧洲钢管研究机构）准则；方法5：全尺寸爆破试验。目前只有方法5适用于超高强度钢管（X90及其以上钢级）。该方法建立在全尺寸爆破试验的基础上，对特定设计与输送流体的管线止裂韧性进行验证。目前，这是唯一在超高强度管线钢管如X100或X120上进行试验的方法。需要更多的实验数据结合理论方法进行分析。

表3 ANSI/API SPECIFICATION 5L 44版/ISO 3183：2007标准
推荐的5种止裂韧性预测方法和应用范围

序号	止裂预测方法	适用范围			
		钢级	输送压力（MPa）	管径 D、壁厚 t（mm）	介质
1	Battelle简化公式	≤X80	≤7.0	$40<D/t<115$	单相气体
2	Battelle双曲线模型	≤X80	≤12.0	$40<D/t<115$	/

续表

序号	止裂预测方法	适用范围			
		钢级	输送压力（MPa）	管径 D、壁厚 t（mm）	介质
3	AISI 公式	≤X70	/	$D\leqslant1219$；$t\leqslant18.3$	单相气体
4	EPRG 指南	/	≤8.0	$D<1430$；$t<25.4$	单相气体
5	全尺寸爆破试验	/	/	/	/

ANSI/API SPECOFOCATON 5L 44 版/ISO3183：2007（Modified）标准规定了 X100 需要满足的基本性能，与真正应用相比还有相当大的差距，因此需要根据实际工程情况制定各种补充技术条件，确定各种技术指标，如拉伸试样类型及尺寸、屈强比、CVN 冲击功值、DWTT 等。

3 X100 管线钢管开发研究的主要问题

3.1 X100 管线钢的断裂控制

延性裂纹的长程扩展是天然气管道失效后果最严重的失效模式之一，往往会造成灾难性的后果，尤其随着输送压力和管线钢强度的升高，风险也变得越来越大。因此，高压输气管道的动态延性断裂控制和止裂预测成为了关系管道安全的最重要的问题。现有的确定钢管延性断裂止裂 CVN 冲击功值的 5 种方法，目前只有全尺寸实物爆破试验法适用于超高强度钢管。

X100 管线钢管现有实物爆破试验数量少、数据分散，不能有效支持止裂韧性预测值（或修正值），按目前实物爆破实验数据（图 1），X100 管线钢管止裂韧性应达到 Battelle 预测值的 2.4 倍。因此 X100 管线钢管的自身止裂问题一直存在争论，一般当设计系数较低时，基本可实现自身止裂，当服役条件很苛刻时（富气、高利用系数、低温）则需要使用外部止裂器。

但全尺寸爆破试验的成本相当昂贵，而且每次试验获得数据较少，试验只针对特殊的管道运行和环境条件，试验结果的普适性较差。为了促进 X100 钢管的工程应用，有必要针对 X100 管线钢展开深入的动态延性断裂研究并开发出新的适用于 X100 管线钢的止裂韧性预测方法。同时，应进行大量全尺寸气体爆破试验，来验证新的止裂预测模型的有效性和 X100 管线钢止裂预测值的准确性。并形成 X100 管线钢全尺寸气体爆破试验数据库，确定合适的修正系数。外部止裂器的结构设计、止裂能力验证、材料优选等都是需要亟待解决的问题。

3.2 高屈强比对稳定塑性变形能力的影响

TMCP 工艺适合批量生产高强度管线钢，但同时屈强比也会随着强度的增加而变大。X100 管线钢管的屈强比可能超过 0.97，因而稳定塑性变形的能力将比低钢级的钢管小。为兼顾经济效益和生产率，钢厂必须优化工艺，生产出低屈强比的高强度钢，增强钢管的塑性变形能力，适应现场冷弯和 100%SMYS 的水压试验的要求。国外通过增加钢管抗变形能力的要求，对应力应变行为做更严格的要求，通过优化 X100 管线钢冶炼及轧制工艺，已经生

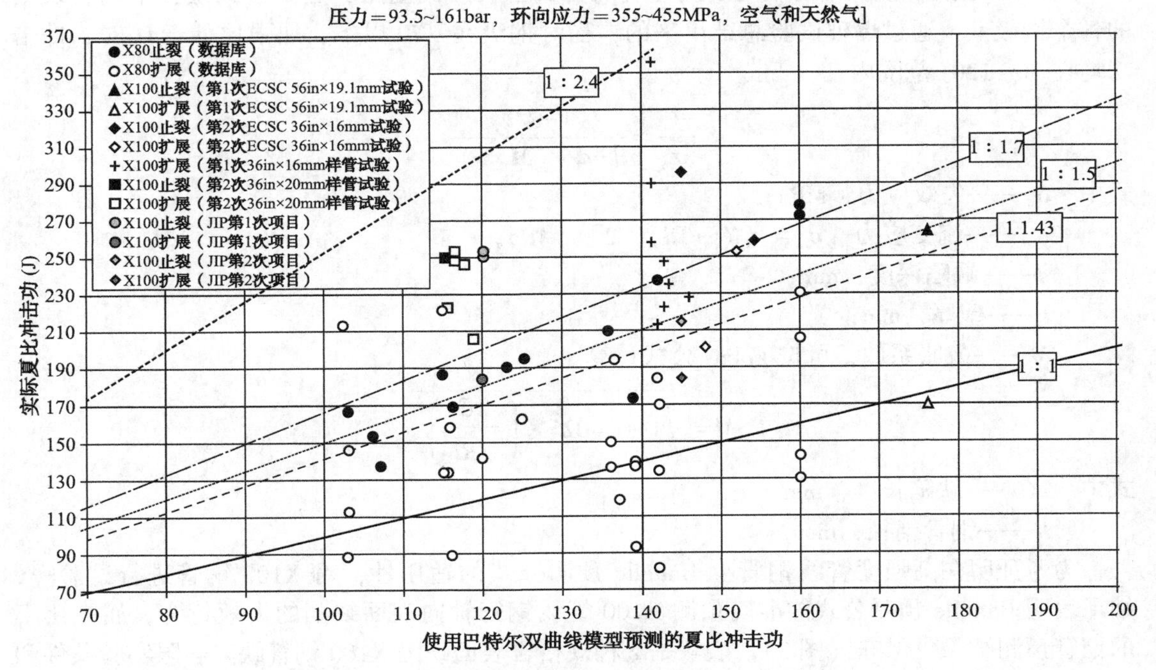

图 1　现有的部分全尺寸实物气体爆破实验结果

产出屈强比不超过 0. 95 的 X100 管线钢管。此外，X100 管线钢管时效试验发现温度会对 X100 的性能产生影响，特别是，经过温度时效后的 X100 钢管的屈强比会显著提高。如何在钢管防腐涂敷过程中对应变时效效应的控制并制定出合理的涂敷方案应引起足够的重视。

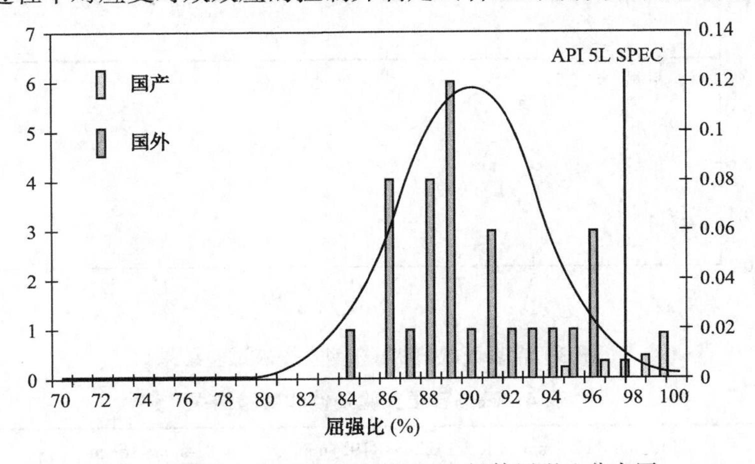

图 2　国产 X100 钢管与国外 X100 钢管屈强比分布图

3. 3　各向异性对钢管变形能力的影响

X100 钢级的各向异性明显，尤其是 UOE 管，横向和纵向的屈服强度差在 100MPa 左右；环向不同位置取样的屈服强度也相差 80 ~ 90MPa，根据国外的研究，当管道变形量较大时各向异性对管道的应力应变行为及变形能力有一定的影响。

3.4 缺陷容限的要求

Battelle 建立的预测钢管轴向裂纹开裂的公式已被广泛应用，这个公式是在早期管线钢的各个钢级上，通过爆破试验总结出来的，当时钢级在 X70 以下，屈强比低于 0.87，冲击功均在 30~120J 范围内。

$$\frac{\sigma_f}{\sigma_o}=\frac{1-d/t}{1-d/(M\cdot t)}$$

式中 σ_f——失效应力，MPa；

σ_o——流变应力 [σ_o=（YS+TS）/2]，MPa；

d——缺陷深度，mm；

t——壁厚，mm；

M——鼓胀系数，可按 Folias 公式计算。

$$M=\sqrt{1+0.4025\cdot\left(\frac{2C}{\sqrt{Rt}}\right)^2}$$

式中 $2C$——缺陷长度，mm；

R——钢管半径，mm。

为了证明在高强度管线钢管上 Battelle 预测公式的适用性，对 X100 钢管进行试验[1]，结果表明 Battelle 预测公式能准确预测 X100 管线钢管轴向表面缺陷的失效行为，屈强比小的钢管预测结果更保守（图 3）。表 4 是 2 种规格含表面缺陷 X100 钢管缺陷容限实验条件和实验结果。

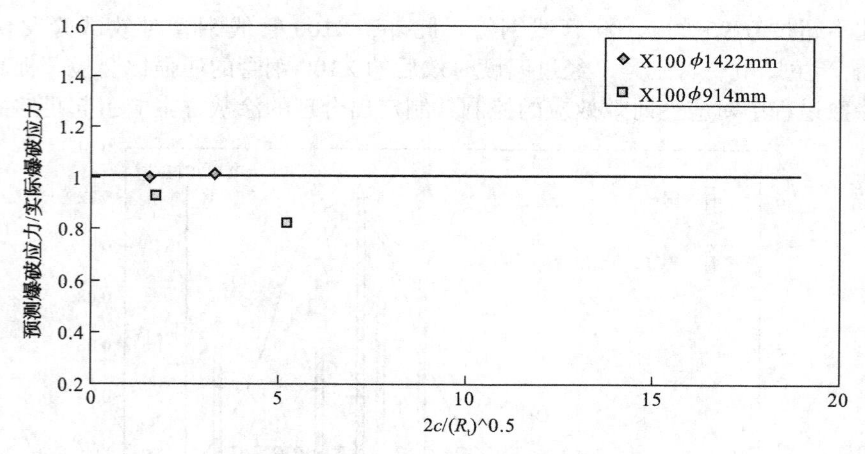

图 3　含表面缺陷 X100 钢管爆破试验结果

表 4　缺陷容限实验条件和实验结果

钢管规格	1422mm×19.1mm	1422mm×19.1mm	914mm×16mm	914mm×16mm
管号	846077	846014	99457	99457
实际壁厚（mm）	19.25	20.1	16.4	16.4
屈服强度（MPa）	740	795	739	739
抗拉强度（MPa）	774	840	813	813
CVN（J）	261	171	253	253

续表

钢管规格	1422mm×19.1mm	1422mm×19.1mm	914mm×16mm	914mm×16mm
屈强比	0.96	0.95	0.91	0.91
缺陷长度 2c（mm）	180	385	150	450
缺陷深度（mm）	10.4	3.8	9	6
缺陷深度比	0.54	0.19	0.55	0.37
计算爆破应力（MPa）	568	723	555	551
实验爆破压力（MPa）	153.5	201.2	214	240.2
实验爆破应力（MPa）	567	712	597	670
爆破后最大环向应变	0.30%	0.30%	0.30%	0.30%

3.5 X100 管线钢管韧脆转变行为

1970 年代进行的大量的实验表明用 DWTT 试验得出的试样剪切面积和韧脆转变温度与全尺寸管线钢管的行为之间有良好的一致性。裂纹扩展的韧脆转变温度可以用 DWTT 试样 85%剪切面积对应的温度来表示，这个要求可以保证管线钢管不发生脆性断裂。

图 4 是 ϕ1422mm×19.1mm X100 管线钢管 DWTT 试验、CVN 试验和 WJ 爆破试验结果的对比[2]，表明用 Battelle 总结的经验方法仍然可以预测 X100 管线钢管的人脆断裂转变行为，尽管因为 TMCP 工艺生产的管线钢上存在的断口分离对断裂面积的测量带来了一定的困难。图 5 是-20℃下 DWTT 试样断口形貌与 West Jefferson 试验钢管端口形貌的图片，这些都表明用 DWTT 试验和 Battelle 总结的 85%SA 对应温度经验方法能准确预测 X100 钢管的韧脆转变温度和断裂行为。

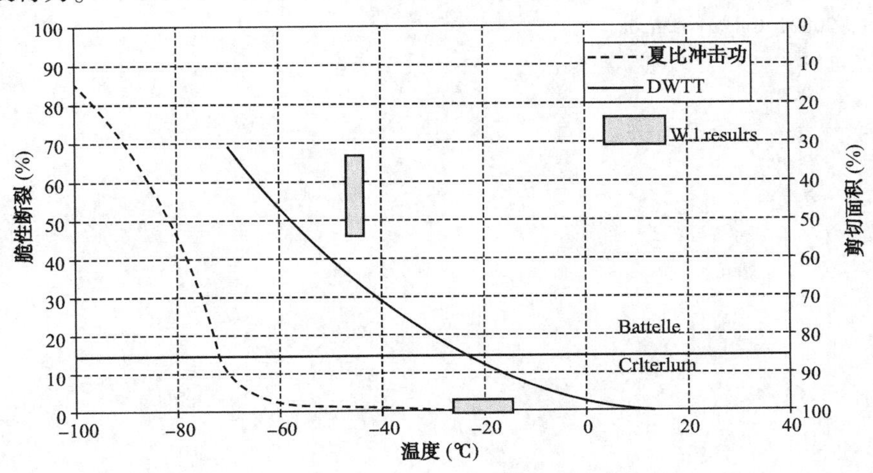

图 4　X100 管线钢管 DWTT 试验、CVN 试验和 WJ 爆破试验结果的对比

4　结论

新一轮干线天然气管线建设将向更大的输送能力发展，单管输送能力将达到（450~500）×$10^8m^3/a$ 的水平。采用更高工作压力和大管径输送是必然的选择。在已有的 X100 管线钢及钢管的研究、开发及应用的基础上，围绕 X100 输气管道的设计方案、X100 管材关键技术指标及技术标准、断裂控制、X100 管道整体服役性能等多方面展开研究，突破超高强

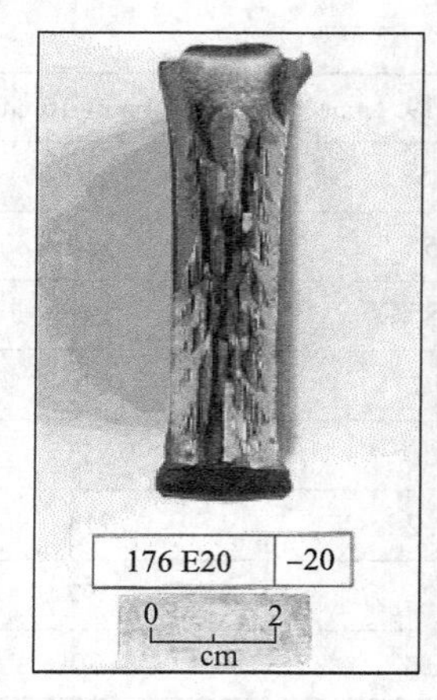

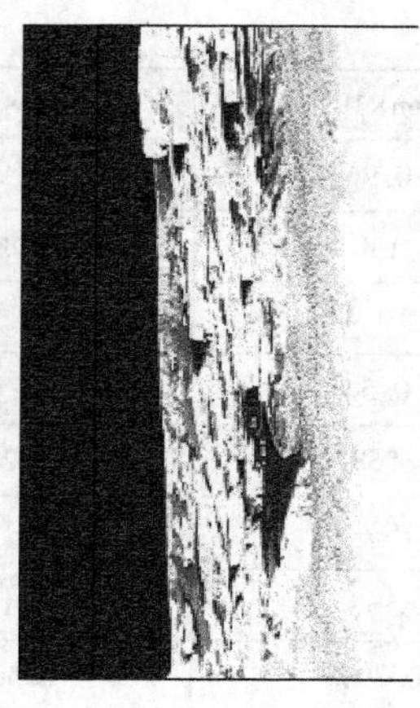

图 5　DWTT 试样断口形貌与 West Jefferson 试验钢管端口形貌对比（-20℃）

度高压输气管道延性动态断裂及止裂预测的瓶颈，形成符合工程需求的 X100 管道延性断裂预测及控制技术；对 X100 管材进行全面的组织、性能和服役行为等方面的全面研究，形成 X100 管材关键性能指标及综合评价技术；通过成型工艺、焊接工艺、焊材开发等研究，形成 X100 管制造技术；最终实现在高性能 X100 高压管线钢管应用方面的突破。

参 考 文 献

［1］Mannucci G，et al. Fracture properties of API X100 gas pipeline steels［J］. Mathematische Annalen，2002，98（1）：422-464.

［2］Barsanti L，et al. Possible use of new materials for high pressure linepipe construction：the experience of snam rete gas and europipe on X100 grade steel［C］//Proceeding of IPC：The international pipeline conference，september，2002，Calgary Canada.

Study on Strain-hardening Properties of High Grade Line Pipes

Ji Lingkang Feng Hui Wang Haitao Zhang Jiming Chen Hongyuan

(Tubular Goods Research Institute of CNPC;
State Key Laboratory for Performance and Structural Safety of Oil Industry Equipment Materials;
Key Laboratory for Oil Tube Engineering of CNPC)

Abstract: The strain-hardening performance and characteristics of pipeline steel material have important influence on the deformation behavior and arrest behavior of the line pipe. In this paper, X70, X80 and X90 longitudinal and spiral SAW pipes with different microstructure characteristics were selected, and the longitudinal and transverse tensile curve and strain - hardening characteristics are analyzed. The influence of line pipe grade, anisotropy, microstructure, and pipe forming method on strain-hardening exponent and tensile curve are stated, and the main factors affecting the strain-hardening exponent were expounded from the macroscopic and microscopic view.

Keywords: Line pipe; Strain - hardening; Grade; Microstructure; Specimen orientation; Forming method; Dual phase microstructure

1 Introduction

Natural gas and coal gas resources in China are mainly concentrated in the western region such as Tarim, Xinjiang field, and the importing natural gas resources are also via the western region such as the gas from middle Asia countries and Siberia region in Russia. All of them need to be transported through pipelines. So natural gas pipeline in our country shall trend to large throughput, high pressure and long-distance[1], resulting in the 1st and 2nd West-East natural gas pipeline more than 8000km built with X80 line pipe and operated with 12MPa pressure. The 3rd West-East natural gas pipeline with the same pipe grade and pressure are being built now. Since 1990, the research and development of high grade line pipe in the world have been rapidly developed. In China, it was characterized by short period for research and development and fast speed for application[2]. In order to ensure the pipeline safety and reliability, for high grade line pipe (especially above X80 grade), the balance of toughness and plasticity needs to be paid more attention in addition to improving strength. In recent years, with the continuous development of pipeline construction technology, on the basis of the traditional Stress-Based-Design, SBD (Strain-Based-Design) technology has been also widely used and developed[3].

The mechanical characteristics, especially the strain – hardening behavior, has an important influence on the behavior of ductile fracture crack arrest and plastic deformation capacity for high strength line pipe. A large number of studies have suggested that the hardening behavior of the pipeline steel with different microstructure characteristics also have obvious differences[4, 5]. For X70 and X80 pipeline steel, with the increasing of strain hardening exponent, yield ratio (Yield ratio is the ratio between yield strength and ultimate strength) decreases, and the critical buckling strain also increases obviously[6]. Meanwhile, the flow stress–strain curve has a close relationship with its ductile fracture crack arrest behavior for high grade pipeline steel[7]. So one of the important problems we need to pay more attention on is the main factors affecting the strain–hardening exponent and the influence rule. The research relating to the strain hardening behavior will provide important technical support for the development and application of high grade line pipe.

2 Test Materials

The test pipes are X70, X80 and X90 line pipe with different microstructure used in different pipeline projects, and all of them are low–carbon micro–alloyed controlled rolling and controlled cooling line pipes. The dimensions and chemical components are shown in Table 1. No. 1 – No. 6 are LSAW pipes with different microstructures manufactured by longitudinal submerged arc weld, and No. 7 is SSAW pipe with microstructure of bainite and a small amount of ferrite manufactured by spiral submerged arc weld. Figure 1 is the optical microstructure of the test pipes.

Table 1 Dimension and chemical component of test pipes

No.	Grade–microstructure*	Pipe type	Dimension		Element (wt%)						
			D (mm)	t (mm)	C	Si	Mn	Mo	Ni	Cu	Nb+V+Ti
1	X70–D	LSAW	1016	21.0	0.052	0.19	1.61	0.10	0.19	0.13	0.11
2	X70–N	LSAW	1016	21.0	0.061	0.23	1.61	0.12	0.17	0.16	0.12
3	X80–D	LSAW	1219	22.0	0.062	0.15	1.82	0.23	0.25	0.25	0.04
4	X80–N	LSAW	1219	22.0	0.044	0.22	1.77	0.001	0.19	0.21	0.10
5	X90–D	LSAW	1219	16.3	0.061	0.25	1.90	0.24	0.21	0.19	0.11
6	X90–N	LSAW	1219	16.3	0.053	0.18	1.82	0.14	0.21	0.18	0.09
7	X80–S	SSAW	1422	21.4	0.042	0.21	1.81	0.14	0.27	0.22	0.12

* Note: D–dual phase (bainite and ferrite), LSAW

N–single phase (bainite), LSAW

S–single phase (bainite and small amount of ferrite), SSAW

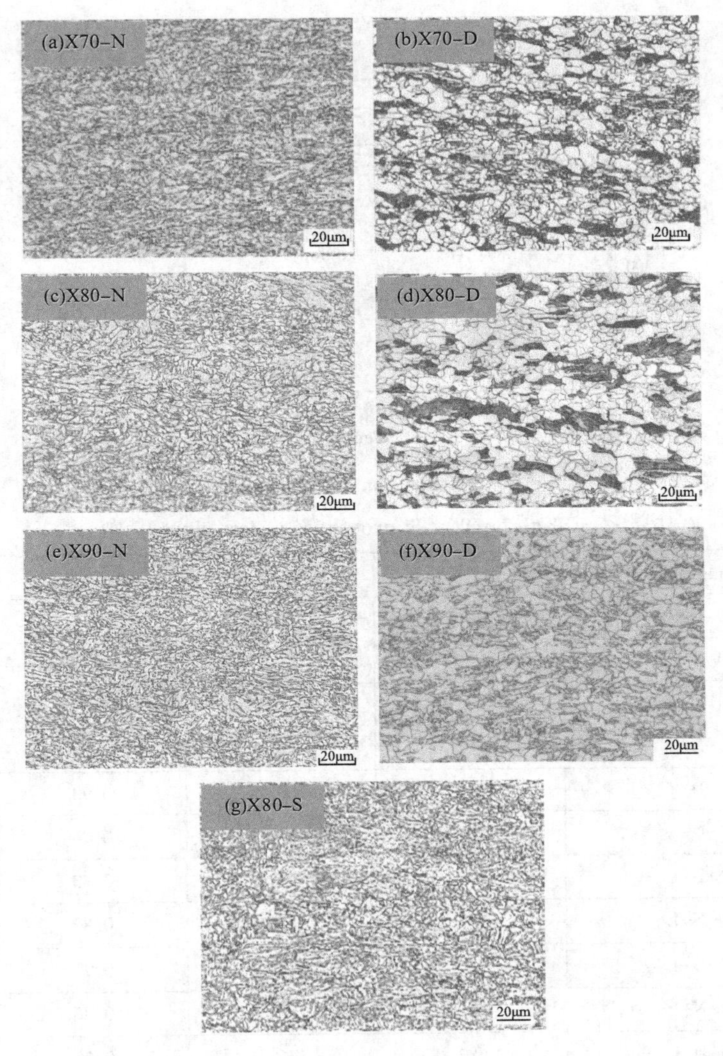

Fig. 1 Microstructure of test pipes

3 Research Methods

3.1 Tensile Tests

Round bar tensile specimens of the pipe metal are sampled at the position 90° from the weld, in both transverse and longitudinal directions. The diameters of the specimens are 12.7mm and the gage length is 50.8mm. The test is load-controlled and executed on a MTS-810 (25t) with loading velocity of 0.03m/s. The ultimate strength (R_m), yield strength ($R_{t0.5}$), total elongation (A), and engineering stress-strain curve are measured experimentally. The test results are summarized in Table 2. Figure 2 shows stress-strain curves for all specimens.

For LSAW pipe with the similar microstructure, along with grade increasing, yield ratio and elongation decrease gradually. The plastic indexes of dual phase are better than that of single phase. The strength in transverse direction is higher than that in longitudinal, but the plasticity is just the opposite.

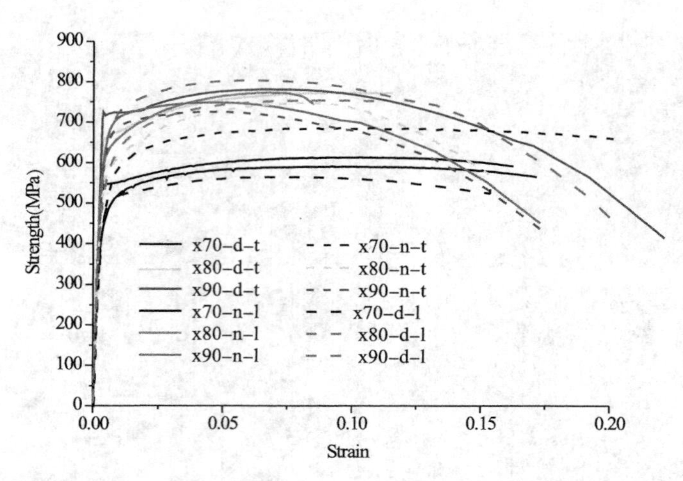

Fig. 2 Stress-strain curver of test pipes

Table2 Tensile properties of test pipes

No.	Grade-structure-direction *	$R_{t0.5}$ (MPa)	R_m (MPa)	$R_{t0.5}/R_m$	A (%)
1	X70-D-T	496	685	0. 72	28. 0
2	X70-N-T	550	690	0. 82	26. 5
3	X80-D-T	577	703	0. 82	27. 0
4	X80-N-T	595	680	0. 88	28. 0
5	X90-D-T	645	691	0. 93	23. 5
6	X90-N-T	763	812	0. 94	22. 0
7	X70-D-L	494	675	0. 73	24. 0
8	X70-N-L	535	650	0. 82	24. 0
9	X80-D-L	557	719	0. 77	26. 5
10	X80-N-L	562	655	0. 86	26. 0
11	X90-D-L	687	769	0. 89	21. 5
12	X90-N-L	636	773	0. 82	21. 0
13	X80-S-T	612	678	0. 90	26. 0
14	X80-S-L	623	695	0. 89	25. 0

* Note: L-longitudinal T-transverse

3. 2 Analysis method of strain-hardening exponent (n)

Strain hardening exponent can be obtained by processing the engineering stress - strain curve[5, 8]. According to engineering requirements, only a section from the yield strength to the tensile strength in the engineering stress strain curve is to be considered. The processing steps are as follows:

According to the following formulas, the true stress - strain curve can be obtained from the engineering stress-strain curve.

$$S = \sigma(1 + \varepsilon) \tag{1}$$

$$e = \ln(1 + \varepsilon) \tag{2}$$

where, S is true stress, e is true strain, σ is engineering stress, and ε is engineering strain. According to the established Hollomon classical equation. $S = Ke^n$,

$$n = \frac{d\ln S}{d\ln e} \tag{3}$$

The strain-hardening exponent (n) is calculated from the slope of the $\ln S$-$\ln e$ curve. Then the n-$\ln e$ curve which will be analyzed can be obtained.

4 Test Result and Analysis

4.1 Strain hardening exponent for different grade line pipe

Figure 3 shows the change of strain hardening exponent along with strain for X70, X80 and X90 line pipe with single phase microstructure. It can be seen that, when strain is smaller than 2% (lne is smaller than −4.0), strain hardening exponent decreases with strain increasing, and the decreasing speed is smaller and smaller gradually. The exponent tends to be stable inthe range of 0.05 ~ 0.10 in the end.

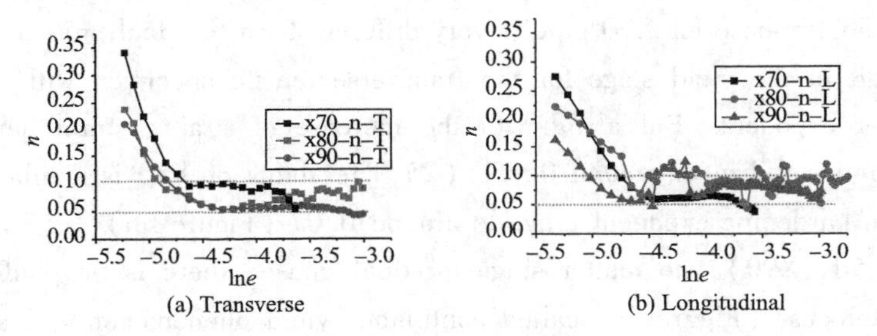

Fig. 3 Strain hardening exponent for different grade line pipe

In terms of comparison of these three grade pipeline steel, with the increase of grade, strain hardening exponent decreased gradually, which illustrates that the effect of strength on strain hardening ability is very obvious. This phenomenon is mainly related to the pipeline steel microstructure type, phase proportion, dislocation density, grain size and so on.

No matter transverse or longitudinal sample, when the engineering strain is below 1% (i. e., lne value is 4.5), the phenomenon is more outstanding. When strain is above 1%, hardening exponent for all the three grade steel gradually tends to converge and fluctuate in the same scope, and the influence of the steel grade is not so obvious, sometimes the hardening exponent for higher grade steel is slightly higher than that for low grade steel, which may be related to the composition of the microstructure.

4.2 Strain hardening exponent for different specimen orientation

Figure 4 shows the change of strain hardening exponent along with strain for X90 and X80 line pipe with single phase or dual phase microstructure. It can be seen that, (1) no matter X90 or X80, strain hardening exponent of longitudinal specimen is higher than that of transverse specimen when the strain is small, especially when engineering strain is less than 1% (Figure 4a and 4b show the single phase n-lne curve). This is because, the pre-strain in the process of pipe forming and expanding improves the yield strength and yield ratio, and consumes the plastic deformation ability of the pipe at the same time, which reduces the strain hardening ability of circumferential.

(2) Figure 4c and 4d show the strain hardening exponent changing curves of dual phase microstructure in longitudinal and transvers direction for X90 and X80 pipe. It can be seen that the transvers n-lne curve of dual phase is also different from longitudinal curve which is similar with the curve of the single phase specimen. Comparing with single phase, for dual phase pipe, longitudinal hardening ability is much higher than transverse's. Especially in the small strain range, transverse deformation hardening exponent is even less than 0.05. Only when the engineering strain is achieved to 1% (lne value is −4.5), strain hardening exponent was increased and stable in the range of 0.05 to 0.10 gradually.

4.3 Strain hardening exponent for different microstructure

Figure 5 shows the change of strain hardening exponent along with strain for single phase and dual phase microstructure in transvers and longitudinal direction. (1) For transverse specimen of X90 pipe with dual phase (Figure 5a), there is obvious yield platform appeared in the small strain stage (Figure 5c), resulting in the low strain hardening exponent, but in the later deforming process, it still has a strong ability of deformation (strain hardening exponent of 0.1 or so). On the contrary, the single phase for X90 pipe is very different form the dual phase. There is no yield platform at the start of the yield stage for the transverse tensile specimen with single phase and leading to a higher exponent. But along with the increase of strain, strain hardening exponent gradually reduce and stabilize to around 0.05. (2) The status of X80 is similar to X90, with a higher stable strain hardening exponent range to around 0.07 (Figure 5b). (3) For longitudinal specimen (Figure 5d, X90), no matter single or dual phase, there is no platform phenomenon during the tensile process (Figure 5c, called continuous yield phenomenon), leading to higher n value for dual phase. The exponent of dual phase below the 2% engineering strain is far higher than that of single phase.

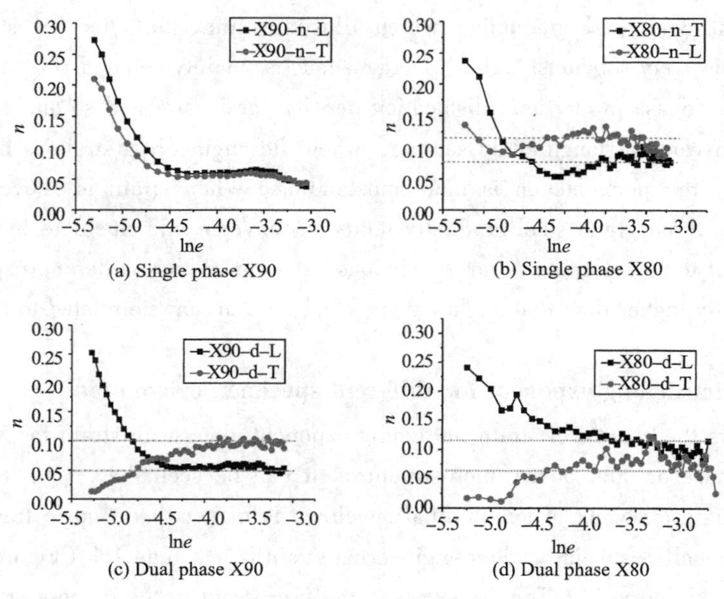

(a) Single phase X90

(b) Single phase X80

(c) Dual phase X90

(d) Dual phase X80

Fig. 4 Strain hardening exponent for different specimen orientation

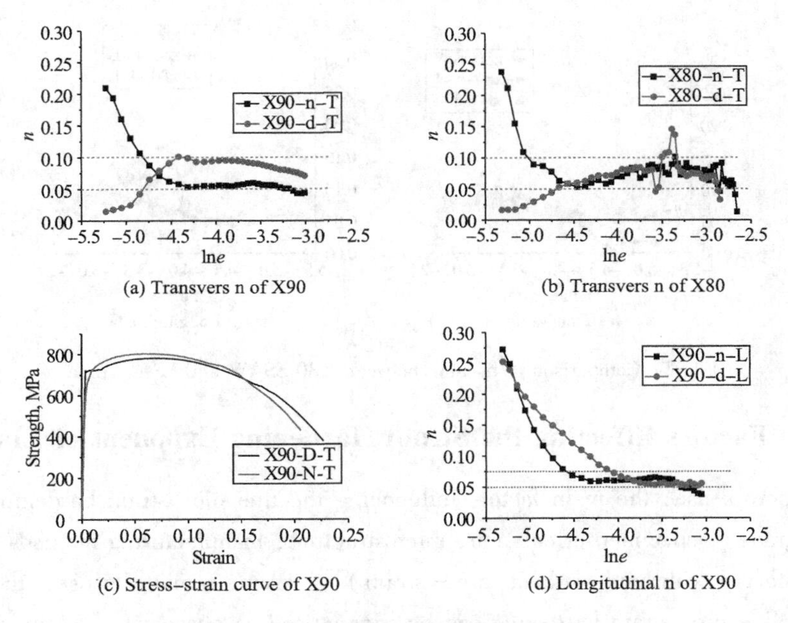

Fig. 5 Strain hardening for different microstructure

4.4 Strain hardening exponent for different forming method

Figure 6 shows strain hardening for spiral submerged arc welding pipe (SSAW). (1) There is no much difference between transverse and longitudinal specimens for SSAW pipe, with the strain hardening exponent ranging from 0.05 to 0.10. This may be due to the angle between the specimen direction and the coil rolling direction. (2) There is the yield platform tendency for the stress-strain curve of SSAW pipe, so the strain hardening exponent is low initial deforming stage.

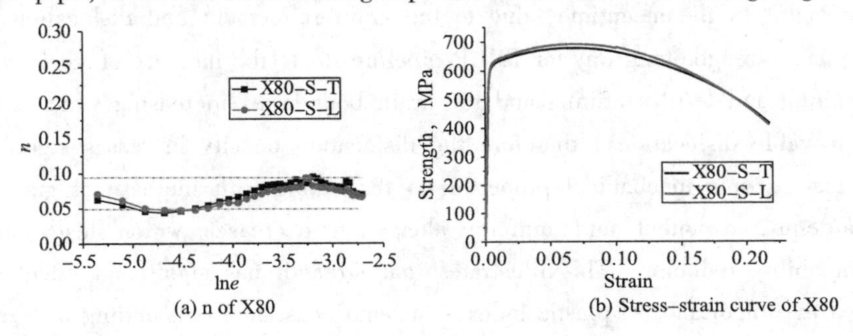

Fig. 6 Strain hardening for SSAW pipe

Figure 7 shows the comparison of strain hardening exponent between SSAW and LSAW pipe, in transverse (i.e. circumferential direction) and longitudinal direction separately. It can be seen that, for transvers specimen (figure 7a), LSAW with single phase has the best strain hardening ability in the small strain range, and when the strain increases to a certain value, the exponents for all pipes trend to a stable value no matter LSAW or SSAW pipe. But longitudinal specimen is different (figure 7b). The exponent of LSAW pipe with dual phase is higher than that of LSAW pipe with single phase or SSAW pipe. Even arrived at large strain, the longitudinal deformability of LSAW pipe can maintain at a high level (The hardening exponent is above 0.1).

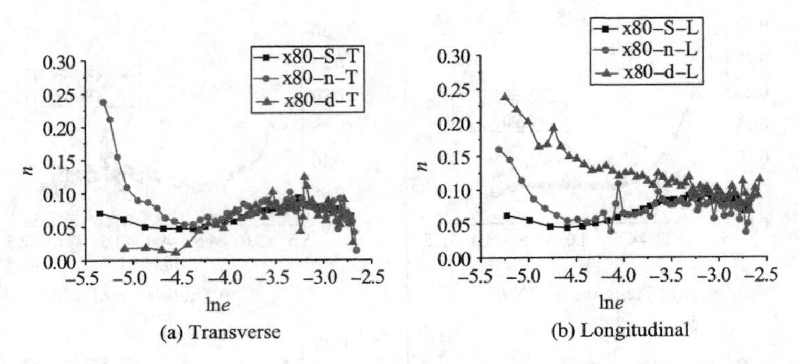

Fig. 7 Comparison of n value between X80 SSAW and LSAW pipe

5 Main Factors Effecting the Strain Hardening Exponent of Line Pipe

From a macro sense, the main factors influencing the line pipe strain hardening are primarily the strength (grade), specimen orientation, microstructure, manufacturing methods of pipe and its deformation history (with and without pre - strain). From a micro sense, the main factors influencing the line pipe strain hardening can be summarized as phase ratio, grain size, dislocation density and the degree of its pile-up and tangling, number of solid solution atoms, precipitation, texture, etc.

5.1 Strength

Generally, with the increase of steel grade, the proportion of massive ferrite in the microstructure, which is soft phase and the important guarantee of high hardening ability of pipeline steel, will gradually decrease, thus the increase of strength can cause the decrease of strain hardening exponent. In the meantime, due to the grain refinement and dislocation strengthening which are the main strengthening way for TMCP pipeline steel, the increase of steel grade will cause the grain of bainite and ferrite refining and the grain boundaries increasing which will hinder the movement of movable dislocations, therefore the dislocation density increases significantly. Main show on the macroscopic mechanical properties is that along with increase of grade, yield ratio increases, hardening exponent and uniform elongation decreases which leads to the plastic deformation capability reducing. This illustrates that strength has significant effect on hardening capability, and the importance of plastic indexes to security is more outstanding with the increase of steel grade, therefore the balance of strength, toughness and plasticity needs to be paid more attention during the application of high grade line pipe.

5.2 Specimen orientation

Strain hardening exponent is closely related to the tensile specimen orientation of pipe. The using of longitudinal and transverse specimen in related standards is different. Transverse specimen mainly describes the circumferential direction mechanical performance of line pipe and ensures its loading ability. Longitudinal specimen mainly describes the axial direction mechanical performance of line pipe and ensures its deformation ability in axial direction. In general, the strain hardening capability in axial direction is far higher than that in circumferential direction for LSAW pipe. The main reason can be attributed to the anisotropy produced during the rolling process of line pipe steel, and the larger plastic strain produced in the process of pipe forming, expansion, hydrostatic test, etc.. Additionally, strain aging phenomenon due to the characteristic of pipeline steel is also the important reason of obvious difference between pipe longitudinal and transverse direction. Because of the angle between the rolling direction and the pipe axial direction, the mechanical behavior of SSAW pipe is different from LSAW pipe.

5.3 Microstructure

The effect of microstructure on pipe strain hardening ability is also obvious. For the kind of pipeline steel with acicular ferrite microstructure i. e. X70, X80, X90, in general, the plastic deformation capability of dual phase with bainite and ferrite microstructure is better than that of single phase with bainite, and the dual phase has higher hardening exponent than the single phase. This is because that in dual phase microstructure, ferrite as the soft phase yields at low stress level and deform firstly, and with the grain internal dislocation multiplication, pile-up and tangling, the deformation resistance increases, showing the higher ability of strain hardening. While in single phase, bainite starts to yield at higher stress level, and its hardening ability is lower than that of dual phase. This point of view is validated well in the longitudinal tensile test of X90 line pipe (figure 5d).

But, transverse tensile specimens do not comply with the law. The transverse stress-strain curve of dual phase appears obvious platform while the single phase S-S curve characterized by the round-house for single phase (chapter 4.3).

The reason is investigated. On the one hand, for dual phase microstructure, the polygonal ferrite with low density dislocation is the important influence factor of yield platform phenomenon appearing in the transverse tensile curve. During the process of plastic deformation, polygonal ferrite firstly deforms, leading to the dislocation multiplication, pile-up and tangle which can make the deformation resistance increase sharply. Meanwhile, some dislocations annihilate, and some pinned dislocation depinning due to the stress, which can make the stress relaxes and the deformation resistance decrease. When the role of these two aspects approximately equal, the S-S curve is characterized by the yield platform on the macro[4] and the hardening exponent is maintained at a lower level. Along with the increase of strain, bainite grains participate into the deforming process, and the dislocation multiplication rate is higher than the rate of annihilation and depinning, which induces the phenomenon of stress increasing continuously and higher strain hardening rate than the yield stage. This is the same as the yield platform phenomenon of S-S curve for the pipeline steel with microstructure of pearlite + ferrite. Due to lack of ferrite phase with low density dislocations, the single phase bainite microstructure does not appear the platform phenomenon in general case.

On the other hand, during the process of manufacture, the pipe bears a large strain along circumferential direction, and the similar strain aging phenomenon appears subsequently. In micro sense, after the deforming in transverse of pipe, lattice appeared slip layer and distortion, and the dissolution ability of solid solution atoms drops which leads to presence the a saturated or oversaturated situation. After a period of aging, the solute atoms precipitate, pinning effect is more apparent which causes the aging phenomenon[9] presenting as platform in transverse tensile curve. Due to the low yield stress for dual phase, the pinning effect of solute atoms presents more fully, therefore its strain aging phenomenon is more obvious than single phase. Additionally, because there is no greater pre-strain conducted on longitudinal specimen, strain aging phenomenon is weak, and without effective precipitation of solid solute atoms, the pinning effect is not obvious, which results in normal dislocation multiplication in microstructure. Therefore, the yield platform is not easy to appear for longitudinal specimen.

5.4 Pipe Forming Method

The pipe forming method is an important factor effecting strain hardening ability. From chapter 4.4, it can be seen that there is significant difference between stress-strain behavior of SSAW pipe and that of LSAW pipe. For longitudinal direction, the strain hardening exponent of dual phase LSAW pipe is obviously higher than that of SSAW pipe. As the raw material of SSAW pipe, the coil is obtained by continuous hot rolling production process in which there is no sufficient relaxation time, so it is not easy to get more polygonal ferrite, and its microstructure is composed by bainite and a small amount of massive ferrite. In the meantime, it is more important that there is a certain angle between the specimen in longitudinal or transverse direction and the coil rolling direction, while the texture in pipeline steel can make a strong strain hardening ability in rolling direction. In this point, SSAW pipe is very different from LSAW pipe, especially the longitudinal hardening exponent of SSAW pipe is obviously lower than that of LSAW pipe, which induces the decrease of deformation resistance ability. This may be the main cause that SSAW pipe cannot be used as the high deformation pipe.

5.5 Factors in Micro Sense

From micro sense, the main factors influencing the pipeline steel deformation hardening can be summarized as phase ratio, grain size and dislocation density, solid solution precipitation atoms, texture, etc.. The higher the soft phase ratio in microstructure of pipeline steel, the stronger the hardening ability in low stress level. While the decrease of grain size, increase of grain boundary, and the precipitation of carbon and nitrogen atoms due to strain-aging, etc. can result in a fast growth rate of the dislocation multiplication, pile-up and tangle, which can make the dislocation density increase, and the deformation resistance increase; meanwhile the texture due to the rolling method of pipeline steel can also make an important effect leading to a strong strain hardening capability in rolling direction.

6 Conclusions

(1) From a macro sense, the main factors influencing the line pipe strain hardening is primarily to the strength (grade), specimen orientation, microstructure, manufacturing methods of pipe and its deformation history (with and without pre-strain).

(2) The microstructure are the natural factors affecting line pipe strain hardening behavior. Phase ratio, grain size and dislocation density, precipitation, texture, etc. will have an effect to the strain hardening behavior of pipeline steel.

Acknowledgements

The authors would like to acknowledge China National Petroleum Corporation (CNPC), Key Lab of Oil Tubular and Environmental Behavior of CNPC for the financial assistance and allowing the publication of the paper.

References

[1] Ji Lingkang, Wang Haitao, Chi Qiang, etc. Research on the Key Technology for Natural gas Pipeline with Large Transportation Capacity, Proceedings of Baosteel Academic Conference 2013, Shanghai, China. June 4~6, 2013: F71-F76.

[2] Ji Lingkang, Huo Chunyong, Feng Yaorong, etc. Research on Key Technology and Production Quality of X80 Line pipe for the 2nd West-East Gas Pipeline, Proceedings of International Seminar on X80 and Higher Grade Line Pipe Steel 2008, Xi' an, China. Jun. 23 ~ 24, 2008: 1-12.

[3] Lingkang Ji, Hongyuan Chen, Chunyong Huo, etc., Key Issues in the Specification of High Strain Line Pipe Used in Strain-Based Designed Districts of the 2nd West to East Pipeline, Proceedings of the 7th International Pipeline Conference, September 29 -October 3, 2008, Calgary, Alberta, Canada.

[4] L. K. Ji, H. L. Li, H. T. Wang, etc., Influence of dual-phase micro structure on the properties of high strength grade line pipe, Journal of Materials Engineering and Performance, 2014 (23): 3867-3874

[5] Ji Lingkang, Li Helin, Zhao Wenzhen, etc. Microstructure and Strain hardening Performance Analysis for X70 High Strain Line Pipe. Journal of Xi' an Jiaotong University, Vol. 46, 2012 (9): 108-113.

[6] Ji Lingkang, Li Helin, Chen Hongyuan, etc. Analysis of Local Buckling Strain of Line Pipe. Chinese Journal of Applied Mechanics, 2012, 29 (6): 758-762

[7] Xian-Kui Zhu. Review of Fracture Control Technology for Gas Transmission Pipelines, Proceedings of the 10th International Pipeline Conference, IPC2014, Calgary, Alberta, Canada Sep. 29 - Oct. 3, 2014.

[8] Lingkang Ji, Xiao Li, Hongyuan Chen, etc. On the Relationship between Yield Ratio, Uniform Elongation, and Strain Hardening Exponent of High Grade Pipeline Steels, Proceedings of The Seventeenth 2007 International Offshore and Polar Engineering Conference, ISOPE-2007, Lisbon, Portugal, July 1~6, 2007: 3007-3012.

[9] Wang Maotang, He Ying, Wang Li, etc. Development and Application of X80 Pipeline Steel in the Second West-East Gas Line. Journal of Electric Welding Machine. 2009, 39: 6-14.

Welding of 2205 Duplex Stainless Steel Natural Gas Pipeline

Li Weiwei Xu Xiaofeng Yang Yang

(Tubular Goods Research Institute of CNPC)

Abstract: There Are Abundant Natural Gas Resources In Western China, But Many Oil And Gas Fields Are Rich In Chloridion, Sulfureted Hydrogen, Carbon Dioxide And Other Corrosive Medium, Which Have Strong Corrosivity To Pipeline. One China Western Gas Field Possesses Abundant Natural Gas With Great Pressure, And The Chloridion Concentration In The Water Separated From Gas Is About 10%, So The Medium Has Great Corrosivity. In Order To Ensure The Safety Of The Pipeline, About 13 km Length Pipeline And The Internal Pipes Of A Gas Treatment Plant That Purifies Gas About 12 Billion Cubic Meter A Year Are Made Of 2205 Duplex Stainless Steel (2205 Dss). 2205 Dss Has Many Characteristics In Welding With Complex Welding Process; And Because Of High Quality Requirements For The Construction Of Natural Gas Pipeline And Restriction Of On - Site Conditions, The Site Welding Is Very Difficult. Focus On Engineering Applications, A Large Number Of Experimental Researches Have Been Carried Out On The Material Microstructure, Properties And Weldability. Finally Weld Joints Which Conform To The Requirements Of Standard Are Obtained, Contributing To The First Large - Scale Application Of This Material In The Field Of Oil And Gas Pipelines. Considering The Engineering Application And The Latest Research Development, The Welding And Key Factors Affecting The Joint Properties Of 2205 Dss Pipes Are Summarized And Analyzed.

Keywords: Duplex stainless Steel; Weldability; Welding key factor; Joint properties

Ferrite-austenitic duplex stainless steel was invented in 1920's, which hadn't been widely used until 1970's because of some questions such as material, welding and so on. As the development of metallurgy technique and the acknowledgement that nitrogen element has an important role in duplex stainless steel, the 2nd generation duplex stainless steel containing nitrogen

---

This work was supported by the Science research and technology development project of China National Petroleum Corporation (04B41101).

Li Weiwei, Xu Xiaofeng, Yang Yang, CNPC Tubular Goods Research Institute, State Key Laboratory of Performance and Structural Safety for Petroleum Tubular Goods and Equipment Materials, Xi'an, China, 710077.
Li Weiwei, Corresponding author, E-mail: liweiwei001@cnpc.com.cn.

had been developed in 1970's and the weldabilty had been improved. Therefore duplex stainless steel started to be widely used in projects from 1980's in some advanced countries and from 1990's in China. 2205 DSS is one of the modern duplex stainless steels and possesses high strength, high toughness and good corrosion resistance, especially good resistance to pitting, crevice corrosion and stress corrosion in chloridion containing environment, and has been applied widely in petroleum and natural gas transportation, ocean and chemistry industries [1].

One gas field in western China possesses abundant natural gas and undertakes about 80% of the total provision for West-East Pipeline Project. Because the medium is very dangerous and the chloridion concentration in the water separated from gas is higher up to 10%, which have great corrosivity to the pipes, so 2205 DSS with good mechanical properties and corrosion resistance had been used in the pipes of the project for the safety and reliability, including a 13 km length pipeline and the internal pipes of a gas treatment plant that purifies gas about 12 billion stere a year. The 2205 DSS consumed was about 5226 tons and occupied about 1/3 of the global annual production at that time[2]. This is the first time for such a large application of 2205 DSS material in China and the world. The project has run over several years safely, which accumulates a wealth of technical results and application experiences for the further application of 2205 DSS.

2205 DSS has many characteristics in welding with complex welding process; and because of high quality requirements for the construction of natural gas pipeline and restriction of on-site conditions, the site welding is very difficult. Based on welding experiments and studies, reasonable welding procedures and strict discipline are developed, which ensure the quality of the project. From the view of engineering application, the welding characteristics of the material and some major welding techniques are introduced in this paper.

1 Properties and weldability of the material

1.1 Material Properties

The typical chemical components of 2205 DSS are given in Table 1, of which the main alloying elements are Cr, Ni, Mo and N. Cr and Mo promote the formation of ferrite, while Ni and N stable austenite. The material is generally delivered in solid solution treatment state, and studies show that this material by solid solution treatment at 1050℃ for 2 hours can obtain best microstructures and performances [3]. In the normal state of delivery, the microstructure is a kind of duplex structure combined about 50% of ferrite with about 50% of the austenite, shown in Figure 1. 2205 DSS combines the advantages of the two microstructures, in which the ferrite microstructure provides high strength while the austenite gives good plasticity and toughness and promotes the formation of fine grain structure with the ferrite microstructure, thus making the material exhibit excellent mechanical properties. The typical mechanical properties of 2205 DSS are given in Table 2.

Table 1 Typical chemical components of 2205 DSS (wt %)

C	Si	Mn	S	P	Cr	Ni	Mo	N
0.023	0.37	1.46	0.001	0.020	22.3	5.78	3.22	0.18

Table2 Typical mechanical properties of 2205 DSS

Tensile strength R_m (MPa)	Yield strength $R_{p0.2}$ (MPa)	Elongation A (%)	Charpy impact absorbed energy at−40℃ K_V (J)	Hardness (HB)
789	564	24	280 290 295	292

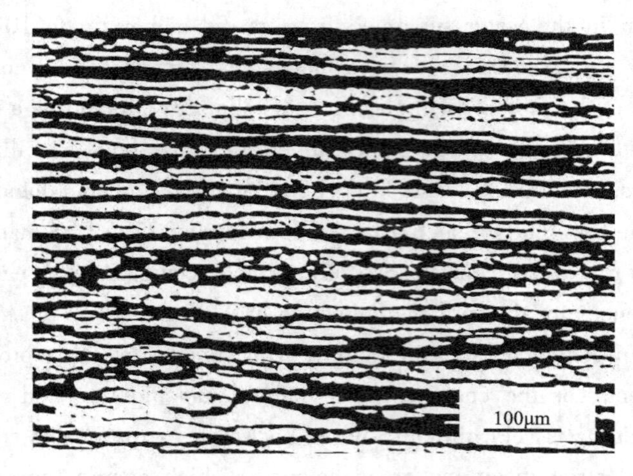

Fig. 1 Typical microstructure of 2205 DSS

The weldability of the early duplex stainless steel was poor and lots of quality accidents happened due to welding problems. Even until now, many people are still worried about the welding of duplex stainless steel. Due to nitrogen alloying technology, sufficient austenite can reform in the heat-affected zone and weld joints with good mechanical properties and corrosion resistance can be obtained; thus, the welding performance of the modern duplex stainless steel has been greatly improved[4-7]. Compared with austenitic stainless steel, 2205 DSS has higher thermal conductivity and smaller linear expansion coefficient, so the hot tearing tendency and deformation are smaller; compared with the low alloy high strength steel, due to the effects of the austenite, the cold cracking tendency is smaller. Overall, the weldability of 2205 DSS is good. It can be welded without preheating and post-weld heat treatment, and it can be welded with the 18-8 type austenitic stainless steel or carbon steel[8, 9].

Its excellent performance relies on an appropriate proportion of two-phase structure; however, the welding thermal cycle has a great influence on the microstructure of the weld. Hence, a major problem for 2205 welding is how to ensure that the welding seam and heat affected zone (HAZ) have basic balanced two-phase structure without the precipitation of intermetallic phases. When the welding heat input is too low, the cooling speed is fast, and then excessive ferrite and nitride may be produced in the weld and HAZ, thereby reducing the corrosion resistance and toughness of welded joints. On the other hand, if the heat input is too high, the cooling speed is too slow, and intermetallic phases may be precipitated in the weld and HAZ, also can make the welded joint corrosion resistance and toughness decrease. Only the right welding process parameters combined with certain technical measures can the weld and HAZ obtain fine microstructure and properties.

Only the process of single side welding with both sides formation can be used in the field welding, andpickling and passivation treatment can not be used for the internal part, then how to get a good back protection and prevent the oxidation of weld and HAZ is another major problem.

1.2 Welding performance influencing factors

1.2.1 Heat input

The welding heat input has a great influence on the microstructure and the proportion of the two phases of the weld and HAZ, and thus has a significant impact on mechanical properties and corrosion resistance of the weld joint. The results of corrosion test and impact test of the weld joint in different heat input in laboratory simulation are as shown in Figure 2.

We can see from Figure 2, in smaller heat input (6.2 kJ/cm), the pitting corrosion rate of weld joint is higher, the corrosion resistance is poorer and pitting may appear, and the impact absorbed energy of the weld and HAZ is also lower. During the heat input range of 8.2 kJ/cm to 15.7 kJ/cm, with the increase of heat input, the corrosion rate increases, the corrosion resistance decreases, and the impact absorbed energy improves, but the combination property is good, meeting the engineering requirements.

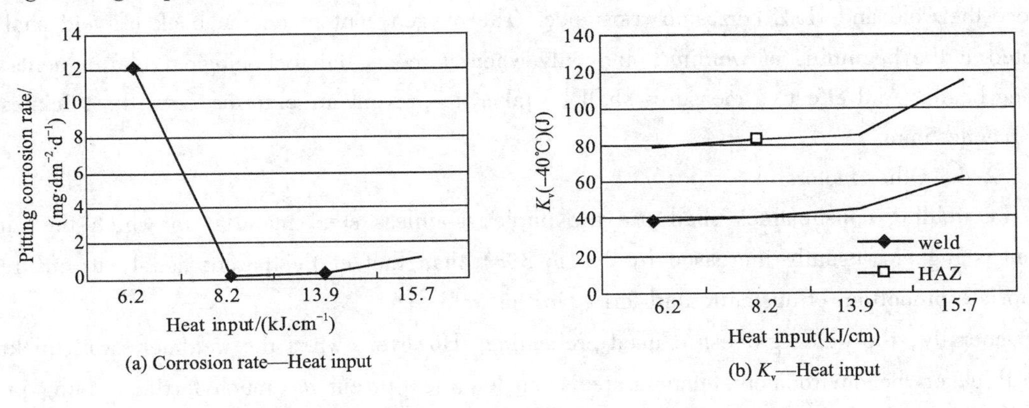

Fig. 2 Varies of corrosion rate and low temperature toughness of the weld joint with the heat input

If the welding heat input is too small, the cooling rate is very fast and the austenite of the weld metal and high temperature HAZ hasn't enough time to fully precipitate, resulting in high content of ferrite and nitride (the solubility of nitrogen in ferrite is low and easy to form nitride), and reducing the corrosion resistance and toughness of the weld joint. Welding in the proper heat input, the austenite of the weld and HAZ transforms fully with proper ratio of two phases, so that the weld joint has a good performance. Within a certain range, with the increase of heat input, the austenite content increases and the toughness of the joint improve. However, with the increase of heat input, the cooling rate becomes lower. If the time from 900℃ to 550℃ stays too long, the likelihood of precipitation of intermetallic phases of the weld or HAZ increases, reducing the corrosion resistance of the joint[10, 11]. Therefore, moderate heat input parameters should be adopted in the welding process, avoid using too low or too high heat input parameters.

1.2.2 Welding method

Many welding methods can apply to 2205 DSS, such as shielded metal arc welding (SMAW), gas tungsten arc welding (TIG) and submerged arc welding (SAW). Theses three methods are used in paper[13] to conduct the welding test on 2205 DSS plates; and the microstructure, mechanical property and corrosive property of weld joints are analyzed. The results indicate that the phase ratio of ferrite in the weld metal and HAZ of the joints is controlled in the range of 30% to 60%. The combination property of TIG joint is the best, and then that of SAW, last the SMAW. The impact absorbed energy at −40℃ of the weld metal by SMAW is the minimum for 37J while the pitting corrosion resistance is the poorest with the weight loss rate at about 9 $dm^{-2}\cdot day^{-1}$, but still meeting the engineering requirements. Restricted by the field welding conditions in the actual construction, TIG is used for backing welding and SMAW is used to fill and cover in the pipe girth welding, which can meet the project quality and schedule requirements.

1.2.3 Back protection

The process of single side welding with both sides formation is used in the field welding, and then effective back gas protection is the premise to ensure the welding quality. Effective back protection tooling shall be used and the purity of the protection gas shall meet the technological requirements. There is a need to add a certain proportion of nitrogen into the back protection gas, to improve the weld and HAZ corrosion resistance. The oxygen content on the back of weld shall be detected at the beginning of welding, and only when it meets the technological requirements can welding begin. And effective measures shall be taken to prevent air entering until the thickness of weld reaches5mm.

1.2.4 Other factors

The welding consumables shall use the duplex stainless steel material in which the nickel content is higher (usually increased by 2% to 3%) than that of the parent metal, to ensure an appropriate proportion of austenite and ferrite in the weld[9].

Generally, the welding doesn't need preheating. However, when the weldment wall thickness is too thick or the environment temperature is too low, to prevent too much ferrite forming in the weld and HAZ due to the rapid cooling rate, if necessary, preheating may be used. To avoid the production of precipitate phase for the slow cooling rate, the interpass temperature between the multi-layers or multi-passes shall be controlled, usually not to exceed 150℃. Multi-pass welding and additional process welds if necessary should be adopted, the following weld having a heat treatment effect on the front, to ensure the balance of the phases of the joint.

It's necessary for the shielding gas to increase a certain percentage of nitrogen to improve the corrosion resistance of the weld. Besides, effective protection from the back is very important for TIG welding. The purity of the gas shall meet the technical requirements and welding shall not start until the oxygen content on the back of the weld meets the process requirements. Effective measures shall be taken to prevent air into the back, and the back gas protection should continue until the effective thickness of the weld reaches5 mm or more.

It should prevent the stainless material from the pollution of carbon steel, copper, low melting point metal or other impurities. Where possible, the stainless steel pipe and carbon steel pipe should be kept separate. Measures shall be taken to prevent spattering, arc blowing, carburizing, local overheating in the welding and cutting.

2 Welding procedure qualification

According to previous studies, the process scheme for 2205 DSS natural gas pipeline is as follows: TIG welding was used for backing welding and the first layer filling welding, then the weld thickness is about 5mm, therefore to ensure good back forming, joint toughness and corrosion resistance, SMAW welding was adopted for the subsequent filling and cover welding in order to improve the welding efficiency. Single V type groove and single side welding with both sides formation were used. For welding materials, filler metal with higher Ni than base metal was used and high purity argon with a certain amount of nitrogen was used for welding torch and back shielding gas protection. Moderate heat input process parameters were used in the welding.

The welding procedure qualifications (WPQ) were processed according to the process specifications developed. The procedure qualification conditions for the field welding of ϕ323 mm× 12.7 mm pipe are given in Table 3, while the test results for the main properties are given in Table 4 and the microstructure are in Figure 3.

Table 3 Main test conditions of WPQ of the girth weld

Welding process	Filler metal	Protect gas	Groove	Position	Heat input E (kJ/cm)
TIG for 2 root pass SMAW for filler and cap pass	AVESTA 2205 AVESTA 2205-PW	Torch: Ar+1.5% N_2 Back of weld: Ar+5% N_2	Single V	45°fixed (6G)	8~20

Table 4 Main results of WPQ of the girth weld

Tensile strength R_m (MPa)	Guided bend	-40℃ CVN K_v/J		Microstructure	Corrosion
		Weld	HAZ		
785 763 785 768	Diameter of the mandrel is 90mm, side bending for 180°. No crack	53 52 45 65 58 67	103 100 100 125 192 95	Weld and HAZ are F+A. No intermetallic precipitations. The F proportion is 35% to 50% in weld and 50% to 65% in HAZ.	No pitting corrosion occurs in specimens' surface after 24h at 22℃, 6%$FeCl_3$ solution. Average corrosion rate of 3 specimens are 0.36, 3.96, 0.00 mg·dm^{-2}·day^{-1}

It can be seen from the results given in Table 4, the tensile strength of weld joints is much higher than the specified lower limit of the base metal (620 MPa), and so the tensile property of the joint is good. The tensile side of the weld joint is intact after bended for 180°, indicating that the ductility of the joints is good. The Charpy impact absorbed energies at −40℃ of the weld metal and HAZ meet the specified requirement in ASTM A923 (no less than 54J for the base metal and HAZ, while no less than 34J for the weld) . The pittting test in 6% $FeCl_3$ solution in accordance with ASTM A923C method shows that the weld joints possess good resistance to chloridion localized corrosion (the specified requirement is no more than 10 mg$\cdot dm^{-2}\cdot day^{-1}$) .

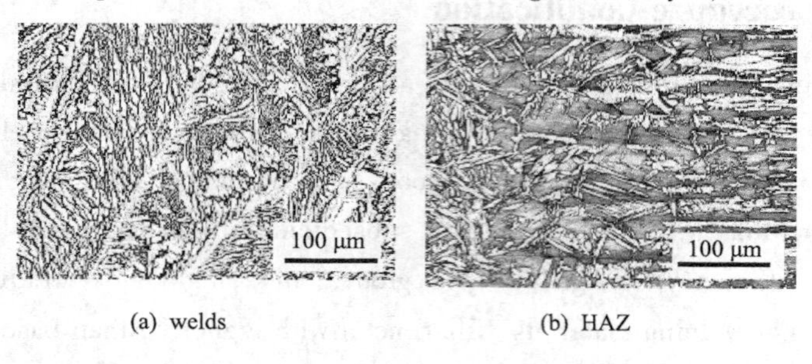

(a) welds (b) HAZ

Fig. 3 Microstructure of weld and HAZ of WPQ

For the weld joint, it's easier to ensure microstructure balance because of adding more austenite formation elements of Ni. However, HAZ is different, especially for the high temperature HAZ near the fusion line, it has a certain degree of difficulty to ensure the microstructure balance and avoid the precipitation of intermetallic phases under field welding condition. From Figure 3, we can see that the weld is a dual phase microstructure of ferrite and austenite without precipitation phase, the ferrite content is 35% ~ 50%, meeting the technical specification requirement (the requirement for ferrite content in weld and HAZ is in the range of 35% ~ 65%), and the austenite phase is relatively much more, being good for the toughness and corrosion resistance.

HAZ also has a dual phase microstructure of ferrite and austenite but the proportion varies large. The ferrite content in the high temperature HAZ near the fusion line arrives at 65% without single ferrite phase, while that away from the weld and HAZ is a bit lower (50% ~ 60%) . The technical specification requirements are satisfied as well as the plasticity, toughness and corrosion resistance. The TEM pictures of morphology for austenite and ferrite microstructure in high temperature HAZ under moderate process parameters are given in Figure 4 we can see that there is no precipitation of intermetallic phase but there are many dislocations.

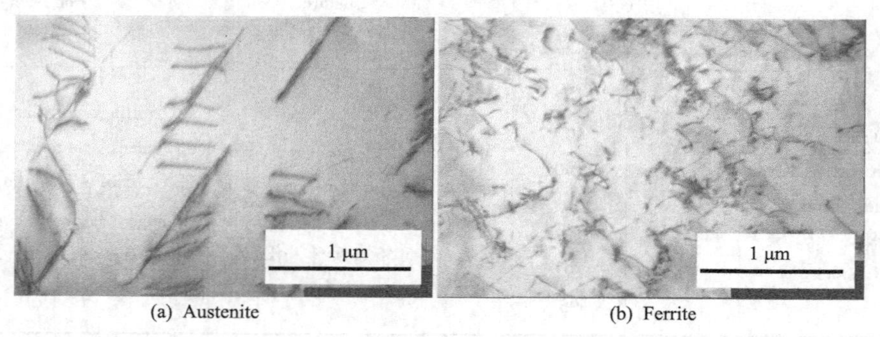

(a) Austenite (b) Ferrite

Fig. 4 TEM picture of morphology for A&F microstructure in high temperature HAZ

The assessment results meet the requirements of relevant standards and excellent weld joint is obtained, indicating that the welding parameters are appropriate and the proposed field welding process scheme and welding process specification for 2205 DSS is correct.

3 Engineering application

For one large - scale CO_2 - bearing gas condensate field in China with high single well production, the transportation pipeline faces the corrosion with high pressure (the maximum working pressure is 13.3 MPa), high temperature (the highest temperature is 80℃), and high chloridion (the chloridion concentration in the water separated from gas is about 10%). According to these characteristics of the gas field, 2205 DSS material with higher mechanical properties and excellent resistance to chloridion and carbon dioxide corrosion is used.

There are many pipe sizes used in the project (the minimumϕ33.4 mm × 3.4 mm, the maximum ϕ508 mm × 19.1 mm). API 1104 is a basic pipeline field welding standard, which is only applicable for carbon steel and low alloy steel material. Based on the experimental studies and analysis, the welding technical requirements for duplex stainless steel are supplemented, and the weld defect inspection is strengthened, and a stricter standard, which is Q/SY TZ 0110-2004 "Specification for construction and acceptance of the welding of 2205 duplex stainless steel", than API 1104 has been developed. 12 welding procedures are developed and the corresponding welding procedure qualifications are processed. According to the results of the qualification, six of the welding specifications are elected for the welding of the project.

In order to guarantee the quality and progress of the project, lots of preparatory work and strict requirements have been done on the technical standards for construction, welder training and assessment, on - site construction discipline and so on. Under the careful organization and coordination of the project department, the pipeline field welding quality met the requirements of Q/SY TZ 0110—2004. After more than five months of hard work of the scientific research, construction and supervision corporations, the project was successfully completed at the end of 2004. Currently, these pipelines run in over ten years in good condition, showing that the construction quality of the pipeline is reliable. This project accumulates a wealth of technical results and application experiences for the further application of 2205 DSS.

4 Conclusions

(1) 2205 DSS has good mechanical properties and corrosion resistance, of which the welding has many characteristics. How to ensure the dual phases balance, avoid the precipitation of intermetallic phases and make the weld back protection are the main issues.

(2) The field welding of 2205 DSS natural gas pipeline is recommended as follows: TIG welding is used for backing welding and the first layer filling welding, while SMAW welding for the subsequent filling and cover welding. Filler metal with higher Ni than base metal is used and high purity argon with a certain amount of nitrogen should be used until the weld thickness is over 5mm. Moderate heat input process parameters are recommended to use in the welding.

(3) During the pipeline filed construction, detailed technical rules and rigorous discipline are developed and implemented strictly in the welding process, ensuring the welding quality of the project.

References

[1] Wu J, Jiang S Z, Han J A, et al. Duplex Stainless Steel. Beijing: Metallurgical Industry Press, 1999. (in Chinese)

[2] Bi Z Y, Zhang J X, Zhang F, et al. 2205/Q235 Duplex stainless steel composite pipe with large diameter used in acid medium. Corrosion & Protection, 2010, 31 (5): 349-352. (in Chinese)

[3] Wei B , Bai Z Q, Yin C X, et al. Effects of solid solution treatment on pitting corrosion behavior of 2205 duplex stainless steel. Transactions of Materials and Heat Treatment, 2009, 30 (4): 73-76. (in Chinese)

[4] Lindblom B E S, Lundquis T B, and Hannerz N. E. Grain growth in HAZ of duplex stainless steels. Scandinavian Journal Metallurgy, 1991, 20: 68-74.

[5] Li S Z, Wiseman R, and Sunter B J. Metal inert gas welding of alloy 2205 duplex stainless steel, Transactions of the China Welding Institution. 1995, 16 (2): 68-73.

[6] Mats L. The welding metallurgy of duplex steel. International Duplex Stainless Steel Conference, Beijing, Oct. 27-28, 2003: 25-39.

[7] Rouault P and Bonnet C. Make the duplex and super duplex welding easier through metallurgical and practical simple recommendations. International Duplex Stainless Steel Conference, Beijing, Oct. 27-28, 2003: 53-58.

[8] Si C Y, Zhou Z F, Qian B N, et al. Welding Handbook--Welding of Materials, Beijing: Mechanical Industry Press, 1992. (in Chinese)

[9] Zhang W Y, Hou S Y. Weldability and welding material of duplex stainless steel. Welding Technology, 2004, 33 (1): 40-42. (in Chinese)

[10] Li J, Wang Y S. Effect on metallographic structure of duplex stainless steel (SAF 2205) by different welding procedure. Pressure Vessel Technology, 2004, 21 (2): 7-11. (in Chinese)

[11] Claes-Ove P, Sven-Ake F. Welding Practice for the Sandvik Duplex Stainless Steels SAF 2304, SAF 2205 and SAF 2507, SANDVIK Steel, Sweden, 1994.

[12] Wang Z Y, Han J , Song H M, et al. Comparative analysis for joint performance of duplex stainless steel by different arc welding methods. Transactions of the China Welding Institution, 2011, 32 (4): 37-40. (in Chinese)

Strain Capacity of Girth Weld Joint Cracked at "Near-Seam Zone"

Chen Hongyuan * Chi Qiang Wang Yalong Niu Jing Yang Fang Ren Jicheng

(1. Tubular Goods Research Institute of CNPC; 2. Xi'an Jiaotong University;
3. State Key Laboratory of Performance and Structural Safety for
Petroleum Tubular Goods and Equipment Materials)

Abstract: Cracking occurs often at "near-seam zone" in the girth weld joint of a ϕ813mm× 14.7mm X70 pipeline in tensile test, which is considered unacceptable for strain-based design pipelines according to some current standards. The tensile strain capacity of girth weld joint for X70 pipelines with "near-seam zone" cracks has thus been studied via the approach of crack driving force and fracture resistance curve. The high strain capacity has been demonstrated by resistance curve tangency approach and curved wide plate test. The results prove that the girth weld joint has considerable fracture resistance which may lead to high strain capacity.

Keywords: Strain-based design; Weld joint; Tensile strain capacity; Single-Edge notch tensile test; Curved wide plate test

1 Introduction

Strain-based approach is the latest progress as a solution to the pipeline design in harsh environment. Ductile fracture under axial tensile strain is one of the most severe limit states. For land pipelines, it is always related to the ground movement such as seismic activities, discontinuous frost soil and mining subsidence. For offshore pipelines, it is always the result of pipeline laying, such as S-lay, J-lay and reeling lay which can produce the strain up to 2%, and the uneven seabed, etc..

The tensile strainlimit of a pipeline depends on the tensile strain capacity of its girth welds joint. The girth welds here refers to the entire weld region, including the weld metal and the heat-affected zone (HAZ). It tends to be the weakest link among the pipeline due to the possible existence of weld defects and regular deteriorative metallurgical and/or mechanical property changes from welding thermal cycles. Consequently, tensile strain capacity is related to the girth welding procedure qualification and flaw acceptance criteria. The welding procedure qualification involves the control of variables to ensure the equivalence of procedure qualification welds, field production

* Corresponding author.

E-mail address: chenhongyuan@ cnpc. com. cn.

welds and the definition and execution of mechanical tests of welds. The flaw acceptance criteria are implemented in field production welds to ensure a certain level of performance. In this case, certain tensile strain capacity is achieved [1], and the weld joint with flaws can ensure its integrity under plastic deformation. These designs are complicated because of the definition of the value of different variables. The variables are always difficult to be defined with interactive influence. When girth weld flaws occur, strain capacity will depend on more material and geometric factors[2]. Numerical studies and pressurized full scale tension (FST) tests have shown that once the threshold level of toughness ensuring ductile failure is achieved, the following variables do influence strain capacity.

- Fracture resistance
- Weld strength mismatch level
- Uniform strain (uEL) capacity of the pipe metal
- Pipe and weld metal strain hardening capacity
- Flaw location (surface or buried) and dimensions (length and depth)
- Flaw depth to wall thickness ratio
- High-low weld misalignment
- …

Although related issues are under research, there is no comprehensively general standard for strain capacity of girth welding joint. According to the current standard in China, the tensile test of weld joint crack on weld, heat affected zone, or "near-seam zone" (which means the softening zone near the weld and HAZ) is not allowed under strain-based design conditions. For the reason of sensitivity to material properties and flaws size, weld joint strain capacity cannot be analyzed quantitatively only by tensile test results for weld.

The resistance-curve approach is a failure criterion by which fracture instability is predicted to occur when the driving force curves exceed the material fracture resistance . The failure mode is assumed to be ductile fracture. The crack driving force, in terms of crack tip opening displacement (CTOD) or J-integral, is derived from finite element analysis (FEA) for various structural geometries (including flaw size) and material properties. The resistance curve (R-curve) is directly measured from test specimens. The failure point or the unstable ductile tearing point is determined by the traditional tangency criteria. There are several organizations pursuing tensile strain capacity prediction by using the tangency approach, two of which are SINTEF[3-5] and ExxonMobil[6-10]. According to the approach, the crack driving forces are presented as a group of curves (iso-strain CTODF curves) of different strain levels. Each iso-strain CTODF curve is expressed as a function of flaw growth.

In the paper, the strain capacity of girth weld joint of ϕ813mm×14.7mm X70 strain-based designed pipeline is studied in detail. The fracture location of weld joint tensile test are always found on near-seam zone as shown in figure 1. The tensile test, hardness test, and single-edge notched tensile tests for related zone are conducted for the description of the mechanical properties of weld joints. Then FE analysis is conducted for the calculation of crack drive force during ductile tearing. Accordingly, the tensile strain capacity is evaluated by tangency approach of CDF curve and CTOD R-curves, which is demonstrated by wide plate tensile tests.

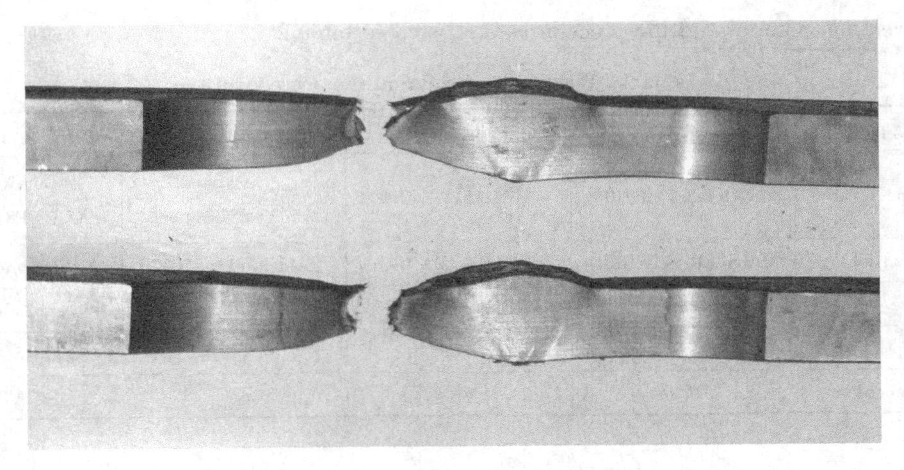

Fig. 1 The specimens cracked on near seam zone

2 Pipe Material and Welding Procedure

2.1 Pipe

The pipe nominal size is ϕ813mm×14. 7mm, the specimens are cut from the same pipe. The chemical compositions of the pipe material are listed in table 1.

Table 1 Chemical compositions of the pipe material

C	Si	Mn	P	S	Mo
0. 052	0. 13	1. 49	0. 0079	0. 0021	0. 17
Cr	Nb	V	Ni	Cu	Pcm
0. 036	0. 051	0. 0041	0. 17	0. 031	0. 15

2.2 Girth weld joint

SMAW has been used for root pass and hot pass, self-shielded flux-core FCAW has been used for fill pass and cap. Extra cap welding has been used for over match of the girth weld joint, as shown in figure 2.

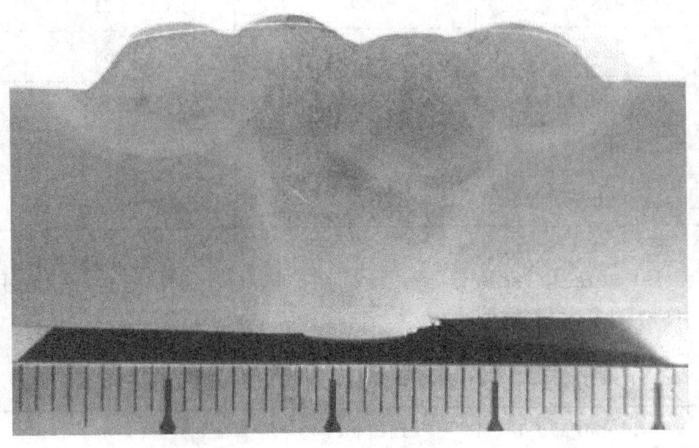

Fig. 2 Macro profile of the girth weld

The welding material and the procedures are listed in table 2.

Table 2 Welding material and the procedures

Welding pass	Root pass	Hot pass	Fill pass	Cap pass
Type	AWS A5. 1 E6010	AWS A5. 1 E6010	AWS A5. 29 E81T8-Ni2	AWS A5. 29 E81T8-Ni2
Brand	BOHLER FOX CEL	BOHLER FOX CEL	Golden bridge JC30	Golden bridge JC30
Size	ϕ4. 0mm	ϕ4. 0mm	ϕ2. 0mm	ϕ2. 0mm
Welding position	5G	5G	5G	5G
Welding procedure	SMAW	SMAW	FCAW-S	FCAW-S

3 Experimental Research

The mechanical performance tests include the tensile test, hardness test, (single-edge notched tension) SENT and curved wide plate (CWP) test.

3. 1 Tensile test

Table 3 shows the longitudinal tensile test results from eight positions along the circumference of the pipe body, including yield strength, tensile strength, the ratio of yield strength to tensile strength (Y/T) and uniform elongations. The pipes used in the strain-based designed pipelines have better strain hardening capacity in longitudinal direction with lower Y/T, higher stress ratios and uniform elongation than common pipes for which the longitudinal tensile properties are not specified.

Table 3 Longitudinal tensile properties of the pipe

Position	YS (MPa)	TS (MPa)	Y/T	UEL (%)
0°~45°	498. 0	672. 4	0. 74	8. 7
45°~90°	518. 3	682. 4	0. 76	7. 5
90°~135°	464. 5	671. 6	0. 69	9. 0
135°~180°	473. 6	669. 6	0. 71	8. 8
180°~225°	459. 9	673. 3	0. 68	9. 1
225°~270°	490. 9	681. 2	0. 72	8. 6
270°~315°	487. 3	685. 6	0. 71	7. 8
315°~0°	496. 5	678. 9	0. 73	8. 0

The all-weld-metal tensile specimens are also tested. Figure 3 and figure 4 show the specimens and one of the stress-strain curves separately. The test results are shown in table 4.

Table 4 Tensile properties of all-weld metal

Position	YS (MPa)	TS (MPa)	Y/T	UEL (%)
All Weld	557. 0	657. 4	0. 85	10. 5

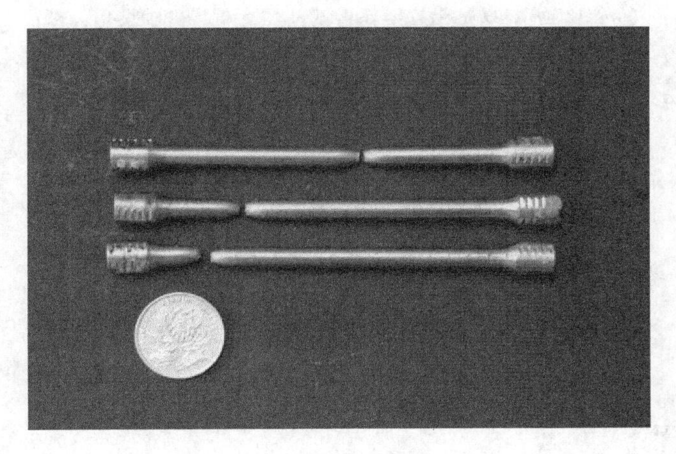

Fig. 3 The all-weld-metal tensile specimens

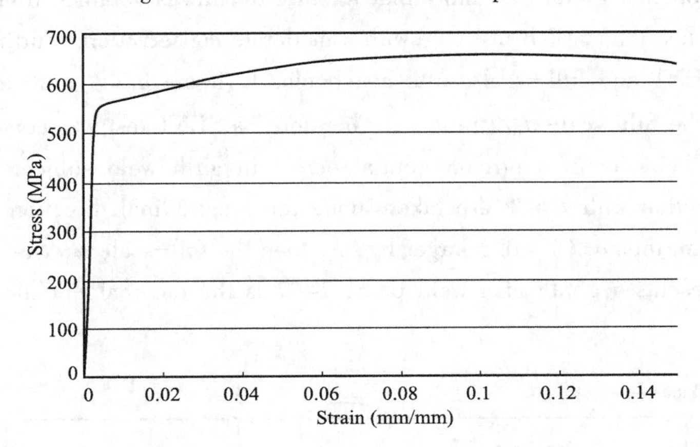

Fig. 4 Engineering stress-strain curves for all-weld-metal

3.2 Hardness test

The Vickers hardness is tested along the inner side, outer side and middle of the specimen. As the hardness value and the distribution shown in figure 5 and figure 6. It shows that the hardness is slightly lower near the all fusion line because of the weld thermal cycles. It would result in the initiation of the crack at "near-seam zone" on weld toe, but the softening are not serious enough so that the crack do not grow along the soften zone, as shown in figure 1.

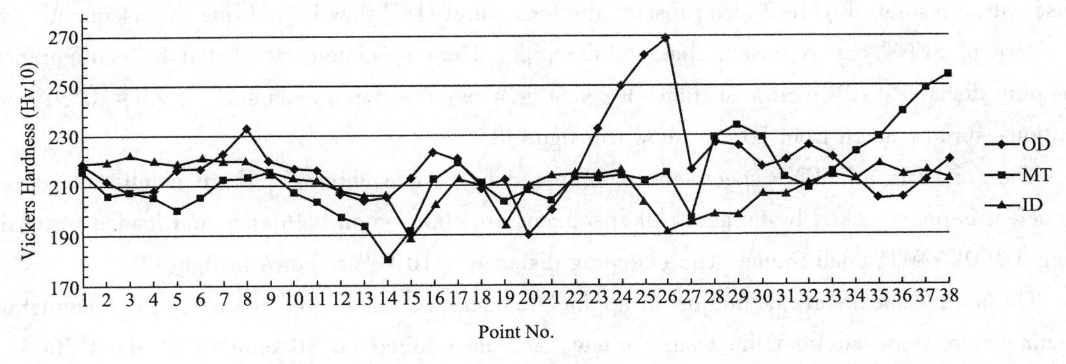

Fig. 5 Vickers hardness along the inner side, outer side and middle of the specimen

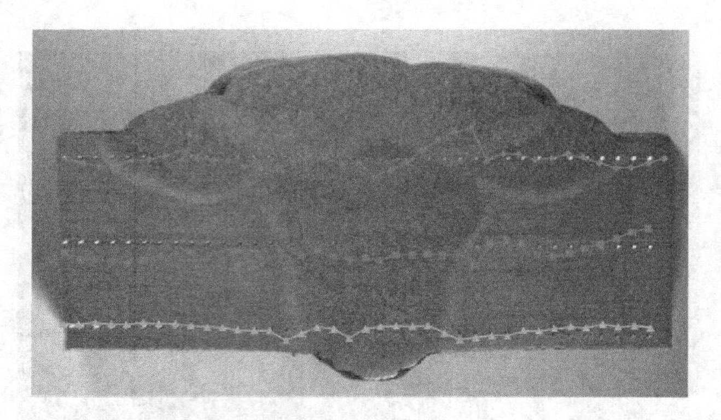

Fig. 6 Hardness distribution on the weld cross section

3.3 SENT test

The related published literatures imply that fracture toughness obtained from SENT test is more similar to full-pipe test than SENB test[11], with a moderate conservation. Further studies show that the R-curves for SENT and full-scale tests match closely if the a_0/W ratio for the SENT test is larger by 0.1 than the full-scale a_0/t ratio[5]. Therefore, a SENT test was conducted to investigate the ductile fracture behavior of a circumferential defect in girth weld subjected to plastic strain. Specimens across section with $B = W$ are taken along the longitudinal direction of the pipe. Initial crack depth is 4.3mm thus $a_0/W = 0.3$ larger by 0.1 than the full-scale a_0/t ratio (3mm/14.7mm = 0.2, the 3mm represents a depth of a weld pass, 14.7 is the nominal thickness of the pipe), as shown in figure 7.

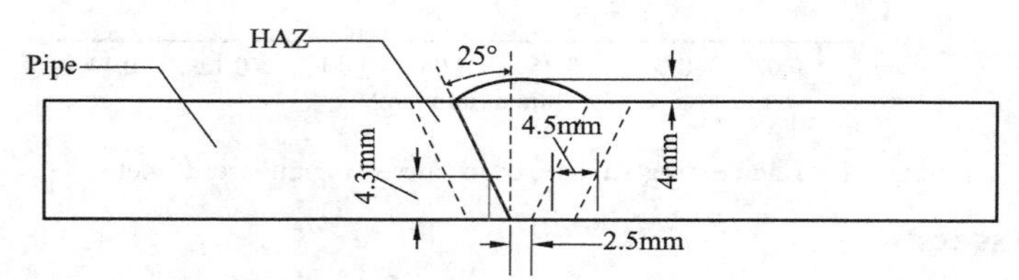

Fig. 7 Location and depth of the flaw

The specimens with flaws on weld metal and on HAZ are all test. For WM specimens, crack location is outside diameter (OD) surface with crack on centerline of the weld metal, and for HAZ specimens, crack location is inside diameter (ID) surface with crack tip on the fusion line for conservative results. Figure 7 also presents the location of HAZ flaw by red line. Consequently, the R-curve of SENT can represent that of full-pipe. Here specimens are listed by configuration: clamping distance: $10W$; cross section = W; side grooves: 5% depth each side; $a_0/W = 0.3$; notch location: surface notch from ID, as shown in figure 8.

The cracks in all SENT specimensare prepared by first machining a sharp notch, after which the defect is pre-cracked by fatigue. All specimens are clamped on both ends and loaded in tension using a 100kN MTS load frame. The clamping distance is $10W$, as shown in figure 9.

The multi-specimens technique is applied to measure crack growth. After test completion, specimens are removed from the testing frame, and heat tinted for 30 minutes at about 300℃ to mark the crack growth tearing zone. In accordance with ASTM 1820, both the initial and final crack

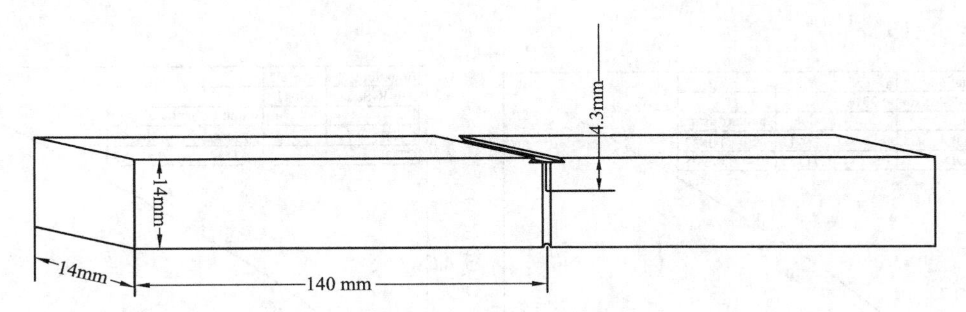

Fig. 8 Configuration of the specimen

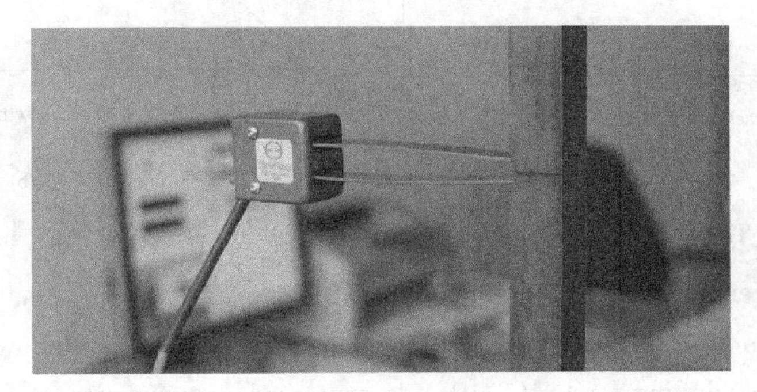

Fig. 9 SENT specimen mounted with a COD gauge

depths are measured at 9 equally spaced points and averaged. Additional details of the SENT testing procedure are beyond the scope of this paper and not to be described here.

Equations (1) have been used to calculate CTOD from the SENT specimens. It should be noted that in this project, single clip gauge was used for measurement. Therefore, the plastic rotation factor $\gamma_p = 1.2$ was used for calculating CTOD-R curve.

$$\delta = \frac{K^2}{2\sigma_{YS}E'} + \frac{\gamma_p(W - a_0)v_p}{\gamma_p(W - a_0) + a + z}$$

The symbols used in equations above are:

δ CTOD

K Stress intensity factor

σ_{YS} Yield strength

E' elastic modulus corrected for constraint conditions $E' = E$ for plane stress

γ_p plastic rotational factor for SENT test

W specimen width

a_0 originalcrack depth

v_p plastic displacement at the crack mouth

a crack depth during the test

z distance of the notch opening clip gage above the surface of the specimen

Finally, the crack extension length (Δa) and the CTOD (δ) for each specimen are pointed and then produce the R-curves for both of weld metal specimen and HAZ (fusion line) specimen, as shown in figure 10. The R-curve formulation is listed below:

$\delta = 1.269\Delta a^{0.7119}$,

$$\delta = 1.215\Delta a^{0.6744},$$

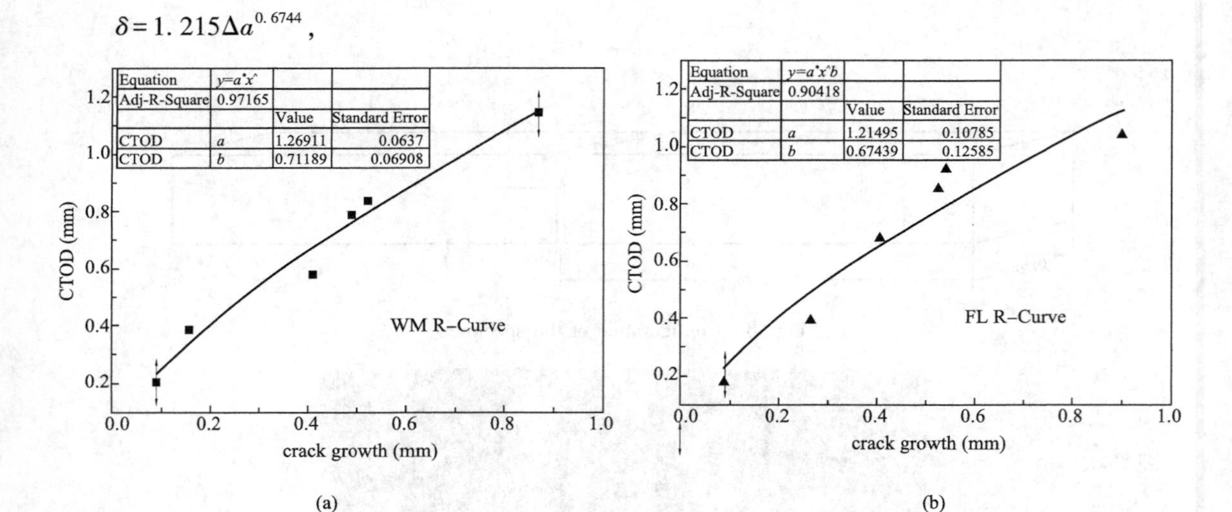

Fig. 10 The R-curves of weld metal (a) and HAZ (b) using fromSENT specimens

3.4 Development of Crack Driving Force Model

The tensile strain modelis built upon two major components: crack driving force relations and limit states. The crack driving force, $CTOD_F$, is expressed as a function of remote strain for any given geometry (including flaw size) and material parameters.

The approach is primarily based on failure prediction by ductile fracture emanating from girth weld defects. When the fracture driving force exceeds the material's ductile tearing resistance (R-curve), failure occurs. The tangency approach is often described by a geometric construction that shows a driving force curve and a material toughness R-curve on a graph that plots a fracture parameter, such as CTOD, versus crack growth.

Finite element analysis for full-pipe tensile model with flaw located in fusion line of girth weld is used to model the crack driving force. Weld strength overmatch is defined as real strain-stress curve, as shown in figure 11. All FEA models have been created using ABAQUS® 6.10. A half of a pipe specimen assuming transversal symmetry, first creates the geometry (cylinder, weld with fusion lines, bevel) which is then meshed for the requirement of fracture analysis, as shown in figure 12.

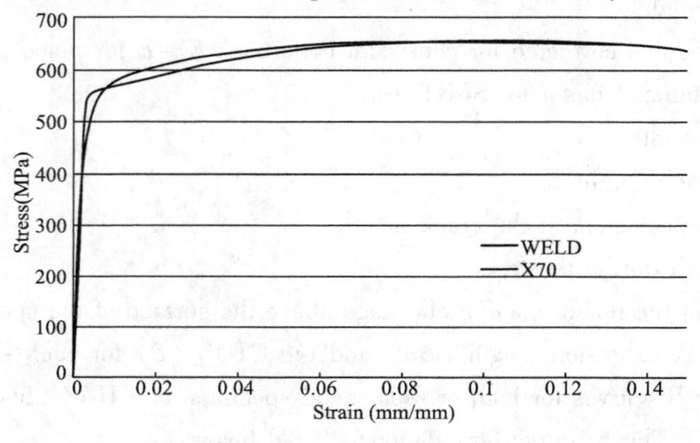

Fig. 11 The strain-stress curves of the pipe and weld metal

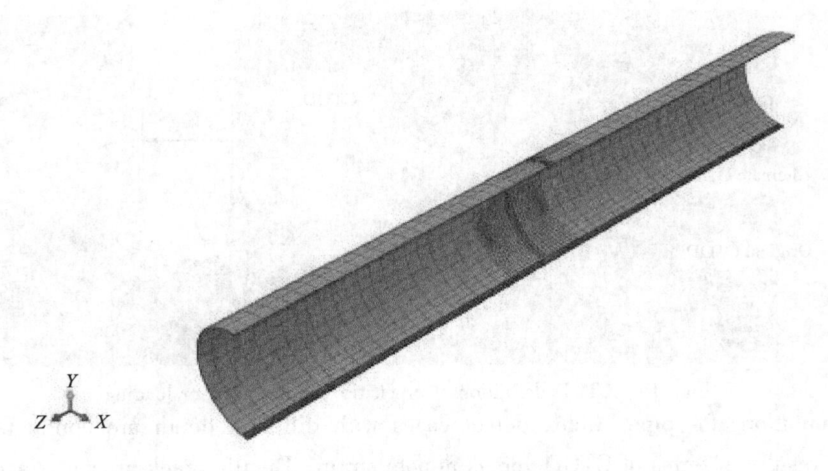

Fig. 12 Symmetry pipe model for FEA

A large-deformation formulation of ABAQUS® standard solver ('nlgeom') is used. All materials are modeled as elastic-plastic, and were assumed to harden isotopically according to the Von Mises yield criterion.

The elements consist of ABAQUS® element type 'C3D8R' (three-dimensional solid linear brick elements with reduced integration) . The analysis process is optimized through a mesh convergence study. An initially blunt defect with a fine spider-web mesh around the defect tip is modeled. The different depth of the flaws are 2, 3, 4, 5, 6 mm with length to depth ratio are 17 according to the defect tolerance 50mm×3mm. The radius of the initial crack tip is 0. 15mm. Figure 13 shows an example model of girth weld with detail of the near-defect mesh.

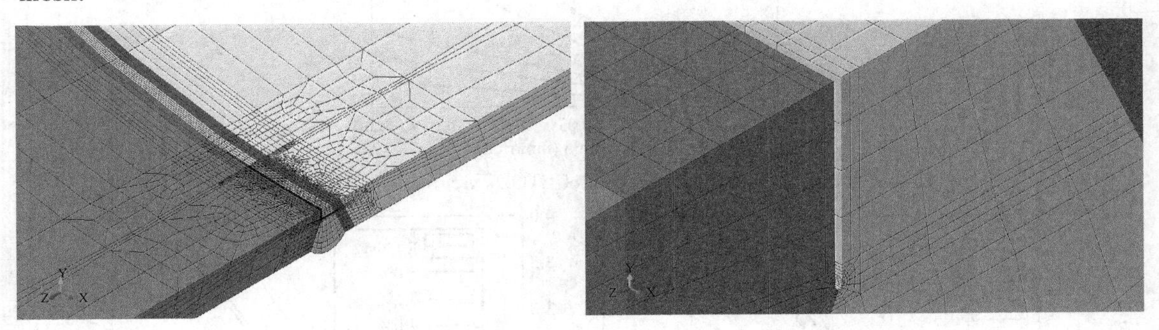

Fig. 13 The mesh of FEA model

In the FEA models, CTOD is defined at a convenient node behind the original crack tip. This position is chosen to avoid crack tip distortions that can occur at the original crack tip during advanced stages of plasticity, as shown in figure 14. Tracking nodes at the crack tip based on the traditional 45° CTOD definitions do not practically consider the deformation noted by the arrow in figure 14[6].

Stable ductile crack keeping growth until the resistance equals the driving force. Once the driving force reaches the tangency point, crack growth proceeds unstably because cracking driving force increases more rapidly than resistance with additional crack growth.

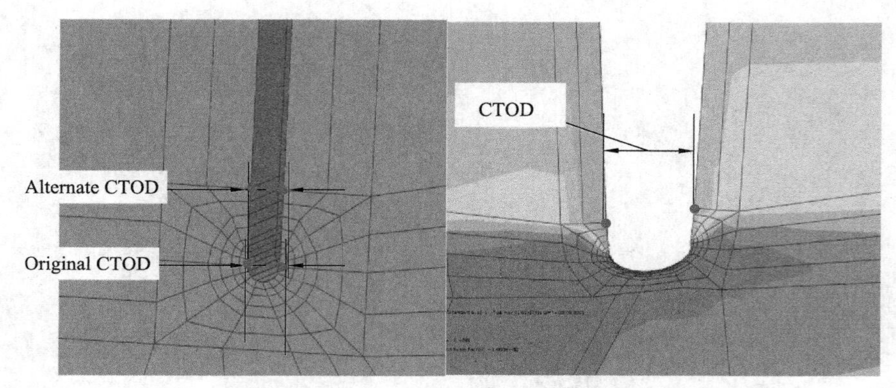

Fig. 14 CTOD definition at crack tip before and after loading

FEA simulations for pipe-weld-defect cases with different depth are conducted to produce driving force results in terms of CTOD versus remote strain. Ductile crack growth is accounted for by building stationary crack models with successively increasing flaw depth. This series of models represents crack status of a growing crack. The CTOD-strain driving force curves are converted into CTOD-crack growth curves to determine tangency by comparison with the tearing resistance R-curve as shown in figure 15. Figure 16 shows that the iso-strain of 5.3% has a tangency point with the R-curve of the SENT test for weld metal flaw and 5.1% for HAZ/FL flaw.

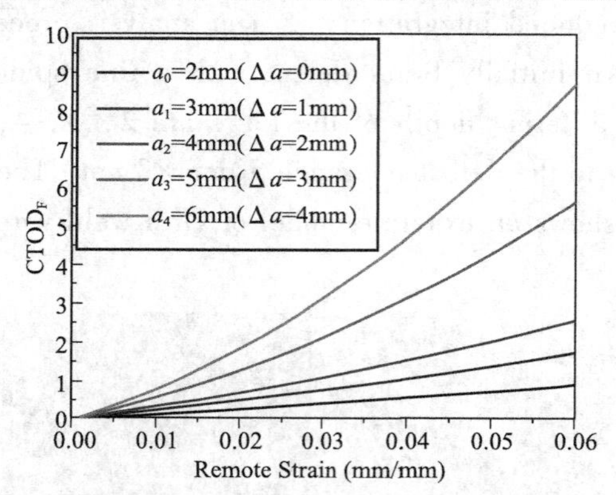

Fig. 15 Schematic drawing of $CTOD_F$ vs. remote strain

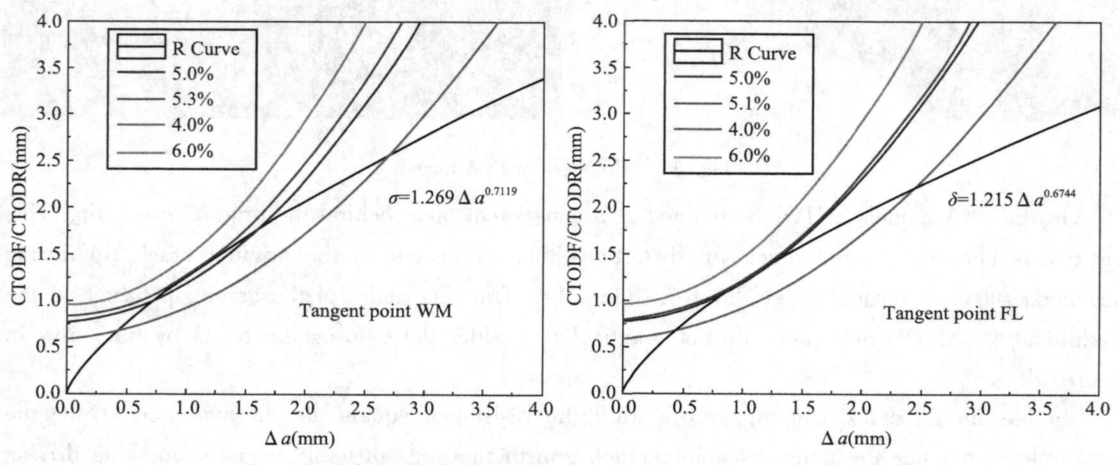

Fig. 16 Schematic drawing of iso-strain curves vs. crack growth

3.5 CWP test

Specimens

For five of the CWP specimens, simulated welding flaws are machined in the middle of the gauge width and outside diameter (OD) of the pipe. The flaws are machined on either the weld centerline (WCL) or the HAZ. Nominal flaw size is 4mm deep by 75mm long beside the standard flaws of 3mm deep by 50mm long. In preparation for flaw installation, the girth weld cap reinforcement of each specimen is removed over a short length (about 10 mm longer than the flaw) at the mi-width of the girth weld.

As shown in figure 17, the total length, gauge length is 1500mm and 1000mm separately, with 300mm width. All specimens are tested at room temperature (approximately 20°C). Two flaw types represent the typical workmanship (50mm×3mm, 3 mm is the average depth of a weld pass and 50 mm is twice the length of a surface breaking or the length of non-surface breaking flaw length typically allowed for every 300mm of weld length under workmanship standards)[12] and the maximum flaws (75mm×4mm) to ensure the pipeline failure by remote yield (with maximum Y/T ≤0.85), which provide in EPRG Tier 2[13, 14]. Flaw machining is initially done by rough cutting a slot in the CWP with a 1.6-mm thick saw blade. This rough notch is cut within approximately 1.5 mm of the final flaw depth. To machine the flaws to the final dimensions, a 70-mm diameter saw blade with a thickness of 0.15 mm is used, which produced a notch with a maximum width of 0.21 mm. Flaw depth is subsequently measured with a feeler gauge at a minimum of nine locations along the flaw depth.

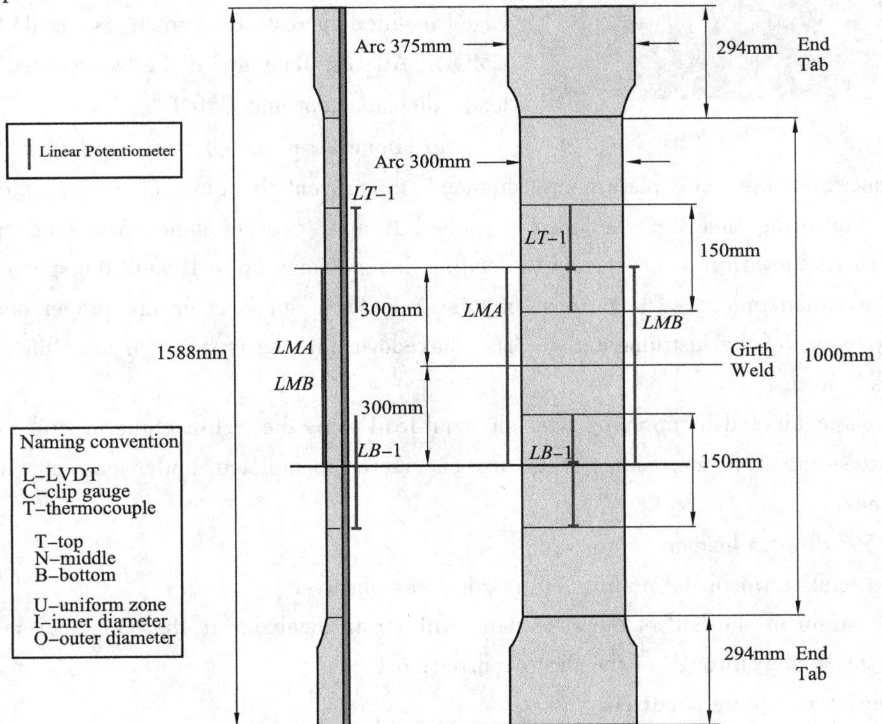

Fig. 17 Instrumentation layout of the specimens

Table 5 Specimens and flaws size

Test No.	Gauge length (mm)	Width in gauge (mm)	Thickness (mm)	Flaws position	Flaw size (mm)
1	1000	300	15.0	N/A	N/A
2	1000	299	15.2	WCL	3×50
3	1000	298	14.9	WCL	4×75
4	1000	299	15.0	HAZ	3×50
5	1000	300	15.1	HAZ	4×75

Fig. 18 Test setup

3.6 Instrument and procedure

The CWP tests are performed in Universal Testing System (UTS), a servo-hydraulic load frame capable of 15 MN of tensile load. The experimental setup, which is conducted by the specimens to a deformation-controlled and axially-oriented tensile load to failure, is shown in figure 18. Specimens are installed in the test frame using high strength bolts through the end plates to the machine base and actuator. Linear variable displacement transducers (LVDTs) which are used to measure overall elongation across the specimen and local elongation over a shorter gauge length centered about the weld. A clip gage mounted across the notch is used to measure CMOD. All test data are digitally recorded, including load, displacement and CMOD.

Specimens are aligned in the test frame and bolted into position, and the end plates are shimmed to prevent bending of the specimen during installation, ensuring that the specimen is pulled from a neutral state. After the specimen is installed, a shakedown run is conducted by loading the specimen up to 10% of the specified nominal yield load and then unload. The purpose of this shakedown run is to ensure proper operation and minimal hysteresis of the instrumentation. The shakedown process is repeated until the specimen is deemed ready to test.

Testingis conducted by applying a tensile axial load using the test machine in stroke control at a rate not exceeding 1.5 mm/min. Tests are considered completed under one of the following circumstances:

- A 5% drop in load was observed;
- 5% total strain in the uniform strain zone was observed;
- 5% strain in one half of the specimen, with strain localizing in the half, was observed;
- A stable pop-through of the flaw occurred; or,
- Specimen rupture occurred.

3.7 Test results

The uniform strain is calculated by averaging the displacement measurements of the two short

(150-mm) linear potentiometers and dividing the result by the average of the gauge lengths. This calculated strain is used to represent the far field remote strain. Gross stresses are calculated by dividing the axial load by the gross cross-sectional area of the gauge section, disregarding the area reduction as a result of the flaw. Failure strain is defined as the uniform strain at peak load, even if the maximum measured uniform strain exceeds this value.

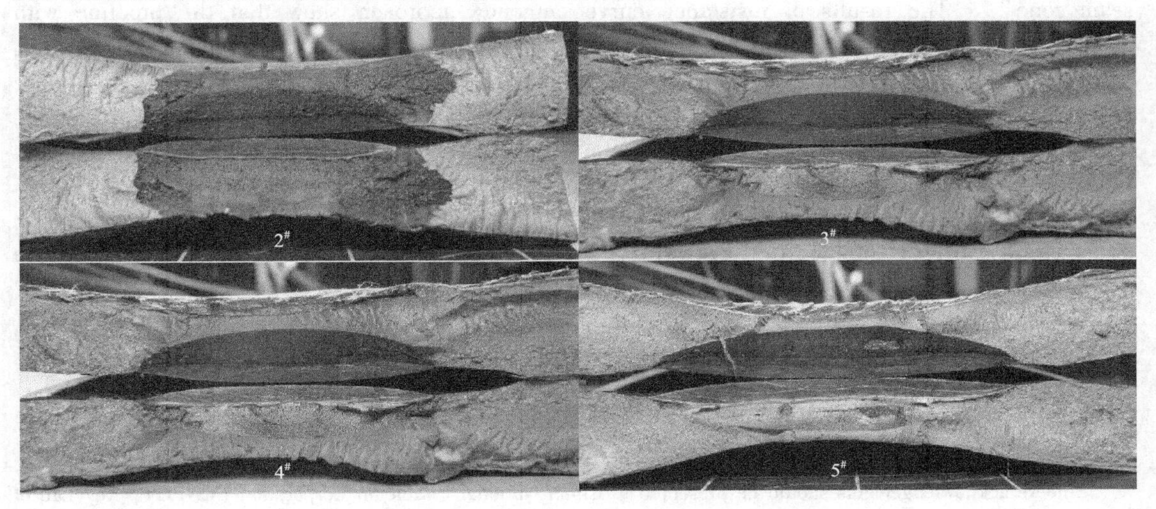

Fig. 19 Fracture morphology of the CWP specimens

As shown in figure 19, to measure the final crack depth, sections containing the flaw are cut from the specimen. These sections extended 150 mm on either side of the flaw in the longitudinal direction and 25 mm on either side of the flaw in the circumferential direction. The overall dimensions of the removed section are 300 by 100 mm.

CWP test results show that the remote strain achieved about 3.98% for specimens with 50mm×3mm WM flaws, meanwhile, 4.73% for HAZ/FL flaws, as $2^{\#}$ specimen and $4^{\#}$ specimen shown in table 6. For larger flaws of 75mm×4mm, remote strain drops down to about 3% strain.

Table 6 Specimens and results of the CWP tests

Test No.	Peak stress (MPa)	Strain (%, LT-1)	Strain (%, LB-1)
1	655	6.26	5.29
2	631	4.07	3.89
3	622	2.71	3.59
4	640	4.86	4.60
5	621	3.43	2.46

Test No.	Remote strain (%)	Total crack growth (mm)	CMOD (mm)
1	5.78	N/A	N/A
2	3.98	10.1	3.30
3	3.15	9.69	2.32
4	4.73	5.41	6.66
5	2.95	5.62	6.18

4 Conclusions

The paper conducts SENT test and FE analysis to demonstrate the strain capacity for an X70 strain-based designed pipeline of which all specimens of girth weld tensile test cracked in "near-seam zone". The results of resistance curve tangency approach show that the pipeline with maximum allowable weld defect can achieve considerable remote strain. Key results and conclusions are summarized below:

(1) Fracture tests are more appropriate than traditional tensile test for girth weld joint of strain-based designed pipelines to demonstrate the strain capacity under ductile fracture.

(2) To quantitative research for specific strain-based pipelines, the crack drive force of flaw under strain can be calculated by FEA. The SENT R-curves of the weld joint with flaws should be conducted for the description of ductile tear behavior.

(3) Although the specimens of girth weld tensile test fractures in "near-seam zone", the tangency approach and the CWP test show considerable plastic strain capacity.

Acknowledgements

Sincere acknowledgements should be presented to China National Petroleum Corporation (CNPC), Key Lab of Oil Tubular and Environmental behavior of CNPC for the support of Petro China Innovation Foundation (2013D-5006-0602) and their agreement of the publication of the paper.

References

[1] Wang Y, Liu M, Song Y. Second Generation Models for Strain-Based Design, Pipeline Research Council International report, PR-350-074509-R01, July 31, 2011.

[2] Hertelé S, O' Dowd N, Minnebruggen K, Denys R, Waele W. Effects of pipe steel heterogeneity on the tensile strain capacity of a flawed pipeline girth weld, Eng. Frac. Mech. 2014, 115 (1): 172-189.

[3] Chiesa M, Nyhus B, Skallerud B, Thaulow C. Efficient fracture assessment of pipelines. A constraint-corrected SENT specimen approach, Eng. Frac. Mech. 2001; 68 (5): 527-547.

[4] Kibey S, Wang X, Minnaar K, Macia M, Fairchild D, Kan W, Ford S, Newbuy B. Tensile Strain Capacity Equations for Strain-based Design of Welded Pipelines, Proc of the 8th Int' l. Pip. Conf., Calgary, Canada, 2010, 1-7.

[5] Fairchild, D, Macia M, Kibey S, Wang X, Krishnan V, Bardi F, Tang H, ChengW. A Multi-Tiered Procedure for Engineering Critical Assessment of Strain-Based Pipelines. Proc. of 21st Int'l Offshore and Polar Eng. Conf., Maui, U. S. A, 2011, 698-705.

[6] Østby E, New strain-based fracture mechanics equations including the effects of biaxial loading, mismatch and misalignment, Proc. of the 24th Int' l Conf. on Offshore Mechanics and Arctic Eng., Halkidiki, Greece, 2005, 649-658.

[7] Sandivk A, Østby E, Naess A, Sigurdsson G, Thaulow C. Fracture control - offshore pipelines: probabilistic fracture assessment of surface cracked ductile pipelines using analytical equations, Proc. of the 24th Int' l Conf. on Offshore Mechanics and Arctic Eng., Halkidiki, Greece, 2005, 641-647.

[8] Nyhus B, Østby E, Knagenhjelm H, Black S, Rostadsand P. Fracture control - offshore pipelines: experimental studies on the effect of crack depth and asymmetric geometries on the ductile tearing resistance, Proc. of the 24th Int' l Conf. on Offshore Mechanics and Arctic Engineering, Halkidiki, Greece, 2005,

731-740.
[9] Minnaar K, Gioielli P, Macia M, Bardi F, Biery N, Kan W. Predictive FEA Modeling of Pressurized Full-scale Tests, Proc. of 17th Int'l Offshore and Polar Eng. Conf., Lisbon, Portugal, 2007, 3114-3120.
[10] Kibey S, Minnaar K, Issa J, Gioielli P. Effect of Misalignment on the Tensile Strain Capacity of Welded Pipelines, Proc. of 18th Int'l Offshore and Polar Eng. Conf., Vancouver, Canada, 2008, 90-95.
[11] Kibey S, Minnaar K, Cheng W, Wang X. Development of a Physics-Based Approach for the Prediction of Strain Capacity of Welded Pipelines, Proc. of 19th Int'l Offshore and Polar Eng. Conf., Osaka, Japan, 2009, 132-137.
[12] Denys R, Hertelé S, Verstraete M. Longitudinal Strain Capacity of GMAW Welded High Niobium (HTP) Grade X80 Steel Pipes, Proc. of HSLP 2010: Int'l Seminar on Application of High Strength Line Pipe, Xi'an, China, 2010, 62-74.
[13] Knauf G, Hopkins P. The EPRG Guidelines on The Assessment of Defects In Transmission Pipeline Girth Welds. 3R International, 35 (10/11), 1996, 620-624.
[14] Denys R, Andrews R, Zarea M, Knauf G. EPRG Tier 2 Guidelines for the Assessment of Defects in Transmission Pipeline Girth Welds, Proc. of the 8th Int'l Pipeline Conf',. Calgary, Canada, 2010, 1-7.

Compressive Strain Capacity of X70 Cold Bend Pipes

Chen Hongyuan Chengshuai Huang Peng Wang Qiang Chi
Jiming Zhang Lingkang Ji

(Tubular Goods Research Institute of CNPC)

Abstract: Cold bending has been widely used in pipeline installation for long time. In general, cold bend pipes are considered not suitable for strain-based design of pipelines crossing harsh areas. However, cases of in-service plastic strains may be observed through the history of pipeline usage due to soil movement on unstable slopes, mining subsidence, and seismic loadings. The full scale bending test is deemed the most accurate ways to estimate the compressive strain capacity of line pipes. It is used in this study to measure the compressive strain capacity of cold bend pipes. In addition, a finite element model of cold bend pipe is constructed using the measured geometry and mechanical properties. The model is used to simulate the full-scale test that characterizes the limit capacity.

Keywords: Cold bend; Bending test; Strain capacity; Compressive strain limit

1 Nomenclature

YS Yield strength
TS Tensile strength
Y/T Yield strength to tensile strength ratio
D Outside diameter
SMYS Specified minimum yield strength
FEA Finite element analysis

2 Introduction

With sharply rising demand of oil and gas, pipeline industry has been experiencing a rapid development in recent years. Meanwhile, the pipelines for strain-based design have also developed. Pipelines in the areas of ground movement from slope instability, seismic sideslip, loess collapse, mining subsidence, frost heave and thaw settlement in discontinuous permafrost can experience displacement-controlled plastic strain. Cold bend pipes which are frequently required along pipeline routes where changes in the elevation of the trench, or the horizontal orientation of the pipeline, are also required by the strain-based design pipelines. These cold bend pipes, like the straight pipes, may experience displacement-controlled plastic strain during the pipeline life.

Consequently, it is of interest to understand the factor that contribute to local buckling for cold bend pipes, and the factor contribute to the material and geometry changes. Current pipeline design codes do not provide methodologies for the prediction of the buckling strain for cold bends, however it is important to predict the strain capacity of a cold bend pipes prior to buckling in order to ensure adequate maintenance of the pipeline.

The paper discusses the strain capacity of L485 SAWL cold bend pipes by TMCP process. The related pipe size are ϕ1016mm×17.5mm. Several issues related to strain capacity such as initial imperfections, material properties are mentioned. Firstly, the effects of plastic strain from the field cold bend process on materials properties are used to FEA model. Secondly, a series point cloud data from cold bend pipe are used to construct a CAD pipe model to present a real pipe FEA model. Finally a full scale bending test was introduced to verify the results of FEA.

3 Background

Because of the plastic deformation during the field bending process, the longitudinal properties of body pipe may change relate to the amount of strain (Chi, et al., 2010). The quantity and symbol are depended on the cold bend process and bending direction. Furthermore, the plastic strain and the curvature are nonuniform due to the stepper bending process. It also has a effects on compressive strain capacity of cold bend pipes. There comes a quantification requirement on the compressive strain capacity to application of cold bend pipes in strain-based design pipelines. The buckling behaviors are investigated by means of full scale bending test and finite element analysis with consideration of mechanical properties changes and shape discontinuous.

In the paper, a ϕ1016mm × 17.5mm X70 cold bend SAWL pipe are investigated to demonstrated the compressive capacity. The pipes are manufactured for strain-based design pipelines of Sino-Burma pipeline by TMCP process (Chen, et al., 2011) with higher strain capacity than traditional pipes. For better understand the contributions of cold bend process on strain capacity, a FEA model are constructed by points cloud from precise laser measurement.

4 Geometry Measurement

Initial geometric imperfections have important effects on the buckling response of pipe segment (Suzuki and Igi, 2007). FEA without geometric imperfection tends to overestimate the critical compressive strain. The geometric imperfections such as change of wall thickness and popple of pipe outer surface should be taken into account in order to predict the compressive strain limit with accuracy. The imperfections which present by wall thickness and outer surface coordinates for more than 1440 points on a section of the cold bend pipe are measured by ultrasonic thickness gauge and by laser tracking system respectively, as shown in figure 1 to figure 3. The radius of the coordinates on pipe outer surface are measured by laser per 30° circumferentially and 50mm longitudinally. The thickness are measured on corresponding point to describe the initial geometric imperfections on inner surface. Figure 4 (a) are measurement results for intrados and extrados while figure 4 (b) show the thickness changes after cold bend and the coordinate change for intrados and extrados. It trend to thicker along the intrados and thinner along the extrados respectively. The points on both outer surface and inner surface are jointed by 240 B-splines separately which construct a line frame of real pipe as shown is figure 5. The purpose of the accurate measurement and CAD modeling is to demonstrate geometric effects from cold bending.

Fig. 1 The measurement of laser tracking system

Fig. 2 The ultrasonic thickness measurement

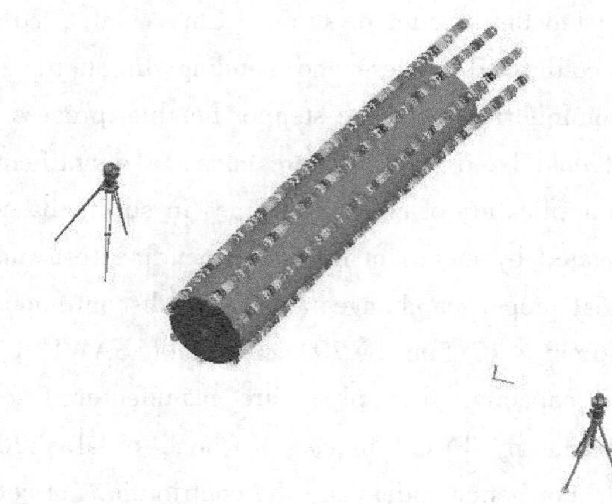

Fig. 3 Data cloud of line pipe

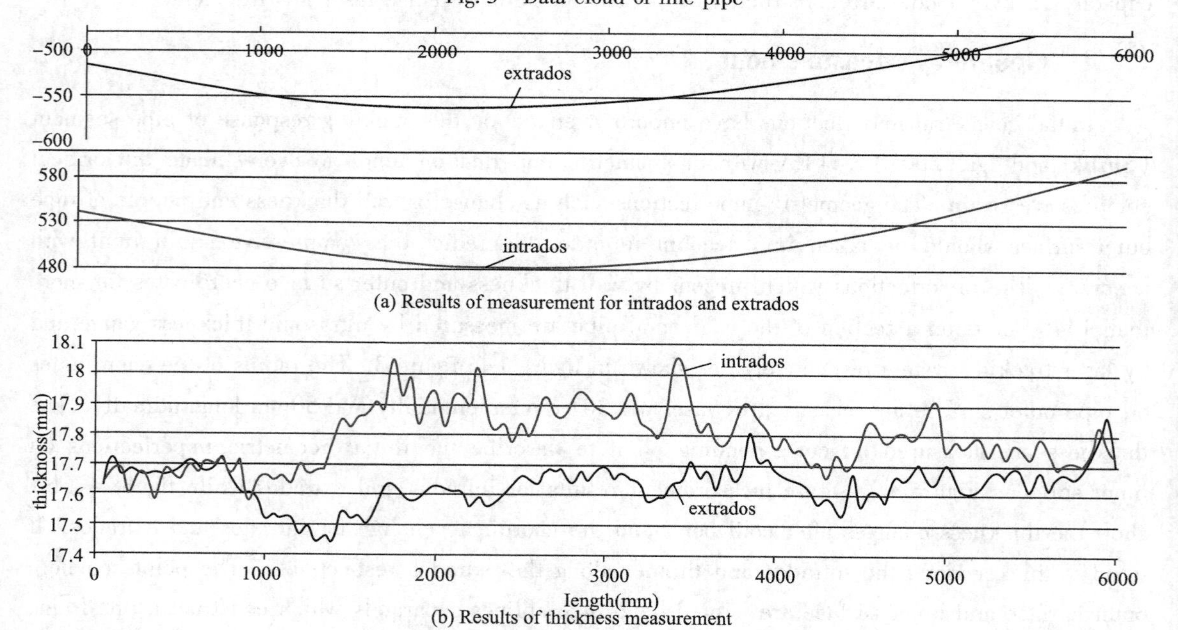

(a) Results of measurement for intrados and extrados

(b) Results of thickness measurement

Fig. 4 Geometric measurement

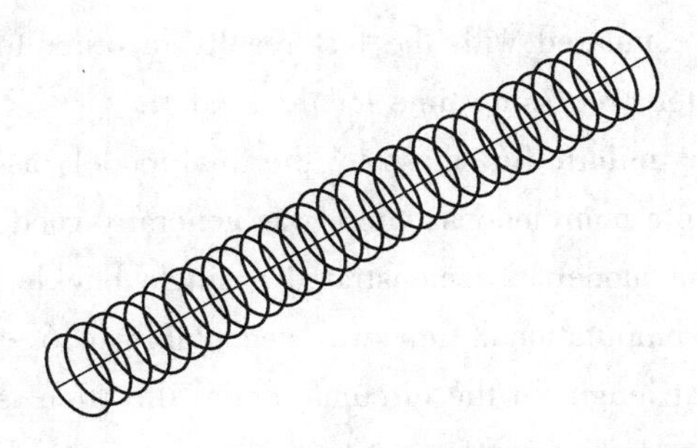

Fig. 5 Line frame of cold bend pipe by splines

The measuring facility is API (Automated Precision Inc.) Track 3 laser tracking system which is a portable, high precision (5μm/m) coordinate measurement system. It combines interferometer based laser optics with state-of-the art servo control technology to measure the position of a target relative to the tracker.

5 FEA Model

A finite element model of cold bend pipe is constructed from geometric imperfection CAD model which is present in figure 5. The surface of the pipe is constructed by the splines. The figure 6 illustrate that the pipe length and moment arms are 10160 mm which is 10 times of pipe outer diameter (OD), for acquiring the results of pure bending deformation. Internal pressure are applied of 10MPa, which is 72% SMYS. The left-side moment arm is assumed to be a rotation-free pin support, and that of the right-side moment arm a slide-free and rotation-free pin support like the bending test facility. Eight-node brick elements were applied to idealize the linepipe, which was divided into 400, 120 and 2 segments in the longitudinal, circumferential and radial directions, respectively. The mesh is shown in figure 7. The FEA software is ABAQUS 6. 10. 1.

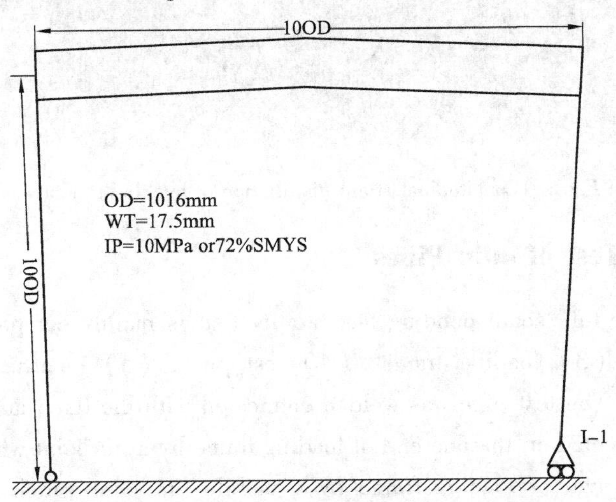

Fig. 6 FEA model of full scale bending test

To understand the effect of the mesh size and distribution on the behavior of the model, a mesh study was conducted. several different meshes were considered for this sensitivity analysis. Finally the mesh considered consisted of a uniform fine mesh. The load - displacement response of the models with the meshes was compared with the test results in order to achieve a suitable balance between accuracy and computer processing time for the model.

It was resolved to use the uniform fine mesh for the final model, as shown in figure 7. Although the mesh contained considerable many nodes, there was generally good agreement with the behavior of the test specimens, and the model is demonstrated a single buckle at the middle of its length. Meanwhile, the duration of computational time was acceptable of 6 ~ 8 hours, depending on the output requested. The element length in the circumferential direction is nearly 25 mm for the pipe specimens. The longitudinal node spacing for the bent segment is nearly 25mm too.

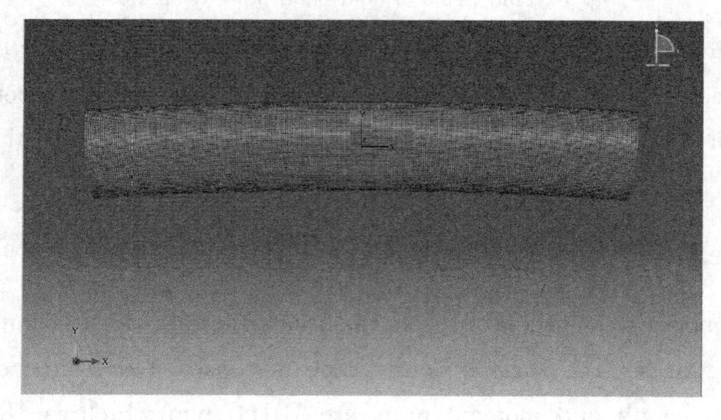

Fig. 7 Mesh of the FEA model

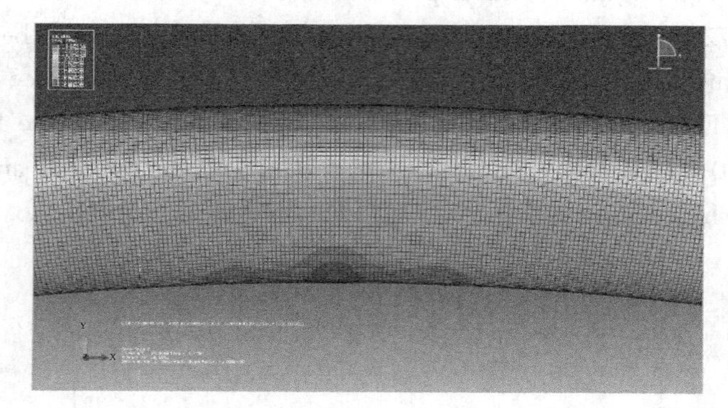

Fig. 8 Longitudinal strain distribution of buckled test pipe

6 Full-scale Test of 40in Pipes

Figure 9 shows the full-scale bending test facility that is mainly composed of (1) two arms, (2) hydro-cylinder, (3) loading frame, (4) test pipe, (5) electric accessories and (6) measuring instruments. The test pipe was welded connected with the fixed arm and the mobile arm. The fixed arm was connected on the one end of loading frame by a pin joint while the mobile arm was connected on the other end of loading frame by a mobile pin joint which can slide in the axial direction of hydro-cylinder. It shows that the facility is located in a low-lying rectangular pit. The test pipe was prepared for almost 8.0m length. The test pipe was filled with water and pressurized to 10Mpa (72%SMYS) during the test. Thereafter the pipe was compressed and bent by the moment from mobile arm and hydro-cylinder until buckling occurred. The maximum load and stroke of hydro-cylinder are 6000 kN and 3m respectively. The length of arms are 6 m thus the maximum moment is 36000 kN · m. The bending load is displacement-controlled and the internal pressure should be adjusted for each displacement increment to keep the 10MPa because the volume changes of pipe during the bending deformation may cause the pressure change. The loading conditions are shown in figure 10.

Fig. 9 Full-scale bending test facility

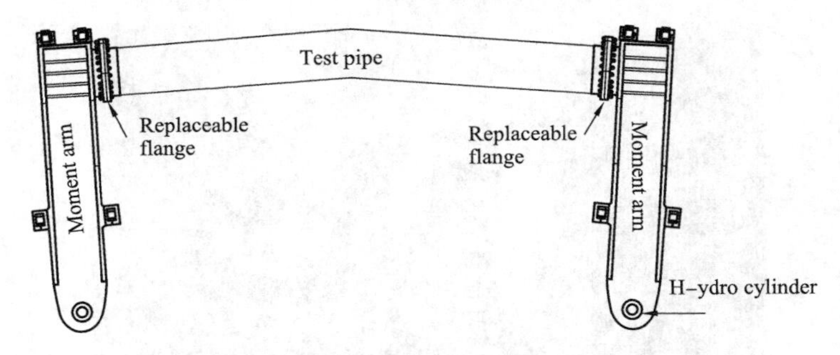

Fig. 10 Sketch of full-scale bending test set-up

The basic parameter of the facility are shown in table 1.

Table 1 Basic parameters of the facility

Item	parameters
Pipe diameter	20~40in
Pipe thickness	Max. 26. 4mm
Pipe length	6~12m
Load capacity	6000kN
Moment Capacity	36000kN
Stroke of hydro-cylinder	3m

7 Test Pipe

A section of 40 in cold bend pipe with 17. 5mm thickness of X70 Grade were prepared for the bending test. The cold bend process are 0. 4°/300mm, as shown in figure 11. The length of the test pipes section was approximately 8m that of 8 times as of the pipe diameter. The pipe was constantly pressurized to 10 MPa (equals to 72% SMYS) .

Table 2 presents the longitudinal tensile properties of 3 specimens from the test pipe including YS, TS, Y/T, and the stress ratios. The stress-strain curves of three longitudinal specimens of the pipe body shows round-house appearance. As for high strain line pipe (Chen, et. al., 2011), the stress ratios of the pipe are listed. It seems higher than most conventional pipes.

Table 2 Longitudinal tensile properties of the pipe

Specimens no.	YS	TS	Y/T
1	556	652	0.85
2	567	661	0.86
3	558	655	0.85
Specimens no.	$R_{t1.5}/R_{t0.5}$	$R_{t2.0}/R_{t1.0}$	$R_{t5.0}/R_{t1.0}$
1	1.11	1.04	1.07
2	1.13	1.05	1.08
3	1.12	1.04	1.07

Fig. 11 Filed cold bend of 40 in pipes

8 Test Result

The strain distribution of intrados and extrados of the test pipe are measured during the test by strain gauges. About 140 strain gauges were used to measure the strain distribution during the bending test as shown in figure 9. The figure 12 shows the strain distribution of intrados and its evolution during the test. The line 0~7 means the compressive strain distribution on the intrados of the pipe at different moment. It shows that the strain concentration occurred on intrados due to buckling. Once the buckle initiated, strain were measured on all strain gauges.

The most important result of the test is buckling strain. It can be present by the average compressive strain on the 2D length which centered around the point of strain concentration. Figure 12 shows that the 2D average strain of buckling is larger than 1.35%. The maximum load are almost 1053kN.

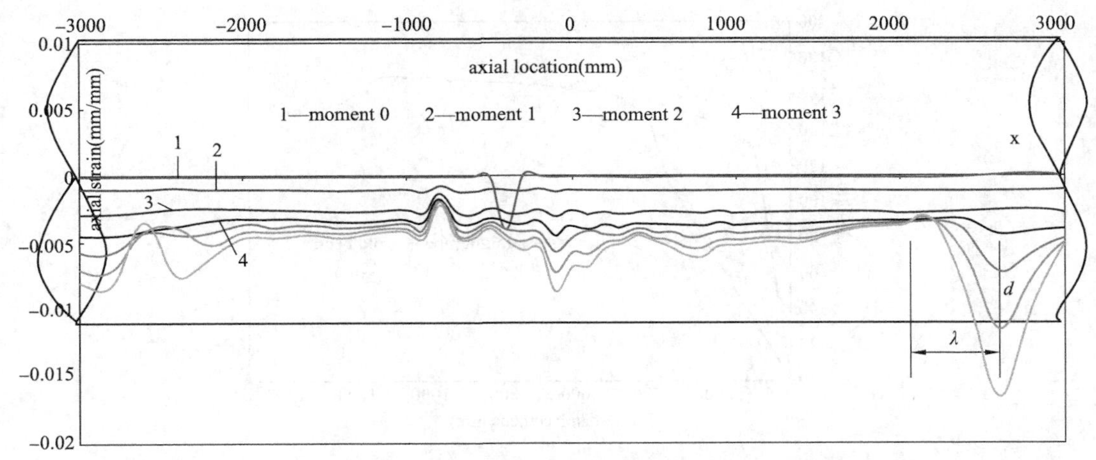

Fig. 12 Strain distribution on intrados during the test

Bending continued until a decrease in the bendingload of approximately 10% was observed (indicating that the buckle is well developed). The figure 13 shows the blister of pipe after test and the comparison with the FEA result. The comparison of the FEA model and test load-displacement responses has demonstrated that the FEA model is adequate in simulating the buckling behavior of a cold bend pipe, as shown in figure 14.

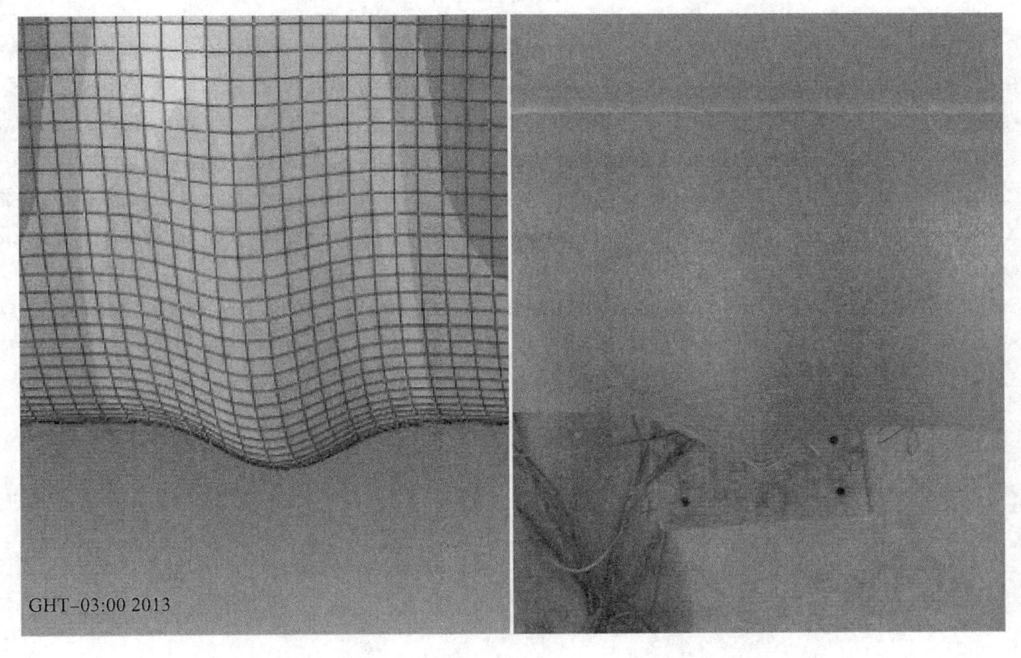

Fig. 13 Comparison of FEA result and post-buckling pipe section

9 Conclusion

A full-scale bending test of 40 in X70 cold bend pipe was conducted to investigate the compressive strain capacity and post-buckling behaviors of cold bend pipe which was cold bent from a pipe manufactured for strain-based design pipelines. The test results agree well with the FEA results. We can conclude as follows based on the research results.

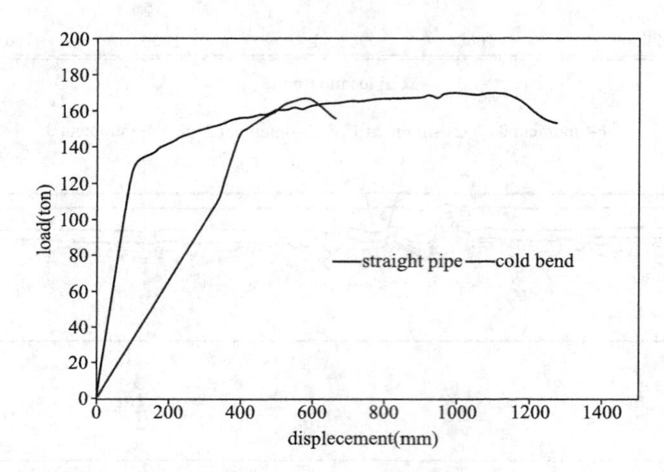

Fig. 14 Comparison of load-displacement curve from FEA and full-scale bending test

Acknowledgements

The authors would like to acknowledge China National Petroleum Corporation (CNPC), Key Lab of Oil Tubular and Environmental behavior of CNPC for the financial assistance and allowing the publication of the paper.

References

[1] Chen H. Y. et. al (2011), "Test Evaluation of High Strain Line Pipe Material," Proceedings of the Twenty-first (2011) International Offshore and Polar Engineering Conference Maui, Hawaii, USA, 4: 581-585.

[2] Chen H. Y. et. al (2012), "Analysis of Bending Test of 40" X70 Line Pipe, "Proceedings of the Twenty-second (2012) International Offshore and Polar Engineering Conference Rhodes, Greece, 4: 571-576.

[3] Chi Q., Ji L. K., Liu Y. L., Wang P (2010), "Investigation on Strain-age of Large Diameter and Thick Wall X80 Grade Cold Bends," Proceedings of the Twentieth (2010) International Offshore and Polar Engineering Conference, Beijing, China, ISOPE, 4: 436-440.

[4] Suzuki N. and Igi S (2007). "Compressive Strain Limits of X80 High-Strain Line Pipes," Proceedings of the Sixteenth (2007) International Offshore and Polar Engineering Conf, Lisbon, Portugal, ISOPE, 4: 3246-3253.

Influence factors of X80 Pipeline Steel Girth Welding with Self-shielded Flux-cored Wire

Qi Lihua Ji Zuoliang Zhang Jiming Wang Yulei Hu Meijuan
Wang Zhenchuan

(Tubular Goods Research Institute of CNPC)

Abstract: For girth weld of high pressure oil and gas transmission pipeline, there is impact toughness values deviation phenomenon with self - shielded flux - cored semi - automatic welding technology. The macro - images, microstructure and mechanical performance of girth welding joint have been investigated by OM, SEM, TEM. The results show that there are several factors of impact toughness unqualified values of weld joints, such as welding heat input, coarse grain zone and a chain of M-A organizations, Al_2O_3 and Zr Precipitates particle sizes and distribution et. al. , which are the main unqualified reasons of welding impact toughness of the semi-automatic self-shielded core butt welding process of X80 pipeline steel.

Keywords: Girth welded joint; Microstructure; Mechanical properties; Coarse grain and M-A mixing zone; Heat input; Precipitates

1 Introduction

The continuously growing demand in natural gas globally over the past 20 years promoted the high speed construction of pipeline in using high strength and large diameter pipeline steel[1-3]. Girth welding faces several new challenges when high - strength and toughness pipeline steel was used, such as, X70, X80 and X90, et. al. [4,5]. It is reported that over 80% high-strength pipeline girth weld was adopting self - shielded flux - cored wire semi - automatic welding method[2, 4-6] in China. Various factors have significant influence on the impact toughness of girth weld[6-13], such as heat input, microstructure of weld joint especially for M-A organizations distribution in the coarse grain zone, and precipitates distribution such as AlN, Al_2O_3 and Zr. Besides the factors mentioned above, technical level of welder, filling pass of welding beam, fluctuation of welding current and voltage, the thickness of each layer also affect the impact toughness of girth weld. Generally, smaller welding heat input results in better impact toughness for girth weld of X80 grade pipeline steel. However, the present work demonstrated that the impact toughness of small heat input is far lower than that of large heat input. Therefore, it is necessary to investigate the influencing factors of girth welding process, it is expected that this work would significantly improve the girth weld quality during pipeline construction, and therefore, reducing the failure risk during pipeline operation [11-13].

2 Experimental

2.1 Two groups test of girth weld

The base material is X80 steel pipe with specification of 1219mm × 18.4mm. Metal powder cored wire with 1.2 mm in diameter was used for root welding, E81T8-Ni2 wire with 2.0 mm in diameter was used for filling and cover welding.

Two groups' tests were carried out. Group-I, three base material samples, such as 1# sample, 2# samples and 3# sample, were adopted by different manufacturers, and welding wire was adopted by the same factory; Group- II, base material was adopted by the same manufacture, and welding wire were fabricated by different manufacturers, Factory - A and Factory - B, Same welding parameters were employed for two groups, as shown in Table 1.

Table 1 welding process parameters of FCAW-S filling cover face with RMD root welding

Weld bead	Welding grade	diameter (mm)	welding current (A)	Voltage (V)	welding speed (cm/min)
Root welding	AWSA5.28 E80C-Ni	1.2	160~170	16~17	25~30
Hot welding	AWSA5.29 E81T8	2.0	200~220	18~20	25
Fill welding	AWSA5.29 E81T8	2.0	200~240	18~22	20~25
cap welding	AWSA5.29 E81T8	2.0	200~230	18~21	18~20

2.2 Chemical element analysis

Chemical element of steel pipe and weld joint were analyzed by spectrometer. The results of Group-I and Group- II are as given in Table 2 and Table 3.

Table 2 The chemical composition of the deposited metal of X80 steel pipe and welding material (wt%)

Element	C	Si	Mn	P	S	Cr	Mo	Ni	Nb	V	Ti	Cu	B	Al
1# sample	0.05	0.19	1.77	0.010	0.001	0.30	0.20	0.13	0.08	0.001	0.010	0.1	0.0001	0.030
2# sample	0.052	0.22	1.89	0.010	0.0018	0.31	0.18	0.048	0.084	0.003	0.014	0.057	0.0003	0.038
3# sample	0.054	0.19	1.82	0.082	0.0025	0.33	0.18	0.05	0.082	0.002	0.015	0.01	0.0002	0.035
E81T8-Ni2	0.042	0.06	1.46	0.0091	0.0035	0.025	0.006	2.43	0.005	0.003	0.0002	0.020	0.0013	1.1294

Table 3 The chemical composition of the deposited metal of X80 steel pipe and welding material (wt%)

Element	C	Si	Mn	P	S	Cr	Mo	Ni	Nb	V	Ti	Cu	B	Al
4# and 5# sample	0.05	0.19	1.77	0.010	0.001	0.30	0.20	0.13	0.08	0.001	0.010	0.1	0.0001	0.030
Welding Factory-A	0.042	0.06	1.46	0.0091	0.0035	0.025	0.006	2.43	0.005	0.003	0.0002	0.020	0.0013	1.1294
Welding Factory-B	0.034	0.18	1.37	0.0107	0.0032	0.027	0.005	1.66	0.003	0.003	0.0015	0.006	0.0014	1.0908

2.3 Mechanical properties

All the samples for mechanical tests of the weld joint were carried out. Tensile test results of Group-I and Group- II are shown in Table 4 and Table 5, respectively. Charpy test results with 3 point and 12 point position of Group-I and Group- II are shown in Fig. 1 (a) and Fig. 2 (c).

2.4 Metallurgical properties

Three samples of Group-I and two samples of Group- II for metallurgical properties of the weld joint were observed, as illustrated in Fig. 1 and Fig. 2.

2.5 Microstructure analysis

In order to investigate the influencing factors of girth welding process, the microstructure of coarse grain zone and M-A constitute of the weld joint was observed by scanning electron microscope (SEM) and transmission electron microscopy (TEM). The precipitate particles of Group - II samples were conducted by the method of carbon membrane extraction, observed by high-resolution transmission electron microscopy (HRTEM).

2.6 Thermal simulation test

Computer finite element modeling of welding temperature field of the weld joint was established according to the girth welding process. According to temperature cycling curve of finite modeling result, weld thermal simulation test was carried out using Gleeble 3500 testing machine, and then mechanical tests was done.

3 Results

3.1 Mechanical properties of Group-I

Macro-images of three girth weld joints prepared at 3 o' clock positions are shown in Fig. 1. Except for root welding, heat welding and cap welding, there are different filling pass in three samples. Samples 1# and 3# are both with 11 filling passes, while Sample 3# is not even, the earlier deposited metal layer was melted partially by the latter deposited metal, the deposited layer boundaries are very hard to identify. Unlike Sample 1# and 3#, Sample 2# only has 3 filling passes. The tensile strength of three samples is given in Table 4, which is between 714 and 740MPa, meeting the API RP 5L standard requirement. However, the values of impact toughness of 3 samples is big different, as shown in Fig. 1 (d)

Table 4 Mechanical performance test results

Sample	Fill welding pass	Standard requirements
1#	11pass	≥625MPa
2#	3pass	
3#	11pass	

The impact toughness of Sample 1# is the highest, mean values of which are 146J and 209J, respectively. That of Sample 3# is the lowest, which are 44 J and 58J, respectively. Welding current and voltage of Samples 1# and 3# are basically the same, each filling pass thickness is also close to each other. But there is big difference of impact toughness values between Samples 1# and 3#.

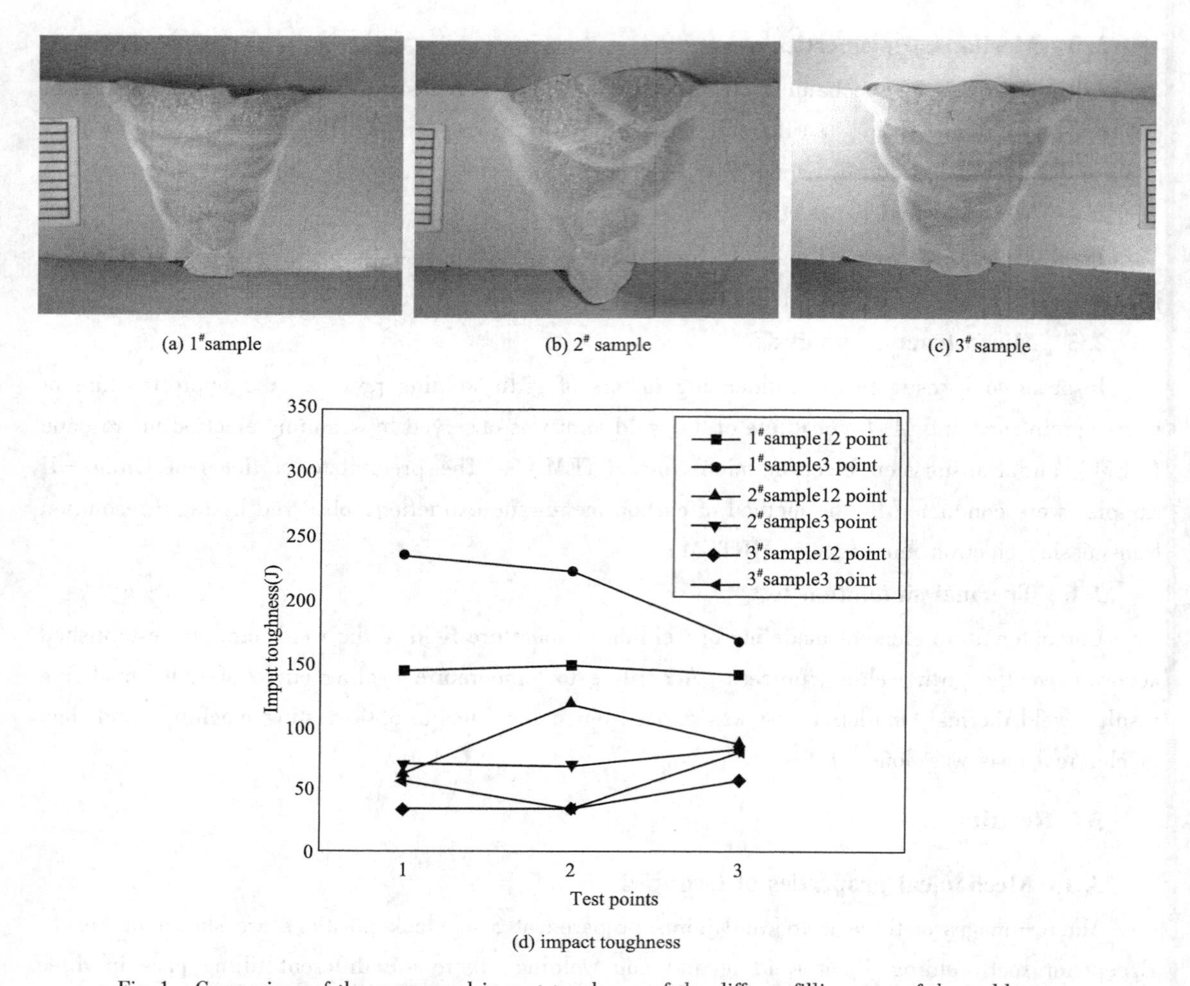

Fig. 1 Comparison of the macro and impact toughness of the differentfilling pass of the weld seam

Compared with Sample 2#, the welding heat input of Sample 2# is the largest, and filling pass thickness is the most thick. However, the impact toughness values of Sample 2# are 90J and 74J, respectively, which is higher than that of Sample 3#, but is lower than that of Sample 1#. Therefore, except for welding current and voltage, it is necessary to analysis the other factors of welding impact toughness by semi-automatic self-protective core welding method.

3.2 Mechanical properties of Group- II

Macro-images of three girth weld joints prepared at 3 o' clock positions are shown in Fig. 2. Except for root welding, heat welding and cap welding, Samples 4# and 5# are both with filling passes. The tensile strength of 2 samples is given in Table 5, which is between 708 and 742MPa[14]. The mean values of impact toughness of Sample 4# are 91J and 215J, respectively, and thats of Sample 5# are 65J and 111J. The highest single value of Sample 5# is 162J, the lowest single value is only 25J, and several single values do not meet the standard requirements. The welding current and voltage of both samples are basically the same, the deposited metal thickness of each pass of the two samples is relatively close, but the impact toughness values of the two samples are quite different.

Table 5 Mechanical performance test results

Sample	Fill welding pass	Standard requirements
4#	12 pass	≥625MPa
5#	12 pass	

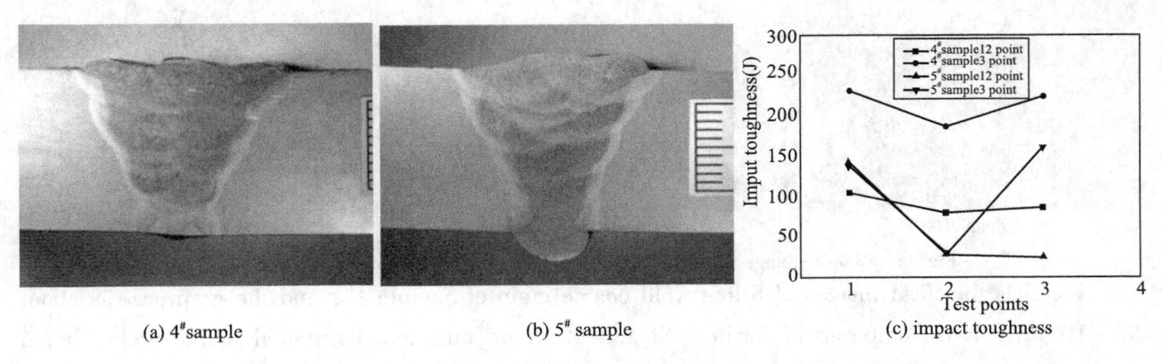

(a) 4#sample (b) 5# sample (c) impact toughness

Fig. 2 Comparison of the macro and impact toughness of thesame filling pass and different wire manufacturer of the weld seam

4 Influencing factors analysis

4.1 Welding heat input

Macro-images and the microstructure schematic diagram of the various regions of girth welding joint are shown in Fig. 3. The regions with arrows in the diagram are the columnar crystal region, the coarse grain region, the coarse grain and the M-A mixed zone respectively. Because of the great influence of coarse grain region and M-A mixed zone on the impact toughness of the welding joint , it will be investigated further.

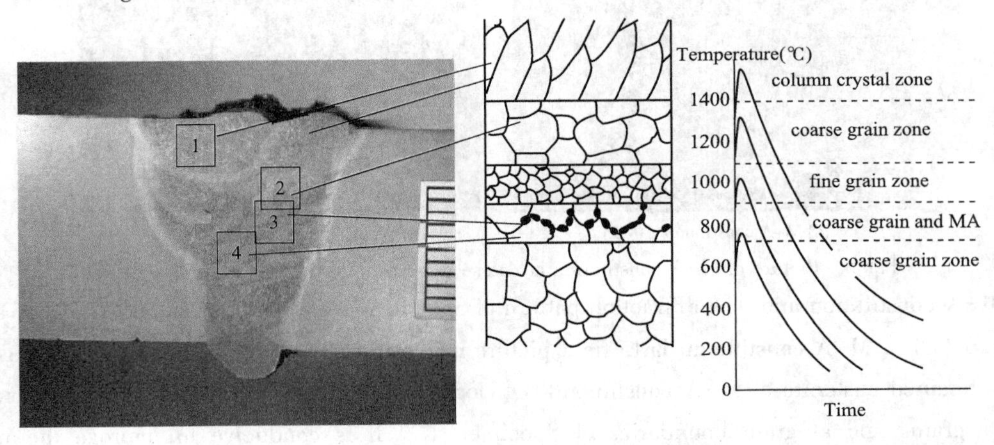

Fig. 3 Macro-images and microstructure schematic diagram of the grain shape of each region

Macro-images of coarse grain area and enlarge images of filling weld are shown in Fig. 4. Grain size of Sample 1# is about 50 ~ 100μm, which is composed of ferrite laths and M-A constituent. That Grain size of Sample 2# is coarser than that of 1# sample, which is mainly composed of lath ferrite and M-A constituent. Compared with Sample 1#, due to the welding heat input increase and filling passes decrease, the deposited metal thickness of each filling pass of

Sample $2^{\#}$ is about 4~5 mm (Fig. 1), and the grain size is significantly coarser than that of Sample $1^{\#}$ and ferrite laths in austenite grain interiors are bunched structure, as shown in Fig. 4 (c). Impact toughness of coarse grain microstructure of Sample $2^{\#}$ is lower.

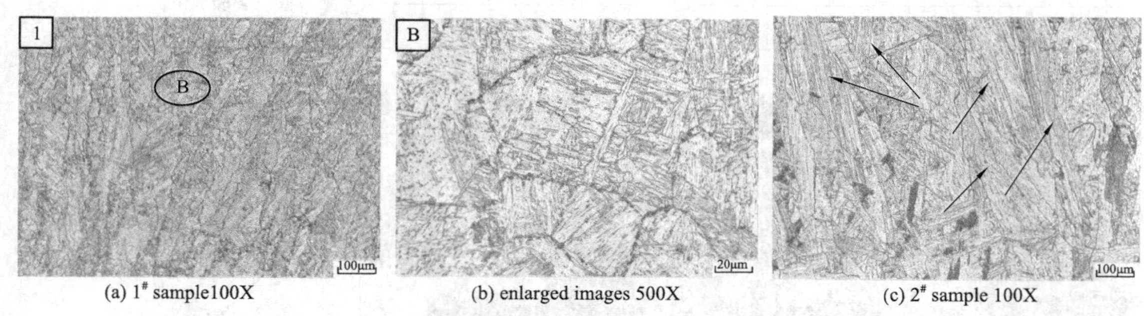

(a) $1^{\#}$ sample100X (b) enlarged images 500X (c) $2^{\#}$ sample 100X

Fig. 4 Macro-image of coarse grain area of Samples $1^{\#}$ and $2^{\#}$ packed layers

Fig. 5 is the SEM images of filling weld coarse grain of Sample $1^{\#}$, and the grain size is about 50 ~ 100μm. It is composed of ferrite lath and M-A organization distributed in the grain, but a chain of M-A constituent distributed at the grain boundary. It is well known that M-A chain weakened interfacial energy at grain boundaries and reduced the impact toughness of the weld. Tiny bright white particles with size of 1 ~ 2μm distributed around grain boundaries are the M-A constituent chain. As we all know, a single tiny distributed in grain or grain boundary is beneficial for grain nucleation, inhibiting grain growth and enhanced microstructure toughness. But for a chain M-A constitution distribution on the grain boundaries, it will be weakened the binding force between the grains, and bring the disadvantages to the toughness.

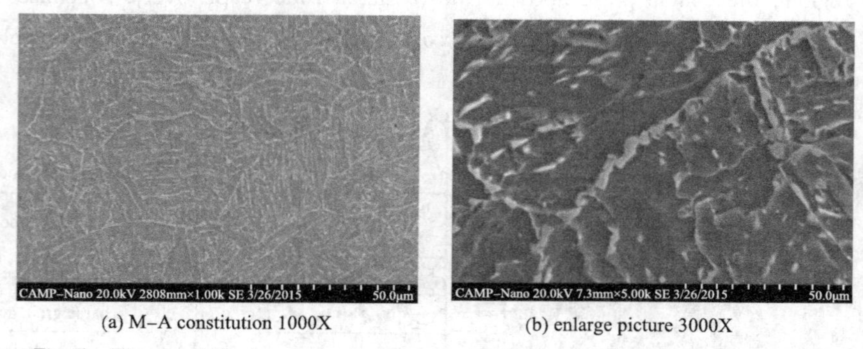

(a) M-A constitution 1000X (b) enlarge picture 3000X

Fig. 5 Picture of M-A constitution at coarse grain zone of the filling welding SEM

M-A constitution and it's diffraction pattern of Sample $1^{\#}$ were observed by HRTEM, as shown in Fig. 6 (a). M-A constituent dark-field picture is shown in Fig. 6 (b), which the bright white part is retained austenite of M-A constituent. Obviously, there is more retained austenite found at internal grains and at grain boundaries of Specimen $1^{\#}$. It is conducive to improve the impact toughness of the weld. According to retained austenite in the M-A constituent, it is beneficial for the high strength of the martensite and the deformation behavior of retaining austenite.

M-A microstructure at the grain boundaries of Sample $2^{\#}$ was investigated by HRTEM, as shown in Fig. 6 (c) and (d), and martensitic diffraction pattern in Fig. 6 (d). Obviously, the dislocation density of the ferrite laths around M-A constituent at grain boundaries is much lower than that of the ferrite grains around martensite interior grains, see Fig. 6 (d). It means that the interface energy of Sample 2# is lower, and the cracks around M-A constituent at grain boundaries are easier to form and expand than that of internal grains. The retained austenite in M - A constitution of sample 2# was not observed both the grains boundary and internal grains.

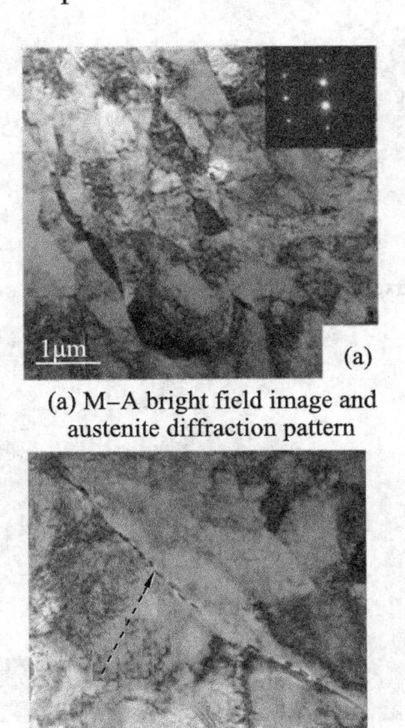

(a) M-A bright field image and austenite diffraction pattern

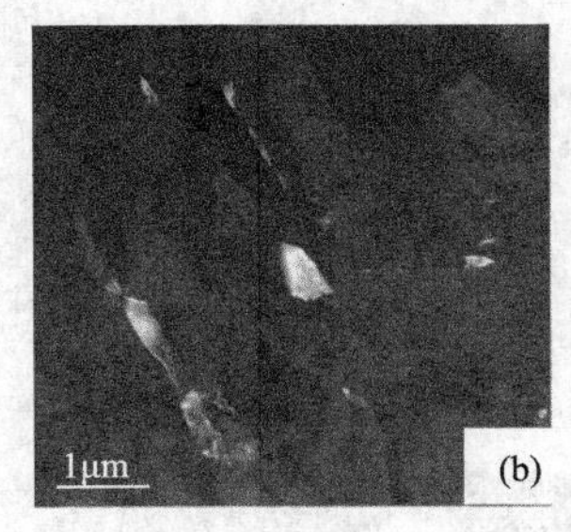

(b) M-A dark field picture of 1#sample

(c) M-A image of 2#sample

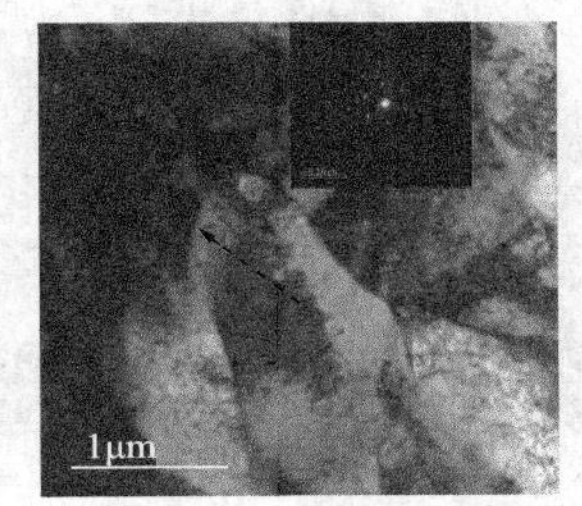

(d) M-A image and martensite diffraction pattern of 2#sample

Fig. 6 Picture of microstructure and diffraction pattern of M-A constituent of 1# and 2# samples TEM

4.2 M-A constitution morphology

Macro-images pictures of coarse grain and M-A mixed area of Samples 1# and 3# are given in Fig. 7. The black boxes in Fig. 7 (c) and (d) are the sampling position of the impact toughness, there are 7 layers coarse grain and M-A mixed zone of Sample 3# and 5 layers coarse grain with M-A mixed zone of Sample 1#, respectively. Although the welding heat input, grain sizes and distribution of two samples are similar, the impact toughness of Sample 1# is about 146~209J, but that of Sample 3# just only 44~58J, it is significantly different, see Fig. 1.

SEM and enlarged images of coarse grain and M-A mixed area at grain boundaries were shown in Fig. 8. Ferrite laths have no obvious direction around M - A constitution due to the high temperature remelting, and lath boundary passivation around the deposited metal. The grain sizes of M-A constitution near the grain boundaries were slightly aggregated and grown. The size of M-A constitution at the boundaries is coarse, 4 ~5μm. By further magnification observation, there are many light fine shape material on the laths as shown in Fig. 8 (c) .

Coarse grain zone and M-A constitution at the grain boundaries were observed by HRTEM of directional sectioning observation, as shown in Fig. 9 (a) and (b), and diffraction pattern of M-A constitution is in Fig. 9 (c) . The size of martensite is about 2~4μm, high carbon martensite, and the distribution is more concentrated. As we all know, martensite with high hardness and low impact toughness shows a brittle phase in the microstructure of welded joint. If it is distributed aggregated, which is easy to form stress concentration, and impact toughness decrease. It will lead to crack formation and extension.

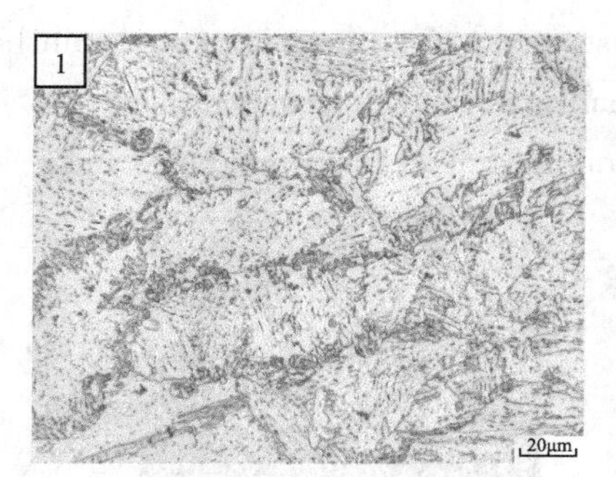

(a) coarse grain and M-A mixed zone of $1^{\#}$ sample

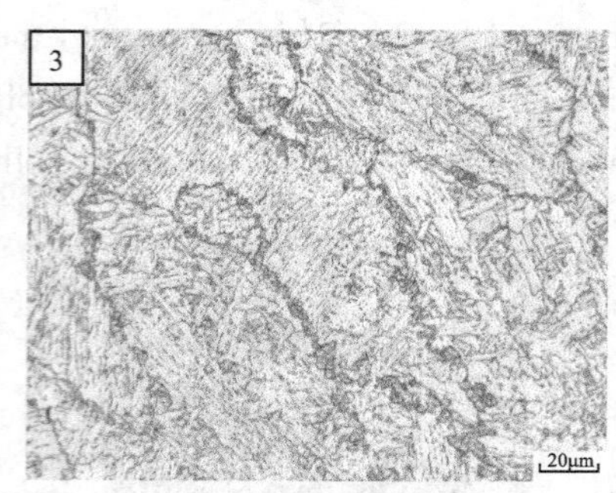

(b) coarse grain and M-A mixed zone of $3^{\#}$ sample

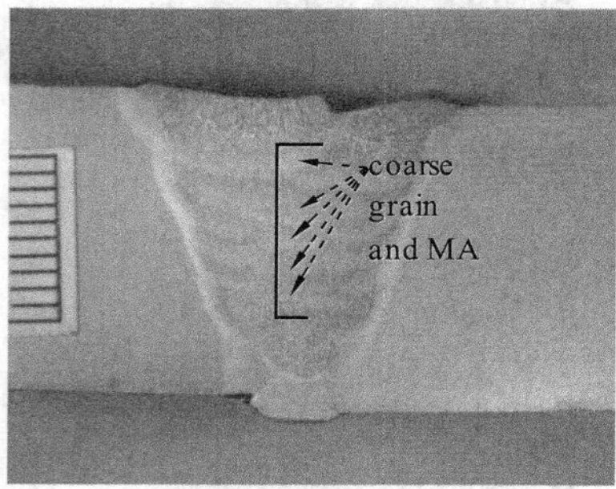

(c) Macro picture of $1^{\#}$ sample

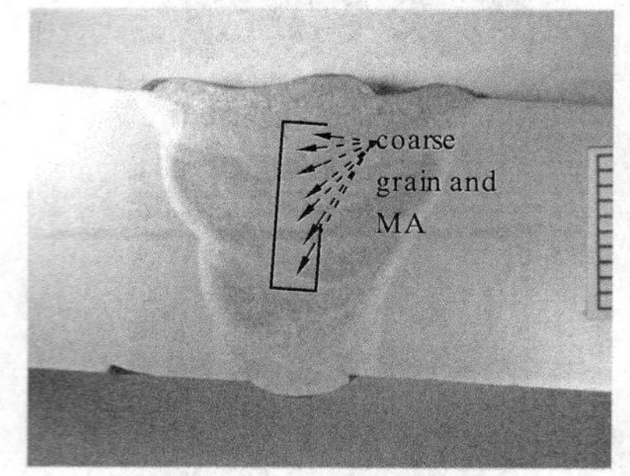

(d) Macro picture of $3^{\#}$ sample

Fig. 7 Metallographic images of comparison of $1^{\#}$ and $3^{\#}$ samples

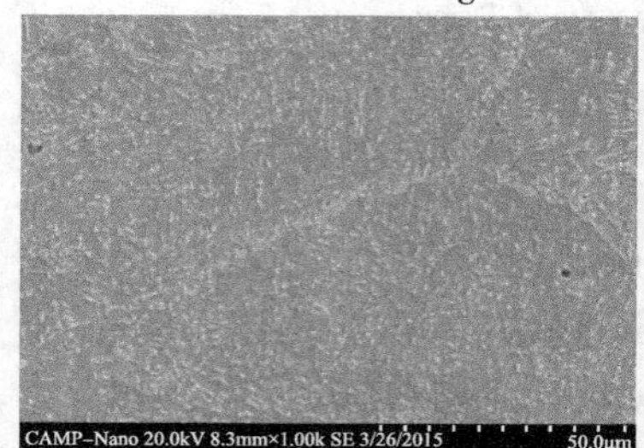

(a) M-A mixed zone 1000X

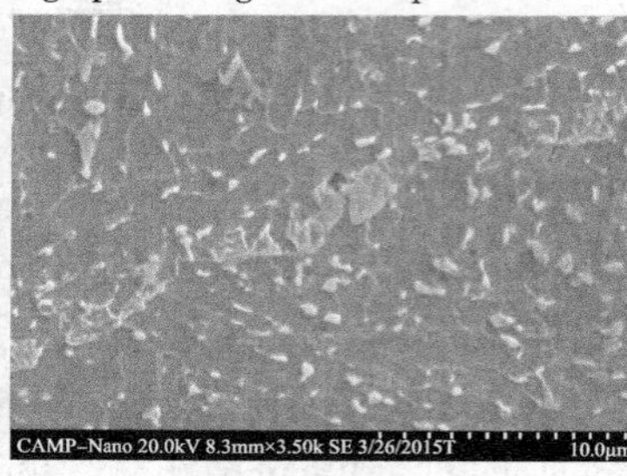

(b) enlarged image 3000X

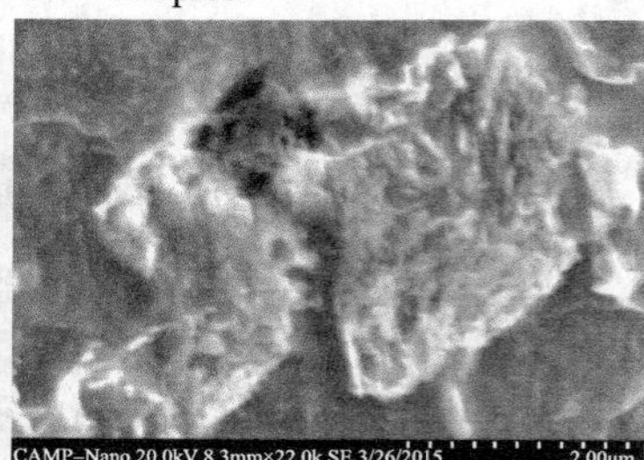

(c) M-A enlarged image 26000X

Fig. 8 The M-A chain organization at grain boundary of coarse grain zone and its magnification SEM

(a) M-A enlarge picture 20000X

(b) M-A enlarge picture 80000X

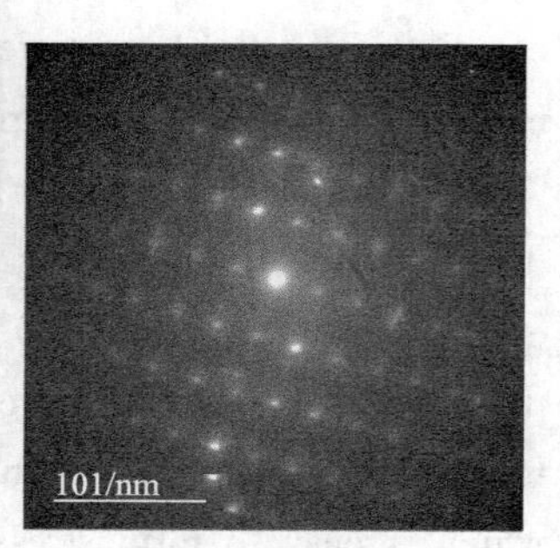

(c) M-A diffraction pattern

Fig. 9 TEM picture of the M-A chain organization and diffraction pattern at grain boundaries

4.3 Precipitates effect

In order to observe the type and shape of precipitates, the pipe body and the weld heat affected zone at two sides of weld joints were separated. The weld area was retained, and the precipitation of the whole weld zone was observed by the method of carbon membrane extraction. Precipitates distributed in the weld area have been shown in Fig. 10. It was obvious, there are a large number of precipitates in the whole weld area, and the particle size of the precipitates is about 100nm ~ 1μm.

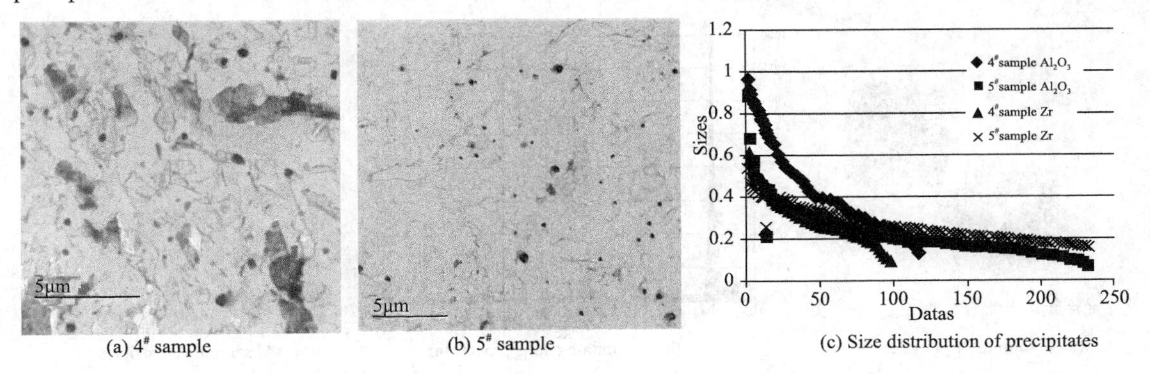

(a) 4# sample (b) 5# sample (c) Size distribution of precipitates

Fig. 10 TEM picture of precipitates distribution of two samples in the weld area

The size distribution of precipitates in the weld area is in Fig. 10 (c) . In the same region, the number of precipitates of Sample 4# is much more than that of Sample 5#. The precipitates sizes of Sample 4# are mostly less than 400 nm, however, nearly 50% of precipitate size of Sample 4# are greater than 400 nm. It is well know that, it is beneficial to the nucleation and grain refinement when the precipitates size is smaller than 200nm, and dispersed in the matrix. It can not only enhance the strength of the matrix, but also increase the toughness. However, the existence of a large number of excessive sizes of precipitates (greater than 400 nm) weakens the matrix toughness and easy to stress concentration under the external force. Furthermore, it will lead to micro crack form, propagation and penetration of each and the formation of macroscopic crack fracture failure.

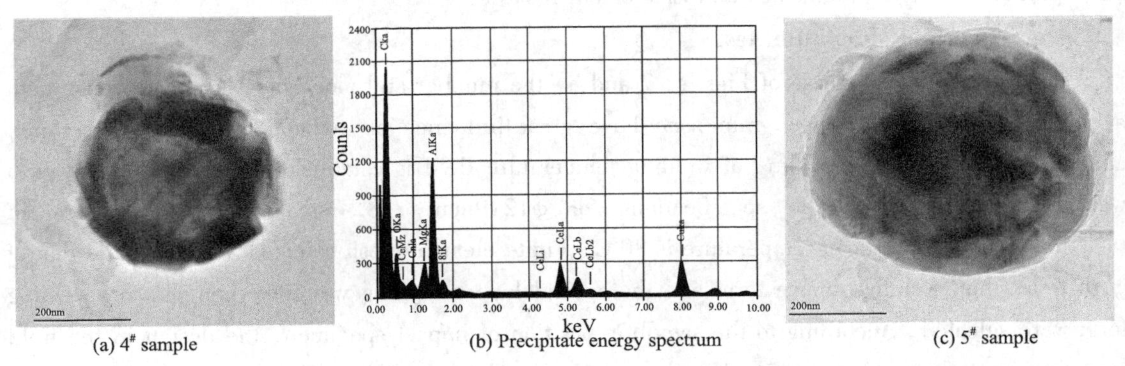

(a) 4# sample (b) Precipitate energy spectrum (c) 5# sample

Fig. 11 Morphology and energy spectrum of the Al_2O_3 precipitates in two samples weld zone

Morphology and energy spectra of Al_2O_3 precipitates were shown in Fig. 11. The particle surface is approximately spherical, and the result of energy spectrum analysis is main components of Al_2O_3. The particles sizes of Samples 4# are about 200nm and smaller particle sizes of Al_2O_3 precipitates. The size of the precipitate of Sample 5# is about 580nm, and the sizes of all Al_2O_3 in the whole observation field are larger than 200nm. The residual Al_2O_3 inclusions formed by flux cored wire welding are in the interior of the welding seam.

Particles morphology of trace alloy elements Zr were shown in Fig. 12, which the number of Zr particles of Sample 4# is larger, additionally, a small amount of Zr particles in the aggregate state are above 500nm, the rest of the size of the majority is about 200nm. The number of Zr particles in the Sample 5# sample is less, and the most in the single crystal form, see Fig. 12 (c).

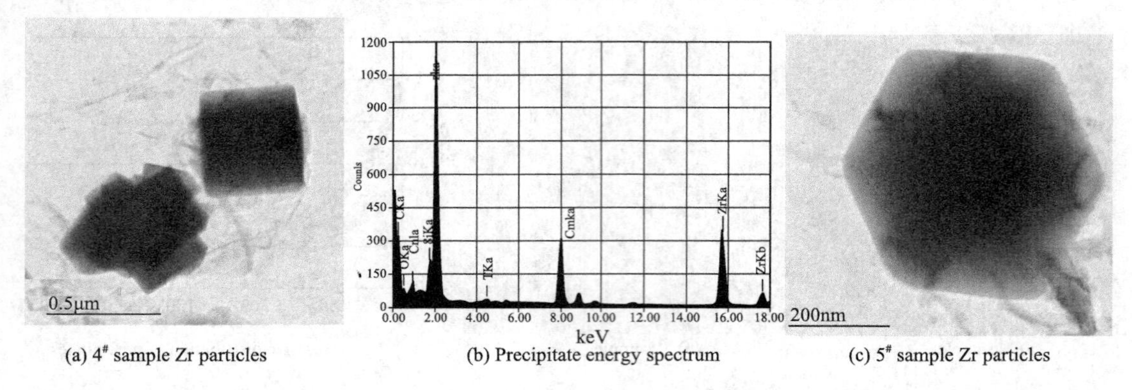

(a) 4# sample Zr particles (b) Precipitate energy spectrum (c) 5# sample Zr particles

Fig. 12 Morphology and energy spectrum of the Zr precipitates in the weld zone

In a word, the larger particles of Sample 4# are the aggregation state of Zr particles, a small amount of Al_2O_3 mixture. On the contrary, the larger particles of Sample 5# are mainly Al_2O_3, and a small amount of Zr precipitates. In order to improve the toughness of the weld, the amount of high temperature-micro alloying element Zr was added to the core. However, excessive residual Al_2O_3 inclusions in the weld are not conducive to the improvement of the overall toughness of the weld. Therefore, in order to improve the toughness of weld, the formula the ratio of flux cored elementary in the wire should be adjusted properly. It is favorable for the welding process of gas - slag protection, increase the fluidity of molten pool on the one hand, on the other hand to avoid excessive large particles of Al_2O_3 inclusions and Zr element residual.

4.4 Thermal simulation test

According to the analysis of Figs. 1, 7 and 8, the number and morphology of M-A constituent at grain boundaries of coarse grain zone have great influence on impact toughness. Therefore, computer finite element modeling of welding temperature field is established according to the girth welding process. Steel pipe specifications for ϕ1219mm × 18.4mm × 200mm, environment temperature is 0, interlayer temperature is 80℃. Finite element mesh selected 3 o'clock position of girth weld, butt welding groove form and the second layer of filling welding when the row welding form were selected. According to the sampling position of impact specimen, the design in the weld center every 0.5mm points to calculate the temperature in the grid, as shown by the yellow line to take the position.

According to temperature cycling curve of M-A constituent at coarse grain zone finite modeling results, weld thermal simulation tests was carried out using Gleeble3500 testing machine, and then mechanical tests was carried out. Impact toughness results after thermal simulation test were shown in Fig. 13 (b). The impact toughness of the weld microstructure after a thermal simulation cycle

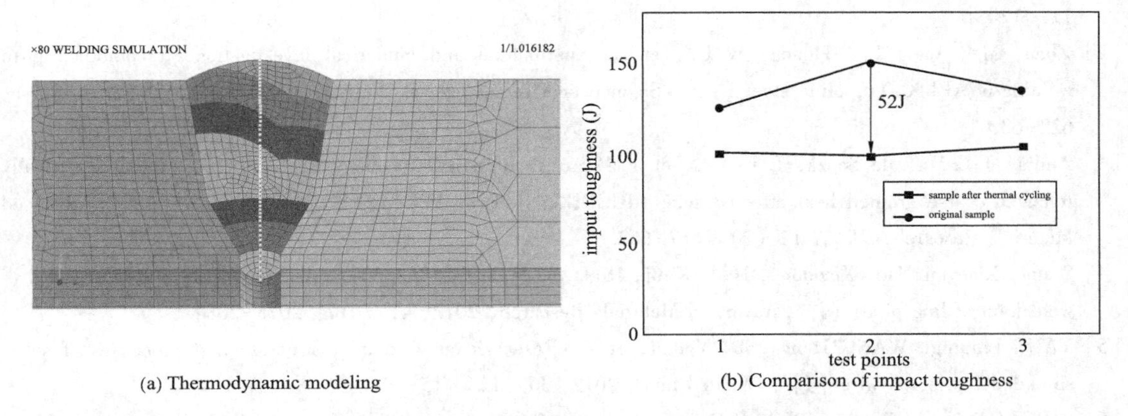

(a) Thermodynamic modeling (b) Comparison of impact toughness

Fig. 13 Mesh division of the thermodynamic modeling of weld temperature field and thermal simulation test

was significantly lower than that of the welding state. The mean reduction value is 35J, the maximum reduction value is 52J. The mixed microstructure of coarse grain and M - A can significantly reduce the impact toughness of the weld.

5 Conclusions

Based on the laboratory test of self-shielded flux-cored wire semi-automatic welding method, the microstructure and mechanical results of girth weld of X80 pipeline were analyzed. The conclusions are as follows:

(1) The tensile strength of the girth weld has little change with different welding heat input and flux cored wire produced by different manufacturer, and the results can meet the standard requirement .

(2) The microstructure of the welded joint is similar to each other when obtained by the same welding heat input, which is composed of ferrite lath and M-A constitution.

(3) The grain sizes of weld joint microstructure were coarser significantly and the deposited metal thickness of filling pass increased due to the heat input increase. Furthermore, there is no retained austenite in M-A constitution observed both in the grain boundaries and internal grains. Therefore, the impact toughness of the weld joints is low relatively.

(4) Under the similar low heat input condition, the microstructure of girth weld with more retained austenite is conducive to improve the impact toughness of the weld joint. Moreover, except for heat input effect, the more M-A constituent exist at grain boundary of coarse grain zone, the lower impact toughness value would be obtained.

(5) Many large sizes residual Al_2O_3 inclusions formed during flux-cored wire welding is adverse to improve the toughness of the girth weld.

Acknowledgements

This work was supported by the Science research and technology development project of CNPC (No. 2014B-3416-0501) and Project of Natural Science Foundation of Shaanxi Province in China (2012GY2-23)

References

[1] LI He-lin. Developing pulse and prospect of oil and gas transmission pipe [J] . Welded Pipe, 2004, 27 (6):

111-120.

[2] Zhao, M., Wei, F., Huang, W. Q., et al. Experimental and numerical investigation on combined girth welding of API X80 pipeline steel [J]. Science and Technology of Welding and Joining, 2015, 20 (7): 622-630

[3] Andia, J. L. M., de Souza, L. F. G. 2, et al. Microstructural and mechanical properties of the intercritically reheated coarse grained heat affected zone (ICCGHAZ) of an API 5L X80 pipeline steel [J]. Advanced Materials Research, 2014, 12 (2): 657-662.

[4] Wang, Xiaoyan, He, Xiaodong, Han, Xinli, et al. Study on FCAW semi-automatic welding procedure of girth weld joint of line pipes [J]. Advanced Materials Research, 2012, 415-417: 2078-2084.

[5] YANG Liuqing, WANG Hong, SUI Yongli, et al. Research on weld microstructure and properties of self-shiekded flux cored wire [J]. Welded pipe, 2012, 36 (12): 15-19.

[6] CHEN Cuixin, LI Wushen, WANG Qingpeng, et al. Microstructure and properties of X80 pipeline steel welded coarse grain zone [J]. Transactions of the china welding institution, 2005, 26 (6): 77-80.

[7] CHEN Cuixin, LI Wushen, WANG Qingpeng, et al. Research on influence factor of impact toughness in coarse grain heat-affected zone [J]. Material engineering, 2005, 5: 22-26.

[8] Gianetto, J. A., Fazeli, F., Chen, Y., et al. Microstructure and toughness of simulated grain coarsened heat affected zones in X80 pipe steels [C]. Proceedings of the Biennial International Pipeline Conference, IPC2014-33254.

[9] Zhenglong, Lei, Caiwang, Tan, Yanbin, Chen, et al. Microstructure and mechanical properties of fiber laser-metal active gas hybrid weld of X80 pipeline steel [S]. Journal of Pressure Vessel Technology, Transactions of the ASME, 2013, 135, 1.

[10] LI Yajuan, LI Wushen, XIE Qi. Research and prediction on cold cracking susceptibility of Nb-Mo X80 pipeline steel [J]. Transactions of the china welding institution, 2010, 31 (5): 105-108.

[11] XU Xueli, XIN Xixian, SHI Kai, et al. Influence of welding thermal cycle on toughness and microstructure in grain-coarsening region of X80 pipeline steel [J]. Transactions of the china welding institution, 2005, 26 (8): 69-72.

[12] MIAO Chengliang, SHANG Chengjia, WANG Xuemin, et al. Microstructure and toughness of HAZ in X80 pipeline steel with high Nb content [J]. ACTA metallugrica sinica, 2010, 46 (5): 541-546.

[13] Technical specification of welding for oil and gas pipeline project Part 1: Mainline welding Q/SY GJX137. 1-2012 [S].

[14] LI Yajuan, LI Wushen, XIE Qi. Research and prediction on coldcracking susceptibility of Nb-Mo X80 pipeline steel [J]. Transactions of the china welding institution, 2010, 31 (5): 105-108.

Microstructure and mechanical property of twinning M/A islands in a High Strength Pipeline Steel

Zhang Jiming Feng Hui Qi Lihua Chen Hongyuan Li Yanhua
Zhang Weiwei Chi Qiang Ji Lingkang

(State Key Laboratory for Performance and Structural Safety of Oil Industry Equipment Materials, Tubular Goods Research Institute of CNPC)

Abstract: Fine microstructure of twinning M/A (Martensite/austenite) islands in a high strength pipeline steel was analyzed by the scanning electron microscope (SEM) and high-resolution transmission electron microscope (HRTEM), and a uniaxial compressive experiment of micro-pillar for twinning M/A islands was conducted in present paper. The experimental results showed that the M/A islands were consisted of retained austenite and nanoscale twins with a size of less than ten nanometers. The nanoscale twins showed a few small blocks in the M/A islands. The twinning M/A islands increased the strength and decreased the toughness of pipeline steel with the increase of M/A contents. The nanoscale twinning M/A island exhibited a higher deformation hardening during the micro-pillar compressive test, and its uniaxial compressive strength could up to 1.35GPa ultrahigh stress level.

Keywords: Pipeline steel; M/A island; Twinning; Micro - pillar; Deformation hardening

1 Introduction

A low alloy pipeline steel is mainly used to the transportation of crude oil and natural gas over a long distance. Line pipes with a combination of high strength and high toughness can obviously improve transportation capability. Recently strength level of line pipes was continuously increased with the increase of diameters and service-pressure of line pipes[1-4]. The unification of high service-pressure and high steel-grades has become the development trend of the line pipes.

High strength pipeline steels for line pipes are commonly manufactured using thermal-mechanical control processing (TMCP) in order to control phase transitions and obtain ultrafine microstructure by the accelerated-cooling process and the addition of niobium and titanium micro-alloy elements. The high strength pipeline steels, especially above X70 grades, commonly represent multiphase microstructure, such as polygonal ferrite (PF)[5], granular bainite (GB)[6], lower bainite (LB)[3], martensite (M)[7], and Martensite/austenite (M/A) islands[8,9]. In addition, a large number of nanoscale precipitates containing niobium and titanium elements distribute in the matrix. An M/A island with a combination of martensite and retained austenite is a phase transformation production of high strength pipeline steels during TMCP processing. For the size of M/A is very small with only a few micrometers, so its mechanical property has not been carefully studied using common research methods until present. But, it should be pointed out that the characteristic microstructure and mechanism of M/A have aroused researchers' interesting[10, 11].

The objective of present paper is to describe the fine structure of M/A islands and reveal formation mechanism in high strength pipeline steels. Furthermore, we conduct micro-pillar compression experiments to study the micromechanical response of M/A islands.

2 Experimental procedure

The material examined in this study is an X100 pipeline steel. Its chemical composition was designed by low carbon and low alloy bearing niobium and titanium elements. The main chemical composition was listed in table 1. The steel plates were rolled to 18.4mm thick plates above the Ac3 temperature, and then the plates were cooled down to below Ac3 temperature holding for about 30 seconds in the air to get a certain amount of ferrite microstructure. Subsequently, the plates were cooled to below the Ms-temperature using accelerated cooling process.

Table 1 Chemical composition of experimental X100 pipeline steel (wt %)

C	Si	Mn	Cr	Mo	Ni	Nb	Ti	Cu	S	P	N
0.06	0.25	1.95	0.3	0.3	0.45	0.06	0.015	0.025	0.0012	0.006	0.0033

Samples for measurement of macro-mechanical properties and microstructure were cut from the plates. Standard tensile specimens with a dimension of ϕ8.9mm diameter and Charpy V-Notch (CVN) impact specimens with a dimension of 10mm×10mm×55mm were prepared, respectively. Tensile tests were carried out at a strain rate of $10^{-3}\,s^{-1}$ on a MTS 810 testing machine with an automatic extensometer at room temperature. Impact toughness tests were performed at 263K. Optical samples were ground on abrasive papers and etched in 4% nital. Optical microstructure observation was performed by field emission scanning electron microscope. Specimens for transmission electron microscope (TEM) were prepared. The specimens were firstly cut into 300μm thick foils, and then were ground on abrasive papers to a thickness of 60μm. Lastly the foil was thinned by the twin-jet polishing at 258K using 5% perchloric acid and 95% ethanol solution as electrolyte. TEM testing was performed on JEM-2100F TEM.

A sample for micro-pillar compression experiment was prepared using a FEI Helios Nanolab650 dual beam FIB. The prepared samples for optical observation were examined in the scanning electron microscope-FIB to determine the location of M/A islands. A micro-pillar was milled using an established top-down method in the dual-beam FIB using 30keV Ga ions with lower currents. The M/A micro-pillar with a diameter of 1.0μm was milled from the bigger M/A island. Micro-pillar compression testing was performed on the Hysitron TI 950 Triboindenter with a 10μm diameter diamond flat punch.

3 Results

3.1 SEM microstructure observation of M/A islands

The SEM images of microstructure for experimental steels are shown in Fig. 1. Fig. 1a illustrates a macroscopic microstructure of consisting of a multiphase mixture including ferrite (F), bainite (B) and M/A islands. The ferrite, which is less than 10μm in size, shows irregular polygonal shape. The bainite displays two morphologies of lath-like bainite (LLB) and granular bainite (GB). The M/A islands uniformly distribute in the matrix with various characteristic morphologies, some M/A islands shows a discontinuously neck-lace with the blocky shape at the ferrite grain boundary (Fig. 1c), and some M/A islands display a strip structure between the laths of LLB (Fig. 1d). In addition, there are some islands showing small particles dispersedly in the GB matrix. The length scale of M/A islands is commonly less than 10μm (Fig. 1b and c).

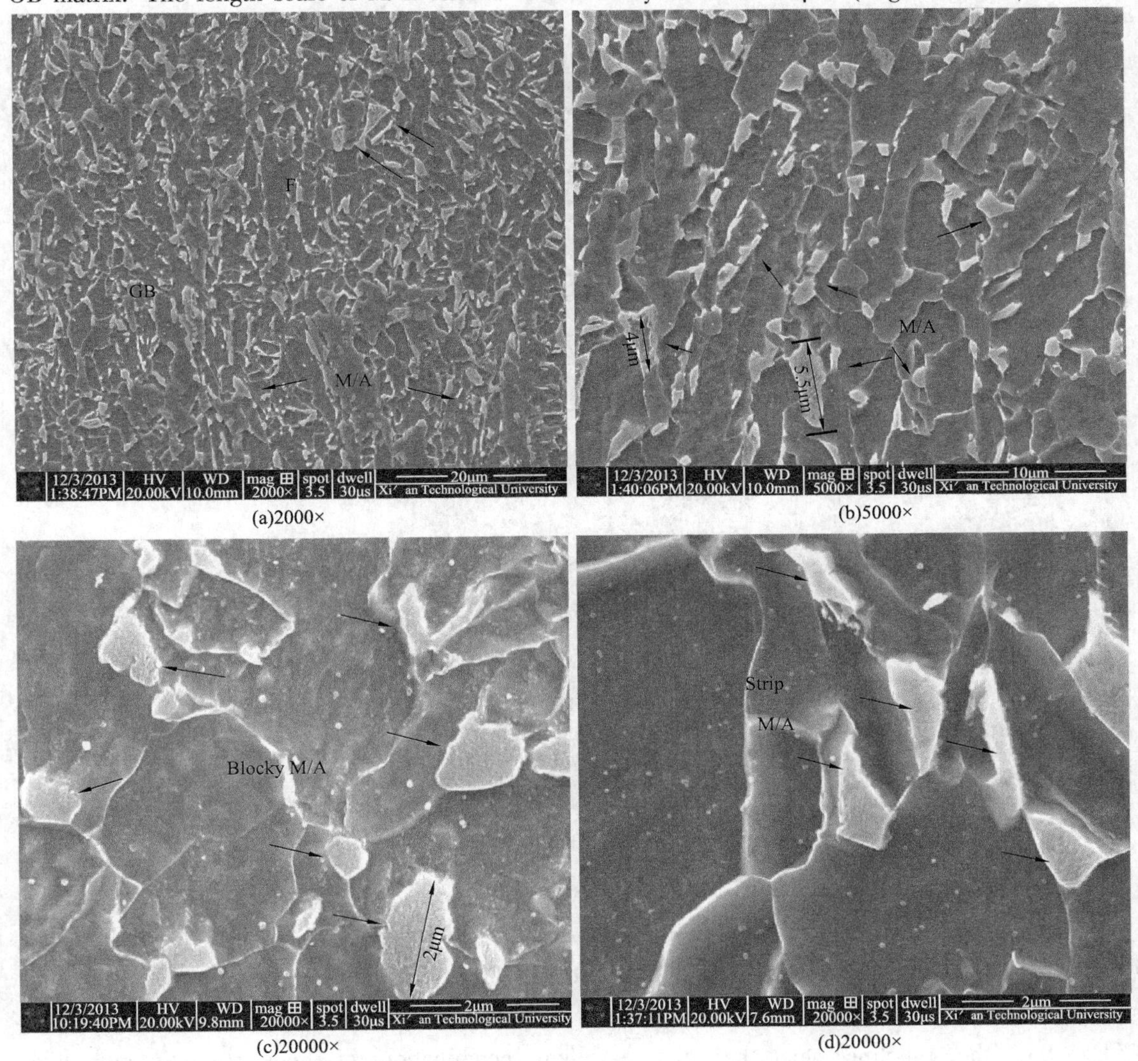

(a)2000×　(b)5000×

(c)20000×　(d)20000×

Fig. 1 SEM M/A morphology of experimental steel with different magnifications, the arrows point to the M/A islands in the figures

3.2 Fine microstructure of M/A islands

Representative TEM micrographs of M/A islands are shown in Fig. 2. Fig. 2 (a) and (b) illustrates the bright field image and dark field image of long strip M/A with the selected area electron diffraction pattern (Fig. 2c), respectively. The long strip M/A island lies in low bainite laths with its long axis orientation parallel with the bainite laths. The M/A island is consisted of retained austenite and twinning martensite. The twinning martensite shows the small discontinuous blocks at the interface between M/A islands and the matrix. These twins are consisted of a large number of micro-twins (MT) with the width of about 5nanometer. These MTs originate in the M/A boundary and not pass through the whole M/A island. Fig. 2 (d) and (e) is the typical TEM morphology of blocky M/A island at the ferrite boundary. This blocky M/A island is consisted of four twinning regions and a retained austenite. The regions of MT located at ferrite grain boundary. Each twinning block is composed of some MT substructures, and the average width size of these twins is less than five nanometers.

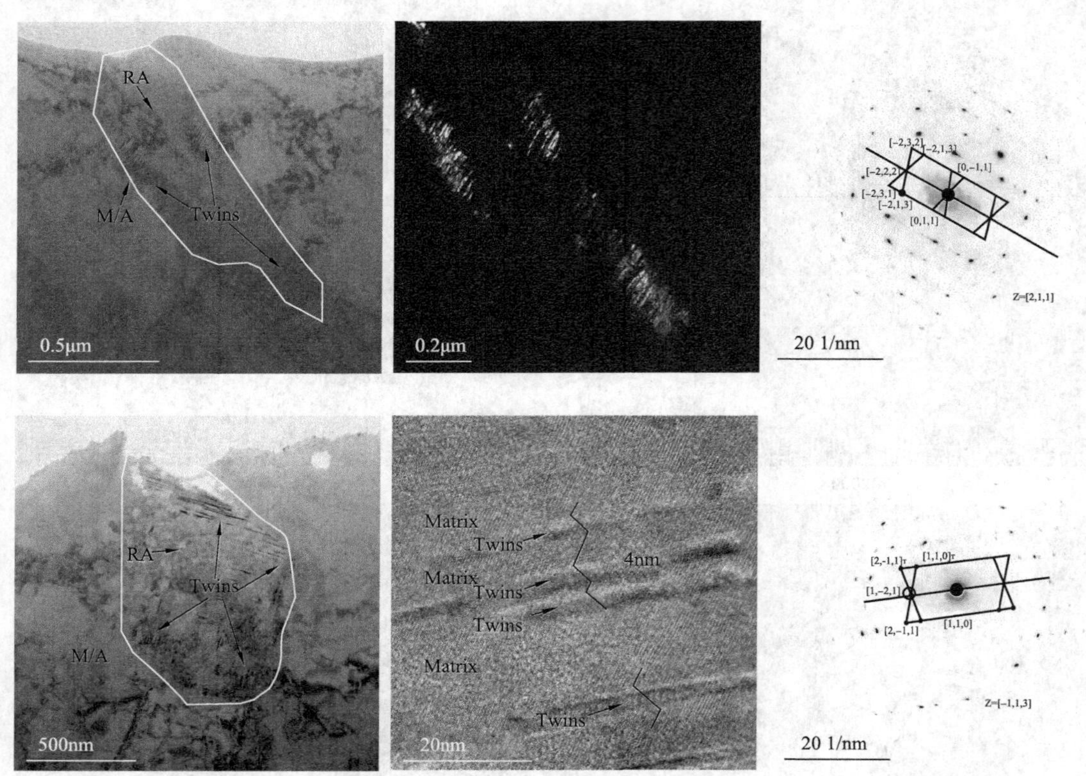

Fig. 2 Representative TEM bright-field images of M/A constituents

3.3 Quantitative analysis of M/A islands

Mechanical properties of X100 pipeline steel were listed in table 2. Ultimate tensile strength (UTS) of X100 pipeline steel was from 846MPa to 1085MPa with the range of 710~815MPa Yield strength (YS). In addition, CVN impact energy of X100 pipeline steel has a range from 202J to 454J. So, the pipeline steel has the excellent combination of high strength and high toughness. Because M/A islands are consisted of body-centered cubic martensitic twins and face-centered cubic structure retained austenite, quantitative analysis of M/A constitutes cannot be obtained using traditional methods of electron backscatter diffraction and X-ray diffraction. So, it is always a difficulty to quantitative statistics of M/A. We found M/A islands appear bright white compared with gray microstructure of ferrite and bainite under the SEM. The main reason is that the M/A islands contained higher carbon and alloy contents in pipeline steel during the recrystallized austenite transformed [12]. According to this feature, the fraction of M/A islands can be quantitatively measured using digital automatic analysis software according to the SEM micrographs of M/A islands. In present experiment, we chose five specimens at different strength levels and gathered randomly ten micrographs under 5000 magnification for each specimen using the SEM. Final fraction of M/A islands is the average of ten measured results. Influence of M/A fractions on mechanical properties of X100 pipeline steel is shown in Fig. 3. It can be seen from Fig. 3 that the volume fractions of M/A island in the steels are from 5.8 percent to 8 percent. With increasing of M/A fraction, the YS of the steels gradually increases. In contrast, the CVN values decrease with the increasing of M/A fraction. We obtained the slopes of two curves by the fit linear method. The slopes of YS and CVN curves are 49.8 ±8 and −93.7 ±12, respectively. This indicates the fraction of M/A islands has the more serious impact on toughness than strength of pipeline steel.

Table 2 Macro-mechanical properties of X100 pipeline steel

Ultimate strength (MPa)	Yield strength (MPa)	Elongation (%)	Charpy-V-notched impact energy (J)
846~1085	710~815	21~25	202~454

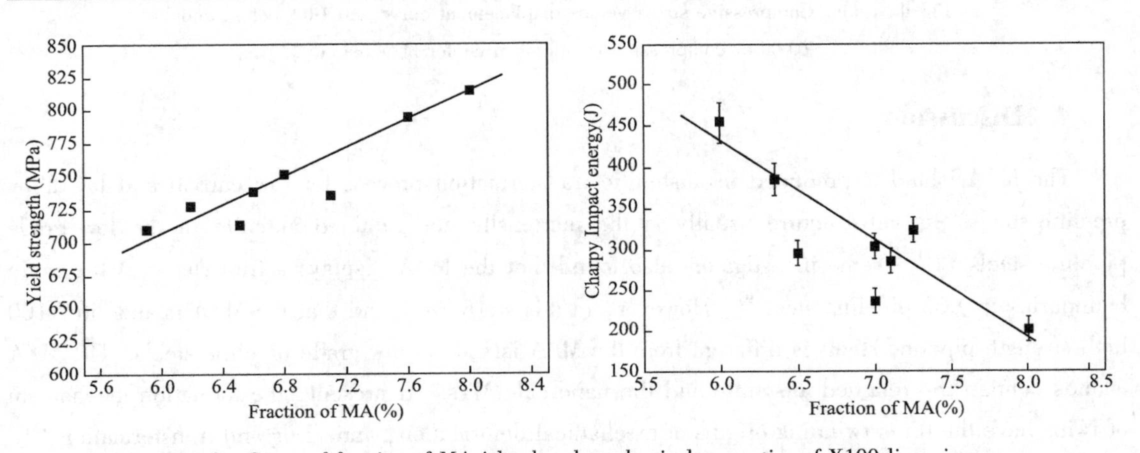

Fig. 3 Curve of fraction of MA island and mechanical properties of X100 line pipe

3.4 Micro-pillar compression response of M/A island

Cylindrical 1mm diameter micro-pillars for the twinning M/A island with a height/diameter aspect ratio of 3 were fabricated using FIB. Nominal constant strain rate is 200μN/s. Compressive stress versus displacement curve of M/A micro-pillar and morphology of before and after compression is shown in Fig. 4. It is clear from the curve that at the initial stage, compressive stress increases quickly, and reached to the yield point at about 0.5μm compressive displacement. It is just the time that a large sliding step appears about at the center of micro-pillar which corresponds to the stress-displacement curve from about 0. 5mm to 1. 0μm displacement stage, it shows that the yield point of twining M/A island is about 200MPa stress. Subsequently compressive curve increases steeply with increasing of strain and finally reaches the 1. 35GPa super-high stress. A large number of small sliding steps are found on the surface of M/A island micro-pillar after compression. These steps are induced by dislocation glide during the compression process [13].

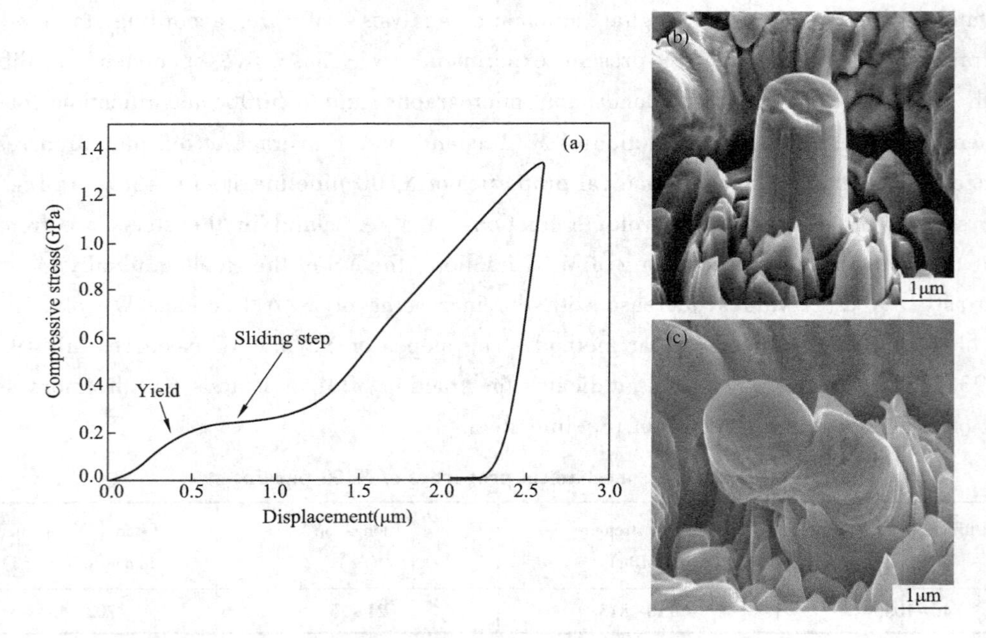

Fig. 4 (a) Compressive stress versus displacement curve and (b) before and (c) after compression of micro-pillar for M/A island

4 Discussion

The M/A island is produced in austenitic transformation process for low carbon and low alloy pipeline steels. Its substructure usually is the martensite and retained autenite in the low grade pipeline steels [4,14], some investigators also found that the M/A displays a film shape at the grain boundaries in X60 pipeline steel[9]. However, in this study we found that the M/A islands in X100 high strength pipeline steels is different from the M/A islands in low grade pipeline steels. The M/A islands contain the retained austenite and the nanoscale MTs. At present, the formation mechanism of twins have the three twinning origins of mechanical deformation, annealing and transformation[15]. In present experiment, These MTs in the M/A islands belong to transformation twins. The nature of transformation twins is that the boundary emits the Shockley partial dislocation during the transformation. Transformation twins form as a result of phase transformation, and produced easily in fcc metals and alloys with high stacking fault energy [16, 17]. The formation of transformation twins in bcc low carbon steels would be unlikely, considering that transformation twins in carbon steels are believed to form when the carbon content exceeds a certain level, such as 0. 2wt% [18]. Therefore, it is very difficult that martensitic twins form in ultralow pipeline steels containing 0. 06% carbon (table 1) . As we all know, the carbon element concentrate in the retained austenite when austenite transforms intoproeutectoid ferrite because the carbon in ferrite has a low solid solubility. So, the retained austenite enriches a large amount of carbon when cooling temperature reaches the martensite transformation points (Ms) . S. V. Subramanian et al [12] have measured the carbon content in M/A island for X100 pipeline steel containing 0. 058wt% carbon using three dimensional atom probe. They found the carbon content in M/A up to 1. 4wt% is twenty-three times as the carbon content in the matrix. So, the carbon-enrichment in M/A is the main reason of twinning M/A formation in the X100 pipeline steel.

The RA has higher toughness and plasticity. However, the twinning martensite displays high strength because the twins are not easy to slip. In the twinning M/A islands, the most fraction is RA and the twins is only small proportion. So, during the micro-pillar compressive process, the face-centered cubic RA with multiple slip systems firstly occurs plastic deformation, and slip system is active at the lower stress level. With the increase of compressive stress, multiple slip systems of RA start and displays a big platform on the curve of stress-displacement. With the further increase of compressive stress, the mobile dislocations in the M/A slip the free surface of the micro-pillar and vanish, which is so called the dislocation starvation mechanism that has been hypothesized and observed on the other experiments [19,20]. The decrease of dislocation density promotes the increase of applied stress and newly generated dislocations. So, the micro-pillar of M/A island quickly deforms hardening. When the applied stress reaches the strength limit of M/A island, the micro-pillar collapsed. In addition, the nanoscale twinning substructure of M/A islands obviously improve the strength of twinning M/A island for the twinning interface inhibits the dislocation slip [21, 22].

5 Conclusion

Fine microstructure of M/A islands in X100 high strength pipeline steel is consisted of retained austenite and nanoscale twins. The twins in M/A island distributes a few small blocks. The existence of M/A is beneficial to improving strength of pipeline steels but deteriorates the toughness of steels with the increase of M/A fraction.

The nanoscale twinning M/A islands show the strong deformation hardening during the compressive test of micro-pillar. The compressive strength of micro-pillar has up to 1. 35GPa although the yield point of twinning M/A begin about 200MPa at low applied stress level.

Acknowledgements

Authors acknowledges support from Xi' an Jiaotong University. Mechanical testing and sample preparation of the micro-pillars were performed in Center for Advancing Materials Performance from the Nanoscale. In addition, we are grateful to Pro. Zhiwei Shan, Dr. Bin Qin and Danli Zhang for their assistance in the conducut of micro-pillar compressive experiments.

References

[1] M. Eskandari, M. A. Mohtadi-Bonab, J. A. Szpunar. Evolution of the microstructure and texture of X70 pipeline steel during cold-rolling and annealing treatments, Materials & Design, 2016, 90 (15): 618-627.

[2] Ping Liang, Xiaogang Li, Cuiwei Du, Xu Chen. Stress corrosion cracking of X80 pipeline steel in simulated alkaline soil solution, Materials & Design, 2009, 30 (5): 1712-1717.

[3] Rahmatollah Ghajar, Giuseppe Mirone, Arash Keshavarz. Ductile failure of X100 pipeline steel- Experiments and fractography, Materials & Design, 2013, 43: 513-525.

[4] Ji-ming ZHANG, Wei-hua SUN, Hao SUN. Mechanical Properties and Microstructure of X120 Grade High Strength Pipeline Steel, Journal of Iron and Steel Research, International, 2010, 17 (10): 63-67.

[5] Jang-Bog Ju, Woo-sik Kim, Jae-il Jang. Variations in DBTT and CTOD within weld heat-affected zone of API X65 pipeline steel, Materials Science and Engineering A, 2012, 546: 258-262.

[6] Xiaoyong Zhang, Huilin Gao, Xueqin Zhang, Yan Yang. Effect of volume fraction of bainite on microstructure and mechanical properties of X80 pipeline steel with excellent deformability, Materials Science and Engineering A, 2012, 531: 84-90.

[7] Xianbo Shi, Wei Yan, Wei Wang, Yiyin Shan, Ke Yang. Novel Cu-bearing high-strength pipeline steels with excellent resistance to hydrogen-induced cracking, Materials & Design, 2016, 92 (5): 300-305.

[8] Nazmul Huda, Abdelbaset R. H. Midawi, James Gianetto, Robert Lazor, Adrian P. Gerlich. Influence of martensite-austenite (MA) on impact toughness of X80 line pipe steels, Materials Science and Engineering A, 2016, 662: 481-491.

[9] Yong Zhong, Furen Xiao, Jingwu Zhang, Yiyin Shan, Wei Wang, Ke Yang. In situ TEM study of the effect of M/A films at grain boundaries on crack propagation in an ultra-fine acicular ferrite pipelinesteel, Acta Materialia, 2006, 54: 435-443.

[10] S. Shanmugam, N. K. Ramisetti, R. D. K. Misra, J. Hartmann, S. G. Jansto. Microstructure and high strength-toughness combination of a new 700 MPa Nb-microalloyed pipeline steel, Materials Science and Engineering A, 2008, 478: 26-37.

[11] Chunming Wang, Xingfang Wu, Jie Liu, Ning'an Xu. Transmission electron microscopy of marensite/austenite island in pipeline steel X70, Materials Science and Engineering A, 2006, 438-440: 267-271.

[12] Y. You, CJ Shang, W. J. Nie, S. Subramanian. Investigation on the microstructure and toughness of coarse grained heat affected zone in X-100 multi-phase pipeline steel with high Nb content, Materials Science & Engineering A, 2012, 558: 692-701.

[13] T Maki, CM Wayman. Transformation Twin Width Variation in Fe-Ni and Fe-Ni-C Martensites, Naure, 2010, 17: 67-94.

[14] Ming-Chun Zhao, Toshihiro Hanamura, Hai Qiu, Ke Yang. Lath boundary thin-film martensite in acicular ferrite ultralow carbon pipeline steels, Materials Science and Engineering A, 2005, 395: 327-332.

[15] K. Poorhaydari, B. M. Patchett, D. G. Ivey. Transformation twins in the weld HAZ of a low-carbon high-strength microalloyed steel, Materials Science and Engineering A, 2006, 435-436: 371-382.

[16] M. J. Szczerba, S. Kopacz, M. S. Szczerba, Experimental studies on detwinning of face-centered cubic deformation twins, Acta Materialia, 2016, 104: 52-61.

[17] S. Xue, Z. Fan, Y. Chen, J. Li, H. Wang, X. Zhang. The formation mechanisms of growth twins in polycrystalline Al with high stacking fault energy, Acta Materialia, 2015, 101: 62-70.

[18] G. R. Spich, W. C. Leslie, Tempering of steel, Metallurgical and Materials Transactions B, 1972, 3: 1043-1054.

[19] Z. W. Shan, RAJA K. Mishra, S. A. Syed Asif, Oden L. Warren, Andrew M. Minor. Mechanical annealing and source-limited deformation in submicrometre-diameter Ni crystals, Nature Materials, 2008, 7: 115-119.

[20] Nix W. D, Yielding and strain hardening of thin metal films on substrates, Scr Mater, 1998, 39: 545-554.

[21] Huang Quan, Yu Dongli, Xu Bo, Hu Wentao, Ma Yanming, Wang Yanbin, Zhao Zhisheng, Wen Bin, He

Julong, Liu Zhongyuan, Tian Yongjun. Nanotwinned diamond with unprecedented hardness and stability, nature, 2014, 510 (7504): 250-253.
[22] Lu K.. Tabilizing nanostructures in metals using grain and twin boundary architectures, nature Reviews Materials, 1, 16019.

Microstructure and properties of HAZ for X100 pipeline steel

Hu Meijuan　Wang Peng　Chi Qiang　Li He　Zhang Weiwei

(Tubular Goods Research Institute of CNPC)

Abstract: In general, the weld thermal cycle results in significant changes in microstructure and mechanical properties of the weld heat affected zone (HAZ). The microstructure, microhardness and low temperature impact toughness of HAZ for X100 pipeline steel were studied by means of welding thermal simulation. Influence of cooling time on the microstructure and properties in coarse-grained heat affected zone (CGHAZ) was investigated. The results illustrated that polygonal ferrite and a small amount of granular bainite were obtained when the cooling time $t_{8/5}$ is larger than 1 500 s. Mainly granular bainite was formed when the cooling time $t_{8/5}$ is in the range of 1 500 s to 100 s. Bainite ferrite was observed when the cooling time is smaller than 60 s. Martensite appeared in the CGHAZ with the 20s cooling time. The value of microhardness in the CGHAZ was higher than that of base metal (BM) when the cooling time $t_{8/5}$ is smaller than 100s. The CVN absorbed energy in the CGHAZ was higher than the value of BM when the cooling time $t_{8/5}$ is smaller than 30s.

Keywords: X100 pipeline steel; CGHAZ; Microstructure; Microhardness; Impact toughness

In oil & gas pipeline industry, there is a continuing push to decrease construction time and associated costs. The use of high strength steel offers major cost benefits for long distance gas transmission pipelines. The cost benefits arise from a combination of reduced materials and construction costs and the ability to operate the pipeline at higher operating pressure thereby reducing the pipeline diameter required for a given throughout[1-3]. Over the last decade, while the use of X80 pipeline steel has become accepted practice in Canada, Europe and China. More attentions have been focused on the application of even higher-grade materials, with particular interest in grade X100 pipeline steel.

Field welding is the key process for pipelineconstruction. Girth welds tend to be the weakest link due to the possible existence of weld defects and microstructure & mechanical property changes of the weld heat-affected zone (HAZ) from welding thermal cycles[4,5]. To date, only a limited amount of welding research has been performed on X100 pipeline steel. The main focus of this work is to understand the microstructure and mechanical property changes that occur in the coarse-grained HAZ (CGHAZ), which is the region that experiences high temperature and then shows remarkable microstructural and properties changes compared with the base metal (BM).

1 Base metal and experimental procedure

1.1 Base metal properties

The base material was obtained from ϕ1 016 mm × 16 mm wall thickness X100 line pipe. Chemical analysis was performed. A summary of analysis results and the calculated carbon equivalent value are found in Table 1. Tensile properties of the grade X100 pipeline steel were measured by extracting round bar specimens (ϕ8.9 mm × 35 mm) transverse to the direction of rolling. Full-size Charpy V-notch (CVN) specimens (ϕ10 mm×10 mm×55 mm) were machined to test the impact toughness. All the results are listed in Table 2. Fig. 1 is the micrograph of the grade X100 pipeline steel, which was made up of primarily oriented acicular/bainitic ferrite with smaller amounts of polygonal ferrite.

Table 1 Chemical compositions of X100 pipeline steel (wt%)

C	Si	Mn	P	S	Cr+Mo+Ni+Cu	Nb+V+Ti	Al	B	N	CE_{Pcm}
0.059	0.023	1.96	0.0078	0.00041	1.32	0.0428	0.0076	0.0003	0.0022	0.22

Note: CE_{Pcm} = C+Si/30+ (Mn+Cu+Cr) /20+Mo/15+Ni/60+V/10+5B.

Table 2 Mechanical properties of X100 pipeline steel

Tensile				CVN absorbed energy (J)				
Tensile Strength (MPa)	Yield Strength (MPa)	Reduction of Area (%)	Yield Strength Ratio (%)	20℃	0	−10℃	−20℃	−40℃
890	875	21	0.98	268	262	252	266	238

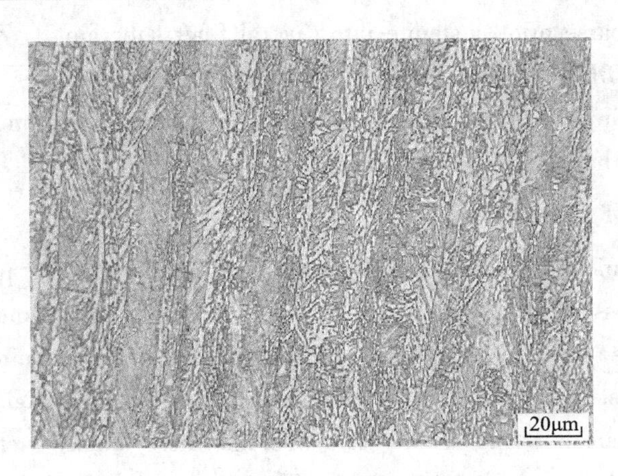

Fig. 1 Microstructure of X100 pipeline steel

1.2 Experimental procedures

CGHAZ simulations were performed using a Gleeble 3500 thermo-mechanical simulator.

Samples of X100 pipeline steel were heated using a 200℃/s heating rate, held at a peak temperature of 1 300 °C for 1s to stabilize the temperature and avoid overshooting. The cooling time from 1 300°C to 900 °C is fixed to 9 s. The cooling time of 800–500 °C ($t_{8/5}$) is typically regarded as the controlling factor for determining HAZ microstructure and properties because it is the temperature where transformation typically occurs. The $t_{8/5}$ can take from a few seconds to several minutes depending upon the heat input and sample dimensions. A series of $t_{8/5}$, which were sequentially 6, 10, 15, 20, 30, 60, 100, 150, 300, 600, 1 000, 3 000 and 6 000 s, were chosen to cover all actual welding procedure.

After subjected to the thermal processing, samples were processed for microstructure observation, microhardness measurement and CVN impact toughness tests. Optical microscopy (OM) was used for microstructural characterization of the CGHAZ. Vickers hardness with a load of 0. 5kg was employed for hardness tests. And the CVN absorbed energy was measured at −20℃.

2 Results and discussion

2. 1 Microstructures of CGHAZ

Fig. 2 presents the representative microstructures of the CGHAZ for X100 pipeline steel after HAZ simulation. There appeared to display only slight differences in CGHAZ prior austenite grain size because of the fixed high temperature dwelling time. The average grain size of coarse prior austenite reached to ~23 μm. The microstructure of CGHAZ with $t_{8/5}$ = 6 000 s was polygonal ferrite (PF) and a small amount of granular bainite (GB) with massive M–A. Mainly granular bainite was obtained in the CGHAZ when the cooling time $t_{8/5}$ is in the range of 1 500 s to 100 s. The difference lied in the morphology and distribution of M–A. With the increase of the cooling rate, the morphology of M–A changed from massive gray–black spots to thin strip form, the M–A distributed in the grain boundary gradually decreased. Direction of the strip M–A in the grain was gradually clear and divided the coarse prior austenite into several finer subgrain[6]. While $t_{8/5}$ is smaller than 60 s, bainite ferrite (BF) was more likely to generate. More slender and smooth baths have nucleated and grown inwards from grain boundaries, resulting in an interlocking structure. Martensite appeared in the CGHAZ with $t_{8/5}$ = 20 s.

2. 2 Properties of CGHAZ

Fig. 3 shows the microhardness variation of the CGHAZ for X100 pipeline steel after HAZ simulation. The hardness of polygonal ferrite is relatively soft. The content of granular bainite increased with a decrease of the cooling time. The M–A in the granular bainite is responsible for the increasing microhardness. The value of microhardness in the CGHAZ is higher than that of BM when the cooling time $t_{8/5}$ is smaller than 100s. The microhardness almost increased linearly with the increase of dislocation density in bainite ferrite. The maximum microhardness of 341 HV in the CGHAZ with $t_{8/5}$ = 20s corresponded to the value expected for a martensitic structure for the carbon level in the steel[7].

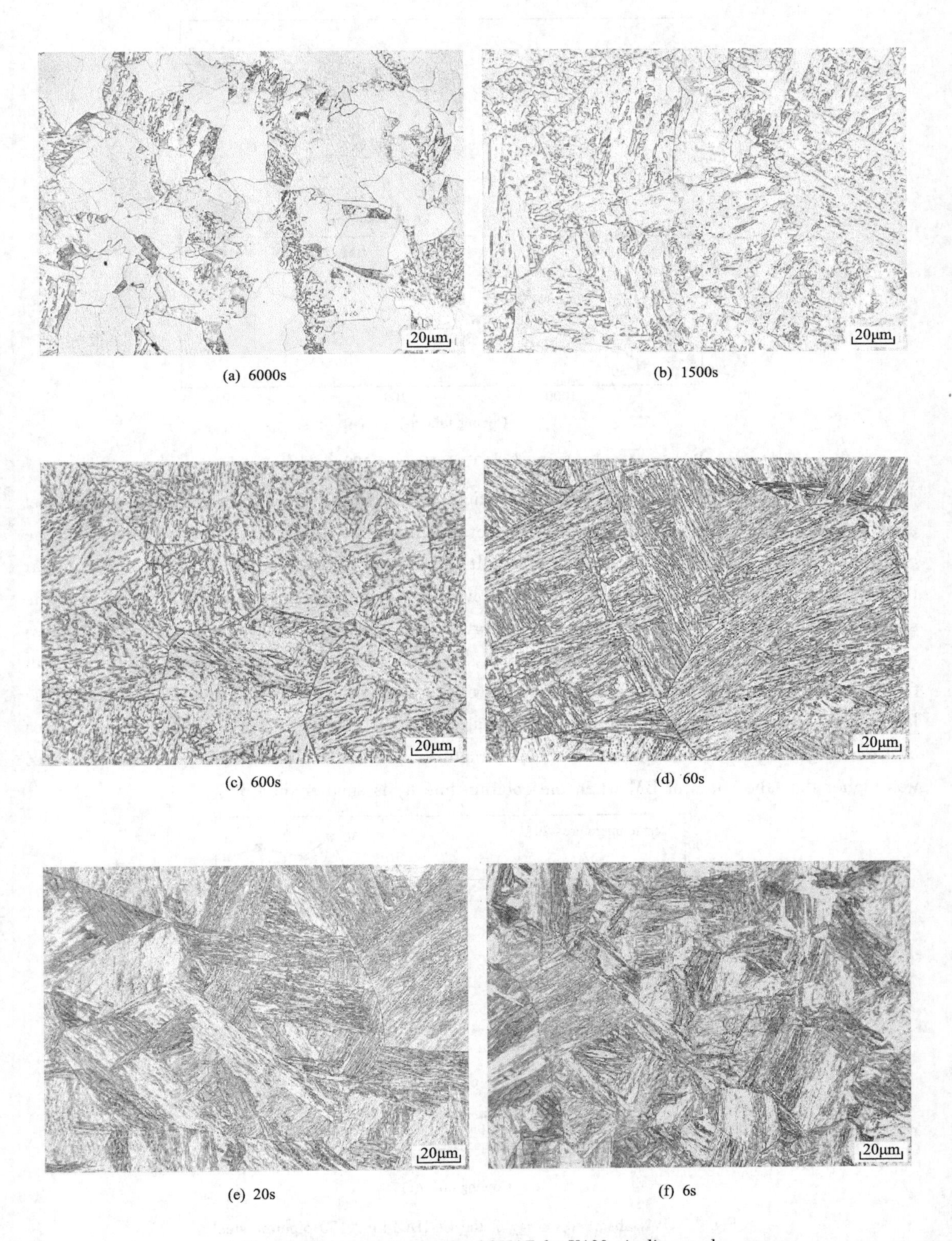

(a) 6000s (b) 1500s

(c) 600s (d) 60s

(e) 20s (f) 6s

Fig. 2 Microstructures of the CGHAZ for X100 pipeline steel

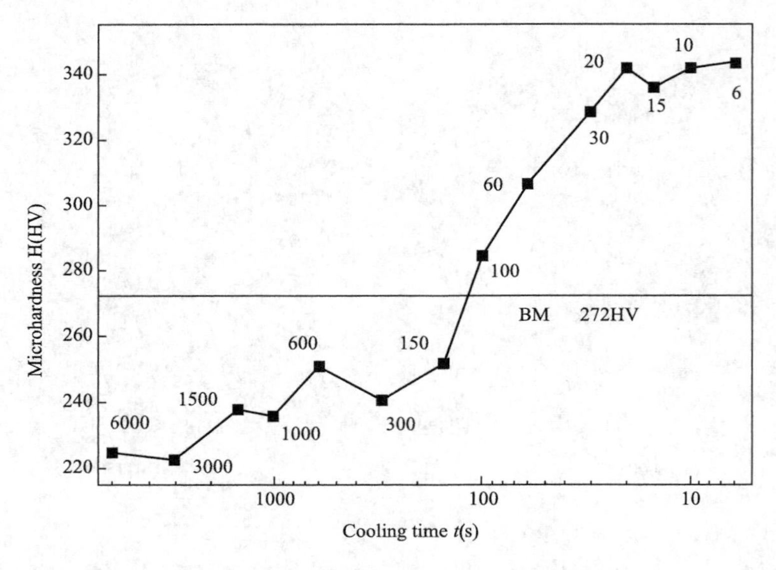

Fig. 3 Microhardness of the CGHAZ for X100 pipeline steel

Fig. 4 shows the CVN absorbed energy variation of the CGHAZ at −20℃ for X100 pipeline steel after HAZ simulation. The low temperature impact toughness of CGHAZ was relatively poor when the cooling time is greater than and equal to 100s. It is believed that the coarse M−A in the granular bainite, which may be the crack source or provide the crack propagation channel, caused the significantly decrease of the CVN absorbed energy in the CGHAZ. The value of impact toughness increases significantly because of the appearance of bainite ferrite when the cooling time $t_{8/5}$ is about 60s. The interlocking structure of bainite ferrite will be more resistant to fracture as compared to the large spanning laths of granular bainite, because a crack front would be forced to make more changes of direction as it propagates through the microstructure. The CVN absorbed energy in the CGHAZ was higher than the value of BM when the cooling time $t_{8/5}$ is smaller than 30s.

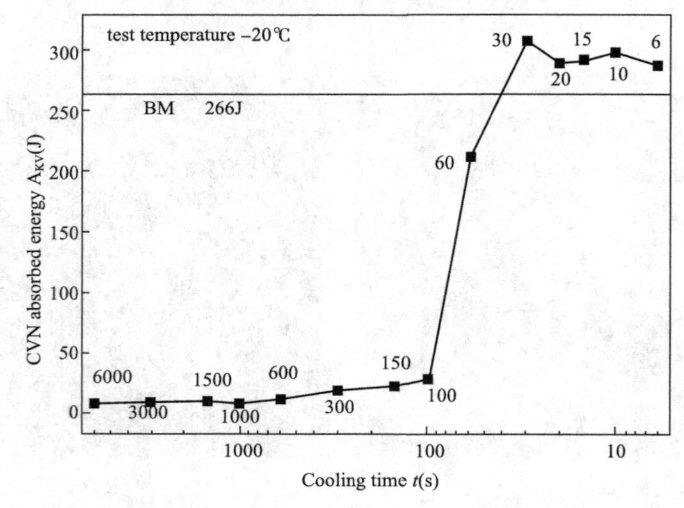

Fig. 4 CVN absorbed energy of the CGHAZ for X100 pipeline steel

3 Conclusions

Microstructural characteristics and mechanical properties of CGHAZ for X100 pipeline steel were examined by welding thermal simulation. Microhardness and CVN absorbed energy variations were explained in terms of the microstructural changes. The results are summarized below:

(1) PF and a small amount of GB were obtained when the cooling time $t_{8/5}$ is larger than 1 500 s. Mainly GB was formed when the cooling time $t_{8/5}$ is in the range of 1 500 s to 100 s. BF was observed when the cooling time is smaller than 60 s. Martensite appeared in the CGHAZ with the 20 s cooling time.

(2) The value of microhardness in the CGHAZ was higher than that of BM when the cooling time $t_{8/5}$ is smaller than 100s.

(3) The CVN absorbed energy in the CGHAZ was higher than the value of BM when the cooling time $t_{8/5}$ is smaller than 30s.

References

[1] Yutaka N, Sadahira Y. High performance steel pipes and tubes securing and exploiting the future demands [J]. NKK Technical Review, 2003 (88): 81-87.

[2] Penniston C, Collins L, Hamad F. Effects of Ti, C and N on weld HAZ toughness of high strength line pipe [C]. Proceedings of IPC2008, Alberta, Canada, 2008: 75-83.

[3] Romera R, Liebeherr M, Güngör O, et al. Development of X100 on coil and first weldability assessment [C]. Proceedings of IPC2010, Alberta, Canada, 2010: 865-871.

[4] Gianetto J, Bowker J, Dorling D, et al. Structure and properties of X80 and X100 pipeline girth welds [C]. Proceedings of IPC2004, Alberta, Canada, 2004: 1485-1497.

[5] Bruce B, Ramirez J, Johnson M, et al. Welding of high strength pipelines [C]. Proceedings of IPC2004, Alberta, Canada, 2004: 1499-1504.

[6] Poorhaydari K, Patchett B, Ivey D. Microstructure/property examination of weld HAZ in grade 100 microalloyed steel [C]. Proceedings of IPC2002, Alberta, Canada, 2002: 2103-2118.

[7] Poorhaydari K, Patchett B, Ivey D. Correlation between microstructure and hardness of the weld HAZ in grade 100 microalloyed steel [C]. Proceedings of IPC2004, Alberta, Canada, 2004: 2739-2751.

Analysis on Aging Sensitive Coefficient of X90 Line Pipe

Li Yanhua Ji Lingkang Chi Qiang Chen Hongyuan Zhang Jiming

(Tubular Goods Research Institute of CNPC, State Key Laboratory for Performance and Structural Safety of Oil Industry Equipment Materials)

Abstract: Tensile and impact toughness test on X90 line pipe with different aging conditions were conducted to study the effect of aging conditions on properties of X90 pipeline steel. The result shows that yield strength and yield ratio of X90 line pipe would increase significantly with the introduction of strain aging, while the impact toughness and tensile strength have little change. The impact toughness of X90 line pipe would decrease with the introduction of 200℃, 5min aging, However, the impact toughness of X90 pipeline would increase with the increase of aging temperature and aging time (250 ℃, 1h). Comprehensive analysis shows that the aging treatment have different effects on different X90 pipeline steel, mainly due to different microstructure and chemical composition of X90 pipeline between manufacturers during the trial production. The result of aging sensitive coefficient analysis shows that the effect of microstructure type is more significant than that of chemical composition.

Key words: X90 line pipe; Strain aging; Microstructure; Chemical composition; Aging sensitive coefficient

1 Introduction

Strain aging refers to the phenomenon that when subjected to plastic deformation and heated to a certain temperature, the strength and hardness of low carbon steel would increase, and the ductility and toughness would decrease[1]. A certain amount of plastic deformation would occur to line pipe during the production process of submerged arc welded pipe. At the same time, local deformation and natural aging would occur when the actual using area of line pipe are earthquake zone or permafrost region where geological conditions is bad. According to the literature[2-6], For every 1% increase in plastic deformation, the ductile-brittle transition temperature of pipeline steel would increase by more than 5 ℃. Three polyethylene or epoxy powder coating should be coated on the surface of the pipe body. During the coating process, the pipe surface needs to be heated to 200 ℃ to 250 ℃, and then strain aging would occur. Strain aging have significant effect on the mechanical properties of line pipe for inducing increase of yield strength and yield ratio, so as to affect the safe use. As a new line pipe product, the research on X90 line pipe is very limited. As a result, it is quite necessary to study the strain aging phenomenon of line pipe.

2 Materials

Taking three X90 SAW pipes as the experimental material, the size of the pipe is ϕ1219mm×16.3mm, and the chemical composition analysis result of the pipe is shown in table 1.

Table 1 Chemical composition analysis result of the X90 SAW pipes (wt %)

Chemical composition	C	Nb	V	Ti	Cu	N
A	0.056	0.08	0.033	0.017	0.25	0.0041
B	0.045	0.098	0.013	0.014	0.24	0.0037
C	0.041	0.085	0.0055	0.015	0.24	0.0040

3 Test results

3.1 Charpy impact test

The relationship between aging conditions and Charpy impact test result of pipe body in transversal direction (10mm×10mm×55mm) is shown in table 2.

Table 2 Charpy impact test result of pipe body in transversal direction corresponding to different aging conditions

Aging conditions	Without aging	Aging at 200℃ for 5 Mins	Aging at 250℃ for 1h
CVN /J (−10℃)	321	298	307
CVN/J (−10℃)	334	307	310
CVN /J (−10℃)	307	285	292

It can be obtained from table 2 that the impact energy would decrease significantly after the introduction of aging at −10℃. While the impact energy would increase slightly with the increase of aging temperature and aging time.

3.2 Tensile properties

The test result of tensile properties of X90 line pipes in transversal direction under different aging conditions (ϕ8.9mm×35mm gauge length) are shown in table 3～table 5.

Table 3 The test result of tensile mechanical properties of sample A in transversal direction under different aging conditions

Aging conditions	Without aging	Aging at 200℃ for 5 Mins	Aging at 250℃ for 1h
$R_{p0.2}$ (MPa)	693	742	736
R_m (MPa)	790	780	770
$R_{p0.2}/R_m$ (%)	88	95	95

Table 4　The test result of tensile mechanical properties of sample B in transversal direction under different aging conditions

Aging conditions	Without aging	Aging at 200℃ for 5 Mins	Aging at 250℃ for 1h
$R_{p0.2}$ (MPa)	636	699	698
R_m (MPa)	746	740	740
$R_{p0.2}/R_m$ (%)	85	94	94

Table 5　The test result of tensile mechanical properties of sample C in transversal direction under different aging conditions

Aging conditions	Without aging	Aging at 200℃ for 5 Mins	Aging at 250℃ for 1h
$R_{p0.2}$ (MPa)	704	792	802
R_m (MPa)	791	811	824
$R_{p0.2}/R_m$ (%)	89	98	97

It can be obtained from table 3～table 5 that the tensile strength would decrease for specimen A and B, while the tensile strength would increase for specimen C. The yield strength and yield ratio would increase significantly after aging, and the rising range of yield strength and yield ratio would decrease with the increase of aging temperature and aging time.

4　Discussion

Strain aging sensitive coefficient is introduced to characterize the strain aging sensitivity of steel. A certain amount of plastic deformation would occur on the X90 line pipe during the forming process, and the coating temperature would reach to 200～250℃ during the anticorrosion process, as a result, the production process of pipes has significant effect on the tensile properties of pipeline steel. There is close connection between the strain aging sensitivity and tensile properties of pipeline steel. The yield strength and yield ratio would increase after strain aging. As a result, the increase percentage of yield strength and yield ratio after the introduction of aging can be adopted to denote the strain aging sensitive coefficient of steel.

Study on the early results of X80 artificial strain aging shows that the yield strength and yield ratio of X80 pipeline steel increased significantly. The results of X90 artificial strain aging shows that the performance of this law is still evident, but the change magnitude of tensile strength is small.

Adoption the thought of GB 4160, the ratio of mechanical properties of pipeline before and after strain aging is introduced to quantify the aging sensitivity of X90 pipeline steel.

Aging sensitivity coefficients based on yield strength according to Equation 1:

$$C_T = (R_{p0.2S} - R_{p0.2}) / R_{p0.2} \times 100\% \tag{1}$$

Aging sensitivity coefficients based on yield ratio according to Equation 2:

$$C_{Y/T} = (R_S - R) / R \times 100\% \tag{2}$$

In the equation, C_T and $C_{Y/T}$ denotes aging sensitivity coefficients based on yield strength and aging sensitivity coefficients based on yield ratio. $R_{p0.2}$ and R the average yield strength and yield ratio before strain aging, $R_{p0.2S}$ and R_S denote the average yield strength and yield ratio after strain aging. The data before and after artificial strain aging is introduced in the equation (1) and (2) to compute the aging sensitivity coefficients based on yield strength and aging sensitivity coefficients based on yield ratio of X90 pipeline steel manufactured by different manufacturers respectively, which are listed in table 6 and table 7.

Table 6 Aging sensitivity coefficients based on yield strength and aging sensitivity coefficients based on yield ratio after 200 ℃, 5min artificial strain aging

Sample information	Sample A	Sample B	Sample C
Aging sensitivity coefficients based on yield strength	0. 09	0. 02	0. 12
Aging sensitivity coefficients based on yield ratio	0. 07	0. 03	0. 10

Table 7 Aging sensitivity coefficients based on yield strength and aging sensitivity coefficients based on yield ratio after 250 ℃, 1h artificial strain aging

Sample information	Sample A	Sample B	Sample C
Aging sensitivity coefficients based on yield strength	0. 09	0. 01	0. 14

The comparison of aging sensitivity coefficients based on yield strength and aging sensitivity coefficients based on yield ratio of X90 pipeline steel manufactured by different manufacturers after 200 ℃, 5min artificial strain aging is shown in Fig. 1.

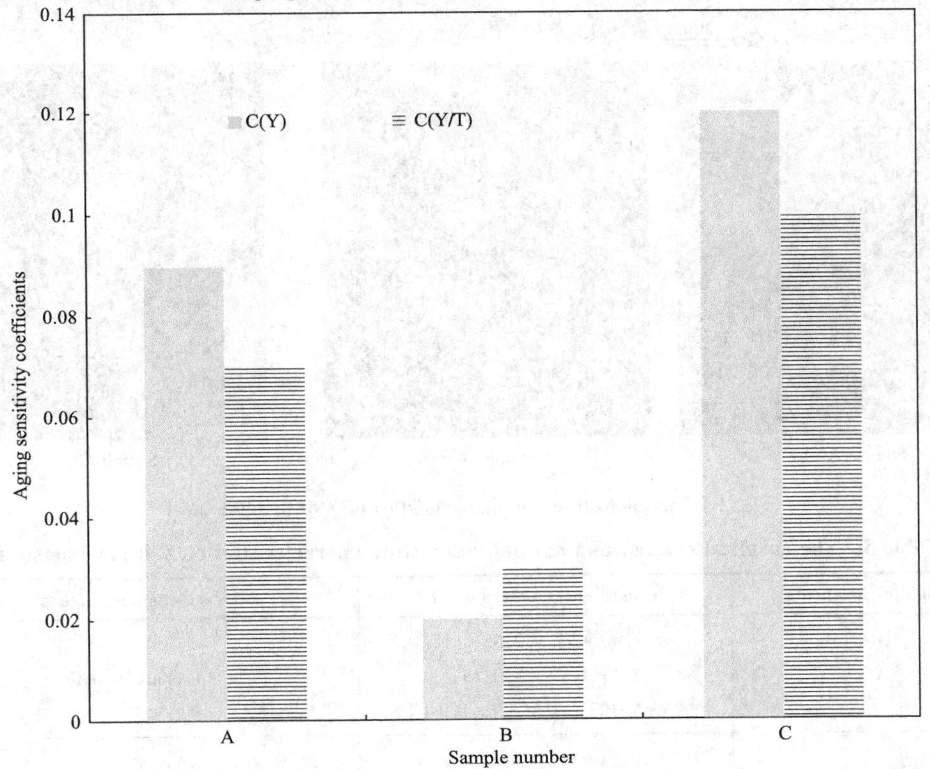

Fig. 1 The comparison of aging sensitivity coefficients based on yield strength and aging sensitivity coefficients based on yield ratio of X90 pipeline steel manufactured by different manufacturers after 200℃, 5min artificial strain aging

The causes of pipeline aging phenomenon is attributed to the segregation of interstitial solid solute C, N near dislocation after short heating process at low temperature so as to stimulate the formation of Cottrell atmosphere which result in dislocation locking. As more and more solute atoms gathered near dislocation, the grain boundary of pipeline steel with high dislocation density would become more brittle, as a result the yield strength and yield ratio of pipeline steel would increase finally. The test results shows that under 200 ℃, 5min aging condition, the sensitivity coefficient of C sample of X90 pipeline steel is maximum, while the sensitivity coefficient of C sample of X90 pipeline steel is minimum, the main reason for different aging sensitivity coefficient of X90 pipeline steel manufactured by different manufacturers should be mainly attributed to different microstructures and chemical compositions of the X90 pipeline steel, as shown in table 8 and Fig. 2.

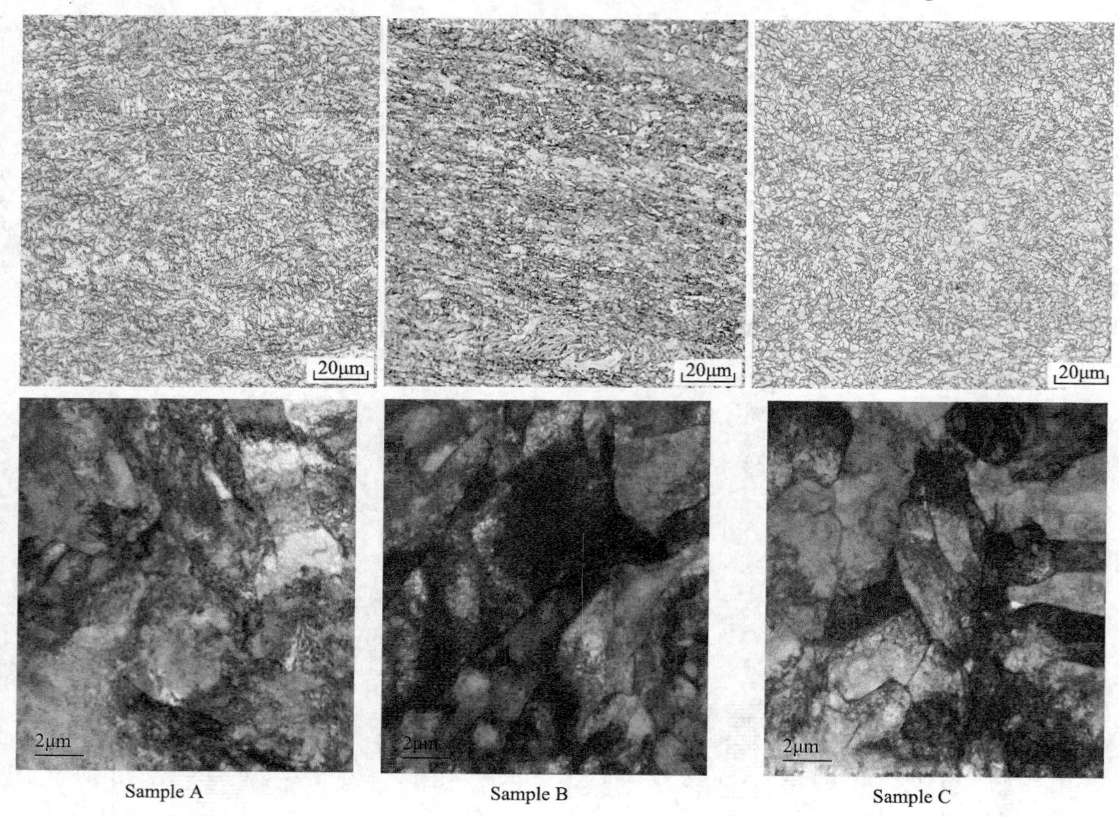

Fig. 2 The microstructure characteristics of X90 pipeline steel

Table 8 The chemical composition and microstructure characteristics of X90 pipeline steel

Sample information	Chemical composition (wt/%)	Microstructure characteristics
A	(C+N) = 0. 06, (V+Nb+Ti) = 0. 13, (V+Nb+Ti) - (C+N) = 0. 07	Granular bainite
B	(C+N) = 0. 05, (V+Nb+Ti) = 0. 13, (V+Nb+Ti) - (C+N) = 0. 08	Granular bainite and MA

Continued

Sample information	Chemical composition (wt/%)	Microstructure characteristics
C	(C+N) = 0.05, (V+Nb+Ti) = 0.11, (V+Nb+Ti) - (C+N) = 0.06	Granular bainite and MA and small amount of PF on inner and outer surfaces, the remaining are granular bainite and MA

Study on the early results of X80 artificial strain aging show that when the microstructures of the X80 pipeline steel is similar, the chemical composition would has significant effect on the aging sensitivity[7-9]. When the C, N content is equal, the aging sensitivity of the pipeline steel would be reduced by increasing the content of V, Nb and Ti, among which, the effect of Nb, Ti content on aging sensitivity is more significant. the main reason should be attributed to these micro- alloying elements would form alloy precipitates with interstitial solid solute C, N, so as to reducing the segregation of interstitial solid solute C, N near dislocation. But the study on aging sensitivity coefficients of X90 pipeline steel shows that the effect of microstructures on aging sensitivity is more significant than chemical composition[10]. Microstructures analysis on the X90 pipeline steel shows that a small amount of polygonal ferrite precipitation appear in the microstructures of sample C, these ferrite precipitation has dissection effect on the original austenite grain, so as to make the bainite lath become short and fine. At the same time, the shear process and the volume expansion during the transformation from austenite to bainite will induce high density dislocation generated on the ferrite around bainite, which effectively reduces the segregation path of C, N atoms to dislocation during the aging process[11,12], as a result, the aging sensitivity of the pipeline steel would exacerbate. In addition, the C content of polygonal ferrite, bainite and MA is quite different between each other, this large concentration gradient is also conducive to promoting the diffusion of C atoms during the aging process, thus further exacerbate the aging sensitivity[13-15]. This may be the main reason for the high aging sensitivity of sample C. Aging pipeline steel under conditions of 250 ℃, 1h also shows a similar result, as shown in Fig. 3.

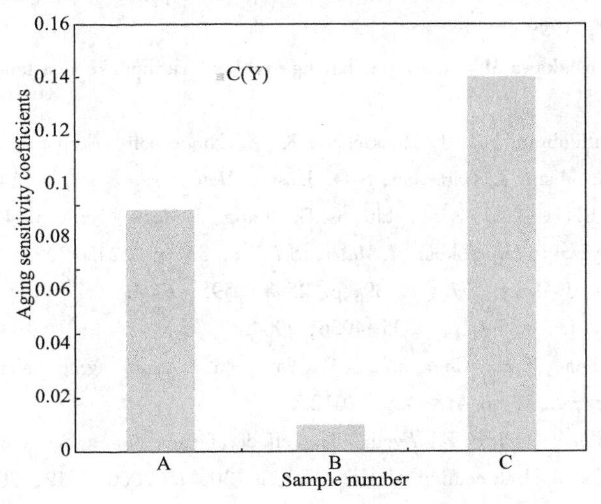

Fig. 3 Aging sensitivity coefficients of X90 pipeline steel under aging conditions of 250 ℃, 1h

5 Conclusions

The impact energy of X90 pipeline steel would decrease significantly after the introduction of 200℃, 5min aging at -10℃, While the impact energy would increase slightly with the increase of aging temperature and aging time.

The tensile strength of X90 pipeline steel would decrease or increase slightly after the introduction of aging. While the yield strength and yield ratio would increase significantly after the introduction of aging, and the rising range of yield strength and yield ratio would decrease with he increase of aging temperature and aging time.

The study on aging sensitivity coefficients of X90 pipeline steel shows that the effect of microstructures on aging sensitivity is more significant than chemical composition. X90 pipeline steel with dual-phase microstructure would has high aging sensitivity coefficient.

Acknowledgements

This work is supported by China National Petroleum Corporation Major Project (2012E-2801-03).

References

[1] GAO Jian-zhong, WANG Chun-fang, WANG Chang-an, etc. Strain Aging Behavior of High Strength Linepipe Steel. *Development and Application of Materials*, v. 24, n. 3, p. 86-90, 2009.

[2] Allen T, Busby J, Meyer M, Petti D. Materials challenges for nuclear systems, *Mater Today*, n. 13, p. 14-23, 2010.

[3] Deodeshmukh VP, Srivastava SK. Effects of short- and long-term thermal exposures on the stability of a Ni-Co-Cr-Si alloy. *Mater. Des.*, n. 31, p. 2501-9, 2010.

[4] Panait CG, Zielińska-Lipiec A, Koziel T, Czyrska-Filemonowicz A, Gourgues-Lorenzon A-F, Bendick W. Evolution of dislocation density, size of subgrains and MX-type precipitates in a P91 steel during creep and during thermal ageing at 600 °C for more than 100000 h. *Mater. Sci. Eng. A*, 527, 4062-4069, 2010.

[5] Sanderson N, Ohm RK, Jacobs M. Study of X100 linepipe costs points to potential savings. *Oli&Gas journal*, v. 3, n. 15, p. 54-57, 1999.

[6] Suzuki N, Kato A., Yoshikawa M. Line pipes having excellent earthquake resistance. *NKK Giho*, n. 167, p. 44-49, 1999.

[7] M. K. Gräf, H. G. Hillenbrand, C. J. Heckmann, K. A. Niederhoff, *Europipe*, n. 35, p. 1-9, 2003.

[8] S. Shanmugam, R. D. K. Misra, J. Hartmann, S. G. Jansto, *Mater. Sci. Eng.*, n. A441, p. 215-229, 2006.

[9] Y. B. Xu, Y. M. Y, B. L. Xiao, Z. Y. Liu, G. D. Wang, J. *Mater. Sci.*, n. 45, p. 2580-2590, 2010.

[10] H. Nakagawa, T. Miyazaki, H. Yokota. J. *Mater. Sci.*, n. 35, p. 2245-2253, 2000.

[11] C. Z. Wang, R. B. Li, J. *Mater. Sci.*, n. 39, p. 2593-2595, 2004.

[12] M. Ozawa, J. *Mater. Sci.*, n. 39, p. 4035-4036, 2004.

[13] W. G. Zhao, M. Chen, S. H. Chen and J. B. Qu. Static strain aging behavior of an X100 pipeline steel. *Mater. Sci. Eng.*, A550, p. 418-420, 2012.

[14] N. Guermazi, K. Elleuch and H. F. Ayedi, The effect of time and aging temperature on structural and mechanical properties of pipeline coating. *Mater. Des.*, n. 30, p. 2006-2012, 2009.

[15] M. S. Rashid. Strain-Aging of Vanadium, Niobiumor Titanium-Strengthened High-Strength Low-Alloy Steels, *Metal. Trans.*, 6A, P. 1265-1267, 1975.

Anisotropic behaviors for X100 high grade pipeline steel under stress constraints

Yang Kun[1,*] Sha Ting[3] Yang Ming[2] Shang Cheng[2] Chi Qiang[1]

(1. Tubular Goods Research Institute of CNPC; 2. Petrochina West Pipeline Company;
3. The No. 771 Institute of Ninth Academy of China Aerospace Science and Technology Corporation)

Abstract: Because of the manufacture process of high grade pipeline steel, the anisotropic behaviors appeared in different directions, including both properties and microstructure. In this paper, mechanical properties and microstructures for X100 high grade pipeline steel were investigated with a series of tests, including not only experiments but also simulation. Tensile tests with DIC (Digital Image Correlation) method was used to get the stress-strain relationship, especially in the process of fracture. SENT (Single Edge Notch Tensile) tests with different notch sizes were used to characterize the fracture resistance anisotropies. The microstructure was characterized by both fracture analysis and inclusion characteristics to infer the relation between voids growth and crack propagation. Moreover, FE simulation of SENT was carried out by complete Gurson model. Finally, the results of tests and simulations were compared to study the effects of stress constraints on crack resistance in different directions.

Keywords: Mechanical properties anisotropies; Microstructure anisotropies; Crack propagation; Inclusion characteristics; X100 high grade pipeline steel; Complete Gurson model; Stress constraints

1 Introduction

Natural gas transported by running pipeline was considered as the most economical and safety way, which had been widely used all over the world. In order to increase the transport capacity of pipeline, the traditional way was to rise the diameter and the grade of pipes. However, safety problem occurred when the grade rise. Ductile crack propagation for long distance pipeline was considered as one of the most dangerous failure types for pipeline steel, and it was easy to occur in high grade pipeline and induced serious accident[1]. It was important to investigate the fracture behavior for high grade pipes (X100).

Full-scale test and simulation method were mainly two ways to study ductile crack propagation

* Corresponding author.

E-mail address: Kunyang073@ cnpc. com. cn (Kun Yang)

for high grade pipes. However, Full-scale test (Full scale burst test) needed to build a pipeline with different toughness pipes, and it was too expensive to carry out and cost too long time. For most of the models, BTC (Battelle two curve method) developed by Battelle institute built a relation between crack resistance and nature gas pressure with a through crack, and the fracture behaviors for the pipeline which was lower than X60 have been predicted successfully. BTC was a semi-empirical method, and it had the assumption of homogeneous material, which did not get the relation between microstructure and properties, and both base metal and inclusions were anisotropic in fact [2-4].

Inclusions were the source of voids, and they may become crack with the rising exterior load. The deformation capacities of inclusions was different from the base metal, they may break away from the base metal and form voids during exterior load. Damage occurred in these places and cracks were produced finally[10-11]. The process of hot roll for pipe manufacture would induce the anisotropic distribution of inclusions, the initial anisotropy had the influence on the growth of voids, and the properties changed finally.

2 Full scale burst test and BTC model

In order to determine the crack arrest toughness of pipeline, full scale burst test was carried on by TGRI (Tubular Goods Research Institute). Fig. 1. (a) and Fig. 1. (b) show the pipes after the burst test. In the test, the pipes with higher toughness were laid on the sides of the pipeline, and the initial crack was made on the lower toughness pipe by shaped explosive in the middle. The crack resistance was rising from the middle to the side for the pipeline, and the toughness indicator could be determined by the crack arrest pipe. CharpyV - notch (CVN) energy was used to characterize the toughness for pipes[5,6]. For the whole process, the toughness of pipes was considered as the resistance for crack propagation and the pressure was considered as the driving force. The balance point of resistance and driving force was the situation of crack arrest.

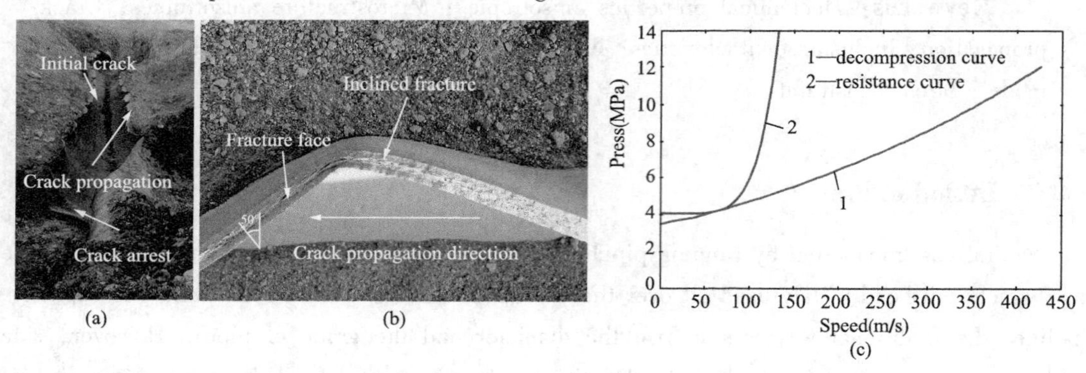

Fig. 1 Full size bust test for X80 nature gas pipeline. (a) The pipes after crack propagation, (b) Inclined fracture on the pipes, (c) BTC curve

BTC used the Eq. 1 to characterize the toughness of pipes.

$$P_a = \frac{4}{3.33\pi} \times \frac{t}{D} \times \sigma_f \times \arccos \exp\left(-\frac{\pi ER}{24\sigma_f^2 \sqrt{Dt/2}}\right) \tag{1}$$

$$V_c = 0.275 \frac{\sigma_f}{\sqrt{R}} \left(\frac{P_d}{P_a - 1}\right)^{1/6}$$

In Eq. 1, P_a was fracture arrest pressure, σ_f was the flow stress, P_d was the dynamic gas pressure, t was the thickness, D was the outside diameter, R was energy density and can be characterized as

$$R = C_v / A_c \tag{2}$$

C_v was the charpy energy, and A_c was the area of sample. The crack arrest situation can be obtained by Fig. 1 (c).

In Fig. 1 (c), fracture arrest pressure could be given by the point of contact between decompression curve and resistance curve. Therefore, R can be calculated and charpy energy can be given, which can be used to predict the crack arrest.

With the rise of grade for pipe, Fig. 2 (a), the flow behavior changed, and the predict results by BTC were inaccurate, which needed to correct, see Fig. 2 (b). The correct factor was lower when the grade of pipes was lower. Therefore, it was limited for BTC to predict the fracture behavior.

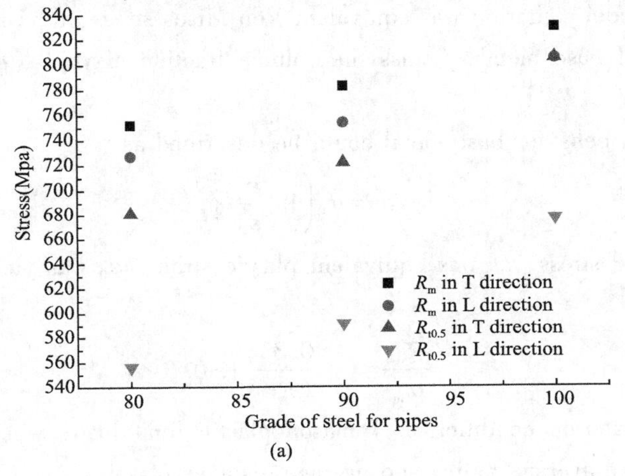

(a)

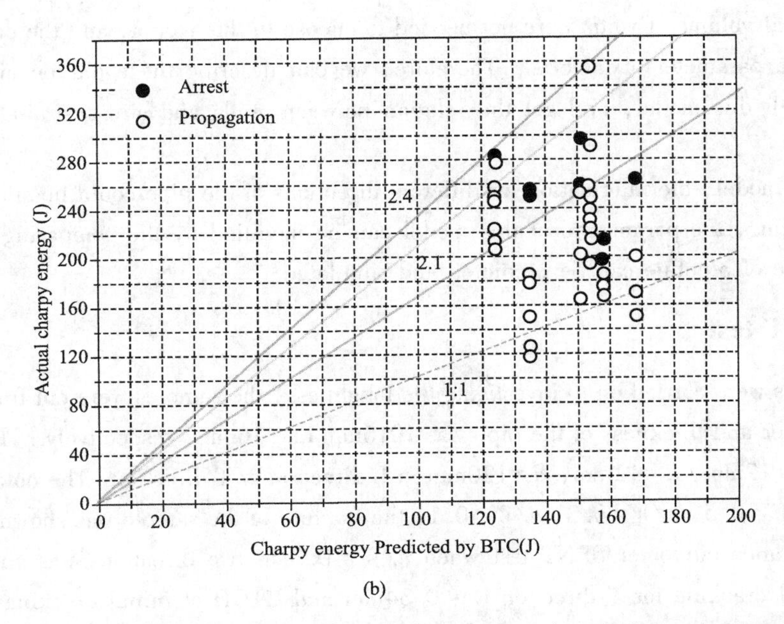

(b)

Fig. 2 (a) Evolution of tensile strength (R_m) and yield strength ($R_{t0.5}$) during the rise of steel grade in different direction in smooth tensile testing, (b) the result of full scale burst test and BTC simulation

3 Improved Gurson model

Gurson model can describe the voids transformation and give the yield function of porous material, which built the relation between microstructures and properties [7,8]. Tvergarrd and Needleman considered the stress field of the voids, the interaction among them, and the properties of material, and built GTN model on the base of Gurson model[9].

GTN model was used to describe the growth of voids and Tomason plastic limit load model was used to describe the coalescence of voids, and the whole process was called the complete Gurson model.

$$\phi\ (q_1,\ \sigma_f,\ f,\ \sigma_m) = \frac{q^2}{\sigma_f^2} + 2q_1 f \cosh\left\{\frac{3q_2\sigma_m}{2\sigma_f}\right\} - 1 - (q_1 f)^2 = 0 \tag{3}$$

Eq. 3 is GTN model, where q was equivalent Von Mises stress, σ_m was the average stress, σ_f was the flow stress of base metal, f wass the volume fraction of voids, q_1 and q_2 was correction factor.

The stress-strain behavior base metal could be described as

$$\sigma_f = \sigma_0\left(1 + \frac{\varepsilon^p}{\varepsilon_0}\right)^n \tag{4}$$

Where σ_0 is yield stress, ε^p was equivalent plastic strain, ε_0 was yield strain, n was stress-intensity factor.

$$\frac{\sigma_1}{\sigma_m} = \frac{0.3}{r/\ (R-r)} + 0.6 \tag{5}$$

Eq. 5 was the criterion condition of Tomason plastic limit load, where σ_1 was the principal stress, R and r was the average radius and average distance of voids.

The critical volume of voids were not needed to choose in this model, and the critical volume of voids was not a parameter of material. Therefore, we can describe the evolution of properties and deformation with this model, and get the relation between voids and stress/strain behavior of the material.

With this model, the information for different directions of the pipe could be simulated by finite element modelling, the parameters of this model can be modified by the comparing with the tests, and the fracture of pipeline can be predicted and simulated.

4 SENT test

SENT tests were carried on to investigate the toughness, the samples were cut from X100 pipes, and the diameter and thickness of the pipe was 1016mm and 16mm, respectively. The size of SENT sample was W=12mm, B=12mm, H=120mm in L direction or T direction. The notch size of SENT sample was a/W=0.5, a/W=0.3, a/W=0.1, the fracture of the sample was shown on Fig. 3, and the crack resistance curve for SENT test when a/W = 0.3 in two directions was shown on Fig. 4. CTOD of initial cracking for T direction was 0.34mm and CTOD of initial cracking for L direction was 0.29mm. Toughness of the material in two directions are different, and it was better for the sample in T direction to block the crack initiation and propagating.

(a) (b) (c) (d)

Fig. 3 Single edge notched tension test with single sample method and the fracture sections for L-sample.
(a) SENT test with double extensometers, (b) fracture sections for the sample when $a/W=0.5$; (c) fracture sections for the sample when $a/W=0.3$; (d) fracture sections for the sample when $a/W=0.1$

5 Microstructure analysis

Inclusions were considered as the source of voids, and they may induce crack with the rising exterior load[10-12]. In order to get the information of initial area of voids, inclusion was investigated by SEM (Fig. 5), and the area fraction in different directions was given in Table 1. The types of inclusion were MnS and Al_2O_3, and the shapes were elongated and equiaxed, respectively. Moreover, the area fractions for inclusion in different direction was anisotropic, which may cause the anisotropic behavior of initial voids. According to the result of inclusion characteristics, the inclusions for X100 are mainly made up of elongated inclusions (MnS) and equiaxed inclusions (Al_2O_3), and the area fraction of equiaxed inclusions was higher.

Table 1 inclusion characteristics for elongated (MnS) and equiaxed particles (Al_2O_3) for the sample, and the scanning area is about 100 mm^2

Type	Section (i)	Area frcation	Number/area (mm^{-2})	Aspect ratio	Size (μm)		Spacing (μm)	
					$\overline{r_J}$	$\overline{r_K}$	$\overline{d_J}$	$\overline{d_K}$
MnS	(L)	0.0000271	13.7	2.1	2.5	1.0	71	65
	(T)	0.0000141	9.2	2.0	1.1	2.3	117	96
	(S)	0.0000419	6.7	2.2	2.9	1.3	161	132
Al_2O_3	(L)	0.0000916	11.2	2.1	3.5	1.5	84	69
	(T)	0.0000774	12.9	2.0	1.8	3.6	97	80
	(S)	0.0000523	11.3	1.9	4.3	2.1	86	79

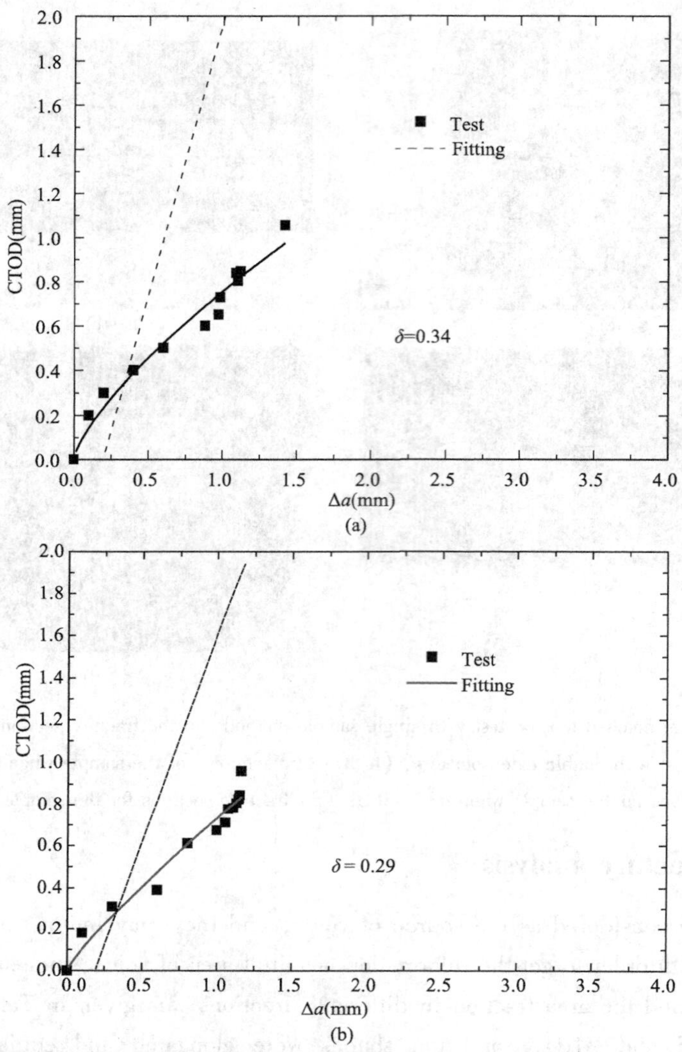

Fig. 4 Crack growth resistance curves for SENT test when $a/W = 0.3$. (a) in T direction, (b) in L direction

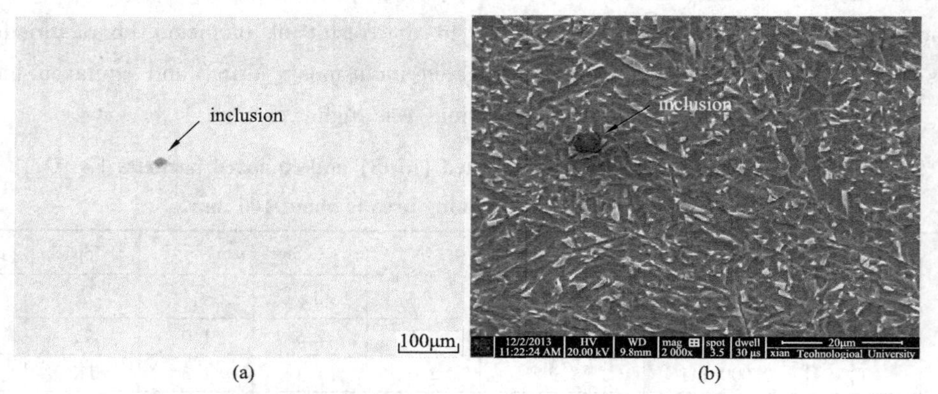

Fig. 5 Inclusion in X100 for OM and SEM. (a) OM, (b) SEM

6 Simulation

To investigate the relation between microstructure and properties for the sample, SENT tests were simulated by the complete Gruson model, and it was carried out by Abaqus with material user subroutine UMAT. The true stress-strain curves in two directions for the simulation were given by the tension test with DIC method, see Fig. 6. Comparing with the traditional way (Axial extensometer), DIC can get the stress-strain relation at a larger strain, especially in the process of crack propagation and fracture. In this study, initial voids area fractions $f_0 = 0.00025$ in T direction and $f_0 = 0.0005$ in L direction. Young's modulus $E = 206\text{GPa}$ and Poisson ratio $\nu = 0.3$. Moreover, yield stress $\sigma_0 = 767\text{MPa}$ in T direction, yield stress $\sigma_0 = 649\text{MPa}$ in L direction, correction factor $q_1 = 1.5$ and $q_2 = 1.0$ in this paper. Moreover, the size of Abaqus SENT model was the same as the test, it was a half model and $W = 12\text{mm}$, $B = 12\text{mm}$, $1/2H = 60\text{mm}$, the initial crack depth of this model was $a/W = 0.5$, 0.3 or 0.1, the simulation results were shown on Fig. 7, Fig. 8, Fig. 9. FE simulation was carried out by ABAQUS software, and the cloud maps for strain in two different directions with different stress constrain were shown in Fig. 7. The P-CMOD curves were shown in Fig. 8, and the max load was higher when the notch size was smaller. The crack resistance curves in two directions when $a/W = 0.3$ were given in Fig. 9. CTOD in T direction was 0.35mm, the value was 0.32mm in L direction, and the result went well with the test result. Moreover, the value of CTOD in T direction was higher than the value in L direction.

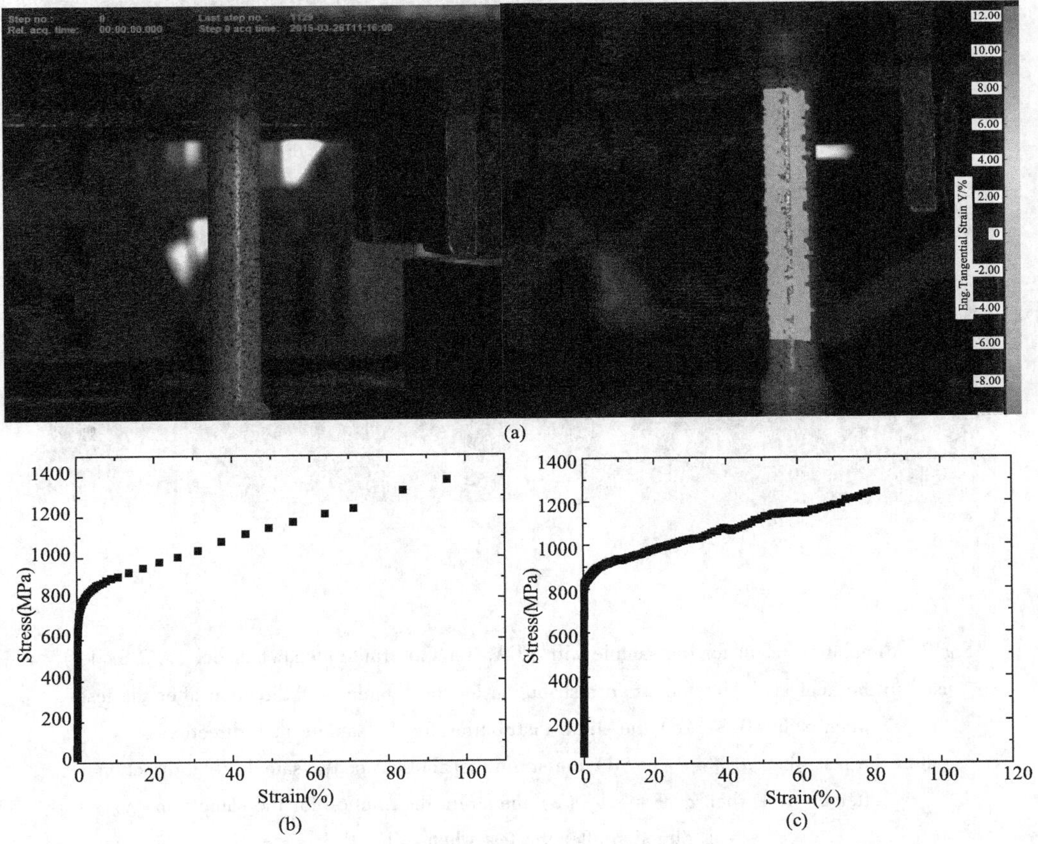

Fig. 6 Tension test for X100 with DIC (digital image correlation) method. (a) two photos from the cameras of DIC, (b) true stress-strain curve forthe sample in T direction, (c) true stress-strain curve for the sample in L direction

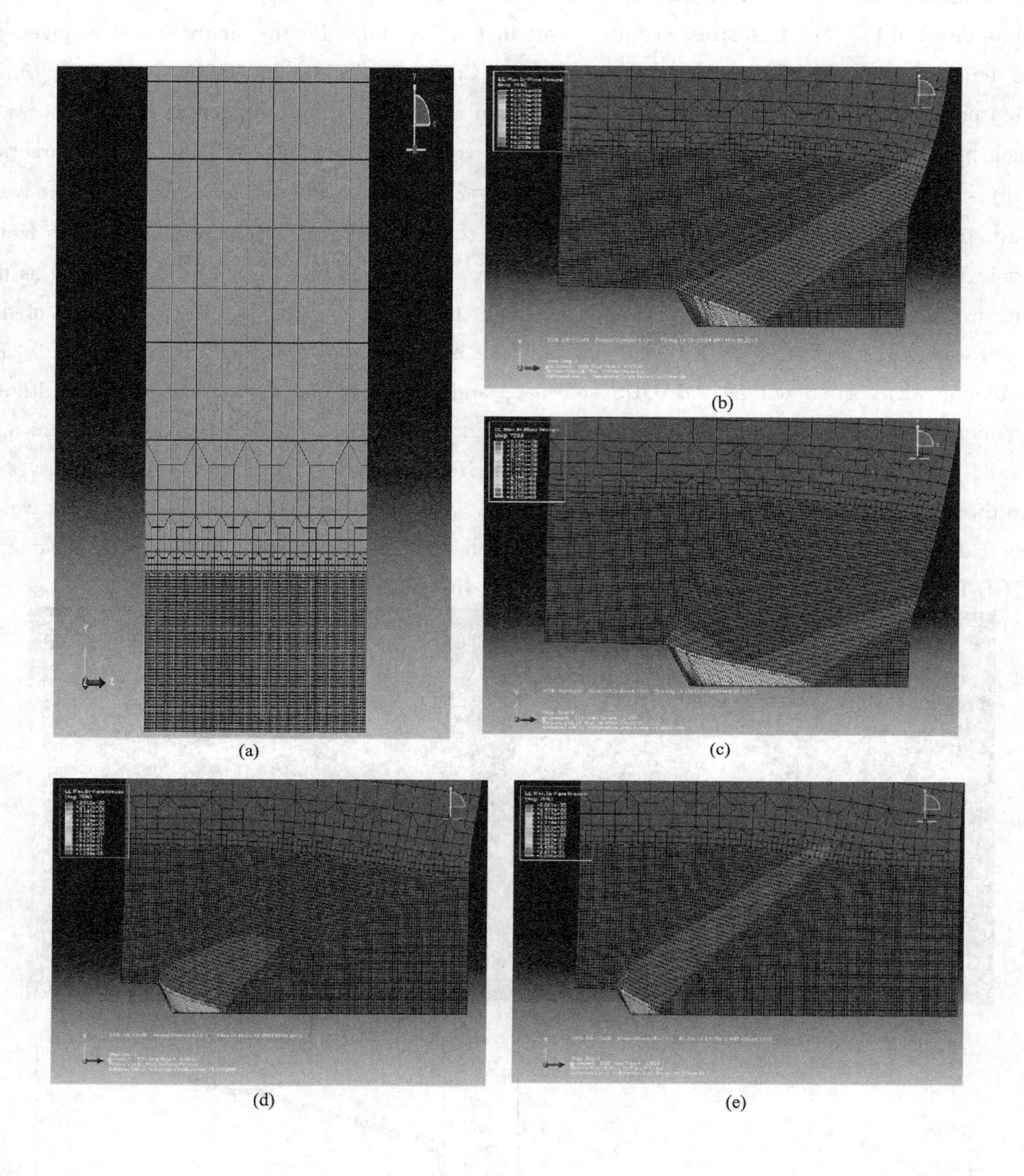

Fig. 7 Simulation result for the sample with FEA. (a) the finite element model (1/2 model) used in the analysis, (b) the strain distribution for the sample in T direction after the test when $a/W=0.3$, (c) the strain distribution for the sample in L direction after the test when $a/W=0.3$; (d) the strain distribution for the sample in T direction after the test when $a/W=0.1$, (e) the strain distribution for the sample in L direction after the test when $a/W=0.1$

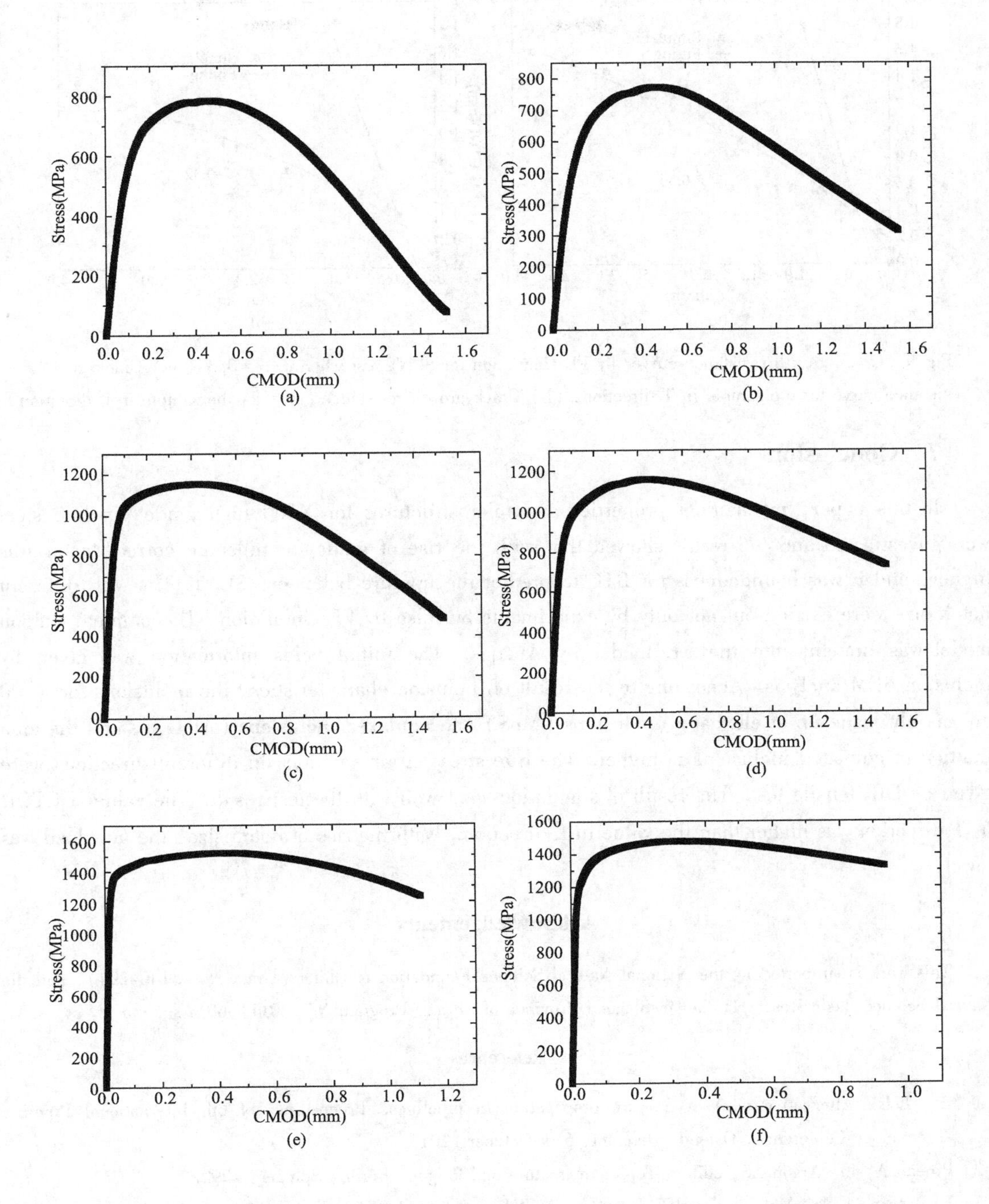

Fig. 8 P-CMOD curves by FE simulation on SENT test. (a) P-CMOD curve for the sample in T direction when $a/W=0.5$; (b) P-CMOD curve for the sample in L direction when $a/W=0.5$; (c) P-CMOD curve for the sample in T direction when $a/W=0.3$; (d) P-V curve for the sample in L direction when $a/W=0.3$; (e) P-CMOD curve for the sample in T direction when $a/W=0.1$; (f) P-CMOD curve for the sample in L direction when $a/W=0.1$.

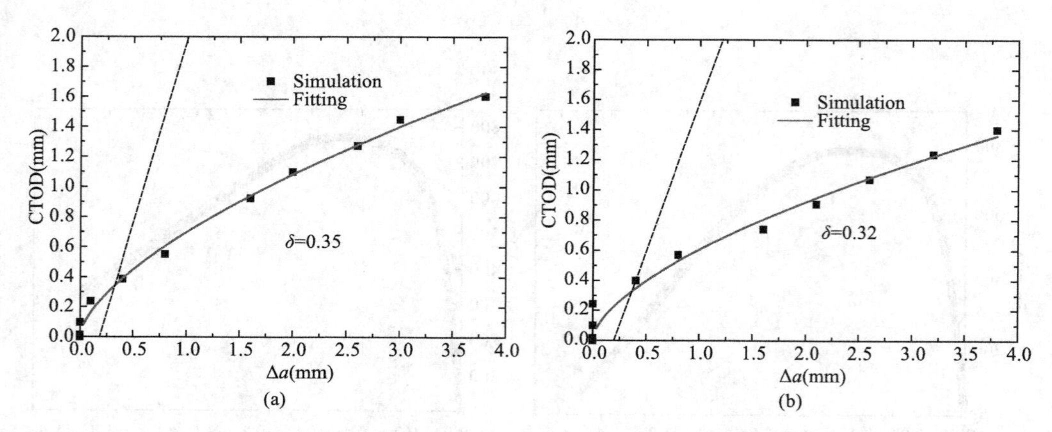

Fig. 9 Crack growth resistance curves by FE simulation for SENT test when $a/W=0.3$. (a) Crack growth resistance curve for the sample in T direction, (b) Crack growth resistance curve for the sample in L direction

7 Conclusion

In this paper, mechanical properties and microstructures for X100 high grade pipeline steel were investigated and the result showed that with the rise of grade for pipe the correct factor was higher, and it was boundedness for BTC to predict the fracture behavior. SENT tests with different notch size were carried out not only by experiments but also by FE simulation. Thecomplete Gurson model was brought into the FE model by Abaqus. The initial voids information was given by inclusion SEM analysis. According to the result of inclusion characteristics, the inclusions for X100 are mainly made up of elongated inclusions (MnS) and equiaxed inclusions (Al_2O_3), and the area fraction of equiaxed inclusions is higher. The true stress strain relations in different directions were given by DIC tensile test. The result of simulation went well with the test result. The value of CTOD in T direction was higher than the value in L direction. With the rise of notch size, the max load was lower.

Acknowledgements

This work is supported by the National Natural Science Foundation of China (Grant No. 51404294), and the Natural Science Basic Research Plan in Shaanxi Province of China (Program No. 2011JQ6017).

References

[1] B. Leis. Alternative view of fracture propagation in pipelines. Proceedings of 6th International Pipeline Technology Conference, Ostend, Belgium, 6-9 October, 2013.

[2] Pineau A. In: Argon AS, editor. Topics in fracture and fatigue. Berlin: Spinger, 1992.

[3] J. Besson, C. N. McCowan and E. Drexler, Modeling flat to slant fracture transition using the computational cell methodology [J]. Engineering Fracture Mechanics, 2013, 104: 80-95.

[4] H. O. Nordhagen, S. Kragset, T. Berstad, A. Morin, C. Dørum, S. T. Munkejord, A new coupled fluid-structure modeling methodology for running ductile fracture [J]. Computation and Structure, 2012, 94: 13-21.

[5] X. L. Yang, Y. B. Xu, X. D. Tan, D. Wu, Influences of crystallography and delamination on anisotropy of Charpy impact toughness in API X100 pipeline steel [J]. Materials Science and Engineering: A, 2014,

607: 53-62.

[6] S. Y. Shin, B. Hwang, S. Kim, S. Lee. Fracture toughness analysis in transition temperature region of API X70 pipeline steels [J]. Materials Science and Engineering: A, 2006, 429: 196-204.

[7] Tvergaard V, On Localization in Ductile Materials Containing Spherical Voids. International Journal of Fracture, 1982, 18 (4): 237-252.

[8] Tvergaard V, Needleman A. Analysis of the cup-cone fracture in a round tension bar. Acta Metall 1984; 32: 157-69.

[9] Zhang ZL, Thaulow C, Ødegård J. A complete Gurson model approach for ductile fracture. Engng Fract Mech 2000, 67: 155-68.

[10] J. Llorca, A. Needleman, S. Suresh. An analysis of the effects of matrix void growth on deformation and ductility in metal-ceramic composites. Acta Metall Mater. 1991, 39: 2317-2335.

[11] M. D. Richards, E. S. Drexler, J. R. Fekete. Aging-induced anisotropy of mechanical properties in steel products: Implications for the measurement of engineering properties [J]. Materials Science and Engineering: A, 2011, 529: 184-191.

[12] A. A. Benzerga, J. Besson, A. Pineau. Anisotropic ductile fracture Part I: experiments [J]. Acta Materialia, 2004, 52: 4623-4638.

X70 高应变钢管冷弯变形及全尺寸弯曲试验研究

王 鹏 陈宏远 池 强

（中国石油集团石油管工程技术研究院；石油管材及装备材料
服役行为与结构安全国家重点实验室）

摘 要： 大口径钢管的现场冷弯常用于改变管道水平方向的。由于其特殊的结构，针对弯曲过程和冷弯管大变形行为的影响进行了实验和分析研究。现场冷弯试验是针对 X70 钢管级高应变钢管，采用 3 种不同的弯曲角度。同时，采用拉伸试验、冲击试验的影响，DWTT 测试冷弯过程对几何和 X70 高应变管的性能进行了分析。对角度约 6°的冷弯管进行了全尺寸弯曲试验。

关键词： 冷弯管；曲率半径；力学性能；高应变钢管

弯管是天然气和石油管道的重要组成部分，其经常用于改变管线的高度，或管道的水平方向的。冷弯管广泛应用于天然气管道中弯曲半径大、弯曲角度小的弯管。冷弯管是由直管采用冷弯机弯曲成型[1]。弯管由于其特殊的结构，被认为是天然气和石油管道中最薄弱的部分．。管道经常受到大变形产生的应力和变形[2]，如土壤沉降、冻胀、热胀冷缩、滑坡、管柱、管道敷设等几种环境负荷。目前，高应变钢管已广泛应用于国内。但冷弯通常是由普通管道，而不采用高应变管。由于现场弯曲过程中的塑性变形，管体的纵向性能发生较大变化[3]。但在这篇文章中，采用了高应变管进行冷弯试验，研究了冷弯过程对 X70 高应变管的影响，并通过全尺弯曲试验研究了冷弯管的大变形行为。

1 钢管的现场冷弯试验

1.1 原材料

现场冷弯实验采用两根直缝埋弧焊高应变管，并在冷弯试验机上完成冷弯试验，高应变钢管为 X70 钢级，尺寸和性能如表 1 所示。

表 1 试验钢管性能和尺寸

编号	直径（mm）	壁厚（mm）	取样方向	屈服强度（MPa）	抗拉强度（MPa）	屈强比
A	1016	17.5	横向	585	710	0.82
			纵向	510	700	0.73
E	1016	17.5	横向	570	705	0.80
			纵向	525	690	0.76

1.2 现场冷弯

冷弯试验中采用不同的曲率半径。A 为直缝埋弧焊高应变管，由于制造工艺的原因，其初始弯曲角为 0.53°。对 B、C 和 D 弯曲参数如表 2 所示。E 是长为 8m 的直缝埋弧焊高应变管，其弯曲参数为0.3°/300mm。E管弯曲参数见表3。在冷弯过程中，弯曲角度由冷弯机的垂直提升位移控制。如果垂直提升位移过大，由于拉伸应力，在外弧测可能发生断裂。由于压应力，在内弧侧可能发生起皱或屈曲。这都可能导致管道产生灾难性事故。因此，在高应变管试验中严格限制弯管工艺，见表2和表3。用数字测角仪测量各弯曲过程的弯曲角度。冷弯管的弯曲角度为1.86°、2.8°、4.16°。和E弯曲角度为6.17°。

表2　冷弯工艺要求

钢管编号	弯曲步	弯曲角度（°）	弯后角度（°）	总角度（°）
A	0		0.53	
B（0.3°/300mm）	1	0.37	0.90	1.86
	2	0.4	1.30	
	3	0.26	1.56	
	4	0.24	1.80	
	5	0.3	2.10	
	6	0.24	2.34	
	7	0.42	2.76	
C（0.5°/300mm）	8	0.42	3.18	2.8
	9	0.42	3.60	
	10	0.45	4.05	
	11	0.48	4.53	
	12	0.48	5.01	
	13	0.29	5.30	
D（0.7°/300mm）	14	0.76	6.06	4.16
	15	0.64	6.70	
	16	0.71	7.41	
	17	0.81	8.22	
	18	0.45	8.67	
	19	0.79	9.46	

表3　E钢管弯曲工艺

钢管编号	弯曲步	弯曲角度（°）	弯后角度（°）	总角度（°）
E（0.3°/300mm）	0		0.67	
	1	0.55	1.22	6.17
	2	0.34	1.56	
	3	0.31	1.87	
	4	0.29	2.16	
	5	0.35	2.51	

续表

钢管编号	弯曲步	弯曲角度（°）	弯后角度（°）	总角度（°）
E（0.3°/300mm）	6	0.33	2.84	6.17
	7	0.36	3.20	
	8	0.30	3.50	
	9	0.53	4.03	
	10	0.23	4.26	
	11	0.39	4.65	
	12	0.40	5.05	
	13	0.45	5.50	
	14	0.43	5.93	
	15	0.39	6.32	
	16	0.52	6.84	

2 试验结果及讨论

2.1 几何形状

对试验前后钢管和冷弯管的椭圆度和壁厚均进行了测量，椭圆度变化如图 1 所示，壁厚变化见图 2。从图中可以看出，冷弯管椭圆度随着弯制曲率半径的增大而增大，而三种弯制工艺下，冷弯管的壁厚变化不明显。

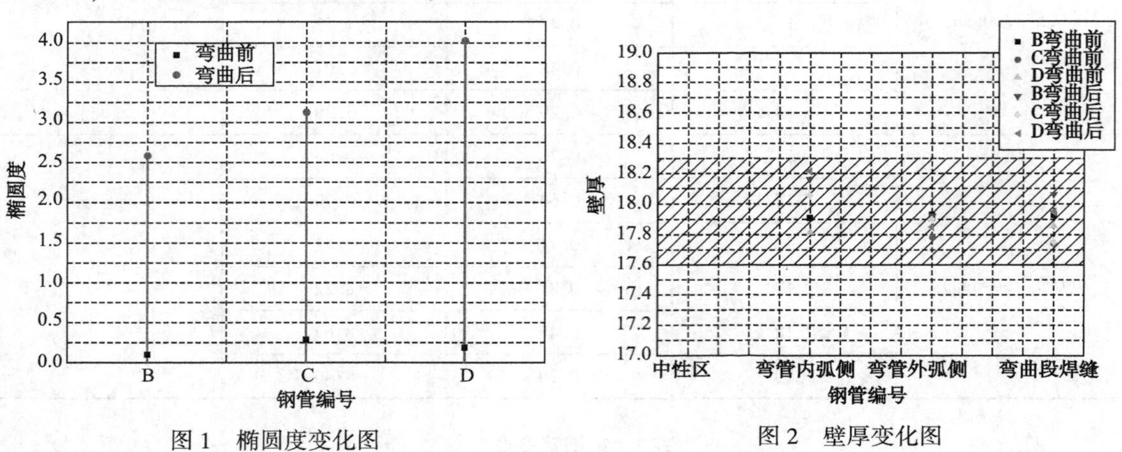

图 1　椭圆度变化图　　图 2　壁厚变化图

2.2 拉伸试验

从冷弯管和原母管上分别取拉伸试样，试样位置分别为中性区、弯管外弧侧、内弧侧和弯曲段焊接接头。试样方向为横向和纵向。试验结果如图 3 和图 4 所示。图 3（a）和图 4（a）横向拉伸试验结果，图 3（b）和图 4（b）为纵向拉伸试验结果。试验研究发现，随着冷弯曲率半径的减小，弯管横向力学性能变化不明显，而冷弯管纵向性能变化明显，具体表现为外弧处屈服强度（$R_{t0.5}$）及屈强比（$R_{t0.5}/R_m$）升高，内弧处强度及屈强比降低，中性区及焊缝变化不明显。这是由于冷弯变形主要是钢管纵向的弯曲变形，而钢管横向的变形很小。

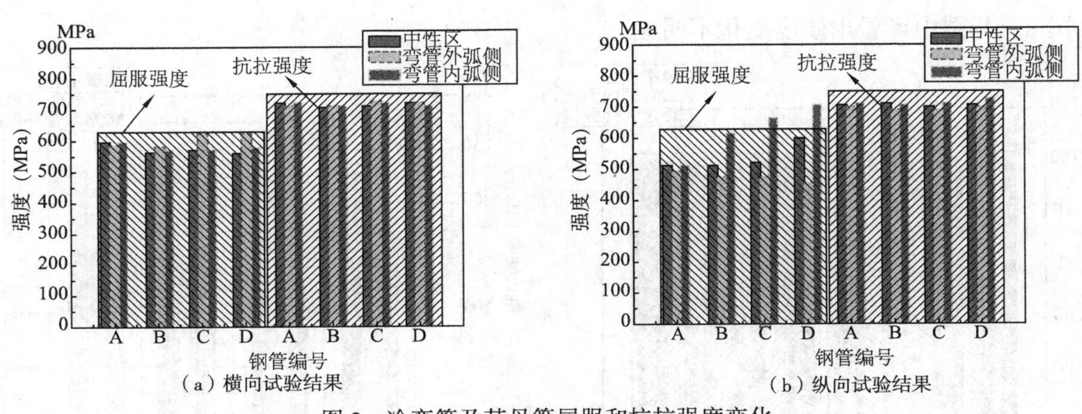

图 3　冷弯管及其母管屈服和抗拉强度变化

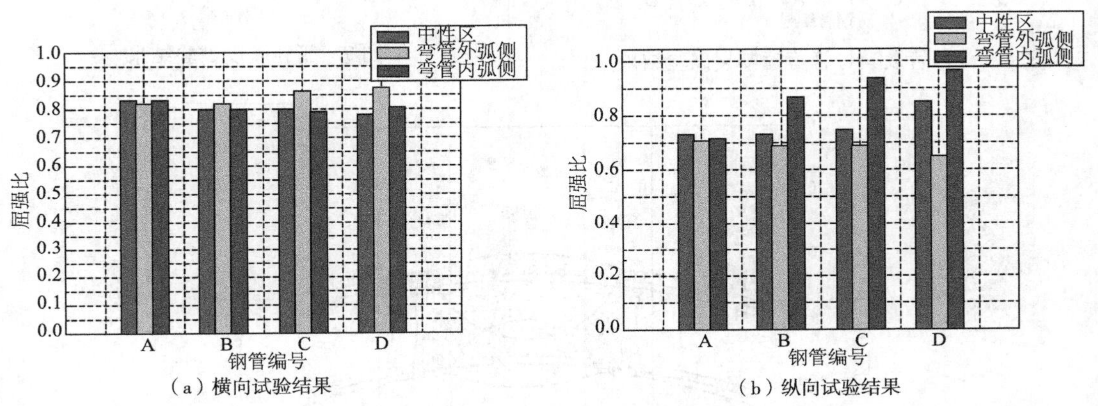

图 4　冷弯管及其母管屈强比变化

2.3　夏比冲击及落锤撕裂试验

从冷弯管和原母管上分别取冲击和落锤试样，试样位置分别为中性区、弯管外弧侧、内弧侧和弯曲段焊缝、热影响区。试样方向为横向和纵向。夏比冲击试验温度为-10℃，试验结果如图 5 所示。图 5（a）为横向试验结果，图 5（b）为纵向试验结果。落锤撕裂试验温度为 0，试验结果如图 6 所示。试验结果表明现场冷弯前后钢管的韧性性能变化不大。

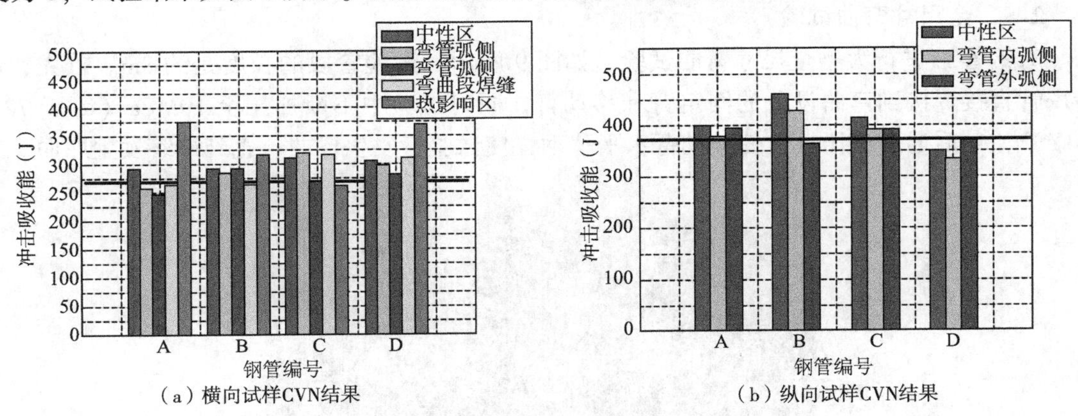

图 5　冷弯管及其母管 CVN 试验结果

2.4　硬度试验

对冷弯管及其母管的横向试样进行了硬度试验，试验结果见图 7。焊接接头的试验结果见图 8。从图中可看出硬度变化不明显。

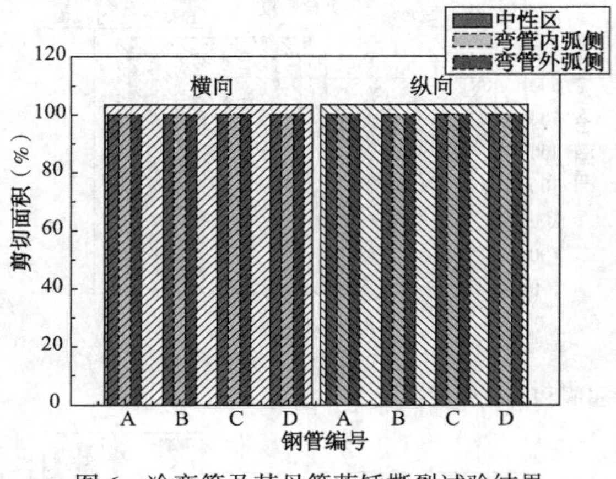

图 6　冷弯管及其母管落锤撕裂试验结果

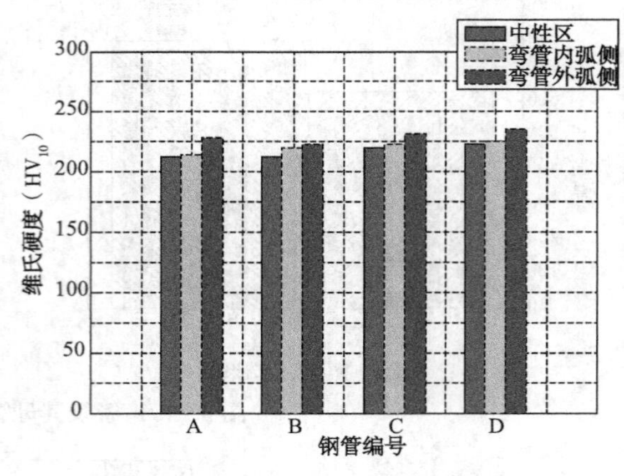

图 7　管体硬度试验结果

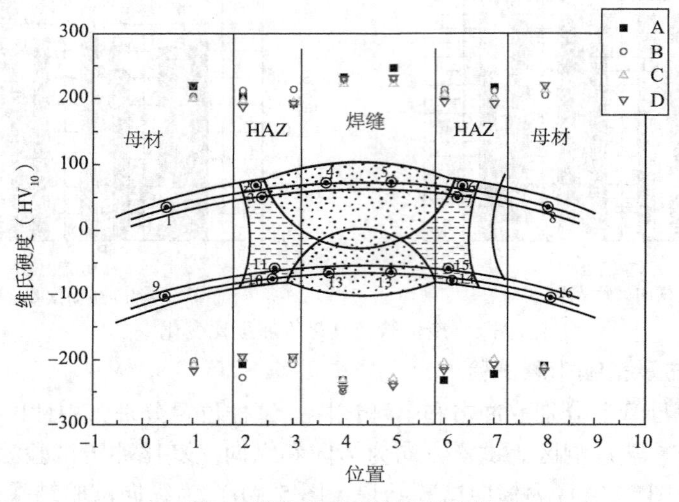

图 8　焊接接头硬度试验结果

2.5　全尺寸弯曲试验

对钢管 E 进行大型全尺寸弯曲试验，如图 9 所示，试验管段的长度是约 8m。首先，试验用管直接焊接到弯曲设备的固定臂和移动臂。同时试验中不断加压至 10MPa（等于 72% SMYS），最后通过液压缸，使移动臂水平移动，使试验管发生弯曲，直到最终发生屈曲。

图 9　全尺寸弯曲试验

在测试过程中，应变和拉伸侧的应变分布由140个应变片收集。在实验过程中的冷弯管（管E）的压缩应变分布如图10所示。图11是一个X70级17.5mm厚度高应变钢管的压缩应变分布。由于其较大的弯曲曲率，图10和图11表明冷弯的集中变形更为明显。在整个钢管压缩应变分布非常均匀。第2步，塑性变形主要集中在位置之间-800~0mm。当弯曲不断，集中塑性变形发生在位置-50mm。在步骤4中，塑性变形发展迅速，最后屈曲的位置在-50mm。

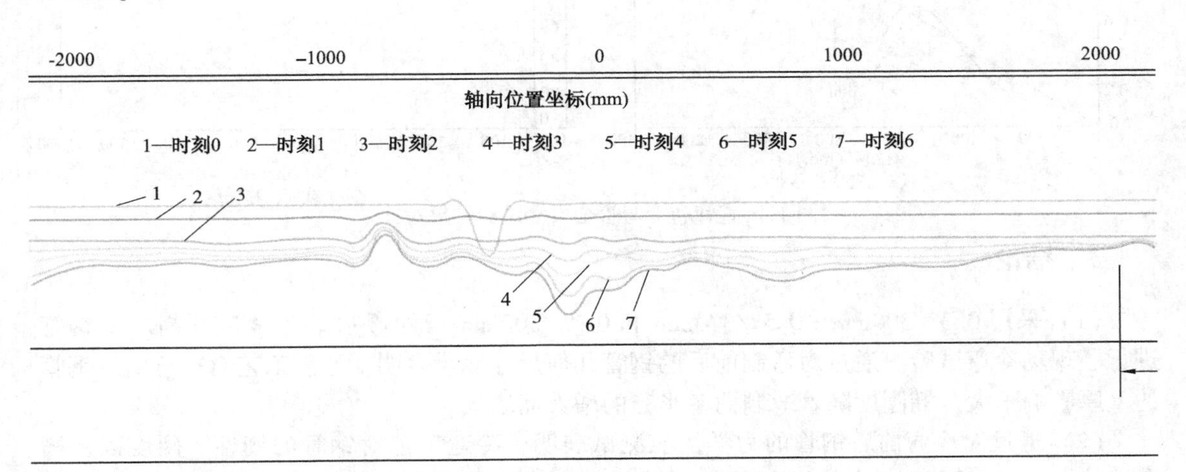

图10　试验过程中内弧侧应变分布

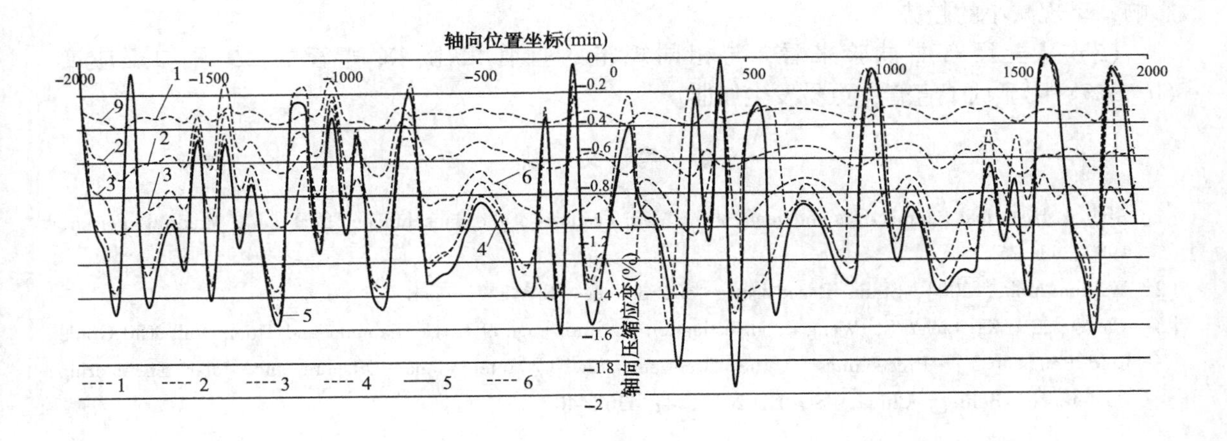

图11　相同规格直管的应变分布

图12所示的冷弯管和钢管的全尺寸弯曲试验比较表明，冷弯管的2*D*平均压缩应变仅为0.73%，小于相同几何形状的直管。由于现场冷弯是一步一步进行的，导致塑性变形的不连续和不均匀，因此与直管相比，冷弯的初始几何缺陷十分显著。但初始几何缺陷对屈曲影响非常显著。

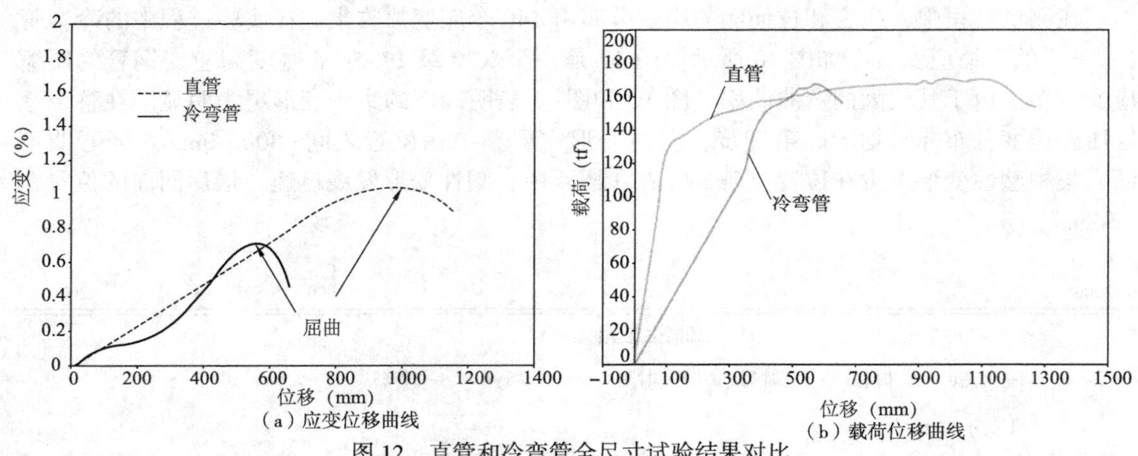

（a）应变位移曲线

（b）载荷位移曲线

图 12　直管和冷弯管全尺寸试验结果对比

3　结论

（1）采用 0.3°/300mm，0.5°/300mm 和 0.7°/300mm 三种弯制工艺对 X70 高应变钢管进行了现场冷弯试验。通过对弯制前后的钢管几何尺寸测量表明，弯制工艺对冷弯前后钢管的壁厚影响不大，椭圆度随着弯制曲率半径的增大而增大。

（2）通过对冷弯前后钢管的力学性能测试表明，冷弯工艺对钢管的韧性、硬度以及横向拉伸性能没有明显的影响。但是，冷弯前后钢管的纵向拉伸性能变化较大，主要表现为，冷弯后钢管的纵向屈服强度及屈强比变化明显。随着弯制曲率半径的增大，由于加工硬化的作用，冷弯管外弧侧的纵向屈服强度和屈强比呈增加趋势。而内弧侧，则由于包申格效应的影响，呈现减小的趋势。

（3）从全尺弯曲试验来看，与相同规格的直管相比，冷弯管的 2*D* 平均压应变（0.73%）明显低于直管，更易发生屈曲。

参　考　文　献

［1］Abel A. Historical Perspectives and some of the Main Features of the Bauschinger Effect［J］. Materials Forum, 1987：11-26.

［2］William Mohr，2003，Strain-Based Design of Pipelines［R］. EWI report.

［3］Chi Q，Ji L K，Liu Y L，Wang P. Investigation on Strain-age of Large Diameter and Thick Wall X80 Grade Cold Bends［C］. Proceedings of the Twentieth（2010）International Offshore and Polar Engineering Conference，Beijing，China，ISOPE，2010，4：436-440.

ϕ1422mm X80 管材技术条件研究及产品开发

张伟卫[1,2] 李 鹤[1,2] 池 强[1,2] 赵新伟[1,2] 霍春勇[1,2]
齐丽华[1,2] 李炎华[1,2] 杨 坤[1,2]

（1. 中国石油集团石油管工程技术研究院；2. 石油管材及装备
材料服役行为与结构安全国家重点实验室）

摘 要：本文结合中俄东线天然气管道工程用 ϕ1422mm、X80 管材技术条件的研究制定过程，对国内外管线钢管技术标准进行了对比分析，同时对制定的 ϕ1422mm X80 管材技术条件中的化学成分、止裂韧性等关键技术指标及制定过程进行了分析探讨，并对 ϕ1422mm X80 管线钢管的开发过程及产品性能进行了介绍。通过生产试制和产品检测，证明技术条件合理有效的解决了化学成分控制、断裂控制、产品焊接稳定性等技术问题，不仅满足工程要求，而且也适应生产情况，可以保障中俄东线天然气管道的本质安全。本工作可为中俄东线建设 ϕ1422mm 天然气管道提供强有力的技术支撑，同时对于其他天然气管道工程技术条件的制定具有重要的指导意义。

关键词：管线钢管；天然气管道；技术条件；技术指标；化学成分；止裂韧性；焊接

近年来，随着天然气需求的日益增加，我国油气管道特别是天然气管道建设进入了一个新的高峰期，大口径、厚壁、高钢级管线钢管成为管道建设的主要选择[1,2]。在“十一五”期间中石油正式立项开展了 ϕ1219mm X80 管线钢前期先导技术研究，取得了一些成果，制定了一系列 ϕ1219mm X80 管材技术条件，并用于西气东输二线等天然气管线建设中。就当时全球已经建成和正在建设的天然气高压长输管道而言，不论钢级、长度、管径、壁厚还是输送压力，西二线工程都堪称世界之最[3]。

2014 年 5 月，中国石油天然气集团公司与俄罗斯天然气工业股份公司正式签署了《中俄东线管道供气购销协议》，约定从 2018 年起，俄罗斯开始通过中俄东线向中国供气。为了满足中俄东线 $380\times10^8m^3/a$ 超大输气量的要求，中国石油通过对 ϕ1422mm、X80 钢级管线钢管应用技术的攻关，形成了第三代大输量天然气管道应用配套技术，并决定在国内 737km 的中俄东线黑河—长岭段干线，首次使用 ϕ1422mm、钢级 X80 钢管。

管线钢的质量是保证管线安全的最基本也是最关键的因素之一，钢管订货技术条件是钢管生产、检验和验收的依据，确定其合理的技术要求对保证管线的安全可靠性、经济性和可行性是非常重要的。受中国石油管道项目经理部的委托，石油管工程技术研究院负责研究、

作者简介：张伟卫，男，1981 年生，高级工程师，硕士，2008 年毕业于北京科技大学并获硕士学位，主要从事输送管与管线材料及标准方面的研究工作。地址：（710077）西安市锦业二路 89 号。电话：（029）81887838。E-mial：zhangweiwei@ cnpc. com. cn。

制定了中俄东线天然气管道工程用 ϕ1422mm、X80 管线钢、钢管系列技术条件。本文对中俄东线天然气管道工程用 ϕ1422mm、X80 管线钢、钢管系列技术条件制定过程中的几个关键问题进行了论述，并对 ϕ1422mm、X80 管线钢及钢管的开发过程及产品性能进行了介绍。

1 国内外相关管线钢标准对比分析

油气输送管道用钢管技术标准从制定方和适用范围进行划分，可分为国际标准、国家标准、行业标准、企业标准等。如 ISO 3183 属国际标准，GB/T 9711 是国家标准，API SPEC 5L 可认为是行业标准，中国石油天然气集团公司发布的 Q/SY 1513 以及中国石油管道建设项目经理部发布的 Q/SY GJX 149—2015 等是企业标准。

目前在我国使用的陆上油气输送钢管基础标准主要有 API SPEC 5L、ISO 3183、GB/T 9711 等。中国石油在吸收国内外技术标准研究成果的基础上，还形成了自己的油气输送钢管技术标准体系，经常使用的通用技术标准有 Q/SY 1513，CDP-S-NGP-PL-006 等，此外根据不同的工程需求，还制定了大量的工程技术条件，如西气东输二线天然气管道工程用管材技术条件、中俄东线天然气管道工程用管材技术条件等。

ISO 3183 是国际标准化组织制定的石油天然气工业管道输送系统用钢管标准，被 GB/T 9711 等同采用。由于 GB/T 9711 的采标修订工作受管理因素限制，更新较 ISO3183 慢，目前等同采用的 ISO3183：2007 版本。API SPEC 5L（管线管规范）是美国石油学会制定的一个被普遍采用的规范。上述标准或规范兼顾了管线钢的技术要求与制造厂实际生产的可行性，但相对管线与制管技术的发展，这些标准或规范中的技术要求显得比较宽松，许多条款仅给出了原则性能要求，具体指标不明确，因此已经很少单独用于管线项目。

目前，世界上大多数石油公司都习惯采用 API SPEC 5L 作为管线钢管采购的基础规范，在该规范基础上，根据当地实际情况或管线工程的具体要求，制订补充技术条件。中国石油的管线钢管通用技术标准 Q/SY 1513、CDP-S-NGP-PL-006 等就是以 API SPEC 5L《管线管规范》为基础，吸收了国内外管线钢和工程经验编制而成，具有很强的实用性和可行性，但由于是通用技术条件，有些指标，如化学成分、夏比冲击功（CVN）等要求比较宽松，需要根据具体工程情况进一步确定。

在具体技术指标方面，GB 9711—2011、ISO3183：2012、API 5L：2012 等标准的最大适用管径包括 1422mm，化学成分指标要求较为宽松，其技术指标要求只满足一般钢管的最基本要求，如要求 C 含量不大于 0.12%，Si 含量不大于 0.45%，其他合金元素范围也非常宽泛[4-6]，没有考虑现场焊接对化学成分的要求，缺乏工程应用指导意义。CVN 要求三个试样最小平均为 54J，仅能满足管体材料不发生启裂失效的最基本要求，不能满足钢管自身止裂要求，因而不能保证长输管道的本质安全。

中国石油企业标准 Q/SY 1513.1—2012《油气输送管道用管材通用技术条件　第 1 部分：埋弧焊管》最大适用管径为 1219mm，最高适用钢级为 X80，理化性能指标基本与 ISO3183、GB/T9711 和 API SPEC 5L：2012 相同，区别在于 Q/SY 1513.1—2012 中碳当量不要求 CE_{IIW} 指标，见表 1。夏比冲击韧性只给出了确定方法（冲击功值要求应按 API Spec 5L 的附录 G 确定），没有给出具体数值，需要根据具体工程进行计算或试验验证确定。

中俄东线天然气管道工程用管材技术条件则是在 Q/SY 1513 基础上，借鉴了 API SPEC 5L：2012 的最新成果，结合中俄东线工程的具体特点，对 ϕ1422mm 管线钢及钢管的各项关键技术指标进行研究攻关，主要针对近年来管线工程的热点问题，确定了更为严格的化学成分指标，计算并验证了钢管的 CVN 值，规定了夹杂物评定标准等，此外，还对管材和板材的试验检验方法和要求进行了优化，并提出了更严格的制造、检验程序和更科学合理的质量控制措施。如化学成分要求 C 不大于 0.07%，Mn 不大于 1.85%，Nb、Mo、Ni 等根据螺旋缝埋弧焊管和直缝埋弧焊管管型的不同分别有不同的要求，有效的解决了现场焊接质量的稳定性问题。关于 CVN 指标，采用了 Battelle 双曲线（BTC）方法进行了理论计算，并利用近年来管道断裂控制技术研究的最新成果，采用 Leis-2、Eiber、TGRC2 等多种修正方法进行修正[7]，同时通过全尺寸气体爆破试验进行了验证。

中俄东线天然气管道工程用 ϕ1422mm X80 管材技术条件还明确了试验样品的加工要求，如板卷力学性能试验样品要求与板卷轧制方向成 20°取样，拉伸试验采用 ϕ12.7mm 的圆棒试样。

此外，通过研究，中俄东线天然气管道工程用 ϕ1422mm X80 管材技术条件还规定了严格的非金属夹杂物验收极限。针对夏比冲击试验中普遍存在的断口分离问题，增加了夏比冲击试样断口分离程度分级方法，针对钢管管端非分层缺陷检测问题，增加了非分层缺陷的检测和验收方法等。

表 1 Q/SY 1513.1—2012 对 X80 管线钢化学成分的要求

钢号	根据熔炼分析和产品分析的最大质量分数（%）									最大碳当量[a]（%）	
	C[b]	Si	Mn[b]	P	S	V	Nb	Ti	其他[i]	CE_{IIW}	CE_{Pcm}
L555 或 X80	0.12[e]	0.45[e]	1.85[e]	0.025	0.015	f	f	f	h	/	0.25

a 根据产品分析。如果碳的质量分数大于 0.12%，则 CE_{IIW} 极限适用；如果碳的质量分数小于等于 0.12%，则 CE_{Pcm} 极限适用。

b 碳含量比规定最大质量分数每降低 0.01%，则允许锰含量比规定最大质量分数增加 0.05%，对于大于等于 L485 或 X70 小于等于 L555 或 X80 的钢级，最大值不应超过 2.00%。

e 除非另有协议。

f 除非另有协议，铌、钒和钛的总含量不应超过 0.15%。

h 除非另有规定，铜的最大含量为 0.50%，镍的最大含量为 1.00%，铬的最大含量为 0.50%，钼的最大含量为 0.50%。

i 除非另有规定，不得有意加入 B，残留 B 含量应≤0.001%。

2 ϕ1422mm X80 管材技术条件制定中的几个关键问题

2.1 化学成分

自西气东输二线管道工程开始，我国 X80 管线钢的生产和应用越来越多，随着钢铁冶金技术的进步，为了降低生产成本，国内各钢铁企业根据自身的特点，开发出了多种合金体系的管线钢，不同钢铁企业生产的管线钢化学成分差别很大，甚至同一企业在不同阶段生产的管线钢的化学成分也有很大的差异[8]。这种化学成分的较大差异，会降低焊接工艺和焊材的适用性，缩小现场焊接的工艺窗口，增加管线焊接的难度，造成焊缝力学性能波动加剧，从而给管道的服役安全带来隐患，对于壁厚 20mm 以上 X80 管线钢，这一问题尤为突出。为了解决这一难题，在中俄东线天然气管道工程用 ϕ1422mm X80 管材技术条件制定过程中，对化学成分指标进行了大量的试验研究工作，目标就是限定中俄东线天然气管道工程用管线钢的化学成分波动范围，制定经济、科学的化学成分指标，从而稳定管线钢质量和现场焊接工艺窗口。

2.1.1 碳、锰、铌

化学成分对管线钢的显微组织、力学性能和焊接性能有着重要的影响。通过研究，决定 ϕ1422mm X80 管线钢采用低 C、Mn 的成分设计，并加入适量的 Mo、Ni、Nb、V、Ti、Cu、Cr 等元素。炼钢时钢材应采用吹氧转炉或电炉冶炼，并进行炉外精炼，并采用热机械控轧工艺（TMCP）生产，最终管线钢的晶粒尺寸达到 10 级以上，从而保证生产出具有良好的强韧性、塑性和焊接性的管线钢。

对西二线等天然气管道工程用 X80 钢管的化学成分及焊接结果进行研究分析发现，管线钢中 C、Mn、Nb 的剧烈波动（图 1~图 3），对焊接性能影响具有较大的影响。在管线钢中 C 是增加钢强度的有效元素，但是它对钢的韧性、塑性和焊接性有负面影响[9]。降低 C 含量可以改善管线钢的韧脆转变温度和焊接性，但 C 含量过低则需要加入更多的其他合金元素来提高管线钢的强度，使冶炼成本提高[10]。综合考虑经济和技术因素，C 含量应控制在 0.05%~0.07%之间。

为保证管线钢中低的 C 含量，避免引起其强度损失，需要在管线钢中加入适量的合金元素，如 Mn、Nb、Mo 等。Mn 的加入引起固溶强化，从而提高管线钢的强度。Mn 在提高强度的同时，还可以提高钢的韧性，但有研究表明 Mn 含量过高会加大控轧钢板的中心偏析，对管线钢的焊接性能造成不利影响[11]。因此，根据板厚和强度的不同要求，管线钢中锰的加入量一般是 1.1%~2.0%。Nb 是管线钢中不可缺少的微合金元素，能通过晶粒细化、沉淀析出强化作用改善钢的强韧性。但有研究表明 Nb 对阻止焊接热影响区晶粒长大和改善热影响区韧性并不十分有效，这是因为在焊接峰值温度下，Nb 的碳、氮化物的热稳定性尚有不足[11]。较低的 Nb 含量，在焊接热循环过程中不能有效抑制热影响区奥氏体晶粒长大，最终导致相变时产生大尺寸的块状 M/A 和粒状贝氏体产物，使韧性恶化。过高的 Nb 含量，在焊接热循环过程中会导致较大尺寸的沉淀析出，同时使晶粒均匀性恶化，也会损害热影响区韧性[12,13]。研究表明，Nb 的加入量一般控制在 0.03%~0.075%比较合理。

通过对 ϕ1422mm X80 管线钢的大量实验研究、工业试制分析和专家组研讨，认为 X80 管线钢的 Mn 含量最高不宜大于 1.85%，Nb 的含量应控制在 0.04~0.08 之间。图 4 和图 5 给出了按最新制定的技术条件工业试制的 ϕ1422mm X80 管线钢管的环焊缝及热影响区在 −10℃下的 CVN 值，可以看出其合格率高达 97%以上。

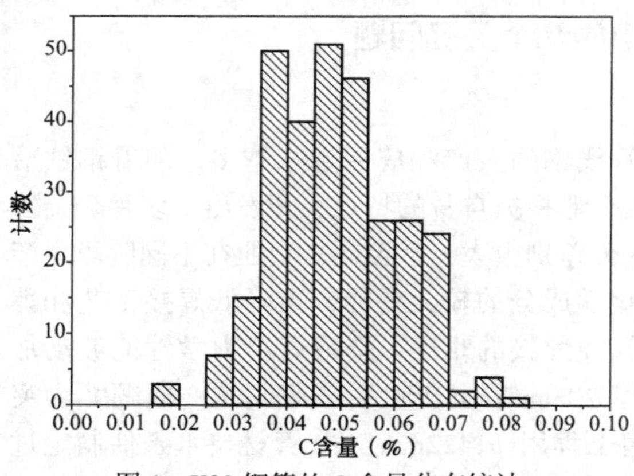

图 1　X80 钢管的 C 含量分布统计

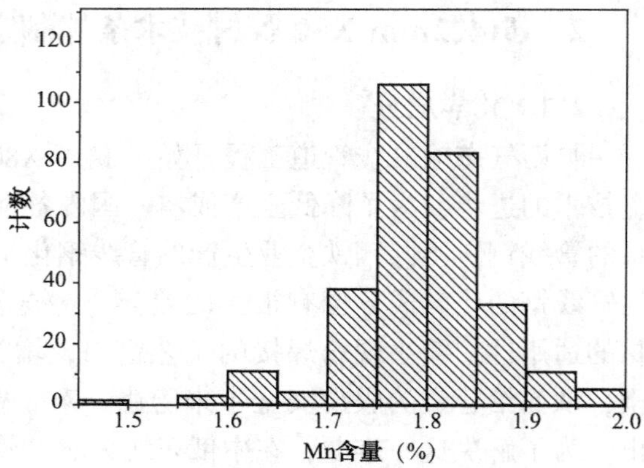

图 2　X80 钢管的 Mn 含量分布统计

2.1.2 其他合金元素

Ti 是强的固 N 元素，在管线钢中可形成细小的高温稳定的 TiN 析出相。这种细小的 TiN

粒子可有效地阻碍再加热时的奥氏体晶粒长大，有助于提高 Nb 在奥氏体中的固溶度，同时对改善焊接热影响区的冲击韧性有明显作用。研究表明 Ti/N 的化学计量比为 3.42 左右，利用 0.02%左右的 Ti 就可以固定钢中 0.006%的 N。管线钢中的 N 含量一般不超过 0.008%，因此技术条件中 Ti 的含量规定控制在 0.025%以下。

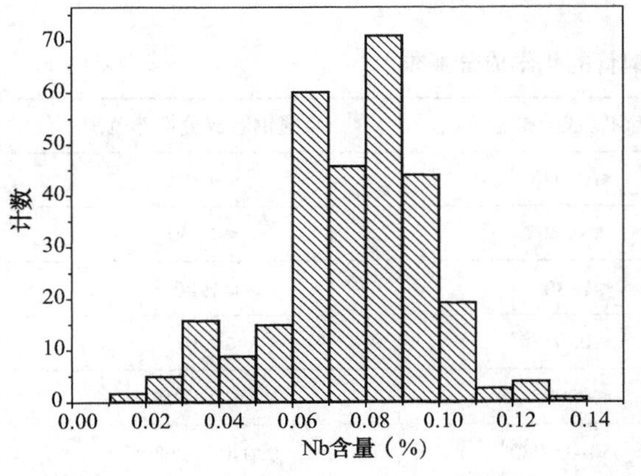

图 3　X80 钢管的 Nb 含量分布统计

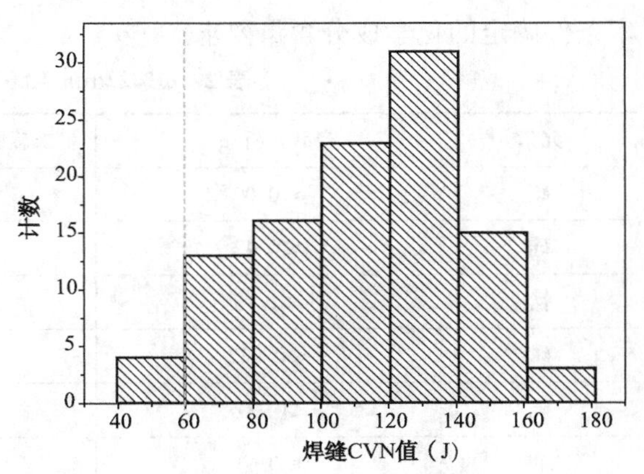

图 4　ϕ1422mm 钢管环焊缝 CVN 分布

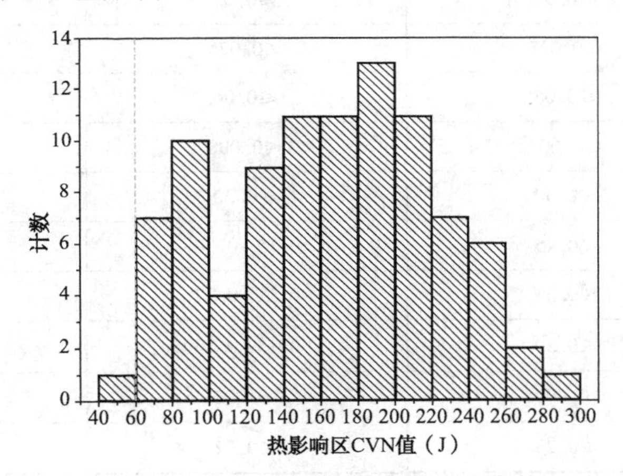

图 5　ϕ1422mm 钢管热影响区 CVN 分布

Cr、Mo 是扩大 γ 相区，推迟 α 相变时先析铁素体形成、促进针状铁素体形成的主要元素，对控制相变组织起重要作用，在一定的冷却条件和终止轧制温度下超低碳管线钢中加入 0.15%~0.35%的 Mo 和低于 0.35%的 Cr 就可获得明显的针状铁素体及贝氏体组织，通过组织的相变强化提高钢的强度。

Cu、Ni 可通过固溶强化作用提高钢的强度，同时 Cu 还可以改善钢的耐蚀性，Ni 的加入主要是改善 Cu 在钢中易引起的热脆性，且对韧性有益。在厚规格管线钢中还可补偿因厚度的增加而引起的强度下降。一般管线钢中铜含量低于 0.30%，镍含量低于 0.5%。

为了更好的稳定产品的理化性能，保证钢管具有良好的现场焊接性，结合国内管线钢生产中合金元素的实际控制能力，ϕ1422mm X80 管材技术条件根据钢管类型对 C、Mn、Nb、Cr、Mo、Ni 的含量进行了约定。通过试验研究，并组织冶金和焊接专家讨论协商，确定管线钢中 C 的含量目标值为 0.060%，Mn 的目标值为 1.75%，Nb 的目标值为 0.06%。直缝钢管中 Ni 目标值为 0.20%，必须加入适量的 Mo，且含量应大于 0.08%。螺旋缝钢管中 Cr、Ni、Mo 的目标值均为 0.20%。考虑到生产控制偏差、检测误差及经济性，ϕ1422mm X80 管材技术条件中规定C含量不大于0.070%，Mn含量不大于1.80%。直缝钢管Nb的含量范围为0.04%~0.08%，Mo的含量范围为0.08%~0.30%，Ni的含量范围为0.10%~0.30%；螺旋缝钢管中Nb的含量范围为0.05%~0.08%，Cr的含量范围为0.15%~0.30%，Mo的含量范围为0.12%~0.27%，Ni的含量范围为0.15%~0.25%。表2给出了ϕ1422mm X80管材技术条件确定的化学成分含量要求。

表2 ϕ1422mm X80钢管的化学成分要求

元素	产品分析max	螺旋缝钢管成分推荐范围	直缝钢管成分推荐范围
碳	≤0.09	≤0.070	≤0.070
硅	≤0.42	≤0.30	≤0.30
锰	≤1.85	≤1.80[c]	≤1.80[c]
磷	≤0.022	≤0.015	≤0.015
硫	≤0.005	≤0.005	≤0.005
铌[a]	≤0.11	0.050~0.080[d]	0.040~0.080[d]
钒[a]	≤0.06	≤0.03	≤0.030
钛[a]	≤0.025	≤0.025	≤0.025
铝	≤0.06	≤0.06	≤0.06
氮	≤0.008	≤0.008	≤0.008
铜	≤0.30	≤0.30	≤0.30
铬	≤0.45	0.15~0.30	≤0.30
钼	≤0.35	0.12~0.27	0.08~0.30
镍	≤0.50	0.15~0.25	0.10~0.30
硼[b]	≤0.0005	≤0.0005	≤0.0005
CE_{Pcm}	≤0.23	≤0.22	≤0.22

a V+Nb+Ti≤0.15%。

b 不得有意加入硼和稀土元素。

c 碳含量比推荐最大含量每减少0.01%时，锰推荐最大含量可增加0.05%，但锰含量不得超过1.85%。

d 碳含量比推荐最大含量每减少0.01%时，铌推荐最大含量可增加0.005%，但铌含量不得超过0.085%。

2.2 止裂韧性

API SPEC 5L：2012和ISO3183：2012中规定的四种止裂韧性计算方法中，只有对BTC计算结果进行修正的方法适用于12MPa、X80、OD1422管道的止裂韧性计算[14]，其中修正系数的确定来源于X80全尺寸爆破试验数据库。目前国际上通用的全尺寸爆破试验数据库如图6所示，由此确定的中俄东线管道工程止裂韧性修正方法为TGRC2，修正系数为1.46。

中俄东线的天然气组分如表3所示，按照中俄东线实际工况ϕ1422mm、壁厚21.4mm、输送压力12MPa、运行温度0进行止裂韧性计算。用BTC方法计算其止裂韧性结果为167.97J，按1.46倍修正后止裂韧性为245J，如表4所示，表4中还给出了Leis-2、Eiber、Wilkowski等方法的修正结果。

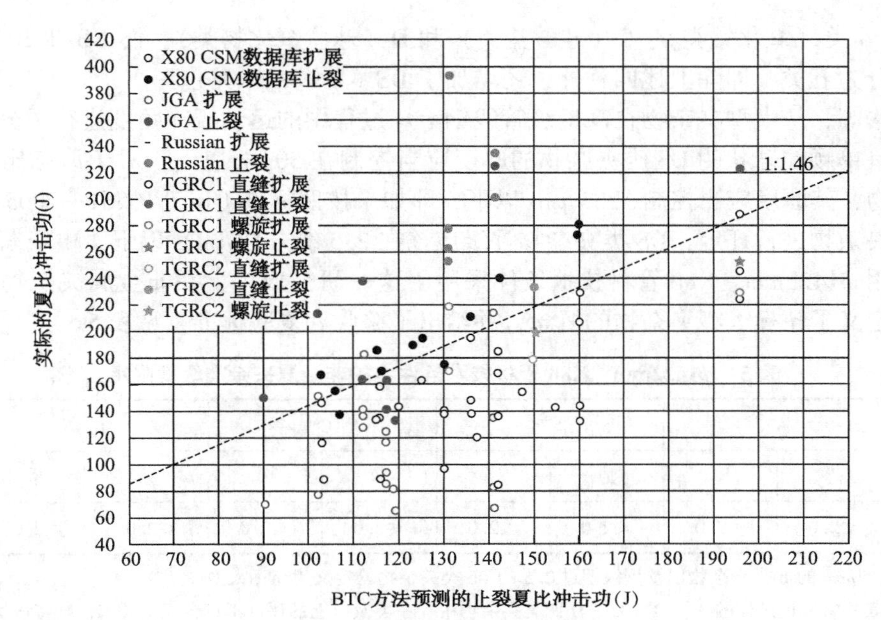

图 6　全尺寸爆破试验数据库

表 3　中俄东线计算用气质组分

组分	C_1	C_2	C_3	C_4	C_5	N_2	CO_2	He	H_2
Mol%	91.41	4.93	0.96	0.41	0.24	1.63	0.06	0.29	0.07

表 4　中俄东线止裂韧性计算结果

计算方法	BTC（J）	BTC+Eiber 修正	BTC+Leis2 修正	BTC+Wilkowski 修正	BTC+TGRC2 修正
CVN 值（J）	167.97	251	250	286	245

由于现有的全尺寸气体爆破实验数据库无法覆盖中俄东线天然气管线 X80 钢级、ϕ1422mm、12MPa 压力下输送富气的设计参数要求。因此，2015 年 12 月在中国石油管道断裂控制试验场，针对中俄东线天然气管道具体的设计参数和服役条件，对 ϕ1422mm、X80 钢管的延性断裂止裂指标进行了全尺寸爆破试验验证，结果表明，采用 BTC 方法计算，并用 TGRC2 方法进行修正后的 ϕ1422mm、X80 钢管止裂韧性指标为 245J 是安全和经济的。

2.3　非金属夹杂物

近年来，许多管道工程使用的高钢级管线钢均在金相检测过程中发现了超尺寸大型夹杂物。管线钢中大型夹杂物的存在会对其力学，焊接，耐腐蚀等性能产生不利影响，进而给油气输送管道的安全运营带来很大的工程风险。为了有效降低管线钢中大型夹杂物的存在给管道输送系统带来的风险，石油管工程技术研究院李炎华等[15]，针对高钢级管线钢中大型夹杂物的特性进行了大量的研究工作，进而为高钢级管线钢中大型夹杂物级别判定标准的制定提供了依据。

目前，国内管线钢夹杂物评判通常采用 ASTM E 45—2005《Standard Test Methods for Determining the Inclusion Content of Steel》和 GB/T 10561—2005《钢中非金属夹杂含量的测定标准评级图显微检验法》。ASTM E45—2005 将夹杂物按形态和分布分为四类，即 A（硫

化物类)、B类（氧化铝类)、C（硅酸盐类）和D（球状氧化物类)；而GB/T 10561—2005将夹杂物分为五类，即除上述四种外，还增加了DS（单颗粒球类)。

李炎华等，从大型夹杂物在高钢级管线钢冶炼过程中的运动规律角度进行了分析，认为对于形态呈单颗粒球状的DS类夹杂物的厚度应当控制在50μm以下，对于形态比小于3的B类夹杂物，其厚度应当控制在33 μm以下，即如果按照标准GB/T 10561—2005对管线钢中的大型夹杂物进行评定，DS类夹杂物评级应该在2.5级（53μm）以下。中俄东线天然气管道工程用ϕ1422mm、X80管材技术条件采用了这一研究成果，在非金属夹杂物级别验收极限中，定义了超标大型夹杂物的概念，并给出了验收和复验标准，见表5。

表5　ϕ1422mm、X80管材技术条件中的非金属夹杂物级别限定

类型	A[a]		B[a,b,c]		C[a]		D[a]		DS[b,c]
系别	薄	厚	薄	厚	薄	厚	薄	厚	—
级别	≤2.0	≤2.0	≤2.0	≤2.0	≤2.0	≤2.0	≤2.0	≤2.0	≤2.5

厚度大于50μm的B类夹杂物以及评级超过2.5的DS类夹杂物均定义为超标大型夹杂物。

a 如果代表一熔炼批试样的A、B、C、D四类夹杂物中有一类及以上的评价不符合规定要求，则将该熔炼批判为不合格。

b 如果评价过程中发现某一视场中同时存在两个或两个以上的同类或不同类超标大型夹杂物，将该熔炼批判为不合格。

c 如果代表一熔炼批钢管的夹杂物检验中发现某一视场中存在单个超标大型夹杂物，则需要在同一熔炼批中再随机抽取两个试样进行复验。如果两个试样的复验结果均符合A、B、C、D四类夹杂物规定要求且未出现超标大型夹杂物，则除原取样不合格的那根钢管外，该熔炼批合格。如果任一个试样的复验结果不符合A、B、C、D四类夹杂物规定要求或出现了超标大型夹杂物，则该熔炼批判为不合格。

2.4　力学性能试样取样位置

西气东输二线以来，油气管道工程用螺旋缝埋弧焊钢管的管径均小于1219mm，为了取样方面，热轧板卷技术条件中力学性能取样位置均要求与板卷轧制方向成30°取样。取样角度与板宽和钢管管径的关系，如公式（1）。

$$\sin\alpha = B/(\pi D)$$

式中　α——螺旋角；

B——板宽；

π——圆周率；

D——钢管直径。

按目前主流热轧板卷产品宽度1500~1600mm计算，对于管径1219mm的螺旋缝埋弧焊管，热轧板卷的取样角度为23.1°~24.7°，对于管径1422mm的螺旋缝埋弧焊管，热轧板卷的取样角度为19.6°~21°。因此对于外径1422mm的螺旋缝埋弧焊管，与板卷轧制方向成20°取力学性能样，更符合实际情况。

图7和图8给出了实际生产的热轧板卷20°、30°位置的力学性能对比图。可以看出与轧制方向夹角20°位置的屈服强度、抗拉强度、DWTT剪切面积高于30°位置，若按与轧制方向成30°位置取样，容易低估热轧板卷的力学性能，造成不必要的浪费。因此在中俄东线天然管道工程用热轧板卷技术条件中力学性能的检测取样位置更改为与轧制方向成20°位置。

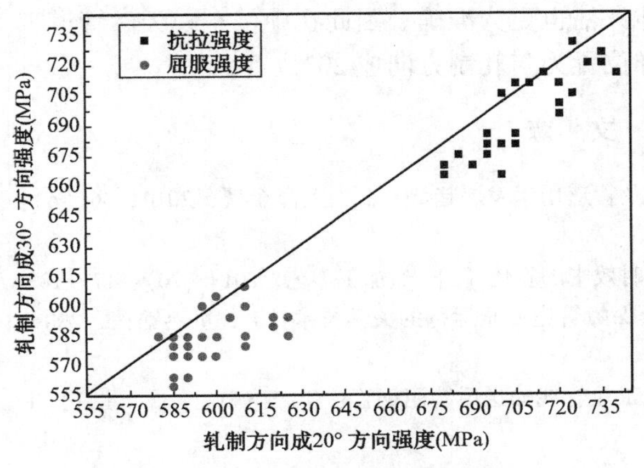

图 7　热轧板卷 20°、30°位置的拉伸性能

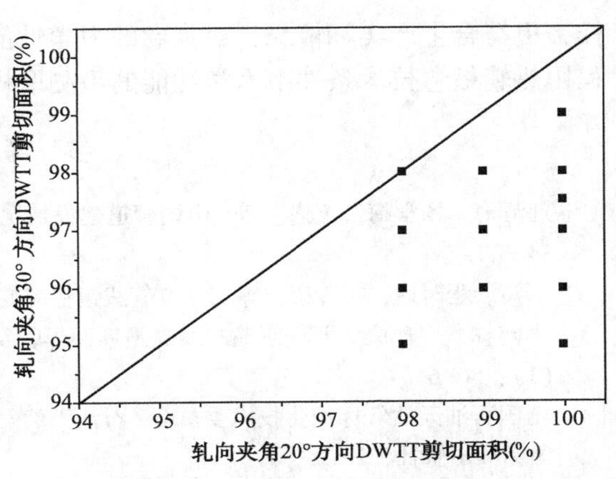

图 8　热轧板卷 20°、30°位置的 DWTT 性能

3　ϕ1422mm、X80 钢级大口径钢管开发

2013 年以来，中国石油组织相关科研单位和国内大型钢铁企业和制管企业，开展了 ϕ1422mm、X80 钢级大口径钢管的联合开发。在研发阶段，共进行单炉产品试制 3 轮，参与生产制造企业 15 家，试制产品 2000 余吨。工业应用阶段，进行了 1 次千吨级小批量试制，参与生产制造企业 8 家，试制产品 6000t。经第三方检测评价表明，试制的 ϕ1422mm X80 钢管的化学成分和力学性能均符合中俄东线天然气管道工程用 ϕ1422mm X80 管材技术条件要求。试制钢管的屈服强度为 595～668MPa，抗拉强度为 677～745MPa，母材 CVN 值为 324～486J，焊缝 CVN 值为 138～232J，热影响区 CVN 值为 172～354J。通过环焊试验证明，所试制的 ϕ1422mm X80 钢管的环焊缝性能均能满足标准要求。

4　结论

（1）中俄东线天然气管道工程用 ϕ1422mm X80 管材技术条件，借鉴了 API SPEC 5L：2012 的最新成果，结合中俄东线工程的具体特点，提出了适合外径 1422mm 管材的化学成分、夹杂物评定、CVN 值、力学性能试验方法等关键技术指标要求，技术条件具有很强的可操作性，即能满足工程要求，也适应生产情况，其研究经验在我国未来的天然气管道工程建设上推广应用。

（2）中俄东线天然气管道工程用 ϕ1422mm X80 管材技术条件规定管线钢采用低 C、Mn 的成分设计，并对添加的 Mo、Ni、Nb、V、Ti、Cu、Cr 等合金元素含量进行了严格的限定，是国内油气管道建设以来对化学成分要求最为严格的工程技术条件。工业试制结果表明，该技术条件制定的化学成分指标符合生产要求，并可有效解决管材的理化性能和焊接性能稳定性问题。

（3）采用 BTC 方法计算，并用 TGRC2 方法进行修正来确定 ϕ1422mm、X80 钢管的止裂韧性指标是安全和经济的，ϕ1422mm、X80 钢管的止裂韧性指标应为 245J。

（4）中俄东线天然气管道工程用 ϕ1422mm、X80 管材技术条件在非金属夹杂物级别验收极限中，采用最新研究成果定义了超标大型夹杂物的概念，并给出了验收和复验标准，有利于提高钢管的力学，焊接，耐腐蚀等性能。

（5）对于 ϕ1422mm 的螺旋缝埋弧焊管用热轧板卷，与板卷轧制方向成 20°取力学性能样，更符合生产实际情况，对板卷的力学性能评估也更为准确。因此在中俄东线天然管道工程用热轧板卷技术条件中力学性能的检测取样位置为与轧制方向成20°位置。

参 考 文 献

[1] 刘清梅，杨学梅，赵谨，等. 中国管道建设情况及管道用钢发展趋势［J］. 上海金属，2014，36（4）：34-37.

[2] 彭涛，程时遐，吉玲康，等. X100管线钢在应变时效中的脆化［J］. 热加工工艺，2013（20）：179-183.

[3] 李鹤林，吉玲康. 西气东输二线高强韧性焊管及保障管道安全运行的关键技术［J］. 世界钢铁，2009，（1）：56-64.

[4] 全国石油天然气标准化技术委员会. 石油天然气工业管线输送系统用钢管：GB/T 9711［S］. 北京：中国标准出版社，2011.

[5] International Organization for Standardization. Petroleum and natural gas industries—Steel pipe for pipeline transportation systems［S］. ISO 3183，2012.

[6] American Petroleum Institute. Specification for Line Pipe. 45th Edition［S］. API SPEC 5L，2012.

[7] 高惠临. 管线钢管韧性的设计和预测［J］. 焊管，2010，33（12）：5-12.

[8] 尚成嘉，王晓香，刘清友，付俊岩. 低碳高铌X80管线钢焊接性及工程实践［J］. 焊管，2012，35（12）：11-18.

[9] Bai Lu，Tong Lige，Ding Hongsheng，Wang Li，Kang Qilan，Bai Shiwu. The Influence of the Chemical Composition of Welding Material Used in Semi-Automatic Welding for Pipeline Steel on Mechanical Properties［C］//ASME 2008 International Manufacturing Science and Engineering Conference，7-10 October 2008，Evanston，Illinois，USA. DOI：10.1115/MSEC_ ICMP2008-72110.

[10] 孙磊磊，郑磊，章传国. 欧洲钢管集团管线管的发展和现状［J］. 世界钢铁，2014，（1）：45-53.

[11] 高惠临. 管线钢与管线钢管［M］. 北京：中国石化出版社，2012：22-27.

[12] Wang BX，Liu XH，Wang GD. Correlation of microstructures and low temperature toughness in low carbon Mn - Mo - Nb pipeline steel［J］. Materials Science and Technology，2013，29（12）：1522-1528. DOI：10.1179/1743284713Y.0000000326.

[13] 缪成亮，尚成嘉，王学敏，等. 高Nb X80管线钢焊接热影响区显微组织与韧性［J］. 金属学报，2010，46（5）：541-546.

[14] 霍春勇，李鹤林. 西气东输二线延性断裂与止裂研究［J］. 金属热处理，2011，36（增刊）：4-9.

[15] 李炎华，吉玲康，池强，等. 高钢级管线钢中大型夹杂物的特性［J］. 管道技术与设备，2013，（1）：4-6.

国内外高性能油气输送管的研发现状

杜　伟　李鹤林

（中国石油集团石油管工程技术研究院）

摘　要：结合高性能油气输送管的发展方向，分别介绍了超高强度钢管、低温环境用钢管、腐蚀环境用管、大应变钢管以及深海油气开发用钢管的需求、成分与力学性能要求、国内外研发与工程应用情况等。与国际领先水平相比，我国高性能输送管的技术及质量还有一定差距。建议国内生产企业和科研院所开展联合攻关，不断提升超高强度钢管、大应变钢管及深海油气开发用钢管的性能稳定性，并研究解决低温环境用钢管韧性指标的确定以及腐蚀环境用管的材料选用等问题。

关键词：超高强度钢管；低温管；耐酸管；大应变钢管；海管

近十年来，我国油气供给与管道建设呈现跨越式发展，管道总里程超过 12×10^4km，跨入中国五大运输体系行列。在全球油气资源方面，非常规油气资源占比更大（约80%），且非常规油气资源几乎未开采，将是未来开发的重点[1,2]。非常规油气资源，一般分布于偏远且地理环境恶劣的区域，这给管道建设及管材选用提出了更高的要求。为保障油气输送管道的安全运行，尽可能降低成本，需着力开展高性能油气输送管的研发与应用，如超高强度钢管、低温环境用钢管、腐蚀环境用管、大应变钢管和深海油气开发用钢管等。

1　超高强度（X90及以上）钢管

提高强度不仅可以减小钢管壁厚和质量，节约钢材成本，而且由于钢管壁厚的减小，还可降低钢管运输成本和焊接工作量，从而大幅降低管道建设的投资成本和运行成本，高钢级钢管的应用已经成为管道工程发展的一个必然趋势[3,4]。

目前，国外的新日铁和欧洲钢管公司均开发了 X100、X120 钢管，国内也正在开展 X90/X100 钢管开发、工程应用关键技术研究，提出了典型高强度油气输送管的化学成分和力学性能（表1和表2）。

表1　典型高强度管线钢的化学成分

钢级	厂家	质量分数（%）					其他	碳当量（%）	
		C	Mn	Mo	Ti	B		CE_{Pcm}	CE_{IIW}
X90	国内A厂	0.059	1.96	0.21	0.013	0.000 2	Ni、Cr、Nb、Cu	0.21	0.54

基金项目：项目来源“集团公司海洋油气开发装备材料的现状及应用发展策略研究”，项目编号 2013D-5003-13。

作者简介：杜伟，男，工程师，1982年生，2008年硕士毕业于北京科技大学材料科学与工程专业，现主要从事石油管的工程应用与应用基础方向的研究工作。地址：陕西省西安市锦业二路89号，710077。电话：15002982056，E-mail：duw002@163.com。

续表

钢级	厂家	质量分数（%）					其他	碳当量（%）	
		C	Mn	Mo	Ti	B		CE_{Pcm}	CE_{IIW}
X90	国内B厂	0.061	1.79	0.0032	0.022	0.00043	Ni、Cr、Nb、Cu	0.19	0.43
X90	JFE	0.064	1.81	0.12	0.018	0.0002	Ni、Cr、Nb、Cu	0.20	0.41
X100	国内C厂	0.054	2.03	0.30	0.014	0.0002	Ni、Cr、Nb、Cu	0.21	—
X100	NSC	0.059	1.96	0.0036	0.014	0.0003	Ni、Cr、Nb、Cu	0.22	0.55
X100	Europipe	0.07	1.90	0.17	0.018	—	Cu、Ni、Nb、V	0.20	0.46
X120	NSC	0.041	1.93	0.32	0.02	0.0012	Cu、Ni、Nb、V	0.21	—
X120	Europipe	0.06	1.91	0.042(Nb)	0.017	0.004（N）	Cu、Ni、Mo、V	0.21	—

表2　典型高强度油气输送管的力学性能

钢级	厂家	壁厚（mm）	屈服（MPa）	抗拉（MPa）	屈强比	延伸率（%）	CVN		DWTT	
							冲击功（J）	温度（℃）	剪切面积（%）	温度（℃）
X90	国内A厂	19.6	643	781	0.82	24	329	-10	85	0
X90	国内B厂	16.3	650	735	0.88	26	309	-10	100	0
X90	JFE	19.6	690	742	0.93	24	310	-10	—	—
X100	国内C厂	17.8	681	781	0.87	22	267	-10	80	0
X100	NSC	16.0	828	834	0.95	20	262	-20	100	0
X100	Europipe	19.1	737	800	0.92	18	200	20	85	20
X120	NSC	19.0	853	945	0.90	31	318	-30	75	-5
X120	Europipe	16.0	843	1128	0.75	14.3	250	-30	—	—

国内钢铁企业在第1轮X90管线钢试制中，采用了较高的合金成分设计方案（A厂），与国外JFE同类产品相比，碳当量更高，合金成本更大。此方案主要通过增加钢中碳和合金元素的含量来提高钢的强度水平，优点是TMCP工艺要求相对较低，缺点是现场焊接性能差。在我国长输管道的现场环焊中，一般采用自保护药芯焊丝半自动焊工艺，其自保护药芯焊丝中的高铝成分设计，会显著恶化焊缝金属组织及性能。因此，第一轮试制的X90钢管未通过自保护药芯焊丝半自动焊环焊工艺评定。在此基础上，调整设计方案，降低合金元素含量，进行了第2轮试制。第2轮试制产品的碳当量（B厂）与JFE产品相差不大，通过了自保护药芯焊丝半自动焊环焊工艺评定。

国内第2轮试制的X90钢管虽然合金成本及整体性能指标与JFE产品相当，但在个别性能方面（如屈服强度等）仍存在不足，说明我国X90等高钢级钢管的生产水平与世界先进水平相比尚有一定差距，因而需要冶金企业、钢管制造厂、施工单位、管材用户和科研院所等机构协调配合、深入研究，逐步提升国内高强度管线钢生产及应用水平。

2　低温环境用钢管

我国新疆油田和大庆油田外输管道冬季最低温度为-34℃或更低[5]。几年前西一线轮南首站的低温液气分离器脆性断裂，造成了严重后果，近年来高钢级三通在试压过程中也频繁出现脆性爆裂（图 1），为此，输送管的低温脆断问题需引起高度重视，并积极开发低温环境用高强度钢管。

(a) 案例1

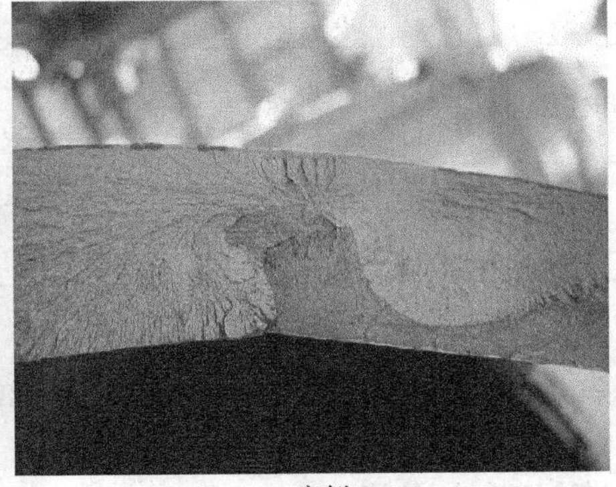
(b) 案例2

图 1　高钢级三通试压脆断实物照片

目前，对于低温环境钢管的断裂韧性要求，不同标准有规定不同。ISO 3183-3：1999[6]（GB 9711. 3—2005）规定夏比冲击功（CVN）≥σ_y（MPa）/10，CVN 试验温度与钢管壁厚有关（尺寸效应，表 3）。2007 年，ISO 3183：2007[7]与 API Spec 5L 44 版[8]合并，降低了 CVN 值的要求，试验温度改为 0。当时专家们的争议很大，现欧洲部分用户仍采用老版本标准中更严苛的规定。

表 3　ISO 3183-3：1999 对 CVN 试验温度的规定

钢管规定壁厚 t（mm）	最低设计试验温度（℃）
t≤20	-10
20<t≤30	-20
t>30	-30

俄罗斯技术规范（OJSC 公司，STT-08. 00-60. 30. 00-KNT-013-1-05）针对 14MPa 下 X70 和 X80 规定，管体 CVN 试验在-40℃下进行，焊缝中心和热影响区 CVN 试验温度为管道正常运行的最低管壁温度，但不能高于-10℃。加拿大 ALLIANCE 管道规范规定，裸露钢管落锤撕裂（DWTT）和 CVN 试验均在-45℃下进行。ASME B31. 8 规定，当管道在低于-45℃的环境下运行时，DWTT 和 CVN 的试验温度为管道服役时所预期的最低金属壁温或更低温度。

针对以上问题，在研究开发低温环境用高强度钢管（表 4[9]）的同时，应该继续深入研究裸露管道 CVN 和 DWTT 试验温度的确定方法（考虑尺寸效应）、CVN 值与 DWTT 判据的确定等问题。

表 4　低温韧性管线钢和抗 HIC 管线钢的化学成分设计

管线钢	质量分数（%）										
	C	Mn①	P	S	Nb①	V①	Al	Ti	Ca	Cu	Ni	ACR②
对低温韧性有较高要求的管线钢	0. 05~0. 09	1. 00~1. 60	≤0. 016	≤0. 005	0. 030~0. 050	0. 010~0. 050	0. 040	0. 008	0. 030	—	—	—

续表

管线钢		质量分数（%）											
		C	Mn①	P	S	Nb①	V①	Al	Ti	Ca	Cu	Ni	ACR②
抗 HIC 管线钢	BP 溶液	0.05~0.09	1.00~1.60	≤0.015	≤0.003	0.030~0.050	0.010~0.050	0.040	0.008	0.030	0.15~0.25	0.15~0.25	≥1.0
	NACE 溶液	0.04~0.05	1.00~1.60	≤0.010	≤0.001	0.030~0.050	0.010~0.050	0.040	0.008	0.030	0.15~0.25	0.15~0.25	≥1.5

①Mn、Nb、V 的含量随钢级和壁厚有所变化；

②$ACR=0.8\frac{F_{Ca_{eff}}}{F_S}$，$F_{Ca_{eff}}=F_{Ca}-(130F_{Ca}+0.18)F_O$，其中 F_S、F_{Ca}、F_O 分别为 S、Ca 和 O 元素的质量分数。

3 腐蚀环境用管

在油气田开发中，未经净化处理的石油天然气，常含有 H_2S、CO_2、Cl^-、H_2O 等腐蚀性介质。为避免管道腐蚀失效，此类油气输送需要采用具有耐腐蚀性能的材料，包括碳钢和低合金钢、不锈钢、镍基/铁镍基合金以及钛合金等。管道的具体选材应根据输送介质的压力、温度及介质中的 H_2S、CO_2 和 Cl^- 含量等确定。根据日本住友金属的选材指南，当 $p_{H_2S}<0.0003MPa$，$p_{CO_2}\leqslant0.02MPa$ 时，选用碳钢和低合金钢；当 $p_{CO_2}>0.02MPa$，$p_{H_2S}<0.003MPa$ 时，选用 13Cr 或超级 13Cr 不锈钢；当 $p_{CO_2}>0.02MPa$，$0.003MPa\leqslant p_{H_2S}<0.01MPa$ 时，选用双相不锈钢；当 $p_{CO_2}>0.02MPa$，$p_{H_2S}\geqslant0.01MPa$ 时，选用镍基或铁镍基合金。NACE MR0175 标准则提供了在含湿硫化氢的油气产品环境下材料的评价和推荐选材作法。

根据 NACE MR0175，API Spec 5L（45 版）附录 H 给出了酸性服役条件 PSL2 钢管（下称“抗酸管”）的技术要求。国外抗酸管的研制较早，欧洲钢管公司抗酸管的销售量已占 30%以上。国外批量供应的抗酸管主要是 X65 钢级，X70 钢级的抗酸管已研制成功，并在墨西哥一条管道上使用。我国抗酸管的研发刚刚起步，部分钢厂开发出了 X65MS、X70MS 耐酸管。

根据抗酸管的基本成分（表 5）和力学性能（表 6）要求，抗酸管的化学成分比低温用钢管要求更加严格，需进一步降低 P、S 含量，并添加 Cu 和 Ni，同时采用 Ca 处理。从成分和整体性能上看，国内研发的抗酸管与国外水平相当，但批量生产时的性能稳定性有待实践检验。

表 5 典型抗酸钢管的化学成分

钢级	厂家	质量分数（%）					其他	碳当量（%）	
		C	Si	Mn	P	S		CE_{IIW}	CE_{Pcm}
X65MS	国内 A 厂	0.039	0.29	1.31	0.011	0.0013	Cr、Cu、Nb、Ti、V	—	0.15
X70MS	国内 A 厂	0.039	0.29	1.31	0.01	0.0012	Cr、Cu、Nb、Ti、V	—	0.15
X65MS	Europipe	0.04	0.28	1.38	0.015	0.0015	Nb、V	0.33	0.13
X70MS	JFE	0.05	0.28	1.13	0.014	0.0005	Mo、Ni、Cr、Cu、Nb	—	0.14
X70MS	Europipe	0.038	0.30	1.43	0.009	0.0005	Mo、Ni、Cr、Cu、Nb、V	0.41	0.17

表 6　典型抗酸钢管的力学性能

钢级	厂家	屈服（MPa）	抗拉（MPa）	屈强比	延伸率（%）	CVN		DWTT		CLR（%）	CSR（%）
						冲击功（J）	温度（℃）	剪切面积（%）	温度（℃）		
X65MS	国内 A 厂	519	595	0.87	40	454	0	100	0	0	0
X70MS	国内 A 厂	521	600	0.87	48	449	0	98	0	0	0
X65MS	Europipe	480	564	0.86	50.0	433	−10	89	0	≤5	≤0.5
X70MS	JFE	531	613	0.87	23.0	373	−10	100	0	0	—
X70MS	Europipe	521	619	0.84	54.2	452	−20	94	0	≤4	≤0.2

当油气介质中的腐蚀性成分含量较高，普通碳钢和低合金钢难以满足耐蚀要求时，需选用不锈钢、镍基/铁镍基合金或钛合金等具有更高耐蚀性能的材料。其中，常用的不锈钢包括奥氏体不锈钢如 316L，超级奥氏体不锈钢如 904L、254SMO，双相不锈钢如 2205、2507，超级双相不锈钢如 2707HD；镍基/铁镍基合金包括 028、G3、625、825 等；常用的钛合金如 TC4（Ti−6Al−4V）等。在实际应用中，因不锈钢和耐蚀合金的价格较高，油气输送管道一般选用以不锈钢或耐蚀合金为衬里的双金属复合管。

双金属复合管是以不锈钢或耐蚀合金为内衬层，提供抗腐蚀能力，以碳钢或低合金钢为外层基管，承受压力。国外在双金属复合管的应用方面早于我国，日、美等国较早开展了双金属复合管的研究，于 1991 年开始使用，随后用量逐年扩大。现行产品标准为美国石油学会制订的 API 5LD—2009《Specification for CRA Clad or Lined Steel Pipe》。目前国内双金属复合管的主要厂家包括西安向阳、浙江久立、上海海隆、新兴铸管等公司，其中以西安向阳的机械复合双金属复合管开发年限最长，供货业绩最大。我国双金属复合管产品基本以机械复合管为主，内衬材料主要是 316L，而以镍基/铁镍基合金、钛合金为内衬的双金属复合管尚处于起步阶段。为此，可以重点关注以耐蚀合金为衬层的双金属复合管的开发及应用。

4　大应变钢管

我国处于地震多发区域，如西二线和西三线管道沿线经过相当长的强震区（地震峰值加速度 0.2g 以上，其中峰值加速度 0.3g 的地段约 96km）和 22 条活动断层。当地震发生时，这些地区的管道将发生较大的位移和变形，必须进行应变控制，要求管道的极限应变（临界屈曲应变）大于设计应变（地震和地质灾害可能给管道造成的最大应变）。为适应大应变环境，管道应采用大应变钢管。大应变钢管具有较低的屈强比、高的形变硬化能力和均匀的伸长率。

大应变钢管（表 7）的制取，目前普遍采用双相钢的技术路线，典型组织类型有 F+B、B+M/A 等。采用双相钢的大应变钢管，最早由日本 NKK 钢铁株式会社提出，并在 NKK 福山工厂试制成功 X65 大应变钢管。目前国外已公开的大应变钢管有日本 JFE 钢铁株式会社（前 NKK 钢铁株式会社与川崎制铁合并）开发的 HIPER 和新日本制铁株式会社的 TOUGH-ACE。欧洲钢管公司也宣称开发了 X100 级别的大应变钢管，并用于 North Central Corridor 管道。我国在2011年中缅油气管道工程项目中首次采用了国产X70大应变钢管。

表7　典型大应变管线钢的力学性能

钢级	组织	纵向拉伸性能				冲击韧性	
		屈服强度（MPa）	抗拉强度（MPa）	屈强比	形变强化指数	冲击功（J）	FATT（℃）
X65	F+B	463	590	0.78	0.16	271	-98
X80	F+B	553	752	0.74	0.21	264	-105
X80	B+M/A	532	702	0.76	0.12	271	-98
X80	B+M/A	581	734	0.79	0.14	264	-105
X100	F+B	651	886	0.73	0.18	210	-143

此外，为准确测定大应变钢管的变形能力，国内还开发了钢管全尺寸弯曲试验系统，其试验能力（管径、钢级及壁厚等参数）是目前国际上同类试验设备中最大的。全尺寸弯曲试验系统的开发为西气东输二线、三线及中缅等管道工程大应变钢管的设计提供了可靠的技术支撑。

5　深海油气开发用钢管

21世纪是海洋的世纪，海洋在国家经济发展及维护国家主权的地位愈加突出。预计2015年末，海洋油气产量占全球油气总产量的比例将分别达到39%和34%。我国海洋油气资源储量巨大，而海上石油资源探明程度约为12.3%（世界平均约为73.0%），天然气资源探明程度约为10.9%（世界平均约为60.5%），探明率远低于世界平均水平。因此，我国海洋油气资源勘探开采潜力巨大。大陆架浅水区域的油气资源勘探开发起步较早，目前需要将开采延伸至海上深水区。

海底管道向深海发展，管道外压的问题逐渐突出，为防止管道发生挤毁事故，深海管道需要应用大厚径比（t/D）钢管，而大t/D钢管使DWTT性能面临更大考验。此外，海底管道在铺设过程中，尤其在使用铺管船铺设时，将承受很大的压缩、拉伸或者弯曲变形，同时，浪、流、平台移动及地质活动亦将造成海底管道在服役过程中发生塑性变形。因此海底管道需采用应变设计，并使用具有一定应变能力的钢管[10]。此外，海洋环境中的浪、流等可能引起涡激振动（VIV），造成钢管的疲劳损伤。

为适应海底管道的安装要求和服役环境，与陆地管线钢相比，海上服役钢管的要求更高：化学成分规定更严格，硫、磷等有害元素含量及碳当量要求更低。在力学性能方面，增加了失效前后的纵向拉伸及CTOD试验要求；屈服强度、屈强比及硬度要求更高。几何尺寸精度要求也更为严苛，尤其对直径、椭圆度和壁厚偏差的要求更为严格。

欧洲钢管集团、日本新日铁住金公司、JFE公司等老牌的冶金制管企业始终在国际上处于管线钢管制造领域的领先地位。其中，欧洲钢管集团在海底厚壁管线管开发方面一直走在世界前列，是近年来世界多个重大海底管道工程的主要供货商。2004年启动的挪威Langeled工程是当时世界上最长的海底管道工程，全长1173km，最大水深1000m，采用了X70管线钢，欧洲钢管集团为供货商（表8）。

表 8 欧洲钢管集团 Langeled 项目 X70 管线钢管的力学性能

钢管规格（mm）		屈服强度（MPa）		抗拉强度（MPa）		屈强比		伸长率（%）		CVN（-30℃）（J）		
外径	壁厚	横向	纵向	横向	纵向	横向	纵向	横向	纵向	母材	HAZ	焊缝
1066	23.3/24.0	522	515	628	603	0.83	0.85	20.8	22.5	209	225	151
1016	29.1/34.0	513	520	623	603	0.82	0.86	22.6	23.9	260	243	125

我国在海底管道工程建设方面起步较晚，但近年来随着海洋油气资源勘探开发的重要性不断提高，对海底管道工程（表 9）的需求迫切。其中，南海荔湾海底管道工程是目前国内钢管应用水深最深、压力最高、壁厚最大的项目，代表了国内海底管道发展的最高水平。渤海石油装备巨龙钢管有限公司（鞍山钢铁集团公司开发板材）、番禺珠江钢管有限公司（武汉钢铁集团公司开发板材）、宝山钢铁集团公司为荔湾项目提供了钢管，实现了 X65、X70 大壁厚海管的国产化。而与国外先进水平相比，国内产品在性能稳定性、尺寸精度等方面尚有一定差距。因此，针对当前各国趋之若鹜的深海油气资源，国内生产厂和科研院需加快进行更高性能深海管道用钢管的研究，实现新产品的研制和性能升级，为深海油气资源的大规模开发做好技术储备。

表 9 近年国内海底管道工程的钢管应用情况

项目名称	材质	规格（mm×mm）	数量（t）	用途	年份
胜利油田埕岛项目	X52	559×12.7	1552	海底油管道	2000
中海油春晓气田群开发项目	X60	711×15.9	68645	海底气管道	2003
番禺/惠州天然气开发项目	X65	508×（14.3~24.4）	58881	海底气管道	2005
中海油乐东气田海底管道工程	X65	610×（13.6~18.7）	27000	海底气管道	2007
南海荔湾海底管道工程	X65	762×30.2 758.8×28.6 765.2×31.8 559×28	140000	海底气管道	2012
	X70	765.2×31.8			

6 结束语

安全性与经济性的兼顾统一是管道工程发展的不变主题，而伴随非常规油气田、偏远油气田的开发，管道工程面临的服役环境日益恶劣。为此，需要加大低成本、高性能油气输送钢管的研发及应用。虽然我国油气管道建设走在了世界前列，但与国际领先水平相比，我国输送管的整体性能和质量还有一定差距。建议国内生产企业和科研院所，对标国外优势产品，针对高性能油气输送管开展联合攻关，同时加强钢管的工程应用和应用基础研究，推动我国管道事业持续、健康发展。

参 考 文 献

[1] SCHWINN V, FLUESS P, BAUER J. Production and progress work of plates for pipes with strength level of X80 and above [C]. Yokohama: International Conference on the Application and Evaluation of High-grade Linepipes in Hostile Environments, 2002: 339-354.

[2] 王志刚．应用学习曲线实现非常规油气规模有效开发［J］．天然气工业，2014，34（6）：1-8.
[3] 张希悉，汪凤，范玉然．高钢级天然气长输管道止裂控制技术现状［J］．油气储运，2014，33（8）：819-824.
[4] 黄维和，郑洪龙，王婷．我国油气管道建设运行管理技术及发展展望［J］．油气储运，2014，33（12）：1259-1262.
[5] 高惠临．管线钢与管线钢管［M］．北京：中国石化出版社，2012：6-7.
[6] International Organization for Standardization. Petroleum and natural gas industries-Steel pipe for pipeline-Technical delivery conditions, Part 3: pipes of requirement class C: ISO 3183-3: 1999 [S]. Geneva: ISO, 1999: 15-16.
[7] International Organization for Standardization. Petroleum and natural gas industries - Steel pipe for pipeline transportation systems: ISO 3183: 2007 [S]. Geneva: ISO, 2007: 30-31.
[8] American Petroleum Institute. Specification for line pipe: API Spec 5L: 2007 [S]. Washington D C: API, 2007: 30-31.
[9] 李鹤林．天然气输送钢管研究与应用中的几个热点问题//李鹤林．石油管工程文集［C］．北京：石油工业出版社，2011：255-265.
[10] 周廉，李鹤林，马朝利，等．中国海洋工程材料发展战略咨询报告［M］．北京：化学工业出版社，2014：247-249.

Study of Thickness Effect on Fracture Toughness of High Grade Pipeline Steel

Zhang Hua　Zhao Xin　Wang Yalong　Li Na

Abstract: The critical fracture toughness decreases when thickness of specimens increases and stress–strain field in crack tip start changing to plane strain state from plane stress state. In this paper, fracture toughness tests were carried. Based on the analysis of stress–strain field in crack tip and fracture toughness test results, a fracture toughness–thickness empirical model was established and the plane strain fracture toughness and critical thickness of X80/X100 were calculated. Then the validity of the empirical model was discussed and verified. The analytical results indicate that the safety of thick wall pipelines is worth of attention.

Keywords: Fracture toughness; Thickness effect; Pipeline steel; Plane strain

1 Introduction

In fracture mechanics, when stress intensity factor in crack tip is equal to the material fracture toughness, crack will start to grow. That means fracture toughness of material can forecast the remaining strength of a component with an initial crack. Although fracture toughness is material inherent attribute, for the same material, different fracture toughness values was determined in different tests as test conditions (temperature, loading rate, et al) and specimen size are different. Of all the influence factors for fracture toughness test, specimen thickness is a most important factor.

Because thereare amounts of oil and gas pipeline in the world, so cracks existing in some pipes body are almost inevitable. Stress in crack tip varies as pipe thickness is different. As thickness increases, stress–strain field in crack tip start changing to plane strain state from plane stress state, which means crack tip is in tension state in all three directions and the plastic zone will be limited in a small scope. So critical fracture toughness will decreases when pipe thickness increases in plane strain state, and brittle fracture is prone to happen, which is much more dangerous compared with ductile fracture.

In Chinese, pipe with high steel grade, large diameter and thick wall have been developing all the time. For pipes with high steel grade and thick wall which have been widely used, studies are focusing on whether fracture toughness has reached plane strain state or not. Fracture toughness tests of X80 pipeline steel that has been widely used and industrial trial–produced X100 pipeline steel with different wall thickness – were conducted and carefully analyzed in this paper. Based on analysis of stress field in crack tip, plane strain fracture toughness forecasting model was built, which is used to forecast whether the pipe thickness meets the plane strain condition or not.

2 Thickness Effect of Fracture Toughness

For component with a crack and external load, stress intensity factor K_I (take I mode crack as example) is a mechanics parameter to describe stress field in crack tip. K_I would increase if external loading increases. When K_I increases to a critical value, the crack in component begins to grow. This critical value is called fracture toughness K_C or K_{IC}, which represents the material's ability to resist unstable propagation of crack. Fracture toughness K_C or K_{IC} is inherent attribute of a material. The difference is that K_C is fracture toughness under plane stress state, which would be influenced by thickness of plate or test specimen. When specimen thickness increased, fracture toughness tends to be a stable and lowest value, which would not be influence by thickness. This value is called K_{IC} or plane strain fracture toughness. K_{IC} is the real material constant, which reflects the material's ability to prevent crack extension.

3 Facture Toughness Tests of Pipeline Steel

3.1 Test Materials

X80 and X100 specimens were separately sampled from ϕ1219mm×26.4mm and ϕ1219mm×23.5mm longitudinal submerged arc welding pipe. Mechanical properties of the X80 and X100 pipes were shown in Table 1.

Table 1 Mechanical properties of test materials

Material	Tensile strength R_m (MPa)	Yield strength $R_{p0.2}$ (MPa)	Elongation (%)
X80	665	556	22
X100	786	717	19

3.2 Test Results

In order to work out the thickest test specimen, CT (compact tensile) specimens were used. For pipelines, axial crack is most dangerous because the largest stress on pipe locate in hoop direction. So the direction of precast crack in specimen was parallel to pipe axial direction. For X80 pipe material, specimen thickness was 25, 22, 20, 18, 16 and 14, 12, 10, 8, 6mm respectively. The thickest X100 specimen was 22mm. There are 3 specimens for each set (grouped by thickness) of sample. Tests were carried out in accordance with the provisions of GB/T 21143—2007[2]. The results are shown in Table 2, in which test results is average value of three specimens.

Table 2 Fracture toughness K_C test results

Thickness (mm)	6	8	10	12	14	16	18	20	22	25
X80 (MPa · $m^{0.5}$)	59.0	57.9	62.3	61.1	61.7	59.8	67.3	65.6	62.8	63.6
X100 (MPa · $m^{0.5}$)	69.5	76.1	75.5	75.1	75.5	78.4	78.6	83.0	81.5	—

Overall, fracture toughness of X100 is bigger than X80. Fracture toughness reaches to the biggest value when the specimen thickness is 18mm for X80 and 20mm for X100. After peak value, fracture toughness begins to gradually decrease with increase of wall thickness. Before reaching to the biggest value, fracture toughness value has a platform which is measured with smaller thickness specimens.

3.3 Fractographic Features Analysis

Fracture SEM photographs of X80 and X100 sample with different thickness are shown in Fig. 1 and Fig. 2. It is found that all fractures are ductile fracture. There are many dimples with different size and non-uniform distribution on the fracture section. So in the thickness range of all the tests, material has not yet to embody the characteristics of embrittlement, and the crack propagation region presents plastic state. A tongue shape crack propagation region existed in fracture section, which is shown in Fig. 3. In the figure, B is the thickness of the original sample, and A is the width of crack propagation in experiment process. Due to constraint effect in thickness direction, central sample material is in three-dimensional tensile stress state. So middle part of the sample begun to crack firstly. With the increase of external load, the region reaching to critical stress intensity factor in thickness direction became larger, but the crack speed in central part of specimen was the biggest. So the fracture section shows a tongue shape.

4 Fracture Toughness-thickness Empirical Model

The shape of crack tip plastic zone is shown in Fig. 4. For stress state in crack tip, the ratio of plastic zone to thickness is an important coefficient. If the values of plastic zone size and specimen thickness are in same order of magnitudes, which means r_p/B tends to be 1, the plane stress state is dominant. In order to make sure most of the material in thickness direction is in stress state of plane strain, r_p/B must be significantly less than 1, which means material near surface in plane stress state in thickness direction just be a small part through thickness section. Tests confirmed that when r_p/B was about 0.025, then stress state when crack happens was typical plane strain[3].

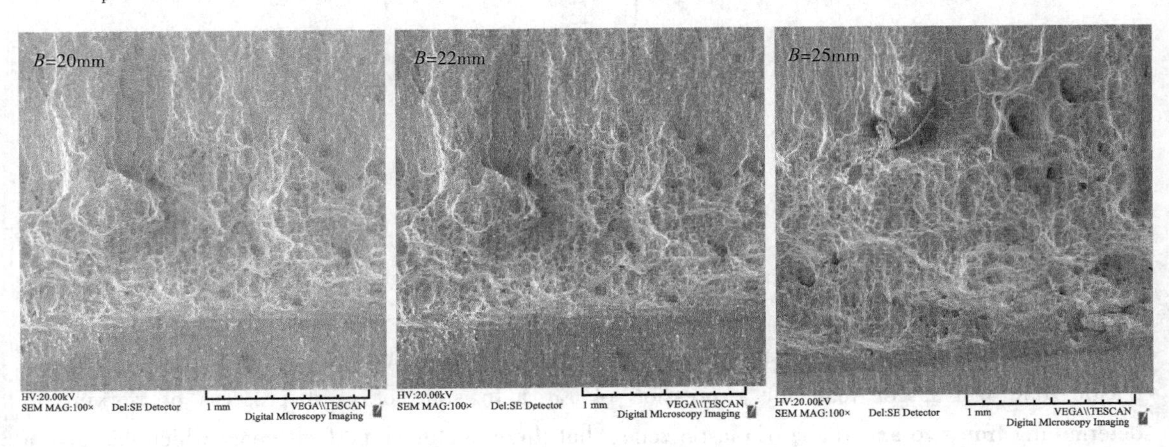

Fig. 1 SEM photographs of X80 fracture section

Based on the elastic solution of plane stress state, crack tip plastic zone radius ($\theta=0$) can be expressed as

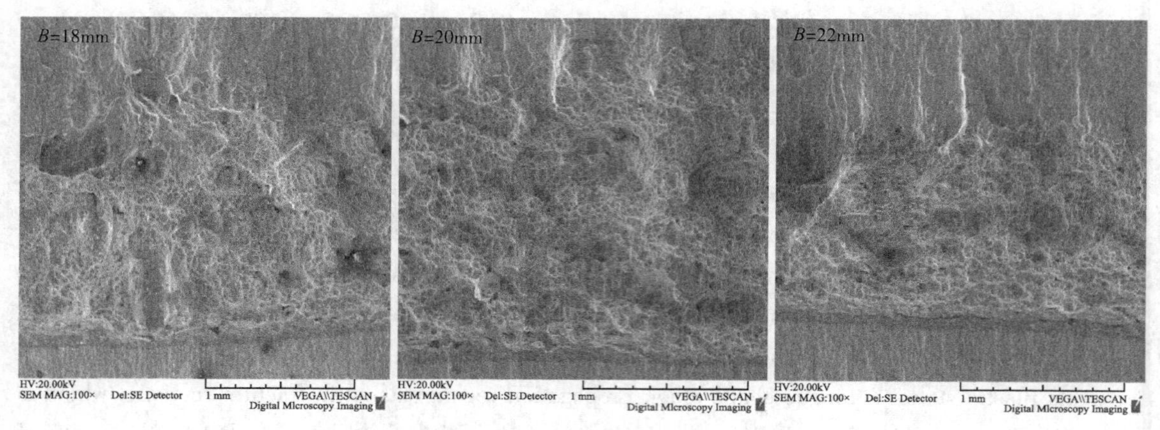

Fig. 2 SEM photographs of X100 fracture section

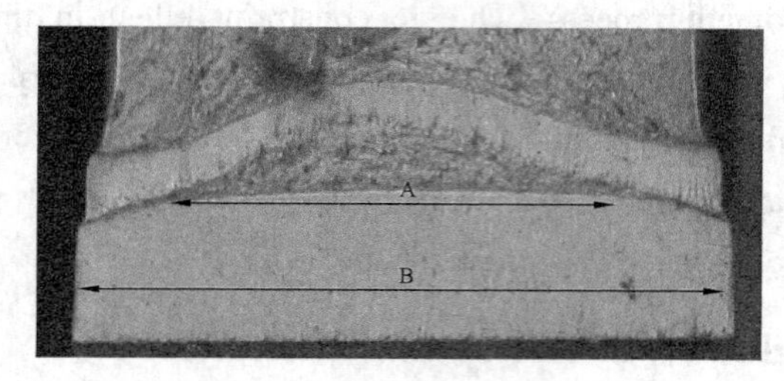

Fig. 3 Tongue shape in fracture section of fracture toughness specimen

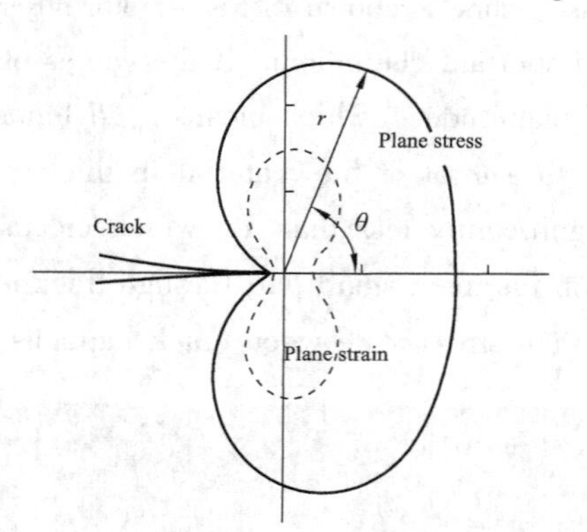

Fig. 4 Plastic zone size in crack tip

$$r_{\mathrm{p}} = \frac{K_{\mathrm{I}}^2}{2\pi\sigma_{\mathrm{ys}}^2} \tag{1}$$

Where r_{p} is the radius of plastic zone size, σ_{ys} is yield strength, K_{I} is stress intensity factor.

An obvious problem in this formula is that when crack tip stress exceeds the material yield strength, the exceeded load must be bore by material outside the hypothesis boundary. So the plastic zone will larger than the expression of Eq. 1. In spite of great progress of working in determining true size and shape of plastic zone, but there is still no perfect theory which can give a satisfactory description to the shape of plastic zone. The biggest radius of plastic zone locates in specimen surface. From surface to specimen center, as crack tip stress field is changing to plane strain state from plane stress state, and the size of plastic zone is also gradually decreasing. For the specimen, the overall plastic zone is smaller than that decided by surface plastic zone radius. So if surface plastic zone is used as specimen plastic zone, it has already amended the size by a correction factor more than 1.

On other hand, the theory of K_{IC} was established on the basis of elastic mechanics. So in situation of fitness for linear elastic fracture mechanics, which means plastic zone is much less than crack size, then plastic correction is not necessary. If plastic zone is not small compared with crack size, then expression of K based on elastic mechanics should be limited to apply. Plastic zone radius determined by linear elastic mechanics and von Mises yield criterion is used to characterize the plastic zone in plane stress state in this paper. As mentioned above, when $r_p/B=0.025$, then the specimen is in stress state of plane strain. That means the plastic zone size radius is only 1/40 of the specimen thickness. In this case, the plastic zone is very small compared with specimen size and Eq. 1 can be used.

In case of $r_p/B=0.025$, plane strain fracture toughness K_{IC} is substituted for K_I in Eq. 1, then the critical thickness of plane strain can be expressed as

$$B_{IC}=6.4\frac{K_{IC}^2}{\sigma_{ys}^2} \tag{2}$$

Where B_{IC} is the critical thickness of plane strain state.

The effect of thickness on fracture toughness was analyzed by Anderson[4] based on existing experimental data. In his study, it was reasonable that K_{IC} linearly decreased with thickness increase. In other studies[5], the relationship between K_C and thickness was expressed as

$$K_C=\xi\times t^{1/2}e^{-kt}+K_{IC}\ (1-e^{-kt}) \tag{3}$$

Where ξ and k were material constants, t was thickness. This equation is applicable to material of TC4 titanium alloy.

Through calculation of Eq. 3, point of fracture toughness – thickness can be got as shown in Fig. 5. Linear pattern was used to fit the points between maximum value of K_C to value close to plane strain fracture toughness, and the fitting curve is shown in Fig. 5. It shows that linear fitting is suitable for this curve segment as the value of correlation coefficient is 0.998.

Butaccording to the Eq. 3, K_C would increase from 0 with thickness increase before reaching to the maximum value. While according to the test data shown in Table 1, the fracture toughness of pipeline steel K_C has not decreased with wall thickness increase before reaching to the maximum value, and there is a platform in the curve. Some material would accord with this law mentioned in the literature[3]. So for pipeline steel, the error is large when fitting the relationship between fracture toughness and thickness.

Experiment data of X80 pipeline steel after fracture toughness reached to the maximum value (data of thicknesslarger than 18mm) were used and fitted by linear pattern, the linear relationship between fracture toughness and specimen thickness can be expressed as

$$K_C=-0.565\times B+76.8 \tag{4}$$

Let $K_C=K_{IC}$, $B=B_{IC}$, then

$$K_{IC}=-0.565\times B_{IC}+76.8 \tag{5}$$

K_{IC} is 48.9MPa · m$^{0.5}$ and B_{IC} is 49.4mm by solving Eq. 2 and Eq. 5. Same method was used to calculate the plane strain fracture toughness and critical thickness of X100. The K_{IC} value is 49MPa · $m^{0.5}$ and B_{IC} is 62. 7mm. Because wall thickness of X100 pipeline steel is thin relatively, and the thickest specimen is only 22mm, so there is only one point after peak point in the fracture toughness-thickness curve. That means the error of this forecasting model for X100 may be big, and the result is for reference only.

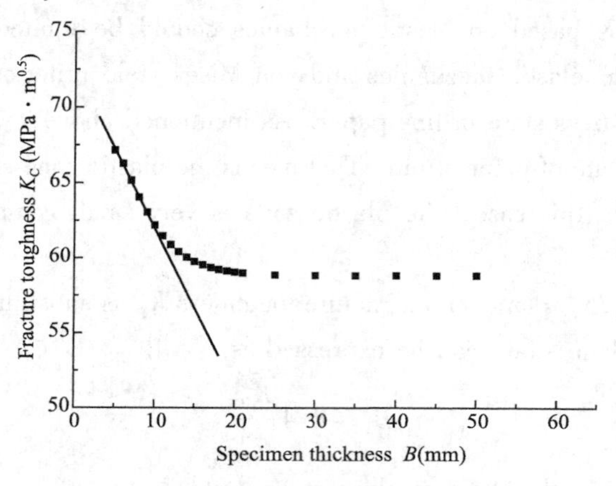

Fig. 5 Linear fitting and validation

5 Analysis and Validation of the Empirical Model

Experiment data of TC4 titanium alloy in literature[5,6] was used to validate this model. K_{IC} test result in the literature was 58. 8MPa · $m^{0.5}$, and the forecasted result with our model is 54. 6MPa · $m^{0.5}$. The error is just 7. 2%. The critical thickness forecasted by this model is 22. 9mm. But this parameter is unknown in the literature. From the forecasted curve (the forecasted points in Fig. 5), it is found that when thickness is more than 20mm, the fracture toughness has slight difference, which means that the stress state in crack tip is nearly plane strain state.

In literature[7], ratio of plane strain width to specimen thickness is an important parameter. K_C gradually tends to K_{IC} when this ratio increases to 100%. As shown in Fig. 3, crack tip does not propagate in the same speed, but faster in central and lower in surface. So the tongue shape formed in fracture section. After analyzing stress-strain field in crack tip, it is found that because central part of the sample was in three-dimensional tensile stress state, so this part begun to crack firstly. At the same time, plastic development is limited and brittle fracture is prone to happen in the central part. As load increases, crack tip width which reaches critical value of fracture strength becomes larger, but central crack propagates in the fastest speed.

The crack front width (A in Fig. 3) was analyzed for both X80 and X100 specimens. When dividing crack width in central crack front divide by the specimen thickness, it is found that for wall thickness larger than 14mm, ratio of crack width to thickness increases when thickness increases. Linear law is obtained as shown in Fig. 6.

If crack front width is used to characterize the width of plane strain, then the specimen is totally in stress state of plane strain when the ratio of crack width to thickness reaches to 1. Although this is impossible, if the ratio reaches a certain value, the specimen can be regarded as be in total plane strain state.

As shown in Fig. 6, it can be seen that when thickness is bigger than 14mm, the linear relationship can be obtained between ratio of crack width to thickness and thickness itself. This ratio increases as thickness increases. When thickness is less than 14mm, no obvious relationship can be found. The possible reason is that in thin specimen test, proportion of plane strain crack is small. In thin specimen crack, Plane stress crack dominates the test, so the crack extension law is different from that of plane strain.

Test results of thickness more than 16mm were used, the linear relationships for both X80 and X100 were obtained between crack width to thickness ratio and thickness. Taking the width-to-thickness ratio up to 100% as criterion of plane strain, it is found that the critical thickness of X80 is 45mm, and that of X100 is 38.7mm.

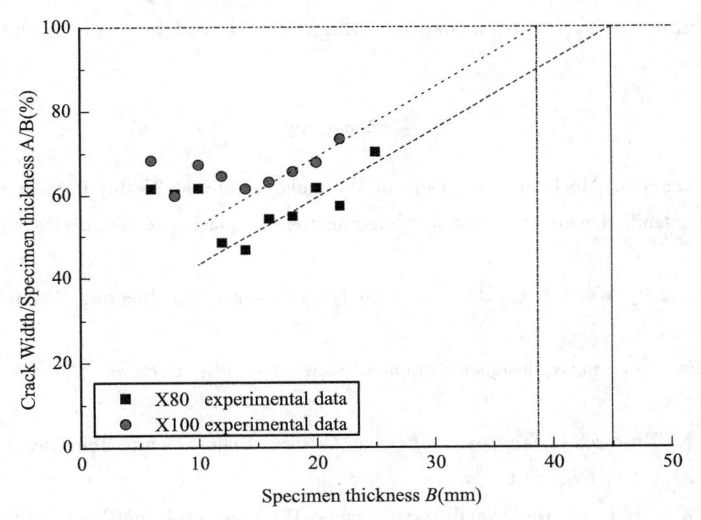

Fig. 6 Relationship of crack width-to-thickness ratio and specimen thickness

For X80, it can be seen that the difference of plane strain critical thickness calculated from our model and from width-to-thickness ratio model in literature[7] is very little (4.7mm). But for X100, the difference is very large. The possible reason is that in our tests, only one point was obtained after peak point in fracture toughness-thickness curve. So the error will be big if the test data is limited when using our model to estimate the plane stress fracture toughness or critical thickness.

Linear fitting was adopted for both the above two kinds of material. But in case of thickness increase, especially the thickness is nearly plane strain condition, the fracture toughness and specimen thickness has not shown a simple linear relationship. Because even if the specimen is thick enough, it cannot completely reach ideal plane strain state, so the ratio of A/B can only be infinitely close to 1, but not to be 1. So calculated critical thickness will be less than actual value, but it can infinitely close to it. From the test results, it can be confirmed that this representation is reasonable, and the model proposed in this paper has been validated.

6 Conclusions and Suggestions

In the second west to east gas pipeline project, the thickness of bends and fittings is more than 50mm. From analysis above, the possibility of the brittle fracture does exist. So the safety of thick wall pipelines is worth of attention. On the other hand, because wall thickness of general X80 pipe is usually not so thick, and the critical thickness of plane strain fracture toughness is nearly 50mm. So characteristics of the plane strain can't appear in actual X80 pipeline and the crack resistance is not determined by K_{IC}. In such cases, the actual fracture toughness is usually higher than K_{IC}, which means that if K_{IC} is used to estimate critical crack size and residual strength of a component, then conservative result will be obtained. Especially when fracture stress intensity factor is much bigger than extension stress intensity factor, the criterion of K_{IC} might be too conservative. So for a specific case, specimen of appropriate size and thickness should be used to determine the fracture toughness.

References

[1] Mc Clintock F, Argon A. Mechanical Behavior of Materials, Addison-Wesley Publishing Company, 1968.

[2] Metallic materials-Unified method of test for determination of quasistatic fracture toughness: GB/T 21143—2007 [S]. 2007.

[3] Brock D,. translated by Wang KR, He MY, Gao H. Elementary Engineering Fracture Mechanics, Beijing Science Press, 1980.

[4] William Anderson, W. Some designer - oriented view on brittle fracture. Battelle Northwest Rept. SA - 2290, 1969.

[5] Yang JY, Zhang X. Research on Thickness Effect on Cracked Plate Fracture Toughness. Journal of Mechanical Strength, 2005, 27 (5): 672-680.

[6] Yang JY, Zhang X, Zhang M. Theory and Application on Thickness Effect on Cracked Plate Fracture Toughness. Journal of Mechanical Engineering, 2005, 41 (11): 32-42.

[7] Marc Andre Meyers, Krishan Kumar Chawla. Mechanical Behavior of Materials, Cambridge University Press, 2009.

X90高强度管线钢预精焊冷裂纹形貌及成因

何小东　霍春勇　路彩虹　仝　珂　宋　娟

（中国石油集团石油管工程技术研究院）

摘　要：采用力学测试、组织分析、扫描电镜和能谱分析的方法，研究了X90高强度管线钢预精焊的焊接裂纹形貌，并分析了裂纹形成原因。结果表明，X90螺旋焊钢管内焊缝上的横向裂纹为延迟冷裂纹，裂纹在焊缝组织中呈穿晶扩展或沿晶界扩展。X90含有较多的Mn、Cr、Mo、Ni、Cu等合金元素，强度高，导致内焊缝的焊接残余应力高于裂纹的临界应力。内焊道维氏硬度比外焊道的高，且焊缝两侧的硬度分布极不对称，造成了内焊道附近的应力集中和分布不平衡。内外焊缝重合区域的扩散氢不易逸出，其含量较高，在气孔、夹渣等“陷阱”处聚集，导致裂纹产生并在断口上形成大量的氢白点。焊缝一次结晶所形成的连续细长的树枝晶晶界为裂纹扩展提供了“通道”。

关键词：X90；高强度管线钢；预精焊；焊接冷裂纹；扩散氢

目前，国内长输油气管道输送用螺旋缝埋弧焊管已突破了国际上的使用禁区，X70、X80高钢级的螺旋缝埋弧焊接钢管已在西气东输等重大管道工程中得到大量的应用，并正在开展X90、X100的研究开发[1~3]。螺旋缝埋弧焊接钢管采用“一步法”成型焊接时，由于焊接位置和较高的焊接速度，导致了外焊缝余高较大或在焊缝中心形成“脊棱”，而在内焊缝形成“马鞍形”。不良的焊缝形状容易造成应力集中，给管道服役带来安全隐患。为了进一步提高螺旋缝焊管的生产效率，改善焊缝质量，先进的预精焊（即“两步法”）工艺在大口径、厚壁、高强度管线钢管制造中得到了大量应用[4,5]。

文献［6，7］指出，对于采用低合金设计、控轧控冷制造的高强度管线钢，虽然碳当量（C_{eq}）较高，但冷裂纹指数（P_{cm}）较小，因此，即使在不预热条件下焊接，X80和X100管线钢对冷裂纹也不敏感．张克修等人[8]分析了X80螺旋焊缝金属出现的横向裂纹和沿熔合线的纵向裂纹，认为是因局部成分偏析和工艺参数不当产生的热裂纹，或是因焊缝中心存在低熔点物质在拉应力作用下产生的横向热裂纹．然而，高强度管线钢焊接时，在焊缝金属上产生冷裂纹也是存在的[9]。由于冷裂纹不是焊接后立即出现，具有“延迟”的特性，随着时间增长逐渐增多和扩展，因而对产品质量和管道的安全运行具有更大的危害。

文中测试研究了X90高强度管线钢预精焊工艺所产生的冷裂纹形貌，并分析了其形成原因，有利促进高强度管线钢焊接质量控制，保障油气管道安全。

基金项目：陕西省自然科学基金资助项目（No. 2011JQ6017）。

作者简介：何小东，男，1970年出生，硕士，高级工程师/国际焊接工程师。主要从事管线钢焊接工艺、材料性能测试及表征。发表论文40余篇．E-mail：xiaodonghe@126.com。

通讯作者：何小东，男，高级工程师．E-mail：xiaodonghe@126.com。

1 试验方法

ϕ1219mm×16.3mm X90 螺旋焊接钢管试制用板卷的化学成分见表 1。钢板两边加工的坡口为不对称的 X 型坡口，焊接工艺采用预精焊。先用 CO_2 气体保护焊对成型的螺旋缝进行预焊，所采用焊丝直径为 4.0mm，牌号为 ER62-G。焊接电流为 850A，焊接电压为 22～24V，气体流量为 80～100L/min，焊接速度为 3m/min。预焊之后，再用多丝埋弧焊接进行内、外焊缝精焊，所用焊丝直径为 4.0mm，牌号为 BHM-11，并配用 YS-SJ105G 焊剂。内外焊缝精焊工艺参数见表 2。

表 1　X90 化学成分（wt%）

C	Si	Mn	P	S	Cr	Mo	Ni	Nb	V	Ti	Cu	B	Al
0.040	0.29	1.92	0.0090	<0.002	0.28	0.36	0.39	0.087	0.037	0.016	0.28	<0.0005	0.027

焊接完成后，对焊缝进行 X 射线探伤和超声波探伤检验未发现裂纹。但是，将样品放置到第二天，在焊缝上出现多处肉眼可见的横向裂纹。截取裂纹试样，利用 MEF4M 金相显微镜、VEGA 扫描电镜和 NSS-300 型能谱仪分析了焊缝的显微组织、裂纹形貌和裂纹断口的成分。为了进一步分析裂纹产生原因，用 UTM 5305 材料试验机和 KB 30BVZ-FA 硬度计分别测试了母材的拉伸性能和焊接接头维氏硬度。

表 2　X90 内外焊道精焊工艺参数

焊道		焊接电流 I（A）	焊接电压 U（V）	焊丝间距 d（mm）	焊接速度 v（m/min）
内焊	前丝	820	32	15	1.5
	中丝	560	34		
	后丝	480	35		
外焊	前丝	1070	31		
	后丝	520	34		

2 试验结果与分析

2.1 裂纹形貌

X90 螺旋焊管焊接完成后放置一段时间，待焊缝金属完全冷却后，检查发现焊缝上的横向裂纹如图 1（a）所示，该裂纹垂直于焊缝，有向两侧母材扩展的趋势。沿焊缝中心用线切割垂直裂纹切开，裂纹在焊缝内部的宏观形貌如图 1（b）所示。从图 1（b）可以看出，裂纹沿壁厚方向，从内焊缝表面，穿越内外焊道重合处向外焊缝扩展，裂纹走向较平直。用金相显微镜观察到的裂纹形貌和附近的显微组织如图 1（c）所示。可以看出，裂纹附近焊缝组织为典型的晶内成核针状铁素体（Intra-granular nucleated Acicular Ferrite，IAF）和粒状贝氏体（Granular Bainitic，GB）的混合组织。

为了进一步观察在深度方向的裂纹形貌，沿裂纹方向分别别从内焊缝到外焊缝取内、中、外三块断口试样，用扫描电镜对断口形貌进行分析，结果如图 2 所示。从图 2 中可以看出，焊缝上的横向裂纹断口平齐，没有明显的韧窝，裂纹沿焊缝柱状晶呈穿晶扩展或沿晶界扩展。裂纹的起源于内焊道近表面部位，启裂之后穿过壁厚中部的内外焊道重合区向外焊道迅速扩展，几乎穿透外表面。由此可以确定 X90 预精焊内焊缝上的横向裂纹为冷裂纹。同时，从图 2 中可以看到，断口上分布有很多圆形的白色“斑点”，并且内外焊缝重合部位的白色“斑点”比内焊缝和外焊缝断口上的“斑点”多。在断口上还分布有圆形或不规则的凹坑。

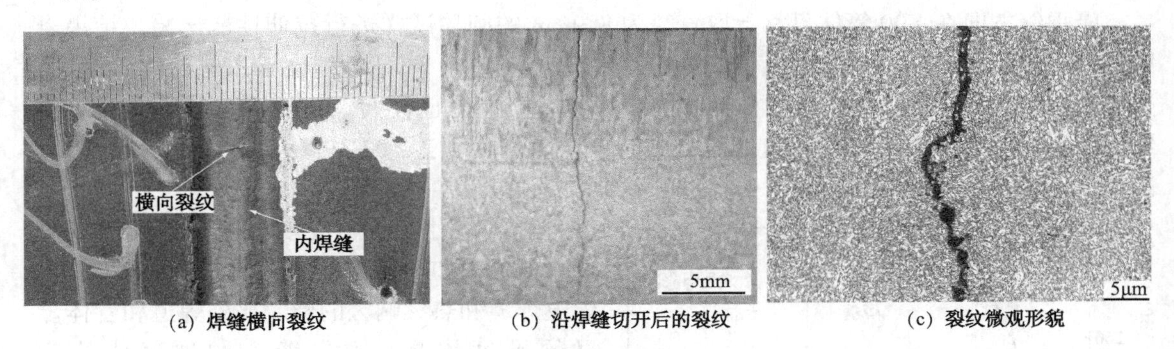

(a) 焊缝横向裂纹　(b) 沿焊缝切开后的裂纹　(c) 裂纹微观形貌

图 1　焊缝横向裂纹形貌

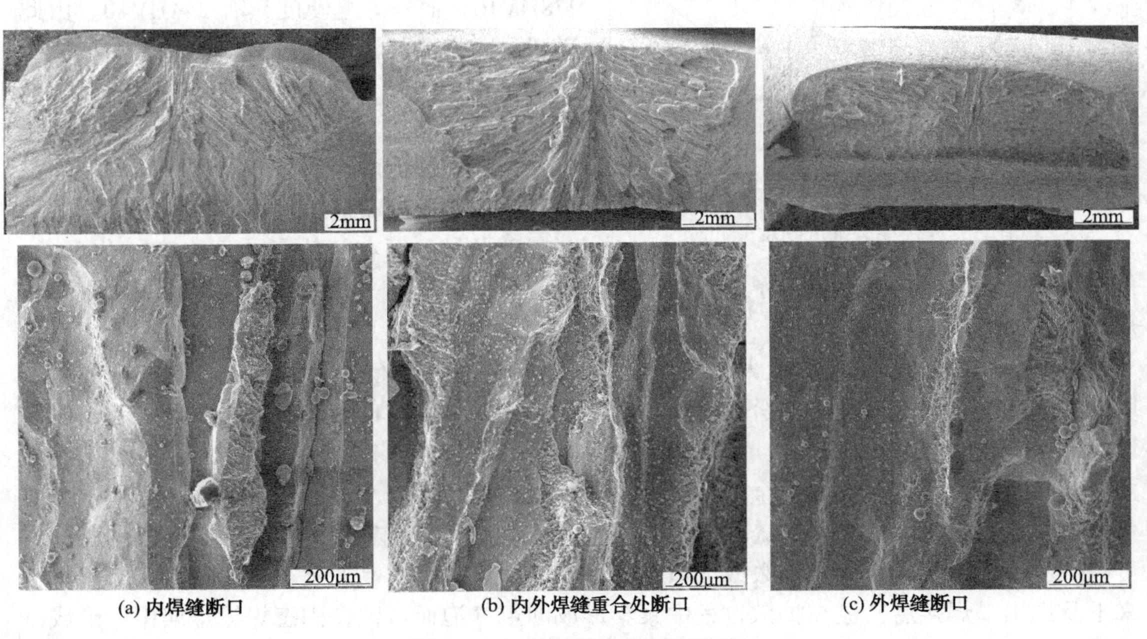

(a) 内焊缝断口　(b) 内外焊缝重合处断口　(c) 外焊缝断口

图 2　X90 焊缝裂纹的断口形貌

2.2　讨论与分析

钢的焊接性可以用碳当量来评价。管线钢碳当量通常采用国际焊接学会(IIW)的 CE_{IIW}[碳含量 w(C)>0.12%适用]和 CE_{Pcm}[碳含量 w(C)≤0.12%适用]来计算，其表达式如下：

$$CE_{IIW}=C+Mn/6+(Cr+Mo+V)/5+(Ni+Cu)/15 \quad (1)$$

$$CE_{Pcm}=C+Si/30+(Mn+Cu+Cr)/20+Ni/60+Mo/15+V/10+5B \quad (2)$$

当 w(C)>0.12%时，用碳当量 CE_{IIW} 作为材料可焊或难于焊接的评价标准。当 w(C)<0.12%时，用 CE_{Pcm} 评价材料的可焊性。此时，认为材料处在易焊区，但实际上 CE_{IIW} 仍对管线钢的焊接有较大影响。研究表明，随碳当量的增加，产生焊接裂纹的临界应力降低。当碳当量大于 0.42 时，管线钢的焊接裂纹的临界应力仅约为 100MPa[10]。

一般地，标准要求 CE_{Pcm} 不超过 0.25%，CE_{IIW} 不超过 0.43%。从表 1 可知，试验用的 X90 板卷的 Mn、Cr、Mo、Ni、Cu 等合金元素成分含量较高。经计算，X90 的 CE_{Pcm} 为 0.21%，CE_{IIW} 为 0.54%. 这表明 X90 管线钢是“易焊”的，但是具有较大的淬硬性，产生焊接裂纹的临界应力较低。

沿焊缝方向在 X90 管体母材上取直径为 8.9mm 的圆棒试样进行拉伸性能试验。结果表明，其屈服强度 $R_{t0.5}$ 为 712MPa，抗拉强度 R_m 为 765MPa。较高的强度导致螺旋焊管成型、焊接过程中其拘束度较大，所产生弹复力和焊接残余应力也较大。有研究表明[11]，对于钢级为 X80 的 ϕ1219×18.4mm 螺旋焊管，其残余应力峰值可达 407MPa，且分布在内焊缝上。因此，对于 X90 高强度管线钢，焊缝上的残余应力会更大。

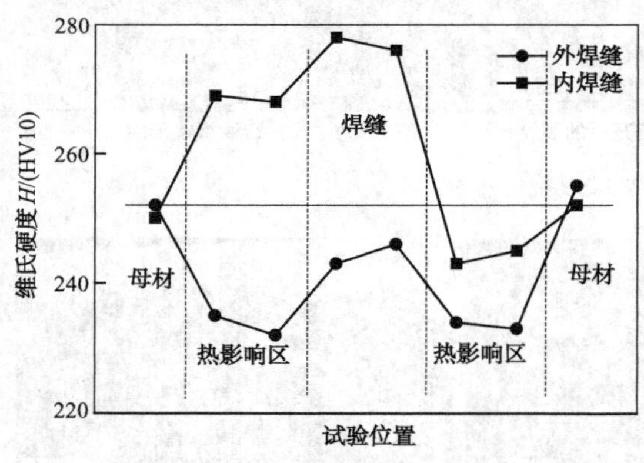

图 3　焊接接头维氏硬度分布

对裂纹附近的焊接接头进行维氏硬度测试，其分布如图 3 所示。从图 3 可以看出，内焊道焊缝和热影响区的硬度比外焊道和管体母材的硬度均高。内焊缝上的硬度值约为 278HV10，而外焊缝硬度只有 244HV10。由此说明，内焊缝的淬硬性较高，为裂纹的形成提供了条件。同时，图 3 也表明，外焊道热影响区的硬度低于母材的硬度 20HV10，存在软化现象，但外焊道两侧的硬度对称分布，而内焊道两侧的硬度分布极不对称，左侧热影响的硬度高于母材，而右侧热影响区也有软化现象，低于母材的硬度，进一步造成了内焊道附近的内应力的分布不平衡和应力集中，促进裂纹的形成和扩展。

对图 2（b）的断口局部放大，用扫描电镜观察断面上分布的圆形白色“斑点”如图 4 所示。这些白色斑点有的密集分布，有的稀疏分布在焊缝断口上。图 5 是白色“斑点”区域的能谱分析结果。图 5 表明，焊缝熔敷金属中除了焊丝本身含有较多的合金元素 Ni 外，主要还有氧元素，说明在焊缝金属中有大量的氧化物存在，如水分、油污、铁锈等，这些氧化物可导致 X90 焊接过程中液态金属吸收氢。焊缝金属中的氢有一部分在熔池结晶过程中可以逸出，但由于焊接工艺参数或坡口尺寸不当，而熔池的结晶速度很快，还有相当多的氢来不及逸出。焊接后，氢在扩散过程中聚集形成的氢气泡而被留在固态焊缝金属中，形成巨大的内应力，致使焊缝开裂，在断口上形成圆形氢白点或“鱼眼”。因此，X90 横向裂纹断口性质属于氢致开裂。

淬硬倾向、扩散氢和焊接接头的应力状态是管线钢焊接时产生冷裂纹的三大主要因素。淬硬性主要取决于碳当量，焊接接头的应力状态与结构的拘束度有关。因此，考虑三大因素时，管线钢的焊接冷裂纹敏感指数可以表示为：

$$P_w = CE_{Pcm} + [H]/A + R_F/B \tag{3}$$

式中 P_w——考虑拘束度和扩散氢含量的冷裂纹敏感指数，%；

CE_{Pcm}——冷裂纹敏感指数，%；

$[H]$——焊缝中扩散氢含量，mL/100g；

R_F——焊缝拘束度，N/mm²；

A、B——常数。

公式（3）表明，当CE_{Pcm}一定时，扩散氢越大，强度越高，约容易产生裂纹。

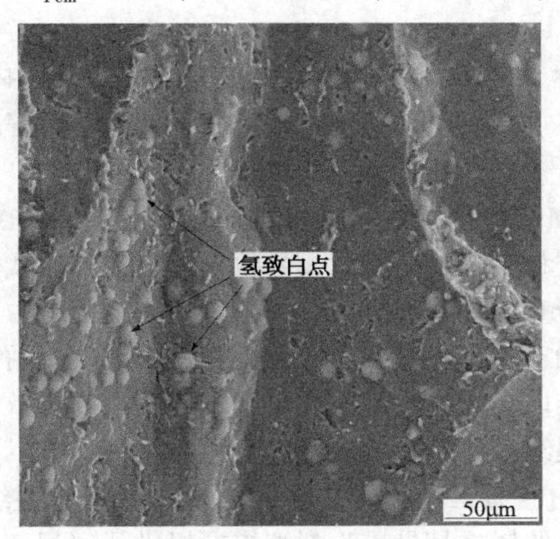

图4　X90焊缝金属中氢聚集形成圆形白色“斑点”

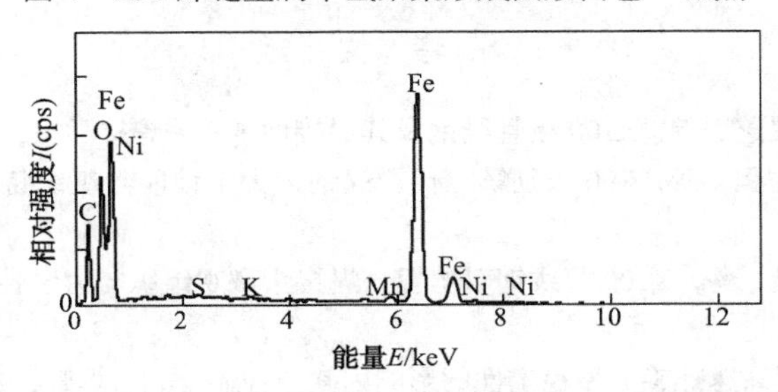

图5　焊缝金属中白色“斑点”区域的能谱分析

实际上，裂纹的形成还与焊接过程中熔池结晶有很大关系。焊缝金属与坡口处形成熔化边界，其温度梯度较大，结晶速度较小，成分过冷度很低，主要形成平面晶。随着远离熔合区边界向焊缝中心过渡时，温度梯度逐渐变小，结晶速度逐渐增大，结晶形态由平面晶向胞状晶、树枝胞状晶发展，一直到焊缝中心形成等轴晶。焊缝金属一次结晶形态对裂纹、夹杂、气孔、耐腐蚀性能都具有严重影响。图6是裂纹附近的焊缝组织。从图6中可以看出，裂纹的走向与焊缝一次结晶的树枝晶晶界方向一致，晶界附近有气孔、夹渣等缺陷，从而形成了图2中裂纹断口上的圆形凹坑或不规则凹坑。焊接缺陷成为扩散氢聚集的“陷阱”，为裂纹的形成提供了条件，而连续细长的晶界为裂纹扩展提供了“通道”。

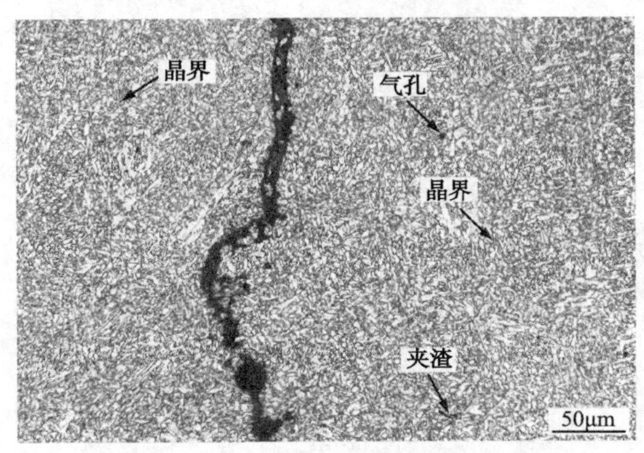

图6　裂纹附近的焊缝组织晶界和缺陷

3 结论

（1）X90 预精焊内焊缝上的横向裂纹为冷裂纹。裂纹起源于内焊缝表面附近，穿越内外焊道重合区向外焊缝扩展。裂纹断口无明显的韧窝，裂纹沿焊缝组织呈穿晶扩展或沿晶界扩展。

（2）X90 含有较高的 Mn、Cr、Mo、Ni、Cu 等合金元素，碳当量较高。淬硬性大。且母材的强度高，成型焊接过程中拘束度较大，导致内焊缝的焊接残余应力较大。内焊道上较高的维氏硬度在焊缝两侧分布极不对称，造成了内焊道附近的应力集中和分布不平衡。

（3）X90 焊接过程中由于含有氢的氧化物浸入，扩散氢在气孔、夹渣等“陷阱”处聚集，在焊缝中部的扩散氢难以逸出，使得内外焊缝重合区域的扩散氢含量比内焊缝和外焊缝表面附近的高，导致焊缝裂纹断口上形成大量的氢白点。

（4）内焊缝具有较高的应力、硬度和扩散氢，为 X90 焊缝裂纹形成提供了条件。焊缝一次结晶所形成的连续细长的树枝晶晶界为裂纹扩展提供了“通道”。

参 考 文 献

[1] 严春妍，李午申，冯灵芝，等. X100 级管线钢及其焊接性［J］. 焊接学报，2007，28（10）：105-108.

[2] 夏佃秀，王学林，李秀程，等. X90 级别第三代管线钢的力学性能与组织特征［J］. 金属学报，2013，49（3）：271-276.

[3] 李继红，杨亮，陈飞绸，等. X100 管线钢双丝埋弧焊接头微观组织与力学性能［J］. 焊接学报，2013，34（10）：27-30.

[4] 刘成坤，陈铭. 预精焊高速螺旋成型稳定性的影响因素与控制［J］. 焊管，2015，38（4）：52-56.

[5] 王凤成，崔晓峰，王国胜，等. 螺旋缝焊管预焊缺陷对精焊质量的影响与控制［J］. 钢管，2012，41（2）：49-52.

[6] 李亚娟，李午申，谢琦. Nb-Mo 系 X80 管线钢焊接冷裂纹敏感性的研究与预测［J］. 焊接学报，2010，31（5）：105-108.

[7] 张君，牛辉. X100 管线钢冷裂纹敏感性研究［J］. 焊管，2011，34（12）：21-26.

[8] 张克修，吴金辉，王树人. X80 螺旋埋弧焊管焊缝横向裂纹产生原因及预防措施［J］. 焊管，2010，33（8）：10-13.

[9] 王高峰，王晓江，梅滨，等. X70 螺旋缝焊管焊缝横向裂纹分析［J］. 焊管，2013，36（3）：39-44.

[10] 高惠临．管线钢—组织 性能 焊接行为［M］. 西安：陕西科技出版社，1995.

[11] 熊庆人，李霄，霍春勇，等. 三种高钢级大直径焊管残余应力分布规律研究［J］. 焊管，2011，34（3）：12-17.

二、油井管与管柱

石油管材及装备材料服役行为与结构安全研究进展及展望

冯耀荣　马秋荣　张冠军

（石油管材及装备材料服役行为与结构安全国家重点实验室，
中国石油集团石油管工程技术研究院）

摘　要：本文论述了加强石油管材及装备材料服役行为与结构安全研究的重要意义、主要研究进展、面临的形势任务、发展目标、研究领域和方向。经过30余年的努力开拓，我国石油管材及装备材料服役行为与结构安全软硬件平台及核心技术体系已基本形成，在高压输气管道断裂与变形控制、在役管道安全风险评价、油气井管柱完整性和适用性评价、石油管材及设备腐蚀与防护、高性能管材应用关键技术等方面取得了重要突破。面对我国油气工业面临的严峻形势和对安全生产提出的新挑战，拟进一步加强“石油管材及装备材料服役行为与结构安全国家重点实验室”建设，围绕油井管与管柱失效预防、输送管与管线安全评价、腐蚀与防护、先进材料及应用技术四大研究方向努力攻关，建立石油管材及装备材料服役行为与结构安全理论和技术体系，力争5到10年内整体技术达到国际先进水平，将实验室建成石油管材及装备材料服役安全科学研究基地、高端人才培养基地、国际合作与学术交流基地、石油工程材料服役安全技术支持中心，有力支撑石油天然气工业发展。

关键词：石油管材；石油装备；服役行为；结构安全；研究方向

鉴于石油管材及装备材料在石油天然气工业中的基础性地位和重要作用以及面临的重大共性理论和工程技术难题，在“中国石油天然气集团公司石油管工程重点实验室”和“陕西省石油管材及装备材料服役行为与结构安全重点实验室”的基础上，国家科技部批复建立了“石油管材及装备材料服役行为与结构安全国家重点实验室”。经过30余年的努力开拓，我国石油管材及装备材料服役行为与结构安全研究已取得了若干重要进展。已初步形成“石油管工程学”“材料服役安全工程学”“石油管安全工程学”“石油管材及装备材料服役行为与结构安全”等新型交叉学科[1-4]。本学科涉及石油天然气工程、材料科学与工程（材料学、材料物理、材料化学、材料加工工程）、冶金工程、机械工程、力学（弹塑性力学、断裂力学、管柱力学、流体力学、岩土力学）、安全科学技术和安全工程、计算机科学与技术、化学、物理等学科，具有多学科交叉的显著特点，因此，本学科的发展与多学科的并行

作者简介：冯耀荣，男，1960年生，毕业于西安交通大学材料科学与工程专业，博士，教授级高工。一直从事石油管材与装备的应用基础研究和重大工程技术支持工作。现任中国石油集团石油管工程技术研究院总工程师，石油管材及装备材料服役行为与结构安全国家重点实验室主任。

中国石油集团应用基础项目“复杂工况气井油套管柱失效控制与完整性技术研究（2014A-4214）”资助。

及耦合交互发展密切相关。本文结合国家重点实验室建设与运行，就加强石油管材及装备材料服役行为与结构安全研究的重要意义、研究进展、发展目标、研究方向等问题进行论述，以促进石油管材及装备材料服役安全学科的发展。

1 加强石油管材及装备材料服役行为与结构安全研究的重要意义

石油管材及装备是石油天然气工业的基础，年耗资达千亿元，石油管材及装备的安全可靠性和使用寿命对石油工业关系重大，其失效会导致巨大经济损失、人员伤亡、环境污染和社会影响，其质量和性能对石油天然气工业采用先进工艺和增产增效有重要影响。

以油气井管柱及其构件为例，约占油气田装备总资产的60%，占建井成本的20%~30%。我国每年消耗油井管约350×10^4t，耗资约250亿元。套管柱的寿命基本上决定了油气井的寿命，油管柱是产油出气的唯一通道，而钻柱是钻井的关键工具。在复杂的拉、压、弯、扭、剪切及复合应力状态和温度及CO_2、H_2S、Cl^-等腐蚀介质的共同作用下，油气井管柱极易发生失效。例如：我国西部深井超深井套管失效导致报废直接经济损失上亿元；川渝地区高含H_2S气井套管泄漏，除造成巨大经济损失外，还会引发大规模人员疏散、伤亡及重大自然环境灾害；西部高温高压气井油管泄漏会导致井筒完整性破坏，造成严重经济损失和安全风险。我国现有油气井约40×10^4口，每年新钻油气井约3×10^4口，套管损坏井已超过3×10^4口，并且每年大约以10%的速度增加。造成的经济损失达数十亿元。

以油气输送管道为例，我国陆上油气长输管道总里程约12×10^4km，原油管道2.3×10^4km，成品油管道2.1×10^4km，天然气管道7.6×10^4km。覆盖31个省区、市和特别行政区，其中60%以上运行已超过20年，管道爆炸着火、断裂、泄漏事故时有发生，年均千公里事故率明显高于美国和欧洲。2013年11月22日发生的黄岛输油管道泄漏爆炸事故，造成62人死亡、136人受伤，直接经济损失7亿元。2014年全国排查出油气管道安全隐患29436处。油气田内部的集输管道，由于复杂的工况条件，失效频繁发生，我国西部某油气田每年发生管道泄漏、断裂、腐蚀穿孔事故超过1000起。近年来，我国高压大口径长距离输送管道建设速度持续提升，预计到2020年我国油气长输管道总里程将达到15×10^4km左右，高钢级管线钢和钢管的大规模应用，安全风险不容乐观。对油气储运的安全生产带来严重挑战。

再以炼化管道和容器为例，目前全国各类化工园区，包括化工聚集区大概有1200多家，炼油设备的腐蚀损失占其产值的6%~7%，因腐蚀进行的大检修占70%，炼1t原油腐蚀控制投入近1美元。炼化管道和容器失效事故时有发生，造成了严重的后果。我国原油加工能力约7×10^8t/a，其中75%以上原油是含硫或高硫原油，其中30%是超高含硫原油。劣质原油和设备老化都给炼化装置的安全生产提出了严峻挑战。

石油天然气工业中的安全事故绝大多数与石油管材及装备材料的服役性能和结构原因有关。随着国家“一带一路”战略和国家“十三五”规划的逐步展开实施，与周边各国和区域的合作不断深化，国际竞争日益激烈，油气需求持续增加，保障国家油气供给安全的难度越来越大。天然气管道高压大输量输送，低温冻土带、地震断裂带等苛刻地质条件，非常规油气资源低成本开发，超深高温高压油气资源开采，极端苛刻腐蚀环境等严酷的服役条件对石油管材及装备的服役性能和安全可靠性提出了更高的要求。加强石油管材及装备材料服役行为与结构安全研究十分必要，也十分重要。

2 我国石油管材及装备材料服役行为与结构安全研究的重要进展

我国石油管材及装备材料服役行为与结构安全的发展起始于石油管材及装备的失效分析，最典型的案例是20世纪60年代四川威远–成都输气管线的失效分析[5]。20世纪70年代后期至80年代，随着国际上含缺陷结构完整性评价技术的发展，失效分析和缺陷评价国内逐步发展起来[6]。20世纪90年代以来，失效分析及预测预防技术和含缺陷结构的适用性评价技术在石油天然气行业逐步开展起来，并得到了较大发展[7,8]。到目前为止，软硬件平台及核心技术体系已基本形成。主要技术进展[4,9-11]表现为：

（1）建立了50000J大摆锤动态断裂试验系统、内压爆破和疲劳试验系统、内压+弯曲复合载荷大变形试验系统，建成全尺寸气体爆破试验场，形成油气输送管道断裂和变形控制试验平台，攻克高钢级大口径管道变形控制技术，提出了确定钢管应变能力的方法，研发了适用于富气组分的天然气减压波分析和高压输气管道止裂预测软件，开展了X80管道实物气体爆破试验，研究形成X80高压输气管道断裂控制及止裂韧性确定技术，研究提出了西气东输二线安全运行参数控制要求和管材止裂韧性要求。推动了X70/X80管线钢和钢管的国产化。为成功建设西气东输管线、西气东输二线、中亚和中缅管线提供了技术支撑。

（2）研究形成油气管道变形和裂纹检测装置，建立了含体积型、裂纹型、几何型、弥散损伤型、机械损伤型缺陷油气管道安全评价和寿命预测方法，开发了工程适用的管道适用性评价软件。针对12种风险因素，建立了天然气管道失效概率计算模型和失效后果的估算模型，研究制定了我国天然气管道的风险可接受推荐准则。研究建立了X80管材的失效评估曲线，提出了基于应变的管道失效评估准则和基于可靠性的失效评估准则。优化管道设计评价技术，将一类地区管道设计系数提高到0.8。基本形成油气管道完整性技术和管理体系。研究成果在20多条在役管道和西气东输、西气东输二线等重大工程中得到应用。

（3）建立了国际先进水平的2500t油井管复合载荷试验系统、轴向+外压复合载荷挤毁试验系统、钻柱构件旋转弯曲疲劳试验系统、实体膨胀管全尺寸实物评价系统等油气井管柱完整性评价试验平台，初步建立非API油井管标准体系，发展了油气井管柱完整性和适用性评价技术。研究形成基于应变的套管柱设计方法、管材选用和适用性评价技术。研究确定了三超气井套管柱失效模式与失效概率计算方法。研究建立了“三超”气井套管密封可靠性设计的极限状态方程、计算程序及判据，系统研究揭示了套管螺纹结构尺寸、材料性能、工作应力等对螺纹密封抗力的影响规律。提出油气井管柱完整性管理流程，形成《油气井管柱完整性管理》等石油天然气行业标准。在塔里木、新疆油气田得到应用。

（4）形成功能强大的油套管实物拉伸+腐蚀试验系统、管材多相流腐蚀试验系统等石油管腐蚀模拟实验平台，系统研究获得了“三超”气井油管腐蚀失效特征及影响因素，揭示了失效规律及机理。研究建立了基于气井全寿命周期的油管选材与评价方法，形成了一套基于井筒全寿命周期完整性的腐蚀选材评价技术。自主研发出多体系超级13Cr酸化缓蚀剂。研究揭示了油气田集输管线的失效机理和原因，初步建立了非金属和复合管材标准体系，形成集输管线选材及评价技术。研究成果为塔里木等油气田安全生产提供了重要技术支撑。

与此同时，国内在石油管材及装备材料服役行为与结构安全人才队伍建设、国内外合作研究、学术技术交流等方面也取得了多项重要进展。为进一步发展石油管材及装备材料服役安全理论和技术奠定了良好的基础。

3 石油管材及装备材料服役行为与结构安全面临的形势和发展目标

石油管材及装备是石油天然气勘探开发、油气储运和炼油化工的重要支撑。随着国内外能源需求的持续攀升，油气资源开发环境日趋恶劣，复杂深层、海洋、极地等各种非常规油气资源动用率日益提高，对石油管材及装备的服役性能和安全保障提出了更高的要求。为此，石油管材及装备材料服役安全学科的发展，要求关联学科并行发展的同时，更加注重多学科的交叉融合、耦合交互，以高性能化、高安全可靠性、长寿命和低成本为导向，突破重大基础理论和关键技术，为石油天然气高效勘探开发、油气储运和炼油化工的顺利进行和安全运行提供强有力的技术支撑与保障。

随着我国"一带一路"战略的实施，能源开发与合作成为重要内容，这为石油管材及装备材料服役安全学科发展提供了空前的发展机遇；而面对国内经济下行、国际油价持续低迷、油气需求不旺等新常态，石油管材及装备材料服役安全学科也面临着诸多新的挑战，同时也肩负着更加重大的责任。"石油管材及装备材料服役行为与结构安全国家重点实验室"的启动建设和运行，为发展石油管材及装备材料服役安全学科提供了重要平台，重点实验室将围绕学科发展前沿和经济社会发展的战略目标，聚集和培养科技人才，研究解决石油管材及装备材料服役行为与结构安全领域重大科学问题和行业关键技术、共性技术，实现高水平、原创性科技成果的突破，提升实验室在国内外的学术地位，增强科技创新能力与竞争实力，为科技进步与经济社会持续发展提供知识、人才储备和技术支撑。

实验室将在已有工作的基础上，逐步发展完善"石油管工程"新学科，建立石油管材及装备材料服役行为与结构安全理论和技术体系。针对油井管与管柱失效预防、输送管与管线安全评价、腐蚀与防护、先进材料及应用技术四大研究方向努力攻关，发展高钢级管道服役安全与失效控制理论，建立适合我国国情的高钢级大口径油气输送管材技术和标准体系；形成复杂工况油气井管柱失效控制及完整性技术，为重点油气田勘探开发提供技术支撑；形成"三超"及严酷腐蚀环境管材腐蚀机理及综合防治技术、炼化管道及设备腐蚀机理与评价技术；解决先进石油管材及装备材料应用技术难题，建立新型管材及装备材料测试与评价核心技术体系。

通过大力推进原始创新，力争 5 到 10 年内整体技术达到国际先进水平，将实验室建成石油管材及装备材料服役安全科学研究基地、高端人才培养基地、国际合作与学术交流基地、石油工程材料服役安全技术支持中心。

同时，要进一步加强石油管材及装备材料服役行为与结构安全国家重点实验室软硬件条件平台建设，在发挥好实验室现有标志性设备作用的同时，逐步配套完善现有实验研究平台设备，提升实验室的研究能力；进一步加强实验室人才队伍建设，吸引和聚集本领域国内外人才，造就一支专业结构合理、老中青结合的高水平创新团队；进一步加强开放交流与产学研合作，加强国际学术技术交流和合作研究，促进实验室应用基础研究与技术应用的有机融合，进一步提升实验室的影响力。

4 石油管材及装备材料服役行为与结构安全的研究领域和方向

"石油管材及装备材料服役行为与结构安全"研究，是从石油管材及装备的服役条件出发，重点在石油管材及装备的力学行为、环境行为、先进材料及其成分/结构—合成/加工—性质与服役性能的关系、石油管材及装备失效控制与预测预防等领域开展研究，揭示石油管材及装备失效的机理和规律，提出失效控制方法，发展全寿命周期的安全可靠性和完整性技术，确保石油管材及装备的长期服役安全。具体研究方向和内容包括：

（1）输送管与管线安全评价技术研究，主要针对我国大口径、长距离、高压力、高钢级管道和储运设施建设和长期安全运行需求，重点研究高钢级管材关键技术指标及表征评价方法、高压大口径天然气管道断裂控制技术、高强度管道变形控制技术、油气管道失效机理与失效控制理论、油气管道安全风险评价技术、油气储运设施完整性技术，建立油气输送管道失效控制理论，形成完整性技术体系。

（2）油井管与管柱失效预防技术研究，主要围绕特殊结构和特殊工艺井，超深、超高温、超高压、超长水平井及高压大排量分段压裂改造对油气井管柱服役性能、安全可靠性、质量和寿命提出的新的更高的要求，重点研究油井管的失效机理和规律、失效预测预防技术、管柱优化设计及适用性评价技术、油气井管柱结构完整性和密封完整性、高性能油井管关键技术指标及表征评价方法，建立油气井管柱失效控制理论，形成完整性技术和标准体系。

（3）石油管材及装备腐蚀与防护技术研究，主要针对高温、高压、严酷腐蚀介质油气田石油管材及装备存在的严重腐蚀问题，重点研究三超及严酷腐蚀环境油套管腐蚀机理及防治技术、高酸性气田管材腐蚀机理及采集系统腐蚀综合防治技术、炼化管道及容器腐蚀行为与评价方法、石油管材及装备腐蚀检测、监测、预测和预防技术，揭示失效机理和规律，提出有效预防措施，确保管材及装备安全。

（4）先进材料及应用技术研究，主要针对复杂工况和特殊服役环境用石油管材及装备，重点研究高性能管线钢管成分/组织/性能/工艺相关性，高强高韧、高抗挤、高抗扭、耐腐蚀、长寿命油井管成分/组织/性能/工艺相关性，高性能非金属管材、复合管材成分/组织/性能/工艺相关性，新型石油装备材料及成分/组织/性能/工艺相关性，发展先进和特殊专用材料应用关键技术，保障其服役安全。

总之，要通过大力推进原始创新，形成比较完善的油气输送管道失效控制技术、高性能管线钢和钢管应用关键技术、油气管道完整性技术、新型钻柱构件材料及安全可靠性技术、非 API 油套管应用及管柱完整性技术、“三超”气井油套管柱腐蚀防护技术、油气田地面管道失效控制及预防技术、新型石油装备材料应用关键技术，形成一批国家标准、发明专利、系列高新技术和产品、高水平学术论文和专著等载体化有形化成果，并加大成果转化推广力度，创造良好经济社会效益。

5 结语

经过多年努力，我国石油管材及装备材料服役行为与结构安全研究已经取得了若干重要进展，基本形成石油管材及装备材料服役行为与结构安全软硬件平台及核心技术体系，在高压输气管道断裂与变形控制、在役管道安全风险评价、油气井管柱完整性和适用性评价、石油管材及设备腐蚀与防护、高性能管材应用关键技术等方面取得了重要突破。鉴于石油管材及装备在石油天然气工业中的重要地位和作用，以及石油管材及装备材料服役行为对结构安全的重要影响，面对我国油气工业面临的严峻形势和对安全生产提出的新挑战，拟进一步加强“石油管材及装备材料服役行为与结构安全国家重点实验室”建设，围绕油井管与管柱失效预防、输送管与管线安全评价、腐蚀与防护、先进材料及应用技术四大研究方向努力攻关，建立石油管材及装备材料服役行为与结构安全理论和技术体系，突破相关核心技术，有力支撑石油天然气工业发展。

参 考 文 献

[1] 李鹤林，张冠军，杜伟．“石油管工程”的内涵及主要研究领域［J］. 石油管材与仪器，2015，1（1）：1-4.

[2] 李鹤林．失效分析与安全生产——“材料服役安全工程学”的建立与实践［J］. 西安石油大学学报，2011，26（1）：1-6.

[3] 冯耀荣，李鹤林，张国正，等. 几起重大装备和器材失效事故的分析及建议［C］//中国机械工业学会. 2006年全国失效分析与安全生产高级研讨会论文集，2006.

[4] 冯耀荣. 石油管工程技术进展及发展展望［R］. 石油管工程技术研究院30年院庆大会学术报告，2011.

[5] 于维华. 某管线试压爆破原因分析［M］//李鹤林，冯耀荣，李平全，等. 石油管材与装备失效分析案例集（一）. 北京：石油工业出版社，2006：409-420.

[6] Milne I.，Ainsworth R. A.，Dowling A. R. and Stewart A. T.：Assessment of the Integrity of Structures Containing Defects［R］，Central Electricity Generating Board Report R/H/R6-Rev. 3，May 1986.

[7] 冯耀荣，张平生，李鹤林. 含缺陷油气管道的完整性与适用性评价［J］. 焊管，1998，21（3）：3-8.

[8] 冯耀荣，陈浩，张劲军，等. 中国石油油气管道技术发展展望［J］. 油气储运，2008，27（3）：1-8.

[9] 冯耀荣，霍春勇，吉玲康，等. 我国管线钢和钢管研究应用新进展及发展展望［J］. 石油管工程，2013，19（6）：1-5.

[10] 冯耀荣，韩礼红，张福祥，等. 油气井管柱完整性技术研究进展与展望［J］. 天然气工业，2014，34（11）：71-81.

[11] 冯耀荣．石油管工程试验平台建设与关键技术创新［C］//中国机械工业学会. 2015年全国失效分析学术会议论文集，2015.

稠油蒸汽吞吐热采井套管柱应变设计方法

韩礼红　谢　斌　王　航　王建军　田志华

（中国石油集团石油管工程技术研究院）

摘　要：本文针对中国新疆油田稠油蒸汽吞吐热采井套管损伤进行了调研分析，对现有管柱设计方法和选材技术进行评估后认为，现有的应力设计方法主要针对钻完井阶段，而套损主要是由于生产阶段热采工艺形成的，预防套损应该采用基于应变的设计方法，并以此为基础进行套管选材。研究表明，套管失效模式主要包括变形、缩颈、断裂、剪切、脱扣和泄漏，这些模式均表明套管是在经历一定的塑性变形后产生失效。前四种失效与材料的纵向塑性变形、应变强化、蠕变、应力松弛、包申格效应有密切关系，这些材料行为是套管应变设计的重要基础。剪切失效主要是由泥岩层吸水膨胀后造成的横向载荷引起，由于生产作业对管柱通径的要求，显著的横向变形是不允许的，预防剪切失效应该从提高螺纹密封性能，提高套管壁厚、刚度等方面进行。基于应变的设计方法主要针对管柱纵向变形进行，其目的是允许材料在安全范围内发生塑性变形，据此，本文以材料均匀延伸率为指标，以临界失效为依据提出了应变设计判据之一，并得到了现场试验的验证。在蒸汽吞吐循环作业中，套管材料实际上处于低周应变疲劳服役状态，据此，本文以油井设计寿命和热循环次数为依据，以材料循环塑性应变临界值为指标，提出了应变设计判据之二。同时，为避免螺纹连接区域应力集中和应变集中效应，提出了强度错配的管材技术指标体系，建立了热采套管选材技术标准，并针对热采工况提出了管柱适用性评价方法。

关键词：热采井；套管；损伤；应变设计

1　背景介绍

稠油是中国石油能源中的主要类型之一，以新疆、辽河、胜利三大油田为主要开采区域。稠油开采主要包括循环蒸汽吞吐、蒸汽驱、火驱等工艺方式，尤其以循环蒸汽吞吐工艺方式采用较多，其涉及的高温井数量远超其他方式，循环变温引发的套损也最显著，是预防套损的重点研究对象。近年来，国内稠油热采井的套损率保持在 20%～30%，局部区块更高，平均单井修井费用超过 100 万元，加上套损井产能下降，热采井套损给油田造成巨大的经济损失。新疆油田的稠油井普遍较浅，井深大多处于 600m 以内，热采井数量接近 2×10^4 口，并且以每年近 2000 口的数量持续增加，由于井筒数量庞大，套损引发的损失尤其显著。近几年，新疆油田已探明可开采稠油储量近 4×10^8t，是未来 20 年开采的重点油藏。开展优化套管柱设计研究，形成适用的套管选材技术，对于预防套损具有重要的经济意义。

2　现场调查

自 2000 年以来，新疆油田热采井套管设计采用两层套管，除表层套管外，技术套管同时作为生产套管，主要采用 ϕ177.80mm 套管，壁厚包括 8.05mm 和 9.17mm 两种，螺纹连接包括 API 偏梯形螺纹及生产厂家由此改进的密封螺纹。采用预应力方式固井。技术套管材质以非 API 的 90H 为主，另有部分非 API 的 80H 套管，均属于低碳 Cr-Mo 系钢种。钢种的标准方面，各制造企业均有自己的标准，没有石油行业内统一标准。对新疆油田的蒸汽吞吐热采井套损情况进行调研后发现，套管的损伤模式主要包括变形、缩颈、断裂、泄漏、脱扣和剪切错断。以 LJ 区为例，截至 2008 年底，已证实套管损坏 362 口井，其中，泄漏 84 口井，变形及断裂共 251 井次。图 1（a）是已修复的 206 口套损井统计结果，其中缩径 94 口，占 46%；断裂脱扣 19 口，占 9.0%；剪切错断 58 口井，占 28%；纵向变形 35 口井，占 17.0%。在 LD 区，从 2000 年至 2008 年期间约 600 口井经历六轮吞吐后，只有 230 口井可以正常生产，370 口井发生了套损。图 1（b）给出了套损模式对比情况，其中，泄漏 152 口井，占 41%；变形与缩颈 144 口，占 39%；剪切错断 74 口，占 20%。另外一个大型套损区是 BZ7 区，从 2000 年开始开发。到 2005 年底，全区投产井数 1732 口，到 2006 年 2 月底，已发现各类套损井 514 口，占投产井数的 30%。尤其是 2001 年底以前完钻井投产的 614 口井套损井数已达 452 口，套损率达到 74%。套损的主要形式是变形缩颈、错断、泄漏，如图 1（c）所示。套损与注汽轮次对比统计分析发现，前 6 轮是套损多发期，套损都是区域性的发生，如图 2 所示。

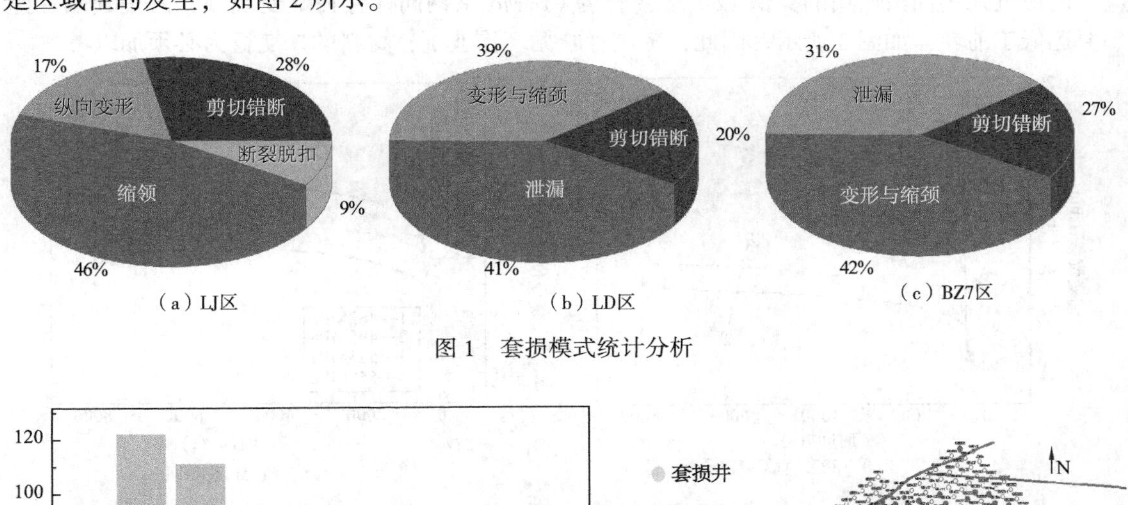

（a）LJ区　（b）LD区　（c）BZ7区

图 1　套损模式统计分析

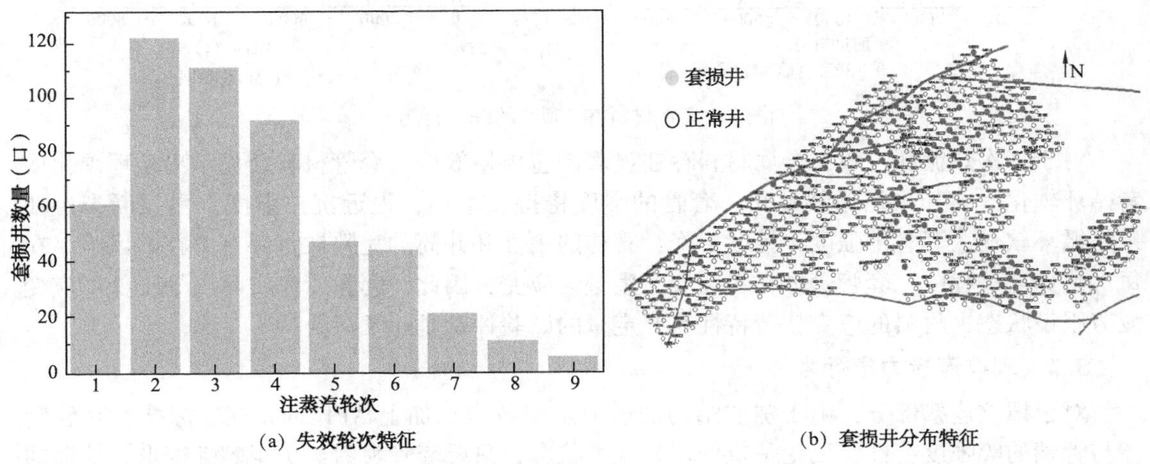

(a) 失效轮次特征　(b) 套损井分布特征

图 2　套损特征

3 套管服役行为及失效分析

3.1 热循环中的材料行为

蒸汽吞吐单次热循环包括注汽、闷井和采油三个阶段。在注汽过程中，井筒受热膨胀，套管的热膨胀系数远高于水泥环及地层，在胶结水泥环与地层的约束下，套管实际承受温度场变化带来的压缩载荷。当温度变化超过180℃左右时，套管管体材料将发生屈服，随后伴随着均匀变形产生形变强化或者软化[1-3]，当变形超出材料的均匀变形能力时，将产生失稳或屈曲。闷井属于持续性的高温，可以和注汽一起看作升温过程。在采油阶段，井筒温度持续下降，同样由于膨胀系数的差异，套管管体将承受拉伸作用，同样伴随着材料的短暂弹性变形、持久性的塑性变形。与注汽闷井阶段不同的是，拉伸状态下，材料超出其均匀变形范围时，将产生明显的缩颈，进而断裂。国内外学者对此力学行为也已经进行了广泛的研究，也取得一些积极的成果[4-6]。

由于持续高温作用，套管材料将显示出不同程度的蠕变现象。一般认为[7]，钢铁材料在超过30%熔点，大约450℃以上时才会有明显的蠕变现象，实际上，对油田现场已经使用多年的套管材料进行的试验结果表明，即使在350℃，普通的N80套管材料也显示出了明显的蠕变效应，而最近几年油田使用的类似L80-2套管的Cr-Mo系钢同样存在蠕变现象，只是其蠕变速率要远低于前者，如图3所示。因此，蒸汽吞吐循环作业下，材料的蠕变行为必须加以考虑。

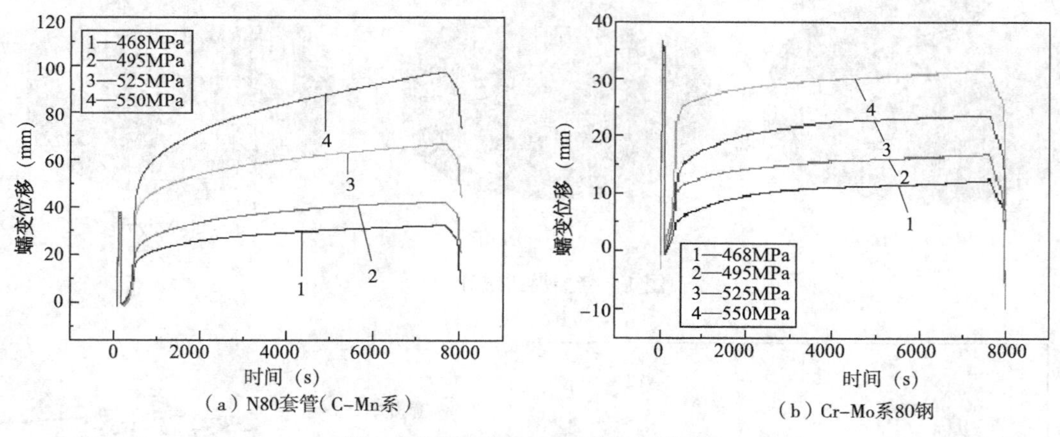

图3 套管材料在350℃的蠕变行为

对于多次热循环过程，金属材料存在显著的包申格效应，套管材料会显示出循环硬化或者循环软化[2]。如果是循环硬化，套管的强度将持续增加，逼近抗拉强度，引发断裂。如果是循环软化，套管的强度将持续下降，管柱的刚性和井筒的完整性将存在潜在的风险。在每次热循环过程中，套管管体材料均会产生塑性应变，因此，套管管体实际上处于低周应变疲劳服役状态，材料的应变疲劳特性需要定量的试验评价是非常必要的。

3.2 螺纹连接力学行为

对于螺纹连接部分，由于螺纹结构的应力集中效应，加上API Spec 5CT标准[8]套管管体与管端的等强度特性，无论是拉伸还是压缩状态，只要套管材料经历了塑性变形，伴随的应变强化效应都会持续性提高管端螺纹根部的应力集中，甚至是应变集中，最终会导致局部应力超过螺纹连接强度，导致管体断裂或脱扣。因此，当套管管体变形时，如何保持管端螺纹连接的安全性，尤其是避免应力集中和额外的塑性变形，甚至应变集中异常重要[2]。

3.3 失效分析

3.3.1 变形、缩颈及断裂

套管的变形是材料在纵向载荷作用下，超过了弹性范围而产生的永久塑性变形特征，当载荷继续增加时，材料将持续发生塑性变形，并在薄弱环节产生剧烈的应变集中，即缩颈现象。缩颈发生后，材料承载的应力将迅速增加，如果达到拉伸强度，将产生断裂。作业现场利用井下铅印方式充分验证了此种失效，如图 4（a，b，c）所示。

3.3.2 脱扣

油田现场的套管柱主要采用偏梯形螺纹连接，这种连接具有高的连接强度。因此，脱扣失效说明作业过程中，螺纹连接部位产生了显著的应力集中超出螺纹连接强度而造成失效。如图 4（d）所示，井下成像分析证实存在此类失效现象。

3.3.3 泄漏

泄漏主要与螺纹的密封性能有关。一般情况下，在低于 200℃ 范围内填充合格的螺纹脂，偏梯形螺纹连接可以具有一定的密封性能。然而，热采井井口注汽温度一般都在 270~350℃之间，目前尚未发现适用的螺纹脂产品可以保证螺纹密封。油田现场经常发生蒸汽泄漏，进而溢出地面的现象，如图 4（e）所示，因此，热采井套管产品需要采用具有气密封性能的螺纹连接。

3.3.4 剪切

如前文所述，剪切主要是地层横向运动诱发的。新疆油田的地质环境中都有不同深度的泥岩夹层，这些泥岩夹层在吸水后将发生膨胀现象，诱发不同地层界面产生显著的横向载荷，进而造成套管的剪切变形，如图 4（f）所示，国外也有类似的看法[3]。这种剪切变形轻者影响井下作业，重者造成井眼报废，因此，显著的剪切变形是不允许的，提高螺纹连接密封性能，进而阻止泥岩吸水膨胀是关键环节。

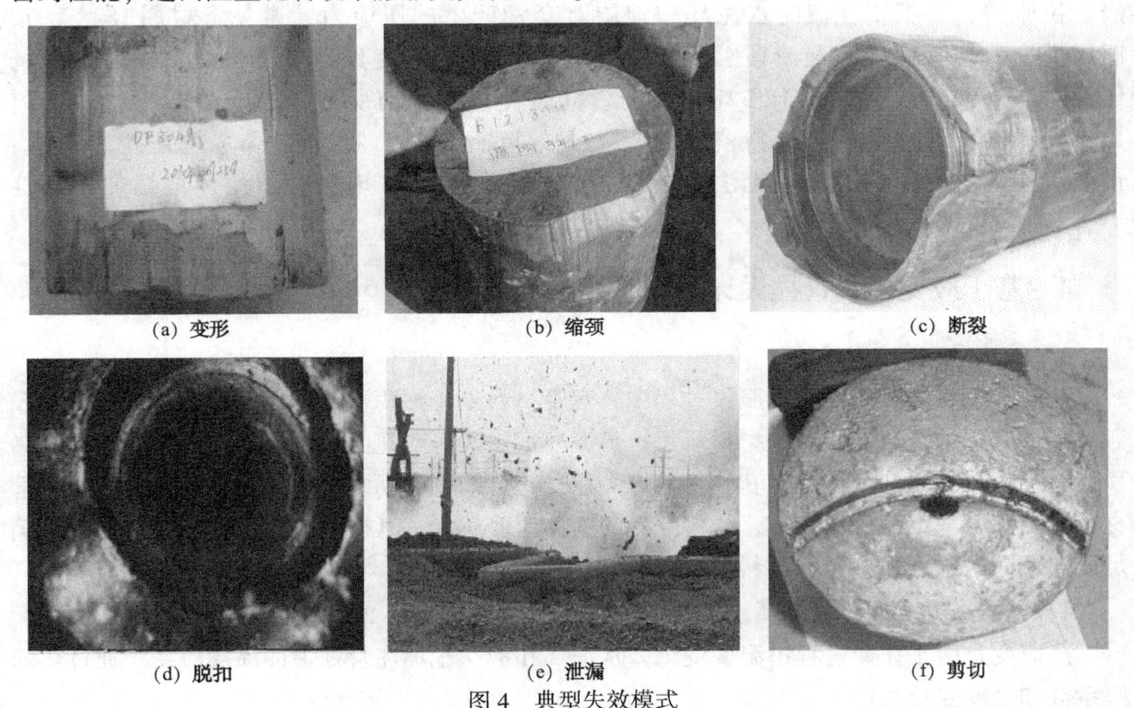

(a) 变形　(b) 缩颈　(c) 断裂

(d) 脱扣　(e) 泄漏　(f) 剪切

图 4　典型失效模式

3.4 提高套管强度的作用分析

有研究认为，热采井套管可以采用提高钢级的方法来避免塑性变形[9,10]。实际上，在接近300℃的温度变化环境下，作业所产生的热应力足以使金属材料发生塑性变形[4]。有学者认为，热采作业过程有伴生 H_2S 产生，提高套管钢级无疑提高了管材发生应力腐蚀开裂的风险[3]。由于 H_2S 应力腐蚀开裂对温度有很强的依赖性，在65℃以上环境很少发生[11]。即便除去此因素的影响，由于高温下金属材料具有应力松弛效应[12]，高的应力是无法保持的，图5中的应力松弛试验结果证明了这一点。

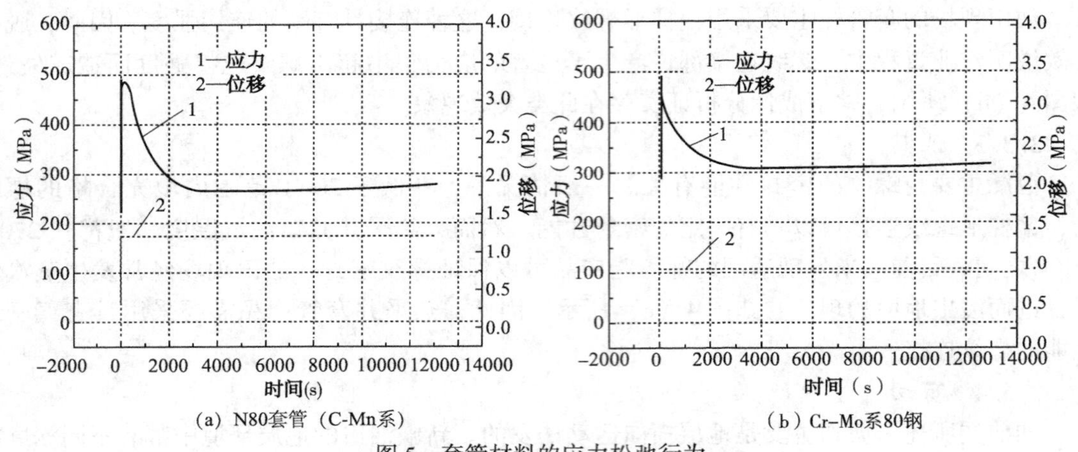

图5 套管材料的应力松弛行为

3.5 预应力固井分析

预应力固井是基于应力交互作用的一种理论模型，该模型认为，在固井时给套管柱施加一定拉应力的同时固井，以便套管—水泥环—地层胶结后，套管柱可以保留一定的残余拉应力，并抵消注汽时水泥环和地层对套管产生的压应力，避免管材屈服[13]。按照该模型，套管柱在采油阶段，所承受的叠加拉应力将使其更易于失效。在油田现场作业时，这种方法更倾向于和高钢级套管同时使用，如前所述，由于高温下材料的应力松弛特性，管柱是无法保持高应力状态的。除此之外，稠油储层往往是砂岩地层，预应力技术所用的地锚往往无法实现和井底的有效结合，现场对于使用地锚的有效性一直持有不同的观点[14,15]。

4 基于应变的套管柱设计

4.1 热采作业中的应变

循环蒸汽吞吐热采井套管服役过程中，套管材料处于累积性的塑性变形和应变疲劳服役条件下。累积性的塑性变形是指材料每次热循环都需经历明显的塑性变形，如果塑性应变累积量超过材料的均匀延伸率，套管材料将失稳，趋于缩颈和断裂失效。在设计寿命内，套管材料需要满足一定的循环次数，材料承受的塑性应变越大，其循环寿命就越短，因此，在满足设计寿命的条件下，套管材料存在一个临界值。服役中的套管主要承受以下几种应变：

4.1.1 热应变 ε_t

热应变是由于作业过程中温度变化，水泥环和地层施加在管柱上的纵向应变，通过热膨胀系数和温度变化来计算。

$$\varepsilon_t = \alpha \cdot \Delta T \tag{1}$$

式中　α——热膨胀系数；

　　ΔT——温度差。

热应变是伴随注汽闷井及采油全过程的应变，随着循环注汽而循环，对材料的应变疲劳失效和均匀变形失效均起主要参数作用，属于设计应变。

4.1.2　蠕变应变 ε_c

蠕变应变是由于套管材料高温下的蠕变速率（$\dot{\varepsilon}$）和持续时间（t）决定。一般注汽闷井周期较短，而采油阶段周期较长，两者产生的蠕变分别为压缩和拉伸变形，一般以较长的采油周期来计算，而蠕变速率需要依据标准 GB/T 2039[16]，在恒温但不同级别应力下，对材料稳态蠕变试验曲线进行数值回归方法获得。

$$\varepsilon_c = \dot{\varepsilon} \cdot t \tag{2}$$

蠕变应变同样随着作业中的热循环而循环产生，对材料的应变疲劳失效和均匀变形失效均起主要参数作用，属于设计应变。

4.1.3　弯曲应变 ε_b

弯曲应变是由于井筒轨迹的狗腿度造成的，受钻井井筒质量影响。在中国，油田钻井时井眼轨迹狗腿度规定在 12°/30m，一般测试结果为（6°～8°）/30m。弯曲应变可以依据 API Spec 5C5 提供的公式计算[17]。

弯曲应变在钻井后即得到确定，是永久性应变，对材料的应变疲劳失效和均匀变形失效均起主要参数作用，属于设计应变。

4.1.4　土壤应变 ε_s

土壤应变是由于稠油开采过程中，储层石油及砂砾排出地面后，引发上覆岩层的压实作用产生的。土壤应变需要借助于数值分析手段，考虑作业周期及地层变化综合计算。土壤应变是随着开采作业逐步累积，对材料的应变疲劳失效和均匀变形失效均起主要参数作用，属于设计应变。

4.1.5　屈曲应变 ε_f

屈曲应变是由于局部水泥环破碎或地层出砂掏空后，管柱失去了水泥环和地层的支撑作用，在纵向压缩载荷作用下管柱屈曲失稳对应的应变量。屈曲应变可以通过理论力学、数值分析或模拟试验来确定，表征管柱失稳失效时的临界应变，对材料均匀变形失效起主要作用，属于设计应变。

4.1.6　剪切应变 ε_{sh}

剪切应变是指由于地层运动或泥岩吸水膨胀诱发的横向载荷作用在套管柱而产生的永久应变。在油田作业中，由于井下作业的需要，管柱需要保持一定的通径要求，因此，明显的剪切变形是不允许的。预防剪切需要从螺纹密封、提高套管局部钢级、壁厚等方面控制。剪切应变属于偶发性事件，不纳入本文应变设计范畴。

4.1.7　均匀延伸率 δ

均匀延伸率属于套管材料的属性，是材料进行均匀拉压塑性变形的极限承载参数，属于许用应变。

4.1.8　应变疲劳极限 ε_x

应变疲劳极限是指套管材料在循环拉压载荷作用下，经历一定循环寿命相对应的临界应变值。循环寿命越长，临界值越低，可依据油井设计寿命通过试验获得。应变疲劳极限属于材料的属性，属于许用应变。

4.2 应变设计安全准则

依据套管材料服役中累积塑性变形不会引起失稳和应变疲劳服役安全，本文提出热采井套管柱基于应变设计的准则，包括：

4.2.1 均匀变形安全准则

套管材料均匀变形过程中，设计应变包括热应变、蠕变应变、弯曲应变、土壤应变及屈曲应变，许用应变为材料的均匀延伸率：

$$\varepsilon_d=\varepsilon_t+\varepsilon_c+\varepsilon_b+\varepsilon_s+\varepsilon_f\leqslant\varepsilon_a=\delta/F \tag{3}$$

4.2.2 应变疲劳安全准则

$$\varepsilon_d=\varepsilon_t+\varepsilon_c+\varepsilon_b+\varepsilon_s\leqslant\varepsilon_a=\varepsilon_x/F \tag{4}$$

式中 ε_d——设计应变；

ε_a——许用应变；

F——安全系数。

4.3 螺纹连接部位设计

基于应变的热采井套管柱设计是针对管体纵向变形进行，不包括横向剪切、挤毁载荷，也不适用于螺纹连接部位。螺纹连接需要具有气密封性能，而气密封性能通过金属材料的过盈配合实现，密封面不允许产生过高的应力和应变集中。同样，螺纹根部的应力集中对螺纹连接的完整性具有重要影响，也不允许过度提高[1,3]。因此，管体在发挥均匀塑性变形能力的时候，不应该显著影响螺纹连接部分的应力分布特征。有研究者试图通过确定螺纹连接的极限应变来建立失效准则[1,18,19]，然而，在当前的 API Spec 5CT 标准规定下，套管管体和管端具有同样的性能指标，同样经历塑性变形，对螺纹连接安全影响显著，因此，套管需要实现管体和管端的强度错配（例如，管端与接箍等强度）或者采用管端加厚处理，以保证管体在整个均匀塑性变形范围内管端及螺纹连接部分都可以保持在屈服强度范围内，保持强度及密封完整性。图 6 给出了该设计的基本原理。

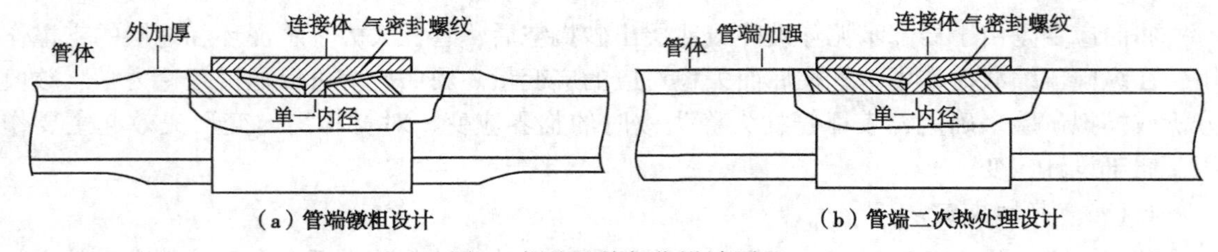

图 6 螺纹连接部位设计原理

5 现场试验案例介绍

本文针对新疆油田的蒸气吞吐热采作业工况，对工业领域的几种管材进行了试验评价，对通过模拟评价的一种套管进行了现场试验，共计 8 口井。迄今为止，试验井已经经历了 5 年，完成 14 轮注汽作业，管柱服役安全，验证了本文提出的设计方法是可行的。试验井用套管柱主要信息如下：

5.1 材料设计与评价

套管材料采用 Cr-Mo 系耐热钢，并含有微合金化元素。材料的热应变采用了 Xie 关于膨胀系数的研究结果[4]。按照最高350℃注汽，降至室温25℃时进入下一循环计算，依据式(1)可得热应变为0.436%。

依据GB/T 2039，在350℃恒温下，采用四个应力水平（图3）测试材料稳态蠕变速率后，对应力-稳态蠕变速率关系按照高斯函数关系进行拟合，确定函数关系式中的常数数值，得到材料的蠕变速率本构关系，见式（5）。

$$\dot{\varepsilon}=(7.40\times10^{-9}\times e^{\frac{\sigma}{91.8}}+1.95\times10^{-6}) \tag{5}$$

式中 $\dot{\varepsilon}$——蠕变速率；

σ——试验应力，须高于热应力。

对实物套管在轴向零位移约束条件下，施加350℃热循环，获得热应力为448MPa。式(5)中的应力采用500MPa保守计算。

依据油田现场每年进行3轮注汽的实际作业条件，考虑管柱服役寿命为10年，共注汽30轮，每轮持续4个月时间，其中，蠕变持续时间为每轮高温注汽闷井3天，共累积90天，应力采用保守的500MPa上限计算，按照最高350℃下的蠕变速率本构方程，即式(5)，可以获得套管材料的蠕变应变为0.285%。

按照现场钻井时，狗腿度上限12°/30m，依据API Spec 5C5标准可计算狗腿度引发的弯曲应变为0.057%。

随着稠油开采，储层砂砾被流体携带返回地面，上覆岩层缓慢压实产生的土壤应变可依据数值分析方法预测。新疆油田循环蒸汽热采环境与加拿大稠油开发环境类似，根据Xie研究预测结果，在十年生产周期后，土壤压实带来的应变约为0.25%[4]。Xie通过现场数据分析后认为，套管柱产生一次屈曲的临界应变为1%~2.5%[20]，因此，该结果可以借鉴用于本文试验井计算方面。

根据式（3）均匀变形安全准则，按照10年设计寿命，可以计算试验井的套管柱设计应变为3.528%。取设计安全系数为2.0，则套管材料的许用应变，即均匀延伸率应不低于7%。本文试验井用套管材料在室温及280℃注汽温度下，均匀延伸率试验值均满足此项要求。材料拉伸试验结果见表1。

表1 材料拉伸试验结果

试样			抗拉强度（MPa）	屈服强度（MPa）	伸长率（%）	均匀延伸率（%）	屈强比
温度	方位	宽度/直径×标距（mm×mm）					
室温	管体纵向	25.4×50	718	625	30.5	9.6	0.870
			718	634	29.0	9.5	0.883
			712	627	29.0	9.1	0.881
		均值	716	628.7	29.5	9.4	0.878
		协议要求	≥655	552~655	—	≥8.0	≤0.90
	接箍纵向	ϕ12.5×50	821	727	22.0	7.8	0.885
			811	717	23.0	7.8	0.884
			822	728	21.5	8.0	0.885
		均值	818	724	22.2	7.87	0.885
		协议要求	—	655~758	—	≥5.0	—

续表

试样			抗拉强度（MPa）	屈服强度（MPa）	伸长率（%）	均匀延伸率（%）	屈强比
温度	方位	宽度/直径×标距（mm×mm）					
280℃	管体纵向	φ5×25	699	556	19.5	6.9	0.795
			689	536	22.0	7.8	0.778
		均值	694	546	20.8	7.35	0.786
	接箍纵向	φ5×25	752	591	21.0	8.9	0.786
			746	604	21.0	7.0	0.809
		均值	749	597.5	21.0	7.95	0.798

根据式（4）应变疲劳准则，套管柱的设计应变为1.028%，作为套管材料的应变疲劳临界值，材料应该保证40次蒸汽循环安全（10年设计寿命）。拉-拉应变疲劳测试结果表明，材料在0.8%塑性应变下可以实现40次循环，如图7所示，这一结果和实际工况是不符合的，后者需要进行拉-压应变疲劳或者轴向零位移条件下的热循环试验值，涉及材料的应变强化、包申格效应、蠕变效应以及应变疲劳寿命评价方法[21]，相关的基础研究还需要开展大量的工作。

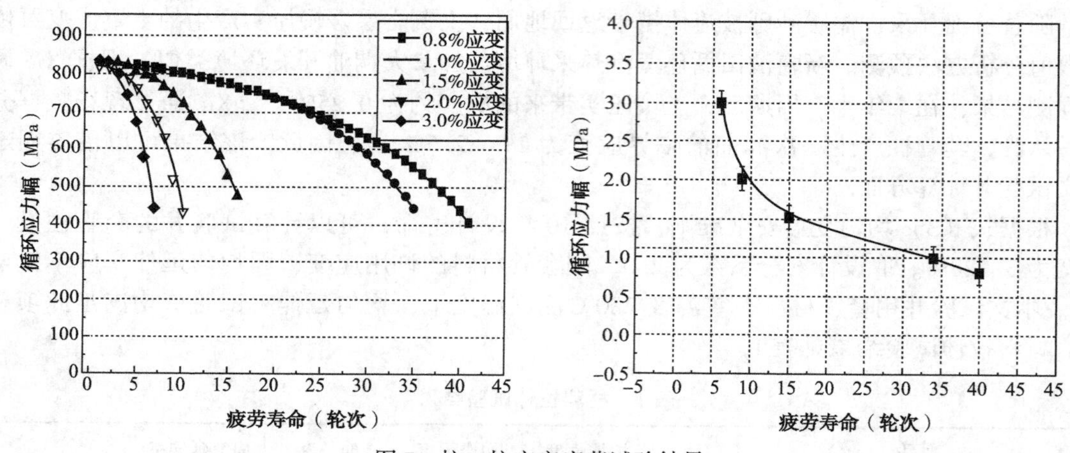

图7　拉—拉应变疲劳试验结果

5.2　适用性评价方法

除套管材料性能及尺寸容差要求之外，热采井套管柱还需要针对蒸汽吞吐工况制定系统的实物性能评价方法，具体包括：

5.2.1　抗黏扣

套管螺纹须保证三上两卸试验中不发生黏扣现象，以保证在现场作业中螺纹保持完整性。本项试验按照一般ISO 13679标准执行。图8（a，b）是本文试验井用套管经历上卸扣试验后的螺纹形貌。

5.2.2　抗内压

套管管体及螺纹连接除满足ISO 13679标准要求之外，还须满足现场注汽压力及井下作业载荷如压裂工况需要。本项技术要求及具体参数由最终用户确定。图8（c）是套管螺纹连接试验段通过抗内压试验后的形貌。

5.2.3　抗外挤

套管管体除满足 ISO 13679 标准外挤性能要求之外，还需要满足最终用户依据现场工况提出的技术要求。图 8（d）是套管管体通过抗外挤试验后的形貌。

5.2.4　拉伸强度

在拉伸强度方面，管体须满足与名义钢级相对应的实物拉伸强度要求，螺纹连接实物还应保证缩颈及断裂发生在管体，而不是螺纹连接处，如图 8（e）是套管柱螺纹连接试样通过实物拉伸试验后的形貌。本项要求是基于应变的设计方法对螺纹连接设计的直接要求。如果断裂发生在螺纹连接部位，说明管体、管端及螺纹的强度错配设计指标还需继续优化。

5.2.5　热循环

热循环是蒸汽吞吐热采井最主要的评价方式，涉及直井、定向井的工况模拟。试验须模拟现场轴向约束环境，即轴向位移为零的条件。施加的恒定试验载荷包括内压和弯曲载荷。热循环从常温至最高注汽温度之间，循环至少 10 次，每次峰值温度需保持恒温至少 5min。在试验循环期间，管柱须保持强度及密封性能的完整性。

由于温度循环的周期较长，热循环加载以前，可以先进行循环拉-压载荷试验进行初步判断。拉-压载荷须高于温度循环引发的热应力。由于拉-压循环试验中，材料不能表现出高温行为，因此，循环拉-压试验不能代替热循环试验。图 8（f）是本文试验井用套管柱在进行工况内压及极限弯曲条件下热循环试验。

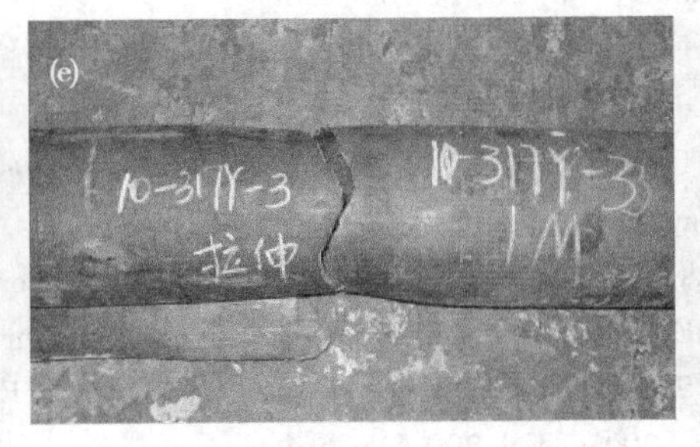

图 8　试验井用套管柱模拟试验

6　结论

稠油蒸汽吞吐热采井套管损伤主要是套管柱在水泥环与地层的轴向约束下，发生塑性变形后引发的失效，套管柱需要采用基于应变的设计方法，主要用来控制材料轴向变形的安全性。

热循环中，套管材料需经历应变强化、蠕变、包申格效应及低周应变疲劳等行为，这些行为特征是建立材料均匀变形安全准则和低周应变疲劳安全准则的重要基础。

套管柱螺纹连接应该采用气密封性螺纹，抑制蒸汽泄漏造成泥岩膨胀，引发地层剪切套管失效。

套管两端推荐采用额外的外加厚或者二次强化处理手段，以保证螺纹连接部分的结构完整性。

热采套管的适用性评价除需进行系统的材料性能评价外，还应通过针对性的工况模拟试验评价。

致　谢

本文获得国家自然科学基金项目（编号：51574278）、中国石油天然气集团公司应用基础研究项目（2011A—4208）和新疆油田公司技术开发项目（GC-JF-2011-57）支持。

参 考 文 献

[1] XIE J, A Study of strain-Based Design Criteria for Thermal Well Casings [C], World Heavy Oil Cinference, Edmonton, Canada, March 2008: 2008-388.

[2] J Nowinma, T Kaiser and B Lepper, Strain-Based Design of Tubulars for Extreme-Service Wells [C], SPE/IADC Drilling Conference & Exhibition, February 2007, Amsterdam (SPE 105717).

[3] J Nowinka and D Dall' Acqua, New standard for Evaluating Casing Connections for Thermal Well Application, SPE/IADC Drilling Conference & Exhibition [C], March 2009, Amsterdam, Netherlands (SPE/IADC 119468).

[4] XIE J, Casing design and analysis for heavy oil wells [C], First World Heavy Oil Conference , Beijing, China, November 2006: 2006-415.

[5] IRP, 2002, Industry recommended Practices for Heavy Oil and Oil Sand Operations [S], Vol. 3.

[6] Trend Kaiser, Post-Yield Material Characterization for Strain-Based Design [J], SPE Journal, March 2009 (SPE 97730).

[7] Zheng X L, Mechanical Behaviors of Engineering Materials [M], Published by Northwestern Polytechnical University Press, China, Nov 2004: 168-180.

[8] API Spec 5CT Standard, Specification for Casing and Tubing [S], Published by America Petroleum Institute, July 2011, Ninth Edition.

[9] Wellhite G P and Dietrich W K, Design Criteria for completion of Steam Injection Wells [J], Journal of Petroleum Technology, January 1967: 15-21.

[10] Holliday G H, Calculation of Allowable Maximum Casing Temperature to Prevent Tension Failures in Thermal Wells [C], ASME Petroleum Mechanical Engineering Conference, Tulsa, Okla, Sept 1969.

[11] R New, Material Selection in the Piping Design for Wet H2S Environment [J], Corrosion & Protection in Petroleum Industry, 2003, Vol. 20 (6): 6-9.

[12] Lepper B, Production Casing Performance in a Thermal Field [C], Petroleum Society of CIM & AOSTRA, 1994: 94-107.

[13] Li Z F, Casing Cementing with Internal Pre-pressurization for Thermal Recovery wells [J], Jounal of Canadian Petroleum Technology, December 2008, Vol. 47 (No. 12).

[14] Zhou M S, Prevention Method and Application for Casing Damage in Unconsolidated sandstone Ultra Heavy Oil Recovery [J], Petroleum Geology and Engineering, 2006, 20 (6): 78-80.

[15] Liu Y X, Fu J T, Lu L H, etc, Analysis for Character and Cause of Casing Failure in Loose Sandstone Reservoir [J], Journal of Shandong Jianzhu University, 2010, 25 (3): 342-346.

[16] 金属拉伸蠕变试验方法：GBT 2039—1997 [S].

[17] ISO 13679 Standard, Petroleum and Natural Gas Industries - Procedures for Testing Casing and Tubing Connections [S], Published by International Organization for Standardization, 2002.

[18] Weiner P D, Wooley G R, Coyne P L and Christman S A, Casing Strain Tests of 13-3/8" N80 Buttess Connection [C], SPE 5598, 50th Annual Fall Meeting of SPE-AIME, Dallas, USA, Oct 1976.

[19] Goodman M A, Desighning Casing and Wellheads for Arctic Service [J], World Oil, 1978.

[20] Wagg B, XIE J, Solanki S and Arndt S, Evaluating Casing Deformation Mechanisms in Primary Heavy Oil Production [C], SPE International Thermal Operations and Heavy Oil Symposium, March 1999, Bakersfield, California, USA.

[21] S Y Hsu, K H searles, Y Liang, etc, Casing Integrity Study for Heavy-Oil Production in cold lake [C], SPE Annual Conference % Exhibition held in Florence, Italy, Sep 2010 (SPE Number 134329).

Strain Based Design and Field Application of Thermal Well Casing String for Cyclic Steam Stimulation Production

Han Lihong Wang Hang Wang Jianjun Zhu Lijuan

(Tubular Goods Research Institute of CNPC, State Key Laboratory of Performance and Structral Safety for Petroleum Tubular Goods and Equipment Material)

Abstract: This paper focuses on the strain based design and field application of thermal well casing for cyclic steam stimulation process. The heavy oil is a main field of China petroleum industry, where the cyclic steam stimulation process has been widely used and playing the dominating role in production. Generally the casing damage rate is 15%~30% under this process, which has resulted in a large economic cost. The high damage rate is mainly caused by the current strength design method, where both the room and elevated yield strength were considered not beyond anticipated field thermal stress. In fact the casing material will serve in cyclic thermal elastic-plastic deformation state and the single yield strength is not sufficient for long term thermal stimulation process.

This paper presented a new strain based design method for thermal casing string. New material parameters were proposed to build the safety evaluation principle throughout the service life. For thermal casing the materialplasticly deforming limit, called as allowable strain, is determined by the homogeneous deforming capacity. For design strain, the material creep rate equation is introduced to evaluate the accumulative elevated strain in long term service together with initial install strain, thermal strain and soil strain. When the design strain is not more than the allowable strain the casing material will safely serve throughout the whole life. In addition the strain fatigue limit is measured to evaluate the material safety for thermal cycle life. According to the factual field engineering environment the gas tight thread joint was adopted to prevent steam leak which otherwise will cause a large transversal stress between different formations and result in slip deformation of casing string.

Before the engineering operation the full scale test procedure was proposed and finished to evaluate the string integrity in multiple thermal cycles. From 2011, this method was used in Xinjiang oil field for eight wells for more than 14 steam cycles and all the experiment wells served well without any damage. Several wells were cyclic measured through multi arms inspector and the data showed that all the casing deformation located in controlled scope except that only one has obvious deformation because poorer cementing quality. The field practice has showed that the new method has better fitness for thermal wells.

Keywords: Thermal well casing; Cyclic steam stimulation; Strain based design; Thermal elastic-plastic; Safety evaluation

1 Introduction

The heavy oil is a main energy type in China petroleum industry, where the cyclic steam stimulation (CSS) process has been widely used and playing the dominating role in production. During exposure to the CSS environment, casing failure occurred regularly. Generally the casing damage rate is 15%~30% under this process, which has resulted in a large economic cost. For example, the additional cost for maintaining and repairing wellbore is about one million RMB dollars per year for each damaged well in Xinjiang oil field. In LJ area there were about 362 wells damaged until 2008, of which 84 wells were due to leakage failure, about 251 wells due to plastic deformation and fracture. Among 206 wells already repaired, 94 wells were in necking mode, 19 wells were in fracture and parting mode, 58 wells were in shear mode and 35 wells were in plastic deformation mode, as illustrated in figure 1a. In LD area there were about 600 wells in the period of production from 2000 to 2008. After six thermal cycles, only 230 wells could operate normally, 370 wells had been seriously damaged. Figure1b presents the failure statistic data, where 152 wells damaged in leakage mode, 144 wells in deformation and necking mode, and 74 wells in shear mode. Another more serious area locates in BZ7 area, which was developed from 2000. Until the end of 2005, about 30 percent of wells among 1732 had been damaged. From 2000 to 2001, there were 614 wells in the stage of production but about 452 wells had been damaged. The damage ratio reached to about 74 percent. The domestic failure modes consisted of deformation plus necking, shear, and leakage, as shown in figure 1c.

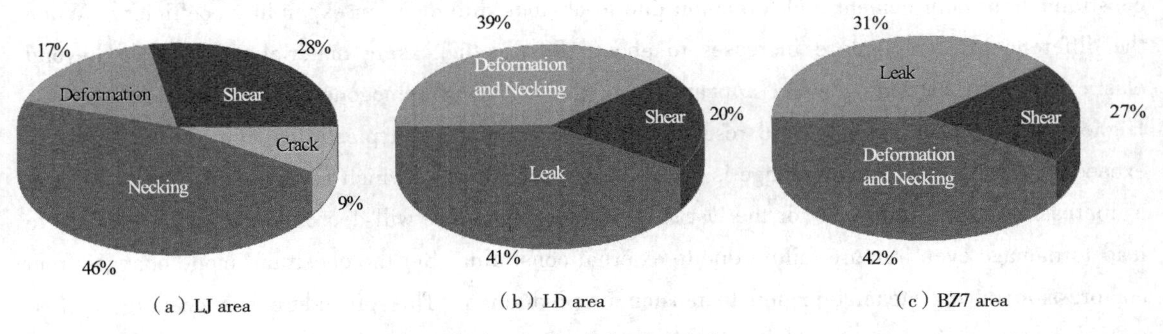

(a) LJ area (b) LD area (c) BZ7 area

Fig. 1 Casing failure statistic

Failure analysis displayed that there are several typical modes as showed in figure 2.

(1) Plastic deformation, necking, crack

Plastic deformation is associated with excessive plastic strain beyond elastic limit and will leads to necking and the following crack when it is beyond the material's homogeneous deforming ability.

(2) Connection parting

This mode is directly related with stress or strain concentration around the thread root.

(3) Shear and thread joint leakage

Shear failure is directly associated with formation movement, which is thought resulted from the steam leakage through thread joint because the current API thread joint has not sealing ability at elevated environment.

(a) Deformation (b) Necking (c) Crack

(d) Connection parting (e) Leakage (f) Shear

Fig. 2 Casing failure modes

CSS process consists of three stages: steam injection, soaking and production. During initial steam injection, the casing will expand and undertake compression resulting from the axial constraint from both cement and formation due to obvious difference in expanding coefficient. When the difference of temperature increases to about 180℃, the casing material would yield beyond elastic limit and hardening effect appear during the following homogeneous deforming scope[1-3]. Higher load or plastic strain would result in ultimate fracture when plastic capacity of material was exceeded. The second soaking stage is still in high temperature, which can be considered as a part of increasing temperature. As for the last stage, the temperature will decrease, which is possible to lead to damage even fracture failure due to external constraint. But the constraint mode changes from compression during steam injection to tension in production. This procedure has been researched widely in the similar opinions[4-6]. So the plastic deformation can be thought as the key failure resource. For service safety the casing string design must consider possible plastic deformation due to process and the material's capability. This means that the current strength design method cannot meet this special process and strain based design method should be applied in CSS production.

2 Field CSS well features

Since 2000, Xinjiang oilfield has adopted both straight and directional wells as the dominant types for CSS process, as shown in figure 3. The reservoir depth is about 500 meter and the dog leg including directional wells is not beyond 12 degree per 30 meters. The casing string contains surface casing about 60 meter depth and production casing until through the hole. The production casing adopts 7in diameter and 8. 05 or 9. 17mm wall thickness and 80ksi or 90ksi grade. The connection traditionally adopts API buttress thread and some gas tight joint in recent years. The CSS process contains 280℃ steam injection for 3 days and 3 days soak time, and then the production begins. In recent years 350℃ steam injection is also being tested to try to increase the production ability. For well completion the production casing will bring an earth anchor to be hoped to seize the earth for preventing bottom casing movement. When the well is completed the surface and production casings were welded together to prevent the axial displacement of production casing string. This paper adopts 80ksi grade as the applied casing with 7in diameter and 8. 05mm wall thickness. The aim steam injection temperature is 280℃ and the material creep tests were done at limit temperature, i. e 350℃.

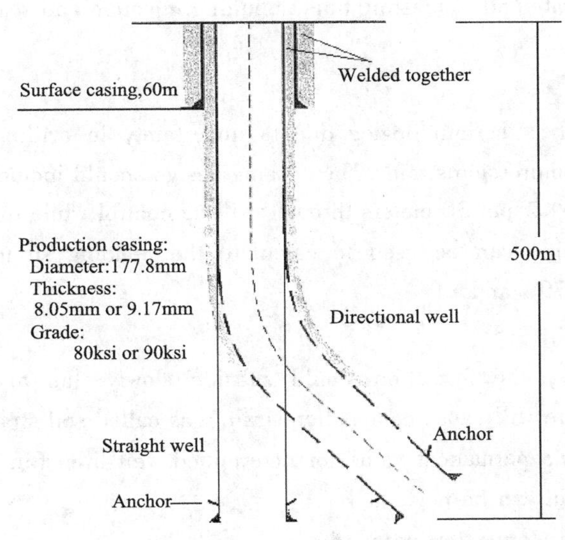

Fig. 3 Oilfield CSS well configure

3 Strain based design for production casing string

In CSS operation, the casing material will undertake fatigue loadingincluding plastic compression and tension periodically when the temperature varied beyond 180℃. For each tension or compression state, the casing material will fail when the cumulated strain exceeds its homogeneous plastic capacity. As a result, necking or fracture would occur. As for the fatigue loading, a critical strain of service cycle exists for the casing material, which must exceed the factual CSS cycle number. In CSS process there are several strains for served casing as following.

3.1 Thermal strain, ε_t

The thermal strain results from axial constraint of cement and formation due to the difference in thermal expanding coefficient. It can be calculated through formulas (1) . This strain belongs to required or design strain which must be satisfied for casing material because it plays an important role in both deformation and fatigue loading.

$$\varepsilon_t = \alpha \cdot \Delta T \tag{1}$$

Where, α is expanding coefficient and ΔT is temperature difference.

3.2 Creep strain, ε_c

Creep strain is caused by elevated environment, related with material's creep rate and lasting time, as described in formulas (2) . For CSS thermal well the recovery phase has longer period than that of both steam injection and soaking, which can be applied as a lasting time. When recovery process begins the hole-bottom temperature will rapidly decrease and the corresponding lasting time can be neglected in calculating creep strain. Creep rate is determined through experiment for specific material. Creep strain has a similar role for both deformation and fatigue loading.

$$\varepsilon_c = \dot{\varepsilon} \cdot t \tag{2}$$

Where, $\dot{\varepsilon}$ is creep rate and t is lasting time including injection and soaking time for well design life.

3.3 Bend strain, ε_b

Any well usually has a certain dogleg due to uncertainty in drilling engineering, inducing permanent tension and compression strain. Then casing design should include these strains. In china this value is limited up to 12°per 30 meters through drilling control while usually between 6°~8°per 30 meters. The limit value can be used to calculate the bending strain according to formulas described in the ISO 13679 standard[7].

3.4 Soil strain, ε_s

During the recovery, the formation would subside slowly due to oil recovery and sand production. This procedure will cause compaction strain, as called soil strain here. It is considered in strain-based design as a permanent value for determined well life. But when the oilfield adopts water reinjection this strain can be neglected.

3.5 Homogeneous elongation rate, δ

This parameter is one important property of casing material and can be measured by tension test. It represents the limit plastic capacity before the stress reaches ultimate strength. When external tensile stress exceeds this limit, necking and fracture failure occur immediately.

3.6 Evaluation criterion for steam injection stage under compressive load

For this criterion, the design or required strain contains thermal strain, creep strain, bending strain and soil strain. All of the cumulated strains described above are required not to exceed the plastic capacity of casing material, i. e. homogeneous elongation rate at either ambient or elevated environment. The equation (3) describes this relationship, where the "A" is the relation coefficient between structural and material strains.

$$\varepsilon_d = \varepsilon_t + \varepsilon_c + \varepsilon_b + \varepsilon_s \leqslant \varepsilon_a = \delta / A \tag{3}$$

3.7 Evaluation criterion for oil recovery stage under tensile load

For this criterion, the design or required strain contains thermal strain, creep strain, bending strain and soil strain. All of the cumulated strains described above are required not to exceed the plastic capacity of casing material, i. e. homogeneous elongation rate at either ambient or elevated environment. This principle proposed a necessary critical value by introducing a security factor, as shown in equation (4).

$$\varepsilon_d = \varepsilon_t + \varepsilon_c + \varepsilon_b - \varepsilon_s \leqslant \varepsilon_a = \delta / A \tag{4}$$

3.8 Evaluation criterion under cyclic tension and compression load

This principle proposed a necessary critical value as shown in equation (5). The n_d is the designed life and the n_a is the material' fatigue life under design strain, which could be determined by experimental method with respect to designed life of oil well, considering ten years operation with three injection-production cycles per year.

$$n_d \leqslant n_a \tag{5}$$

4 Casing fitness evaluation

The casing is a new developed material for strain based design and a simple gas tight joint was used together, as shown in table1 and figure 4. The ingot was heat rolled and experienced strict heat treatment process.

Table 1 Chemical composition of new casing material

Element	C	Si	Mn	P	S	Cr	Mo	Ni	Nb	V	Ti	Cu
Body	0.17	0.24	0.98	0.011	0.003	0.99	0.33	0.06	0.02	0.03	0.01	0.21
Couple	0.19	0.24	0.96	0.011	0.003	0.98	0.32	0.06	0.02	0.03	0.01	0.21

Fig. 4 New developed casing and gas tight joint

4.1 Casing deforming capability evaluation

The material's homogenous elongation rate is 9.6% and 7.8% at ambient and 280℃ respectably through tension tests. The strain relation coefficient between casing and material can be deduced to be 1.55 through full scale and material samples as shown in figure 5 test data. This means the casing design strain for 10 years production life is not beyond 6.2% and 5.0% at ambient and 280℃ respectably.

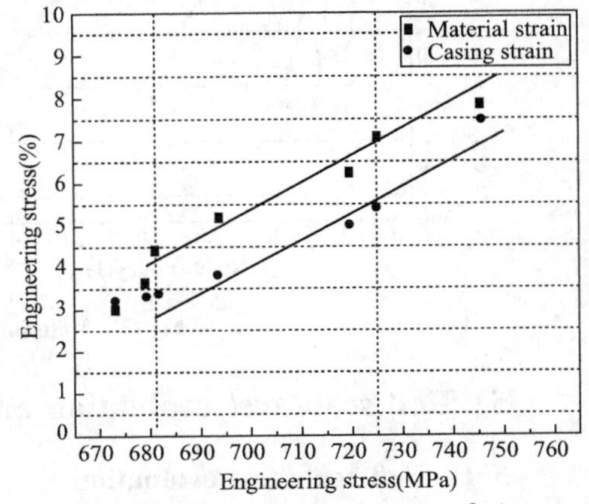

Fig. 5 Material and casing strains relation

Considering the limit steam injection environment, 350℃ the thermal strain for CSS production is 0.436

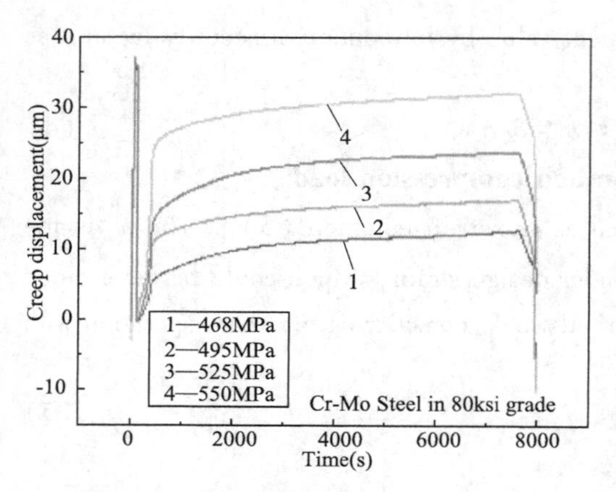

Fig. 6 Material creep behavior at different stress

according to formula (1) and the creep strain is 0. 285 through the material creep rate equation achieved through tests as shown in figure 6, where the steam injection and soaking time were considered as creep time according to field process 3 cycle per year and 10year life. Due to the temperature decreases continually and rapidly in recovery period and the material creep rate becomes lower and lower and so the tension creep strain adopts the same strain with that in steam injection and soaking period. According to ISO 13679[7] the bend strain can be calculated to be 0. 057 considering the most serious dog leg degree, 12 degree per 30 meters. There was a similar environment with Canada field in Xinjiang oil field and a soil strain about 0. 25% was adopted here for ten years well life according to related research results[4]. If the oilfield can continually inject water to balance the production layer pressure the soil strain can be neglected here. For different plastic strains the low cycle fatigue behavior was tested as shown in figure 7, which showed that the casing material has a critical strain about 1. 3 for 10 years' CSS design life.

So according to the evaluation equation (3), the accumulated compress strain is 1. 028 needed for multi cycle injection period during 10 years' production life. The material's homogenous elongation rate can better meet this requirement whatever is the ambient or 280℃ high temperature. Similarly the accumulated tension strain is 0. 528 according to the evaluation equation (4), and the material has a higher safety factor than steam injection. As to the fatigue strain the higher one of both is 1. 028 which is also lower the materials limit strain for 30 cycles CSS operation. According to the evaluation (5) the material can better meet the strain fatigue requirement for 10 year design life.

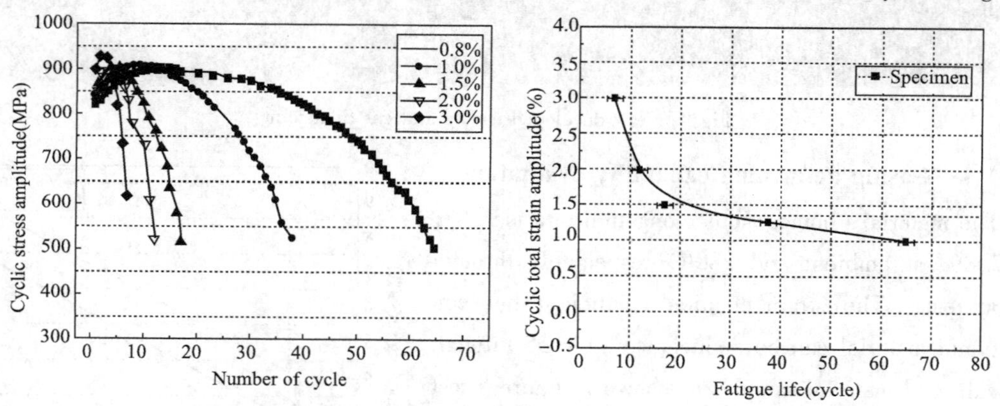

Fig. 7 Material low cycle strain fatigue behavior

5 Full scale test evaluation and field application

5. 1 Full scale test evaluation

Full scale tests were performed for the casing applied in thermal well, including galling, tension, collapse and sealing resistance and multi thermal cycle simulation. Figure 8 presented typical morphologies after full scale tests, which were conducted according to ISO 13679 Standards.

For strain-based design method, the fracture surface after tension test usually locate in pipe body instead of connection, which had been verified as showed in figure 8 (e) and proved the efficiency of this new design.

Thermal cycle tests were performed for 10 cycles from ambient to elevated environment with a holding time about 5 minutes at each peak temperature, as shown in figure 8 (c). In addition, directional wells were also simulated through transverse loading. The experimental results demonstrated that all tested casings display better performance with respect to the whole integrity.

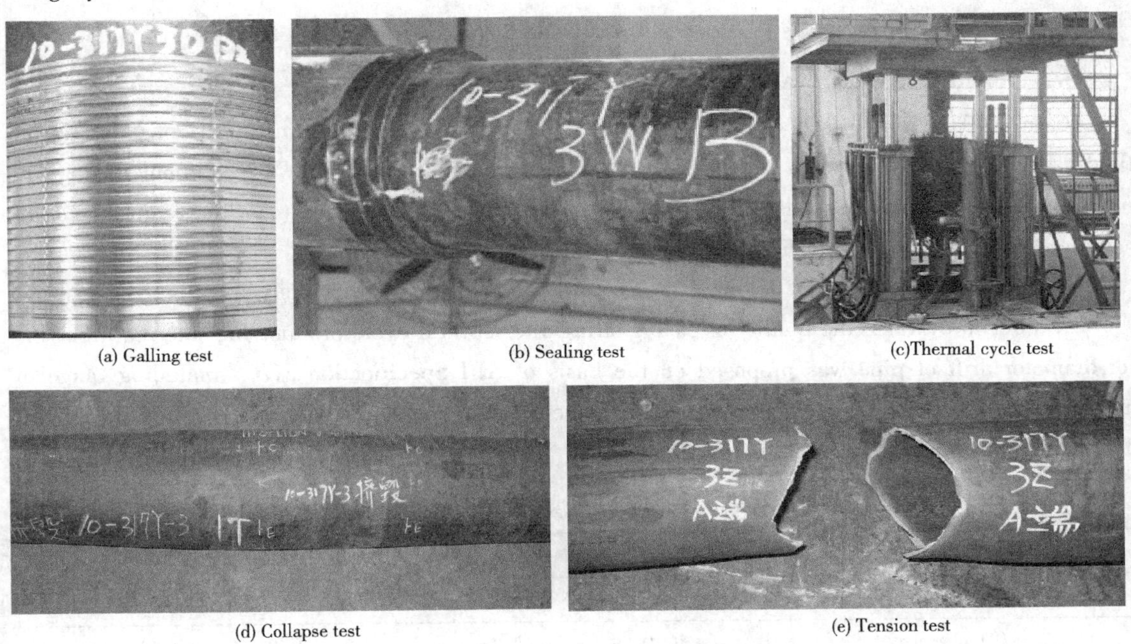

(a) Galling test (b) Sealing test (c)Thermal cycle test

(d) Collapse test (e) Tension test

Fig. 8 Full scale test evaluation

5.2 Analysis of well cementation without pre-tension

The theory of well cementation with pre-tensile is based on the constant property of material, which widely used by oil field company currently. It is considered that pre-tensile can counteract compression stress during cyclic steam stimulation, and then casing would be in the state of elastic stress[8]. It is believed that formation lithology is an important fact associated with the effect of well cementation with pre-tensile, as for sandsone, the anchor would be difficult to fix. In addition, the experimental result demonstrated that stress relaxation effect has direct influence on pre-tensile at temperature of steam injection. Figure 9 showed that the pre-tensile stress would be relaxed thoroughly. As a result, the pre-tensile stress is then no effect for casing string in thermal well. Therefore, there are still different opinions for this pre-tensile stress technology among oil field users now[9-11]. Based upon these experiment results the pre-tension cementation was removed from field tests.

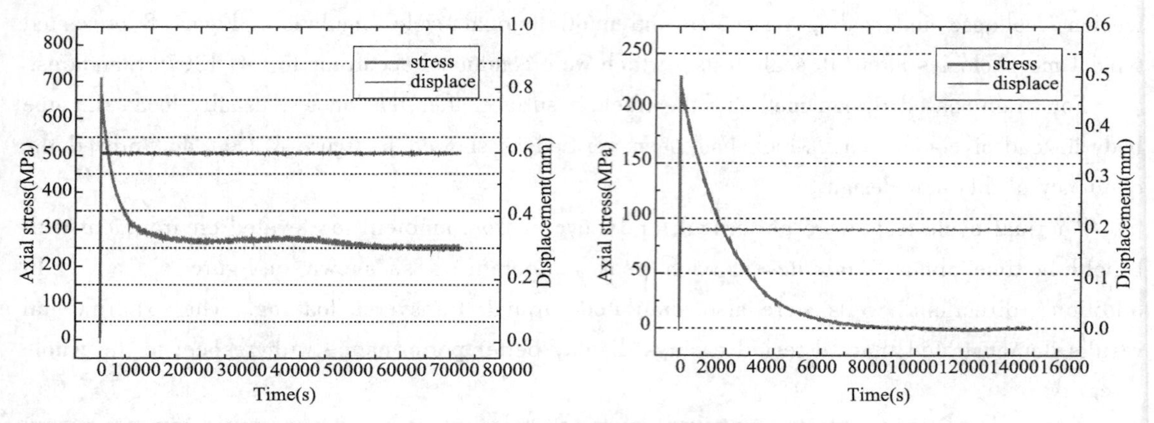

Fig. 9 Stress relaxation curves of thermal casing at temperature 350℃

5. 3 Field test wells evaluation

From 2011, strain based design for casing string was used in 8 thermal wells of Xinjiang oil field. After 14 cycles of steam injection, casing string has still serviced well, without damage in eight experimental thermal wells. The strain based design for casing string meet the requirements of security criterions in both plastic capacity and strain fatigue limit.

5. 3. 1 Requirements for diameter drift of pipe and deformation capacity

In order to evaluate the quality of casing string after several cycle production, the requirements for diameter drift of pipe was proposed on the basis of API Specification 5CT, including range of admission variation for inside diameter, drift diameter, thickness deviation and logging deviation. The result was summarized as following table 2.

Table 2 Comparison of inside diameter in casing string in thermal well

Type	casing in experimental well	casing in convention well for comparing
Normal size	177. 8mm×8. 05mm	177. 8mm×9. 19mm
OD tolerance	(−0. 5%~+1. 0%) D	(−0. 5%~+1. 0%) D
Thickness tolerance	−10%	−10%
Drift diameter	158. 52mm	156. 24mm
Logging accuracy	±0. 64mm	±0. 64mm
Admission ID range	158. 52~166. 38mm	156. 24~164. 32mm

Note: logging accuracy for 40-arm image logging

5. 3. 2 Logging data analysis

The inside diameter of casing string was measured by means of 40 − arm image logging for experimental well. These casing string based on strain design have been serviced 14 cycles in thermal wells. The data was summarized in table 3. These results indicated that the range of variation of inside diameter was still meet the requirement for diameter drift of pipe. Then, the assessment of these casing strings will be in state of safety for experimental thermal well. Meanwhile, the convention casing damage reaches to 4. 2% in the same production zone on the basis of date from operators.

Table 3 Variation range of inside diameter for casing string in tested thermal well

Number	Inside diameter range (mm)	Assessment
1#	159. 20~166. 30	normal
2#	159. 80~165. 40	normal
3#	159. 90~164. 30	normal

5. 3. 3 Comparison of casing string between test and conventional well

The effect of casing string in experimental thermal was compared with conventional well in table 4. It is worth noting that the same production zone was selected, as well as the similar steam injection process and cycle. Furthermore, the pipe thickness of casing is 8. 05mm for experimental thermal well, but 9. 19mm for conventional thermal well. In other word, the pipe thickness reduced significantly for casing string based on strain design.

Table 4 Over proof data between test and convention thermal well

Type	Number of well	Negative over proof (mm)	Positive over proof (mm)
Test well	001	0	0
Test well	002	0	0
Test well	003	0	0
Convention well	001	-3. 2	+5. 4
Convention well	002	-1. 3	+0. 4

Note: the location of casing string above perforation area in wellbore.

5. 3. 4 Exception analysis

For 8 field test wells only one site from one well was found to be has bigger deformation as shown in figure 10, but still in production now. This site is about 60 meters deep nearby the surface casing bottom. Cement quality inspection showed that this site has poorer cementing quality, which was thought to result in obvious compression deformation. This negative case showed a suitable cementing quality is necessary for casing string safety.

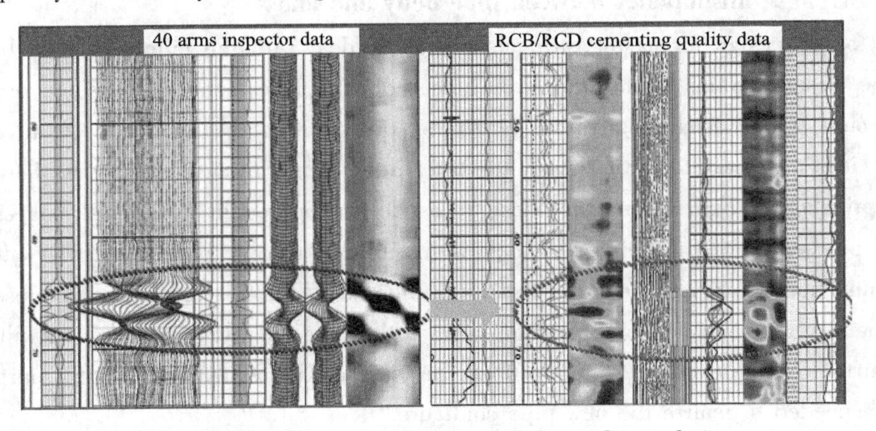

Fig. 10 Negative case and poor cementing quality analysis

6 Further consideration

6.1 Cyclic hardening or softening

For CSS wells the Bauschinger effect should be considered because it will result in cyclic hardening or softening. This effect will leads to structure integrity failure when softening occurs or decreased life when hardening occurs. Figure 11 showed the rough test result for material used in field test. From this result it can be found that the stress will continuously increase and it is possible to reach the ultimate strength of material. From the evaluation results described above it seems that the fatigure strain limit is the dominant factor for a well design life. As the stress increases during different CSS cycle the fatigure strain limit will be smaller and smaller. When the stress reaches ultimate strength the material has no longer deforming capability which measns the well life will end. For different casing products it is easy to meet the initial deforming capacity but hard to meet the fatigue life especially in the case of cyclic hardening effect exists. So the long term service behavior for materials should be carefully studied to enhance the new design method.

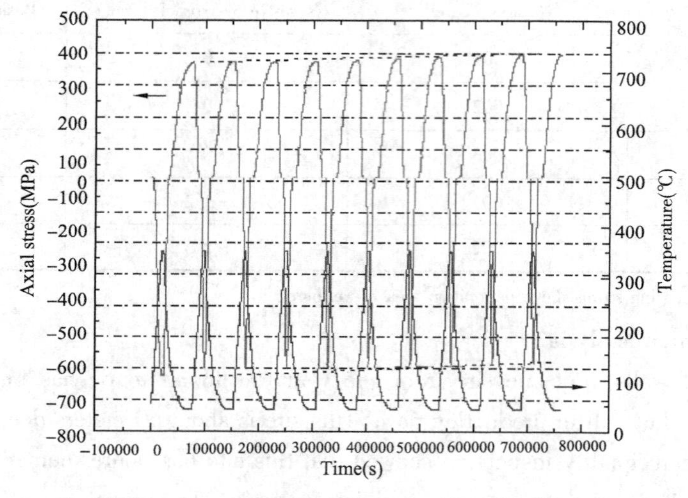

Fig. 11 Cyclic hardening effect of casing material

6.2 Strength mismatches between pipe body and ends

Strain based design for casing string is mainly to deal with the failure of pipe body, which controls the design strain in the range of homogenous deformation. Otherwise, pipe end and coupling would be in danger state due to stress concentration during cyclic steam stimulation. Therefore, it is necessary to further study the thread connection of casing string in thermal well. A new conception of strength mismatches design has been proposed in figure 12, suggesting higher strength in pipe end and coupling, in comparison with that of pipe body, which results in elastic state in connection and plastic deformation only in pipe body. For this concept either upset or second heat treatment process should be applied to enhance the pipe ends property, which will completely alter the current casing configure different from the current API regulation. For this purpose a lot of work is still needed to realize the new pipe configure.

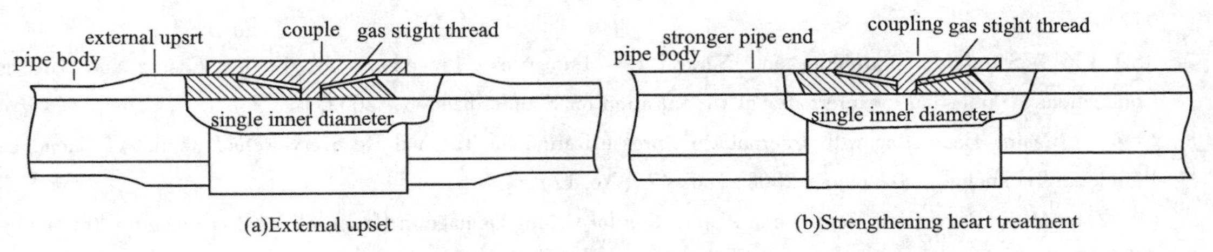

Fig. 12 Recommended designs of pipe ends for connections

7 Conclusions

The strain based design and field application of thermal well casing for cyclic steam stimulation process were investigated in the present work. The casing damages in cyclic steam stimulation wells are caused mainly by extra plastic deformation under tension and compression loads during thermal cycle.

Strain based design for casing string in thermal well was proposed to prevent axial failures. New material parameters were proposed to build the safety evaluation principle throughout the service life. For thermal casing the material deforming limit, called as allowable strain, is determined by the homogeneous deforming capacity. For design strain, the material creep rate equation is introduced to evaluate the accumulative elevated strain in long term service together with initial install strain, thermal strain and soil strain. According to this design, the casing material is required to be evaluated according to security criterions in both plastic capacity and strain fatigue limit. Before the engineering operation, the full scale test procedure was proposed to evaluate the string integrity during thermal cycles.

Strain based design for casing string has been used in eight experimental wells in Xinjiang oil field since 2011. The experimental well serves well without causing damage after 14 cycles. The field application has showed that the new method has better fitness for casing string in thermal wells.

Acknowledgement

The authors would like to thank the financial supports from China National Science Foundation Project (51574278).

References

[1] XIE J, A Study of strain-Based Design Criteria for Thermal Well Casings, World Heavy Oil Cinference, Edmonton, Canada, March 2008, 2008, 388.

[2] J Nowinma, T Kaiser and B Lepper, Strain-Based Design of Tubulars for Extreme-Service Wells, SPE/IADC Drilling Conference & Exhibition, February 2007, Amsterdam (SPE 105717).

[3] J Nowinka and D Dall'Acqua, New standard for Evaluating Casing Connections for Thermal Well Application, SPE/IADC Drilling Conference & Exhibition, March 2009, Amsterdam, Netherlands (SPE/IADC 119468).

[4] XIE J, Casing design and analysis for heavy oil wells, First World Heavy Oil Cinference, Beijing, China, November 2006, 2006, 415.

[5] IRP, 2002, Industry recommended Practices for Heavy Oil and Oil Sand Operations, Vol. 3.

[6] Trend Kaiser, Post-Yield Material Characterization for Strain-Based Design, SPE Journal, March 2009 (SPE

97730).

[7] ISO 13679 Standard, Petroleum and Natural Gas Industries - Procedures for Testing Casing and Tubing Connections, Published by International Organization for Standardization, 2002.

[8] Li Z F, Casing Cementing with Internal Pre-pressurization for Thermal Recovery wells, Jounal of Canadian Petroleum Technology, December 2008, Vol. 47 (No. 12).

[9] Zhou M S, Prevention Method and Application for Casing Damage in Unconsolidated sandstone Ultra Heavy Oil Recovery, Petroleum Geology and Engineering, 2006, 20 (6): 78-80.

[10] Liu Y X, Fu J T, Lu L H, etc, Analysis for Character and Cause of Casing Failure in Loose Sandstone Reservoir, Journal of Shandong Jianzhu University, 2010, 25 (3): 342-346.

[11] S Y Hsu, K H searles, Y Liang, etc, Casing Integrity Study for Heavy-Oil Production in cold lake, SPE Annual Conference & Exhibition held in Florence, Italy, Sep 2010 (SPE 134329).

一种抽油杆用渗铝表面改性空冷贝氏体钢组织与力学性能研究

冯　春

（中国石油集团石油管工程技术研究院；石油管材及装备材料服役行为与结构安全国家重点实验室）

摘　要： 钢铁构件表面进行渗铝处理之后可以大幅度的提高其耐腐蚀性能等而得到广泛的关注。本文重点探讨了非调质渗铝型贝氏体抽油杆的研发。利用 Mn 系贝氏体钢空冷自硬的优势，经过合理的成分设计，进而调控贝氏体钢的相变行为和关键相变点，以满足渗铝之后空冷至室温（或适当回火之后）即可达到抽油杆的性能要求，避免了渗铝之后二次调质处理对渗层的破坏。本文设计了两种 20Mn2SiCrMoV 贝氏体抽油杆，选择不同的温度进行快速渗铝，最终可以达到抗拉强度为 1057MPa，屈服强度为 755MPa，延伸率为 20%，冲击韧性 A_{KV} 为 50~70J 的综合性能，基本达到 HL 级抽油杆的性能要求。渗层总厚度平均约为 100μm，包括纯铝层和 Al—Fe 金属间化合物层，其中化合物层的厚度平均为 50μm，而且化合物层与基体结合紧密。贝氏体抽油杆渗铝工艺简单，避免了渗铝之后二次调质处理对渗铝层的破坏，具有一定的应用前景。

关键词： 贝氏体；抽油杆；渗铝；显微组织；力学性能

众所周知，国内外每年钢铁材料因腐蚀、氧化、磨损造成的损失均在万亿元以上。钢铁构件表面进行渗铝（Al）处理之后可以大幅度的提高其耐腐蚀性能等而得到广泛的关注[1-3]。抽油杆在水、CO_2、H_2S 等腐蚀介质的作用下，腐蚀失效十分严重[4]，因此对抽油杆进行渗铝处理具有十分重要的现实意义。不同于其他的低强度级别的钢构件只要求防腐性能，HL 级别抽油杆在使用过程中还需要承受很大的循环载荷，因此抽油杆表面渗铝在提高其抗腐蚀、氧化、磨损性能的同时，不能降低其力学性能指标，特别是拉伸强度、冲击韧性和疲劳性能等[5]。HL 级别抽油杆的力学性能指标不仅与渗铝工艺有关，还与原材料、热处理工艺等密切相关。

前期的工作表明，采用传统的 HL 级抽油杆钢（如 20CrMoA、25CrMoA、30CrMoA 等）渗铝之后，为了保证强度，必须再进行热处理，通常有两种方式：一是渗铝之后直接淬火至室温，然后高温回火，但是直接淬火会导致渗铝层与基体开裂，而且力学性能不稳定，例如部分试样强度不足或超标；二是渗铝之后二次调质处理，但是会造成分渗铝层有熔化、龟

基金项目：陕西省青年科技新星项目（2016KJXX-60）。

作者简介：冯 春（1980—），男，新疆乌鲁木齐人，高级工程师，博士，主要从事金属结构材料研究，特别涉及石油管材领域先进材料的研究开发、适用性评价及油田现场应用。发表论文 40 余篇。联系电话：029-81887663；E-mail：fc80x@ sina. com。

裂、脱落等现象，不能有效地发挥渗铝层的优势。因此开发新型的渗铝抽油杆材料势在必行。

清华大学方鸿生教授等发明的 Mn 系空冷贝氏体钢，以 Mn—Si—Cr 为主要合金元素，合金设计简单、成本较低，而且利用 Mn 元素的“拖拽和类拖拽效应”提高了钢的贝氏体淬透性，可以在空冷条件下获得贝氏体组织，保证钢的强韧性匹配[6,7]。例如唐山某厂利用该技术生产的 HL 级粒状贝氏体高强度抽油杆用钢可代替 20CrMo 调质钢，空冷性能可达到 $\sigma_{0.2}\geqslant750\sim850$MPa，$\sigma_b\geqslant900\sim1100$MPa，$\delta_5\geqslant14\%$，$\psi\geqslant45\%$，$A_{ku}\geqslant90$J/cm^2，采用低碳空冷粒状贝氏体钢生产的 HL 级抽油杆已用于大庆、胜利、克拉玛依等油田[7]。鉴于此，本文利用 Mn 系空冷贝氏体钢空冷自硬的特点，开发免调质渗铝抽油杆，即渗铝之后直接空冷至室温，避免渗铝之后重新淬火或调质，在保证渗铝层完好无损的情况下，基体的力学性能达到 HL 级抽油杆的性能要求。

1　实验材料及工艺

本文设计了两种不同成分系列的低碳贝氏体钢，实验钢的成分如表 1 所示。两种实验钢均以 Mn-Si-Cr 为主要合金元素，选择不同的 Si 元素含量以调控钢的相变点，进而满足渗铝温度的要求。实验钢经 50kg 的真空感应炉冶炼后，锻造成直径为 25mm 的棒材（与抽油杆直径相当），在热处理之前，将锻造的钢材进行退火处理，利用 DIL-805L 淬火膨胀仪测定实验钢的关键相变点，具体工艺是以 0.05℃/s 加热至 950℃，保温 5min 之后，然后以 10℃/s 冷却至室温，结果如图 1 所示，其中 B1#钢的 Ac_1、Ac_3和 M_s 点分别是 720℃、830℃和 382℃；B2#钢的 Ac_1、Ac_3和 M_s 点分别是 740℃、890℃和 378℃。由此可见 Si 对 Ac_3具有显著影响，但是对 Ac_1和 M_s 点影响不大。

表 1　实验钢的化学成分（wt%）

No.	C	Si	Mn	Cr	Mo	Fe
B-1	0.18~0.22	0.3	适量	0.4	0.28	Bal.
B-2	0.18~0.22	1.5	适量	0.4	0.28	Bal.

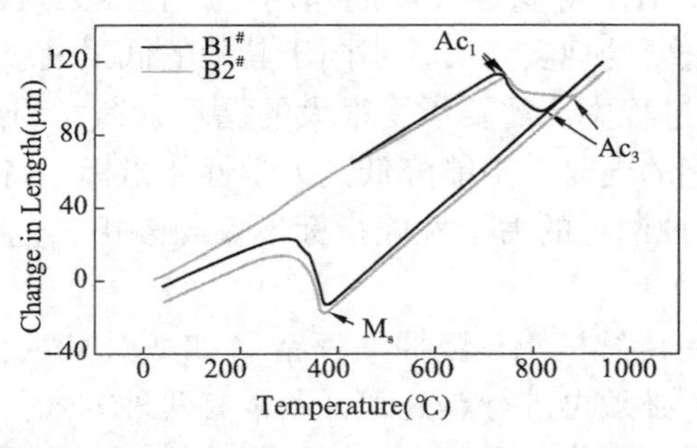

图 1　两种实验钢（B1#和 B2#）的膨胀曲线及关键相变点

首先将棒材在 920℃保温 1h 之后空冷至室温，然后通过除油→水洗→除锈→助渗→烘干→渗铝→空冷至室温，其中 B1 的渗铝工艺参数为：渗铝温度选择 820℃，渗铝时间为 5min，渗铝之后选择直接空冷至室温。B2 的渗铝工艺选择两种：（1）与 B1 相同，即渗铝温度选择 820℃，渗铝时间为 5min，（2）渗铝温度选择为 740℃，渗铝时间为 10min。

将经过渗铝处理之后的贝氏体棒材按照标准 GB/T 228—2002 机加工成标距为 50 mm、

直径为 12.5mm 的棒拉伸试样，在 MST 万能拉伸试验机上进行拉伸实验。冲击试样加工成 $10\times10\times55mm^3$ 的 Charpy V 型冲击试样（GB/T 299—2007）。显微组织经 2%硝酸—乙醇溶液浸蚀，采用扫描电镜（SEM；Zeiss EVO18，20 kV）进行观察，渗铝层采用 EDS 表征。

2 实验结果

2.1 力学性能分析

表 2 是实验钢经过渗铝处理后的力学性能指标，根据标准《SY/T 5029—2006：抽油杆》的要求，其中 B1#实验钢渗铝后综合力学性能基本达到了 HL 级别抽油杆的性能需求，而 B2#钢经 820℃ 渗铝 5min 之后，虽然抗拉强度达到了要求，但是屈服强度偏低，而经 740℃渗铝 10min 之后，抗拉强度和屈服强度均不能达到性能要求。由图 1 可知，B1#的 Ac_3 点为 830℃，渗铝温度为 820℃，接近 Ac_3 温度，钢在渗铝过程中重新奥氏体化，当温度接近 Ac_3 时，奥氏体的含量也接近 100%。而 B2#的 Ac_3 为 890℃，渗铝温度为 820℃时，仍在两相区，组织为铁素体+奥氏体。为了更好地分析力学性能的变化规律，本文对渗铝之后的试样进行显微组织表征。

表 2 渗铝后实验钢的力学性能

No.	Alumetizing process	Tensile strength（MPa）	Yield strength（MPa）	Elongation（%）	Impact toughness（J）
B-1	820℃@5min	1057	755	20	50~70
B-2	820℃@5min	1120	650	15	—
B-2	740℃@10min	790	610	21	—

2.2 显微组织分析

图 2 是不同实验钢经不同的渗铝工艺处理之后的显微组织的 SEM 照片。图 2（a）是 B1#钢经过渗铝之后的显微组织，由图可见，显微组织以粒状贝氏体为主，其中贝氏体铁素体（bainitic ferrite，BF）的宽度为 1~3μm，马氏体/奥氏体（martensite/austenite，M/A）岛的尺寸平均为 1μm，M/A 岛均匀地分布在贝氏体铁素体基体上。本课题的研究表明，粒状贝氏体的组织参量，即 M/A 岛的形状、数量、尺寸和分布，是性能的决定因素，一般而言，随小岛总量增加，小岛弦长及岛间距减小，强度增加；随小岛总量减少，小岛弦长减小，岛间距增加，韧性提高；在一定成分范围内，适当控制小岛数量及小岛尺寸，可以得到强韧性良好配合的力学性能。经适当温度回火后可进一步改善钢的强韧性。本文经过合理的成分设计，在空冷条件下获得了适量 M/A 岛比例和尺寸的粒状贝氏体组织，因而得到了良好的强韧性匹配[8]。

图 2（b）是 B2#钢经过 820℃ 渗铝 5min 之后的显微组织，虽然形貌上与图 2（a）相似，但是该组织是由铁素体+M/A 岛组成，由于渗铝温度（820℃）介于 B2 钢的 Ac_1 和 Ac_3 之间，处于 α/γ 两相区之间，在渗铝温度下保温 5min 的过程中，形成了铁素体+逆转变奥氏体，在后续的冷却过程中，铁素体保留下来，而逆转变奥氏体转变成 M/A 岛。这与粒状贝氏体不同，粒状贝氏体中的贝氏体铁素体和 M/A 岛均是在渗铝之后空冷过程中由奥氏体转变而来的。所以与贝氏体铁素体不同，这种在 α/γ 两相区形成的铁素体位错密度较低，强度不足，这也是造成 B2#钢渗铝之后屈服强度偏低的主要原因。

图 2（c）是 B2#钢经过 740℃ 渗铝 10min 之后的显微组织，该组织以回火马氏体/贝氏

体为主，其中马氏体/贝氏体板条依旧模糊可见，部分板条已经发生了合并，少量的开始回复，碳化物在贝氏体/马氏体板条之间析出。这是由于渗铝温度在 Ac_1 线附近，原有的基体组织（贝/马复相组织）发生了回火。在回火过程中，位错密度降低、马氏体中过饱和的碳以碳化物形式析出，因此强度偏低。

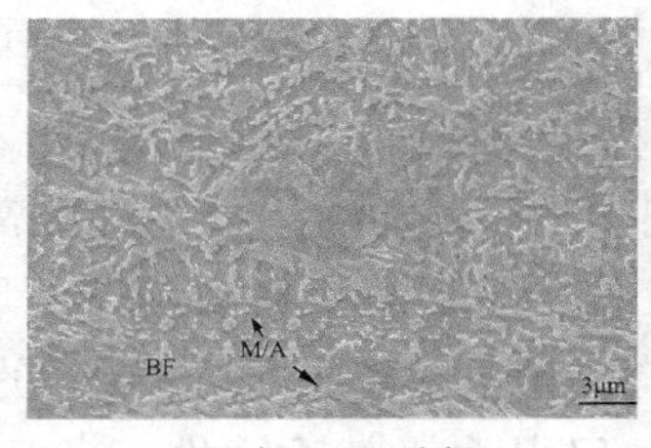

(a) B1钢 820℃渗铝 5min

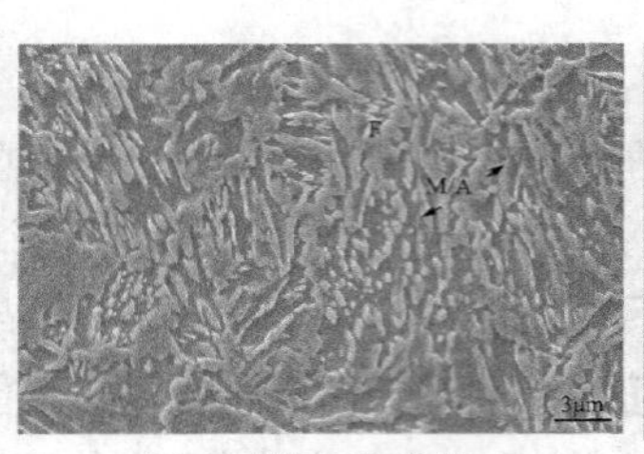

(b) B2钢 820℃渗铝 5min

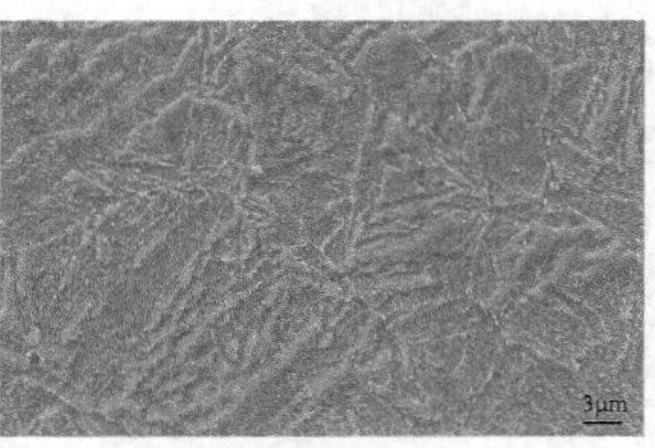

(c) B2钢 740℃渗铝 10min

图 2　实验钢经渗铝处理之后的显微组织

2.3　渗铝层表征

一般认为，在渗铝过程中，液态 Al 与固态铁之间发生浸润和扩散，进而形成冶金结合，是获得致密而有效的渗铝层的关键。因此，渗铝层由表层向里，一般分为纯 Al 层和 Fe—Al 金属间化合物层。根据 Fe—Al 相图（图 3）可知[9]，随着 Al 浓度的增加，在 Fe—Al 固溶体之间的金属间化合物可以依次是，β1（Fe_3Al）和 β2（FeAl）$FeAl_2$（ζ 相）、Fe_2Al_5（η 相）和 $FeAl_3$（θ 相）等。渗铝层的组成还与渗铝的温度和时间密切相关。

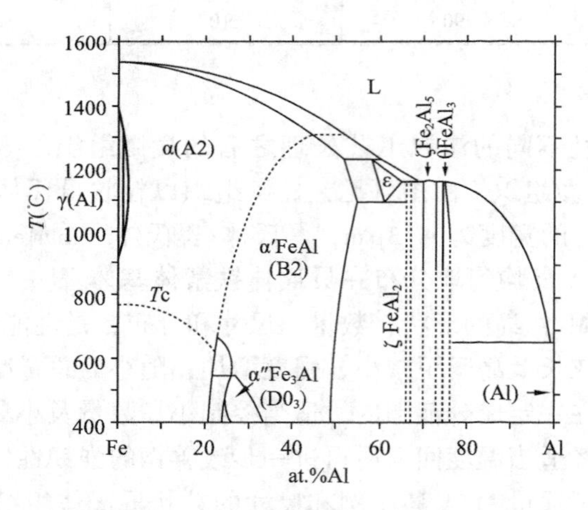

图 3　Fe-Al 二元相图（Kattner and Burton）

图 4 是 B1#钢的渗铝层 SEM 照片和 EDS 能谱分析的结果。图 4（a）是渗铝层的宏观图，可见，渗铝层的总厚度在 70~170μm 之间，平均约为 100μm。图 4（c）和（d）显示了 Al 和 Fe 元素沿渗铝层由表及里的线扫描结果。由线扫描结果可以看出，Al 含量沿渗铝层由表及里逐渐降低，其中表层基本全为 Al 元素，为纯 Al 层，其厚度约为 20μm。往里是合金层，其中 Al 和 Fe 的原子百分比为 2~3 之间，介于 Fe_2Al_5 与 $FeAl_3$ 之间，其中合金层的厚度约为 50μm。图 4（e）和（f）分别是 Al 和 Fe 元素的面分布图，由图可见，最外层基本全为 Al 元素，往里是 Al 和 Fe 元素，再往里即是基体。从面分布图可见，纯铝层存在很多空洞，而合金层十分致密。

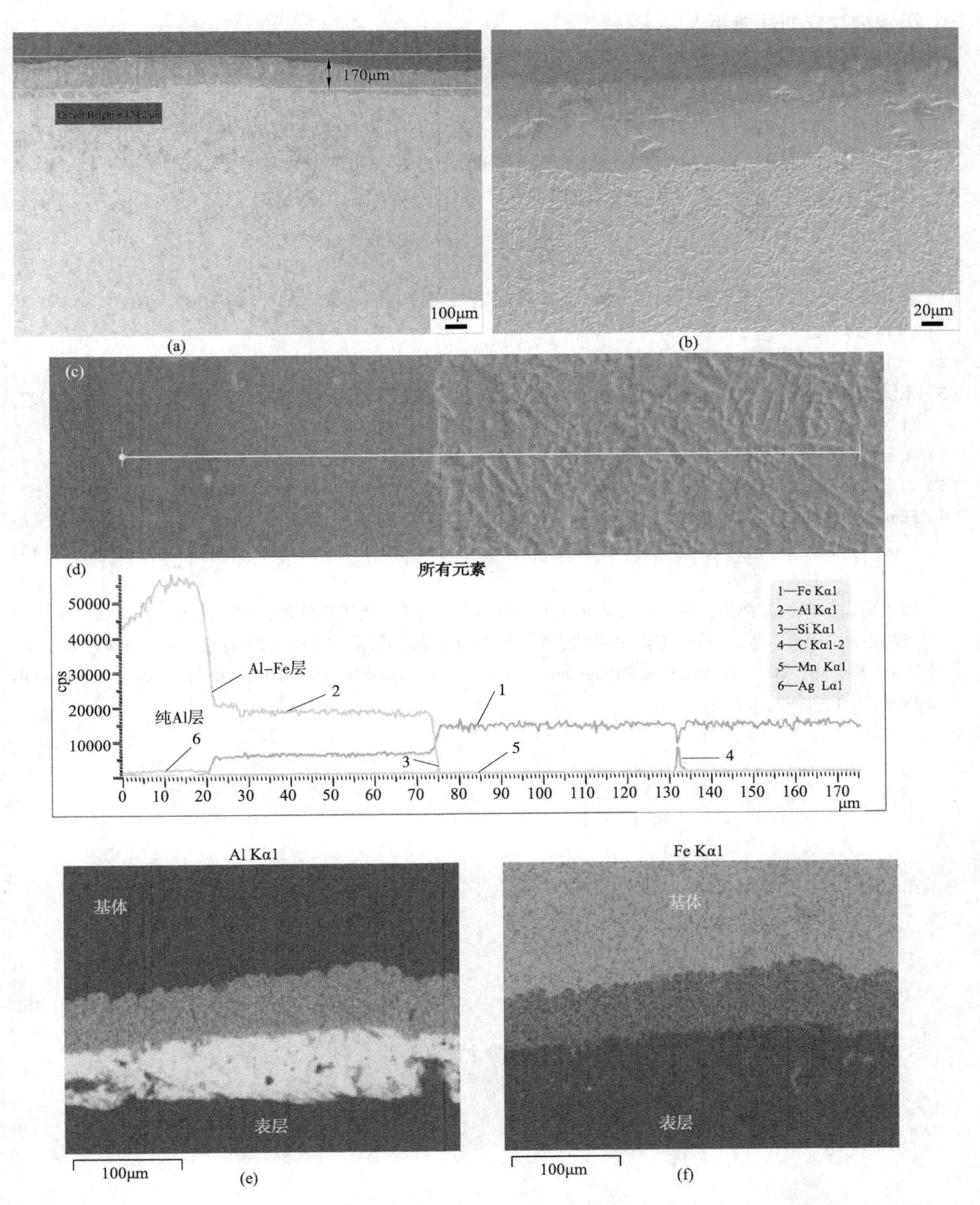

图 4 渗铝层表征，（a-b）渗铝层的 SEM 图，（c-d）渗铝层的线扫描图，（e-f）渗铝层的面扫描图

3 结论

（1）Mn 系贝氏体钢具有空冷自硬特征，经渗铝后空冷至室温，可以获得抗拉强度为 1057MPa，屈服强度为 755MPa，延伸率为 20%，冲击韧性为 50~70J 的综合性能，基本达到

HL 级别抽油杆的性能要求。

（2）经过合理的成分设计，调控贝氏体钢的 Ac_1 和 Ac_3 点，在 820℃ 渗铝 5min 之后，B1#钢的显微组织为典型的粒状贝氏体组织。

（3）通过对渗铝层的表征发现，渗铝层的总厚度平均约为 100μm，包括纯 Al 层和 Fe—Al 金属间化合物层，Fe—Al 金属间化合物层平均约为 50μm，致密无空洞，与基体结合良好，渗铝之后空冷至室温，免去了调质环节，避免了对渗铝层的破坏。

参 考 文 献

[1] 罗吉媛，俞敦义，杨继林．渗铝钢在 Na_2S 溶液中的腐蚀行为研究［J］，材料开发与应用，2000，15（4）：14-25.

[2] 宋世，刘顺华，李长茂，等．钢材的连续热浸镀铝［J］，金属热处理，2005，5：15-18.

[3] 张洪斌，黄永昌，潘健武．钢材热浸镀层的腐蚀性能研究及其耐蚀性比较［J］，全面腐蚀控制，1997，11（4）：1-7.

[4] 纪云岭，张敬武，张丽．油田腐蚀与防护技术［M］，北京：石油工业出版社，2006，1-2.

[5] 边勇俊，石伟，劳金越，等. 20CrMo 钢抽油杆断裂原因分析［J］，金属热处理，2011，50（9）：106-108.

[6] Fang H S，Zheng Y K，Chen X Y，et al. Development of a serious of New air cooled bainitic steels in China［A］. Geoffrey Tither，ZhangShouhua. HSLA Steels，Processing，Properties and Applications，Beijing：TMS，1992：119-125.

[7] 刘东雨，徐鸿，方鸿生，等. 我国低碳贝氏体钢的发展［J］. 金属热处理，2005，20（2）：12-19.

[8] 方鸿生，白秉哲，等．粒状贝氏体和粒状组织的形态与相变［J］. 金属学报，1986，22（4）：A283-287.

[9] U. R. Kattner，B. P. Burton Phase Diagrams of Binary Iron Alloys H. Okamoto ASM International，Materials Park，OH，1993，12-13.

稠油热采井套管柱应变设计方法

王建军[1]　杨尚谕[1]　薛承文[2]　韩礼红[1]　王　航[1]

（1. 中国石油集团石油管工程技术研究院，石油管材及装备材料服役行为与结构安全国家重点实验室；2. 中国石油新疆油田公司工程技术研究院）

摘　要：为解决稠油热采井不断出现的套管损坏现象，改变传统的管柱强度设计方法。基于弹塑性力学理论，在满足套管柱强度设计的基础上，建立了套管柱应变计算模型；通过对比套管材料应变与结构应变，借助 Rambery-Osgood 模型得到了应变安全系数最小取值，当套管屈服应力范围介于560~750MPa 之间，应变安全系数最小值取 1.8，进一步提出了套管柱应变设计的理论判据，最终形成套管柱应变设计方法。该设计方法在西部油田 8 口稠油热采井 ϕ177.8mm×8.05mm TG80H 特殊螺纹套管柱设计中得到应用，生产 4 轮次后测井显示套管柱未出现变形和泄漏现象，经过 14 轮次后，套管服役性能依然良好，有效延长了套管安全生产周期。

关键词：稠油；热采井；套管柱；应变设计；设计准则

国内在稠油热采井管柱设计方面，一直采用传统的强度设计方法（应力设计方法）[1-3]，该方法是以弹性力学理论为基础，确保管材不会发生屈服现象。这种方法主要考虑管材的强度指标，满足钻完井工程需求，并考虑热采过程中的热应力[4-8]，满足高温条件下管材不会发生屈服现象。强度设计方法在我国的实际生产作业中，又通过提高管材强度余量、施加预拉应力技术[9-11]，抵消注汽中的管柱压缩载荷，提高管柱服役安全性，得到了进一步发展。而实际上，国内稠油热采套管的变形、缩径、剪切、断裂等套损形式充分说明了管材在服役中的确发生了塑性变形[12]，而正是不同的塑性变形造成了管材永久变形，甚至断裂。在实际套损案例中，螺纹接头的脱扣现象充分说明，管体在此之前就已经发生了较为明显的永久变形，造成不可逆损伤。现有的套管柱设计方法均考虑的是钻完井工况，对后期生产运行工况没有更多的考虑，而实际上热采井的危险工况是在注蒸汽过程中发生的，套管变形损坏也多发生在该期间。因此，单纯以强度为主要指标来设计管柱是不能满足油田现场作业需求的。

为控制稠油热采井套损率，部分国外学者也提出稠油热采井应采用应变设计[13-14]，但均仅提出了应变设计的概念，对应变设计的方法和理论没有明确说明。笔者利用弹塑性力学理论和油田应用实际，提出套管柱应变设计基本原理，建立了套管柱应变设计模型，并通过

基金项目：国家自然科学基金项目（51574278），中国石油天然气集团公司资助项目（2015E-4006），陕西省自然科学基础研究计划资助项目（2013KJXX-07）。

作者简介：王建军（1979—），男，博士，高级工程师，国家注册安全评价师，现从事油气开发工程与管柱力学研究。地址：陕西西安市锦业二路 89 号（710077），电话：029-81887677，E-mail：wg_ j_ jun @ 163. com。

套管材料应变与结构应变对比给出了应变安全系数最小取值，明确应变设计的判据，确定了热采井套管柱应变设计方法。

1 热采井套管柱应变设计基本原理

稠油热采井的套管柱设计方法包括套管柱强度设计和应变设计[15]，其基本原理示意图如图1所示。即在进行热采井套管柱应变设计时，首先应进行套管柱强度设计，然后再进行套管柱应变设计。强度设计是为了使套管柱满足稠油热采井钻完井过程要求，应变设计是为了使套管柱满足稠油热采生产过程要求。

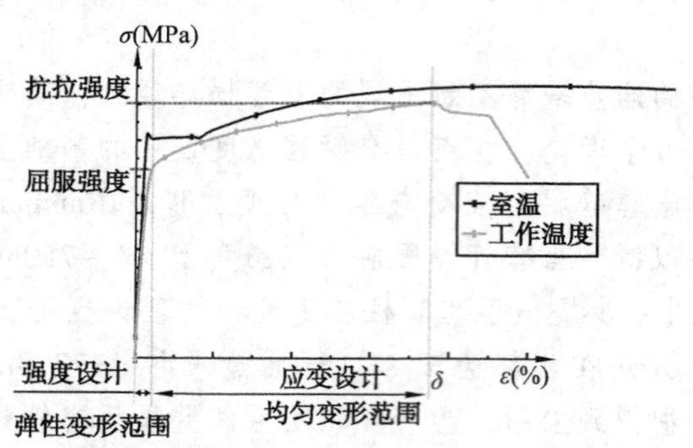

图1 热采井套管柱应变设计原理示意图

因此，稠油热采井套管柱应变设计遵循2个设计准则，一是强度设计准则，二是应变设计准则。套管柱强度设计准则已近完善和成熟，主要以SY/T 5724标准[3]中强度设计为主，下面重点阐述套管柱应变计算方法和设计准则。

2 热采井套管柱应变计算模型

2.1 套管轴向应力

对于稠油热采井，套管柱轴向应力来源主要有以下四个方面：

(1) 由井内套管柱自重而产生的初始轴向应力。采用压力面积法[2]进行计算，见式(1)。

$$\sigma_a = \{0.00981\int_h^{L_c}\rho_s\pi(r_2^2 - r_1^2)\cos\theta dz + \pi[p_i(L_c)r_1^2 - p_o(L_c)r_2^2]\}/[\pi(r_2^2 - r_1^2)] \tag{1}$$

(2) 由管柱内外压力泊松效应[16]而产生附加轴向应力。对于平面应变问题，由广义虎克定律[17]可推导式(2)。

$$\sigma_{zp} = \mu(\sigma_r + \sigma_\theta) \tag{2}$$

其中，

$$\sigma_r = \frac{p_{pi}r_1^2 - p_{po}r_2^2}{r_2^2 - r_1^2} - \frac{(p_{pi} - p_{po})\ r_1^2 r_2^2}{(r_2^2 - r_1^2)\ r^2} \tag{3}$$

$$\sigma_\theta = \frac{p_{pi}r_1^2 - p_{po}r_2^2}{r_2^2 - r_1^2} + \frac{(p_{pi} - p_{po})\ r_1^2 r_2^2}{(r_2^2 - r_1^2)\ r^2} \tag{4}$$

将式(3)和式(4)代入式(2)后，有：

$$\sigma_{zp} = 2\mu\frac{p_{pi}r_1^2 - p_{po}r_2^2}{r_2^2 - r_1^2} \tag{5}$$

（3）由弯曲井段带来的弯曲应力[18]，见式（6）。

$$\sigma_b = 0.060156 D_{leg}(D-2t) \tag{6}$$

（4）因温度变化而产生的热应力[19]，见式（7）。

$$\sigma_T = E_t \alpha \Delta T \tag{7}$$

综合上述轴向应力计算，进行线性叠加，获取最大的总轴向应力，见式（8）。

$$\sigma_z = \sigma_a + \sigma_b + \sigma_{zp} + \sigma_T \tag{8}$$

式中 σ_z——总轴向应力，MPa；

σ_a——初始轴向应力，MPa；

σ_b——弯曲载荷产生的轴向应力，MPa；

σ_{zp}——内外压力产生的附加轴向应力，MPa；

σ_T——热应力，MPa；

σ_r——套管径向应力，MPa；

σ_θ——套管周向应力，MPa；

L_c——套管下入深度，m；

ρ_s——套管质量密度，g/cm^3；

$p_i(L_c)$——套管下端的内压力，MPa；

$p_o(L_c)$——套管下端的外压力，MPa；

r_1——套管内半径，mm；

r_2——套管外半径，mm；

r——套管任一点半径，mm；

μ——套管泊松比，一般取 0.3；

θ——井斜角，(°)；

D_{leg}——弯曲度或狗腿度，°/30m；

t——套管管体名义壁厚，mm；

D——套管管体名义外径，mm；

E_t——注汽温度下套管弹性模量，MPa；

α——线性热膨胀系数，1/℃；

ΔT——套管温度变化，℃；

p_{pi}——套管内压力，MPa；

p_{po}——套管外压力，MPa。

2.2 套管柱轴向应变

轴向应变主要是指由式（8）轴向应力产生的应变。由图 1 知，应变设计中是允许轴向应力超出弹性阶段，在塑性阶段内变化。而在稠油热采井中后期作业中总轴向应力也往往大于套管屈服强度，套管进入塑性变形阶段。因此，对于轴向应变的计算要综合考虑弹性应变和塑性应变，依据线性强化弹塑性力学模型[20]，则有：

当 $\sigma_z \leqslant f_{ymnt}$ 时，套管柱变形在弹性范围内，其轴向应变计算见式（9）。

$$\varepsilon_z = \frac{\sigma_z}{E_t} \tag{9}$$

当 $\sigma_z > f_{ymnt}$ 时，套管柱变形在塑性范围内，其轴向应变计算见式（10）。

$$\varepsilon_z = \frac{f_{ymnt}}{E_t} + \frac{\sigma_z - f_{ymnt}}{E_{pt}} \tag{10}$$

式中 ε_z——套管柱轴向应变,%;

f_{ymnt}——注汽温度下套管最小屈服强度，MPa;

E_{pt}——注汽温度下套管塑性模量或切线模量，MPa。

2.3 套管柱蠕变应变

蠕变应变是与时间密切相关，套管柱的蠕变应变计算公式如下：

$$\varepsilon_c = \sum_{n=1}^{n} \dot{\varepsilon}_c \cdot t_h \tag{11}$$

式中 ε_c——蠕变应变量,%;

$\dot{\varepsilon}_c$——套管蠕变速率（可采用 GB/T 2039 标准[21]规定试验获得）,%/s;

t_h——每轮次注汽时间，s;

n——注汽轮次。

2.4 套管柱累计应变

依据稠油热采井地质环境和作业工况，按式（12）计算服役周期内的套管柱累计应变，即套管柱工作应变。

$$\varepsilon_\Sigma = \varepsilon_z + \varepsilon_c \tag{12}$$

3 热采井套管柱应变设计准则

3.1 应变设计准则

套管柱应变设计准则，见式（13）

$$\varepsilon_\Sigma \leqslant [\varepsilon] = \frac{\delta}{S_s} \tag{13}$$

式中 ε_Σ——套管柱工作应变,%;

$[\varepsilon]$——套管柱许用应变,%;

δ——套管材料均匀延伸率,%;

S_s——应变安全系数。

3.2 应变安全系数

在应变设计原理示意图 1 以及设计准则式（13）中，均以套管材料均匀延伸率 δ 为最终判据。但 δ 由几何尺寸均匀且标准的套管材料棒状试样试验获得，而套管本身实际尺寸是不均一的，其变形曲线与材料变形存有差异[22]。因此，若以 δ 作为套管柱应变设计的判据，则必须清楚套管材料变形与套管柱结构变形之间的关系，即应变安全系数 S_s 取值至关重要。下面我们通过套管材料单轴拉伸试验和全尺寸套管实物单轴拉伸试验，分别获取材料应力—应变关系曲线和套管结构应力—应变关系曲线，并进行比较分析。

对于套管的非线性应力—应变曲线关系，可以采用 Rambery-Osgood 模型[20]描述其弹塑性特性，其 Rambery-Osgood 模型如下式：

$$\varepsilon = \frac{\sigma}{E} + \left(\frac{\sigma}{A}\right)^n \tag{14}$$

式中 ε——应变,%;

σ——应力，MPa；

E——弹性模量，取 2.07×10^5 MPa；

A 和 n——材料常数，通过试验数据拟合获得。

依据 GB/T 228 标准[23]和 ISO 13679 标准[18]，分别获取 ϕ177.8mm×8.05mm TG80H 热采套管的材料工程应力—应变曲线和结构工程应力—应变曲线，并利用真实应力—应变曲线计算方法[88]，分别转化为材料真实应力—应变曲线和结构真实应力—应变曲线，如图 2 和图 3 所示。选取取图 2 和图 3 中套管均匀塑性变形阶段的应力、应变数据，利用式（14）对套管材料应变和结构应变数据点分别拟合，拟合曲线如图 4 和图 5 所示。

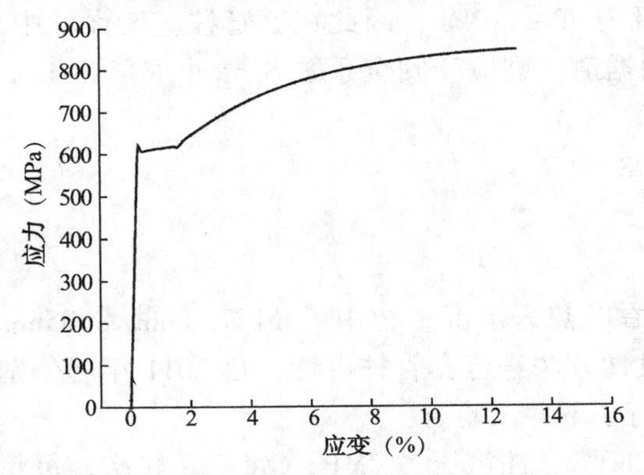

图 2　热采套管材料真实应力—应变曲线

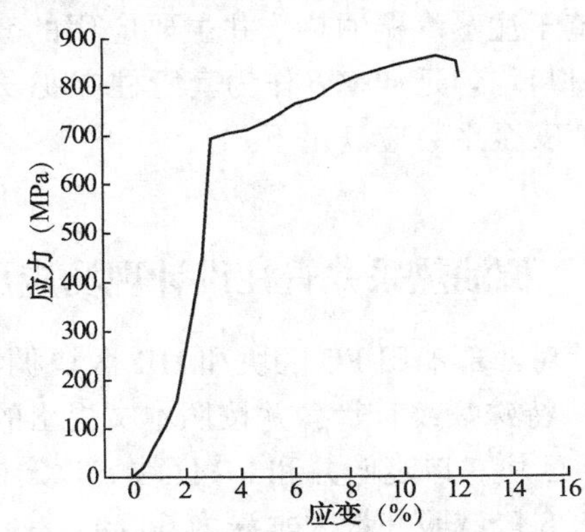

图 3　热采套管结构真实应力—应变曲线

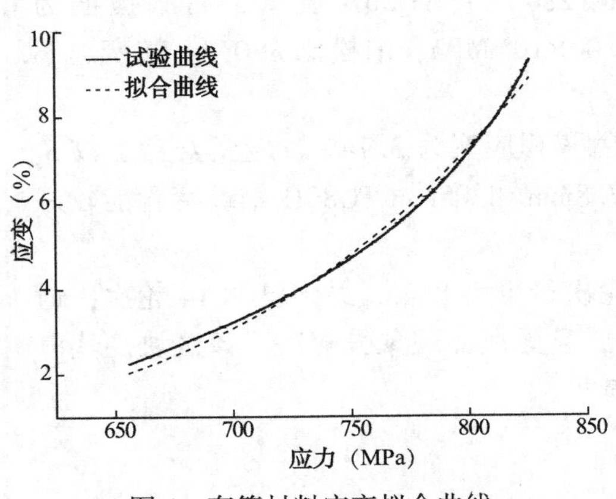

图 4　套管材料应变拟合曲线

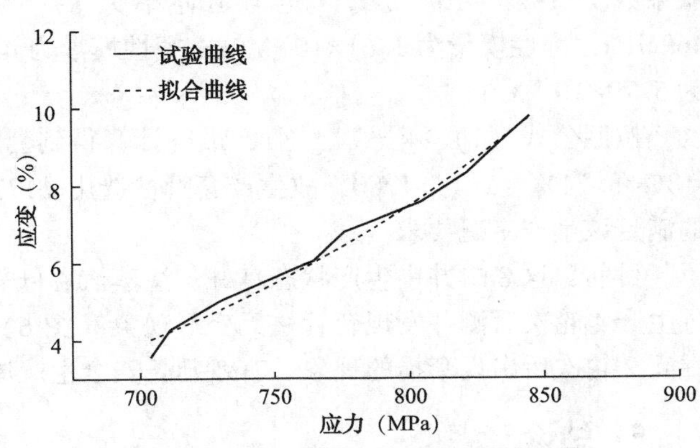

图 5　套管结构应变拟合曲线

由图 4 和图 5，获得的拟合公式分别如下：

（1）套管材料应变拟合公式见式（15），拟合相关系数 $R^2=0.994$。

$$\varepsilon_{\mathrm{m}}=\frac{\sigma}{E}+\left(\frac{\sigma}{586.97}\right)^{6.39} \tag{15}$$

（2）套管结构应变拟合公式见式（16），拟合相关系数 $R^2=0.983$。

$$\varepsilon_{\mathrm{s}}=\frac{\sigma}{E}+\left(\frac{\sigma}{527.86}\right)^{4.84} \tag{16}$$

比较式（15）和式（16），可知套管材料应变与实物套管结构应变之间的关系为：

$$\varepsilon_{\mathrm{m}}=k\varepsilon_{\mathrm{s}} \tag{17}$$

其中，

$$k=\frac{\left(\dfrac{\sigma}{E}+\left(\dfrac{\sigma}{527.86}\right)^{4.84}\right)}{\left(\dfrac{\sigma}{E}+\left(\dfrac{\sigma}{586.97}\right)^{6.39}\right)} \tag{18}$$

式中 ε_m——套管材料应变,%;

ε_s——套管结构应变,%;

k——材料常数。

在均匀塑性变形阶段对应的应力范围 $\sigma=560\sim750$MPa，由式（18）可求得 $k=1.2\sim1.8$。

鉴于注蒸汽稠油热采井主要应用的套管钢级为80~110ksi，因此在套管柱应变设计中，若以材料均匀延伸率 δ 作为套管柱应变设计的判据值，则应变安全系数 S_s 最小取值为1.8，即应变安全系数应满足下式：

$$S_s \geqslant 1.8 \tag{19}$$

4 稠油热采井管柱设计现场应用

针对西部油田FC区块和HD区块稠油蒸汽吞吐热采工况，选用了 ϕ177.8mm×8.05mm TG80H特殊螺纹套管，并按照本文提出的应变设计方法进行套管柱设计，自2011年起分别在HD区块3口定向井和FC区块5口定向井进行了下井试验。

该8口试验井最大垂深为600m，注汽温度280℃，注汽压力7MPa，每年2轮次，每轮次注汽焖井时间15d，设计寿命6年。TG80H套管已知参数：质量密度为7.85g/cm^3，线膨胀系数为 1.21×10^{-5} 1/℃，均匀延伸率为8%。在280℃下TG80H套管，屈服强度为469MPa，弹性模量为 1.61×10^5 MPa，塑性模量为 4.00×10^3 MPa，泊松比为0.3，蠕变速率为 5.79×10^{-8}%/s。

按照公式（1）~公式（13），最终计算得到套管累积应变为3.74%，应变安全系数 S_s =8%/3.74% =2.14>1.80，故生产套管柱选用 ϕ177.8mm×8.05mm TG80H钢级套管能较好的满足该生产工况要求。

目前，这8口井中生产状态良好，注蒸汽吞吐轮次最少为8轮，最多已达14轮次，通过生产4轮次后测井发现套管柱无变形现象（图6），且无泄漏现象，解决了该区块注蒸汽吞吐2轮次后出现套损的现象，有效延长安全生产周期。

5 结论

（1）结合弹塑性力学理论与材料蠕变本构关系，建立了热采井套管柱应变计算模型，最终形成了热采井套管柱应变设计方法，该方法允许套管在可控范围内变形，具有能有效延长套管使用寿命、降低套损率等特点。

（2）利用套管材料拉伸试验和全尺寸实物拉伸试验，借助Rambery-Osgood模型，形成了套管柱应变设计安全系数的确定方法，该方法考虑了套管尺寸的不均匀性和不同热采工况，使应变设计安全系数取值更符合实际。

（3）热采井套管柱应变设计方法，较有限元数值计算方法更方便油田设计人员使用，在现场8口井获得了成功应用，有效延长了套管安全生产周期。

（4）热采井套管柱应变设计方法，最终被纳入到中华人民共和国石油天然气行业标准《SY/T 6952.1—2014 基于应变设计的热采井套管柱 第1部分：设计方法》中。

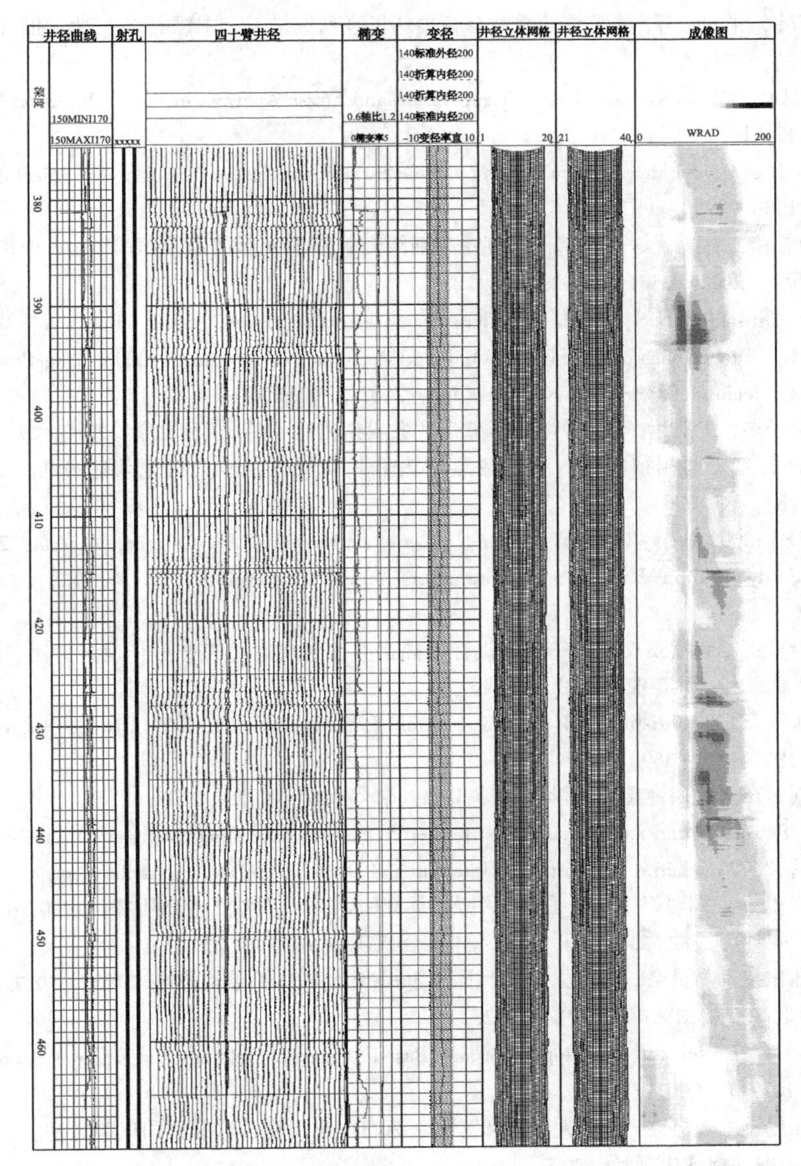

图6　HD区块1井生产4轮次后375~470m井段四十臂井径测井解释成果图

参　考　文　献

[1] 张锐. 稠油热采技术［M］. 北京：石油工业出版社，1999：1-5.

[2] 韩志勇．关于“套管柱三轴抗拉强度公式”的讨论［J］. 中国石油大学学报（自然科学版），2011，35（4）：77-80.

[3] 石油钻井工程专业标准化委员会．中华人民共和国石油天然气行业标准 套管柱结构与强度设计 SY/T 5724—2008［S］. 北京：石油工业出版社，2008.

[4] 王兆会，高宝奎，高德利．注汽井套管热应力计算方法对比分析［J］. 天然气工业，2005，25（3）：93-95.

[5] 王建军，冯耀荣，闫相祯，等. 高温下高钢级套管柱设计中的强度折减系数［J］. 北京科技大学学报，2011，33（7）：883-887.

[6] 崔孝秉，曹玲，张宏，等. 注蒸汽热采井套管损坏机理研究［J］. 中国石油大学学报（自然科学版），1997，21（3）：57-64.
[7] WU Jiang，MARTIN E. Knauss. Casing Temperature and Stress Analysis in Steam-Injection Wells［C］. SPE 103882，2006：1-7.
[8] HECTOR Rodriguez. Improved Calculation of Casing Strains and Stresses in a Steam Injector Well Simulator［C］. SPE 21080，1990：1-11.
[9] 王建军，闫相祯，林凯，等. 内壁椭圆度对高钢级套管挤毁变形的影响试验［J］. 中国石油大学学报（自然科学版），2011，35（2）：123-126.
[10] Smith R J，Alinsangan N S，Talebi S. Microseismic response of well casing failures at a thermal heavy oil operation［C］. Proceedings of the SPE/ISRM Rock Mechanics in Petroleum Engineering Conference. Irving，Texas，USA：Secioty of Detroleum Engineers Inc，2002：390-396.
[11] 阳鑫军，李子丰，王兆运，等. 热采井预膨胀固井模拟实验［J］. 工程力学，2010，27（6）：223-227.
[12] 王建军，韩礼红，闫相祯，等. 稠油注蒸汽热采井套管柱预应力松弛效应分析［J］. 石油机械，2013，41（8）：65-67.
[13] Enform. IRP3-2002 Heavy Oil And Oil Sands Operations Industry［S］. Calgary，Alberta，2002.
[14] J. XIE. A Study of Strain-Based Design Criteria for Thermal Well Casings［C］. WHOC，PAPER 2008-388，2008：1-9.
[15] 石油管材专业标准化委员会. 基于应变设计的热采井套管柱　第 1 部分：设计方法：SY/T 6952. 1—2014［S］. 北京：石油工业出版社，2014.
[16] G. R. Wooley，S. A. Christman，J. G. Crose. Strain Limit Design of 13 3/8 in N-80 Buttress Casing［C］. SPE6061，1977：355-359.
[17] 徐秉业，刘信声. 应用弹塑性力学［M］. 北京：清华大学出版社，2008.
[18] ISO13679，Petroleum and natural gas industries — Procedures for testing casing and tubing connections［S］. first edition，2002. Geneva，Switzerland：International Organization for Standardization.
[19] 钻井手册（甲方）编写组. 钻井手册（甲方）［M］. 北京：石油工业出版社，1990.
[20] 陈慧发等. 弹性与塑性力学［M］. 北京：中国建筑工业出版社，2003.
[21] 全国钢标准化技术委员会. 中华人民共和国国家标准 金属材料单轴拉伸蠕变试验方法：GB/T 2039—2012［S］. 北京：中国标准出版社，2012.
[22] J. Nowinka，T. Kaiser and B. Lepper. Strain-Based Design of Tubulars for Extreme Service Wells［C］. SPE/IADC 105717，2007：1-7.
[23] 全国钢标准化技术委员会. 中华人民共和国国家标准　金属材料室温拉伸试验方法：GB/T 228—2002［S］. 北京：中国标准出版社，2002.

地下储气库套管和油管腐蚀选材分析

王建军　李方坡

（中国石油集团石油管工程技术研究院，石油管材及装备材料服役行为与
结构安全国家重点实验室，中国石油集团石油管工程重点实验室）

摘　要：针对国内储气库注入气介质一样而选材差异较大的问题，从现有选材标准或方法，到生产应用实际，指出造成该差异的主要原因是标准的不适用。通过分析现场应用和标准使用情况，提出对于储气库管柱选材应考虑低含水工况和材质间匹配性。同时，经高温高压釜腐蚀试验和电化学腐蚀试验分析，指出储气库生产套管和油管的腐蚀选材应区别对待，并建立了电化学腐蚀管材匹配选用图版，可指导储气库井管柱选材，并纳入了中国石油天然气行业标准。

关键词：地下储气库；套管；油管；汽相；腐蚀；电位

因地下储气库设计寿命30年以上[1]，且注入气中含2%左右的CO_2，这要求储气库管柱具有较好的耐蚀性，尤其是生产套管和油管。笔者通过中国石油6座储气库的调研发现，在井深、运行压力、注采气量差异不大且注入气介质一样的情况下，各储气库套管和油管选用的材质差异比较大，既有普通碳钢和抗硫材质，又有超级13铬材质，主要包括L80、L80 13Cr、95S、P110、110 Cr13、110 Cr13S等材质。

笔者通过详细分析储气库井下环境和现有标准[2-10]，发现主要是由于标准的不适用造成此类选材的差异，提出储气库管柱选材应考虑低含水工况和材质匹配性，并进行了气液两相下高温高压腐蚀试验和电化学试验，研究建立了储气库套管和油管选材方法。

1　储气库管柱使用环境

调研国内主要储气库环境工况（表1），发现国内储气库气藏地层水中Cl^-含量不超过10000mg/L，水型主要为$NaHCO_3$，因注入气中CO_2含量在1.89%～2.18%之间，CO_2分压主要在0.5～0.8MPa，部分井CO_2分压高达1.16MPa。此外，储气库注入的是经处理后的干燥气体，但在采出时因受气藏地层水影响，采出的气体低含水（10000m^3气约含1m^3水）。

基金项目：中国石油天然气集团公司资助项目（项目批准号2014F-1501、2015E-4006），陕西省自然科学基础研究计划资助项目（项目批准号2013KJXX-07）。

作者简介：王建军（1979—），男，博士，高级工程师，国家注册安全评价师，现从事油气开发工程与管柱力学研究。地址：陕西西安市锦业二路89号（710077），电话：029-81887677，E-mail：wg_ j_ jun @163.com。

表 1　国内主要储气库环境工况与选材情况

储气库		XC	HK	BN	SK
气藏埋深（m）		2782	3585	2900	3300~5000
运行压力（MPa）		11.7~28	18~34	13~31	19~48.5
地层温度（℃）		65	92.5	116	110~157
注入气 CO_2含量（mol%）		1.8909	1.89	2.48	2.37
地层水	Cl^-（mg/L）	少量凝析水 干气气藏	9974	1170~5000	3456
	总矿化度（mg/L）		17800	6800~13300	7630
	水型		$Na_2SO_4/NaHCO_3$	$NaHCO_3$	$NaHCO_3$
选材特点		生产套管柱和注采管柱主要考虑原始地层含有 H_2S，以抗硫管材为主，分别选用 95SS 和 80S 钢级	生产套管柱回接采用 Q125HC 高抗挤套管，尾管悬挂采用改良型 13Cr 套管；注采管柱采用改良型 13Cr 油管	生产套管柱采用 P110 套管；注采管柱采用 L80 油管	生产套管柱采用 P110 套管；注采管柱采用 L80 13Cr 油管

2　腐蚀选材标准

基本的油管和套管腐蚀选材标准，主要是 ISO15156—3 和 GB/T 20972.2（ISO 15156—2，MOD）[2,3]，但是这些标准着重于含 H_2S 环境的选材。依据腐蚀介质不同，石油管材专业标准化技术委员会制定了针对性的选材标准，具体有 SY/T 6857.1、Q/SY—TGRC 2、Q/SY—TGRC 3、Q/SY—TGRC 18 等标准[5-8]，尤其是 Q/SY—TGRC 18 标准主要针对含 CO_2环境选材。

在 Q/SY—TGRC 18 标准中规定，当 0.21 MPa$<CO_2$分压$<$1.0 MPa 时属于严重腐蚀，应选用碳钢与加注缓蚀剂协同作用，或者直接选用普通 13Cr 钢，如 L80-13Cr 钢等；当 1.0 MPa$<CO_2$分压$<$7.0MPa 时属于极严重腐蚀，应选用改良型 13Cr 钢（Cr13M）或超级 13Cr 钢（Cr13S），或选用普通 13Cr 钢与与加注缓蚀剂协同作用。

依据上述标准方法和生产厂家选材推荐方法，并结合现场应用实际情况，各储气库确定了选用的套管和油管，从普通碳钢到超级 13Cr 钢都有选用（表 1）。主要是因现有标准或方法中 CO_2腐蚀条件均是 100%液体环境，在建储气库均是依据现有标准或方法进行选择，造成在注气介质相同而材质选择各异。也就是说，对于枯竭式气藏，在低含水率的情况下，这种选材方法是否合适，还有待试验验证。

此外，对于生产套管上部使用碳钢管（如 Q125HC），下部使用耐蚀合金管（如超级 13Cr），这种组合材质在井下环境会否产生电化学腐蚀，仍需要慎重评价其匹配性。

3　高温高压釜腐蚀试验

分析表 1 各储气库环境工况，现提取主要且通用的工况试验条件，进行汽液两相条件下的动态高温高压釜腐蚀试验。试验条件：CO_2 分压 0.8MPa，Cl^- 浓度 10000mg/L，水型 $NaHCO_3$，试验温度 90℃，流速 2m/s，试验周期 168h。试验结果见表 2 和表 3。

通过分析表 2 和表 3 结果，可知 13Cr 材质较好地适用于 CO_2上述汽/液相腐蚀环境，腐蚀速率：L80 13Cr ≈ 13Cr110 >110 Cr13S；其他材质均较好地适用于 CO_2 上述汽相腐蚀环境，其腐蚀速率均小于 SY/T 5329 标准[9]规定的 0.076mm/a。

通过国内储气库用系列管材汽液两相的高温高压釜腐蚀试验，认识到生产套管和油管的腐蚀选材应分别对待，生产套管依据 100%液体环境（现有标准）选材，油管应按照低含水工况选材。

表 2　国内储气库油套管材质汽相下动态腐蚀试验结果

材质	平均年腐蚀速率（mm/a）	参照 NACE RP 0775 规定[9]
Q125HC	0.0543	中度腐蚀
95S	0.0467	中度腐蚀
P110	0.0463	中度腐蚀
L80	0.0282	中度腐蚀
110 Cr13M	0.0034	轻度腐蚀
L80 13Cr	0.0020	轻度腐蚀
110 Cr13S	0.0002	轻度腐蚀

表 3　国内储气库油套管材质液相下动态腐蚀试验结果

材质	平均年腐蚀速率（mm/a）	参照 NACE RP 0775 规定
95S	8.3054	极严重腐蚀
Q125HC	6.8185	极严重腐蚀
P110	2.2370	极严重腐蚀
L80	1.7981	极严重腐蚀
110 Cr13M	0.0052	轻度腐蚀
L80 13Cr	0.0031	轻度腐蚀
110 Cr13S	0.0007	轻度腐蚀

4　电化学腐蚀试验

对国内储气库所使用的 L80、95S、P110、Q125 以及 11Cr、L80 13Cr、Cr13M、Cr13S 不同钢级和材质的套管、工具等随机抽样，并在 Cl^- 浓度 10000mg/L 和 $NaHCO_3$ 水型的地层水环境介质中且常温下的测定自腐蚀电位，并对试验结果进行统计（图 1）。

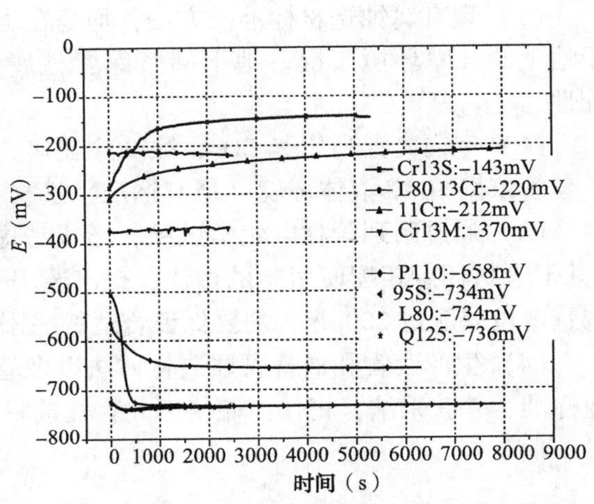

图 1　不同钢级、材质的管材腐蚀电位测定曲线

分析试验结果（图 1），可发现管材的腐蚀电位主要分布在 2 个区域，11Cr、13Cr 等耐蚀合金管材主要位于高电位区域，L80、P110 等低合金碳钢管材主要位移低电位区域，如图 2 所示。

为寻找不同材质间合适的匹配关系，选

取低电位材质 P110 钢级作为阳极，选取高电位材质 L80 13Cr 钢级作为阴极，组成电偶对，在上述地层水介质中进行电偶电流曲线测定，试验结果见图 3。

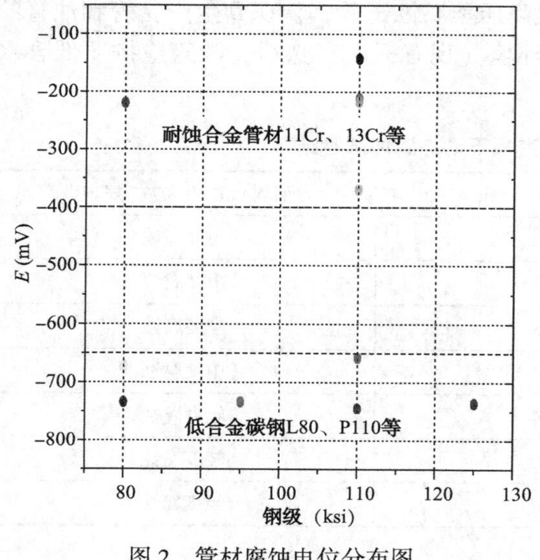

图 2　管材腐蚀电位分布图

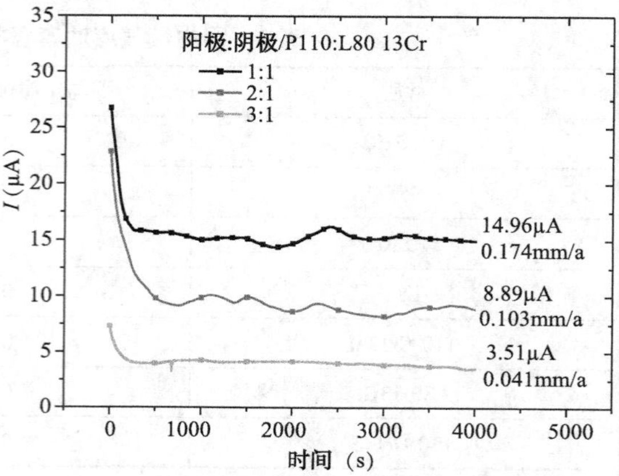

图 3　地层水中电偶电流曲线测定

根据电化学腐蚀速率计算方法[11]，依据图 3 曲线可计算获得相应的腐蚀速率，结果见图 3，可知为避免材质间产生电化学腐蚀，阳极/阴极的面积比必须大于 3，腐蚀速率值小于 0.076mm/a。

通过对 2 种不同电位区域的材质组成电偶对，在地层水和环空保护液环境中进行多次电化学腐蚀试验，发现 2 种材质电位差超过 200mV 需要注意电化学腐蚀，在材质匹配上要求在同一空间内，低电位与高电位材质的面积比至少要大于 1∶1，在地层水环境中面积比要求在 3∶1 以上，才可能保证腐蚀速率值小于 0.076mm/a（SY/T 5329 标准规定）。

因此，进行管柱选材时，应尽可能依据图 2 在同一电位区域内选择使用，若要跨区域，需要保证合适的阴阳面积比。

5　结论

（1）现有腐蚀选材标准或方法，均是在 100%液体环境下确定管材选用的，不完全适用于地下储气库环境工况。地下储气库管柱腐蚀选材应考虑低含水工况和不同构件材质匹配性。

（2）室内高温高压釜 CO_2腐蚀试验证实，生产套管和油管的腐蚀选材应分别对待，生产套管依据 100%液体环境（现有标准）选材，油管应按照低含水工况选材。

（3）通过系列管材电化学试验，形成考虑自腐蚀电位差的管材匹配选用图版，改变了“井下管材任意匹配使用”的做法，提出选用要求：在同一电位区域内选材，若要跨电位区域且电位差超过 200mV，需要保证合适的阴阳面积比。

（4）生产套管和油管腐蚀选材方法以及管材匹配选用图版，已提交中国石油天然气行业标准，并获采纳，将作为储气库井管柱选材依据。

参　考　文　献

[1] 王建军. 地下储气库注采管柱密封试验研究［J］. 石油机械，2014，42（11）：170-173.

[2] ISO15156-3，Petroleum and natural gas industries. Materials for use in H_2S-containing environments in oil and gas production. Part 3：Cracking-resistant CRAs（corrosion-resistant alloys）and other alloys [S]. 2015. Geneva，Switzerland：International Organization for Standardization.

[3] 全国石油钻采设备和工具标准化技术委员会. 石油天然气工业 油气开采中用于含硫化氢环境的材料 第2部分：抗开裂碳钢、低合金钢和铸铁（ISO 15156-2-2003，MOD）：GB/T 20972.2—2008 [S]. 北京：中国标准出版社出版社，2009：6-9.

[4] ISO15156-2，Petroleum and natural gas industries. Materials for use in H_2S-containing environments in oil and gas production. Part 2：Cracking-resistant carbon and low-alloy steels，and the use of cast irons [S]. 2015. Geneva，Switzerland：International Organization for Standardization.

[5] 石油管材专业标准化技术委员会. 石油天然气工业特殊环境用油井管 第1部分：含 H_2S 油气田环境下碳钢和低合金钢油管和套管选用推荐做法：SY/T 6857.1—2012 [S]. 北京：石油工业出版社，2012：7-9.

[6] 中国石油天然气集团公司管材研究所. 含 H_2S 油气田环境下碳钢和低合金油管和套管选用推荐作法：Q/SY TGRC 2—2009 [S]. 西安：中国石油天然气集团公司管材研究所，2009.

[7] 中国石油天然气集团公司管材研究所. 耐蚀合金套管和油管：Q/SY TGRC 3—2009 [S]. 西安：中国石油天然气集团公司管材研究所，2009.

[8] 中国石油天然气集团公司管材研究所. 含 CO_2 腐蚀环境中套管和油管选用推荐作法：Q/SY TGRC 18—2009 [S]. 西安：中国石油天然气集团公司管材研究所，2009.

[9] 油气田开发专业标准化技术委员会. 碎屑岩油藏注水水质指标和分析方法：SY/T 5329—2012 [S]. 北京：石油工业出版社，2012：1-3.

[10] NACE RP 0775，Preparation，Installation，Analysis，and Interpretation of Corrosion Coupons in Oilfield Operations [S]. 2005. Houston TX：NACE International.

[11] 严密林，李鹤林，邓洪达，等. G3 油管与 SM80SS 套管在 CO_2 环境中的电偶腐蚀行为研究 [J]. 天然气工业，2009，29（2）：111-112.

A Study of Technical Specifications for High Grade Casing in Deep Wells

Wang Jiandong[1] Wang Huan[2] Mao Yuncai[3]
Wei Fengqi[3] Ma Yong[4] Li Yufei[4] Zhou Lang[4]

(1. Tubular Goods Reseach Institute of CNPC; 2. China Petroleum Materials Corporation; 3. China Petroleum Exploration and Production Company; 4. Petrockina Southwest Oil and Gas Field Company)

Abstract: In Western China oilfield where the well depths are more than 6000m TVD, steel grades higher than 140V are often used for the intermediate and production casing strings. No industry standard has been established for such high grade casing, although there are some obvious shortcomings associated with the higher grade casing, such as sensitivity to defects and working environment, prone to fracture. This paper presents a discussion on the technical specifications of the high strength casing steels, based on the analysis of the fracture failure at three couplings for a ϕ339. 7mm×13. 06mm 140V grade intermediate casing.

Investigation was conducted using failure assessment diagram (FAD) on the minimum transverse charpy impact energy required to prevent instable crack propagation with various defects and effective wall thicknesses of the pipe body and coupling. This investigation was based on comparison of similar products from various manufactures with the analysis on physical and chemical properties as well as microstructure of the casing materials used in the well. The analysis results were found to be consistent with the full scale burst testing results.

The technical specifications for the 140V grade casing steel were proposed based on the results of this research. (1) To meet NDT L2 level and to keep more than 90% effective wall thickness, the minimum transverse impact energy of the pipe body and coupling should be greater than 80J. (2) To reduce the ductile-brittle transition temperature by maintain range of yield strength965 ~ 1103MPa. It was found that all casings meeting these requirements have not experienced any fracturing during the subsequent application.

1 Introduction

The deep well and ultra deep well is increasing with more difficult for exploration and development. The high grade casing of out of API 5CT rang was applied more and more.

In Western China oilfield where the well depths are more than 6000m TVD, steel grades higher than 140V are often used for the intermediate and production casing strings. No industry standard has been established for such high grade casing, although there are some obvious shortcomings associated with the higher grade casing, such as sensitivity to defects and working environment, prone to fracture.

This paper presents a discussion on the technical specifications of the high strength casing steels, based on actual failure accidentsin oil field.

The casing of ϕ339. 7mm×13. 06mm 140V was used for intermediate casing strings of 4500m depth. Three locations were seriously blocked and the scrap iron was returned in the process of drilling plug after secondary cementing. The Primary cementing segment casing string fracture was found by electromagnetic logging. The mechanical and chemical as well as physical performance was tested. The test sample was taken from remaining unused casing on the well site. The casing order technical conditions were analyzed, based on test results and theoretical analysis.

In order to determine whether casing material mechanics performance indicators meet the practical application.

2 Study on the material performance by test

Material tensile properties and charpy impact energy have significant influence for casing used. The casing material the well and other similar products from various manufacture was analyzed by comparative. The test results were shown in table 1. The ductile brittle transition temperature of charpy impact energy and the shear area ratio were shown in figure 1, 2.

Table1 Tensile properties of 140V grade of casing

item	performance	sample quantity	mean	SD	COV	max	min
A	yield strength	21	1091 (966~1172)	21. 63	0. 02	1123	1031
	Tensile strength	21	1164 ≥1034	15. 23	0. 01	1184	1135
	elongation	21	26. 29	0. 72	0. 03	27. 00	25. 00
B	yield strength	9	1033 (966~1138)	9	0. 01	1050	1023
	Tensile strength	9	1112 ≥1034	14	0. 01	1130	1094
	elongation	9	27	2	0. 09	29	24

Notes: A-the well casing, B-similar products. The following is the same.

By comparison analysis of mechanical performance tests result: (1) the range of yield strength has a significant effect on impact energy with the decrease of yield strength to increase. (2) the ductile brittle transition temperature of charpy impact energy, the well casing was rang of −20~−10℃, The other similar product was rang of −40~−30℃, based on FATT50.

3 Analysis of impact energy of preventing cracking

Casing was damaged in the process of transportation, loading and unloading. The coupling

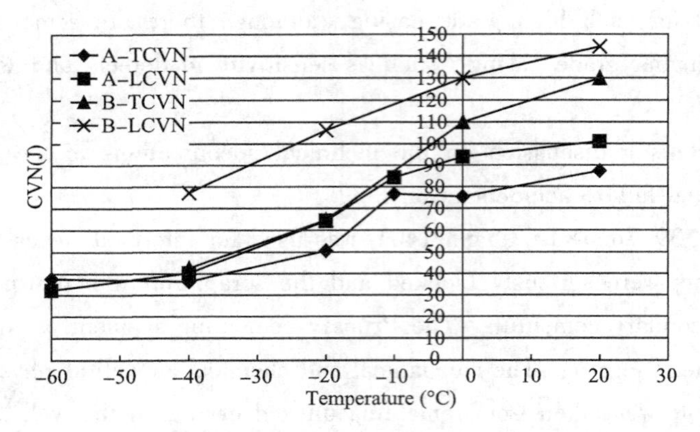

Fig. 1 Impact energy with temperature changes

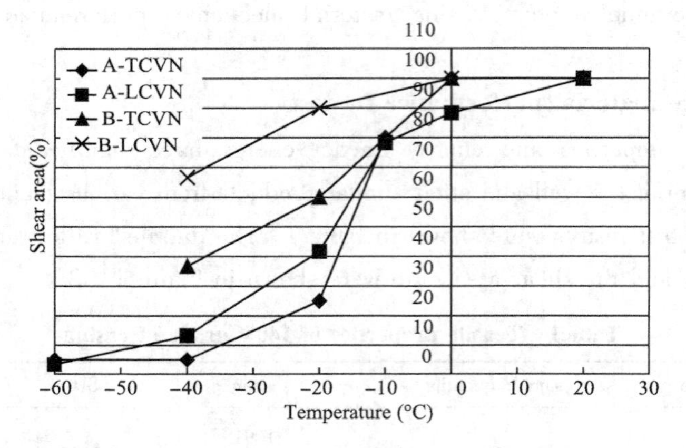

Fig. 2 Shear area with temperature changes

defect was undetected. The above things were prone to failure of casing of downhole. The damage form for pipe body and coupling longitudinal defect was shown in figure. 3.

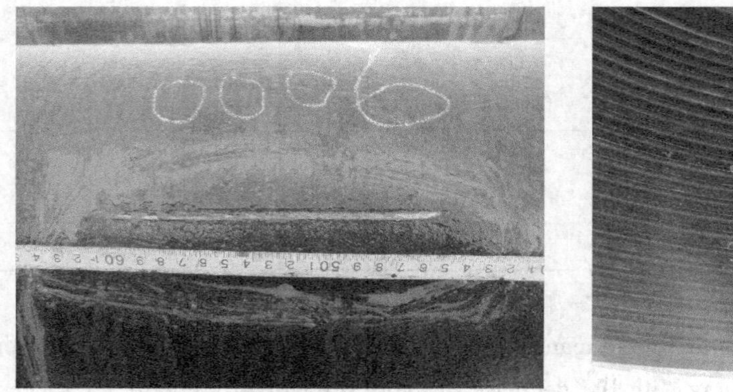

Fig. 3 Damage macro morphology of tube and coupling

Therefore, damage tolerance and impact energy of preventing cracking were needed study. The failure assessment diagram was the most effective methods. Based on fracture limit state equation in ISO 10400 standards, Fracture toughness of prevent crack propagation of crack-like imperfection on the outer surface of the pipe was analyzed.

$$(1-0.14L_r^2)(0.3+0.7\exp[-0.65L_r^6]) = [p_{iF}(D/2)^2(\pi a)^{1/2}]/[((D/2)^2-(D/2-K_{wall}t)^2K_{Imat}]\times \{2G_0-2G_1[a/(D/2-K_{wall}t)]+3G_2[a/(D/2-K_{wall}t)]^2-4G_3[a/(D/2-K_{wall})]^3+5G_4[a/(D/2)-K_{wall}t)]^4\}$$

$$L_r=\sqrt{3}/2(p_{iF}/f_{ymn})[(d_{wall}/2+a)/(K_{wall}t-a)]$$

Equation1 is empirical for CVN of calculation.

$$(K_{Imat}/f_{ymn})^2=5\ (CVN/f_{ymn}-0.1) \tag{1}$$

The transverse impact energy was needed to analysis of calculation as shown in figure 4, 5 for different of the effective thickness and NDT level.

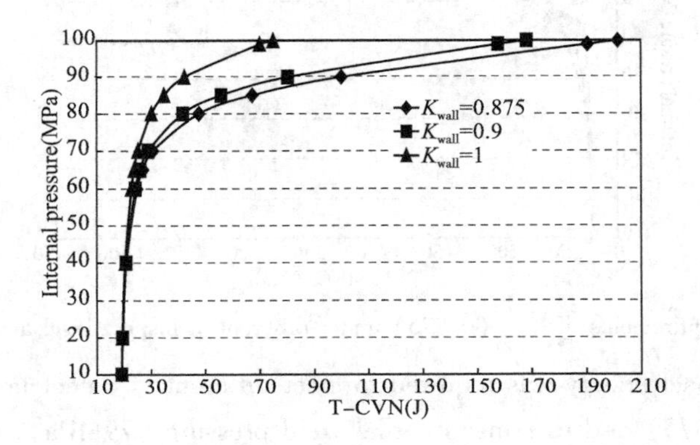

Fig. 4 Level L2 with different of the effective thickness

Internal yield pressure equation 2

$$p_{iYAPI}=[2f_y\ (K_{wall}t)\ /D] \tag{2}$$

When yield strength was range of 966 ~ 1172MPa and coefficient of effective thickness was range of 0.875 ~ 1, The transverse impact energy of tube with different of internal yield pressure was shown in Table 2. When extremer internal yield pressure was 90MPa, impact energy with different of effective thickness was shown in Table 3. With internal yield pressure increasing, sensitivity of effective thickness was significantly increased. The impact energy was needed increasing with effective thickness decreasing. When coefficient of effective thickness was less than 90%, the impact energy was needed more than 1 times under extreme internal yield pressure.

Table 2 The transverse impact energy with internal yield pressure changes

Item	f_y (MPa)	K_{wall}	Internal yield pressure (MPa)	NDE Level	T-CVN (J)
1	966	0.875	65	L2	27
2	966	0.9	67	L2	28
3	966	1.0	75	L2	29
4	1172	0.875	79	L2	50
5	1172	0.9	81	L2	40
6	1172	1.0	90	L2	43

Table 3 The transverse impact energy with effective thickness change

internal yield pressure (MPa)	K_{wall}	NDE Level	T-CVN (J)	Increment (%)
90	0.875	L2	100	133
	0.9	L2	80	86
	1.0	L2	43	0

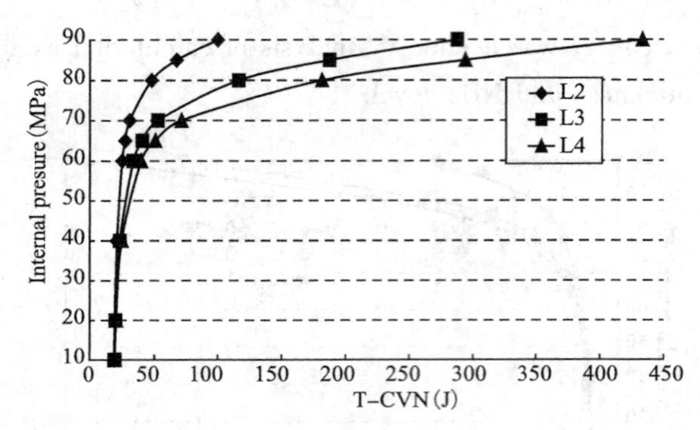

Fig. 5 T-CVN requirements of K_{wall} (0.875) under different of inspect level and internal pressure

The transverse impact energy was required to meet different of defect level under coefficient of effective thickness (0.875) and extremer internal yield pressure (79MPa).

Table 4 Impact energy requirements with different of defect level

Internal Yield Pressure (MPa)	K_{wall}	NDE Level	T-CVN (J)	Increment (%)
79	0.875	L2	48	0
		L3	116	142
		L4	181	277

Thedefect depth was allowed to cracking in casing under internal yield pressure 65 MPa and transverse impact energy 60 J as well as different effective thickness. It was shown in table 5.

Table 5 Allowed defect depth in casing

Internal yield Presser (MPa)	T-CVN (J)	K_{wall}	Defect depth (%)
65	60	0.875	14
		0.9	15.5
		1.0	21.5

The impact energy requirement was improved with improvement of defect tolerance under same effective thickness and internal pressure. The impact energy increased more than 1 times in L2 level above. The defect tolerance was improved with improvement of effective thickness under same internal pressure and impact energy.

Hoop stress distribution of premium connection was analyzed by FEA under make-up and internal pressure loading. It was shown in figure 6. The hoop tensile stress in coupling inside surface was greater than outer surface. Under internal pressure loading, hoop tensile stress was changed on internal pressure. It was shown figure 7. The coupling damage in thread nearby end was found on site. It was shown in figure 3. The impact energy of preventing cracking propagation was analyzed based on width of bearing face.

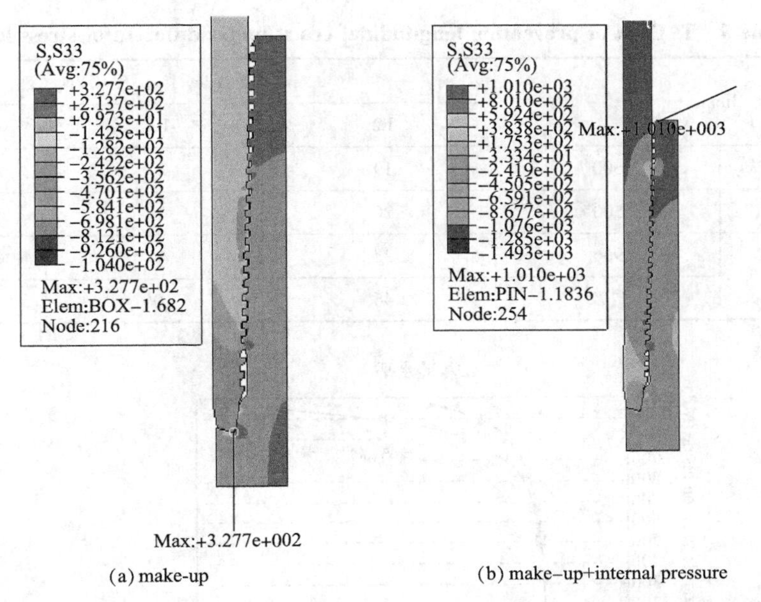

(a) make-up (b) make−up+internal pressure

Fig. 6 Hoop tensile stress distribution of premium connection under make−up and internal pressure

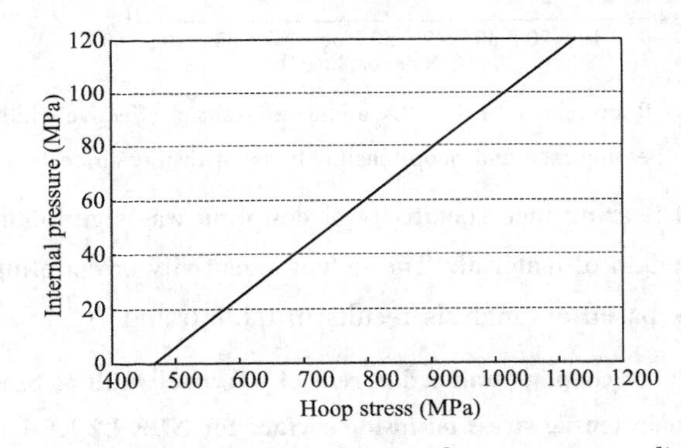

Fig. 7 Hoop tensile stress changed with internal pressure on coupling a location

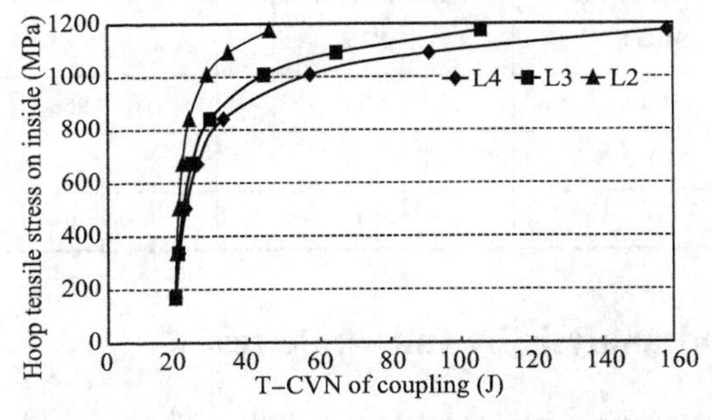

Fig. 8 Requirement of transverse impact energy under different of inspect level and inside hoop stress

The impact energy to prevent cracking from coupling end was shown in table 6. Thedefect sensitivity of coupling was poor under state of low stress level less than nominal yield strength of material. The defect sensitivity of coupling and requirement of transverse impact energy was significantly improved with stress level increasing.

Table 6 T-CVN of preventing longitudinal crack under different of stress level

Item		Inspect of level and T-CVN (J)		
		L2	L3	L4
Residual stress (MPa)	300	19	20	21
Stress of make-up (MPa)	500	20	21	22
YSM (MPa)	966	28	44	57
	1172	46	106	158

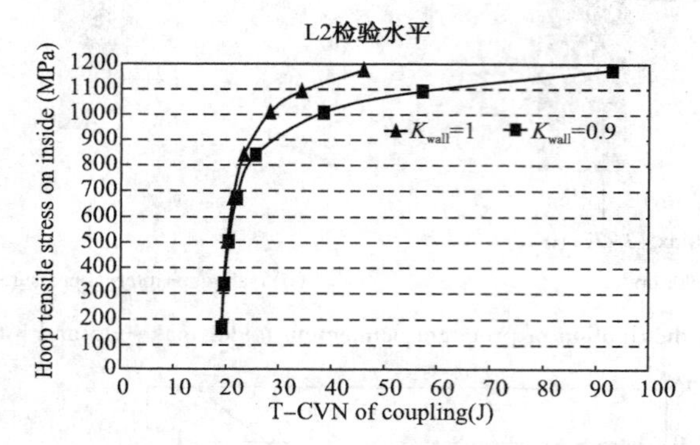

Fig. 9 Requirement of T-CVN under different of effective width of bearing face and hoop tensile stress on inside surface

The effective width of bearing face sensitivity of coupling was significantly improved with Stress level greater than yield strength of material. The defect sensitivity of coupling was much greater than the effective wall thickness based on analysis results in table 6 and 7.

Table 7 T-CVN of coupling under different of effective width of bearing face and hoop tensile stress on inside surface for NDE L2 level

Item	Residual stress (MPa)	Stress of make-up (MPa)	YSM (MPa)	
	300	500	966	1172
$K_{wall}=1$ T-CVN (J)	19	20	28	46
$K_{wall}=0.9$ T-CVN (J)	20	21	40	92

4 Verification and analysis by full-scale test

The error of theoretical analysis was validated by full-scale physical test. The three sample of number $5^{\#}\sim7^{\#}$ was selected. Field end thread make up were used by maximum and optimum as well as minimum torque. The internal pressure burst test was performed. Test result was showed in table8, 9 and macro was showed in figure 10.

Table 8　Internal pressure burst test for tube

Test No.	Burs value (MPa)	Minimum wall thickness (mm)	NDE level	T-CVN measured (J)	T-CVN calculated For internal pressure100MPa (J)
$5^{\#}$	101. 4	13. 19	L2	90	
$6^{\#}$	101. 1	13. 10	L2	91	75
$7^{\#}$	101. 9	13. 06	L2	70	

Table 9　Internal pressure burst test for coupling

Test No	Burs value (MPa)	width ofbearing face in coupling (mm)	NDE level	T-CVNmeasured (J)	T-CVN calculated For internal pressure120MPa (J)
$5^{\#}$	101. 4	9. 6	L2		
$6^{\#}$	101. 1	10	L2	75	75
$7^{\#}$	101. 9	9. 6	L2		

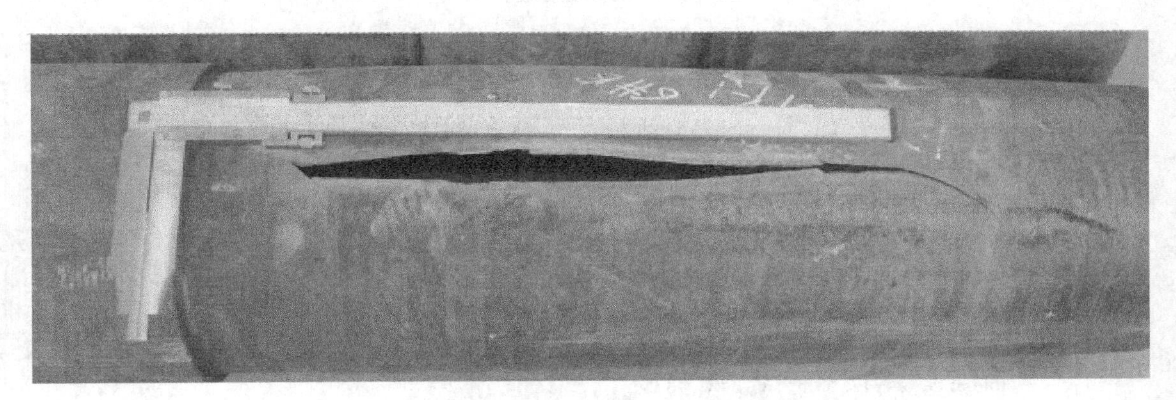

Fig. 10　Macro for internal pressure burst test

The Internal pressure failure test phenomenon was tube burst. The fracture macroscopic had obvious plastic deformation and showed ductile fracture. When the tube had NDE level for L2, Effective thickness of more than 100%, and T-CVN of more than 70J, burst pressure was more than 100MPa. The test value and calculated value of absolute error was only 5J and relative error less than 7%. when internal pressure was 120MPa in casing, coupling happened crack on bear face location and effective thickness of 90% by calculate. The burst occurred in tube was explained by the analysis of calculation. Test results and calculation analysis proved that FAD method has high reliability and fit for high grade casing defect analysis.

5　Conclusions and recommendations

The following conclusions and recommendations was obtained by 339. 7mm × 13. 06mm 140 grade of casing material mechanics performance analysis and full-scale test.

(1) The material yield strength should be controlled in the range of 966 ~ 1100MPa to meet the requirements of ductile brittle transition at low temperature.

(2) The transverse impact energy was greater than 40J at low temperature (-40℃) to meet the needs of winter operation.

(3) The transverse impact energy of pipe and coupling were greater than 80J at NDE level L2 and effective thickness of more than 90% to meet extremity internal yield pressure of casing 90MPa.

References

[1] Specification for Casing and Tubing, API Specification 5CT, 2010.

[2] ISO/TR 10400 Petroleum and natural gas industries-Equations and calculations for the properties of casing, tubing, drill pipe and line pipe used as casing or tubing, 2007.

[3] API 579-1 Fitness-For-Service, 2007.

[4] Roberts, R. and Newton, C. Interpretive Report on Small Scale Test Correlations with KIC Data [S]. Welding Research Council, England, Bulletin265 (Feb. 1981).

[5] P. W. Moore, SPE, Grant Prideco, J. G. Maldonado, InterCorr International, Inc. Review of the Recent API PRAC Project Attempt to Evaluate the ISO DIS 10400 Failure Assessment Diagram Through Full Scale Testing of Pipe Containing Surface Imperfections (SPE 97578).

Nomenclature

L_r——the load ratio;

a——for a limit state equation, the maximum actual depth of a crack - like imperfection; for a designequation, the maximum depth of a crack - like imperfection that could likely pass the manufacturer's inspection system;

d_{wall}——the inside diameter;

D——the specified pipe outside diameter;

f_{ymn}——the specified minimum yield strength;

K_{Imat}——the fracture toughness of the material in a particular environment;

K_{wall}——the factor to account for the specified manufacturing tolerance of the pipe wall. For example, for a tolerance of -12. 5 %, K_{wall} =0, 875;

L_r——the load ratio;

p_{iF}——the internal pressure at fracture;

p_{iR}——the internal pressure at ductile rupture of an end-capped pipe;

p_{iRa}——p_{iR} adjusted for axial load and external pressure;

t——is the specified pipe wall thickness;

T-CVN——transverse charpy V-notch impact test minimum absorbed energy;

L-CVN——longitudinal charpy V-notch impact test minimum absorbed energy;

NDE——non-destructive examination;

L2 (4) ——acceptance (inspection) level.

Cracking Mechanism of the Interface of Precipitation in High Strength Drill Pipe during Sulfide Stress Corrosion

Wang Hang[1,2] Han Lihong[1,2] Lu Caihong[1,2] Yang Shangyu[1,2] Tian Tao[1,2] Jiang Long[1,2]

(1. Tubular Goods Research Institute of China National Petroleum Corporation;
2. State Key Laboratory of performance and structural safety for petroleum tubular goods and equipment Materials)

Abstract: Precipitation strengthening enhanced steel grade, but induced sulfide stress corrosion crack (SSCC). So it is difficult to be compatible between strength and resistance to SSCC. Cracking mechanism of the interface of precipitation was studied in high strength drill pipe via uniaxial tensile, NACE-A method, In-situ TEM and SEM observations. The results showed that the strength maintained above 120ksi while impact toughness enhanced to 196J as tempering temperature increased. The specimen at 580℃ fractured but the one at 625℃ remained well during NACE-A method test. The microscopic observation indicated that the straightness crack propagation path occurred along the interface of rod-like precipitation in the former. Whereas, the curve path appearred as the morphology of precipitation evolved into spherical during in-situ TEM. Corrosion was observed at grain boundary firstly. Particularly, the void initiated and evolved into micro-crack in the interface of rod-like precipitation, which resulted in intergranular fracture finally. Cracking mechanism of the interface of rod-like precipitation could be attributed to the discrepancy of inharmonious deformation between precipitation and the matrix, as well as the dilution of alloy element.

Keywords: Precipitation; Strengthening; Toughness; Crack propagation; Sulfide stress corrosion

1 Introduction

Chuan Yu and Tarim Tazhong area are both strategy oil and gas field in china. Their work condition, such as well depth and high concentration of sulfide, restricted the exploration and development. Currently, the steel grade of anti-sulfide drill pipe available is not more than 105ksi, which does not meet requirement of stress design. Moreover, it is demanded to be fine match

Corresponding author. Tel: +86 29 88726206. E-mail: wanghang008@ cnpc. com. cn.

between strength and toughness, and satisfy the fracture criterion of leakage before fracture [1].

The steel grade can be enhanced by means of precipitation strengthening. The high strength low alloy steel had been developed successfully through nano-sized carbide in the ferritic matrix[2]. Its strength reached to 780MPa, of which approximately 300MPa was considered to be from nano-sized caribide due to coherency/semi-coherency crystal relationship between nano-sized carbide and the ferritic matrix. Meanwhile, fracture toughness also depended on microstructural feature[3,4], specifically carbide size and distribution. The investigation indicated that multiple coupling of precipitation size and morphology enhanced the strength. In addition, the impact toughness also improved as volume fraction of spherical precipitation increased[5].

The sensitivity to SSCC increased evidently as strength enhanced for drill pipe. Therefore, the hardness of traditional anti-sulfide drill pipe was limited to below HRC22. The related study indicated that the FeS film was formed in the surface, the hydrogen from corrosion was absorbed and induced crack (HIC), then came into being hydrogen bubble in the surface as well, which resulted in SSCC for both carbon and low alloy steel under sulfide environment. So far, the useful process was added alloy element (Cr, Mo and Ni) to enhance the resistance to SSCC. In addition, homogeneous and spherical carbide interior grain when addition of 0.1wt% Nb[6], as well as nano-sized carbide (10~20nm) through addition of V element [7].

So far, the investigation was still concentrated on theoretical analysis, such as hydrogen embrittlement and bubble. The microstructural evidence was absented, specifically, cracking mechanism of the interface of precipitation during SSCC. Therefore, the relationship between strength and resistance to SSCC was unclear for high strength drill pipe.

2 Experimental

Drill pipe with high steel grade was selected, on the basis of chemical composition of anti-sulfide drill pipe, including Cr, Mo, and Ni. Moreover, micro-alloy element such as Nb, V and Ti was added, as follow in Table 1.

Table 1 Chemical composition of drill pipe with sulfur resistance

C	Si	Mn	P	S	Cr	Mo	Ni	Nb-V-Ti	Cu	Al
0.20	0.21	1.00	0.007	0.0007	0.91	0.23	0.069	≤0.012	0.043	0.017

Testing material was melted in vacuum induction furnace, forged at 800~1100℃, followed by quenching at 900℃ for 60 minutes, then tempering at 580℃ and 625℃ for 60 minutes, respectively. Mechanical properties were measured using Instron 1195 electro-hydraulic service test machine.

The specimen for in-situ TEM was cut from longitudinal cross section with size of 3mm×9mm using wire-electrode cutting. Its thickness was reduced from 0.5mm to 60μm through mechanical friction, then two-jet thinning. The micro-crack was regarded as primary main crack, which was introduced in centre thin zone of the specimen during two-jet operation, and then observed the fracture process. In-situ TEM was performed using JEM-200CX transmission electron microscope (TEM) operating at 200kV.

SSC test was conducted using A method according to NACE TM0177 standard, Specifically, uniaxial tensile loading to lowerlimit of 85% yield strength for 720 hour, identification whether crack or not. The tensile specimen was prepared according standard, its surface roughness was required to be not more than 7 grade, i. e. degree of finishing 0. 8μm. H_2S gas was pumped into solution cyclically to ensure in state of saturation. The feature of crack propagation after SSC was observed on OSL-4100 laser scanning confocal microscope. Meanwhile, fracture morphology was observed on TESCAN VEGA Ⅱ scanning electron microscope (SEM).

3 Results

3.1 Uniaxial tensile and sulfide stress corrosion

Mechanical properties was showed in Table2 for drill pipe at different tempering temperature, indicating the steel grade above 120ksi as tempering temperature increased to 625℃.

The specimens were observed to fracture within 720 hour during NACE-A method test for drill pipe at 580℃, of which the first one remained 17 hour, the second 18 hour, and the third only 14 hour, as shown in Figure 1. Whereas, as for drill pipe tempering at 625℃, its specimen remained well after 720 hour during NACE-A method test. as shown in Figure 2. These results indicated that the latter specimen had much higher resistance to SSCC.

Table 2 Uniaxial tensile property of drill pipe at different tempering temperature

Number	Tempering temper (℃)	Yield strength (MPa)	Tensile strength (MPa)	Elongation (%)	Impact toughness (J)
1#	580	876	979	22. 1	175
2#	625	834	899	26. 0	196

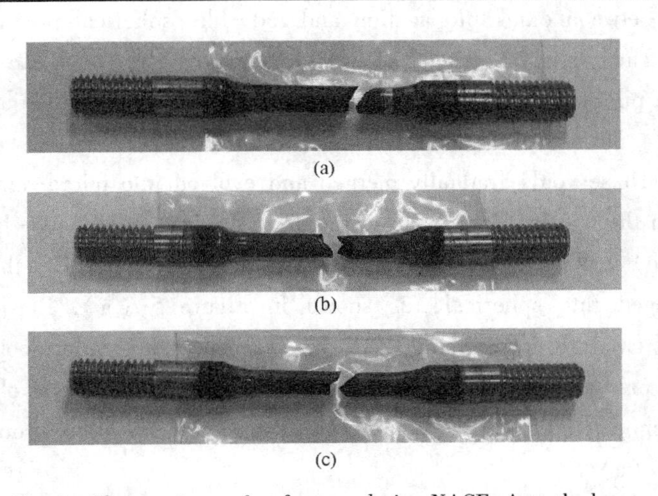

Fig. 1 The specimen after fracture during NACE-A method test

3.2 Precipitation Feature and In-situ TEM Observation

The microstructure and precipitation were observed in high strength drill pipe, as shown in Figure 3. The typical microstructure was tempered sorbite at 580℃, accompanying with rod-like precipitation [Figure 3 (a)]. It is worthy of note that grain boundary was almost covered by strip and rod-like

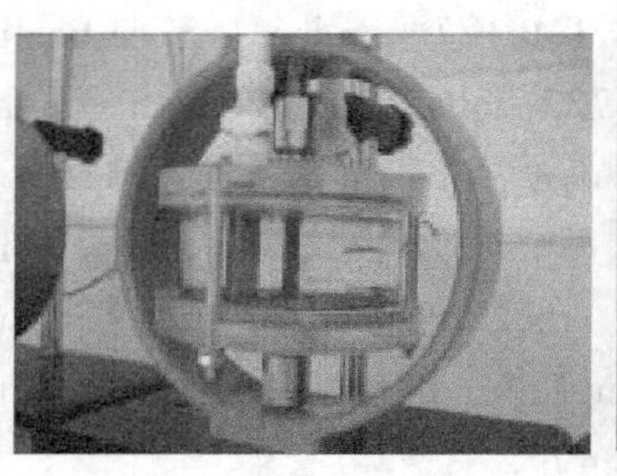

Fig. 2 The specimen remaining well after 720h during NACE-A method test

precipitation. As temperature increased to 625℃, the morphology of typical precipitation evolved from rod-like to spherical, which distributed homogeneously. On closer examination, the precipitation was characterized by isolate and discontinuous each other [Figure 3 (b)].

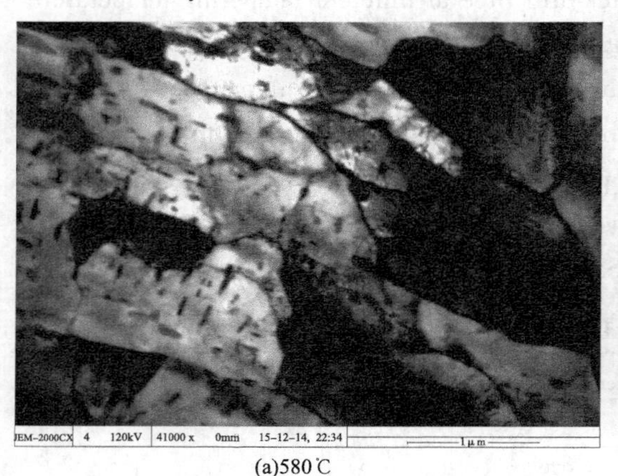

(a)580℃

(b)625℃

Fig. 3 The morphology of precipitation in drill pipe at different temperature.

The interaction between crack propagation and rod-like/spherical precipitations was studied using in-situ TEM. The primary crack were showed in Figure 4 (a). it can be seen that a large number of dislocation pileup occurred along the interface of rod-like precipitation, which introduced stress concentration, and then initiated void at this local region, as shown in Figure 4 (b) indicating by arrow. These voids gradually merged and evolved into micro-crack, then connected with primary crack in the matrix, leading to cracking of the interface of rod-like precipitation, as shown in Figure 4 (c). When tempering temperature increased to 625℃, the typical morphology of precipitation changed into spherical, as shown in Figure 5 (a). Plastic deformation zone occurred at the tip of crack during in-situ TEM. It is seen that dislocation loop was released at the interface of spherical precipitation, as shown in Figure 5 (b). The void was observed in the plastic deformation zone, connected with primary crack, and formed curve crack propagation path finally, as shown in Figure 5 (c).

3.3 Fracture morphology

Based on SEM observation, the fracture morphology was showed in Figure 6. The source region of crack initiation located close to surface of specimen, The region of crack propagation was characterized by flat and smooth, as shown in Figure 6 (a). The specimen fractured in the form of intergranular mode, accompanying with cleavage mode in local area, as shown in Figure 6 (b).

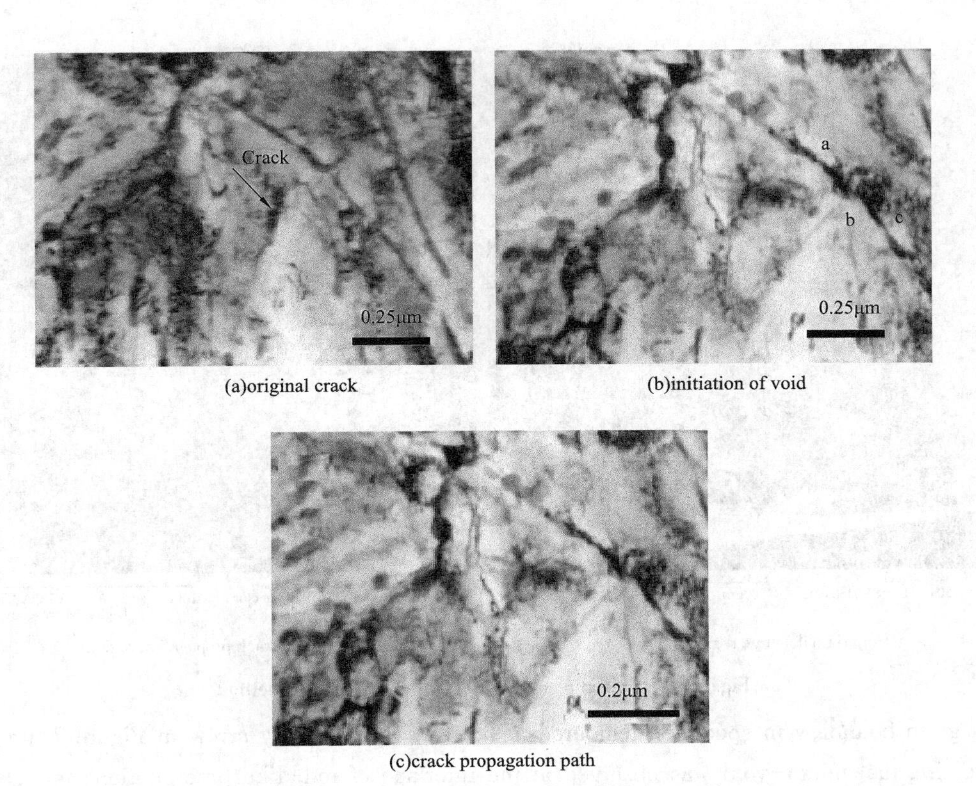

(a)original crack (b)initiation of void

(c)crack propagation path

Fig. 4 The crack propagation in drill pipe tempering at 580℃ during in-situ TEM

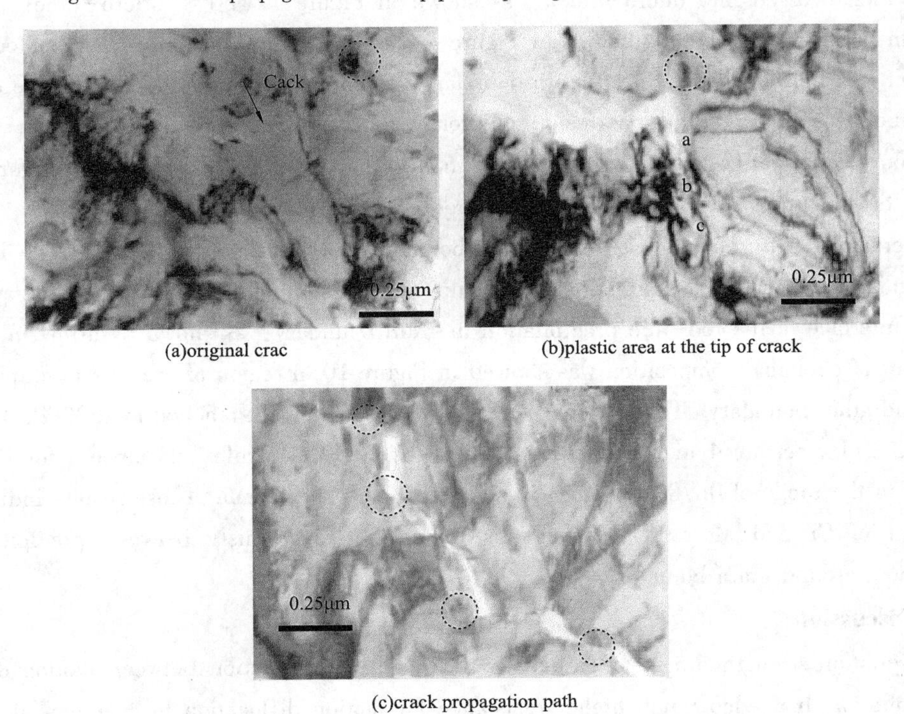

(a)original crac (b)plastic area at the tip of crack

(c)crack propagation path

Fig. 5 The crack propagation in drill pipe tempering at 625℃ during in-situ TEM

3.4 Microstructure feature during SSCC

The SEM observations were showed in Figure7, 8. A number of micro-voids were observed

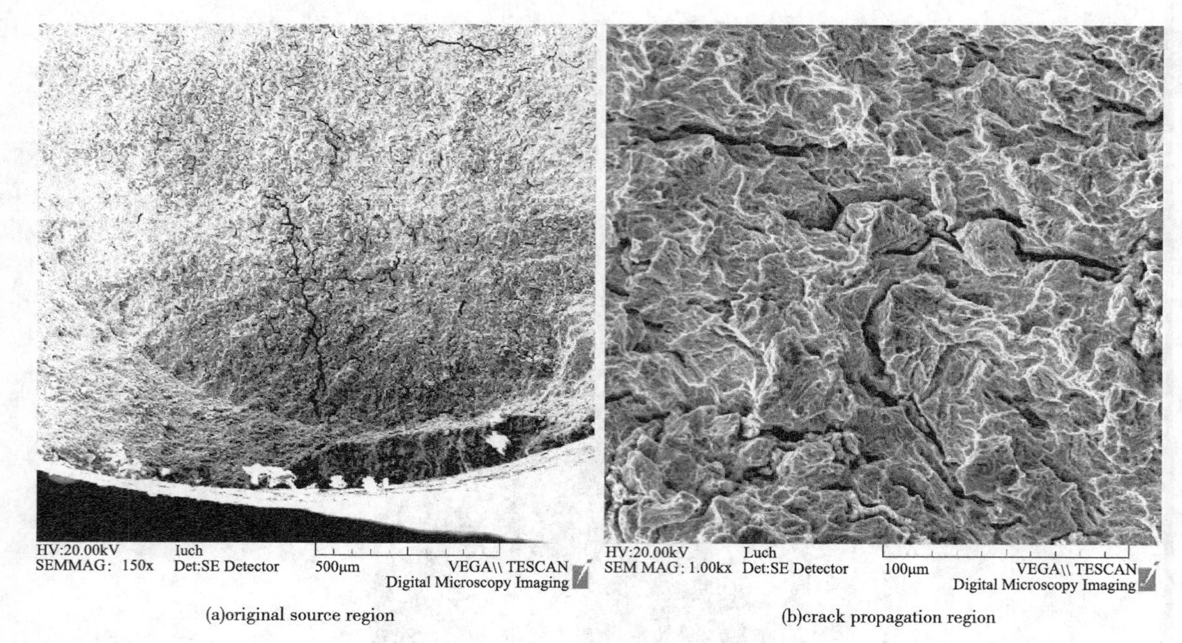

(a)original source region

(b)crack propagation region

Fig. 6 The fracture morphology after NACE-A method

around grain boundary in specimen tempered at 580℃, as shown by arrow in Figure 7 (a). It is worth noting that micro-void was observed at the interface of rod-like precipitation, merged each other and then evolved into micro-crack, as shown in Figure 7 (b). Micro-crack propagated along grain boundary finally, as shown in Figure 7 (c). Meanwhile, the laser confocal optical observation was showed in Figure 8. Grain boundary was damaged gradually, formed the defect and became thickness under sulfide corrosion environment, as shown in Figure 8 (a). On closer examination, corrosion was observed at grain boundary firstly, as shown by arrow in Figure 8 (b), indicating the sensitivity of microstructure to sulfide stress corrosion environment.

The cracking feature was observed at grain boundary, which occurred among triple junction of grain boundary. On closer examination, rod-like precipitation appeared, and crack propagated along the interface of the rod-like precipitation in grain boundary, as shown by arrow in Figure 9. The content of chemical composition was showed in Figure 10 in region of precipitation and adjacent area around grain boundary. The content of Cr and Mo alloy element reached to 1.30%, 0.91% for spectrum 3 and spectrum 4 in precipitations, respectively. Meanwhile, its amount for Cr element decreased to the range of 0.78 to 0.72% in spectrum1 and spectrum5. These results indicated that the content of Cr and Mo alloy element distributed heterogeneously between precipitation and adjacent area around grain boundary.

4 Discussion

Precipitation strengthening is regarded as the result of interaction between motion dislocation and precipitation. It needed much higher resistance for motion dislocation to overcome the obstacle from precipitation, and then induced the enhancement of strength. Precipitation strengthening is considered to depend on the hardness, morphology and its distribution for precipitation. Zhu et al[8] pointed out that the value of exponent q decreased from 2.0 to 1.0 as the strength of spherical

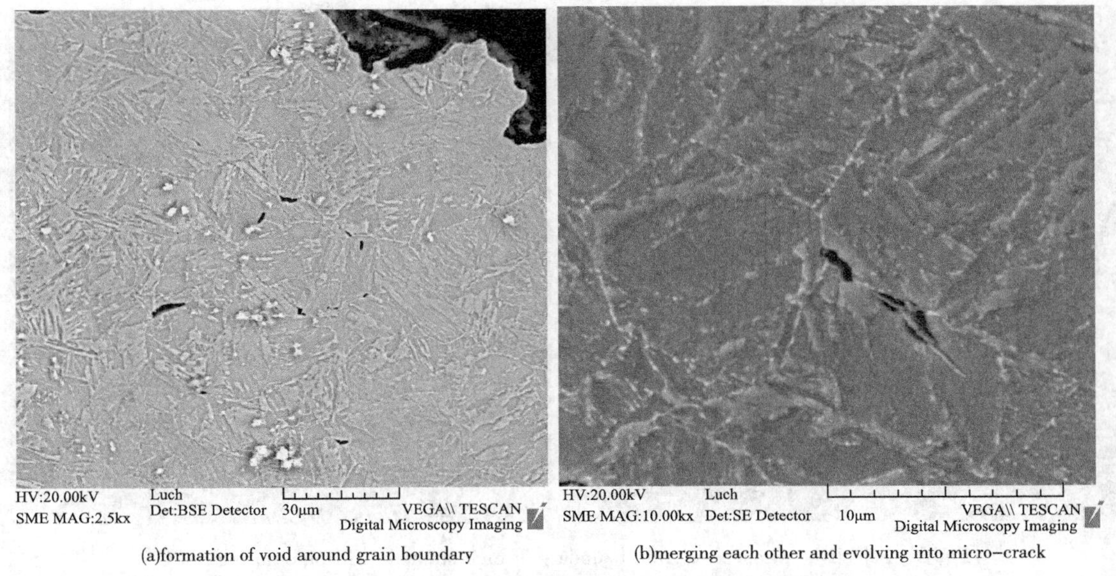

(a)formation of void around grain boundary (b)merging each other and evolving into micro-crack

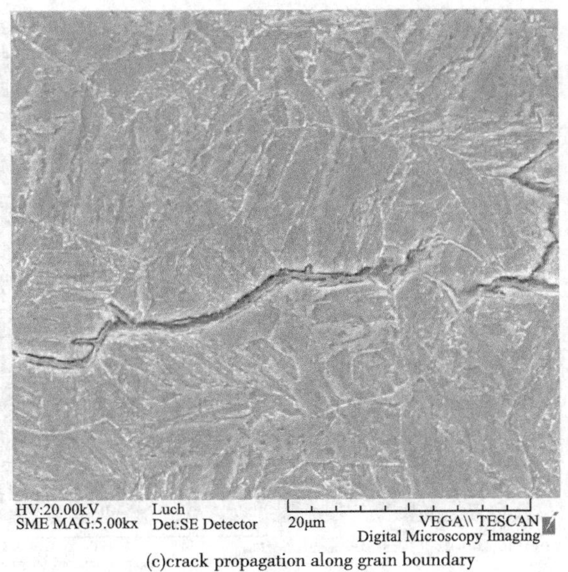

(c)crack propagation along grain boundary

Fig. 7 The cracking process during SSCC

particle declined in the system including two types of shearable and unshearable particle. As for two types of particle, their strengthening effect much higher than that induced by single type. The investigation results demonstrated that precipitation possessed two side effect for fracture toughness in aging aluminium alloy[9], one side the fracture toughness increased due to the enhancement of the resistance to deformation, another side the ductile and fracture toughness were reduced by the localization of slip deformation. Multiscale precipitations with spherical, rod-like and acicular morphology occurred in two-stage aging aluminium alloy. As a result, yield strength increased, as well as fracture toughness. The anomaly relationship between strength and toughness could be attributed to the competition with each other, including the interaction mechanism between precipitation and dislocation, the inharmonious deformation mechanism between precipitation and

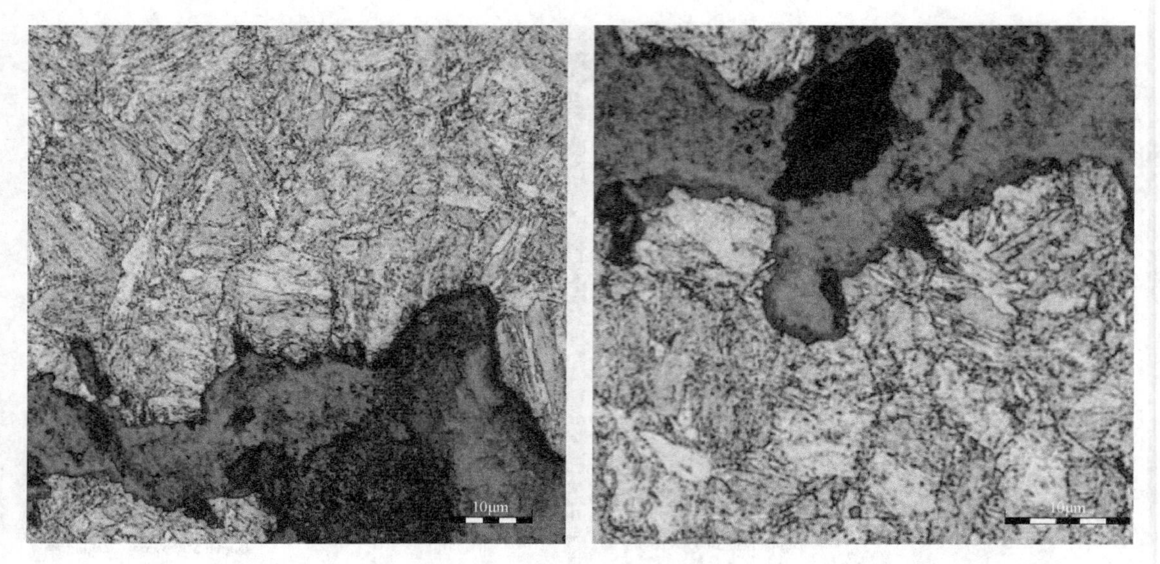

Fig. 8 Selective corrosion at grain boundary in high strength drill pipe during SSCC

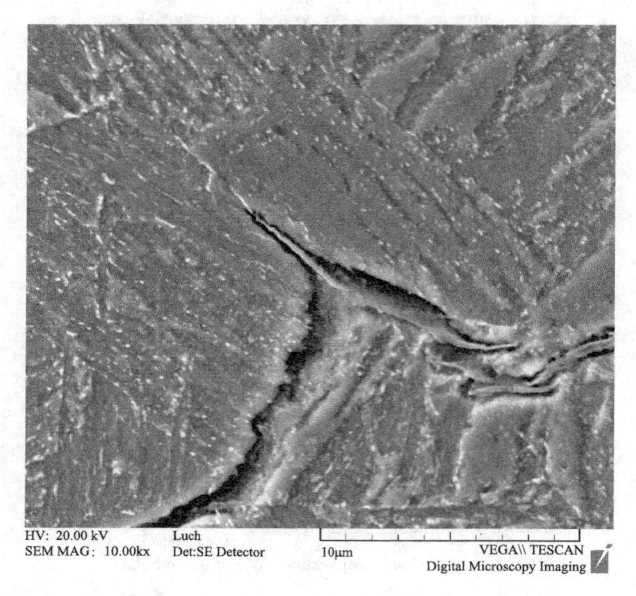

Fig. 9 Cracking feature of the interface of rod-like precipitation at grain boundary during SSCC.

the matrix.

The both deformation capacity and fracture resistance of alloy were considered to depend on two mechanisms: (1) interaction between dislocation and precipitation; (2) deformation harmony between precipitation and the matrix. The related investigation demonstrated that the strengthening effect from the interaction between spherical precipitation and dislocation was higher than that between rod-like precipitation and dislocation, then caused local stress concentration more easily. On the other hand, plastic deformation was accommodated only through rotation for precipitation due to body stiffness, resulted in inharmonious deformation with the matrix, which induced geometry necessary dislocation (GND)[10,11]. As for spherical precipitation, the density of GND (ρ_s^G) related with shear strain, as well as volume fraction of precipitation.

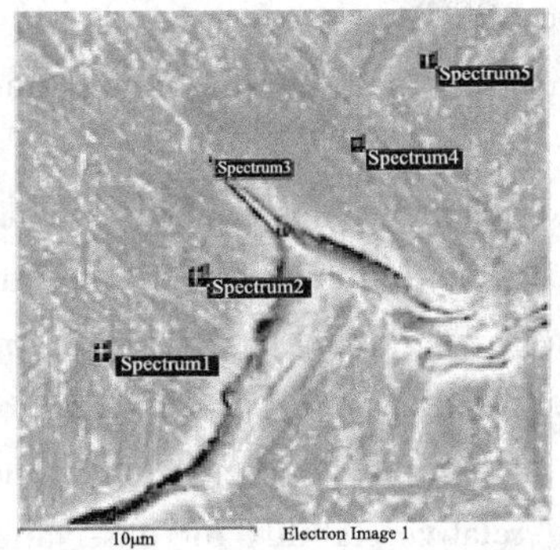

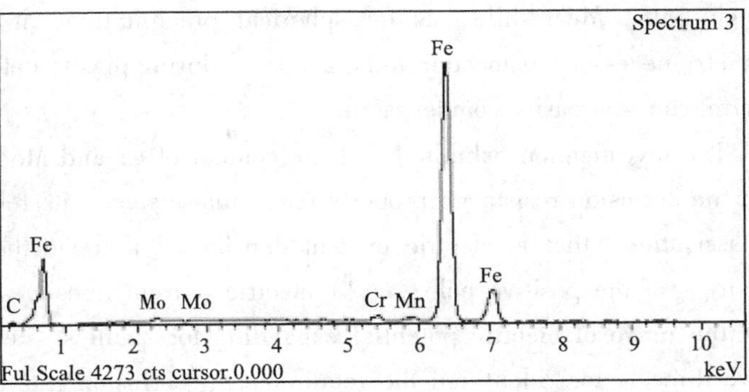

Processing option: All elements analyses(Normalized)(weight%)

Spectrum	In stats.	C	O	S	Cr	Mn	Fe	Mo	Total
Spectrum 1	Yes	5.37	7.20	1.86	0.78	0.95	83.84		100
Spectrum 2	Yes	2.50	5.57	2.71	0.89	1.01	87.32		100
Spectrum 3	Yes	3.29			1.04	1.00	93.75	0.91	100
Spectrum 4	Yes	4.21		0.83	1.30	1.32	92.35		100
Spectrum 5	Yes	7.99			0.72	0.97	89.91		100

Fig. 10　The EDS analysis of chemical composition in precipitation and adjacent area around grain boundary

$$\rho_s^G = \frac{4f_s\gamma}{br_s}$$

As for rod-like precipitation, the density of GND ($\rho_{r/n}^G$) can be formulated as follow[10]:

$$\rho_{r/n}^G = \frac{4\gamma}{b\lambda_{r/n}}$$

$\lambda_{r/n}$ represents the distance between particles, $\lambda_{r/n} = \min(\lambda_r, \lambda_n)$.

The effect of precipitation morphology on inharmonious deformation between precipitation and the matrix can be evaluated in term of the ratio of $\rho_{r/n}^G$ and ρ_s^G ($\rho_{r/n}^G/\rho_s^G$). Yuan[12] etc pointed out that the density of GND induced by rod-like precipitation ($\rho_{r/n}^G$) was higher one order of magnitude than that by spherical precipitation (ρ_s^G) when the particle distance ($\lambda_{r/n}$) was 100nm approximately. These results implied that it is needed to activated much more GND to compensate the inharmonious deformation between rod - like precipitation and the matrix during plastic deformation, in comparison with spherical precipitation, as shown in Figure 11.

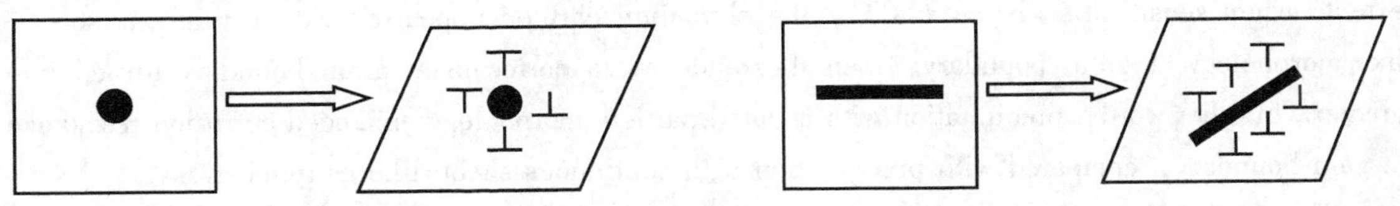

Fig. 11　The model of inharmonous deformation between spherical/rod-like precipitation and the matrix

When the dominant mechanism was deformation incompatibility between precipitation and the matrix, it needed to activate much higher density of GND to compensate the inharmonious deformation for rod-like precipitation. Therefore, void initiated and evolved into micro-crack due to dislocation pileup, and then resulted in crack propagation in the form of intergranular, forming straight path. Meanwhile, as for spherical precipitation, it was much lower for the density of geometry necessary dislocation to be activated during plastic deformation, and then the inharmonious deformation was easily compensated.

The investigation exhibited that the content of Cr and Mo alloy elements was closely associated with the corrosion resistance property for stainless steel, its related factors included: (1) be liable to passivation, that is electric current density i_{cc} of the cathodic greater than passivation current density i_c of the positive pole; (2) electric current density after passivation (i_p) becoming low; (3) the range of electric potential expanding for stainless steel in the state of passivation. These above three factors identified the requirement, corrosion rate and relative stability for passivation. As for the content of Cr, this has an obvious effect on above parameters. With increasing in the content of Cr, the critical electric current density (i_c) decreased during passivation, and the electric current density after passivation (i_p) declined, as well as the Flade potential (E_F). In other word, the corrosion resistance property improved with addition of Cr element. A large number of experimental results showed that both Cr and Mo elements were the most effective ones to increase the pit corrosion resistance for stainless steel, which made the breakdown potential toward positive shift in 3% NaCl solution at 22℃.

The physical process included three stages. Firstly, HS^- ionization was adsorpted above the the surface of oxidation film electrode with compact structure, then the original oxidation film became reduction continuously as sulfide film of Fe was produced under the reaction between HS^- and the surface film, finally the oxidation film in surface of the electrode destroyed thoroughly, and produced the sulfide file with loose structure. As for drill pipe steel, the diffusion rate of carbon atom toward grain boundary was faster than that of Cr element during tempering. Consequently, the carbide of $(CrFe)_{23}C_6$ precipitated at grain boundary, which resulted in the lacking of Cr element in adjacent domain around grain boundary. Every side mean width was reported to be approximately $(1.5 \sim 2.0) \times 10^{-5}$ cm for dilution zone of Cr element [13]. The corrosion accelerated at grain boundary due to the formation of micro-battery between lager cathode and smaller anode, when the content of Cr element decreased to the below value of that required to passivation. It is worth noting that corrosion resistance at intergranular was associated with the morphology of precipitation at grain boundary. When sensitization above 730℃, the chromium carbide occurred in the form of isolated particle at grain boundary, then the tendency to corrosion at intergranular became less in certain extent. when sensitization below 650℃, the chromium carbide appearred with continuous sheet-like morphology at grain boundary, then the tendency to corrosion at grain boundary turned into greater. In other word, precipitation with isolated particle morphology enhanced corrosion resistance at gain boundary, compared with precipitation with continuous sheet-like morphology.

The precipitation of strip and rod-like carbide at grain boundary induced the dilution of Cr and Mo element in adjacent domain around grain boundary, which reduced the resistance to sulfide stress corrosion crack at intergranular. Moreover, it was much more difficulty to compensate the inharmonious deformation between rod-like precipitation and the matrix, compared with spherical precipitation, then induced the local stress concentration and finally leaded to crack at interface of rod-like precipitation more easily.

5 Conclusions

(1) The strength remained above 120ksi while impact toughness enhanced to 196J for drill pipe as tempering temperatureincreased. Meanwhile, the morphology of precipitation evolved from rod-like to spherical feature.

(2) The former specimen tempering at 580℃ fractured below 720h, while the latter at 625℃ remained well after 720h for drill pipe during NACE-A method test, indicating much higher resistance to SSCC for the latter specimen.

(3) In-situ TEM and SEM observations showed that the void initiated and merged into microcrack, then formed straightness crack propagation path in the interface of rod-like precipitation.

(4) Cracking mechanism of the interface of precipitation could be attributed to the discrepancy of inharmonious deformation between precipitation and the matrix, as well as the dilution of alloy element.

Acknowledgements

The authors would like to thank the scientific research special project of quality inspection public welfare industry (No. 201510205-03) for its financial support.

References

[1] Li Helin, Han Lihong, Zhang Wenli. Demand for and development of hi-performance OCTG [J]. Steel Pipe, 2009, 38 (1): 1-9.

[2] Funakawa Y, Shiozaki T, Tomita K, Yamamoto T, Maeda E. Development of high strength hot-rolled sheet steel consisting of ferrite and nanometer-sized carbides [J]. ISIJ international, 2004 (44): 1945-1951.

[3] Kim B C, Lee S, Kim N J, Lee D Y. In situ fracture observations on tempered martensite embrittlement in an AISI 4340 steel [J]. Metallurgical and Materials Transaction A, 1991, 22 (A): 1889-1892.

[4] Lee S, Kim S, Wang B H, Lee B S, Lee C G. Effect of carbide distribution on the fracture toughness in the transition temperature region of an SA 508 steel [J]. Acta materialia, 2002 (50): 4755-4762.

[5] Wang Hang, Han Lihong, Hu Feng, Feng Yaorong, Sun Jun. Effect of tempering temperature on precipitate and mechanical properties of an anti-sulfur drill-pipe steel in H_2S containing environments [J]. Transactions of Materials and Heat Treatment, 2012, 33 (3): 88-93.

[6] Grobner P J, Sponseller D L, Diesburg D E. Effect of molybdenum content on the sulfide stress cracking resistance of AISI 4130-type steel with 0. 035% Cb [J]. Corrosion, 1978 (40): 1-21.

[7] Ravi T, Ramaswamy V, Namboodhiri T K G. Effect of molybdenum on the resistance to H2S of high sulphur microalloyed steels [J]. Materials Science and Engineering. A, 1993 (169): 111-118.

[8] Zhu A W, Csontos A, Starke E A. Computer experiment on superposition of strengthening effects of different particles [J]. Acta Materialia, 1999 (47): 1713-1721.

[9] Hahn G T, Rosenfield A R. Metallurgical factors affecting fracture toughness of aluminum alloys [J]. Metallurgical Transactions A, 1975 (6): 653-668.

[10] Ashby MF. Deformation of plastically non-homogeneous materials [J]. Philosophical Magazine, 1970, 21 (170): 399-&.

[11] Russell KC, Ashby MF. Slip in aluminum crystals containing strong, plate-like particles [J]. Acta

Metallurgica, 1970, 18 (8): 891.

[12] Yuan S P, Liu G, Wang R H, Sun J, Chen K H. Effect of precipitate morphology evolution on the strength-toughness relationship in Al-Mg-Si alloys [J]. Scripta Materialia, 2009 (60): 1109-1112.

[13] Yoshino Y, Minozaki Y. Sulfide stress cracking resistance of low-alloy nickel steels. Corrosion, 1986 (42): 222-232.

基于可靠性的S135钻杆疲劳裂纹萌生寿命预测研究

李方坡[1,2]　王建军[1,2]　韩礼红[1,2]

（1. 中国石油集团石油管工程技术研究院；
2. 石油管材及装备材料服役行为与结构安全国家重点实验室）

摘　要：本文基于可靠性理论对S135钻杆疲劳裂纹萌生寿命预测进行了系统研究。结果表明，服役过程中，S135钻杆管体部位承受的应力载荷明显大于接头螺纹连接部位。在应力比为-1条件下，S135钻杆的疲劳强度约为500MPa；在置信度为95%，误差为5%条件下，应力水平分别为660MPa、620MPa、580MPa和540MPa时的疲劳裂纹萌生寿命分布符合正态分布规律，随着应力水平的降低，疲劳裂纹萌生寿命对数分布概率密度函数峰值逐渐降低，离散性逐渐增强。计算获得了50%，90%，99%和99.9%可靠度条件下S135钻杆的疲劳裂纹萌生寿命预测方程。

关键词：S135钻杆；疲劳寿命；裂纹萌生寿命；寿命预测；可靠性

随着深层油气资源的钻探开发，S135钻杆以其优异的服役性能获得广泛应用，已成为目前油气钻井过程中的最主要工具。S135钻杆在服役过程中承受复杂的交变载荷，极易发生疲劳失效事故，钻杆疲劳失效尤其是断裂失效的发生会造成严重的经济和社会损失，有统计表明，平均每例钻杆断裂失效造成的经济损失达25万英镑[1]。S135钻杆的疲劳寿命由疲劳裂纹萌生寿命和疲劳裂纹扩展寿命两部分组成，其中疲劳裂纹萌生寿命占疲劳寿命的绝大部分比例[2]，是决定钻杆疲劳寿命的关键。与其他性能指标不同的是，钻杆的疲劳寿命具有很大的离散性，由于疲劳寿命离散性表述自身的复杂性，使得目前关于S135钻杆疲劳寿命预测的主要方法均未重点考虑疲劳寿命的离散性，这也成为目前钻杆疲劳寿命预测结果与实际偏差较大的主要原因之一[3,4]。

可靠性是指产品在规定条件和规定时间内完成规定功能的能力，其关键指标是可靠度[7]。针对金属材料疲劳寿命离散性的研究发现，多数金属材料疲劳寿命的分布符合一定规律，具有概率学特征。为了描述材料疲劳寿命的分布规律，国内外学者通过引入可靠性的方法建立了疲劳可靠性理论。近年来，有国外学者基于可靠性理论对钻杆疲劳裂纹无损检测可靠性开展了卓有成效的研究[5,6]，而对于S135钻杆疲劳寿命预测可靠性方面的研究还鲜见报道。有鉴于此，本文基于可靠性理论对S135钻杆疲劳裂纹萌生寿命预测进行研究，以期建立S135钻杆疲劳裂纹萌生寿命预测方法。

基金项目：中国石油天然气集团科学研究与技术开发项目（2016A-3905）。

作者简介：李方坡（1982.10），男，工学博士，高级工程师，2008年毕业于中国石油大学（华东），主要从事石油管材失效机理及服役安全性研究。E-mail：lifangpo@cnpc.com.cn，Tel：029-81887678。

1　试验方法及材料

钻杆材料化学成分（%）为 0.27C，0.25Si，0.81Mn，0.93Cr，0.38Mo，Fe 其余，材料金相组织为回火索氏体，晶粒度等级为 8.5 级。采用德国 Zwick 高频疲劳试验机测试钻杆疲劳寿命。疲劳试样沿钻杆纵向截取，加载部位直径为 4.0mm，应力比为-1，以 10^7 周次为作为疲劳试验终止条件。采用 VEGA Ⅱ 扫描电子显微镜对疲劳试样的微观形貌进行分析。

2　S135 钻杆应力载荷分析和疲劳应力的确定

2.1　S135 钻杆服役过程中的应力载荷分析

钻井作业过程中，钻杆的受力状态与钻井工艺及井眼轨迹等多方面因素有关，概括起来，作用在钻杆上的载荷包括拉伸载荷、扭矩载荷、钻井液压力、弯矩载荷及振动载荷等，其中拉伸载荷、弯曲载荷和振动载荷是钻杆承受的主要载荷，也是导致钻杆疲劳失效的主要因素，而钻井液压力和扭矩载荷形成的应力载荷则通常远小于前者。由井下钻柱振动所诱发的振动载荷是一个极其复杂的系统，不确定性非常大，为了提高计算结果的实用性，依据 SY/T 6719—2008 标准[7]，井下振动引起的交变载荷取 45kN。

Lubinski 等人的研究结果表明，钻杆的疲劳损伤主要是由于钻杆在井斜角变化位置旋转造成的。井眼轨迹变化所形成的大曲率主要出现在水平井中[8]。目前我国水平井作业过程中应用的 S135 钻杆的主要规格尺寸如表 1 所示，为了便于表示，将五种规格的钻杆分别记为 A，B，C，D 和 E。

表 1　S135 钻杆结构尺寸参数

<table>
<tr><th>代号</th><th>管体外径（mm）</th><th>壁厚（mm）</th><th>接头外径（mm）</th><th>接头内径（mm）</th><th>螺纹扣型</th><th>螺纹中径（mm）</th><th>锥度（mm/mm）</th><th>牙底削平高度（mm）</th><th>扩锥孔直径（mm）</th></tr>
<tr><td>A</td><td rowspan="3">127.0</td><td>9.19</td><td>168.3</td><td>69.9</td><td>NC50</td><td>128.0592</td><td>1/6</td><td>0.9652</td><td>134.94</td></tr>
<tr><td>B</td><td>9.19</td><td>184.2</td><td>88.9</td><td>5-1/2FH</td><td rowspan="2">142.0114</td><td rowspan="2">1/6</td><td rowspan="2">0.635</td><td rowspan="2">150.02</td></tr>
<tr><td>C</td><td>12.7</td><td>184.2</td><td>82.6</td><td>5-1/2FH</td></tr>
<tr><td>D</td><td rowspan="2">139.7</td><td>9.17</td><td>190.5</td><td>76.2</td><td>5-1/2FH</td><td rowspan="2">142.0114</td><td rowspan="2">1/6</td><td rowspan="2">0.635</td><td rowspan="2">150.02</td></tr>
<tr><td>E</td><td>10.54</td><td>190.5</td><td>76.2</td><td>5-1/2FH</td></tr>
</table>

依据我国目前水平井作业情况，选取典型工况条件对钻杆服役过程中的载荷进行计算分析。钻柱浮重取 20~160t，计算不同钻柱浮重作用下五种规格 S135 钻杆管体部位及接头螺纹部位承受的拉应力如图 1 所示，由图可见，随着浮重的增大，钻杆管体和接头螺纹部位（紧邻扭矩台肩 15.875mm 位置）承受的拉应力载荷逐渐增大，且管体部位承受的拉应力载荷明显大于接头螺纹部位。井眼曲率取（5°~90°）/30m 范围，钻柱浮重为 50t 条件下五种不同规格 S135 钻杆管体和接头螺纹部位承受的弯曲应力如图 2 所示。由图 2 可见，随着井眼曲率的增大，钻杆管体和接头螺纹部位承受的弯曲应力逐渐增大，且钻杆管体部位承受的弯曲应力远大于接头螺纹部位。分析认为，这主要由于钻杆接头部位的刚度远大于管体部位

所造成，这也正是钻杆管体部位的失效数量明显高于接头部位的主要原因。

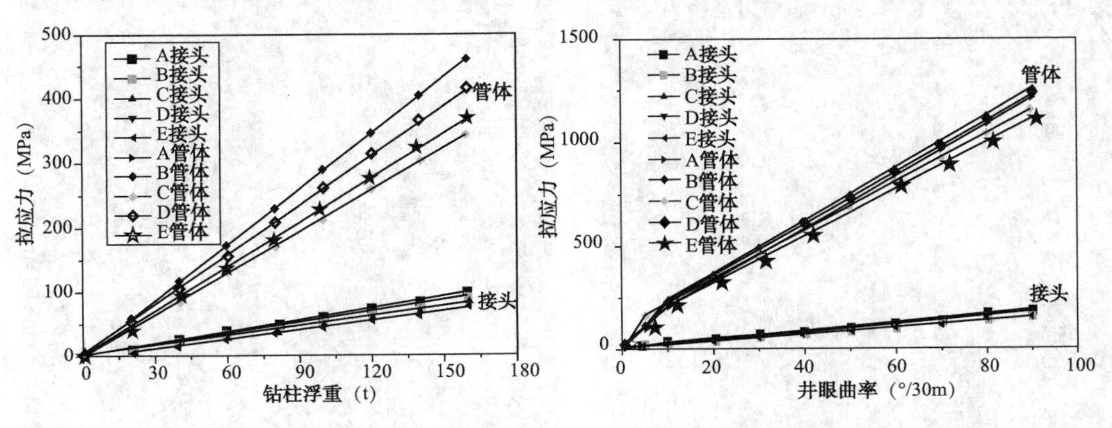

图1　钻柱浮重形成的拉应力　　图2　井眼曲率形成的弯曲应力

2.2　S135钻杆疲劳应力水平的确定

采用单点试验法对S135钻杆管体的疲劳寿命进行试验测试，以10^7周次为作为试验终止条件，在对数坐标系内对试验结果进行分析，钻杆疲劳寿命试验结果如图3所示。由图3所示结果判断，S135钻杆材料的疲劳强度约为500MPa，对应力幅不低于520MPa时的疲劳寿命试验结果进行曲线拟合，拟合方程如式（1）所示。

$$\lg N=43.7165-13.7655\lg S_a \quad (1)$$

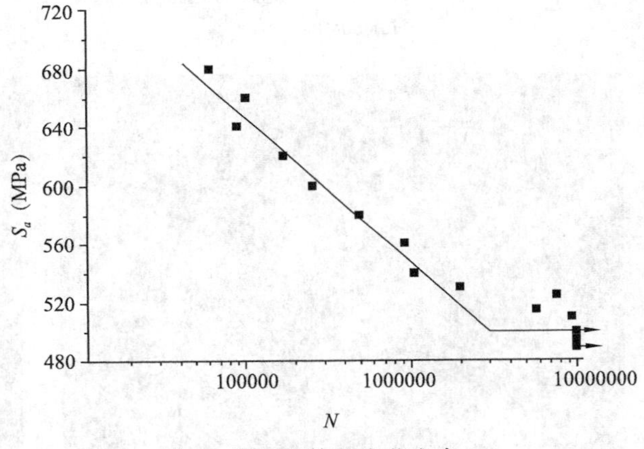

图3　钻杆管体的疲劳寿命

对不同应力水平下的疲劳试样断口形貌进行分析发现，试样断口可分为明显的裂纹萌生区、扩展区和瞬断区，随着应力幅的增加，疲劳裂纹扩展区面积变化不明显。裂纹扩展面可见明显收敛于试样外表面裂纹源位置的放射状扩展纹路，裂纹源区主要为解理断裂形貌，裂纹稳定扩展区可见规则排布的疲劳条带形貌，660MPa、620MPa、580MPa及540MPa四种典型应力幅的试样疲劳条带形貌如图4所示，试样的疲劳辉纹间距较为均匀，无明显变化。监测疲劳试验过程发现，疲劳裂纹一旦形成宏观尺寸，疲劳试验的频率和应力载荷会在极短的时间内（2~5s）发生剧烈变化，试样发生断裂。分析认为，这主要是由于疲劳试样尺寸较小，在拉应力载荷作用下，裂纹一旦形成便会迅速扩展，直至发生断裂，不同应力幅下的疲劳裂纹扩展区差别不明显，也正是基于此，本实验用试样的疲劳寿命可视为钻杆疲劳裂纹的萌生寿命。

试验结果表明，S135钻杆的疲劳裂纹萌生寿命具有一定的离散性，为了系统研究钻杆疲劳裂纹萌生寿命的离散性，预测疲劳裂纹的萌生寿命，下面针对660MPa、620MPa、580MPa及540MPa四种不同应力幅下的S135钻杆疲劳裂纹萌生寿命可靠性进行分析研究。

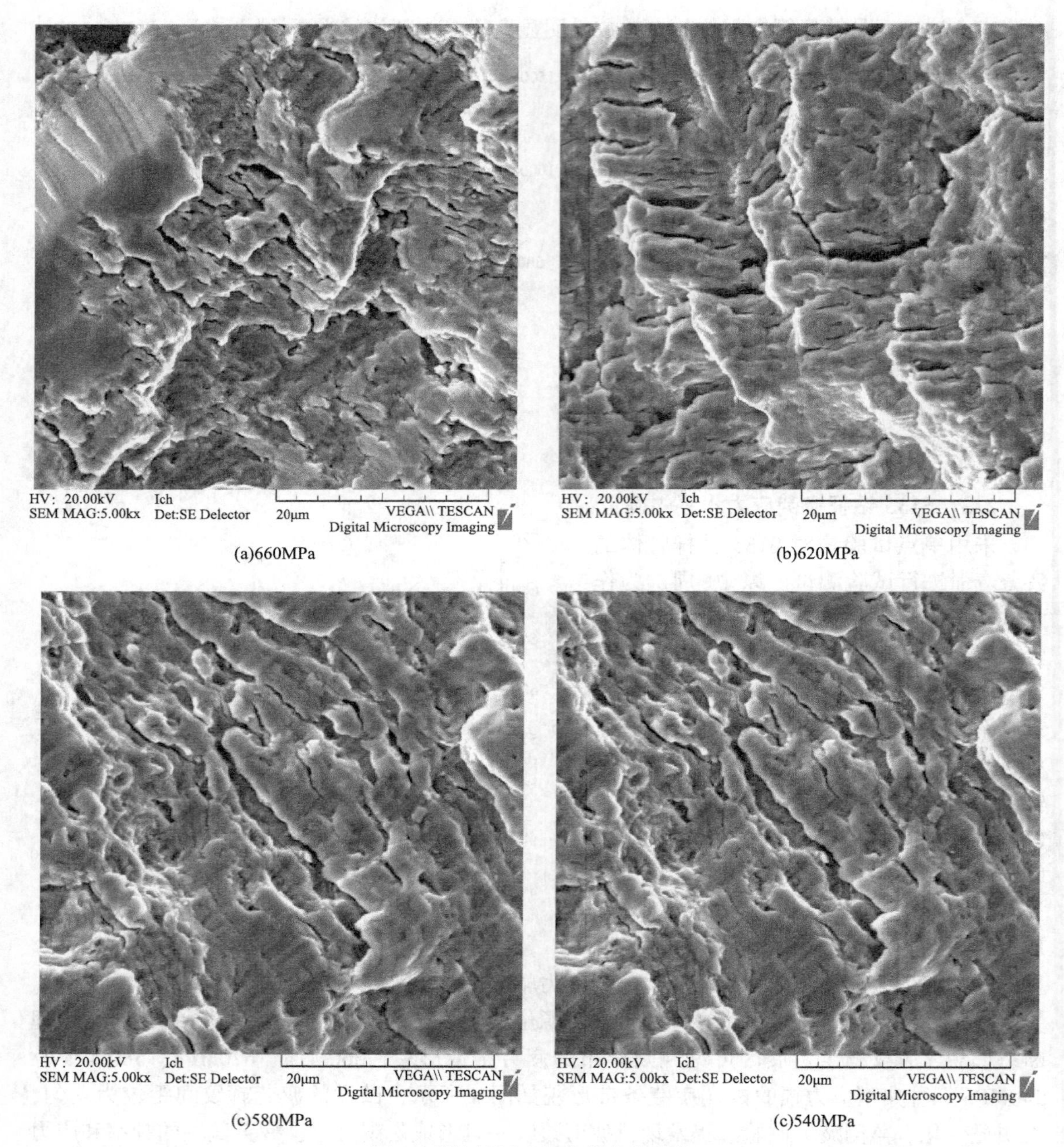

(a)660MPa (b)620MPa (c)580MPa (c)540MPa

图 4 裂纹扩展区的疲劳条带形貌

3 S135 钻杆疲劳裂纹萌生寿命可靠性分析

3.1 可靠性分析试样数量的确定

依据数理统计理论，对于钻杆材料疲劳寿命母体数据而言，母体平均值记为 u ，作为从母体中抽取的含有 n 个疲劳试验数值的子样，其平均值记为 $\bar{x}$ 和标准差记为 s ，则由 t 分布理论，u 的区间估计式为：

$$\bar{x} - t_{r}\frac{s}{\sqrt{n}} < u < \bar{x} + t_{r}\frac{s}{\sqrt{n}} \tag{2}$$

即：

$$-\frac{st_r}{\bar{x}\sqrt{n}} < \frac{u-\bar{x}}{\bar{x}} < \frac{st_r}{\bar{x}\sqrt{n}} \tag{3}$$

其中 $\frac{u-\bar{x}}{\bar{x}}$ 表示子样平均值 $\bar{x}$ 与母体平均值 u 的相对误差，令 δ 为相对误差极限，即：

$$\delta = \frac{st_r}{\bar{x}\sqrt{n}} \tag{4}$$

其中 $\frac{s}{\bar{x}}$ 为变异系数，考虑实际需要，在本文的研究中，取置信度为95%，误差限度为5%，计算不同变异系数条件下至少需要的疲劳试样数量如表2所示。

表2　变异系数与试样数量

变异系数	最少试样数量	变异系数	最少试样数量
小于0.0201	3	0.0541~0.0598	8
0.0201~0.0314	4	0.0598~0.0650	9
0.0314~0.0403	5	0.0650~0.0699	10
0.0403~0.0476	6	0.0699~0.0744	11
0.0476~0.0541	7		

3.2　疲劳裂纹萌生寿命可靠性分析

依据2.2的试验结果，选择高于疲劳强度的660MPa、620MPa、580MPa和540MPa应力幅进行成组疲劳试验，其中660MPa和620MPa应力水平下各完成疲劳试验试样10件，580MPa和540MPa应力水平下各完成疲劳试验试样12件。试验结果表明，疲劳试样均断裂于中部薄弱部位，试验结果有效，四种不同应力水平下钻杆疲劳裂纹萌生寿命试验结果如图5所示。

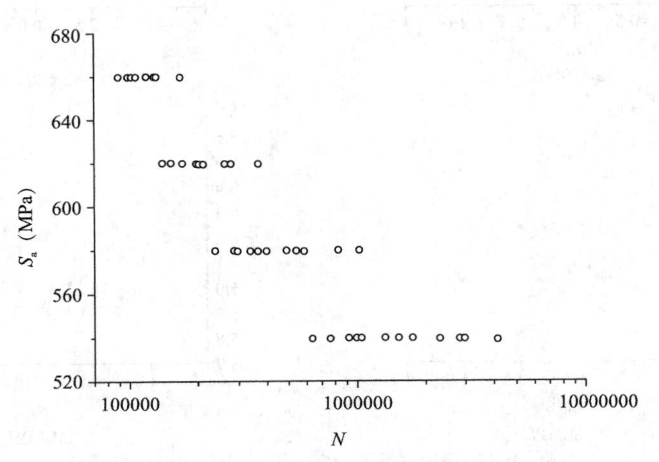

图5　不同应力水平下的钻杆疲劳裂纹萌生寿命

表 3　线性相关参数

应力载荷（MPa）	变异系数	线性相关系数（r）	拟合的直线系数（a）	拟合的直线系数（b）
660	0.0182	0.9589	5.0544	0.0919
620	0.0289	0.9759	5.3152	0.1536
580	0.0387	0.9855	5.6543	0.2190
540	0.0486	0.9893	6.1774	0.3004

分别计算 4 种应力水平下疲劳裂纹萌生寿命的可靠度估计值 p_i 及其对应的标准正态偏量 u_{pi}，u_{pi} 与疲劳裂纹萌生寿命对数值 $\lg N_i(x_i)$ 之间的数据关系如图 6 所示。由图可见，4 种应力水平下的 $u_{pi}-\lg N_i(x_i)$ 数据均近似分布在一条直线上，基于最小二乘法分别拟合计算 $u_{pi}-\lg N_i(x_i)$ 线性相关直线参数如表 3 所示。$n=10$ 和 12 时的线性相关系数起码值分别为 0.632 和 0.576，4 种应力水平下 u_{pi} 与 $\lg N_i(x_i)$ 之间的相关系数均明显大于线性相关系数起码值，且为正值，故可以认为四种应力水平下的 u_{pi} 与 $\lg N_i(x_i)$ 线性相关，由此可以判断，四种试验应力水平下的 S135 钻杆疲劳裂纹萌生寿命对数值符合正态分布规律。计算四种应力水平下疲劳裂纹萌生寿命试验结果的变异系数如表 3 所示，对比表 1 可知，四种应力水平下 S135 钻杆疲劳试样数量均满足置信度为 95%和误差限度 5%情况下最少试样数量的要求。分别计算四种应力水平下的正态分布概率密度函数，如图 7 所示。由图可见，随着应力幅水平的降低，概率密度函数的峰值逐渐降低，宽度逐渐增大，S135 钻杆疲劳裂纹萌生寿命的离散性逐渐增强。由上述试验结果分别计算四种应力幅下可靠度分别为 50%、90%、99% 和 99.9%时的 S135 钻杆疲劳裂纹萌生寿命对数值，结果如表 4 所示。

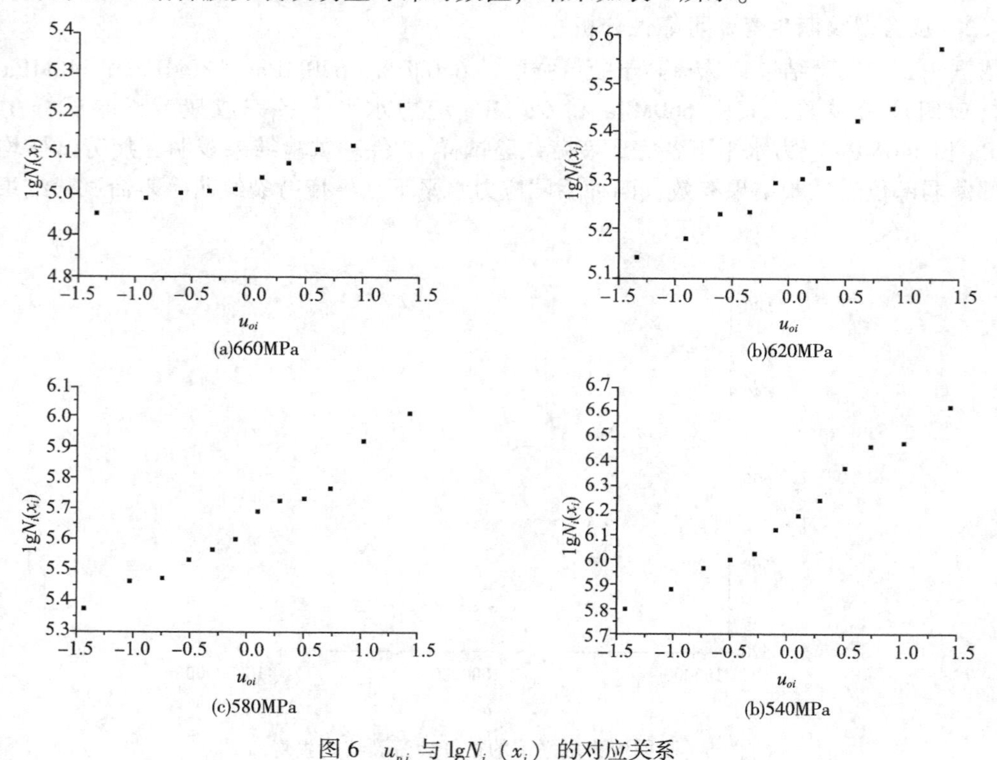

图 6　u_{pi} 与 $\lg N_i$（x_i）的对应关系

表 4　不同可靠度下的疲劳裂纹萌生寿命对数值

可靠度 \ 应力幅	660MPa	620MPa	580MPa	540MPa
50%	5.0544	5.3152	5.6543	6.1774
90%	4.9365	5.1183	5.3735	5.7923
99%	4.8406	4.9579	5.1449	5.4787
99.9%	4.7704	4.8406	4.9776	5.2492

采用最小二乘法对 $\lg S_a$ 与 $\lg N$ 之间进行线性拟合，50%、90%、99%和 99.9%可靠度下的 S135 钻杆疲劳裂纹萌生寿命方程分别 $\lg N = 41.1446 - 12.8178\lg S_a$、$\lg N = 32.4234 - 9.7655\lg S_a$、$\lg N = 25.3196 - 7.2793\lg S_a$ 和 $\lg N = 20.1150 - 5.4577\lg S_a$，如图 8 所示。对于 $n=4$ 时，线性相关系数 r 的起码值为 0.950，计算四种不同可靠度下的 $\lg S_a$ 与 $\lg N$ 线性相关性系数 r 分别为−0.9916，−0.9875，−0.9802 和−0.9682，其绝对值均大于 0.950，且为负值，所以可以认为 $\lg S_a$ 与 $\lg N$ 线性负相关，而且，随着可靠度的增加，$\lg S_a$ 与 $\lg N$ 拟合直线的线性相关系数绝对值逐渐减小。分析认为，这主要是由于不同应力幅下的疲劳裂纹萌生寿命概率密度函数不同所致，高可靠度下的疲劳裂纹萌生寿命是利用直线延伸推算得到的数据，与真实数值的偏离相对增大，从而导致高可靠度下疲劳裂纹萌生寿命线性相关系数相应有所降低。

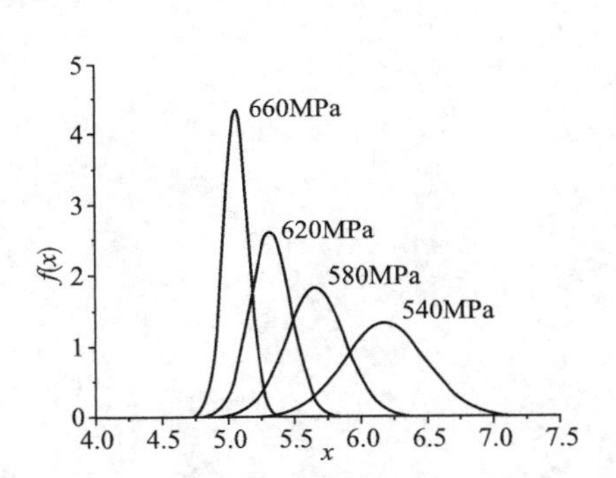

图 7　钻杆疲劳寿命对数值的概率密度分布曲线

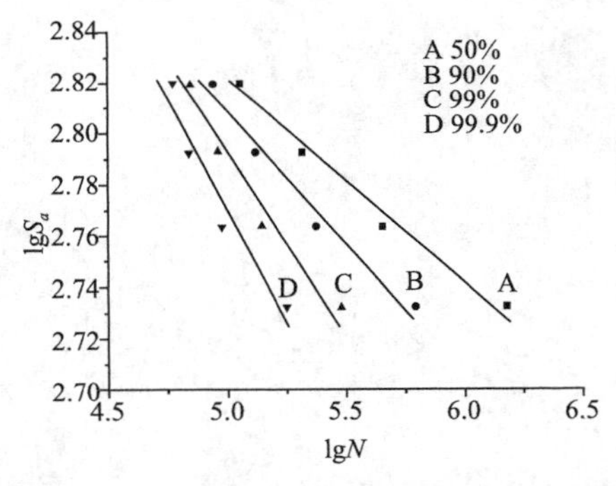

图 8　不同可靠度的疲劳裂纹萌生寿命

4　结论

（1）钻井作业过程中，S135 钻杆管体部位承受的弯曲及拉伸应力载荷均明显大于接头螺纹连接部位，钻杆管体在应力比为−1 条件下的疲劳强度约为 500MPa。

（2）应力幅分别为 660MPa、620MPa、580MPa 和 540MPa 条件下的 S135 钻杆疲劳裂纹萌生寿命分布符合正态分布，且随着应力水平的降低，疲劳裂纹萌生寿命分布概率密度函数的峰值逐渐降低，疲劳裂纹萌生寿命的离散性逐渐增强。

（3）50%，90%，99%和 99.9%可靠度下的 S135 钻杆疲劳裂纹萌生寿命预测方程分别为 $\lg N = 41.1446 - 12.8178\lg S_a$，$\lg N = 32.4234 - 9.7655\lg S_a$，$\lg N = 25.3196 - 7.2793\lg S_a$ 和 $\lg N = 20.1150 - 5.4577\lg S_a$。

参　考　文　献

[1] Macdonald K A, Bjune J V. Failure Analysis of Drillstrings [J]. Engineering Failure Analysis, 2007, 30

(14)：1641-1666.
[2] 崔忠圻，谭耀春．金属学与热处理［M］．北京：机械工业出版社，2008.
[3] Miscow G F，Netto T A. Techniques to characterize fatigue behaviour of full size drill pipes and smallscale samples［J］. International Journal of Fatigue，2004，26：575-584.
[4] 李方坡，王勇．钻杆疲劳寿命预测技术的研究现状与发展［J］．材料导报，2015，29（6）：88-91.
[5] Grondin G. Y，Kulak G Y. Kulak. Fatigue Testing of Drillpipe［J］. Drilling&Completion，1994，6：95-102.
[6] Angelo Ligrone，Giovanni Botto，Angelo Calderoni. Reliability Methods Applied to Drilling Operation［C］// 1995 SPE/IADC drilling conference，Amsterdam，1995：227-234.
[7] 含缺陷钻杆的适用性评价方法：SY/T 6719—2008［S］．中华人民共和国石油天然气行业标准．
[8] J. J 阿扎，G. 罗埃罗．萨莫埃尔．钻井工程手册［M］．北京：石油工业出版社，2011.

钢管折叠缺陷案例分析

王　鹏　冯　春　王新虎　韩礼红

（中国石油集团石油管工程技术研究院）

摘　要：折叠是一种常见的钢管热加工缺陷。本文基于两个典型的折叠缺陷钢管失效分析案例，简明介绍了其共有的失效特征，通过对折叠缺陷部位承压爆裂过程的数值模拟分析阐述了其对钢管服役性能的影响。

关键词：钢管；折叠；失效分析；数值模拟

钢管是石油天然气行业中大量使用的重要设施，管材性能的好坏直接决定了其服役的安全可靠性与否。折叠是一种常见的钢管热加工缺陷，严重影响钢管产品质量。作者通过两个钢管折叠典型案例分析，总结其共有的失效特征，并运用扩展有限元法（XFEM）[1,2] 对径向深度 2mm 折叠的 ϕ60mm×7mm 无缝钢管爆裂实例进行了数值模拟分析，阐述了折叠对钢管服役性能的影响。

1　外壁折叠案例分析

某 ϕ88. 9mm×6. 45mm 110 Ksi 钢级油管外表面修磨处发现线性缺欠，该缺欠基本位于修磨区域中央，为平行于管体轴向的细线状，目测连续长度约 20mm 并具有一定的深度，形貌如图 1a 所示。按照 API Spec 5CT 标准 L2 检验等级的超声波检测发现该部位存在回波显示，经磁粉检测其存在线性缺欠磁痕显示（图 1b），缺欠轴向长度约 18mm。

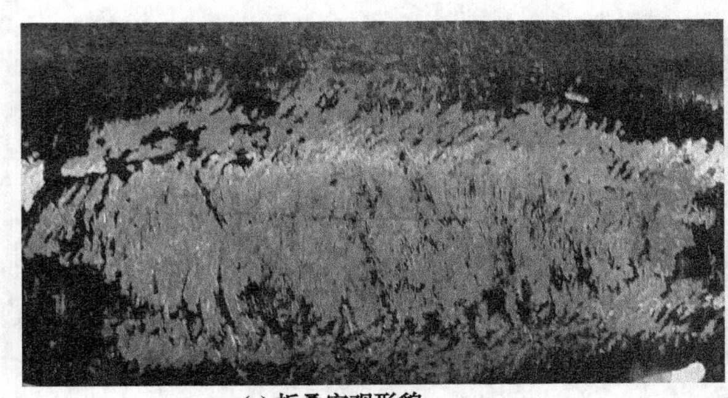

(a) 折叠宏观形貌

(b) 折叠磁痕显示

图 1　油管外壁折叠宏观形貌

对缺欠及附近的微观形貌、金相组织及能谱分析显示（图 2）：缺欠裂纹特征明显，起始于管体外表面，具有多个分支，裂纹内填充灰色物质，裂纹周边组织脱碳明显，裂纹最大深度为 0. 305mm（名义壁厚的 4. 7%），裂纹内部灰色填充物富含氧元素，判断其主要为金

作者简介：王鹏（1980—），男，高级工程师，博士，主要从事石油管失效分析与安全评价技术研究。地址：陕西省西安市锦业二路 89 号，E-mail：wangpeng008@ cnpc. com. cn。

属氧化物。根据以上各种特征判断该为轧制折叠类缺欠，其裂纹附近脱碳严重，裂纹尖端不甚尖锐，推测缺欠形成于轧制管坯，轧制过程进一步破坏了该区域组织连续性，接近5%名义壁厚的裂纹深度势必造成油管承载能力大大降低。

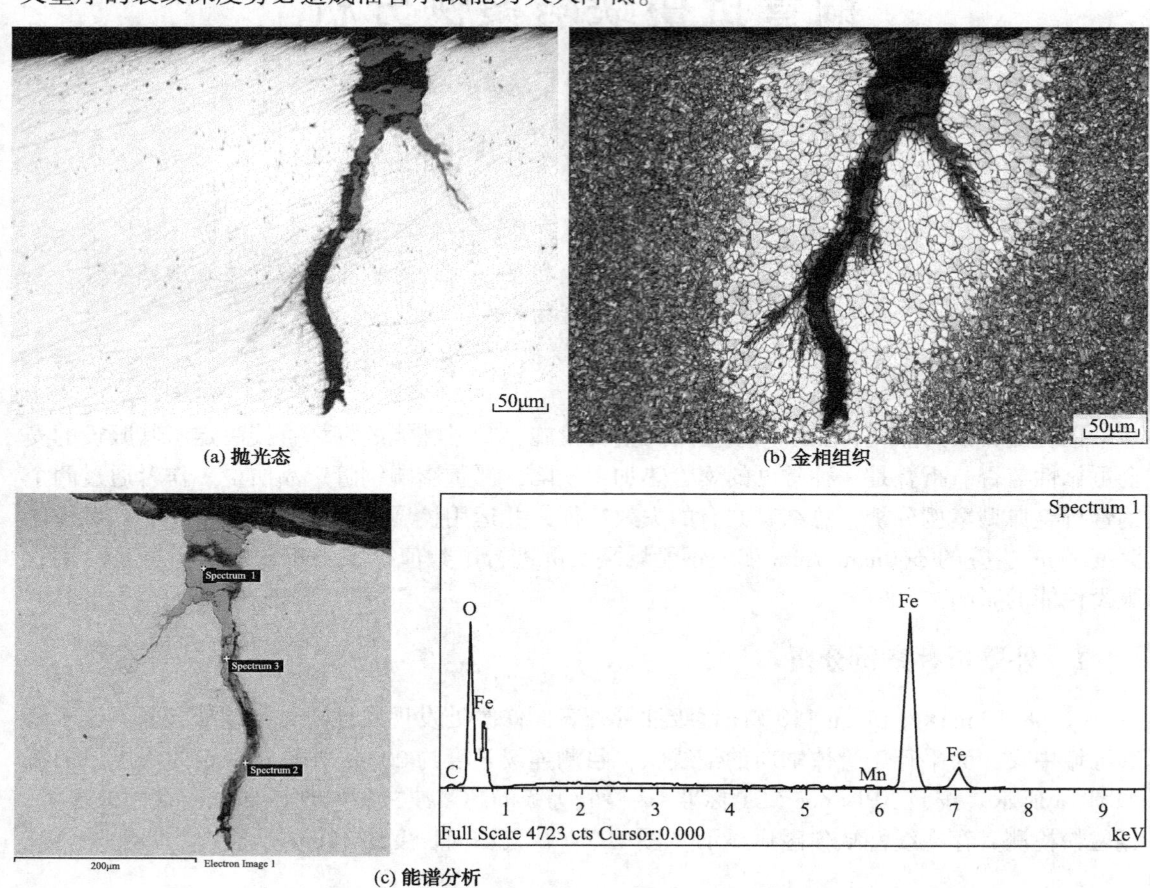

(a) 抛光态

(b) 金相组织

(c) 能谱分析

图2　油管外壁缺欠微观形貌、金相组织及能谱分析

2　内壁折叠爆裂案例分析

某石油公司的管线压力试验中内压加至约35MPa时，ϕ60mm×7mm 20G锅炉钢管发生爆裂（图3）。参考API 5C3标准中有关光管在内压作用下的屈服强度计算公式，按照GB 5310—1995标准规定材料屈服强度下限245MPa，估算管体可承受内压为50.02MPa，爆裂试验内压远低于此值。

图3　爆裂的无缝钢管

经过理化检验和分析发现在管体开裂断口面处存在纵向带状脱碳区域（图4a），其长度至少大于66mm、径向深度约为2mm，该脱碳区域位于断口靠近管体内壁的带状台阶状区域外侧（图4b），由这些特征判断管体存在纵向呈细线状分布的内表面折叠缺陷，爆裂即在此位置[3]。热轧无缝钢管的内折缺陷一直是影响钢管一次合格率和成材率指标的主要原因之一，主要在穿孔过程中产生，其产生的主要原因与连铸管坯质

量有关，也与穿孔工艺制度合理与否有关[4]。该内折缺陷不仅大大降低了管体的有效承载壁厚，同时缺陷尖端又会造成此处的应力集中。

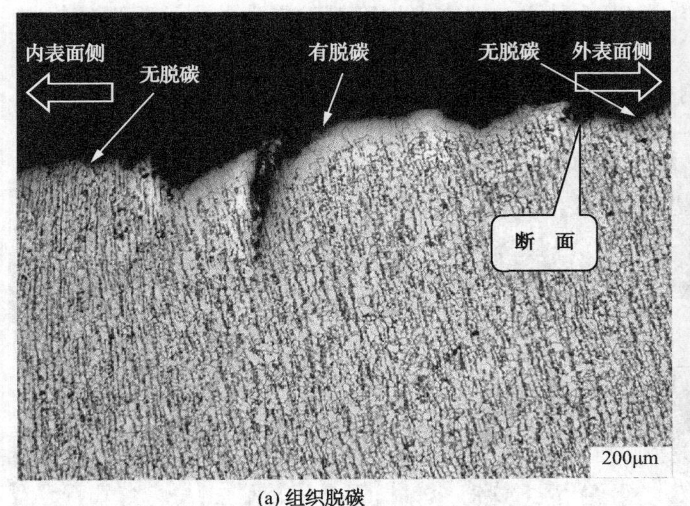

(a) 组织脱碳

(b) 断面宏观形貌

图 4　爆裂断口分析

3　内壁折叠爆裂数值模拟

金属材料断裂是一种复杂的强不连续力学问题，其复杂性由几何界面处的位移不连续性和端部的奇异性引起物体内部物理界面的脱粘或起裂。于 1999 年由以美国西北大学 Belytschko 教授为代表的研究组首先提出的扩展有限元法（XFEM）继承了 CFEM 的框架，划分网格时不需要考虑结构内部的几何或物理界面，克服了裂纹尖端等高应力和应变集中等问题所带来的计算困难。

借助 ABAQUS/Standard 商用有限元软件平台，基于 XFEM 对无缝钢管在内压作用下的整个爆裂过程进行模拟分析。通过设置损伤起始和扩展的判据等相关参数实现裂纹扩展计算，模拟裂纹起始时，选取最大主应力损伤准则作为判据，即

$$f = \left\{ \frac{\langle \sigma_{\max} \rangle}{\sigma^{o}_{\max}} \right\} \tag{1}$$

式中　$\sigma^{o}_{\max}$ ——最大主应力；

$\langle\rangle$ ——当材料处于纯压缩应力状态时，不会发生起裂。当最大主应力比值 f 大于某值时，认为裂纹起始。

模拟裂纹扩展时，选取基于能量的混合型幂指数损伤演化法则作为判据，即

$$f = \left(\frac{G_{\mathrm{I}}}{G_{\mathrm{IC}}} \right)^{am} + \left(\frac{G_{\mathrm{II}}}{G_{\mathrm{IIC}}} \right)^{an} + \left(\frac{G_{\mathrm{III}}}{G_{\mathrm{IIIC}}} \right)^{ao} \tag{2}$$

式中　G_{I}、G_{II} 和 G_{III} ——分别表示 Ⅰ、Ⅱ 和 Ⅲ 型能量释放率；

G_{IC}、G_{IIC} 和 G_{IIIC} ——分别表示 Ⅰ、Ⅱ 和 Ⅲ 型临界能量释放率。

建立平面应变计算模型，在管体内壁上构造深 2mm 的原始缺陷，管体施加内压为：从 0 增至 35MPa，材料参数选取试验检测值[3]。采用四节点四边形单元，单元数为 1875，节点数为 2020，其中对裂纹区域的网格进行了加密以保证计算精度。

图 5 为数值模拟得到的有缺陷无缝钢管在施加 35MPa 内压作用下的裂纹起裂和扩展过

程以及材料内部的等效应力分布演变。可以看出，在内压的作用下，管体内应力分布极不均匀，在原始缺陷尖端有明显的应力集中。当管体内压从0加至约7MPa时，承受较大应力的缺陷尖端就逐步产生了细微的起始裂纹，随后裂纹尖端应力加剧严重集中，裂纹附近的材料受到周向拉应力和剪应力的综合作用，裂纹逐步扩展延伸，管体截面不断削弱，内压升至约35MPa最终导致管体完全断裂。折叠爆裂处未见塑性变形，低应力脆断特征明显。

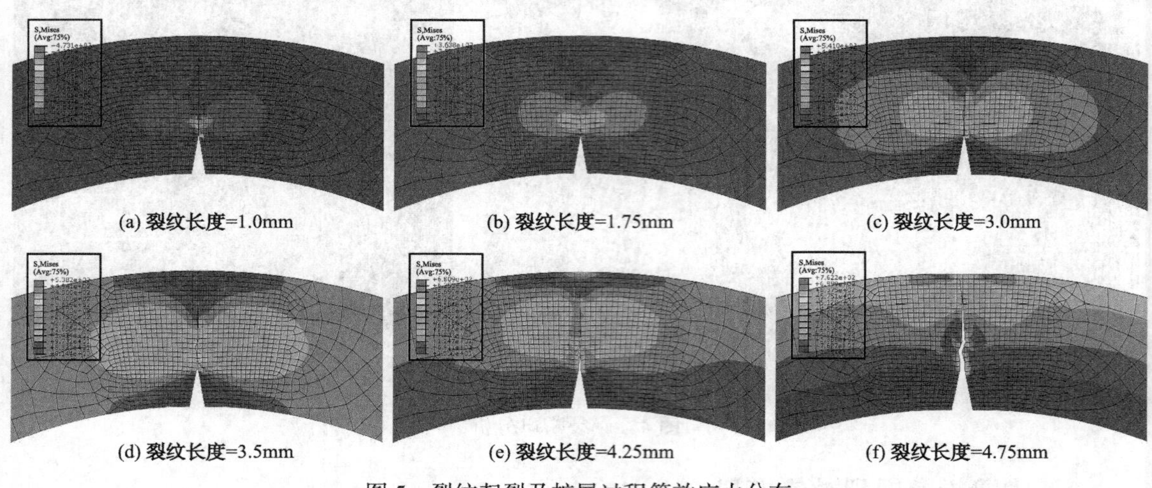

(a) 裂纹长度=1.0mm (b) 裂纹长度=1.75mm (c) 裂纹长度=3.0mm

(d) 裂纹长度=3.5mm (e) 裂纹长度=4.25mm (f) 裂纹长度=4.75mm

图5 裂纹起裂及扩展过程等效应力分布

4 结论

本文基于钢管折叠分析案例介绍了该类缺陷典型特征，并运用XFEM模拟技术对折叠缺陷无缝钢管纵向爆裂实例进行了有限元分析，为相关工程技术人员提供借鉴参考。

参考文献

[1] Moes N, Dolbow J, Belytschko T. A finite element method for crack growth without remeshing [J]. International journal for numerical methods in engineering, 1999, 46: 131-150.

[2] 李录贤，王铁军. 扩展有限元法（XFEM）及其应用 [J]. 力学进展，2005，35（1）：5-20.

[3] 王鹏，李方坡，路彩虹，等. 高压锅炉无缝钢管爆裂原因分析 [J]. 理化检验—物理分册，2011，47（6）：375-378.

[4] 周晓锋. 热轧无缝钢管内折缺陷分析 [J]. 钢管，2009，38（10）：48-51.

Migration of Variable Density Proppant Particles in Hydraulic Fracture in Coal-Bed Methane Reservoir

Yang Shangyu[1] Han Lihong[1] Wang Jianjun[1] Feng Yaorong[1], Wu Xingru[2]

(1. Tubular Goods Research Institute of CNPC, Key Lab for Petroleum Tubular Goods Engineering of CNPC; 2. University of Oklahoma)

Abstract: The well performance in Coal-Bed Methane (CBM) reservoir is highly dependent on the conductivity of hydraulic fracture created to increase well production rate and contact area with the reservoir. Aimed at improving the effective fracture length of hydraulic fracture in CBM reservoir and the seam proppant concentration, the objective of this paper is to investigate the migration of variable density proppant particles in hydraulic fracture and seams using numerical modeling and simulation. In the proposed model, the interaction among fracture proppant particles distributed in a pseudo fluid model is taken into consideration. Comparison between the model calculation and field data shows that the proposed model is accurate and reliable in applications. Additionally, we carried out the sensitivity study on the viscosity of fracking fluid, fracture area, proppant density, injection rate and other factors. It is shown that, with the condition of a confining pressure of 69 MPa and an ambient temperature of 90℃, the crush value of nutshell proppant is less than 2%, which meets the field application requirements. Additionally, with the increase of both fracturing fluid viscosity and injection rate, the propped effective fracture length increases, and lead to a more uniform distribution of proppants. The increase in proppant particle diameter, inversely, results in a decrease in the effective fracture length. As for variable-density proppant, the effective crack support length would be longer than the single ceramic proppant, and the particle distribution of the case with variable density is more uniform.

Keywords: Coalbed Methane; Variable density proppant; Proppant migration; Hydraulic fracturing; Effective fracture length

1 Introduction

With the increasing demand on oil and gas and the decreasing production of conventional fossil fuels in China, as an important resource of natural gas, Coal Bed Methane (CBM) reservoir development is becoming more and more prominent in China's energy structure (Zou et al. 2012). In the last 20 years, nearly 10,000 CBM wells had been drilled in China. However, many of these wells have tremendous challenges such as low productivity facing engineers and geoscientists (Yan et al. 2010, Rickards, Brannon, and Wood 2006, Wang and Zhang 2008). The surveillance data from those CBM wells shows that most of proppant particles in the hydraulic fracture were deposited in a range of 0~40 m from the wellbore their fractures. The insufficient propped length of fracture is one of the main reasons for low production of CBM wells. The objective of this paper is to study the deposition and distribution of the variable density proppant particles along the hydraulic fracture.

Proppant distribution in the hydraulic fracture and fracture network has been studied experimentally from late 1950s and theoretically and multiple models are available in literature (Kern, Perkins, and Wyant 1959, Wang et al. 2003, Sahai, Miskimins, and Olson 2014, Tong and Mohanty 2016). However, most of these studies are either for tight formation fracture or for proppants of uniform density using stoke's law to determine the proppant distribution. Generally, using uniform proppant usually results in a high concentration of proppant in the near wellbore region and little of no proppant in the region from the wellbore. This distribution can yield a short effective or propped length, which would lead a well with low productivity (Zhang et al. 2016, Senthamaraikkannan, Gates, and Prasad 2016). The experiments in literature (Liu and Sharma 2005, Neto and Kotousov 2013) show that, the velocity of proppant particles depends on the ratio of proppant particle diameter to local crack width. When the ratio is close to a unit, the velocity of proppant particles drops down sharply along the fracture. The migration of proppant particles has been studied by two parallel plates instead of crack walls (Rahman and Rahman 2010, Shiozawa and McClure 2016), however, the influence of crack width on the migrating velocity hasn't been considered. In this paper, we study the proppants of mixing ultra-low density proppant with tradition sand to get a desirable propped fracture.

The ultra-low density proppant was proposed to increase the sand-laden distance and extend the effective propped length (Kang et al. 2016). Therefore, the fracture conductivity can be enhanced. However, the ultra-low density proppant is expensive for large quantity of applications. In this paper, variable-density proppants are proposed to solve the problem of short sand-laden distance within the fracture, in which the mixed high-density proppant and ultra-low density proppant are both used to hydraulically fracture. The migration process of variable density proppant within the fracture is studied. The experiments of ultra-low density proppant on compressive properties and crushing rate are conducted. This method could increase the effective support length of fracture and improve the CBM well's productivity.

2 Mechanical properties of low-density nutshell proppant

Figure 1 shows two different nutshell proppant samples that are prepared for field test in a CBM field in China; in which (a) Ultra-Light Weight Proppant (ULW1) has a density0.8 g/cm^3 (apparent density 1.20 g/cm^3), and (b) ULW2 with a density 1.25 g/cm^3 (apparent density 1.75 g/cm^3). When the low-density proppant is used to prop hydraulic fractures, we shall test whether it can withstand the formation closure pressure to prop the fracture open by following the crush-testing procedures (API RP-56 1983) established by American Petroleum Institute (API). The test is measured in crush value which is defined as the weight percentage of the damaged proppant in the test (Palisch et al. 2009). The compressive strength and crush value of proppant with grain sizes of #20 and #40 are measured under the temperatures of 20℃ and 90℃. Tests showed that at a closure pressure of 100 MPa, after 10 minutes of testing, the crush values of ULW1 proppant at 20℃ are 1.41%, 1.33%, 1.59% and 1.41%; respectively. The crush values at 90℃ are 1.47%, 1.64%, 1.93% and 1.85%; respectively. From the stress-strain relationship curves of proppant particles as shown in Figures 2 through 5, the elastic modulus of the nutshell proppant at 20℃ is determined to be 172.42MPa and at 90℃ is 137.93MPa. In addition, with the increase of temperature, the ULW1 crush value rises while the elastic modulus reduces.

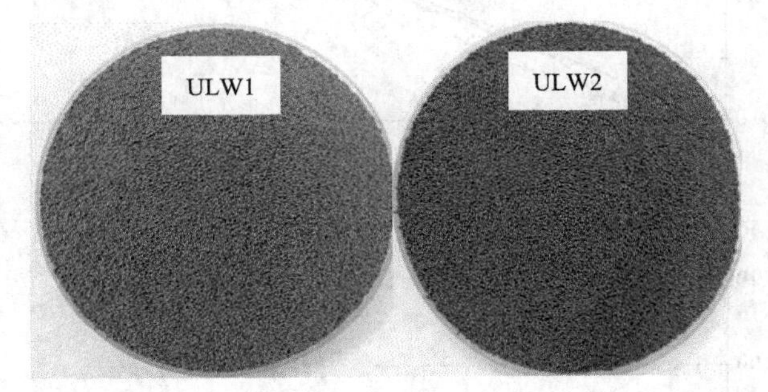

Fig. 1 Samples of nutshell proppants with different densities

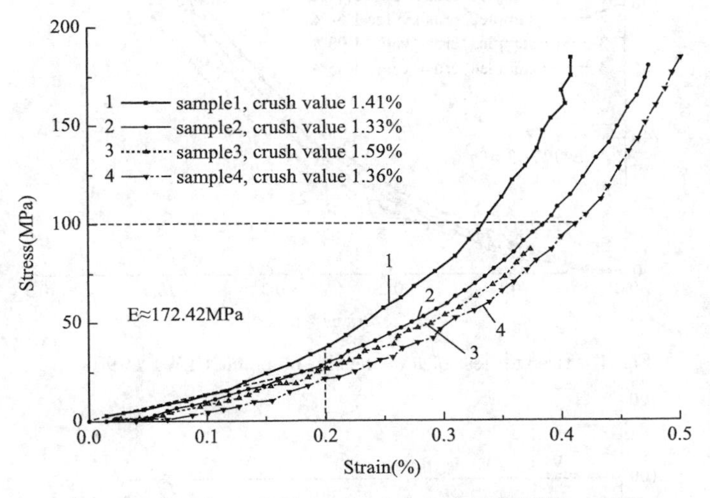

Fig. 2 Strength test of low-density proppant ULW1 at 20℃

Three groups of ULW2 proppant are selected to test their compression prosperity. Tests showed that under a 100MPa closure pressure and after 10 minutes of screening, the elastic modulus of the proppant is 344.83MPa and crush values of ULW2 proppant at 20℃ are 4.02%, 6.38% and 4.02%, respectively. However, at the temperature of 90℃, the crush values respectively are 5.29%, 7.90% and 7.32%, and the elastic modulus of the proppant turns to 275.86MPa, as shown in Figure 5. Compared with ULW1, the crush values of ULW2 are larger with a 100MPa closure pressure. While with the closure pressure reduced to 69MPa, the largest crush value of the three samples groups turns out to be 2.0% at 90℃. The compressive property and crush value can meet the filed requirements on of CBM hydraulic fracturing proppant.

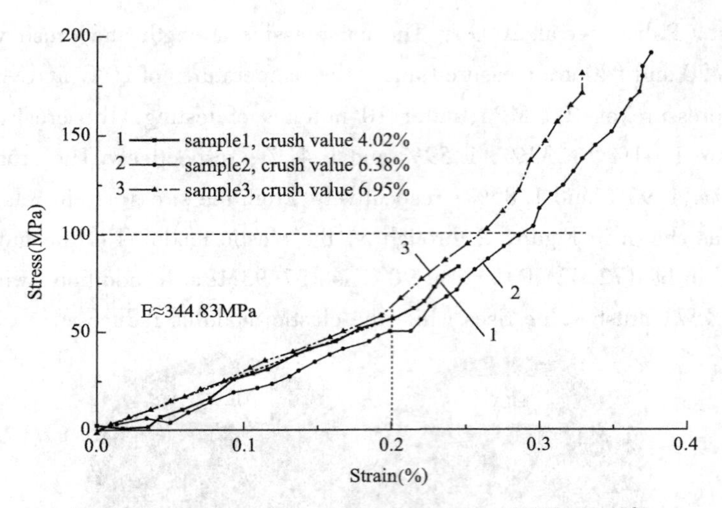

Fig. 3 Strength test of low-density proppant ULW2 at 20℃

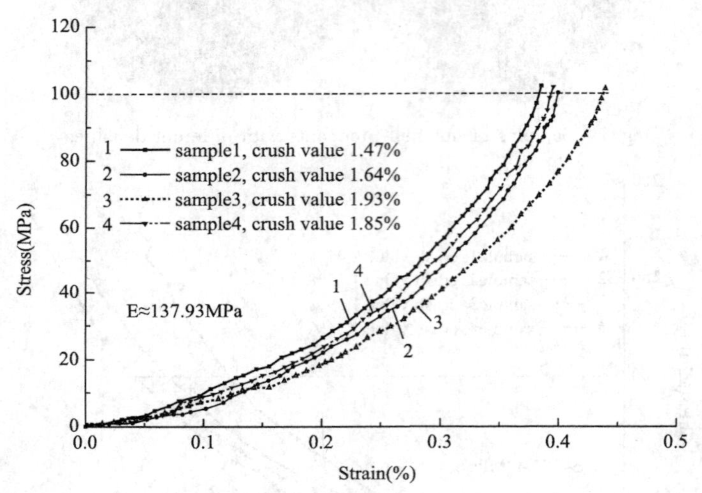

Fig. 4 Strength test of low-density proppant ULW1 at 90℃

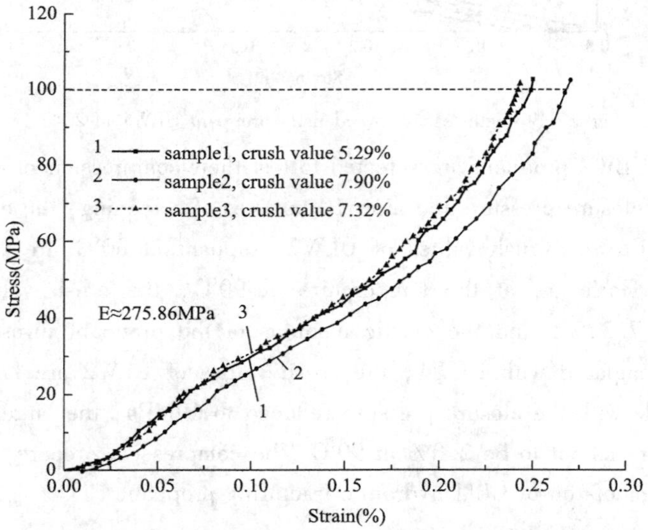

Fig. 5 Strength test of low-density proppant ULW2 at 90℃

3 Modeling variable density proppant deposition in fracture

To derive the equivalent fracture width for fractures propped with variable density proppants, we used a simplified semi-empirical pseudo fluid model (Zhou et al. 2016) to consider the interaction between the suspended particles in the fracking fluid along the fracture. The pseudo fluid model is based on equivalent fracture width which is adopted to simulate the depositing mechanism of the variable density proppant particles. As the proppant particles flowing in a hydraulic fracture, the fracture width is changing because of an additional fluid drag-force caused by the nonzero proppant concentration. The variation of fracture width w_c can be determined using Eq. (1):

$$\frac{1}{w_c^2} = 1.411\left(\frac{1}{d_p^2} - \frac{1}{w^2}\right)c^{0.8} \tag{1}$$

In which w is the fracture width as shown in Figure 6. The equivalent width of hydraulic fracture w_{eff} can be expressed as Eq. (2):

$$\frac{1}{w_{eff}^2} = \frac{1}{w^2} + \frac{1}{w_c^2} \tag{2}$$

Where, w_{eff} refers to the corresponding equivalent width of fracture based on Pseudo Fluid Model. As shown by Figure 6, the shape of major fracture is thin and long when the fracture width change (Z axis) is assumed to be uniform. Neglecting the pressure gradient in the z-direction, the governing equation of the fracturing fluid can be formulated as follows using a two-dimensional fluid model (Guo et al. 2015):

$$-\frac{\partial(w_{eff}V_x)}{\partial x} - \frac{\partial(w_{eff}V_y)}{\partial y} - q_L = \frac{\partial w_{eff}}{\partial t} \tag{3}$$

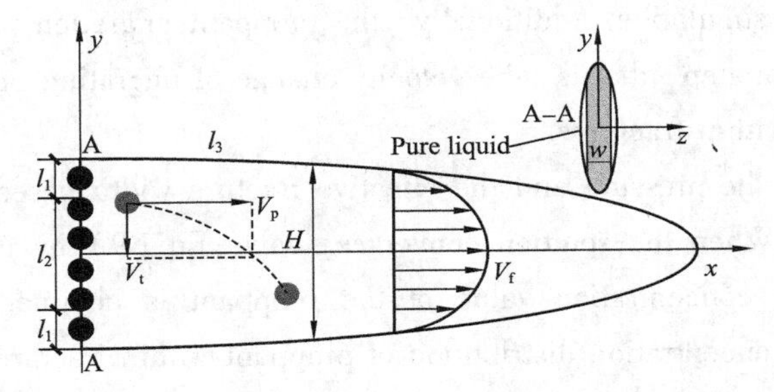

Fig. 6 Motion scheme of the proppant particle in a fracture (l_1, l_2 and l_3 are boundaries)

The mass conservation equation of the proppant carrying liquid within fracture is:

$$\frac{\partial(\rho w_{eff})}{\partial t} + \nabla\cdot(\rho\vec{q}) = -\rho_f q_l \tag{4}$$

The mass conservation equation of thevariable density proppant particles is:

$$\frac{\partial(c\rho_p w_{eff})}{\partial t} + \nabla\cdot(c\rho_p\vec{q_p}) = 0 \tag{5}$$

For a proppant-laden liquid, the proppant particle moving velocity is smaller than that the carrying fluid velocity because of higher density of proppant and viscous drag force between them. In order to precisely calculate the proppant concentration in the fracture, the transport velocity of the proppant should be analyzed. The transport velocity of proppant particles can be obtained from Eq. (6):

$$\vec{q}_p = k_{wc} V_x \vec{i} + (V_y + V_t) \vec{j} \tag{6}$$

$$k_{wc} = \frac{V_p}{V_f} \tag{7}$$

The modified settling velocity of particles is calculated as follows:

$$V_t = V_s \cdot f_{Re} \cdot f_c \cdot f_w \cdot f_T \tag{8}$$

Combined Eq. (4) and (5) with Eq. (6), the transport equation of the variable density proppant within induced cracks is obtained as:

$$w_{eff} \frac{\partial c}{\partial t} - (1 - c)\frac{\partial w_{eff}}{\partial t} - \frac{\partial}{\partial x}\left(\left((1 - c) + c\frac{\rho_p}{\rho_f}(1 - k_{wc})\right) w_{eff} \cdot V_x\right) -$$

$$\frac{\partial}{\partial y}\left((1 - c) w_{eff} \cdot V_y - cV_t \frac{\rho_p}{\rho_f}\right) = q_l \tag{9}$$

The corresponding boundary conditions (Figure 6) of Eq. (9) are:

For boundary of l_2:

$$c = c_p \tag{10}$$

For the boundary between l_1 and l_3:

$$\partial c / \mathrm{d}n = 0 \tag{11}$$

The initial condition is:

$$c(x,\ y,\ 0) = 0 \tag{12}$$

It is difficult to simultaneously work out the fluid pressure, the fracture width and the proppant concentration within hydraulic fracture as they are interdependent to each other. While calculating the velocity field in fractures, the influences of fluid pressure and fracture width on the proppant migration are ignored in simulation. Additionally, the proppant migration process is assumed to be quasi-steady at each time step, that is, the velocity change of migrating proppant does not directly affect the fluid velocity within fractures.

Iterative calculation the pressure and the effective fracture width based on the initial proppant concentration is needed. When the solution converges, solve Eq. (9) by the fracture size and the fluid velocity. Then, the concentration value of the proppant is obtained. Convergence of above algorithm will yield the concentration distribution of proppant within fractures.

4 Case analysis

Table 1 shows the key parameters of the coal reservoir of interest based on collected core samplesfrom a CBM well. With these parameters, we studied the fracturing process and proppant distributions.

Table 1 Fracturing parameters of a coal reservoir in China

Fracturing parameter	Value	Fracturing parameter	Value
Perforation depth (m)	1217.0~1224.0	Shut-in pressure (MPa)	14.7
Fracturing fluid type	water	Closure pressure (MPa)	10.8
Elastic modulus (GPa)	27	Porosity (%)	11.7

Continued

Fracturing parameter	Value	Fracturing parameter	Value
Poisson's ratio	0. 2	Permeability (mD)	1. 0
Fracture pressure (MPa)	30. 1		

Using an injection rate of 6. 5 m^3/min with an average proppant content of 12% (the highest proppant ratio of 25% ~ 30%) and fracturing fluid viscosity of 10 mPa. s, we obtained the simulated fracture geometry for this well as shown in Figure 7. The total injected fluid volume, proppant, pad fluid and displacement fluid are respectively 390m^3, 46. 8m^3, 303. 3 m^3 and 17. 9 m^3. Figure 7 shows the effective half-length of the crack is 70. 6 m, the maximum fracture width is 9. 54 mm, the upper fracture height is 9. 05 m and the lower fracture height is 3. 96 m. Compared with the micro-seismic data, the simulated result has about 3. 06% deviation. Stress data shows that the stress difference between the pay zone and the bed rock is less than 3MPa, and the difference between the cap rock and pay zone is about 6 MPa. The differences of stress attribute to the uneven fracture width.

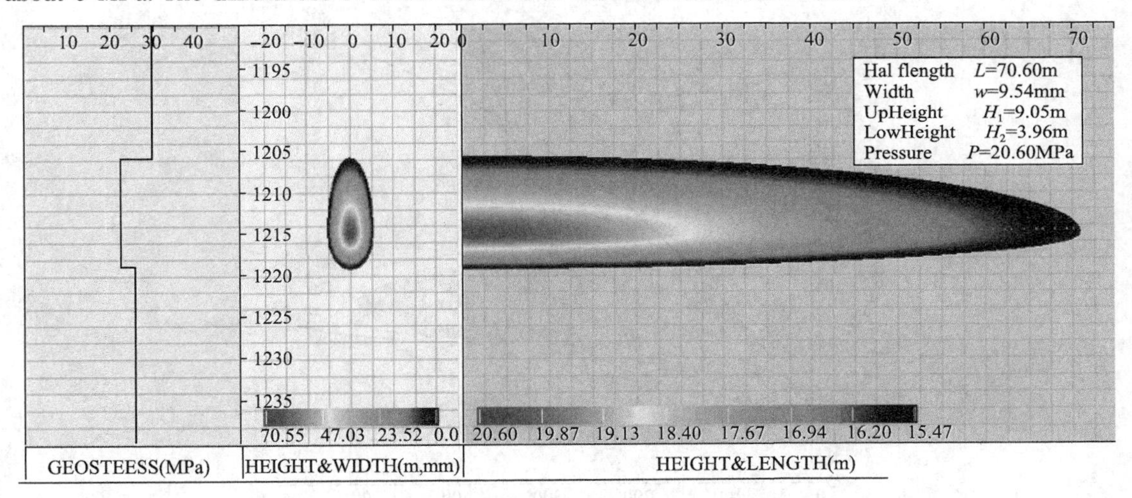

Fig. 7 Simulated fracture geometry for the CBM well

Figure 8 shows the effective half-lengths of the induced fractures for proppants with different densities. The effective propped half - length reaches to 44. 5 m by using the variable density proppant, which is 19. 5 m longer than that of ceramic proppants. Besides, the particle distribution will be more uniform that may increase the well productivity in CBM reservoirs. Figure 9 shows the comparison between the data from simulation and the field data, and it shows that the simulated result are acceptable by using the proposed model for this well fracturing design.

Figure 10 shows contour maps of the proppant concentrations of the cases with and without the wall effect by assuming the same injection parameters. The length of the fracture is reduced from 47. 67 m to 30. 85 m, and the height is reduced from 10. 41m to 9. 32 m due to the influence of the fractured wall. With the increase of fracturing fluid viscosity, the height of propped fracture increases. The fracture wall decreased the drag force of proppant particles. Thus, the average velocity of proppant would be larger than that of fracturing fluid, leading to increases in both of fracture's effective propped length and height, as shown in Figure 8.

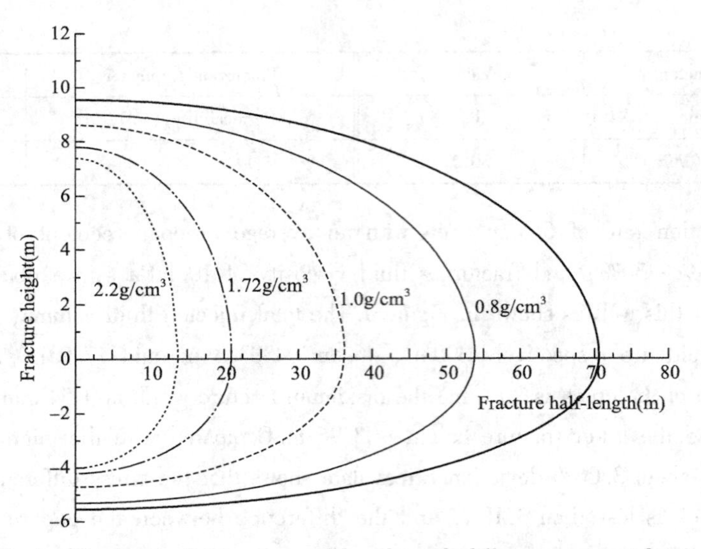

Fig. 8 Simulated effective propped lengths with different proppant density

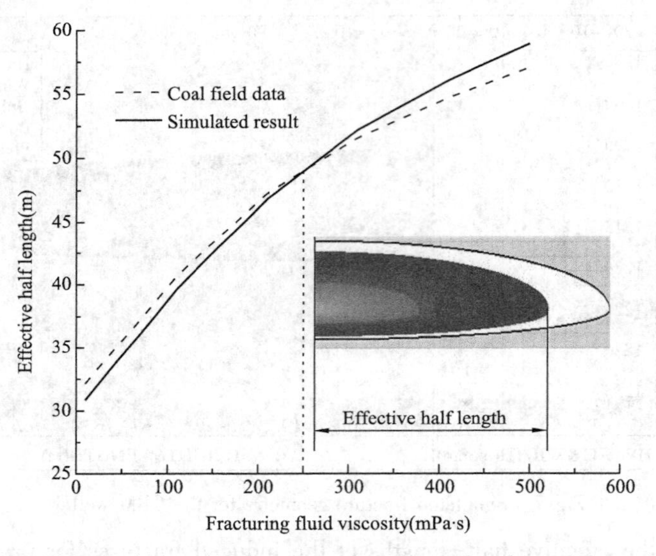

Fig. 9 Comparison of field data with the simulated result on the effective lengths at different fracturing fluid viscosities

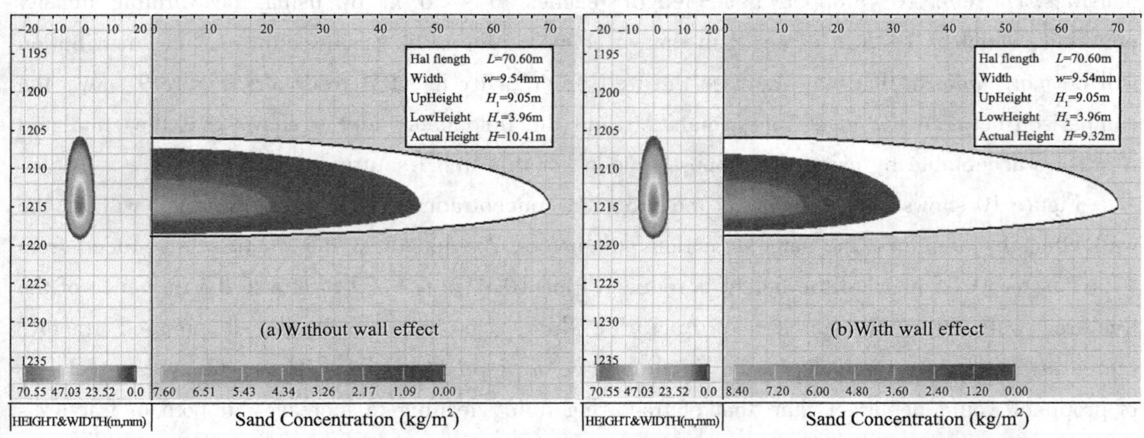

Fig. 10 Wall effect on proppant distribution and fracture geometry (The used fracturing fluid viscosity is 10 mPa · s)

Figure 11 shows the effect of proppant size on the fracture geometry and proppant distribution. The increase of the proppant particle size may cause an increment of both fracturing fluid drag force and proppant resistance. Since the velocity component of the proppant particles decreases along the x direction of fracture, the propped length drops sharply when larger proppant (40/70) is used.

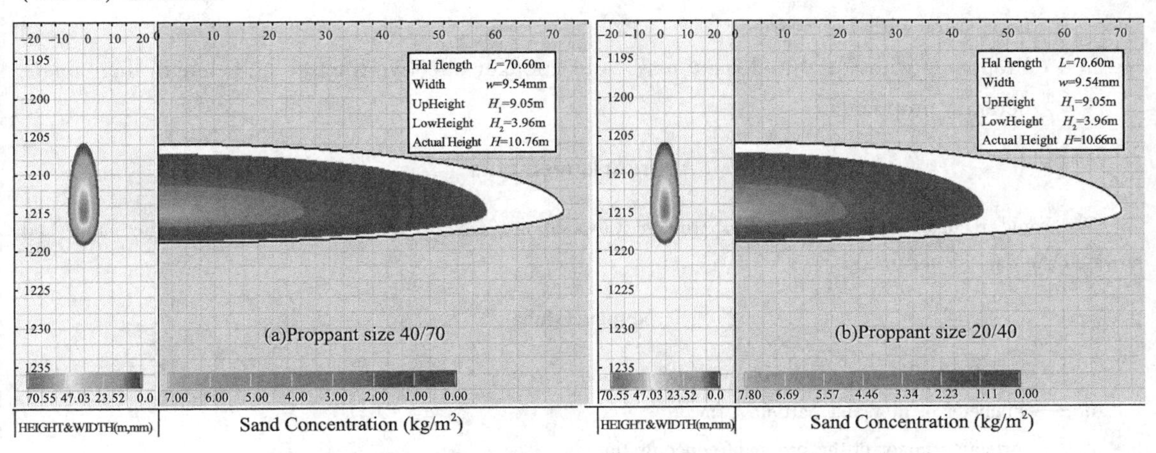

Fig. 11 Proppant concentration map for the cases of different proppant size
(The used fracturing fluid viscosity is 500 mPa · S)

Figure 12 show the influence of the injection rate on proppant concentration distribution within fractures. By doubling the injection rate will lead to an increase of the pressure dropping speed along the fracture. The effective support length is 1.41 times longer than the conventional one.

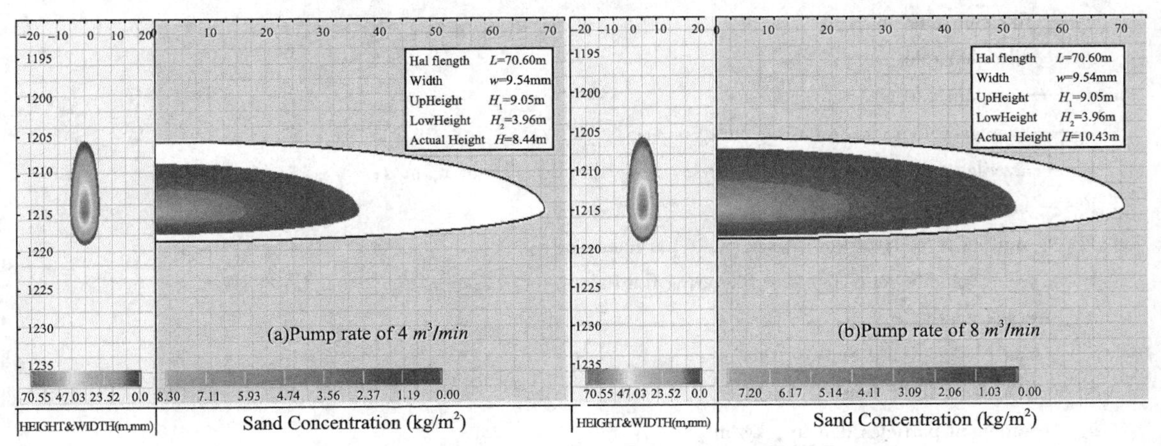

Fig. 12 The effect of pump rate on the proppant distribution.

5 Conclusions

Following API crushing test procedures on a variety of nutshell proppants, the crushing values of the selected proppants are less than 2%, which indicates that the proppants with variable density meets the operating requirements in propping hydraulic fractures. From above analysis, with the selected proppants, the following conclusions can be drawn.

(1) We proposed and developed a model to calculate the proppant distribution along fracture

and the effective length of propped fracture length.

(2) Computation result is validated against field microseismic data and the error is acceptable for fracture designing purposes.

(3) Using the developed model, sensitivity study shows that smaller propants and higher fracturing bump rate are favorable in creating longer effective fracture high length. This is consistent with literature.

(4) Proppants with variable density may yield a longer effective fracture half-length than single ceramic proppants.

Acknowledgements

We would recognize China National Petroleum Corporation (CNPC) for allowing publishing the paper and provided the data for this study.

Nomenclature

c——sand-laden fluid proppant volume concentration;

dp——diameter of proppant particles, m;

f_c——correction factor of the proppant concentration;

f_{Re}——correction factor of the inertial effect;

f_T——correction factor of turbulent fluctuations;

f_w——correction factor of the wall effect;

k_{wc}——average velocity ratio of proppant and fracturing fluid along the crack;

q——fracturing fluid flow, m/s;

q_l——filtration rate, m/s;

q_L——filtration rate, m/s;

q_p——proppant particles rate, m/s;

V_f——average velocity of fracturing fluid, m/s;

V_p——average velocity of proppant particles along the direction of x, m/s;

V_s——settling velocity, m/s;

V_t——settling velocity of the revised Stoke particles, m/s;

Vx, Vy——velocity components of fracturing fluid along the direction of x, y, m/s;

w——crack width, m;

ρ——sand-laden liquid density, kg/m^3;

ρ_f——fracturing fluid density, kg/m^3;

ρ_p——proppant particles density, kg/m^3。

References

[1] Guo Tiankui, Zhang Shicheng, Zou Yushi, and Xiao Bo. 2015. "Numerical simulation of hydraulic fracture propagation in shale gas reservoir." *Journal of Natural Gas Science and Engineering* 26: 847-856. doi: 10.1016/j.jngse.2015.07.024.

[2] Kang Yili, Huang Fansheng, You Lijun, Li Xiangchen, and Gao Bo. 2016. "Impact of fracturing fluid on multi-scale mass transport in coalbed methane reservoirs." *International Journal of Coal Geology* 154: 123-135. doi: 10.1016/j.coal.2016.01.003.

[3] Kern, L. R., T. K. Perkins, and R. E. Wyant. 1959. "The mechanics of sand movement in fracturing."

Journal of Petroleum Technology 11 (07): 55-57. doi: http: //dx. doi. org/10. 2118/1108-G.

[4] Yajun Liu, and Mukul M Sharma. 2005. "Effect of fracture width and fluid rheology on proppant settling and retardation: an experimental study." SPE Annual Technical Conference and Exhibition.

[5] Neto, Luiz Bortolan, and Andrei Kotousov. 2013. "Residual opening of hydraulic fractures filled with compressible proppant." *International Journal of Rock Mechanics and Mining Sciences* 61: 223-230. doi: 10. 1016/j. ijrmms. 2013. 02. 012.

[6] Palisch, Terrence T, Robert John Duenckel, Mark Aaron Chapman, Scott Woolfolk, and Michael C Vincent. 2009. "How to use and misuse proppant crush tests—exposing the top 10 myths." *SPE Production & Operations* 25 (03): 345-354. doi: http: //dx. doi. org/10. 2118/119242-PA.

[7] Rahman, MM, and MK Rahman. 2010. "A review of hydraulic fracture models and development of an improved pseudo-3D model for stimulating tight oil/gas sand." *Energy Sources, Part A: Recovery, Utilization, and Environmental Effects* 32 (15): 1416-1436. doi: 10. 1080/15567030903060523.

[8] Rickards, Allan R, Harold D Brannon, and William D Wood. 2006. "High strength, ultralightweight proppant lends new dimensions to hydraulic fracturing applications." *SPE Production & Operations* 21 (02): 212-221. doi: http: //dx. doi. org/10. 2118/84308-PA.

[9] Sahai, Rakshit, Jennifer L Miskimins, and Karen E Olson. 2014. "Laboratory results of proppant transport in complex fracture systems." SPE Hydraulic Fracturing Technology Conference, The Woodlands, Texas, USA.

[10] Senthamaraikkannan, Gouthami, Ian Gates, and Vinay Prasad. 2016. "Development of a multiscale microbial kinetics coupled gas transport model for the simulation of biogenic coalbed methane production." *Fuel* 167: 188-198. doi: doi: 10. 1016/j. fuel. 2015. 11. 038.

[11] Shiozawa, Sogo, and Mark McClure. 2016. "Simulation of proppant transport with gravitational settling and fracture closure in a three-dimensional hydraulic fracturing simulator." *Journal of Petroleum Science and Engineering* 138: 298-314. doi: 10. 1016/j. petrol. 2016. 01. 002.

[12] Songyang Tong, and Kishore K Mohanty. 2016. "Proppant transport study in fractures with intersections." *Fuel* 181: 463-477. doi: 10. 1016/j. fuel. 2016. 04. 144.

[13] Wang J, D. D. Joseph, N. A. Patankar, M. Conway, and R. D. Barree. 2003. "Bi-power law correlations for sediment transport in pressure driven channel flows." *International journal of multiphase flow* 29 (3): 475-494. doi: 10. 1016/S0301-9322 (02) 00152-0.

[14] Lei Wang, and Shicheng Zhang. 2008. "Influence of the backflow velocity of fracturing on the backflow volume and distribution of proppant in fractures." *PGRE* 15 (1): 101-102.

[15] Xiangzhen Yan, Yan-tao Zhang, Tong-tao Wang, and Xiu-juan Yang. 2010. "Permitted build-up rate of completion strings in multi-branch CBM well [J]." *Journal of China Coal Society* 5 (35): 787-791.

[16] Qiaofei Zhang, Li Yakun, Chai Ruijuan, Zhao Guofeng, Liu Ye, and Lu Yong. 2016. "Low-temperature active, oscillation-free PdNi (alloy) /Ni-foam catalyst with enhanced heat transfer for coalbed methane deoxygenation via catalytic combustion." *Applied Catalysis B: Environmental* 187: 238-248. doi: 10. 1016/j. apcatb. 2016. 01. 041.

[17] Desheng Zhou, Zheng Peng, He Pei, and Peng Jiao. 2016. "Hydraulic fracture propagation direction during volume fracturing in unconventional reservoirs." *Journal of Petroleum Science and Engineering* 141: 82-89. doi: 10. 1016/j. petrol. 2016. 01. 028.

[18] Caineng Zou, Zhu Rukai, Wu Songtao, Yang Zhi, Tao Shizhen, Yuan Xuanjun, Hou Lianhua, Yang Hua, Xu Chunchun, and Li Denghua. 2012. "Types, characteristics, genesis and prospects of conventional and unconventional hydrocarbon accumulations: taking tight oil and tight gas in China as an instance." *Acta Petrolei Sinica* 33 (2): 173-187.

一种 N80 1 类油管的脆性断裂机理

路彩虹　冯　春　韩礼红　朱丽娟　蒋　龙　朱丽霞

（中国石油集团石油管工程技术研究院，石油管材及装备材料服役行为与结构安全国家重点实验）

摘　要：N80 1 类油管属于非调质经济型低碳低合金钢管。对发生脆性断裂的 N80 1 类油管进行化学成分、力学性能、微观组织、断口形貌及微区能谱等分析，判断油管断裂机理为：由于钢管终轧后或者正火处理时冷却速度较快，钒的碳氮化合物析出不充分，较多的钒元素仍固溶于奥氏体中，提高局部淬透性，促进马氏体转变，尤其是内表面型变量较大的情况下，马氏体带状组织偏析较明显，严重降低了油管韧性，导致油管在使用过程中发生脆性断裂。将失效油管重新加热到 850℃，通过控制冷却速度消除马氏体相后，得到细小均匀的珠光体和铁素体组织，油管强韧性良好。

关键词：N80 1 类油管；脆性断裂；马氏体带状组织；成分偏析

N80 1 类非调质油管属于经济型中碳微合金高强韧钢，是非调质油井管用钢中最高钢级，其生产方法是在中碳含锰钢中添加微合金元素，通过热轧后控制冷却速度来获得细小铁素体+珠光体和弥散析出的 V 碳氮氧化物，即可满足需要的良好的强韧性[1-3]，该生产方法简化了生产流程，降低了生产成本，提高了生产效益，在油气田得到广泛应用。由于 API Spec 5CT 标准[4]没有对 N80 1 类油管管体冲击韧性进行强制要求，导致有些生产厂仅追求化学成分和高强度满足要求，而忽略了韧性的匹配，使该类油管在使用过程中发生脆性断裂，本文通过化学成分、力学性能、金相显微组织及断口分析等试验方法对该类油管脆性断裂机理进行研究，并提出改进措施。

1　试验过程及结果

1.1　化学成分分析

试验材料为油田现场发生脆性断裂的 N80 1 类油管，规格为 ϕ88. 90mm×6. 45mm，在断口附近取样，按照标准 ASTM A751—14a 规定的试验方法，用 ARL 4460 直读光谱仪对试样进行化学成分分析，结果见表 1，满足 API Spec 5CT 标准要求。

作者简介：路彩虹，工程师，长期从事石油管材的科研及失效分析工作。联系电话：18966842473。Email：lucaihong@ cnpc. com. cn。

表1　断裂油管化学成分结果（wt×10^{-2}）

样品	C	Si	Mn	P	S	Cr	Mo	Ni	Nb	V	Ti	Cu	B	Al
管体	0.36	0.31	1.63	0.017	0.0022	0.049	0.0034	0.033	0.0005	0.15	0.0027	0.082	0.0005	0.020
API Spec 5CT 标准要求	/	/	/	≤0.030	≤0.030									

1.2　力学性能试验

在管体上取纵向拉伸和冲击试样，分别按照 ASTM A370—15 和 ASTM E23—12c 标准规定的试验方法，在 UTM 5305 材料试验机和 PIT302D 冲击试验机上进行拉伸和冲击试验，结果见表2和表3，可见油管抗拉和屈服强度满足 API Spec 5CT 标准要求，但伸长率和冲击吸收能均低于标准要求。

表2　拉伸性能试验结果

试样			试验结果			
位置	取向	宽度×标距（mm×mm）	加载下的总伸长率（%）	屈服强度（MPa）	抗拉强度（MPa）	伸长率（%）
管体	纵向	19.1×50	0.5	773	1067	15
			0.5	751	1056	16
API Spec 5CT 标准要求			0.5	552~758	≥689	≥20

表3　夏比冲击试验结果

试样		试验温度（℃）	试验结果			
			吸收能量（J）			
取向	规格（mm×mm）		单个值			平均值
纵向	5×10×55V 形缺口	0				
API Spec 5CT 标准要求		0±3	≥27 7≤仅允许1个值<11			≥27

1.3　金相分析

在断口附近取金相试样，用 MeF4A 金相显微镜及图像分析系统对试样进行组织，晶粒度，夹杂物分析，结果为A、B、C、D类非金属夹杂物均未超标，晶粒度6.0级，内表面组织（图1）为珠光体+网状铁素体+带状马氏体+少量贝氏体，壁厚中心组织（图2）为珠光体++网状铁素体+少量马氏体+少量贝氏体，外表面组织（图3）为珠光体++网状铁素体+少量马氏体+少量贝氏体，横截面内壁侧有1/2壁厚组织中存在马氏体带状组织偏析，见图4。

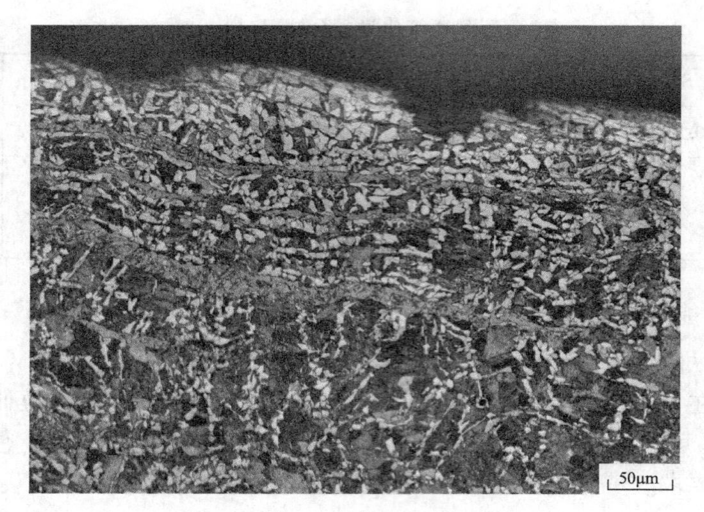

图 1　内表面组织

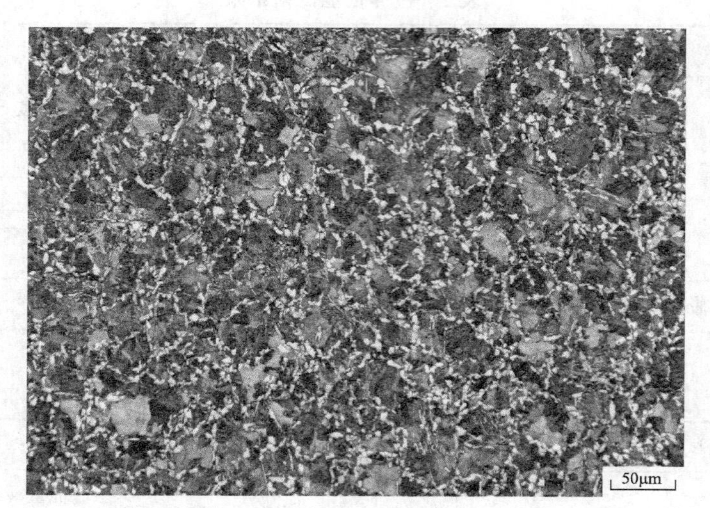

图 2　中心组织

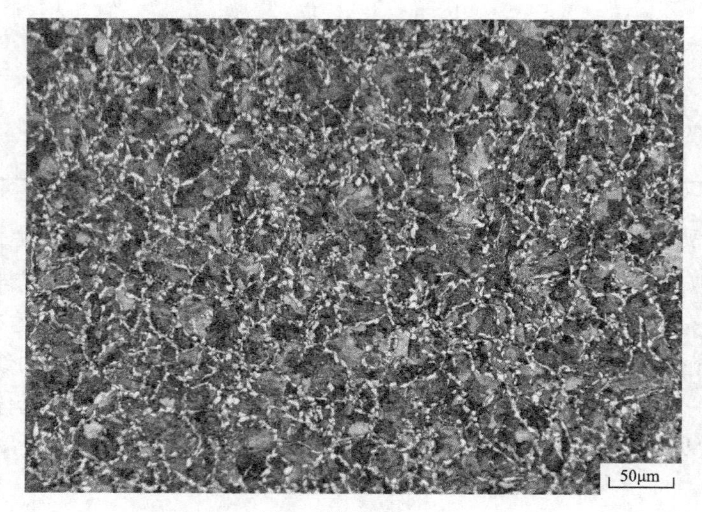

图 3　外表面组织

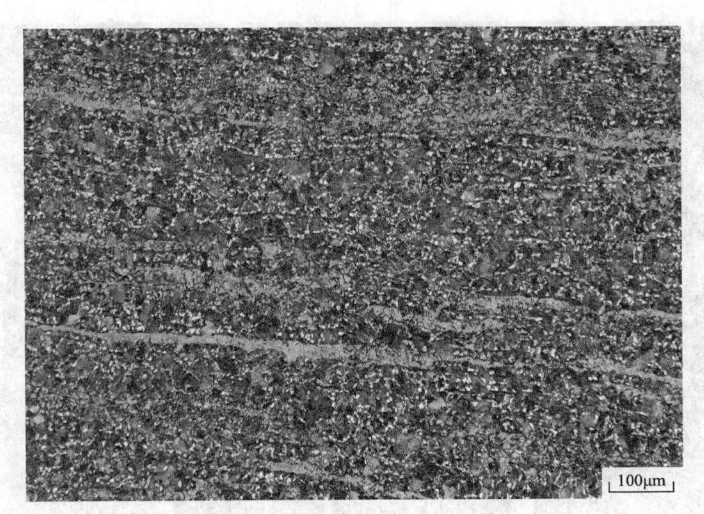

图 4　内表面带状组织偏析

1.4　断口形貌及能谱分析

断裂油管宏观断口形貌见图 5，整个断口表面较平整，断裂“人字纹”收敛于内表面，为典型的起裂于内壁的脆性断口。在源区附近取断口试样在 TESCAN VEGAII 型扫描电子显微镜下进行观察，整个断口为准解理脆性断裂特征见图 6。对源区组织偏析处进行微区能谱分析，结果见图 7 和表 4，可见马氏体组织内 V 和 Mn 元素含量相对较高，而 C 含量相对较少。

图 5　油管断口宏观形貌

表 4　源区微区能谱分析结果

能谱	C	Si	V	Mn	Fe	总和
能谱 1	3.03	0.43	0.34	2.17	94.04	100.00
能谱 2	3.32	0.37		1.88	94.43	100.00

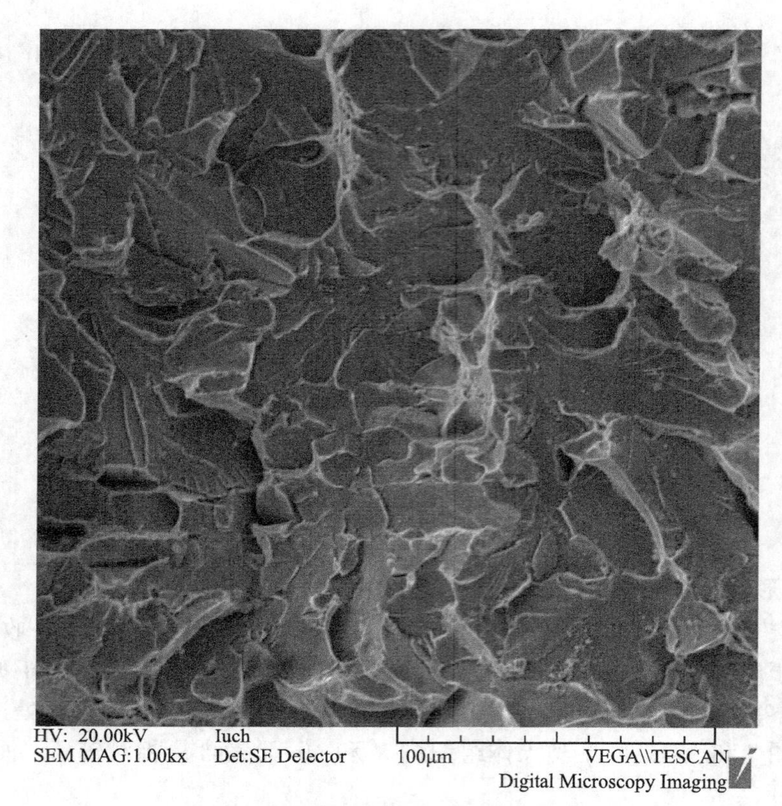

图 6　油管断口微观形貌

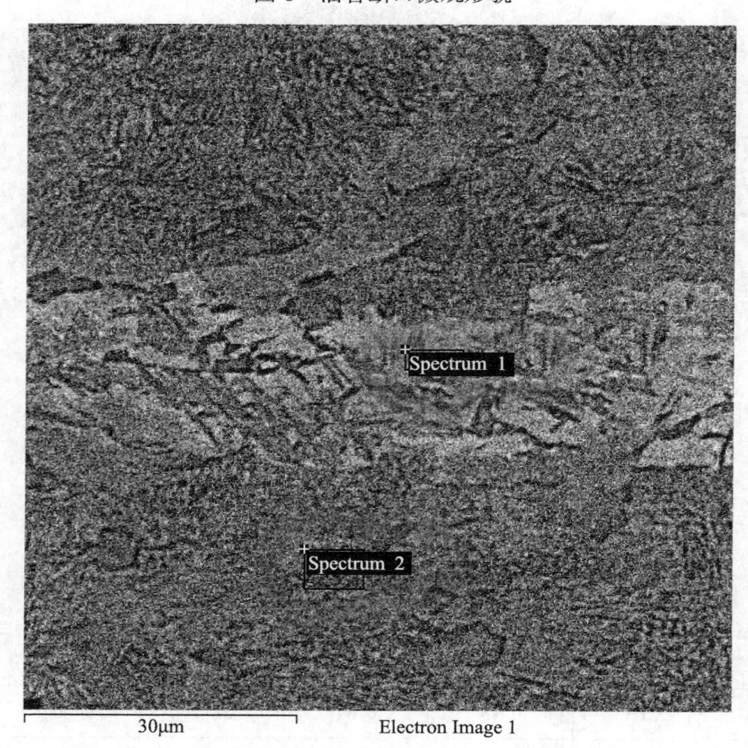

图 7　源区微区能谱分析位置

2 分析与讨论

以上试验结果表明，该油管的化学成分、非金属夹杂物及拉伸强度、屈服强度都满足标准要求，但伸长率和冲击吸收能低于标准要求，晶粒度较正常油管的8~9级较粗大，且组织中出现异常马氏体和贝氏体组织，尤其是靠近内表面一侧，存在严重马氏体带状组织偏析，较多的V元素存在于马氏体相内，相比正常的细小铁素体+珠光体和弥散析出的V的碳氮化合物微粒，韧性明显降低。断口为典型的脆性准解理形貌，且起裂于油管内表面马氏体带状组织较多区域，可判断油管发生脆性断裂的主要原因是硬脆相马氏体带状组织的存在。马氏体带状组织存在的主要原因是终轧温度和冷却速度。

在快冷速条件下[5,6]，比如冷却速度大于2.3℃/s时，微合金碳氮化物的析出受到抑制，此时形变对析出的诱导作用显著，若形变发生在适宜钒的碳氮化物析出的温度区间，则相变前析出的大量微细相可以成为 γ/α 相变的形核核心，促进铁素体和珠光体转变，若形变温度低于钒的碳氮化物的析出温度（如880℃），则钒、碳仍大量固溶于奥氏体中，固溶于奥氏体中的V还可提高淬透性，促进贝氏体转变，如果冷却速度大于10℃/s时则发生马氏体转变，使韧性显著降低。

而在慢冷区，比如冷却速度小于2.3℃/s时，如果终轧温度较低，发生不完全再结晶，沿奥氏体边界及内部变形带上生成新相的形核率增加有利于铁素体沿晶界和晶内析出，加之试样在高温区的停留时间相对较短以及微细碳氮化物的弥散析出，有效地抑制了晶粒的长大。晶内针状铁素体的存在可提高钢的韧性，但在工业生产中，为了获得铁素体、珠光体组织良好的强韧性配合，铁素体含量应尽量控制，同时要避免出现贝氏体和马氏体组织，终轧温度应控制在880~1000℃范围内，冷却速度控制在0.7~2.3℃/s。

基于以上分析，可判断该油管的终轧温度高于1000℃，冷却速度大于10℃/s，在较高温度下使得奥氏体晶粒度偏粗大，网状铁素体析出较多，同时由于冷却速度过快，V的碳氮化合物析出不充分，对晶粒长大的钉扎作用减小，使得较多V元素固溶于奥氏体中，提高淬透性，促进马氏体转变，尤其是内表面型变量较大的情况下，马氏体带状组织偏析较明显，显著降低韧性，导致油管在承受拉应力作用下发生脆性断裂。

3 结论

（1）失效的N801类油管得化学成分、非金属夹杂物、拉伸强度、屈服强度等性能满足API Spec 5CT标准要求，但其延伸率。冲击韧性都低于标准要求。

（2）非调质N80 1类油管断裂机理是油管终轧后或者正火处理后冷却速度过快，V的碳氮化合物析出不充分，对晶粒长大的钉扎作用减小，使得较多V元素固溶于奥氏体中，提高淬透性，促进马氏体转变，尤其是内表面型变量较大的情况下，马氏体带状组织偏析较明显，显著降低韧性，导致油管在使用过程中，承受拉应力作用时，从内表面起裂，发生脆性断裂。

（3）建议对失效油管进行重新热处理，加热温度900℃，保温30min，然后以0.7 ~ 1.0℃/s的冷却速率进行冷却，即可获得细小的珠光体和铁素体组织，和弥散析出的V的碳氮化合物微粒，以保证油管的良好的强度和韧性。

参 考 文 献

[1] 鞠艳美，张伟，李法兴．油井管用非调质钢36Mn2V冶金质量控制技术［J]，中国冶金，2012，22

(12)：17-21.
[2] 刘雅政，刘照，徐进桥．非调质 N80 石油套管轧制工艺优化的试验研究［J］．钢铁，2006，41（7）：41-53.
[3] 方剑，谢凯意，李阳华．V 的碳氮化合物析出对 36Mn2V 非调质钢组织性能的影响［J］，武汉科技大学学报，2012，35（2）：81-84.
[4] 套管和油管规范：API Spec 5CT［S］，2012，1.
[5] 卢忠山，王福明，张博．不同冷却速度下 36Mn2V 钢坯高温塑性及碳氮化物的析出［J］．材料热处理学报，2011，32（8）：118-121.
[6] 吕文涛，黄长虹．冷却速度对36Mn2V 制 N80-1 钢级油管性能的影响［J］．材料热处理技术，2011，9：164-166.

G105 钢制钻杆腐蚀失效机理分析

朱丽娟　冯　春　袁军涛　路彩虹

（中国石油集团石油管工程技术研究院，石油管材及装备材料服役
行为与结构安全国家重点实验）

摘　要：某 ϕ127mm G105 钢质钻杆表面在服役过程中发生了严重的局部腐蚀。通过力学性能分析、金相分析、宏观及微观腐蚀形貌观察，并结合能谱（EDS）和 X 射线衍射（XRD）等方法对该钻杆进行了失效分析，确定了该钻杆发生失效的原因及机理。结果表明，钻杆管体表面发生严重了点蚀且呈条带状分布；在钻杆管体和接头的过渡区也发生了严重的点蚀；引起点蚀的主要原因是氧腐蚀、Ca^{2+} 等离子引起的垢下腐蚀；氯离子的存在加速了钻杆的局部腐蚀。

关键词：钻杆；氧腐蚀；点蚀；失效分析

在钻井工程中，钻具腐蚀是普遍存在的问题。钻具自身的化学性质和金相结构决定了该类铁基合金是一类容易被腐蚀的合金，并随着钻井向高速、深井、苛刻腐蚀环境方向发展而日趋严重。某公司在淮北地区进行钻井作业用的 ϕ127mm 5in G105 内加厚钻杆在服役过程中发生了严重的局部腐蚀。该批钻杆服役两年多时间，闲置时间超过一年。现场观察发现部分钻杆表面有大量的腐蚀坑点，点蚀异常严重。钻井时采用钾）和钠基钻井泥浆，泥浆主要含 O、Si、Ca 和 Al，同时含 Fe、K、Mg、Na、Cl、C 等元素。订货技术协议要求该批次钻杆符合 API SPEC 5DP—2010 标准。

本文主要对上述失效的 5in G105 钻杆的成分组织，力学性能和腐蚀形貌进行分析，对腐蚀产物的成分和分布进行测定，并结合失效理论，对该钻杆腐蚀原因进行了分析。

1　结果与讨论

1.1　宏观腐蚀形貌

失效钻杆管体的宏观形貌如图 1（a）和 1（b）所示。钻杆管体外表面布满了大小不一、深浅各异的点蚀坑，腐蚀区域布满红褐色的腐蚀产物，腐蚀较为严重。管体表面存在沿轴向分布，具有一定宽度，点蚀严重的区域，呈条带状分布，如图 1（a）所示；该区域内点蚀坑尺寸和密度较大，点蚀深度较深。如图 1（b）所示，钻杆接头只是在局部区域出现了少量点蚀坑，点蚀坑的尺寸较小；在钻杆管体与接头的过渡区，出现了严重的点蚀。

基金项目：中国石油集团大型油气田及煤层气开发项目，碳酸盐岩、火成岩及酸性气藏高效安全钻井技术课题（2011ZX05021-002）资助。

作者简介：朱丽娟，女，博士，2013 年毕业于中国科学院金属研究所，腐蚀科学与防护专业，现就职于中国石油集团石油管工程技术研究院，主要从事石油管材腐蚀与防护的研究工作。联系方式：（029）81887609，18392995830，zhulijuan1986@ cnpc. com. cn。

因此，失效钻杆的腐蚀是以点蚀为主。一般来说，点蚀通常在腐蚀介质静滞条件下容易发生。该失效钻杆服役时间2年多，闲置时间一年以上，约占使用时间的50%。钻杆出井时，因重力作用，钻杆表面残留的钻井液或井下腐蚀介质易在钻杆接头和管体的过渡区滞留。如果钻杆在存放前未及时冲洗或存放期间无防腐措施，则因重力作用，在横向放置的钻杆外表面底部，钻井液或井下腐蚀介质滞留较多。在图1（a）所示钻杆外表面的另一面，还有另一明显的腐蚀条带；这是钻杆多次使用，存放时放置的位置不同的结果。在这些腐蚀介质的长期作用下，钻杆表面底部，以及钻杆接头和管体的过渡区将发生严重点蚀。韩勇等[1]的研究工作也曾得出类似的结论。

钻杆管体剩余壁厚测量结果表明，钻杆管体最小剩余壁厚仅为5.32mm，约为名义壁厚的41.9%，远远低于API SPEC 5DP—2010标准中缺欠为87.5%最小规定壁厚的要求，该钻杆已失效。按使用时间两年半进行简单估算，该钻杆最大点蚀速率达2.95mm/a，点蚀异常严重。

力学性能进行测试结果表明，该钻杆的力学性能均符合API SPEC 5DP—2010标准要求。

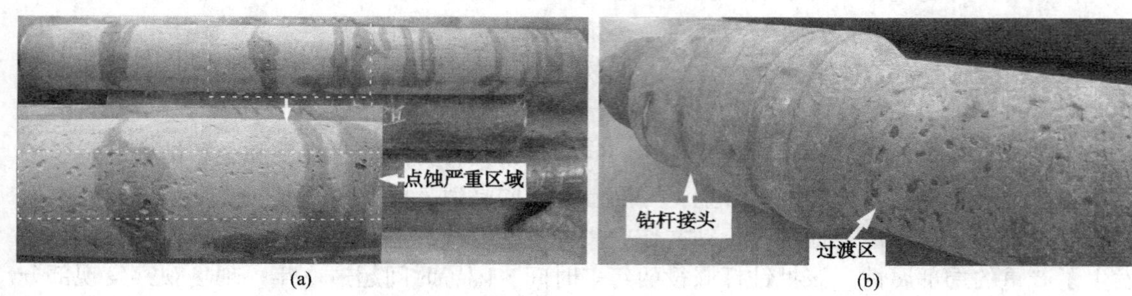

图1　G105钻杆宏观形貌

1.2　化学成分分析

钻杆化学成分符合API Spec 5DP-2010标准要求（表1）。从表中可以看出，不论是提高碳钢耐均匀腐蚀的Cr元素[2]，还是提高碳钢耐点蚀能力的Mo，Ni元素[3]，在钻杆接头中的含量均高于钻杆管体中的含量。在含Cl^-的环境中，Mo能促使铁基合金尤其是不锈钢表面形成钝化膜，阻止铁基合金发生点蚀[4,5]。Chapetti等的研究表明，添加镍能有效提高双相不锈钢的耐点蚀性能和焊缝的耐腐蚀疲劳性能[6]。Zhou等研究表明添加镍能使低碳微合金钢表面生成致密的锈层，从而提高其耐腐蚀性能[7]。Mn元素是良好的脱氧剂和脱硫剂，但是锰含量增高，在钢中会形成MnS夹杂物，诱发点蚀，从而减弱钢的耐腐蚀能力[8,9]；钻杆接头中Mn元素的含量低于钻杆管体中的含量。因此，化学成分差异是钻杆管体较接头腐蚀更为严重的主要原因。

表1　钻杆化学成分（wt%）

元素	C	Si	Mn	P	S	Cr	Mo	Ni	V	Ti
钻杆	0.23	0.22	1.10	0.0083	0.0027	0.84	0.16	0.028	0.0048	0.0020
钻具接头	0.37	0.22	0.91	0.0073	0.0045	1.09	0.30	0.085	0.0054	0.0024
API Spec 5DP—2010	/	/	/	≤0.020	≤0.015	/	/	/	/	/

1.3 金相检验

钻杆管体和接头的非金属夹杂物及晶粒度均符合 API Spec 5DP—2010 标准要求。钻杆管体局部腐蚀坑底有垂直于轴向的裂纹（图 2），裂纹起源于点蚀坑底部，尖部较细；裂纹及腐蚀坑周围组织未见异常（图 2b）。钻杆在使用过程中受力极其复杂，壁厚减薄的点蚀坑处存在应力集中效应，在腐蚀介质和应力作用下，引起腐蚀疲劳致使点蚀坑底部裂纹萌生。

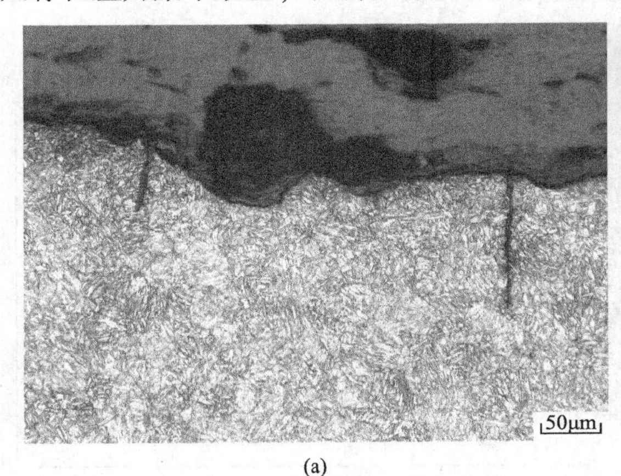

(a)

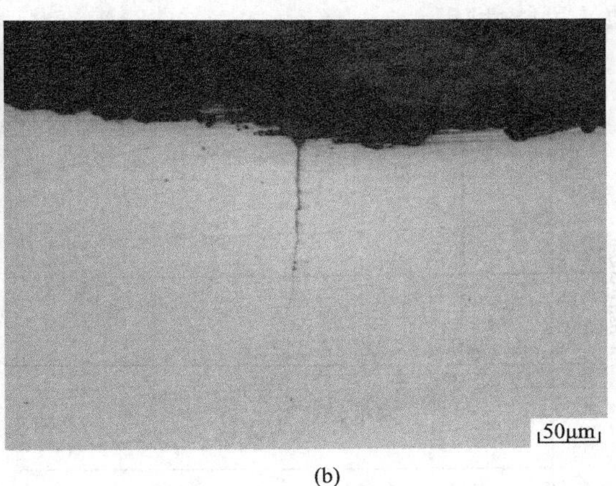

(b)

图 2　钻杆管体外表面腐蚀坑底裂纹形貌（a）及裂纹周围组织形貌（b）

1.4 腐蚀产物分析

1.4.1 表面形貌及能谱分析

腐蚀失效钻杆管体表面形貌及能谱分析结果如图 3 和表 2 所示。结果表明，管体表面腐蚀产物存在开裂和剥落现象，表面腐蚀产物以 O、Fe、C 元素为主，局部区域富含 O、Ca、Si、C、Fe、Mg 等元素。腐蚀失效钻杆接头表面形貌及能谱分析结果与钻杆管体类似。

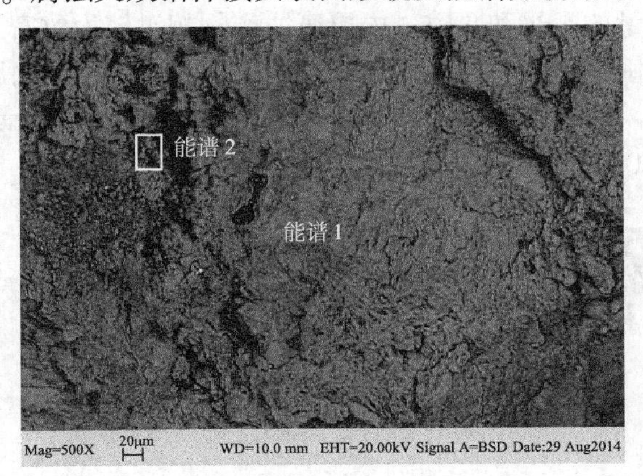

图 3　管体腐蚀产物表面形貌及谱图

表 2　管体腐蚀产物能谱结果（wt%）

元素	C	O	Na	Mg	Al	Si	K	Ca	Ti	Cr	Mn	Fe	Mo
能谱 1	7.01	33.14	/	/	0.40	1.22	/	0.40	0.38	1.53	0.47	54.69	0.77
能谱 2	9.20	48.57	0.43	3.51	1.88	9.81	0.37	18.13	1.38	/	1	6.73	/

1.4.2 腐蚀产物物相分析

从钻杆管体表面取块状样品，对其进行物相分析，结果表明，管体表面的物相主要为 $CaCO_3$，SiO_2（图 4a）。该结果验证了钻井液和腐蚀环境中含大量的 Ca 和 Si 元素。从失效钻杆管体表面刮取腐蚀产物，制成粉末样进行物相分析，结果表明，腐蚀产物主要为α-FeO(OH)，Fe_2O_3，$Fe_{0.98}O$，Fe_3O_4（图 4b）。因腐蚀产物中不含 $FeCO_3$ 等物相，可以排除 CO_2 腐蚀的可能性。

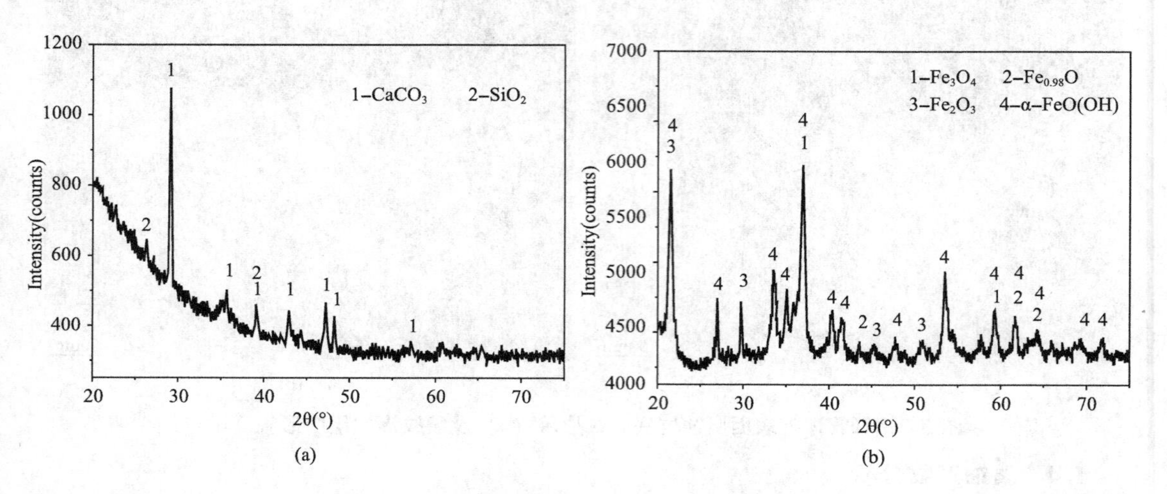

图 4 钻杆管体表面物相

（a）和腐蚀产物；（b）物相分析结果

1.4.3 截面形貌及能谱分析

失效钻杆管体腐蚀坑内的腐蚀产物层厚度超过 1mm。腐蚀产物大致可分为两层：外层腐蚀产物疏松，存在大量的裂纹和孔隙；内层腐蚀产物相对较为致密，局部区域存在微裂纹和少量孔隙，腐蚀坑底垂直于轴向的裂纹形貌如图 5 所示，对裂纹内腐蚀产物进行分析，能谱分析结果如表 3 所示。裂纹内的腐蚀产物以 Fe、O、C、Cr、Mn 元素为主，图 5 所示能谱 1 处，S 和 Cl 含量分别 0.43 和 0.45 wt%。

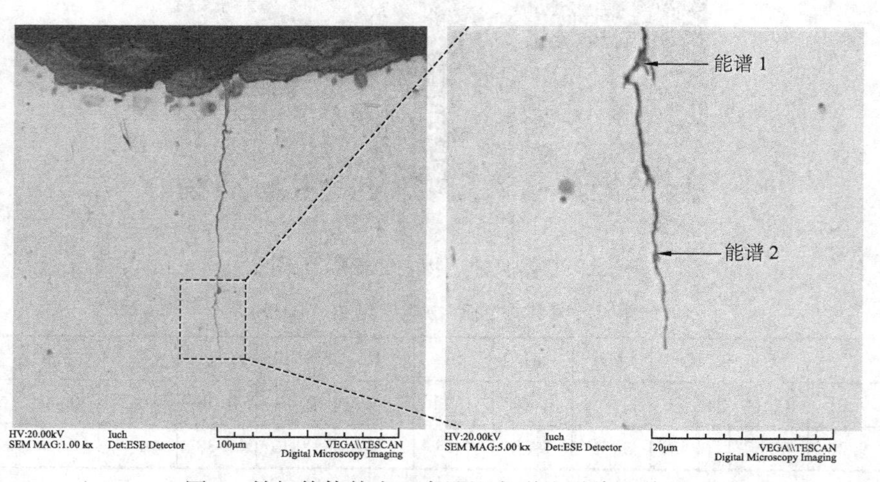

图 5 钻杆管体外表面腐蚀坑底裂纹形貌及谱图

表 3　钻杆管体裂纹内腐蚀产物能谱分析结果（wt%）

元素	C	O	S	Cl	Cr	Mn	Fe
能谱 1	4.51	19.24	0.43	0.45	1.15	1.00	73.21
能谱 2	6.35	13.29	0.39	0.35	0.88	1.13	77.62

1.5　腐蚀机理分析

综合上述分析，钻杆表面腐蚀产物以 O、Fe、C 为主。局部区域，富含 Ca 元素，最高值达 27.20wt%，同时含一定量的 Si、Al 和 Mg 等元素，及少量的 S。腐蚀产物主要为α-FeO(OH)、Fe_2O_3、$Fe_{0.98}O$ 和 Fe_3O_4，排除了 CO_2腐蚀的可能性。虽然腐蚀产物含少量的 S，但是 XRD 分析并未发现 FeS_{1-X}；钻杆管体表面最小剩余壁厚仅为 5.32 mm，并且在腐蚀坑底部存在裂纹的情况下，钻杆服役时间超过两年，却未发生断裂。因此，可排除硫化物应力腐蚀的可能性，所检测到的 S 元素很可能来自工况环境中的 SO_4^{2-}，与钻杆的腐蚀失效并无直接关系。

排除了硫化物应力腐蚀和 CO_2腐蚀的可能性，结合 XRD 分析的结果，可判断钻杆腐蚀主要是由氧腐蚀引起的。这是因为α-FeO(OH) 主要是在有氧或其他氧化剂存在的环境中产生的。研究表明，即使氧的浓度非常低，对碳钢的腐蚀仍有显著的影响[10]，且腐蚀速率随溶解氧含量增加而增加[11,12]。钻井液中存在游离态氧，在 G105 钻杆表面极易发生吸氧腐蚀[13]，即

阳极反应 $$Fe-2e \longrightarrow Fe^{2+} \tag{1}$$

阴极反应 $$O_2+2H_2O+4e \longrightarrow 4OH^- \tag{2}$$

总反应式 $$2Fe+O_2+2H_2O \longrightarrow 2Fe^{2+}+4OH^- \tag{3}$$

Fe^{2+}随后水解成α-FeO(OH)

$$4Fe^{2+}+O_2+6H_2O \longrightarrow 4FeO(OH)+8H^+ \tag{4}$$

FeO(OH) 失水后形成红棕色的 Fe_2O_3，氧化产物下方继续氧化，生成 Fe_3O_4，$Fe_{0.98}O$ 等腐蚀产物。

此外，工况环境中含大量的 Ca^{2+}、Mg^{2+}等元素，Ca^{2+}、Mg^{2+}的存在使结垢倾向增大，形成以 $CaCO_3$为主的碳酸盐结垢。碳酸盐结垢和腐蚀产物在钻杆表面沉积形成垢层，构成缝隙腐蚀的条件，从而诱发垢下点蚀。并且，垢层在钻杆表面不同区域覆盖度不同，不同覆盖度的区域之间就形成了具有很强自催化作用的腐蚀电偶，从而加速了钢材表面的局部腐蚀[14]。

但是，钻杆的最大点蚀速率达 2.95 mm/a，点蚀异常严重。这是因为，环境介质和钻井液中含有 Cl 离子，Cl 离子能诱发并加速点蚀。能谱分析结果表明，腐蚀坑底部裂纹中腐蚀产物的 Cl 离子含量达 0.45wt%。一般认为 Cl^-的存在会破坏钢表面的腐蚀产物膜，阻碍其形成，甚至会促进产物膜下钢的点蚀。研究表明，溶液中 Cl^-含量增加时，碳钢的腐蚀速率增加[15]。Liu 等的研究表明，在油气井的环境中，Cl^-能加速 CO_2腐蚀的阳极反应，破坏腐蚀产物膜并改变腐蚀产物膜的形貌[16]。Cl^-诱发局部腐蚀进而导致腐蚀穿孔的机理如下[17]：活性 Cl^-优先吸附在钢管内表面的缺陷（非金属夹杂或砂眼等）处吸附而诱发腐蚀；Cl^-的存在致使钢管内表面钝化膜在组织结构上发生改变并加速钝化膜溶解；尺寸较小的 Cl^-极易穿过垢层的疏松区域或缺陷处到达金属表面造成垢下腐蚀，甚至导致管材穿孔。

另外，煤层水可引起电偶腐蚀。该批次钻杆用在淮北的煤矿井中，淮北主要出产焦煤，焦煤的煤化度较高，仅次于无烟煤。煤化度越高，其电位越高。当煤层水与相对电位较低的金属接触，分散于煤层水中的固相煤与金属接触，在介质中形成微电偶作用，加速金属的腐

蚀[18]。该批失效钻杆点蚀速率大于一般情况下由垢下腐蚀引起的点蚀速率，这很可能与煤层水导致的电偶腐蚀有关。关于煤层水引起钻杆的电偶腐蚀还有待进一步研究。

因此，钻杆腐蚀主要是由极强的腐蚀性介质作用引起的，腐蚀介质包括 O_2，Ca^{2+}，Mg^{2+}，Cl^-等。腐蚀机理为氧腐蚀，Ca^{2+}、Mg^{2+}等离子引起的垢下腐蚀；Cl^-的存在加速了钻杆的局部腐蚀。

2 结论

（1）G105 钻杆的化学成分、金相组织、拉伸性能和冲击性能符合 API SPEC 5DP—2010 标准要求。

（2）G105 钻杆管体表面发生严重点蚀且呈条带状分布；在钻杆管体和接头的过渡区也发生了严重的点蚀；引起点蚀的主要原因是氧腐蚀、Ca^{2+}等离子引起的垢下腐蚀。Cl^-的存在加速了钻杆的局部腐蚀。

参 考 文 献

[1] 韩勇，冯耀荣，李鹤林 . G105 钻杆管体刺穿原因分析［J］. 石油机械，1990，18（1）：37.

[2] Wu H B，Liu L F，Wang L D，Liu Y T. Influence of chromium on mechanical properties and CO_2/H_2S corrosion behavior of P110 grade tube steel［J］. Journal of Iron and Steel Research，International，2014，21（1）：76.

[3] 张旭昀，高明浩，徐子怡，等 . Ni，Mo 和 Cu 添加对 13Cr 不锈钢组织和抗 CO_2腐蚀性能的影响［J］. 材料工程，2013，8：36.

[4] Ameer M A，Fekry A M，Heakal F E. Electrochemical behaviour of passive films on molybdenum-containing austenitic stainless steels in aqueous solutions［J］. Electrochimica Acta，2004，50：43.

[5] Pardo A，Merino M C，Coy A E，Viejo F，Arrabal R，Matykin E. Pitting corrosion behaviour of austenitic stainless steels - combining effects of Mn and Mo additions［J］. Corrosion Science，2008，50：1796.

[6] Távara S A，Chapetti M D，Otegui J L，Manfredi C. Influence of nickel on the susceptibility to corrosion fatigue of duplex stainless steel welds［J］. International Journal of Fatigue，2001，23：619.

[7] Zhou Y L，Chen J，Xu Y，Liu Z Y. Effects of Cr，Ni and Cu on the Corrosion Behavior of Low Carbon Microalloying Steel in a Cl^containing Environment［J］. J. Mater. Sci. Technol.，2013，29（2）：168.

[8] 王丽萍，董文学，郭二军，等 . Mn 含量对 00Cr22Ni5Mo3MnX 双相不锈钢组织和性能的影响［J］. 热加工工艺，2013，42（2）：63.

[9] Tobler W J，Virtanen S. Effect of Mo species on metastable pitting of Fe18Cr alloys—A current transient analysis［J］. Corrosion Science，2006，48：1587.

[10] Rosli N R，Choi Y S，Young D. Impact of Oxygen in CO_2 Corrosion of Mild Steel［J］. Corrosion，N 2014，9-13 March，San Antonio，Texas，USA.

[11] Jones D A. Principles and Prevention of Corrosion，2nd ed.（Upper Saddle River，NJ：Prentice Hall，1996）.

[12] Uhlig H H，Triadis H H，Stern M. Effect of oxygen，chlorides，and calcium ion on corrosion inhibition of iron by polyphosphates［J］. Electrochem. Soc.，102（1995）：59.

[13] 马桂君，杜敏，刘福国，等 . G105 钻具钢在含有溶解氧条件下的腐蚀规律［J］. 中国腐蚀与防护学报，2008，28（2）：108.

[14] 谢飞，吴明，张越，等 . 辽河油田注水管线结垢腐蚀原因分析及阻垢缓蚀剂应用试验［J］. 石油炼制与化工，2011，42（9）：92.

[15] 万里平，孟英峰，王存新，等 . 西部油田油管腐蚀结垢机理研究［J］. 中国腐蚀与防护学报，2007，27（4）：247.

[16] Q. Y. Liu, L. J. Ma, S. W. Zhou. Effects of chloride content on CO_2 corrosion of Carbon steel in simulated oil and gas well environments [J]. Corrosion Science, 2014, 84: 165.
[17] 崔志峰，韩一纯，庄力健，等. 在 Cl^- 环境下金属腐蚀行为和机理 [J]. 石油化工腐蚀与防护，2011，28 (4): 1.
[18] 孙智，孙岚，欧雪梅，等. 金属在电导性煤层中的腐蚀 [J]. 煤炭学报，1994，19 (6): 605.

10Cr3Mo 钢与 N80 钢的高温力学行为

魏文澜[1,2,3]　韩礼红[1,2]　王建国[4]　王　航[1,2]　冯耀荣[1,2]

(1. 中国石油集团石油管工程技术研究院；2. 石油管材及装备材料服役行为与结构安全重点实验室；3. 西安交通大学材料科学与工程学院；4. 新疆油田公司风城油田作业区)

摘　要：10Cr3Mo 钢是以 N80 钢为基础加入了 Cr、Mo、Ni 等合金元素，对相同热处理工艺下的这两种套管钢进行了高温力学性能试验，并对其进行显微组织结构观察。通过蠕变试验，推得了这两种钢的蠕变本构方程。对两种钢进行室温、高温拉伸试验以及室温冲击试验结果表明，在提高 Cr、Mo 等元素的含量后，10Cr3Mo 钢的抗拉强度仍维持原有钢级水平，屈强比略有增加，伸长率稍有下降，冲击性能有所降低，但高温抗蠕变能力显著提升。使用 10Cr3Mo 钢作为 80 钢级的热采井套管将能有效减少因为蠕变因素导致的热采井失效问题。

关键词：套管；力学性能；显微组织；高温拉伸；蠕变

在石油资源中，稠油占有很重要的一部分。开采稠油的方式主要为循环蒸汽吞吐，其过程分为注气、焖井、采油三个过程[1,2]。注气阶段由井口通入高温高压热蒸汽，井内迅速升温；焖井阶段保持高温高压，加热井下周围油层；采油阶段开始开采经高温软化的油层，不再加热，井下温度逐渐降低[3]。经历这三个过程成为一个热循环，在一口油井的开采过程中需要经历数十甚至上百个热循环过程。

由于其循环变温的特殊工况，导致井中套管的失效频繁发生。在对发生失效的热采井的分析研究中发现，套管主要以断裂形式的失效为主[4,5]。一般来说，套管的失效主要是因为地层运动产生的剪力发生错断，但热采井主要以拉伸断裂为主，其原因在于热采井中高温影响下材料力学性能发生变化，并存在蠕变等一系列问题。

目前热采井套管用钢仍以 N80 钢为主，针对 350℃ 的中温耐热钢仍在研究阶段。10Cr3Mo 钢是以 N80 钢为基础，采用和 N80 钢相同的热处理工艺，加入 Cr、Mo 等合金元素，同时加入微量合金元素 Nb、V、Ti，专门针对高温热采环境设计的套管用钢[6]。本文选用 N80 钢和 10Cr3Mo 钢进行高温力学性能研究，对这两种钢的高温力学性能分别进行了 350℃ 下的高温拉伸和蠕变试验研究。

1　试验材料及方法

1.1　试验材料

N80 钢是常见的套管钢级，10Cr3Mo 钢中提高的 Cr、Mo 和 Ni 的含量，具体化学成分如

基金项目：国家自然科学基金（51574278）。

作者简介：魏文澜（1988—），男，博士研究生，主要研究方向为材料服役安全理论，联系电话：13991371781，E-mail：weiwenalnnds@163.com。

表1所示。试验材料从成型好的套管上取下，沿管壁轴线方向取圆棒拉伸和蠕变试样。

表1　N80钢和10Cr3Mo钢的化学成分（wt%）

钢号	C	Si	Mn	P	S	Cr	Mo	Ni	V	Ti
N80	0.26	0.24	1.27	0.015	0.0076	0.04	0.01	0.003	0.006	0.003
10Cr3Mo	0.17	0.24	0.98	0.011	0.0034	0.99	0.33	0.059	0.029	0.013

1.2　显微组织分析

两种套管钢均沿管壁轴线方向取样品，抛光后用4%的硝酸酒精溶液侵蚀，并在光学显微镜下做金相分析。之后对金相试样进行深腐蚀，在扫描电镜下观察其微观结构。

1.3　蠕变试验和高温拉伸试验

蠕变试验根据GB/T 2039—2012《金属材料 单轴拉伸蠕变试验方法》取圆棒试样，试样直径为5mm。根据热采井工况，焖井阶段温度恒定很长一段时间，此时井下的温度同350℃，因此选取350℃为蠕变试验温度。选用应力松弛过程中的4个应力水平作为试验拉伸载荷，对于N80钢，分别为350MPa、400MPa、450MPa、500MPa；对于10Cr3Mo钢，分别为475MPa、500MPa、550MPa、525MPa。在进行蠕变试验时，先开启保温箱，对试样进行加热，将试验机调整为试样保护模式。温度到达350℃时，保温0.5h，然后开始蠕变试验，分别按照选取的应力水平进行加载，在进行到蠕变第二阶段明显出现并且速率稳定时终止试验。

高温拉伸试验根据GB/T 228.1—2010《金属材料 拉伸试验 第1部分：室温试验方法》取圆棒拉伸试样，由于套管管壁很薄，拉伸试样直径ϕ5mm。根据热采井工况以及之前现场的数据取样，拉伸试验的温度选取270℃，拉伸速率为0.5mm/min[7]。在进行高温拉伸试验时，先开启保温箱，对试样进行加热。此时，将试验机调整为试样保护模式（试验机会自动调整位移，防止试样出现载荷），以此来避免由于热膨胀导致的额外载荷，影响试验结果。在达到270℃时，保温0.5h，然后开始拉伸试验，直至试样断裂为止。

2　结果和讨论

2.1　显微组织

N80钢采用调质处理，其工艺为淬火后在650℃回火处理，得到回火索氏体，其中包含少量的上贝氏体[8]。图1（a）显示，其基本都为回火索氏体，其中密集的灰黑色区域为针状铁素体和第二相，细小的黑点是析出物，有极少的粒状的白色组织为粒状贝氏体。10Cr3Mo钢提高了Cr、Mo、Ni的含量，加入微量合金元素Nb、V、Ti，其热处理工艺与N80钢相同，显微组织如图1（b）所示。可以看出10Cr3Mo钢晶粒尺寸明显较大，存在明显的晶界，由于Mo、Ni含量的影响，产生较多的粒状贝氏体，灰黑色区域为针状铁素体和第二相，析出物较N80钢少。

扫描电镜下，N80钢呈现片层状组织，是典型的回火索氏体，白色点状物为析出物，如图1（c）。而在10Cr3Mo钢中，并未出现密集的片层装组织，有较多的粒状组织，如图1（d）。在扫描电镜下，发现10Cr3Mo钢中不仅有和N80一样的析出物，而且存在尺寸较大的碳化物，其直径为2~3μm。

分别对两种钢进行硬度测试，采用洛氏硬度，按照ASTM E18进行硬度试验，N80钢的

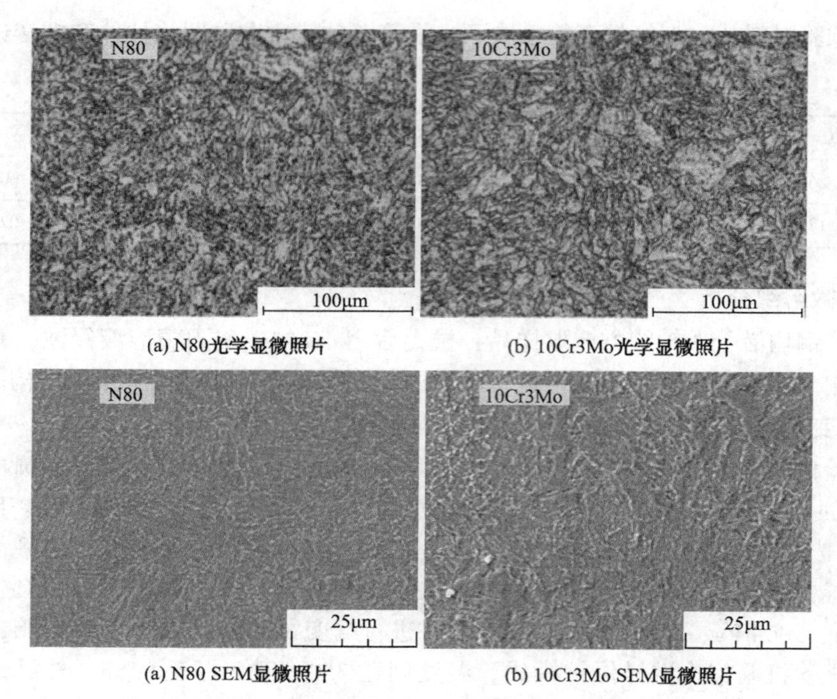

(a) N80光学显微照片　(b) 10Cr3Mo光学显微照片

(a) N80 SEM显微照片　(b) 10Cr3Mo SEM显微照片

图1　N80钢和10Cr3Mo钢的显微组织

平均硬度为20.6 HRC，10Cr3Mo钢的平均硬度为18.1 HRC。虽然两种钢采用了同种热处理工艺，但由于10Cr3Mo钢中合金元素的变化，导致产生组织较N80钢更为粗大，N80钢中密集的片层组织使其具有密集的位错分布[7]，因此产生较大的硬度。相对10Cr3Mo钢中组织较为粗大，硬度较低[6]。

2.2　高温力学性能

一般认为，钢铁材料在超过0.3T_m，大约450℃以上时才会有明显的蠕变现象。实际上，对油田现场已经使用多年的套管材料进行的试验结果表明，即使在350℃，普通套管材料也显示出了明显的蠕变效应[9]。在350℃的试验温度下，在25h时蠕变第二阶段已经趋于稳定，两种钢的蠕变试验结果分别见图2。

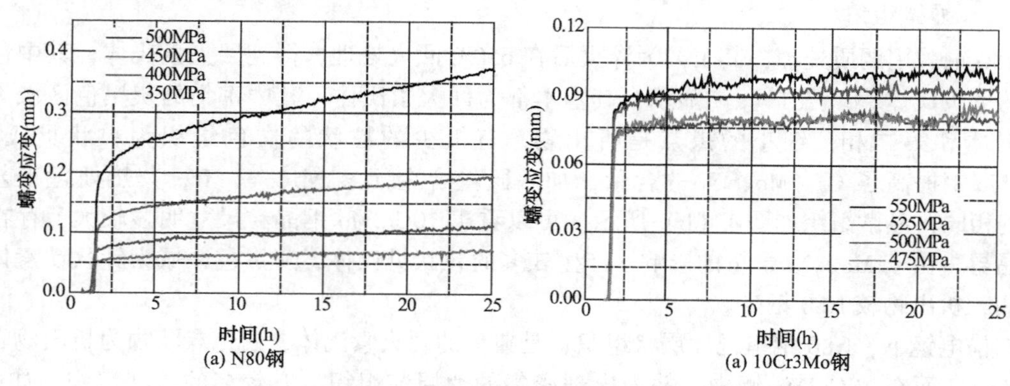

(a) N80钢　(a) 10Cr3Mo钢

图2　350 ℃下N80钢和10Cr3Mo钢的蠕变曲线

参照E. N. da C. Andrade理论，根据蠕变的时效机制，在图2（a）中，5 h之前蠕变过程处于第一阶段，5 h之后处于第二阶段，图中的4条曲线分别对应不同载荷下的蠕变—时间曲线，蠕变变形程度随载荷的增加而增加。低于 350MPa 时，材料的蠕变形变量较小，在热采应用中可以忽略不计，而高于 500MPa 时，材料的蠕变应变量急剧增加对整体应变有很大影响。

蠕变速率与载荷的本构方程可以通过 Monkman-Grant 关系式[10,11]来拟合，其表达形式为：

$$(\dot{\varepsilon})^{\alpha}t=C \tag{1}$$

式中 ε——应变；

t——时间；

α——常数；

C——常数。

拟合本构方程结果如下：

$$\dot{\varepsilon}=2.92\times10^{-10}\times e^{\frac{\sigma}{58.12}}+3.96\times10^{-8} \tag{2}$$

而在图 2（b）中，5h 之前 10Cr3Mo 钢的蠕变过程处于第一阶段，5h 之后处于第二阶段。10Cr3Mo 钢的应力幅明显要高于 N80 钢的应力幅。且在 550MPa 时，仍未出现很大的蠕变速率，可以很直观地看出，其蠕变速率明显小于 N80 钢在 500MPa 的蠕变速率。其拟合本构方程结果如下：

$$\dot{\varepsilon}=5.10\times10^{-18}\times e^{\frac{\sigma}{22.14}}+1.75\times10^{-7} \tag{3}$$

两种套管钢在 350℃下的蠕变速率—应力曲线如图 3 所示。N80 钢的蠕变速率明显高于 10Cr3Mo 钢，也就是说，N80 钢的抗蠕变性能较差。在应力幅小于 500MPa 时蠕变速率并未明显增加，而高于 500MPa 时，N80 钢的蠕变速率开始出现显著上升，而 10Cr3Mo 钢的蠕变速率几乎未出现较大变化。

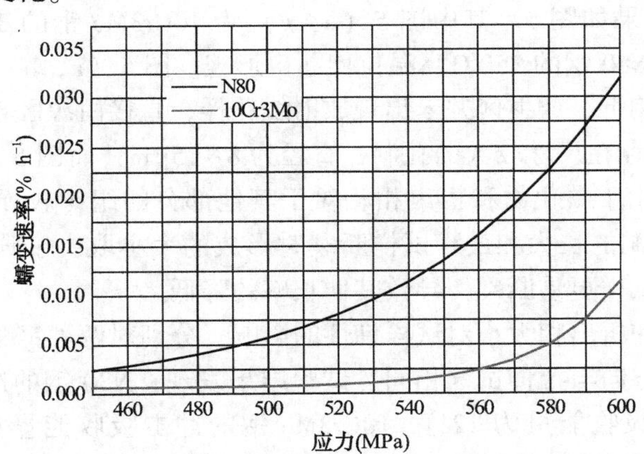

图 3 试验钢的蠕变速率—应力曲线

在加入了 Cr、Mo 等合金元素后，抗蠕变性能得到了明显提升，由于 10Cr3Mo 钢的晶粒较 N80 钢更为粗大，位错分布少于 N80 钢，其强度可能会有所降低，因此对其拉伸性能进行试验研究。为了与高温性能对照分析，对两种钢在室温下同样做了拉伸试验。室温拉伸试验的结果见图 4（a），高温拉伸试验的结果见图 4（b），试样的各项拉伸性能见表 2。

表 2 拉伸试验结果

试验温度（℃）	钢号	屈服强度（MPa）	抗拉强度（MPa）	伸长率（%）	屈强比
室温	N80	562	679	24.1	0.828
	10Cr3Mo	575	693	22.3	0.830

续表

试验温度（℃）	钢号	屈服强度（MPa）	抗拉强度（MPa）	伸长率（%）	屈强比
350	N80	504	655	20.2	0.769
	10Cr3Mo	521	652	19.2	0.799

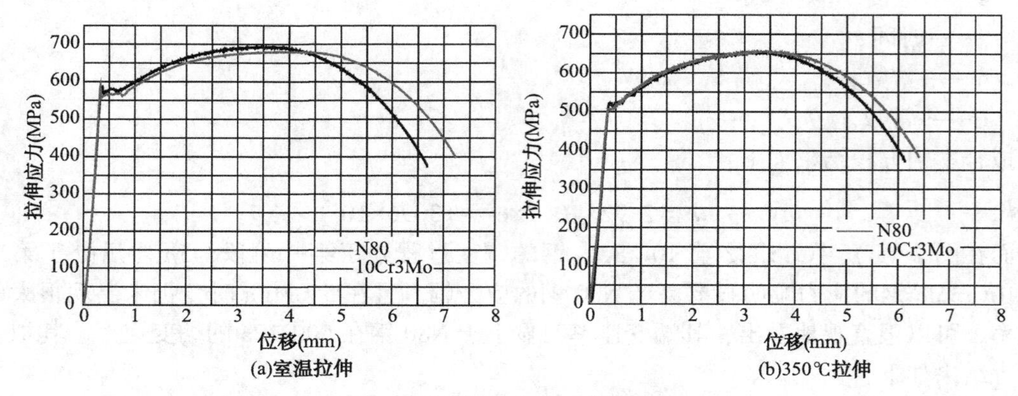

图 4　试验钢的应力—位移曲线

由图 4 中可见，在室温时，N80 钢的屈服强度和抗拉强度略小于 10Cr3Mo 钢，N80 钢的伸长率较大，而在 350℃高温下，两种钢的屈服强度均出现了 10%的降低，抗拉强度降低了 5%，而且伸长率呈现出一定的降低。

由于两种钢晶粒尺寸差别较大，且 10Cr3Mo 钢尺寸较大的碳化物，对拉伸试样的断口进行了 SEM 观察，结果如图 5。其中图 5（a，c）为 10Cr3Mo 钢的 350℃高温拉伸断口形貌，图 5（b，d）为 N80 钢的 350℃高温拉伸断口形貌。图 5（a，b）分别为断口源区的形貌。在高温作用的影响下，放射区并未出现放射线花样，直接由源区转为剪切唇区。可以看出，图 5（a）中普遍存在尺寸较大的孔洞，直径为 8~15μm；而图 5（b）中孔洞则明显较小，直径小于 5μm。由于碳化物和析出相相对于其他部分的组织较硬，在应力作用下试样内部出现孔洞[12]，孔洞的长大和最终试样断裂时形成的大小取决于碳化物和析出相尺寸的大小。较大的孔洞使试样更易断裂，导致其伸长率的降低。

为了确认碳化物和析出相大小对试样韧性的影响，分别对两种钢分别做了室温下的冲击试验，冲击试样选取方向与拉伸试样相同。试验结果发现，N80 钢的冲击吸收能量为 192J，而 10Cr3Mo 钢的冲击吸收能量为 121J。10Cr3Mo 钢的冲击吸收能量相较于 N80 钢减少了 37%，说明在相同的热处理工艺下增加了 Cr、Mo 等元素后，晶粒尺寸的增大和形成的碳化物影响了材料脆性，使韧性下降，脆性增加，更容易发生脆断，导致其均匀伸长率下降。

3　结论

（1）10Cr3Mo 钢的微观组织晶粒较 N80 钢更大，同时伴随出现 2~3μm 的碳化物。断口和试样组织的 SEM 观察表明，碳化物和析出相尺寸的大小决定了试样内部孔洞的尺寸的大小，较大的孔洞尺寸使试样伸长率降低，导致脆性增大。

（2）室温和高温拉伸试验表明，10Cr3Mo 钢在提高 Cr、Mo、Ni 含量之后，抗拉强度仍维持原有钢级水平，屈强比略有增加，伸长率稍有下降。

（3）350℃蠕变试验表明，在温度小于 $0.3T_m$ 时，仍然存在着显著的蠕变现象，

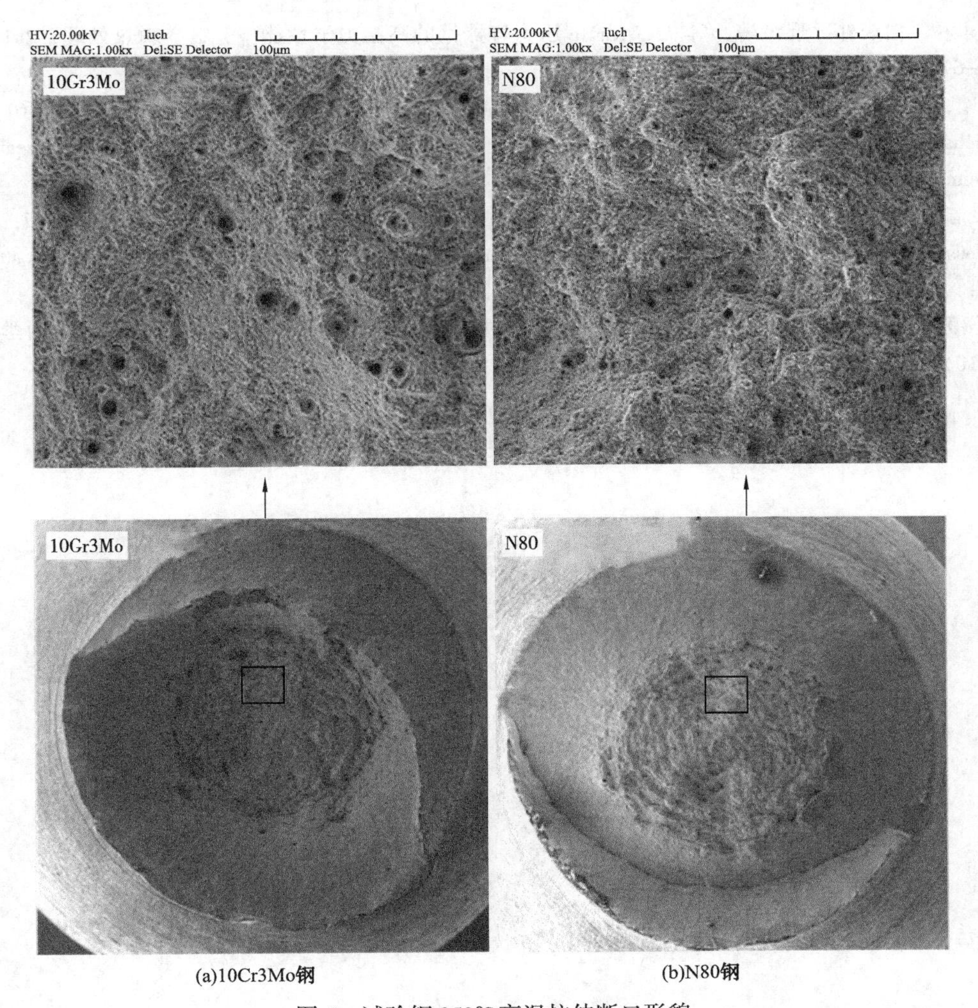

(a)10Cr3Mo钢 (b)N80钢

图 5 试验钢 350℃高温拉伸断口形貌

10Cr3Mo 钢较 N80 具有良好的抗蠕变性能。在应力幅为 550MPa 时，10Cr3Mo 钢的蠕变速率未有明显提升，而 N80 钢的蠕变速率显著增加。使用 10Cr3Mo 钢作为 80 钢级的热采井套管将能有效减少因为蠕变因素导致的热采井失效问题。

参 考 文 献

[1] 王大为，周耐强，牟凯．稠油热采技术现状及发展趋势［J］. 西部探矿工程，2008，12：129-131.

[2] 刘佳．概论稠油热采技术的现状及发展前景［J］. 化工管理，2013（10）：251-252.

[3] 罗全民，张清军，罗晓惠，等．稠油热采高效驱油技术应用研究［J］. 石油天然气学报，2010（3）：361-363.

[4] 崔孝秉，曹玲，张宏，等．注蒸汽热采井套管损坏机理研究［J］. 石油大学学报（自然科学版），1997（3）：59-66.

[5] 赵洪山．稠油热采井套管柱损坏机理及预防措施研究［D］. 青岛：中国石油大学（华东），2007：12-18.

[6] 王航，王建军，韩礼红，等．中温低合金耐热钢蠕变行为及其微观机理［J］. 材料热处理学报，2014（4）：131-136.

[7] 麻永林，白庆伟，邢淑清，等.12NiCrMo压力容器钢的高温力学性能［J］. 金属热处理，2015，08：63-67.
[8] 牛靖，董俊明，薛锦，等. 石油套管钢N80的显微组织分析［J］. 焊管，2002（1）：15-17，61.
[9] Michael E. Kassner, Kamia Smith. Low temperature creep plasticity [J]. Journal of Materials Research and Technology, 2014, 3 (3): 280-288.
[10] 樊志菁，王起江，单爱党，等. 热处理对P9钢蠕变行为的影响［J］. 金属热处理，2013，05：1-5.
[11] Hassan Osman Ali, Mohd Nasir Tamin. Modified Monkman-Grant relationship for austenitic stainless steel foils [J]. Journal of Nuclear Materials, 2013, 433 (1-3): 74-79.
[12] 姜薇，李亚智，束一秀，等. 基于微孔贯通细观损伤模型的金属韧性断裂分析［J］. 工程力学，2014（10）：27-32.

Assessment of Wellbore Integrity of Offshore drilling in Well Testing and Production

Wang Yanbin, Gao Deli, Fang Jun

(China University of Petroleum)

Abstract: The mechanical model and control equations have been established to analyze the characteristics of casing-cement-formation of offshore drilling in well testing and production under the combination action of temperature and non-uniform in-situ stress. The equations have been solved by basic equations of plane strain and the stress distributions of each layer medium have been obtained. Meanwhile, the safety factor of casing in service has been figured out with consideration of the influence of temperature on casing mechanical properties. On this basis, the variations of casing radial stress and its degree of non-uniform with cement elastic modulus, Poisson's ratio, casing grade and temperature have also been discussed. Results show that, under the combination action of these two kinds of external forces, the radial stress on each contact surface is non-uniform. With the increase of temperature, the casing radial stress reduces while the degree of non-uniform and the Von Mises stress increases which indicates the ability of casing to resist the collapse load declines. High cement elastic modulus and low cement Poisson's ratio can improve the ability to resist extrusion loading and is good for wellbore integrity.

Keywords: Offshore drilling; Wellbore temperature; Non-uniform in-situ stress; Wellbore integrity

1 Introduction

With the continuous development ofoffshore drilling technology, people pay more attention to the issue of wellbore integrity. According to the wellbore integrity investigation of 106 offshore wells with various years of development and production made by Petroleum Safety Authority (PSA), 18% wells had the problem of wellbore integrity and 7% among them was forced to shut down which caused major damage to environment and economy (Birgit Vignes et al., 2008; 2010).

The researches about wellbore integrity of offshore drilling are mainly focused on the issue of casing collapse and damage caused by the increase of casing annulus trapped pressure resulted from wellbore temperature increment during well testing and production. A. Admas (1991) has established the analysis model and figured out the trapped pressure for the first time. A. Admas

Corresponding author: Yanbin Wang; Tel: +86 (10) 89733702; E-mail: wyb576219861@126.com

(1994) has improved the analysis model and obtained the calculation results which are more consistent with actual situation. A. S. Halal et al. (1993) have established a new analysis model and discussed the influence of liquid temperature, pressure, coefficient of thermal expansion on trapped pressure. G. Robello Samuel et al. (1999) have analyzed the influence of temperature on additional load of uncemented casing and provided guidance for casing selection and well structure design. However, the common methods used to calculate the trapped pressure are established by A. Admas and A. S. Halal. P. Oudeman et al. (2006) have found that the main reason for affecting prediction accuracy of annulus pressure is the parameter selection of annulus fluid through field investigation. In addition to the theoretical calculation, many scholars have focused on the prevention measures of the trapped pressure. C. P. Leach et al. (1993) have summarized several ways to take precautions against additional load of uncemented casing, which include use of heavyweight and/or high yield casing, full height cementing, cement shortfall (such as, leaving the cement short of the previous casing shoe) and providing a leak path or bleed port. Wang et al. (2007) have presented several methods to take precautions against additional load caused by temperature build-up and put forward to reduce the thermal expansion pressure through injecting compressible fluid to uncemented casing annulus and established corresponding calculation model based on these methods. Practices have proved the most effective measures to prevent casing trapped pressure in offshore well are installation of rupture discs and usage of compressible foam (Roger Williamson et al., 2003; Vargo et al., 2003; Azzola J H, et al., 2007).

High temperature fluid can not only cause the trapped pressure, it can result in additional thermal stress in casing and formation in offshore well. If the wellbore temperature increment is significant, the additional stress caused by thermal expansion of casing-cement- formation cannot be ignored, which will threat the wellbore integrity and is another important reason for casing damage. S. S. Rahman et al. (1995) have deduced the stress distribution of multilayer medium based on theory of axisymmetric of thermal elastic mechanics, which have laid the foundation for later study. However, there are several minor errors in his model. Wu et al. (2006) have calculated the stress distribution of casing and cement ring under the in high temperature and indicated that the study of casing-cement-formation had contribution to analyze the casing failure mechanism. Li et al. (Li et al., 2006; Zhang et al., 2008; Li et al., 2009) have established the interaction model of casing and elastic formation based on thermal elastic mechanics theory and deduced the thermal stress and thermal displacement of casing and formation according to the fitting function of temperature. However, the influence of temperature on casing mechanical properties is ignored.

In actual condition, in well testing or production of offshore well, casing is under the action of non-uniform extrusion load from in-situ stress on the one hand, on the other hand, if the temperature increment is remarkable, casing-cement-formation will generate additional thermal stress. The casing deformation and stress state are codetermined by these two kinds of stresses. However, in previous researches, the combination action of the two kinds of stresses has been ignored which indicates that the calculation results may have a larger degree of deviation with actual situations. In order for more accurate assessment for wellbore integrity of offshore well in testing and production, the mechanical model and control equations have been established with consideration of the combination action of temperature and non - uniform in - situ stress in this paper. The stress distributions of each layer medium have been obtained and the safety factor of casing in service has been figured out with consideration of the influence of temperature on casing mechanical properties. Meanwhile, some discussions about casing radial stress have been presented. The research has some reference value to the wellbore integrity of offshore well.

The remainder of the paper is organized as follows. The mechanical model to analyze the stress distribution of casing-cement-formation is established in Section 2. The thermal stress analysis of the system is shown in Section 3. The mechanical response of the system under non-uniform in-situ stress is presented in Section 4. A case study, calculation results and discussions are given in Section 5. The corresponding conclusions are drawn in Section 6.

2 Mechanical model

In offshore well, subsea test tree and X - mas tree are used in well testing and production. During well testing or production, reservoir thermal fluid flow up to the wellhead through oil tube and heating the oil tube, casing, cement and formation, which will result in thermal displacement and stress in casing. At the same time, casing is subjected non-uniform extrusion load due to the inequality of in-situ stress in the two horizontal directions ($\sigma_H \neq \sigma_h$), which will also induce displacement and stress in casing. The real mechanical characteristic is determined by the two kinds of effects together. In order for convenient calculation and formula derivation, the following assumptions are used to stipulate the present formulation:

- Casing, cement andformation are linear elastic material and isotropic;
- Casing, cement and formation are completely consolidated with good contact;
- All medium are ideal annulus cross section with uniformity of thickness;
- The wellbore temperaturetransmission to casing, cement and formation is homogeneous and radial.

Accordingly, the establishment of mechanical model of casing - cement - formation under the combination action of non - uniform load and temperature and the establishment of the coordinate system are shown in Fig. 1.

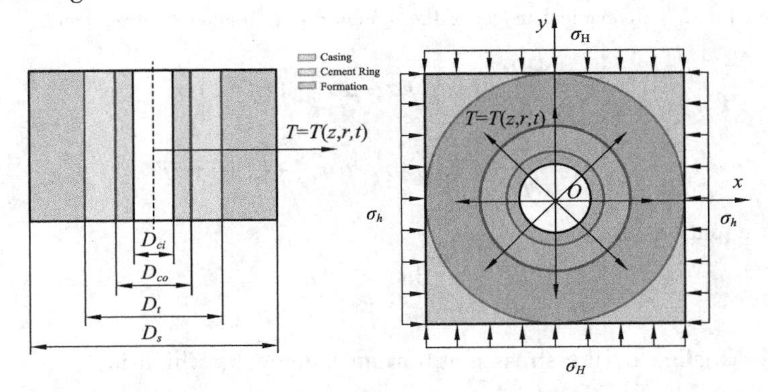

Fig. 1 Mechanical model for assessment of the characteristic of casing-cement-formation

Because of the gravity of the subsea X - mas tree and good consolidation properties of the system, the longitudinal deformation of casing is restricted and the mechanical model can be simplified to a plane strain problem according to the theory of elasticity.

As shown in Fig. 1, σ_H is the maximum horizontal principal stress, σ_h is the minimum horizontal principal stress, $T=T(r)$ is the temperature increment, D_{ci} is the I. D. of casing, D_{co} is the O. D. of casing, D_t is the O. D. of cement and D_s is the O. D. of formation. In order to reduce the influence of boundary conditions on the combined system, the O. D. of formation has to be at least 20 times than the O. D. of cement according to the Saint-Venant principle.

Since the system is linear elastic and isotropic, the total response of the system can be obtained by linear superposition of the results under the above two kinds of external forces, which are:

$$u(r)=u_q(r)+\mu_T(r) \tag{1}$$

$$\sigma(r)=\sigma_q(r)+\sigma_T(r) \tag{2}$$

Where, $u(r)$, $u_q(r)$ and $\mu_T(r)$ are the total radial displacement, radial displacement caused by non-uniform in-situ stress and the radial displacement resulted from temperature increment. $\sigma(r)$, $\sigma_q(r)$ and $\sigma_T(r)$ are the total radial stress, radial stress caused by non-uniform in-situ stress and the radial stress resulted from temperature increment.

3 System responses due to temperature increment

3.1 Mechanical model

As shown in Fig. 2, the system responses due to temperature increment belongs to axisymmetric plane strain problem. So, according to the thermoelasticity (S. P. Timoshenko et al., 1990), the stress-strain relations of casing-cement-formation can be represented by:

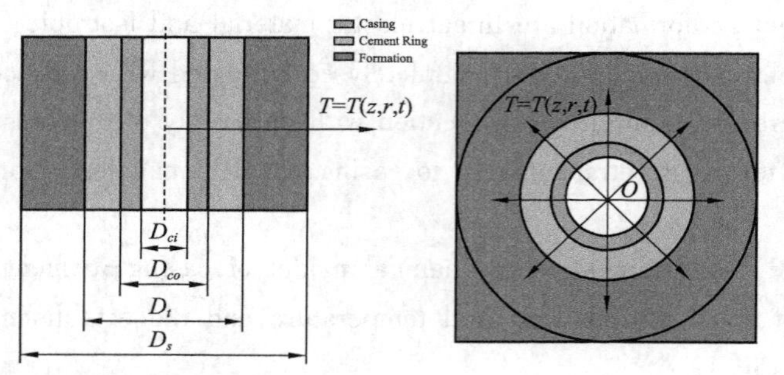

Fig. 2 Mechanical model of the system due to temperature increment

$$\begin{cases} \varepsilon_r=\dfrac{1+\mu}{E}[(1-\mu)\sigma_r-\mu\sigma_\theta]+(1+\mu)\alpha T \\ \varepsilon_\theta=\dfrac{1+\mu}{E}[(1-\mu)\sigma_\theta-\mu\sigma_r]+(1+\mu)\alpha T \end{cases} \tag{3}$$

Thegeometric equations are:

$$\varepsilon_r=\frac{\mathrm{d}u}{\mathrm{d}r},\ \varepsilon_\theta=\frac{u}{r} \tag{4}$$

σ_r and σ_θ are satisfied by the stress equilibrium equations, which is:

$$\frac{\mathrm{d}\sigma_r}{\mathrm{d}r}+\frac{\sigma_r-\sigma_\theta}{r}=0 \tag{5}$$

Where, σ_r is the radial stress, MPa; σ_θ is the circumferential stress, MPa; ε_r is the radial strain; ε_θ is the circumferential strain; $u = u(r)$ is the radial displacement, m; E is the elastic modulus, MPa; μ is the Poisson's ratio and α is the coefficient of thermal expansion, 1/℃.

Substituting Eq. 3 and Eq. 4 into Eq. 5, one obtains:

$$\frac{d^2u}{dr^2}+\frac{1}{r}\frac{du}{dr}-\frac{u}{r^2}=\frac{1+\mu}{1-\mu}\alpha\frac{dT}{dr} \tag{6}$$

After solving Eq. 6, the thermal displacement of casing-cement-formation can be figured out, which is:

$$u(r) = \frac{1+\mu}{1-\mu}\frac{\alpha}{r}\int_{r_{ci}}^{r} T(r)\, r dr + C_1 r + \frac{C_2}{r} \tag{7}$$

Where, r_{ci} is the casing radius, m; r is the radial displacement from casing inner wall to the integral point, m.

Substituting Eq. 7 into Eq. 4 and Eq. 3 successively, the thermal stress of casing-cement-formation can be obtained, which are:

$$\begin{cases} \sigma_r = \dfrac{\alpha E}{(\mu-1)r^2}\displaystyle\int_{r_{ci}}^{r} T(r)\, r dr + \dfrac{E}{1+\mu}\left(\dfrac{C_1}{1-2\mu} - \dfrac{C_2}{r^2}\right) \\ \sigma_\theta = \dfrac{\alpha E}{(1-\mu)}\left(\dfrac{1}{r^2}\displaystyle\int_{r_{ci}}^{r} T(r)\, r dr - T(r)\right) + \dfrac{E}{1+\mu}\left(\dfrac{C_1}{1-2\mu} + \dfrac{C_2}{r^2}\right) \\ \sigma_z = \dfrac{\alpha E T(r)}{\mu - 1} + \dfrac{2\mu E C_1}{(1+\mu)(1-2\mu)} \end{cases} \tag{8}$$

3.2 Determination of unknown parameters

As shown in Eq. 8, the temperature increment $T(r)$ needed to be determined. The temperature in casing can be regarded as constant since the casing belongs to thin-walled tubes and has good thermal conductivity. So, the temperature of casing equals to that of testing liquid. According to the research results by Yu et al. (2008), the radial temperature variation of cement and formation obey exponential decay laws, which is:

$$T(r) = T_c \exp(r_{co} - r) \tag{9}$$

Where, T_c is the temperature increment od casing, ℃.

In addition, the integration constants of C_1 and C_2 are also unknown. What's more, the integration constants of casing, cement and formation are different. Suppose thatC_1^c, C_2^c, C_1^t, C_2^t, C_1^fandC_2^f stand for the integration constants of casing, cement and formation respectively. Therefore, there are 6 unknown parameters to be determined and 6 equations are needed. According to the boundary system conditions, the determination process the of equations are shown as following:

- Boundary conditions: the thermal stresses on casing inner wall and that on formation infinity are 0, which are:

$$\sigma_r^c\big|_{r=r_{ci}} = 0,\quad \sigma_r^f\big|_{r\to\infty} = 0 \tag{10}$$

- Continuity conditions: the radial stress and displacement of casing equal to that of cement on their contact interface. The radial stress and displacement of cement equal to that of formation on their contact interface, which are:

$$
\begin{cases}
\sigma_r^c \big|_{r=r_{co}} = \sigma_r^t \big|_{r=r_{co}}, \quad u_r^c \big|_{r=r_{co}} = \sigma_r^t \big|_{r=r_{co}} \\
\sigma_r^t \big|_{r=r_t} = \sigma_r^f \big|_{r=r_t}, \quad u_r^t \big|_{r=r_{ct}} = \sigma_r^f \big|_{r=r_t}
\end{cases}
\tag{11}
$$

Where, σ_r^c, σ_r^t and σ_r^f are radial stress of casing, cement and formation respectively. u_r^c, u_r^t and u_r^f are radial displacement of casing, cement and formation respectively. r_{co} and r_t are the external radius of casing and cement.

Substituting Eq. 7 and Eq. 8 into Eq. 10 and Eq. 11, the 6 unknown parameters can be determined uniquely. The detail of calculation results are shown in Appendix A.

4 System responses due to non-uniform in-situ stress

4.1 Mechanical model

The mechanical model of system under non-uniform in-situ stress is shown in Fig. 3. Usually, σ_H and σ_h are described in rectangular coordinate system. So, we need to transform them into expressions in polar coordinate. According to the coordinate transformation (Yang, 2004), one obtains:

$$
\begin{cases}
\sigma_r = -\dfrac{1}{2}(\sigma_H + \sigma_h) - \dfrac{1}{2}(\sigma_H - \sigma_h)\cos 2\theta \\
\tau_{r\theta} = \dfrac{1}{2}(\sigma_H - \sigma_h)\sin 2\theta
\end{cases}
\tag{12}
$$

If let $\sigma = \dfrac{1}{2}(\sigma_H + \sigma_h)$ and $s = \dfrac{1}{2}(\sigma_H - \sigma_h)$

Thus,

$$
\begin{cases}
\sigma_r = -\sigma - s\cos 2\theta \\
\tau_{r\theta} = s\sin 2\theta
\end{cases}
\tag{13}
$$

Therefore, the system responses due to non-uniform in-situ stress can be divided into two parts: one is the response caused by uniform compressive stress (σ); the other is the response induced by non-uniform compressive stress and shear stress ($s\cos 2\theta$ and $s\sin 2\theta$), which is:

$$
\begin{cases}
\sigma_r = -\sigma - s\cos 2\theta \\
\tau_{r\theta} = s\sin 2\theta
\end{cases}
=
\begin{cases}
\sigma'_r = -\sigma \\
\tau'_{r\theta} = 0
\end{cases}
+
\begin{cases}
\sigma''_r = -s\cos 2\theta \\
\tau''_{r\theta} = s\sin 2\theta
\end{cases}
\tag{14}
$$

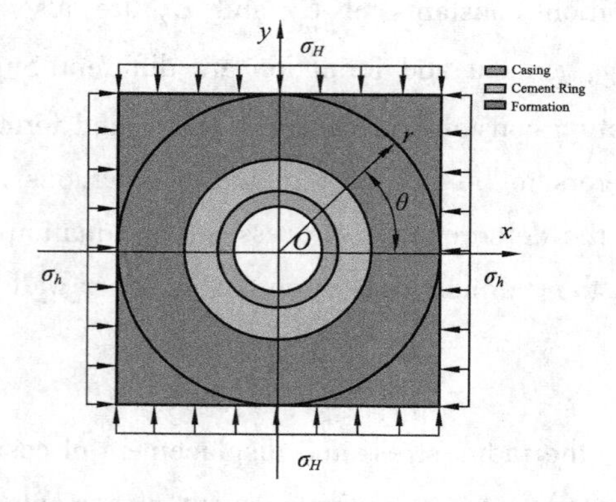

Fig. 3 Mechanical model of the system due to non-uniform in-situ stress

4.2 System response caused by σ

Suppose $q_1$and q_2 are the contact radial stress acting on the interface of casing-cement and cement-formation respectively, the system response caused by uniform compressive stress can be calculated by:

$$\begin{cases} q_1 = \dfrac{(1-\mu_f)(\sigma_H+\sigma_h)}{\dfrac{k_s}{m_2 k_c}k_1+k_2} \\ \dfrac{q_2}{q_1} = \dfrac{k_3+k_4}{k_5} \end{cases} \tag{15}$$

Where, the solusions of parameters in Eq. 15 are presented in Appendis B.

4.3 System response caused by s cos 2θ and s sin 2θ

4.3.1 Determination of stress expression

The system boundary conditions under the action of s cos 2θ and s sin 2θ are:

$$\begin{cases} \sigma_r \big|_{r=a} = -s \cdot \cos 2\theta \\ \tau_{r\theta} \big|_{r=a} = s \cdot \sin 2\theta \\ \sigma_r \big|_{r=a_0} = 0 \\ \tau_{r\theta} \big|_{r=a_0} = 0 \end{cases} \tag{16}$$

According to the stress boundary conditions, the stress function can be written as:

$$\varphi(r, \theta) = f(r) \cos 2\theta \tag{17}$$

Substituting Eq. 17 into thestress harmonistic equations and solving, one obtains:

$$f(r) = Hr^4 + Ir^2 + G + Kr^{-2} \tag{18}$$

Then, substituting Eq. 18 into the basic equation in polar coordinate, the stress distribution of casing, cement and formation can be represented by:

$$\begin{cases} \sigma_r = -(2I+4Gr^{-2}+6Kr^{-4}) \cos 2\theta \\ \sigma_\theta = (12Hr^2+2I+6Kr^{-4}) \cos 2\theta \\ \tau_{r\theta} = (6Hr^2+2I-2Gr^{-2}-6Kr^{-4}) \sin 2\theta \end{cases} \tag{19}$$

Where, H, I, G and K are unknown number related to the mechanical parameters and geometric parameters of the system and they are different for casing, cement and formation. If let the subscript $i=1$, 2, 3 stands for casing, cement and formation respectively, one obtains:

$$\begin{cases} (\sigma_r)_i = -(2I_i+4G_i r^{-2}+6K_i r^{-4}) \cos 2\theta \\ (\sigma_\theta)_i = (12H_i r^2+2I_i+6K_i r^{-4}) \cos 2\theta \\ (\tau_{r\theta})_i = (6H_i r^2+2I_i-2G_i r^{-2}-6K_i r^{-4}) \sin 2\theta \end{cases} \tag{20}$$

4.3.2 Determination of unknown number

There are 12 unknown parameters and 12 equations are needed. The determination process of the 12 equations is shown as follows.

- The radialstress and shear stress on casing inner wall are 0, which are:

$$\begin{cases} (\sigma_r)_1 \big|_{r=r_{ci}} = 0 \\ (\tau_{r\theta})_1 \big|_{r=r_{ci}} = 0 \end{cases} \tag{21}$$

• The radial stress, shear stress, radial displacement, circumferential displacement of casing equal to that of cement on their contact interface, which are:

$$\begin{cases}(\sigma_r)_1|_{r=r_{co}}=(\sigma_r)_2|_{r=r_{co}},(\tau_{r\theta})_1|_{r=r_{co}}=(\tau_{r\theta})_2|_{r=r_{co}}\\(u_r)_1|_{r=r_{co}}=(u_r)_2|_{r=r_{co}},\ (u_\theta)_1|_{r=r_{co}}=(u_\theta)_2|_{r=r_{co}}\end{cases}\tag{22}$$

• The radial stress, shear stress, radial displacement, circumferential displacement of cement equal to that of formation on their contact interface, which are:

$$\begin{cases}(\sigma_r)_2|_{r=t}=(\sigma_r)_3|_{r=t},\ (\tau_{r\theta})_2|_{r=t}=(\tau_{r\theta})_3|_{r=t}\\(u_r)_2|_{r=t}=(u_r)_3|_{r=t},\ (u_\theta)_2|_{r=t}=(u_\theta)_3|_{r=t}\end{cases}\tag{23}$$

• The stress boundary conditions on formation are satisfied by:

$$\begin{cases}(\sigma_r)_3|_{r=r_s}=-s\cdot\cos 2\theta\\(\tau_{r\theta})_3|_{r=r_s}=s\cdot\sin 2\theta\end{cases}\tag{24}$$

8 equations can be listed through substituting the 8 stress conditions out of 12 conditions mentioned above in to Eq. 20. 4 out of 12 conditions mentioned above are related to the system displacement and 4 equations can be listed after displacement analysis.

4.3.3 Displacement analysis

Thestrain-displacement relationship in polar coordinate is:

$$\begin{cases}\varepsilon_r=\dfrac{\partial u_r}{\partial r}\\\varepsilon_\theta=\dfrac{1}{r}\dfrac{\partial u_\theta}{\partial\theta}+\dfrac{u_r}{r}\\\gamma_{r\theta}=\dfrac{\partial u_\theta}{\partial r}+\dfrac{1}{r}\dfrac{\partial u_r}{\partial\theta}-\dfrac{u_\theta}{r}\end{cases}\tag{25}$$

The system radial displacement can be obtained by integrating with respect to r from the first expression in Eq. 25, which is:

$$u_r=\frac{1+\mu}{E}\left[-4H\mu r^3-2Ir+4G(1-\mu)r^{-1}+2Kr^{-3}\right]\cos 2\theta\tag{26}$$

The system circumferential displacement can be obtained by integrating with respect to θ from the second expression in Eq. 25, which is:

$$u_\theta=\frac{1+\mu}{E}\left[2H(3-2\mu)r^3+2Ir+2G(2\mu-1)r^{-1}+2Kr^{-3}\right]\sin 2\theta\tag{27}$$

4 displacement equationscan be obtained by substituting Eq. 26 and Eq. 27 into Eq. 22 and Eq. 23 and 12 equations can be listed by adding 8 equations mentioned above. That is to say the 12 unknown number can be determined uniquely. The system response caused by $s\cos 2\theta$ and $s\sin 2\theta$ can be figured out by substituting the 12 parameters into Eq. 20.

5 Case study and discussion

5.1 Case study

The well structure is shown in Fig. 4. The temperature of testing liquid is 250° C, σ_H is 34.5MPa and σ_h is 28.3MPa. Other calculation parameters of casing , cement and formation is shown in Table 1.

Table 1 Calculation parameters of casing, cement and formation

	Elastic modulus (MPa)	Poisson's ratio	I. D. (mm)	O. D. (mm)	Coefficient of thermal expansion $/10^{-6}/°C^{-1}$
Casing	148180	0.3	250.19	273.05	11.7
Cement	20000	0.15	273.05	335	10.3
Formation	10500	0.23	335	30000	10.3

The radial stress of each layer medium on contact interface under the two kinds of external forces are shown in Fig. 5

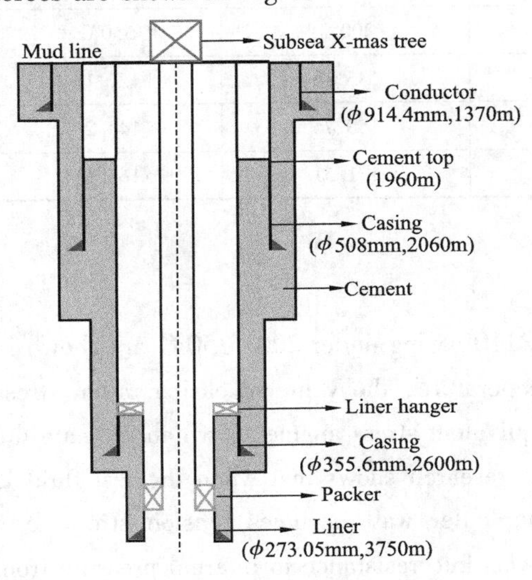

Fig. 4 Well structure

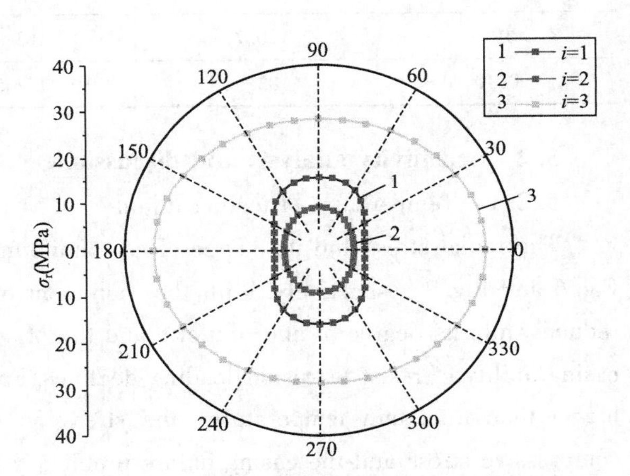

Fig. 5 Radial stressof each layer medium

In Fig. 5, $i=1$, 2, 3stand for the contact interface of casing-cement, cement-formation and formation boundary respectively. As can be seen from Fig. 5, the radial stress distribution shape on contact interface is ellipse under the combination of temperature and non-uniform stress. The main reason for the phenomenon is the existence of non-uniform compressive stress and shear stress. In this example, the direction of the major axis of the stress ellipse on formation boundary coincides with the direction of σ_h, while that of casing-cement and cement-formation coincide with the direction of σ_H. In aspect of the value stress, the stress on formation boundary is the largest among the three one while the stress on cement-formation is the smallest.

If the non-uniform degree of the radial stress can be defined as:

$$\lambda = \frac{\sigma_{r\max}}{\sigma_{r\min}} \tag{28}$$

Thus, in this example, the non-uniform degrees of the radial stress of casing, cement and formation are 1.67, 1.26 and1.22 respectively. That is to say the non-uniform degree of the radial stress on casing is the largest among them which indicates that the casing ability of resisting collapse load will dramatically reduce. According to the fourth strength theory, the Von Mises equivalent stress on casing inner wall is 297MPa. The minimum yield strength for N80, T95, P110 casing are 552MPa, 655MPa and 758MPa and the safety factor of casing in service are1. 86, 2. 21, 2. 55 accordingly. Due to temperature increment, the thermal stress can cover a certain part of the radial compression caused by non-uniform in-situ stress, which make casing tend to safer. However, with the increase of temperature, the elastic modulus declines which will make the casing tend to be extrusion failure. The variations of casing elastic modulus with temperature are shown in Table 2. Besides, the casing safety factor can also be affected by cement elastic modulus, Poisson's ratio and the casing grade and so on.

Table 2 Elastic modulus of casing under different temperatures

Casing	Elastic modulus (MPa)			
	200℃	250℃	300℃	350℃
N80	161. 18	148. 18	135. 18	122. 18
T95	170. 75	163. 25	155. 75	148. 25
P110	183. 51	177. 51	171. 51	165. 51

5. 2 Sensitivity analysis and discussion

5. 2. 1 Temperature of testing liquid

The radial stress and its degree of non-uniform of P110 casing under 200~350℃ are shown in Fig. 6 and Fig. 7 respectively. With the increment of temperature, the value of casing radial stress reduce while its degree of non-uniform and the Mises equivalent stress augment, which indicate the casing ability of resist extrusion loading declines. Further research shows that when the test fluid is higher than a certain temperature, the stress on casing outer wall becomes tension stress form compressive stress and the casing failure problem transforms into resistance to internal pressure from resistance to collapse load accordingly. Under the condition of 350℃, the radial stress of N80, T95, P110 casing is shown in Fig. 8, which shows that the higher the casing grade is , the greater the radial stress is.

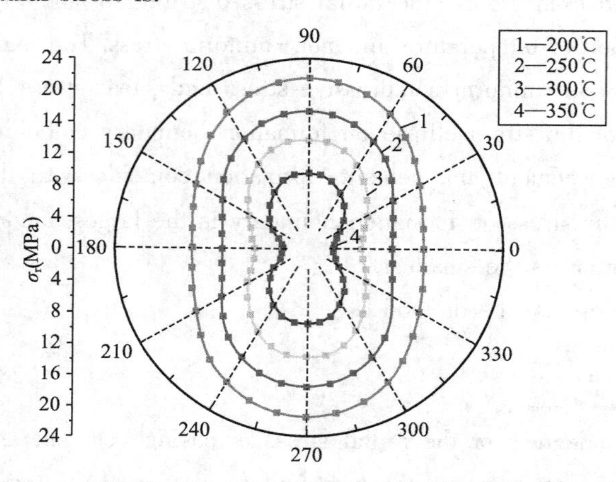

Fig. 6 Influence of temperature on radial stress of P110 casing

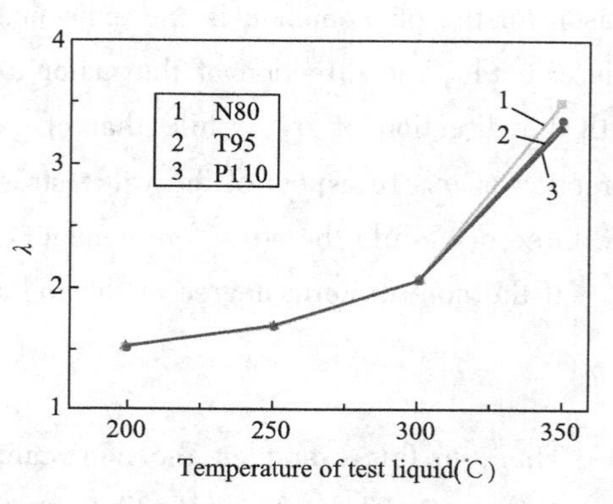

Fig. 7 Influence of temperature on degree of non-uniform

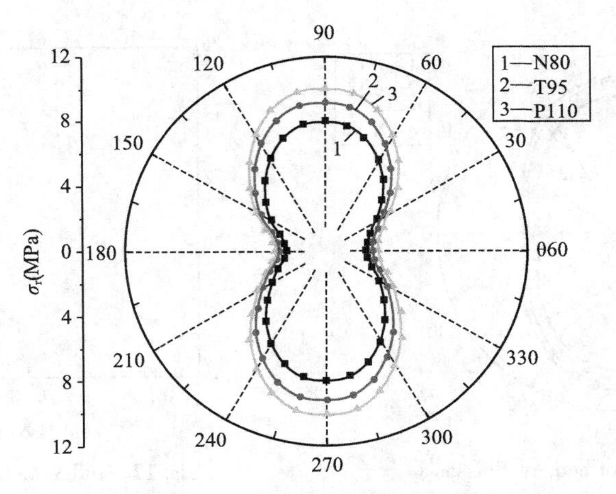

Fig. 8 Influence of casing grade on radial stress

5. 2. 2 Cement elastic modulus

The variations of radial stress and its degree of non-uniform of N80 casing with cement elastic modulus ranging from 10MPa to 25MPa are shown in Fig. 9 and Fig. 10. With the increase of cement elastic modulus, the value of casing radial stress declines while the degree of non-uniform rises gradually. However, the casing ability to resist collapse load tends to be enhanced by the comprehensive effect of high cement elastic modulus. Therefore, higher cement elastic modulus can improve the wellbore integrity.

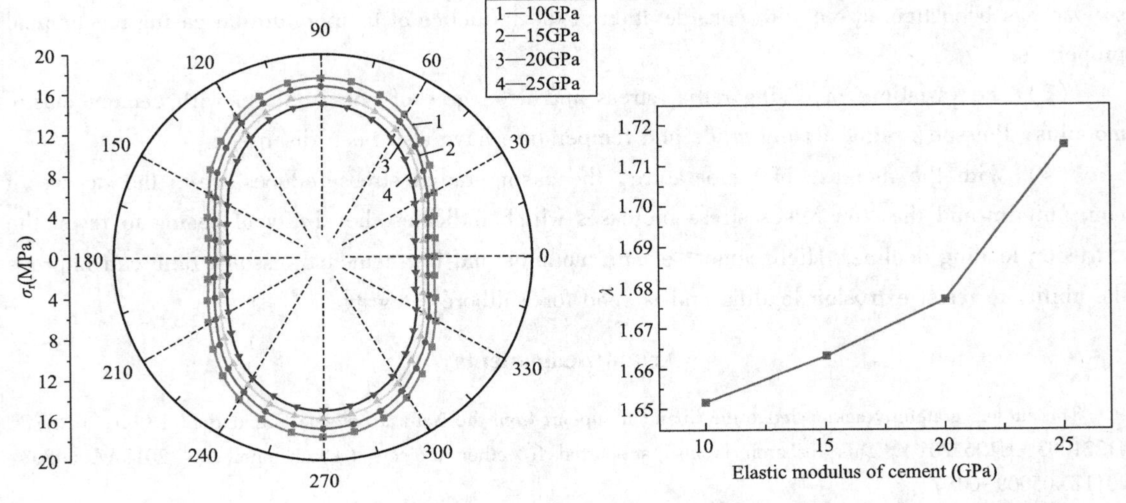

Fig. 9 Influence of cement elastic modulus on radial stress

Fig. 10 Influence of cement elastic modulus on degree of non-uniform

5. 2. 3 Cement Poisson's ratio

The variations of radial stress and its degree of non-uniform of N80 casing with cement Poisson's ratio ranging from 0. 15 to 0. 24 are shown in Fig. 11 and Fig. 12. With the increase of Poisson's ratio, the value of casing radial stress declines while the degree of non-uniform rises gradually. The casing ability to resist collapse load tends to be enhanced by the comprehensive effect of low cement Poisson's ratio. Therefore, lower cement Poisson's ratio is good for the wellbore integrity.

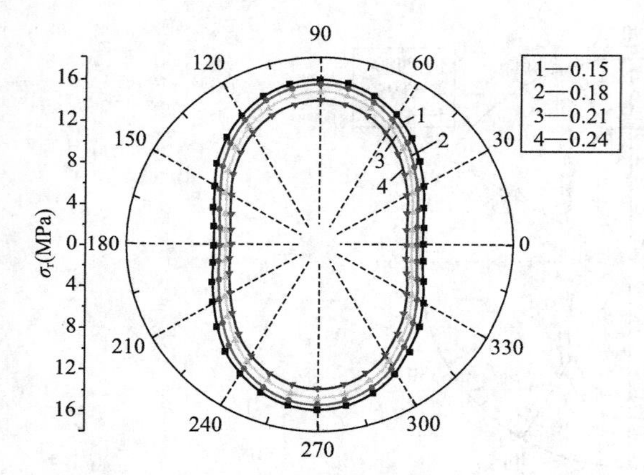

Fig. 11 Influence of cement Poisson's ratio on radial stress

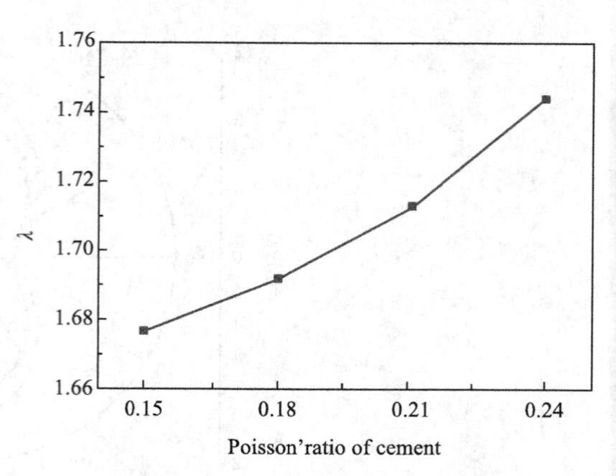

Fig. 12 Influence of cement Poisson's ratio on degree of non-uniform

6 Conclusions

(1) A mechanical model and control equations have been established to analyze the characteristics of casing-cement-formation of offshore drilling in well testing and production under the combination action of temperature and non-uniform in-situ stress.

(2) The equations have been solved by basic equations of plane strain and the stress distributions of each layer medium have been obtained. On this basic, the safety factor of casing in service has been figured out with consideration of the influence of temperature on casing mechanical properties.

(3) The variations of casing radial stress and its degree of non-uniform with cement elastic modulus, Poisson's ratio, casing grade and temperature have also been discussed.

(4) With the increase of temperature, the casing radial stress reduces while the degree of non-uniform and the Von Mises stress increases which indicates the ability of casing to resist the extrusion loading declines. High cement elastic modulus and low cement Poisson's ratio can improve the ability to resist extrusion loading and is good for wellbore integrity.

Acknowledgements

The authors gratefully acknowledge the financial support from the Natural Science Foundation of China (NSFC, 51221003, U1262201). This research is also supported by other projects (Grant numbers: 2013AA064803, 2011ZX05009-005).

Appendix A

Theintegration constants of 6 unknown parameters are shown as follows:

$$C_1^t = (CE+BF)/(AE+BD) \tag{A-1}$$

$$C_2^t = (AF-CD)/(AE+BD) \tag{A-2}$$

$$C_1^c = \frac{r_{co}^2 C_1^t + C_2^t - \dfrac{\alpha_c T_c (1+\mu_c)(r_{co}^2 - r_{ci}^2)}{2(1-\mu_c)}}{r_{co}^2 + \dfrac{r_{ci}^2}{1-2\mu_c}} \tag{A-3}$$

$$C_2^c=\frac{r_{ci}^2}{1-2\mu_c}C_1^c \tag{A-4}$$

$$C_1^f=0 \tag{A-5}$$

$$C_1^f=\frac{(1+\mu_t)\alpha_t T_c \exp(r_{co})}{\mu_t-1}[\exp(-r_t)(1+r_t)-\exp(-r_{co})(1+r_{co})]+C_1^t r_t^2+C_2^t \tag{A-6}$$

$$A=\frac{E_t}{(\mu_t+1)(1-2\mu_t)}\left[r_{co}+\frac{r_{ci}^2}{r_{co}(1-2\mu_c)}\right]-\frac{E_c(r_{co}^2-r_{ci}^2)}{(1+\mu_c)(1-2\mu_c)r_{co}} \tag{A-7}$$

$$B=\frac{E_t}{(\mu_t+1)r_{co}^2}\left[r_{co}+\frac{r_{ci}^2}{r_{co}(1-2\mu_c)}\right]+\frac{E_c\left(1-\frac{r_{ci}^2}{r_{co}^2}\right)}{(1+\mu_c)(1-2\mu_c)r_{co}} \tag{A-8}$$

$$C=\frac{-\alpha_c E_c T_c}{2(1-\mu_c)}\left(1-\frac{r_{ci}^2}{r_{co}^2}\right)\left[r_{co}+\frac{r_{co}}{(1-2\mu_c)}\right] \tag{A-9}$$

$$D=\left[\frac{E_t}{(1-2\mu_t)(1+\mu_t)}+\frac{E_f}{1+\mu_f}\right]r_t^2 \tag{A-10}$$

$$E=\frac{E_f}{1+\mu_f}-\frac{E_t}{1+\mu_t} \tag{A-11}$$

$$F=\frac{\alpha_t T_c \exp(r_{co})}{(1-\mu_t)}[\exp(-r_t)(1+r_t)-\exp(-r_{co})(1+r_{co})]\left[\frac{E_f(1+\mu_t)}{1+\mu_f}-E_t\right] \tag{A-12}$$

Where, μ_c, μ_t and μ_f are Poisson's ration of casing, cement and formation respectively. E_c, E_t and E_f are elastic modulus of casing, cement and formation respectively, MPa. α_c and α_t are coefficient of thermal expansion of casing and cement respectively, 1/°C

Appendix B

The specifics of Eq. 15 are shown as follows:

$$k_1=\frac{[1+\mu_f]E_t}{(1+\mu_t)E_f}\frac{1-m_2^2}{2(1-\mu_t)}+\frac{(1-2\mu_t)+m_2^2}{2(1-\mu_t)} \tag{B-1}$$

$$k_2=(1-2\mu_t)\frac{(1+\mu_t)E_f}{(1+\mu_f)E_t}\frac{1-m_2^2}{2(1-\mu_t)}+\frac{(1-2\mu_t)m_2^2+1}{2(1-\mu_t)} \tag{B-2}$$

$$k_3=\frac{(1-2u_c+m_1^2)(1+u_c)r_{co}}{(1-m_1^2)E_c} \tag{B-3}$$

$$k_4=\frac{[(1-2u_t)m_2^2+1](1+\mu_t)r_{co}}{(1-m_2^2)E_t} \tag{B-4}$$

$$k_5=\frac{2(1-\mu_t)(1+\mu_t)r_{co}}{(1-m_2^2)E_t} \tag{B-5}$$

$$k_s=\frac{E_f}{(1+\mu_f)r_t} \tag{B-6}$$

$$k_c=\frac{(1-m_1^2)E_c}{[(1-2\mu_c)+m_1^2](1+\mu_c)r_{co}} \tag{B-7}$$

$$m_1=\frac{r_{co}}{r_{ci}} \tag{B-8}$$

$$m_2=\frac{r_t}{r_{co}} \tag{B-9}$$

References

[1] A. Adams, 1991. How to Design for Annulus Fluid Heat-Up. SPE Annual Technical Conference and Exhibition, Dallas, Texas, SPE-22871.

[2] A. Adams and A. MacEachran, 1994. Impact on Casing Design of Thermal Expansion of Fluids in Confined Annuli . SPE Drilling & Completion, SPE 29229: 210-216.

[3] A. S. Halal and R. F. Mitchell, 1994. Casing design for trapped annulus pressure buildup. SPE 25694: 107-113.

[4] Azzola J H, et al. , 2007. Application of Vacuum-insulated Tubing to Mitigate Annular Pressure Buildup. SPE Drilling & Completion, SPE 90232: 46-51.

[5] Birgit Vignes et al. , 2008. Injection Wells with Integrity Challenges on the NCS. SPE 118101.

[6] Birgit Vignes et al. , 2010. Well-Integrity Issues Offshore Norway. SPE 112535: 145-150.

[7] C. P. Leachand A. J. Adams, 1993. A New Method for the Relief of Annular Heat-Up Pressures. SPE Production Operations Symposium. Oklahoma City, Oklahoma, SPE 25497: 819-825.

[8] G. Robello Samuel and Adolfo Gonzales, 1999. Optimization of Multistring Casing Design with Wellhead Growth. SPE Annual Technical Conference and Exhibition, Houston, Texas, SPE 56762.

[9] Li Jing et al. , 2009. Theoretical Solution of Thermal Stress for Casing - Cement - Formation Coupling System. Journal of the China University of Petroleum (Edition of Natural Science), 33 (02): 63-69.

[10] Li Zhiming ans Yin Youquan, 2006. Oil and Water Wells Outside Casing Extrusion Force Calculation and Mechanical Foundation. Beijing: Petroleum Industry Press: 97-116.

[11] P. Oudeman and M. Kerem, 2006. Transient Behavior of Annular Pressure Buildup in HP/HT Wells. SPE Drilling & Completion, SPE 88735: 234-241.

[12] Roger Williamson et al. , 2003. Control of Contained Annulus Fluid Pressure Buildup. SPE/IADC Drilling Conference, Amsterdam, The Netherlands, SPE/IADC 79875.

[13] S. P. Timoshenko and J. N. Goodier, 1990. Theory of Elasticity (Third Edition), Higher Education Press: 396-410.

[14] S. S. Rahman and G. V. Chilingarian, 1995. Casing Design Theory and Practice. Elsevier Science Ltd.

[15] Vargo et al. , 2003. Practical and Successful Prevention of Annular Pressure Build Upon the Marlin Project. SPE Drilling & Completion, SPE 85113: 228-234.

[16] Wang shuping et al. , 2007. Reducing Casing Additional Load Caused by Temperature. Journal of Southwest Petroleum University, 29 (06): 149-152.

[17] Wang Shuping et al. , 2007. A Model for Prevention of Casing Additional Laod From High Temperature. Natur. Gas Ind, 27 (09): 84-86.

[18] Wu Jiang and M. E. Knauss, 2006. Casing Temperature and Stress Analysis in Steam - Injection Wells. International Oil & Gas Conference and Exhibition. Beijing, China, SPE 103882.

[19] Yang Guitong, 2004. Induction to Elasticity and Plasticity. Beijing: Tsinghua Unversity Press: 115-120.

[20] Yu et al. , 2008. Numerical simulation and industrial application of the casing production stress inside steam injection wells. Mechanics in Engineering, 30 (1): 66-69.

[21] Zhang Yonggui, 2008. Casing Strength Theoretical and Experimental Study of Thermal Recovery Steam Injection Wells. Qinhuangdao: Yanshan Unicersity.

深水钻井管柱力学与设计控制技术研究新进展*

高德利　王宴滨

（中国石油大学石油工程教育部重点实验室）

摘　要： 深水钻井作业主要包括导管喷射安装、表层套管井段钻井、水下防喷器组和深水钻井隔水管安装及后续钻井等四个主要作业环节，涉及导管、钻井隔水管、送入管柱等三类管柱系统。与陆地及浅水近海钻井不同，由于深水钻井工况的独特性，管柱在作业过程中产生复杂的力学行为，严重影响深水钻井的安全高效作业。因此，开展深水钻井管柱力学与设计控制技术研究，对于推动深水钻井科技进步具有重要意义。

深水导管喷射安装技术是适应深水钻井的特殊要求而发展起来的一种浅层作业技术，也是深水钻井程序的第一步。作业过程涉及导管和送入管柱两类管柱系统，主要目的在于建立安全稳定的水下井口，为后续的钻井作业奠定基础。例如，送入管柱的力学行为分析与优化设计研究、水下井口的管土相互作用与导管承载能力研究等，对实现水下井口安全稳定的目标具有重要意义。本文从工程应用与技术研发两个方面，对涉及其中的送入管柱强度设计与校核、导管喷射安装工艺和导管承载能力等三个方面的研究进展进行了综述与展望。认为深水导管喷射安装的未来研究将侧重于极限工况下导管的入泥深度与承载力计算、喷射钻进参数优化、导管喷射安装风险评估与可靠性预测，以及深水导管喷射安装模拟实验等内容。

深水钻井隔水管是连接浮式钻井平台与水下井口的重要设备，可提供钻井液循环通道、支持辅助管线、引导钻具、下放与回收防喷器组等。深水钻井隔水管在整个钻井作业过程中涉及安装、正常钻进、回收与紧急撤离等作业过程。由于波流联合作用力的动态效应，深水钻井隔水管在服役期间会产生轴向拉伸、横向弯曲、耦合振动等一系列复杂力学行为，给深水钻井安全作业带来巨大挑战。因此，对深水钻井隔水管力学行为进行研究，确保其安全可靠性，是深水钻井研究的关键问题之一。本文着眼于深水钻井隔水管的顶张力控制、纵横弯曲变形、横向振动特性、纵向振动特性、耦合振动特性及涡激振动特性等主要力学问题，从载荷计算、控制方程、边界条件及求解方法等方面入手，总结了深水钻井隔水管系统在力学与设计控制技术方面取得的新进展，对目前研究中仍然存在的问题进行了剖析和探讨。认为在以后的工作中，应在深水钻井隔水管安装作业窗口分析预测、隔水管涡激振动响应与抑制、隔水管疲劳寿命计算与评估，以及隔水管力学行为模拟实验等方面加强

* 基金项目：国家自然科学基金创新研究群体项目和重点项目（项目批准号：51521063 和 U1262201）。

作者简介：高德利（1958—），男，中国石油大学（北京）石油与天然气工程国家重点学科负责人、石油工程教育部重点实验室主任、中国科学院院士。E-mail：gaodeli@cast.org.cn；wyb576219861@126.com。

研究。

在深水井筒整个寿命期间，最大限度地使井筒中地层流体处于有效控制的安全运行状态，防止浅层气和浅水流入侵，提高固井质量，避免水下套管柱变形甚至挤毁等，对于提高深水油气井筒的完整性具有重要的实际意义。本文以深水井筒的温度分布规律、套管环空压力变化及套管应力分布等研究为主，对深水井筒完整性预测和预防研究进行了综述，主要内容包括：地层非稳态传热、套管环空循环温度分布、密闭环空内流体升温膨胀引起的附加载荷和预防措施、多层套管柱环空压力计算、套管-水泥环-地层系统热力耦合响应等。分析认为充分考虑深水钻井特殊工艺与环境约束条件、建立适用于深水井身结构与套管柱优化设计方法、开展深水井筒完整性风险评估与设计控制技术研发将是未来关注的重点。

开展深水钻井管柱力学模拟实验研究，获取相关的有效数据，对于提高深水管柱力学与设计控制研究水平具有必要性。本文建立了“深水管柱力学模拟实验系统”，并对其结构组成、操作方法、技术参数及主要功能等进行了详细说明，介绍了深水钻井隔水管力学行为模拟实验与疲劳寿命测试方面取得的新进展。

本文对深水导管喷射安装技术、深水钻井隔水管系统力学与设计控制技术以及深水井筒完整性预测和预防研究等方面的新进展进行了综述和展望，对指导今后的深水钻井管柱力学与设计控制技术研究具有参考价值。

关键词：深水钻井；管柱力学；设计控制技术；深水导管；海洋钻井隔水管

深水钻井工程是深水条件下海洋油气工程作业的关键环节之一。与近海浅水钻井不同，深水钻井必须面对更为复杂的海洋深水环境和地层条件，面临“入地、下海”的双重挑战，需要采用浮式钻井作业平台，建立安全稳定的水下井口与钻井系统，使用特殊的深水管柱（包括导管、钻井隔水管、套管柱及送入管柱等）、水下智能控制系统等，是一项复杂的系统工程，具有高科技、高投入及高风险等基本特征。

随着世界能源需求的增长及海洋油气勘探开发技术的不断进步，人们越来越重视海洋油气资源的勘探开发[1-3]。深水钻井管柱是实施深水钻井不可或缺的基本工具，各种管柱系统在服役过程受到海洋环境载荷的作用，表现出复杂的力学行为，并且其服役条件随着水深的增加而更加恶劣[4-8]。对深水管柱系统进行力学预测分析并形成相应的控制方法，对确保深水钻井安全高效作业具有重要的实际意义。

本文介绍了包括深水导管、送入管柱、钻井隔水管、水下井口在内的主要深水钻井管柱的力学研究进展及其存在的问题，以及今后的发展方向，以期为深水钻井管柱力学深入研究与深水钻井技术发展提供有益参考。

1 深水钻井特点与关键问题

由于深水钻井环境、装备及部分钻井技术与陆地及浅水近海有很大不同，因而在深水钻井作业流程及其所需技术装备上具有特殊性。一般情况下，深水钻井平台（船）拖航到位并完成定位和浅层钻井的工程作业过程如图 1 所示。

深水钻井作业主要包括导管喷射安装、表层套管井段钻井、水下防喷器组及深水隔水管安装及后续钻井等四个主要作业环节[9-12]。

深水导管喷射安装是适应深水钻井的特殊要求而发展起来的一种浅层作业技术[13,14]。

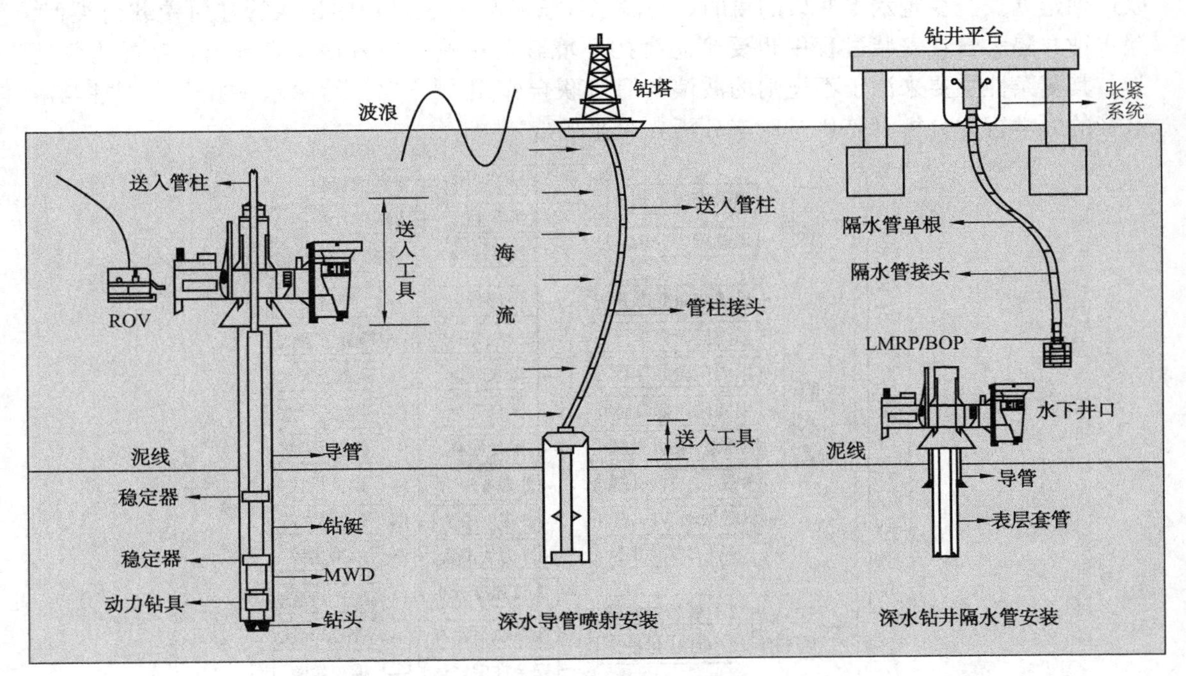

图 1　深水钻井作业流程

在喷射安装过程中，送入管柱通过导管送入工具连接喷射管柱，而喷射管柱主要有两部分组成：一部分为喷射导管本身，另一部分由钻头（一般为牙轮钻头）、动力马达、稳定器、钻铤及低压井口头等组成[15-17]。导管喷射安装的主要目的在于使导管与周围土层形成足够支撑水下井口的承载力，作业过程涉及的管柱系统包括送入管柱与导管，其中送入管柱的合理设计对导管喷射安装具有决定性的作用[18-20]。由于深水钻井导管喷射安装技术在钻井时间与经济成本方面具有较大优势，因而被广泛应用于深水及超深水钻井工程中。在深水喷射安装过程中，需要对钻压、排量、送入管柱受力变形等多个参数进行严格控制，以防止扩孔、卡套管、导管下沉等工程事故的发生[21,22]。

当表层套管固井后，接着就要进行隔水管和防喷器组安装作业[23,24]。由于深水钻井隔水管系统是连接海底井口与水面钻井装置的重要设备，因此隔水管安装作业的主要目的是实现水下井口与 LMRP/BOP（Lower Marine Riser Package/Blowout Preventer）的精确连接，为后续的钻完井作业提供循环钻井液通道及必要的控制设施，其合理的安装操作亦将直接关系到深水钻井作业的顺利进行[25-27]。根据深水钻井作业规范，无论是隔水管各组件的强度和性能，还是安装作业程序，都必须符合相关标准[28-32]，同时在安装过程中应对隔水管的力学行为进行详细分析及预测和控制，以便保证安装作业的安全可靠性。

与陆地及浅水近海钻井不同，在深水钻井过程中，管柱力学行为具有独特性和复杂性，如图 2 所示。深水钻井作业的特点主要体现在深水钻井导管喷射安装和深水钻井隔水管安装两大作业过程，涉及导管、隔水管及送入管柱等三类管柱，其中深水导管喷射安装过程涉及送入管柱和导管，其主要目的在于建立安全稳定的水下井口，为后续的钻井作业提供基础。深水钻井导管承载能力是导管喷射安装过程的研究重点，但深水导管需要通过送入管柱下放到海底，送入管柱的设计合理与否亦直接影响深水导管喷射安装作业的成败，因此需要以深水导管喷射安装为目标对送入管柱进行优化设计，并对导管承载能力及工艺过程进行控制，以达到建立安全稳定水下井口的目的。与深水导管喷射安装过程中送入管柱所受载荷类似，深水钻井隔水管在安装过程中也受到恶劣的环境载荷影响，二者均会表现出复杂的力学行为，其复杂性主要来源于不规则的波浪与海流联合作用力[33-35]，给深水钻井送入管柱与隔水管的力学行为分析（特别是动态分析）带来挑战。

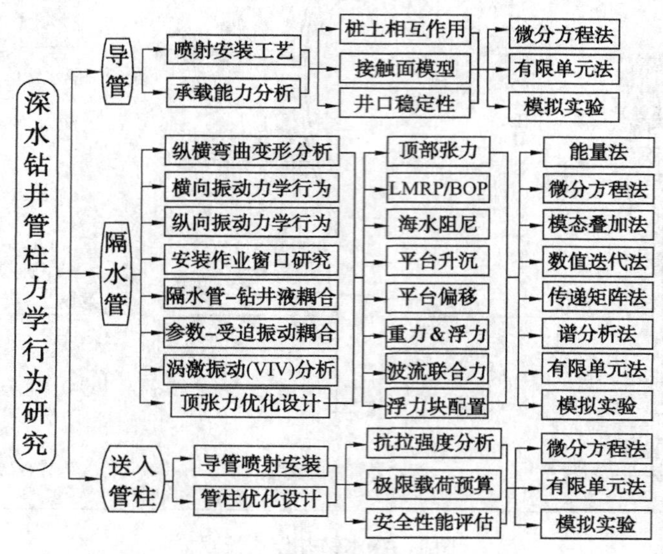

图 2　深水钻井管柱系统力学问题

2　深水导管喷射安装技术研究与实践

对深水导管喷射安装作业的研究，目前主要集中在送入管柱的强度设计与校核、导管喷射安装工艺及导管承载能力研究三个方面，基本涵盖了深水导管喷射安装过程的关键环节。

2.1　工程应用

墨西哥湾是全球深水钻井活动最频繁的海域之一，也是最早使用深水导管喷射安装技术的海域。在 20 世纪 60 年代，SHELL 在没有使用喷射钻头与动力马达的情况下，只借助一个喷射接头，将外径29.5 in（英寸）、壁厚 1.0 in 的导管安装在泥线以下 100 ft（英尺）深度，由于作业过程中底部钻具组合配置不合理，导致钻井液从导管外排除，对地层形成了较大的冲刷作用，使得导管承载力遭到较大破坏[36]，其工艺过程原理如图 3（a）所示。从 20 世纪 70 年代开始，随着钻井技术的不断进步，井下动力钻具性能的提高与各种新型导管送入工具的使用，使得导管喷射安装技术得以较大改进[37,38]。井下动力钻具可以直接驱动牙轮钻头旋转，喷射出的钻井液可以从底部钻具组合与导管的环空返到海底，极大降低了对地层土壤的干扰。到 20 世纪 90 年代，深水导管喷射安装技术已经成为全球深水及超深水钻井的首选[39-41]，其工艺过程原理如图 3（b）示。1994 年，HALLIBURTON 公司首次将深水导管喷射安装技术引入中国南海，采用深水作业模式开发了流花 11-1 油田[42-44]，喷射安装的深水导管为外径 762.0 mm（30.0 in），壁厚 25.4 mm（1.0 in），这是深水导管喷射安装技术在我国首次成功使用，作业时效比常规安装导管方法提高了 3 倍多。近几年来，随着国内海上油气勘探开发活动的逐渐增多，深水导管喷射安装技术在我国的深水及超深水区域得到了广泛的应用。LH29-2-A 井、LW6-1-B 井、BY13-2-C 井及 LW21-1-D 井是我国首批自营深水井[45]，也是国内自主设计建造的第六代深水半潜式平台 HYSY981 承钻的第一批井，这 4

口井的导管（ϕ914.4mm）全部采用喷射安装工艺，并且取得了良好效果，后期作业无井口下沉现象，为我国后续深水导管喷射安装作业积累了宝贵的经验。

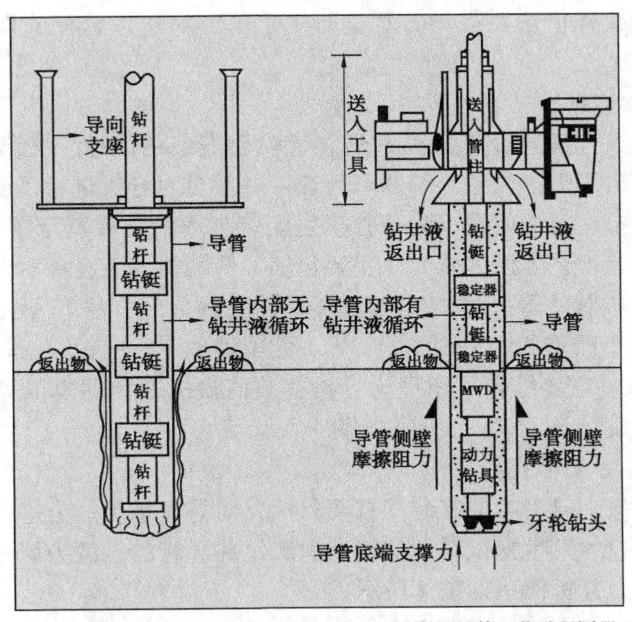

(a)早期喷射下导管工艺过程原理　(b)现在喷射下导管工艺过程原理

图3　深水导管喷射安装技术开发

2.2　技术研究

送入管柱的强度设计与校核、喷射导管串的工艺参数设计及深水钻井导管承载力，是深水钻井导管喷射安装最为重要的三个方面，直接关系到喷射安装作业的成败。

2.2.1　送入管柱的强度设计与校核

国外学者首先在送入管柱的强度设计与校核方面进行了相关研究，早期的研究主要集中在以钻柱的抗静拉力设计为基础的送入管柱设计[46-49]，但此方法与常规的钻柱设计方法类似，未涉及送入管柱的强度设计，也没有很好地体现送入管柱的作业特点。到目前为止，送入管柱的设计方法主要包含安全系数法、抗卡瓦挤毁系数法及拉力余量法[50-52]。安全系数法主要考虑起下钻过程中的井壁摩阻力、流体黏滞阻力以及加速度产生的动载荷；拉力余量法主要考虑管柱遇卡、解卡的情况，由于送入管柱需要同时满足导管喷射安装和表层井眼钻进的作业要求，上述设计方法中，卡瓦挤毁是制约送入管柱作业能力的主要因素，计算方法主要是将送入管柱所受外载等效为轴向载荷[53]。同时使用高强度钻杆[54,55]是提高送入管柱作业能力的另一种方式，为钻杆配备新型接头、在送入管柱接头区域进行加厚处理[56,57]均可减少管柱所受卡瓦外挤力，从而提高送入管柱的抗拉强度。实际设计过程中分别采用以上三种方法计算各段钻杆的最大允许使用长度，并选取其中的最小值作为设计结果。此外，送入管柱在作业过程中会受到来自海洋环境的恶劣载荷，在横向波流联合作用力及平台纵向升沉运动影响下，送入管柱会发生横向振动与轴向振动，产生较大弯曲变形，严重影响导管喷射安装的控制精度。为确保导管喷射安装的顺利进行，需要对其复杂力学行为进行准确的预测与控制。目前对于送入管柱的力学行为研究大多集中在轴向振动对其载荷的影响[58]，而对涉及纵横耦合振动的研究则较少。国内相关学者[59-63]根据深水钻井导管喷射安装作业特点，详细阐述了送入管柱在操作过程中受到的载荷，系统研究了各种环境载荷、送入管柱结构及浮式平台偏移等对钻柱轴向力的影响，并据此建立了管柱的静力学分析模型、控制方程及求解方法，并用加权残值法对控制方程进行了求解，对送入管柱的设计安装提供了理论指导。

2.2.2 喷射导管串的工艺参数设计

喷射导管串的设计合理与否直接关系到导管喷射安装的成败，文献[64，65]讨论了导管喷射安装方法，指出了喷射安装导管的可行性，并着重对导管进行了受力分析，阐述了导管喷射安装过程中需要注意的问题，为导管喷射安装技术的发展奠定了基础。国内从21世纪初开始导管喷射安装工艺方面的研究，主要包括导管喷射安装在深水及超深水钻井中的有效性及实用性探讨，喷射导管尺寸的选择、入泥深度及导管强度与稳定性的优化分析[66]。导管喷射安装过程需要严格控制包括钻压[67]、机械钻速[68]、钻井液排量[69,70]及钻头伸出量[71]在内的关键参数，需要综合采用理论分析、数值模拟及物理模拟等方法对喷射导管串的工艺参数进行优化设计。

2.2.3 深水钻井导管承载力

足够的承载力是深水导管喷射安装的首要目标，对导管承载能力进行研究可为防止井口失稳、下沉等复杂事故的发生提供科学依据。深水钻井导管的承载力属于桩土相互作用研究范畴，导管的承载能力分析模型如图4所示。

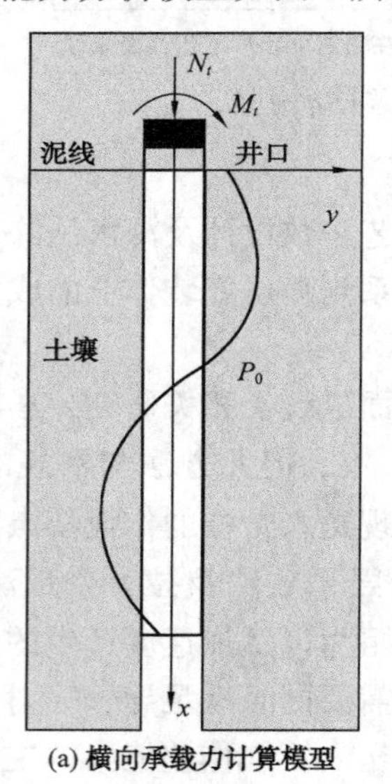

(a) 横向承载力计算模型

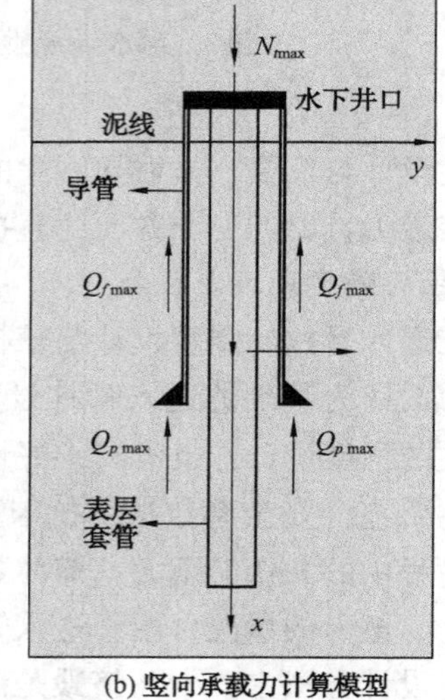

(b) 竖向承载力计算模型

图4 深水钻井导管承载能力分析模型

从20世纪90年代初开始，国外相关公司针对喷射安装导管的工程实践，详细讨论了根据地层土壤剪切强度计算导管承载力的方法，为后续导管设计及承载力计算奠定了基础[9]。在通常情况下，导管的承载力以API RP-2A给出的P-Y曲线为计算依据[72]，计算时需要考虑导管的瞬时承载力[73]及不同海底土层性质对承载力的影响[74]。另外，不同的作业工况、导管承受的外载荷不同、准确计算不同工况下导管的承载能力及各种作业载荷和环境载荷对承载能力的影响，对保证水下井口安全稳定性具有重要意义。21世纪初国内相关学者开始了喷射安装导管承载力方面的研究，在这方面比较典型的研究有：文献[75，76]给出了基于地层破裂压力与岩土力学的深水导管极限承载力计算方法；文献[77-79]介绍了土力学和桩基理论在深水导管喷射安装中的应用，着重讨论了导管下入深度计算方法及时间效应对导管承载力的影响；文献[80]应用 ABAQUS 有限元软件，通过编写相应的本构模型有限元程序，对不同桩土接触面模型下导管的承载能力进行了数值模拟。深水导管喷射安装的未来研究，将侧重于极限工况下导管的入泥深度与承载力计算、喷射钻进参数优化、导管喷射安装风险评估与可靠性预测，以及深水导管喷射安装模拟实验等方面。

3 深水钻井隔水管力学行为研究

深水钻井隔水管是水下井口与浮式钻井设备之间最重要也是最脆弱的连接设备，起到提供钻井液循环通道、支持辅助管线、引导钻具、下放与回收防喷器组等重要作用。深水钻井隔水管在整个深水钻井作业过程中涉及安装、正常钻进、回收与紧急撤离等作业过程。在整个作业过程中，隔水管要受到来自重力、顶张力、海流力及波浪力等复杂载荷的联合作用，还要承受来自不同作业工况载荷，产生拉伸、弯曲、振动等复杂力学行为，会出现挤毁、屈曲、疲劳损伤甚至断裂等失效情况，严重威胁深水钻井安全。因此，对深水钻井隔水管力学行为进行研究，确保其安全可靠性，是深水钻井工程必须考虑的重要问题之一。

3.1 深水钻井隔水管顶张力分析

深水钻井隔水管在安装过程中底部连接防喷器组和隔水管底部组合，顶部通过张紧系统与浮式钻井设备相连，隔水管底部受到 LMRP/BOP 重力作用产生的拉力，顶部受到张紧器产生的近似恒定拉力，隔水管在两端拉力作用下产生轴向拉伸变形。顶张力对隔水管力学行为的控制体现在两个方面，一方面控制隔水管的横向弯曲变形，顶张力越大，隔水管横向弯曲变形越小，但作用在隔水管上的 Von Mises 应力越大；另一方面顶部的超张力使隔水管底部受拉，防止底部管柱屈曲失稳。目前由于对顶张力的认识不统一，从而产生了不同的计算方法：API 算法和底部残余张力算法。两种计算方法均基于有效张力理论[81,82]，其中 API 算法规定，顶张力的配置应保证隔水管各部分的有效张力大于零，即应保证隔水管在水中的稳定性；底部残余张力算法认为隔水管底部的残余张力要满足对 LMRP 的过提要求，以真实轴向力计算隔水管的底部残余张力。

3.2 深水钻井隔水管纵横弯曲变形分析

深水钻井隔水管在轴向拉力与横向波流力联合作用下产生纵横弯曲变形，如果把横向波流作用力视为静态力，那么隔水管受力变形即为静力学问题，通常在隔水管静力分析时假设波流的传播方向一致，取最危险的工况进行分析，其力学分析模型如图 5 所示。

对于隔水管系统的受力变形及强度问题，国外学者进行了大量研究，早期采用二维静力学分析方法，用弹性力学计算隔水管的变形，并进行强度分析，其变形控制方程如下[83]：

$$EI\frac{d^4y}{dx^4} - T(x)\frac{d^2y}{dx^2} - w\frac{dy}{dx} = F(x) \tag{1}$$

式中 EI——隔水管柱截面抗弯模量，N·m^2；

$T(x)$——隔水管轴向力沿水深分布，N；

w——单位长度隔水管柱的重量，N/m；

$F(x)$——横向波流作用力沿水深的分布，N/m。

对于正常钻进过程而言，隔水管上下两端为球铰固定，并且具有旋转刚度，此时式(1)的边界条件为：

$$y\big|_{x=0}=S,\ EI\frac{\partial^2 y}{\partial x^2}\bigg|_{x=0}=k_t\frac{\partial y}{\partial x}\bigg|_{x=0};\ y\big|_{x=L}=0,\ EI\frac{\partial^2 y}{\partial x^2}\bigg|_{x=L}=k_b\frac{\partial y}{\partial x}\bigg|_{x=L} \tag{2}$$

式中　S——隔水管顶部偏移位移，m；

k_t，k_b——隔水管上下两球铰的旋转刚度，N·m/rad。

对于安装及回收与紧急撤离过程而言，隔水管上端为固定端，下端为自由端，此时式（1）的边界条件为：

$$y\big|_{x=0}=0,\ EI\frac{\partial^2 y}{\partial x^2}\bigg|_{x=0}=0;\ EI\frac{\partial^2 y}{\partial x^2}\bigg|_{x=L}=0,\ EI\frac{\partial^3 y}{\partial x^3}\bigg|_{x=L}=F(x)\big|_{x=L} \tag{3}$$

基于上述变形控制方程和边界条件，可求得隔水管在不同作业模式下的变形状态，然后根据材料力学物理关系可对隔水管的内力与强度进行计算和分析。式（1）的求解，通常采用有限差分法[84]、加权残值法[85]等数值解法。

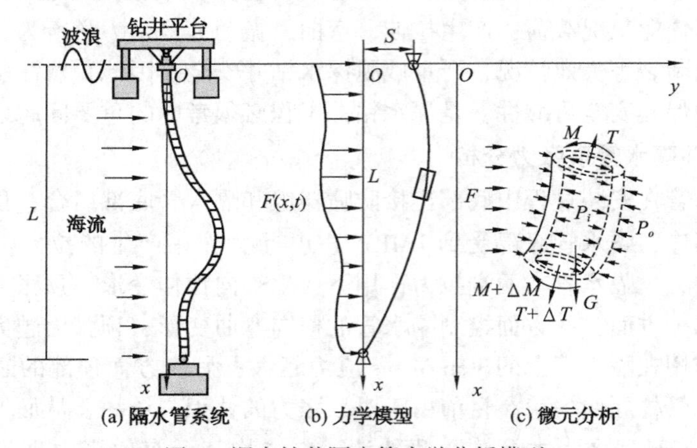

图5　深水钻井隔水管力学分析模型

3.3　深水钻井隔水管横向振动分析

深水隔水管静力分析方法对于浅水隔水管具有一定的适用性，但随着水深增加，动态因素对隔水管的影响越来越显著。在前期的研究中考虑水质点的速度是由周期的稳态正弦波产生，即用一个正弦稳态解法来求与此同周期的隔水管横向振动位移，通过调和函数变换将力学模型简化，但由于模型简化和求解过程烦琐，这种解析法并不实用。后来在文献[86]中给出了一种分析隔水管动态受力的数值求解方法，其数学模型如下：

$$\frac{\partial^2}{\partial x}\left(EI\frac{\partial^2 y}{\partial x^2}\right)-\frac{\partial}{\partial x}\left[T(x)\frac{\partial y}{\partial x}\right]-w\frac{\partial^2 y}{\partial t^2}=F(x,\ t) \tag{4}$$

动态分析与静态分析的最大区别在于前者将横向波流作用力视为动态，但精确计算波流联合作用力也是隔水管动态分析的最大难点。考虑到深水隔水管本身运动的影响，横向动态波流联合作用力可用修正的莫里森方程计算[35]，即：

$$F(x,\ t)=\frac{\pi}{4}C_m\rho_w D^2 a_r-\frac{\pi}{4}(C_m-1)\rho_w D^2\frac{\partial^2 y}{\partial t^2}+\frac{1}{2}C_D\rho_w D\left(v_r+v_c-\frac{\partial y}{\partial t}\right)\left|v_r+v_c-\frac{\partial y}{\partial t}\right| \tag{5}$$

式中　C_m——附连水质量系数，一般在0.93~2.30之间取值，对于隔水管等圆形横截面的构件一般取2.00；

C_D——阻力系数，一般在0.4~1.6之间取值，其值与雷诺数相关；

ρ_w——海水密度，kg/m^3；

D——隔水管外径，m；

v_r——波浪水质点水平速度，m/s；

v_c——海流速度，m/s；

a_r——波浪水质点水平加速度，m/s^2。

同样，对于正常的钻进条件下的隔水管而言，其上下两端仍可看作球铰固定，并且具有旋转刚度；对于安装过程及回收与紧急撤离过程而言，隔水管上端为固定端，下端为自由端。将对应的边界条件方程与式（4）联立求解，其求解方法一般包括模态叠加法[87]、泛函分析法[88]、有限差分法[89]和有限单元法[90]等。

随着研究的不断深入，隔水管的三维动力分析研究越来越多，通过综合运用流体力学理论、波动理论等，考虑海流、波浪等海洋环境载荷的动态作用[91]，以及在外载作用下的小应变大变形和轴向力对隔水管动力学行为的影响[92]，同时引入非线性理论[93]，对隔水管的三维变形、弯曲载荷、振型和频率、动态位移和应力等进行研究，使计算结果更加符合实际情况。同时，有关学者根据随机振动理论计算了随机波浪和规则波浪下船体运动对隔水管非线性动力响应的影响，并对船体平均偏移及波频响应和低频响应的周期和幅值对隔水管动力响应的影响进行了研究[94]，采用模态分析的离散化方法，分析了隔水管在随机载荷波浪力作用下的横向随机振动问题，给出了隔水管横向随机振动位移相关函数和均方位移的计算公式，为深水隔水管的随机响应计算和工程设计提供了一定的理论基础。

3.4 深水钻井隔水管纵向振动分析

动态的波浪载荷一方面使隔水管产生横向振动，另一方面通过钻井平台的升沉运动作用在隔水管顶部引起隔水管纵向振动，对隔水管的轴向力及横向振动特性产生影响。随着水深增加，隔水管纵向振动固有频率增大，若平台升沉振动频率接近隔水管纵向振动的固有频率就会引发共振，产生极大的振动载荷，甚至使隔水管发生破坏。即使轴向振动载荷不超过隔水管的应力屈服极限，隔水管在轴向交变载荷的作用下也可能发生疲劳破坏。在对隔水管进行纵向振动分析时，通常将其看成具有无限多自由度的连续系统[95]，考虑海水阻尼力作用下的纵向振动变形微分控制方程如下：

$$EA\frac{\partial^2 y}{\partial x^2} - H\frac{\partial y}{\partial t} = w\frac{\partial^2 y}{\partial t^2} \tag{6}$$

式中　A——隔水管的横截面积，m^2；

H——微元段单位长度的线性阻尼系数，$N \cdot s/m^2$。

目前，隔水管纵向振动研究主要侧重于隔水管安装及紧急回收与撤离两个阶段，分别对应两种悬挂方式：软悬挂与硬悬挂。“软悬挂”是指隔水管顶部连接在钻井平台升沉补偿器上[96]，“硬悬挂”是指隔水管顶端直接坐在钻井平台卡瓦上[97]，两种悬挂方式如图6所示。

“紧急避台”是隔水管设计中的重要环节，相关学者通过采用理论分析、数值模拟等综合方法对两种悬挂模式下的隔水管纵向振动力学行为进行了大量研究，主要包括悬挂条件下隔水管轴向动力放大系数[98]、轴向振动位移与应力[98]、固有频率[99]、隔水管避台撤离管理策略[100]、安装作业窗口[101]、浮力块配置的影响[102]、隔水管寿命管理[103]等内容。通过研究发现，两种悬挂模式下隔水管的纵向振动力学特性有较大差别，具体来讲，软悬挂操作主要受上球铰转角与伸缩节冲程的限制，硬悬挂操作主要受隔水管过度张力波动的限制，相对于硬悬挂模式，软悬挂模式比较安全可靠。推荐采用隔水管软悬挂模式实施避台风撤离。如果不具备软悬挂条件，则可部分回收隔水管，而将剩余的部分隔水管采用硬悬挂模式实施避台风撤离，也是一种经济可行的技术方案。

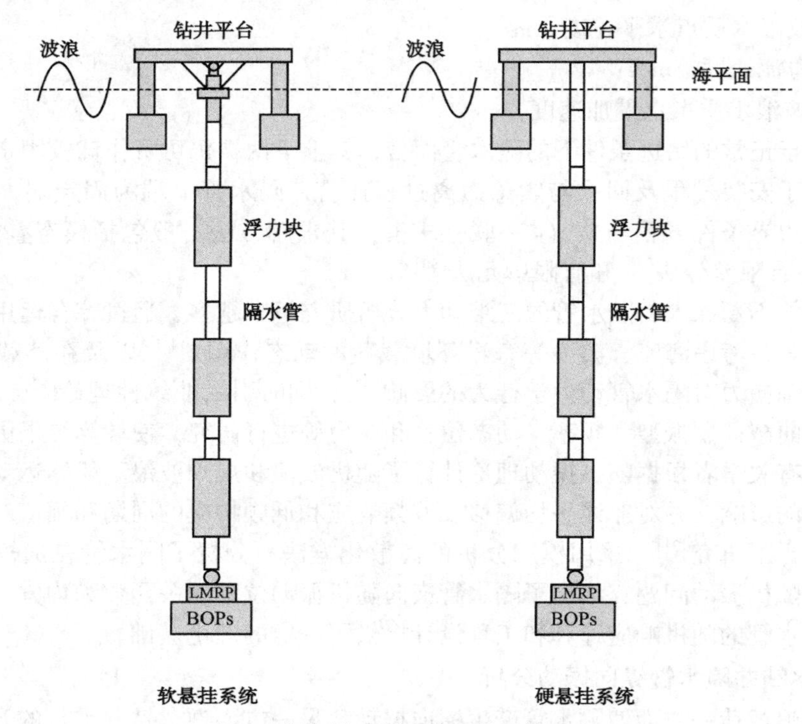

图 6　深水钻井隔水管两种悬挂方式

3.5　深水钻井隔水管耦合振动分析

深水钻井隔水管的耦合振动主要包括两种：第一种是隔水管与钻井液的耦合振动，第二种是隔水管参数激励与强迫激励的耦合振动。前者是指在深水钻井作业过程中，位于隔水管与钻柱环空之间的钻井液自下而上由水下井口返至钻井平台，上返的钻井液与横向振动的隔水管二者产生耦合振动[104]；所谓“强迫激励”是指隔水管在动态波流联合作用下的受迫振动，所谓“参数激励”是指由于波浪动态作用导致钻井平台升沉运动，从而对隔水管顶部施加位移时程激励，在这两种激励共同作用下发生上述第二种隔水管耦合振动[105]。

与隔水管的横向振动相比较，第一种耦合振动方程中增加了因隔水管内流体向上流动产生的水平牵连惯性力、相对惯性力和科氏惯性力的影响因素，其控制方程如下[104]：

$$EI\frac{\partial^4 y}{\partial x^4} + (m_d v^2 - T)\frac{\partial^2 y}{\partial x^2} + (m_c + m_d)\frac{\partial^2 y}{\partial t^2} + 2m_d v\frac{\partial^2 y}{\partial x \partial t} = F(x,\ t) \tag{7}$$

式中　m_d——单位长度隔水管内钻井液的质量，kg；

v——钻井液上返流速，m/s；

$m_d v^2 \dfrac{\partial^2 y}{\partial x^2}$——水平牵连惯性力，N；

$m_d \dfrac{\partial^2 y}{\partial t^2}$——相对惯性力的水平分量，N；

$2m_d v \dfrac{\partial^2 y}{\partial x \partial t}$——科式惯性力，N。

由于钻井液的影响，隔水管横向振动的最大位移与波高呈正相关，与顶张力、环空钻井液排量呈负相关；当由隔水管壁厚变化引起的隔水管横向振动固有频率与波浪横向振动频率相接近时，横向振动位移急剧增大；耦合振动条件下隔水管的振动特性不仅取决于隔水管本身的力学与几何特性，而且与隔水管所处的外部环境因素密切相关[106]。因此，在隔水管的设计与管理过程中，全面掌握外部因素对于正确评估隔水管疲劳寿命、防止隔水管共振、延长隔水管服役期限等具有重要的实际意义。

第二种隔水管耦合振动控制方程如下：

$$EI\frac{\partial^4 y(x,\ t)}{\partial x^4} - [T(x) + S\cos\Omega t]\frac{\partial^2 y(x,\ t)}{\partial x^2} + c\frac{\partial y(x,\ t)}{\partial t} + m\frac{\partial^2 y(x,\ t)}{\partial t^2} = F_y(x,\ t) \quad (8)$$

式中 $T(x)$——隔水管沿水深变化的轴向力，N；

S——隔水管动态张力振动幅值，N；

Ω——隔水管顶端激振力频率，rad/s。

这时，垂直于水流方向的波流联合作用力可分为两部分：一是由于涡街泄放过程产生的涡激升力 $F_L(x,\ t)$，二是由于隔水管 y 向运动而产生的流体阻尼力 $F_r(x,\ t)$，即：

$$F_y(x,\ t) = F_L(x,\ t) - F_r(x,\ t) \quad (9)$$

式（8）可采用龙格—库塔法[105,107]进行求解，通过采用摄动法可将控制方程转化为马蒂厄方程，并据此分析隔水管参数激励的不稳定区，进而得到隔水管发生参激共振的顶部激励频率。以此频率为基础可对隔水管参激振动最危险的工况（参激共振）进行分析，讨论振动的模态响应历程，以及弯矩和剪力的变化。需要说明的是，虽然简支隔水管模型是多自由度系统，但由于各自由度之间不耦合，因此可用单自由度系统方法来分析[105]。波流联合作用下隔水管耦合振动响应的非线性特征更为显著，相应的幅值变化及相位和振动响应频率的变化更为复杂，海流单独作用下隔水管总剪力主要由第一阶模态谐振引起，波流联合作用下各阶振动模态对总剪力均有影响，在计算波流联合作用下隔水管剪力时宜多考虑前几阶振型。

3.6 深水钻井隔水管涡激振动分析

从流体力学角度上来说，在一定的恒定流速作用下的任何非流线型物体，其两侧均会交替地产生脱离结构物表面的旋涡。对于海洋工程上普遍采用的圆柱形断面结构物，这种交替发放的旋涡又会在柱体上生成周期性变化的顺流向和横流向脉动压力。对于深水钻井中的送入管柱及隔水管而言，脉动流体力将引发管柱的周期性振动，这种规律性的柱状体振动反过来又会改变其尾流的旋涡发放形态，这种流体-结构物相互作用的问题被称作涡激振动[35]（Vortex-Induced Vibration，简称“VIV”）。目前对隔水管涡激振动方面的研究主要集中在涡激动力响应计算、数值模拟分析、涡激抑制和涡激疲劳损伤预测等方面。

由于深水海域的海流速度通常比浅水海域要高，隔水管长度的增加降低了隔水管本身的固有频率，从而降低了激励 VIV 的流速阈值，因此深水条件下隔水管的涡激振动问题更加严重。目前隔水管的 VIV 研究主要以尾流振子模型[108-110]为基础，通过建立隔水管涡激振动问题的数学与力学模型来阐述隔水管的涡激振动机理，主要考虑旋涡形成[111]、泄放规律[112]、尾流结构特征[113]、升曳力系数[114]等影响因素。在 VIV 数值模拟方面通过借助 CFD 软件模拟洲涡发放过程，对隔水管在海流流场、海浪流场以及海流与海浪的混合流场中进行流—固耦合分析，将所得结果与理论分析及实验数据进行对比，进一步加深对涡激振动发生时隔水管受力、振幅及旋涡脱落等方面的认识。目前常用的 VIV 模拟软件包括 Ansys+CFX，Fluent+Abaqus，Adina，COMSOL Multiphysics，SHEAR 7 等，但目前在流固耦合作用机理、结构动态模型、获取全尺寸实验数据、实验数据解释及预测技术等方面仍存在较大困难[115]。

涡激振动会导致隔水管产生疲劳损伤，影响其使用寿命，因此相关学者将疲劳可靠性理论应用在海洋工程中，以计算构件和系统的可靠度。疲劳分析、断裂力学、随机载荷响应分析和以概率为基础的设计和分析方法的发展，为分析预测隔水管疲劳寿命奠定了良好的理论基础[116]。在深水钻井过程中，波流、风载、现场作业载荷等对隔水管的力学行为影响非常复杂，尤其是隔水管柱连接点及焊接部位的疲劳寿命预测精度，受到环境载荷、隔水管整体响应、疲劳强度及损伤积累等不确定性因素的严重制约。因此，关于隔水管疲劳寿命计算方法提出了雨流计数法、S—N 曲线法和 Miner 线性累积损伤法[115]等多种方法，不同方法侧重点不同。例如，断裂力学方法与传统的 S—N 曲线相比较，前者通过考虑裂纹扩展模拟及裂纹非稳定性来确定构件裂纹的临界尺寸，而后者表征的是在一定的循环特征条件下标准试件的疲劳强度与疲劳寿命之间的关系曲线。可见不同的评价方法对应不同的要求，对深水钻井隔水管系统的疲劳评估，应根据具体的结构和环境情况，选取适当的评估方法[117]。在实际工程中，可通过安装 VIV 抑制装置[118]来抑制隔水管的涡激振动，以提高其疲劳使用寿命。

4　深水井筒完整性研究

“油气井完整性”指在井眼整个寿命期间，应用技术、操作和组织措施最大限度地减少地层流体不可控流动的风险，以确保油气井始终处于安全可靠的状态[119]。在深水钻井过程中，浅层气与浅水流侵入井筒、深水固井失败及井筒环空圈闭压力升高等均会引发严重的井筒完整性问题。据挪威石油安全管理局（PSA）对海上 106 口不同开发年限和生产类别的井进行井筒完整性调查发现，18%存在井筒完整性问题，其中 7%因井筒不完整而被迫关井，对环境和经济造成了重大损失[120,121]。技术套管固井后，水下井筒会受到诸如软泥页岩膨胀、地层蠕动及非均匀地应力等复杂海洋地质环境的影响，套管柱载荷分析面临较大挑战[122]。同时，在生产及测试过程中，由于温度升高使各层套管环空之间的流体膨胀，导致环空压力升高及井口抬升[123,124]，对深水井筒完整性构成较大威胁。因此，深入探讨深水钻井套管载荷分布特征，分析掌握诸多因素对套管强度及井身结构设计的影响规律[125,126]，对于提高深水钻井作业和水下井筒完整性具有重要的理论和工程意义。目前关于深水井筒完整性方面的研究，主要侧重于井筒温度分布规律预测，多层套管柱环空压力计算，环空圈闭压力增加导致套管破坏等。

4.1　井筒温度分布规律研究

井筒温度分布规律是深水钻井套管设计的重要组成部分，对于分析深水钻井工况下套管的内外压力和轴向力分布特征并进行强度设计具有重要的实际意义。

准确分析地层的非稳态传热是计算井筒温度分布规律的基础。国外学者首先把井筒内的流体传热与地层瞬态导热相结合，先后提出了井筒温度预测方法、综合传热系数及总传热系数的计算公式[127-129]。后续井筒与地层间传热规律的研究主要集中在提高公式计算精度与适用范围方面[130-132]，但由于地层非稳态传热的复杂性，目前仍需要在公式计算精度方面进行深入研究。早期的井筒循环温度分布计算主要依靠 API 简易估算法[133]，后来相关学者对固井过程中的井筒温度进行了数值模拟，并提出了瞬态和拟稳态条件下预测井筒循环温度的方法及拟稳态条件下的解析解[134]。由于井筒循环温度受到井深、套管、井眼尺寸、泵速、时间、钻井液参数、储层物理参数等诸多因素的综合影响，通常将温度预测模型中地层的影响转化成热量随时间变化的边界进行分析，并且忽略井筒内钻井液的轴向热传导，基于上述研究方法，先后得到了井壁温度预测方法[135]、地层传热拟合公式[136]和不同循环情况下钻井液循环温度的解析解[137]。国内相关学者通过建立循环和静止过程中井内温度分布的预测模型，对影响井筒温度的因素进行了敏感性分析，并利用矩阵变换将二维传热问题简化为一维传热问题，从而提高了求解速度[135,136]。但总体上讲，目前我国关于深水钻井循环温度预测的研究还未能充分考虑深水井筒的特点，也不能准确预测套管环空的循环温度分布，一些理论和技术问题仍需进一步探索研究。

4.2 深水井筒套管环空压力分析

水下防喷器安装完成后，由于井口没有泄压阀，在深水井测试和采油期间由于温度变化引起的密闭环空内流体升温膨胀会使套管环空产生附加载荷[138]，严重威胁深水管柱的安全服役寿命与井筒完整性。

国内外对陆地和浅水高温高压井因温度升高引起的附加载荷进行了研究。文献[139]首先给出了多层套管柱附加载荷计算模型，并通过与单层套管计算结果进行对比分析，发现由于温度变化产生的多层自由段套管环空压力会产生较大的套管附加载荷，在套管设计时应予以考虑。一般情况下，套管环空压力受流体温度、压力、热膨胀系数和压缩系数的影响，工程应用中通常根据经验来选取模型的某些参数值，严重降低了模型的计算精度。目前计算自由套管段附加载荷的模型基本上是沿用 Helal 和 Admas 给出的方法，但 Oudeman 通过现场试验发现现有的套管环空压力计算模型高估了套管环空压力值，并认为环空流体参数取值是影响模型精度的原因[140]。除了研究理论计算以外，Leach 和 Adams 总结了五种预防自由套管段附加载荷的方法，包括：改变水泥返高、设计井口卸压通道、提高套管强度、填充可破裂泡沫球和注入压缩流体[141]。同时，该文献也详细讨论了填充可破裂泡沫球的控制方法，还讨论了填充可压缩低密度流体的预防方法，这两种方法操作简单、成本低，在室内和现场试验中得到了良好效果。国内针对深水钻井作业特点，考虑隔水管段井筒传热、钻井液增注、套管及其环空等因素的影响，建立了套管环空循环温度预测模型[142]。通过计算分析发现，隔水管增注对海水段井筒温度影响较大，对于地层段井筒温度影响较小，其影响程度随增注量的增大而增加；套管及其环空流体对地层段井筒内流体起保温作用，与生产过程相比，钻井过程中井筒环空流体温度更低，与套管环空流体的温度差更大。同时，考虑管柱与环空流体的相互作用，建立了多层套管柱环空压力计算模型，从理论上解释了深水套管环空圈闭压力产生附加载荷的机理。今后仍需要进一步针对不同深水钻完井作业工况，考虑温度、力学、流体热膨胀与压缩等诸多效应，建立深水井筒温差产生附加载荷的计算模型，并形成相应的设计控制新技术，指导工程实践。

4.3 深水井筒套管应力分布研究

深水井筒产出的高温流体不仅会引起套管环空附加压力，还会引起套管与地层之间的附加载荷，而套管的热膨胀还会加剧套管环空附加压力，因而有必要研究套管-水泥环-地层系统热膨胀产生的载荷与位移。国外早期研究了温度对套管强度的影响[143]，随着研究的深入，人们意识到当近井地带井筒温度升值较大时，套管-水泥环-地层系统同时膨胀形成的附加应力会造成严重的井筒完整性破坏。国外学者 Rahman 在热弹性力学的轴对称理论基础上，基于前人在温度场方面的研究成果，推导了多层介质的应力分布计算模型，为后来的研究打下了基础[144]。目前温度载荷下套管的受力问题在国内的研究也比较活跃，一些学者在热弹性力学理论的基础上建立了套管与弹性地层的相互作用模型，并根据拟合的温度函数推导了套管与地层耦合系统的热应力和热位移计算公式[145,146]。其中，文献[147]分析了系统的热应力和热位移径向分布规律，以及套管温升、弹性模量、热膨胀系数、壁厚、水泥环和地层的弹性模量和泊松比等参数对套管热应力和热位移的影响；文献[148]将岩层系统分成冷区和热区，并讨论了冷热地层刚度对套管载荷的影响，但求解过程中温度分布函数是假设的，没有利用井筒传热方面的研究成果。还有部分学者通过数值方法研究了套管在高温井及热采井中的破坏机理与预防方法[149-151]。总体上讲，目前主要通过数值方法来模拟和求解套管应力，但现有的理论计算公式难以充分借鉴井筒温度分布的研究成果，且忽略了套管内压的影响，今后仍需在此方面进行深入探索。

5 深水管柱力学模拟实验方法研究

深水钻井管柱在服役状态下所受的外载荷包括平台偏移、横向波流力、轴向拉压、内压外挤等，会产生磨损、断裂、挤毁等不同形式的失效形式。除了理论分析以外，通过试验方法模拟分析深水管柱力学行为，对于提高深水管柱力学的综合研究水平具有必要性。然而，外压施加一直是深水钻井隔水管模拟实验过程中的难点，外压加载一方面要求设备具有足够的空间以安装模拟试件，另一方面又要求实验装备具有较大的承压能力。经调研，巴西里约热内卢大学装备有 1 套海洋深水模拟实验装置，可提供 10MPa 的水压，最大模拟水深仅为 1000m，而目前第 5 代和第 6 代钻井平台（船）的钻井水深为 3000m（最大水压为 30MPa）。为此，中国石油大学（北京）相关学科研制了“深水管柱力学模拟实验装置”，科研提供 30MPa 的额定工作压力，能够有效模拟 3000m 的钻井作业水深，并且能够对试验管柱试件施加内外压力、轴向力及横向力[152]，主要模拟实验设备如图 7 所示。

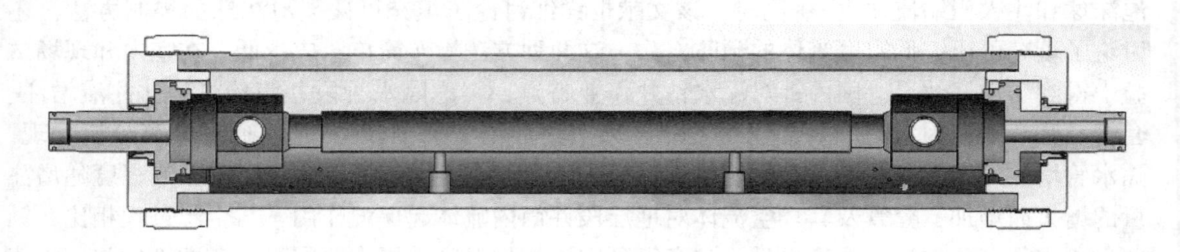

图 7　深水管柱力学模拟实验系统

该模拟实验装置主要有承压主缸筒、液压作用器、液压伺服控制系统、数据采集系统及相应的管路与控制线等组成，承压主缸筒主要由主缸筒、轴向活塞、活塞杆、端部卡箍及相应的密封元件等组成，是系统承压的主体组成部分；液压作用器主要由水—油伺服增压作用缸及伺服控制阀组成，可以输出试验要求所需要的压力和流量；液压伺服控制系统主要由伺服系统控制柜及相应的软件组成，向液压伺服阀发出指令，控制伺服阀的动作及幅度；数据采集系统主要由压力、应变、位移等数据采集仪器组成，将采集得到的信号存入电脑以备后续处理。模拟实验试件通过两端的连接头及连接销与活塞杆相连，实现轴向加载，为了配合不同试验的需要，系统配备了一套液压比例控制系统和一套液压伺服控制系统、一套静态数据采集仪器及一套动态数据采集仪器。另外，系统配备了电控可移动的拆装小车与拆装吊车，可以方便地对实验装置和模拟试件进行拆装，其主要技术参数如表 1 和表 2 所示。

表 1　模拟实验系统的主要结构参数

主缸筒				端部活塞		
外径（mm）	内径（mm）	长度（mm）	耐压值（MPa）	直径（mm）	活塞杆直径（mm）	行程（mm）
1500	1200	8000	30	1000	400	220

表 2　液压控制系统的主要参数

设备	油缸内径（mm）	活塞杆径（mm）	测控行程（mm）	最大速度（m/s）	油水增压比	输出最高水压（MPa）	限定工作水压（MPa）	阀压力等级（MPa）
比例增压泵	180	120	700	0.5	2.25	56	30	31.5
伺服增压缸	180	120	700	1.0	1.65	46	30	28
横向增压缸	140	80	100	—	—	25	25	28

采用水下应变测试技术[153]，该实验系统可以完成深水管柱的抗外挤、抗内压、轴向拉压、横向弯曲、组合实验、内外压下的旋转弯曲及弯曲疲劳、管材 S—N 曲线测定及深水设备的密封性能测试等系列实验。

另外，文献[154]推导了光纤光栅的中心波长与隔水管最大应力和弯矩之间的关系，通过将光纤光栅传感器沿隔水管周向布置，可以实现对隔水管应力的实时监测，该种类型传感器的最大优点是可免受海水腐蚀；文献[155]介绍了基于钻井和海洋工况耦合作用下的深水钻井隔水管力学性能实验，通过利用光纤光栅传感器对隔水管的力学特性进行测量，发现了剪切流作用下深水钻井隔水管的“三分之一效应”；文献[156]介绍了一种可以对深水钻井工况下隔水管进行振动特性模拟的实验装置及实验方法，利用该实验装置可以考察不同长度隔水管在不同顶张力、钻井液密度、排量及转速下的隔水管振动特性的影响；文献[157]针对海洋深水钻井隔水管在疲劳实验过程中的加载问题，提出了一种疲劳实验载荷的计算方法，并用该方法计算了试样在各种应力幅下进行疲劳实验时的内压数据，可作为深水钻井隔水管疲劳实验外载荷的计算依据；文献[158]针对海洋隔水管疲劳实验问题，特别对轴向拉伸疲劳实验和共振弯曲疲劳实验，阐述了不同试验方法的试验原理、试验机模型、优缺点及使用情况，最后确定了隔水管材料和结构疲劳实验方法，以及相应的实验参数，可为我国隔水管疲劳实验机设计与疲劳实验研究提供参考。

6　结束语

（1）深水钻井管柱力学研究涉及内容较多，研究方法也比较先进，并取得了重要研究进展，对深水钻井理论创新与技术进步产生了重要的推动作用。

（2）水下井口是深水油气工程关键建设内容之一。为提高水下井口的稳定性和安全可靠性，需要准确获取海底岩土层的力学参数，进而对深水导管入泥深度、喷射安装工艺及桩土相互作用规律等进行详细的系统研究，不断提高相应的设计控制技术水平。

（3）深水钻井隔水管安装是深水钻井作业的关键环节之一。为确定更为合理的深水隔水管安装作业窗口，需要考虑环境载荷与作业因素的综合影响，对深水钻井隔水管安装过程中的力学行为进行理论分析及数值和物理模拟。

（4）深水钻井隔水管的涡激振动和疲劳寿命评估，事关隔水管的安全服役寿命，在流固耦合机理、结构动态模型、全尺寸模拟实验及预测模型等方面仍需开展深入研究，并根据具体的结构和环境情况，选取相适应的的计算方法对其疲劳寿命进行综合评估。

（5）深水井筒完整性科学评价与设计控制问题十分复杂，在井筒温度、环空压力及套管应力等方面仍需进行深入研究。

（6）模拟实验是进行深水管柱力学研究的必要手段之一，需要根据不同要求进行管柱力学模拟实验设计，获取有效的模拟实验数据，以期对相关理论模型进行验证或修正。

参 考 文 献

[1] P. A. Watson, A. W. Iyoho, R. A. Meize, et al. Management Issues and Technical Experience in Deep-water and Ultra-Deep-water Drilling [C]. Offshore Technology Conference, 2-5 May, Houston, Texas, OTC17119, 2005.

[2] J. Shaughnessy, W. Daugherty, T. Graff, et al. More Ultra-deep-water Drilling Problems [C]. SPE/IADC Drilling Conference, 20-22 February, Amsterdam, Netherlands, SPE 105792, 2007.

[3] P. A. Charlez, A. Simondin. A Collection of Innovative Answers to Solve the Main Problematic Encountered When Drilling Deep Water Prospects [C]. Offshore Technology Conference, 5-8 May, Houston, Texas, OTC 15234, 2003.

[4] J. C. Cunha. Innovative Design for Deep-water Exploratory Well [C]. IADC/SPE Drilling Conference, 2-4 March, Dallas, Texas, SPE 87154, 2004.

[5] 杨进，曹式敬．深水石油钻井技术现状及发展趋势［J］. 石油钻采工艺，2008，30（2）：10-13.

[6] 张晓东，王海娟．深水钻井技术进展与展望［J］. 天然气工业，2010，30（9）：46-48.

[7] 王友华，王文海，蒋兴迅．南海深水钻井作业面临的挑战和对策［J］. 石油钻探技术，2011，39（2）：50-55.

[8] 杨金华．全球深水钻井现状与前景［J］. 石油科技论坛，2014，1：46-50.

[9] R. D. Beck, C. W. Jackson, T. K. Hamilton. Reliable Deep-water Structural Casing Installation Using Controlled Jetting [C]. SPE Annual Technical Conference and Exhibition, 6-9 October, Dallas, Texas, SPE 22542, 1991.

[10] G. W. King, I. J. Solomon, The Instrumentation of the Conductor of a Subsea Well in the North Sea To Measure the Installed Conditions and Behavior under Load [J]. SPE Drilling & Completion, 1995, 10 (4): 265-270.

[11] P. Sparrevik. Suction Pile Technology and Installation in Deep Waters [C]. Offshore Technology Conference, 6-9 May, Houston, Texas, OTC14241, 2002.

[12] J. Choe, H. C. J. Wold. Unconventional Method of Conductor Installation to Solve Shallow Water Flow Problems [C]. SPE Annual Technical Conference and Exhibition, 5-8 October, San Antonio, Texas, SPE 38625, 1997.

[13] R. V. Noort, R. Murray, J. Wise, et al. Conductor Pre-Installation, Deep-water Brazil [C]. Offshore Technology Conference, 4-7 May, Houston, Texas, OTC 20005, 2009.

[14] B. Mackenzie, M. Francis, I. Garrett, et al. Conductor Jetting Experiences in Deep-water Offshore Ghana-An Investigation into Geotechnical and Operational Influences on Success and Establishment of Future Best Practice [C]. Offshore Site Investigation and Geotechnics: Integrated Technologies - Presentand Future, 12 - 14 September, London, UK, SUT-OSIG-12-13, 2012.
[15] 黄小龙，刘正礼，陈建兵．深水结构导管喷射钻井技术研究［J］．科技创新导报，2003，1：88.
[16] 陈彬，刘正礼，罗俊丰，等．南海深水钻井表层导管喷射作业实践［J］．石油天然气学报（江汉石油学院学报），2014，36（9）：109-111.
[17] R. J. Adams. Proven Landing String Design for Ultra-deep-water Application [J]. World Oil, 2001, 222 (7): 73-79.
[18] J. W. Breihan, J. A. Altermann, M. J. Jellison. Landing Tubulars Design, Manufacture, Inspection and Use Issues [C]. SPE/IADC Drilling Conference, 27 February - 1 March, Amsterdam, Netherlands, SPE 67723, 2001.
[19] A. Cantrell, C. Beiriger, D. Everage, et al. Design and Qualification of Critical Landing String Assemblies for Deep-water [C]. IADC/SPE Drilling Conference, 4-6 March, Orlando, Florida, USA, SPE 112787, 2008.
[20] B. Simpson, M. L. Payne, M. J. Jellison, et al. 2000000-lbf Landing String Developments: Novel Slipless Technology Extends the Deep-water Operating Envelope [J]. SPE Drilling and Completion, 2005, 20 (2): 109-122.
[21] 路宝平．深水钻井关键技术与装备［M］．北京：中国石化出版社，2014.
[22] 陈建兵．深水探井钻井工程设计方法［M］．北京：石油工业出版社，2014.
[23] O. Egeland, T. Wiik, B. J. Natvig, et al. Dynamic Analysis of Marine Risers [C]. Offshore South East Asia Show, 9-12 February, Singapore, SPE10420, 1982.
[24] R. A. Khana, A. Kaurb, S. P. Singhc, et al. Nonlinear Dynamic Analysis of Marine Risers under Random Loads for Deep-water Fields in Indian Offshore [J]. Procedia Engineering, 2006, 14: 1334-1342.
[25] 畅元江，陈国明，许亮斌，等．深水顶部张紧钻井隔水管非线性静力分析［J］．中国海上油气，2007，19（3）：203-207.
[26] 张炜，高德利，范春英．钻井隔水管挤毁分析［J］．钻采工艺，2010，33（4）：74-76.
[27] 鞠少栋，畅元江，陈国明，等．超深水钻井作业隔水管顶张力确定方法［J］．海洋工程，2011，29（1）：100-104.
[28] American Petroleum Institute. API RP 2A-WSD-2000, Recommended Practice for Planning, Designing and Constructing Fixed Offshore Platforms-working Stress Design [S]. Washington, D. C, 2000.
[29] American Petroleum Institute. API RP 2A-LRFD-1993: Recommended Practice for Planning, Designing and Constructing Fixed Offshore Platforms-Load and Resistance Factor Design [S], Washington, D. C, 1993.
[30] API BULL 5C3. Bulletin on Formulas and Calculations for Casing, Tubing, Drill Pipe and Line Pipe Properties [S], Washington, D. C, 1994.
[31] DET NORSKE VERITAS. Offshore Standard DNV-OS-F101 Submarine Pipeline System [S], DNV, 2000.
[32] American Petroleum Institute. API RP 16Q 93: Recommended Practice for Design, Selection, Operation and Maintenance of Marine Drilling Riser Systems [S], Washington, D. C, 1993.
[33] 朱艳蓉．海洋工程波浪力学［M］．天津：天津大学出版社，1991.
[34] 方华灿．海洋石油钻采设备理论基础［M］．北京：石油工业出版社，1984.
[35] 吕苗荣．石油工程管柱力学［M］．北京：中国石化出版社，2012.
[36] T. J. Akers. Jetting of Structural Casing in Deep-water Environments: Job Design and Operational Practice [J]. SPE Drilling & Completion, 2008, 23 (1): 29-40.
[37] J. Yang, S. J. Liu, J. L Zhou, et al. Research of Conductor Setting Depth Using Jetting in the Surface of Deep-water [C]. International Oil and Gas Conference and Exhibition in China, 8-10 June, Beijing, China,

SPE130523, 2010.
[38] M. F. Pereira, A. de Silvio, J. C. Ruiz. Deep-water Conductor Pre-Installation for First TLWP in Brazil [C]. OTC Brazil, 29-31 October, Rio de Janeiro, Brazil, OTC24291, 2013.
[39] 叶庆志，王建华．喷射导管安装后实时承载力分析 [J]. 低温建筑技术，2014，1：112-114.
[40] 袁光宇．我国海上钻井隔水导管使用现状及发展趋势 [J]. 长江大学学报（自然科学版），2012，9（7）：102-103，123.
[41] 易先忠，钟守炎，水运震．井下动力钻具的现状与发展 [J]. 石油机械，1995，23（11）：53-57.
[42] J. D. Hughes, R. A. Coleman, R. P. Herrmann, et al. Batch Drilling and Positioning of Subsea Wells in the South China Sea [C]. International Meeting on Petroleum Engineering, 14-17 November, Beijing, China, SPE 29909, 1995.
[43] R. P. Herrmann, R. A. Coleman, J. D. Hughes, et al. Liuhua 11-1 Development-Subsea Conductor Installation in the South China Sea [C]. Offshore Technology Conference, 6-9 May, Houston, Texas, OTC 8174, 1996.
[44] 林广辉．随钻下套管技术在我国南海油田的首次应用 [J]. 中国海上油气（工程），1996，8（1）：53-58.
[45] 周建良，杨进，严德，等．深水表层导管下入方式适应性分析 [J]. 长江大学学报（自科版），2013，10（2）：66-69.
[46] A. J. Dick, R. Cassells. Developing HP/HT Landing String Technologies [C]. Offshore Technology Conference, 3-6 May, Houston, Texas, OTC16208, 2004.
[47] F. Amezaga, J. R Rials, K. Heidecke. Landing String Slip System: State-of-the-Art Design to Minimise Pipe Crushing Problems [C]. SPE Asia Pacific Oil & Gas Conference and Exhibition, 14-16 October, Adelaide, Australia, SPE 171408, 2014.
[48] M. J. Jellison, M. L. Payne, J. S. Shepard, et al. Next Generation Drill Pipe for Extended Reach, Deep-water and Ultra-deep Drilling [C]. Offshore Technology Conference, 5-8 May, Houston, Texas, OTC 15327, 2003.
[49] A. Cantrell, C. Beiriger, D. Everage, et al. Design and Qualification of Critical Landing String Assemblies for Deep-water [C]. IADC/SPE Drilling Conference, 4-6 March, Orlando, Florida, USA, SPE 112787, 2008.
[50]《钻井手册》编写组．钻井手册 [M]. 北京：石油工业出版社，1990.
[51] 董星亮，曹式敬，唐海雄，等．海洋钻井手册 [M]. 北京：石油工业出版社，1996.
[52] American Petroleum Institute. API RP 7G: Recommended Practice for Drill Stem Design and Operating Limits [S]. American Petroleum Institute. Washington D. C., 1998.
[53] S. Udaya B., P. Mike, S. P. V., et al. Advanced Slip-crushing Considerations for Deep-water Drilling [J]. SPE Drilling and Completion, 2002, 17 (4): 210-223.
[54] S. Burnie, P. Michael L., J. Michael J., et al. 2000000-lbf Landing String Developments: Novel Slipless Technology Extends the Deep-water Operating Envelope [J]. SPE Drilling and Completion, 2005, 20 (2): 109-122.
[55] J. N. Brock, R. B. Chandler, M. J. Jellison, et al. 2 Million-lbf Slip-Based Landing String System Pushes the Limit of Deep-water Casing Running [C]. Offshore Technology Conference, 30 April-3 May, Houston, Texas, U. S. A, OTC 18496, 2007.
[56] D. W. Bradford, M. L. Payne, D. E. Schultz, et al. Defining the Limits of Tubular Handling Equipment at Extreme Tension Loadings [J]. SPE Drilling & Completion, 2009, 24 (1): 71-88.
[57] S. D. Everage, N. J. Zheng, S. E. Ellis. Evaluation of Heave-induced Dynamic Loading on Deep-water Landing Strings [J]. SPE Drilling & Completion, 2005, 20 (4): 230-238.
[58] N. J. Zheng, J. M. Baker, S. D. Everage. Further Consideration of Heave-Induced Dynamic Loading on Deep-water Landing Strings [C]. SPE/IADC Drilling Conference, 23-25 February, Amsterdam, Netherlands, SPE

92309, 2005.
[59] 高德利，张辉．无隔水管深水钻井作业管柱的力学分析［J］. 科技导报，2012，30（4）：37-42.
[60] H. Zhang, D. L. Gao, H. X. Tang. Landing string design and strength check in ultra-deep-water condition. Journal of Natural Gas Science and Engineering, 2010, 2 (4): 178-182.
[61] 张辉．深水导管设计与安装力学行为研究［D］. 北京：中国石油大学（北京），2010.
[62] 张辉，高德利，唐海雄．喷射安装导管作业中喷射管串力学分析［J］. 西南石油大学学报（自然科学版），2009，31（6）：148-151.
[63] 张辉，高德利，唐海雄，等．深水导管喷射安装过程中管柱力学分析［J］. 石油学报，2010，31（3）：516-520.
[64] B. Zhou, J. Yang, Y. J. Xu, et al. Experimental Research on Structural Casing Soaking Time in Deep-water Drilling [C]. SPE Deep-water Drilling and Completions Conference, 10-11 September, Galveston, Texas, USA, SPE 170317, 2014.
[65] P. Jeanjean. Innovative Design Method for Deep-water Surface Casing [C]. SPE Annual Technical Conference and Exhibition, 29 September-2 October, San Antonio, Texas, SPE 77357, 2002.
[66] 刘正礼，唐海雄，王跃曾，等．深水喷射导管实用设计方法［J］. 长江大学学报，2010，07（1）：189-191.
[67] 付英军，姜伟，朱荣东．深水表层导管安装方法及风险控制技术研究［J］. 石油天然气学报（江汉石油学院学报），2011，33（6）：153-157.
[68] 刘书杰，杨进，周建良，等．深水海底浅层喷射钻进过程中钻压与钻速关系［J］. 石油钻采工艺，2011，33（1）：12-15.
[69] 汪顺文，杨进，刘正礼，等．深水表层导管喷射钻进机理研究［J］. 石油天然气学报，2012，34（8）：157-160.
[70] 周建良．深水表层导管喷射钻进过程中钻井液排量优化研究［J］. 中国海上油气，2012，24（4）：50-52.
[71] 杨进，周波，刘书杰，等．深水喷射下表层导管合理钻头伸出量计算［J］. 石油勘探与开发，2013，40（3）：367-370.
[72]. P. Jeanjean. Re-Assessment of P-Y Curves for Soft Clays from Centrifuge Testing and Finite Element Modeling [C]. Offshore Technology Conference, 4-7 May, Houston, Texas, OTC 20158, 2009.
[73] G. W. King, I. J. Solomon. The Instrumentation of the Conductor of a Subsea Well in the North Sea to Measure the Installed Conditions and Behavior under Load [C]. Offshore Technology Conference, 3-6 May, Houston, Texas, OTC 7232, 1993.
[74] T. G. Evans, S. Feyereisen, G. Rheaume. Axial Capacities of Jetted Well Conductors in Angola [C]. Offshore Site Investigation and Geotechnics "Diversity and Sustainability"; Proceedings of an International Conference, 26-28 November, London, UK, OSIG-02-325, 2002.
[75] 杨进，彭苏萍，周建，等．海上钻井隔水导管最小入泥深度研究［J］. 石油钻采工艺，2002，24（2）：79-81.
[76] 杨进．海上钻井隔水导管极限承载力计算［J］. 石油钻采工艺，2003，25（5）：28-30.
[77] 苏堪华，管志川，苏义脑．深水钻井导管喷射下入深度确定方法［J］. 中国石油大学学报（自然科学版），2008，32（4）：47-50.
[78] 唐海雄，罗俊丰，叶吉华，等．南海超深水喷射钻井导管入泥深度设计方法［J］. 石油天然气学报（江汉石油学院学报），2011，33（3）：147-151.
[79] 管志川，苏堪华，苏义脑．深水钻井导管和表层套管横向承载能力分析［J］. 石油学报，2009，30（2）：285-290.
[80] 王宴滨，高德利，房军．考虑不同桩土接触模型的深水钻井导管承载能力数值分析［J］. 中国海上油

气，2014，26（5）：76-81.

[81] C. P. Sparks. The Influence of Tension, Pressure and Weight on Pipe and Riser Deformations [J]. Journal of Energy Resources Technology, 1984, 106 (1): 46-54.

[82] J. L. Thorogood, A. S. Train, A. J. Adams. Deep water Riser System Design and Management [C]. IADC/SPE Drilling Conference, 3-6 March, Dallas, Texas, IADC/SPE 39295, 1998.

[83] Y. B. Wang, D. L. Gao, J. Fang. Static Analysis of Deep-water Marine Riser Subjected to Both Axial and Lateral Forces in Its Installation [J]. Journal of Natural Gas Science and Engineering, 2014, 19: 84-90.

[84] 李妍，吴艳新，高德利．深水钻井隔水管纵横弯曲变形解析［J］．石油矿场机械，2011，40（7）：21-24.

[85] A Ertas, T. J. Kozik. Numeric Solution Techniques for Dynamic Analysis of Marine Riser [J]. Journal of Energy Resources technology, 1987, 109 (1): 1-5.

[86] 贾星兰，方华灿．海洋钻井隔水管的动力响应［J］．石油机械，1995，23（8）：18-22，28.

[87] Y. B. Wang, D. L. Gao, J. Fang. Study on Lateral Vibration Analysis of Marine Riser in Installation via Vibrational Approach [J]. Journal of Natural Gas Science and Engineering, 2015, 22: 523-529.

[88] 畅元江．深水钻井隔水管设计方法及其应用研究［D］．东营：中国石油大学（华东），2008.

[89] 石晓兵，陈平．三维载荷对海洋深水钻井隔水管强度的影响分析［J］．天然气工业，2004（12）：86-88.

[90] 石晓兵，郭昭学，聂荣国，等．海洋深水钻井隔水管动力分析［J］．天然气工业，2003，（S1）：81-83.

[91] 周俊昌．海洋深水钻井隔水管系统分析［D］．四川：西南石油学院，2001.

[92] 王林，石晓兵，聂荣国，等．轴向载荷对海洋深水钻井隔水管力学特性的影响分析［J］．石油矿场机械，2004，（1）：30-32.

[93] 王腾，张修占，朱为全．平台运动下深水钻井隔水管非线性动力响应研究［J］．海洋工程，2008，26（3）：21-26.

[94] 李军强，方同．海洋钻井隔水管随机振动的理论分析［J］．石油机械，2000，28（8）：47-49.

[95] Y. B. Wang, D. L. Gao, J. Fang. Axial Dynamic Analysis of Marine Riser in Installation [J]. Journal of Natural Gas Science and Engineering, 2014, 21: 112-117.

[96] B. D. Ambrose, F. Grealish, K. Whooley. Soft Hangoff Method for Drilling Risers in Ultra Deep-water [C]. Offshore Technology Conference, 30 April-3 May, Houston, Texas, OTC 13186, 2001.

[97] 张炜，高德利．深水钻井隔水管脱开模式下纵向动态行为研究［J］．石油钻探技术，2010，38（4）：7-9.

[98] 孙友义，陈国明，畅元江．下放或回收作业状态下超深水钻井隔水管轴向动力分析［J］．中国海上油气，2009，21（2）：116-119.

[99] 高沣，高玉平，张啸斐，等．海洋钻井隔水管悬挂状态下轴向动力特性比对研究［J］．中国造船，2014，55（2）：114-121.

[100] 孙友义，陈国明，畅元江，等．超深水隔水管悬挂动力分析与避台风策略探讨［J］．中国海洋平台，2009，24（2）：29-32.

[101] Y. B. Wang, D. L. Gao, J. Fang. Mechanical Behavior Analysis for the Determination of Riser Installation Window in Deep-water Drilling [J]. Journal of Natural Gas Science and Engineering, 2015, 24: 317-323.

[102] 王宴滨，高德利，房军．浮力块对深水钻井隔水管安装过程性能的影响［J］．石油机械，2015，43（7）：47-50.

[103] 彭朋．深水钻井隔水管寿命管理技术研究［D］．东营：中国石油大学（华东），2009.

[104] 王宴滨，高德利，房军．海洋钻井隔水管—钻井液横向耦合振动特性［J］．石油钻采工艺，2015，37（1）：25-29.

[105] Y. B. Wang, D. L. Gao, J. Fang. Coupled Dynamic Analysis of Deep-water Drilling Riser under Combined

Forcing and Parametric Excitation [J]. Journal of Natural Gas Science and Engineering, 2015, 27: 1739-1747.

[106] 刘清友，周守为，姜伟，等．基于钻井工况和海洋环境耦合作用下的隔水管动力学模型 [J]. 天然气工业，2013，33 (12)：6-12.

[107] H. Park, D. Jung, C. Piao. 3-D Numerical Analysis of a Long Slender Marine Structure under Combined Axial and Lateral Excitation [C]. The Ninth International Offshore and Polar Engineering Conference, 30 May-4 June, Brest, France, ISOPE-I-99-159, 1999.

[108] J. F. Beattie, L. P. Brown, B. F. Webb. Lift and Drag Forces on a Submerged Circular Cylinder [C]. Offshore Technology Conference, 19-21 April, Houston, Texas, OTC 1358, 1971.

[109] T. Sarpkaya. Hydrodynamic Lift and Drag on Rough Circular Cylinders [C]. Offshore Technology Conference, 6-9 May, Houston, Texas, OTC 6518, 1991.

[110] 盛磊祥．海洋管状结构涡激振动流体动力学分析 [D]. 东营：中国石油大学（华东），2009.

[111] 林海花．隔水管涡激动力响应及疲劳损伤可靠性分析 [D]. 大连：大连理工大学，2008.

[112] 赵鹏良．隔水管及附属整流罩涡激振动的流固耦合模拟研究 [D]. 上海：上海交通大学，2011.

[113] 赵卓茂，王嘉松，谷斐．附属管对钻井隔水管涡激振动流动控制的研究 [J]. 水动力学研究与进展 A 辑，2012，27 (4)：401-408.

[114] J. K. Vandiver. Research Challenges in the Vortex-Induced Vibration Prediction of Marine Risers [C]. Offshore Technology Conference, 4-7 May, Houston, Texas, OTC 8698, 1998.

[115] 时米波．深水钻井隔水管系统疲劳可靠性分析 [D]. 东营：中国石油大学（华东），2008.

[116] 谢彬，段梦兰，秦太验，等．海洋深水立管的疲劳断裂与可靠性评估研究进展 [J]. 石油学报，2004，25 (3)：95-99.

[117] R. K. Lubbad, S. Løset, O. T. Gudmestad, A. Tørum, G. Moe. Vortex Induced Vibrations of Slender Marine Risers-Effects of Round-Sectioned Helical Strakes [C]. The Seventeenth International Offshore and Polar Engineering Conference, 1-6 July, Lisbon, Portugal, ISOPE-I-07-037, 2007.

[118] 左金菊．深水钻井隔水管涡激振动机理及其抑制研究 [D]. 成都：西南石油大学，2015.

[119] Norsok Stanardd-010. Well Integrity in Drilling and Well Operations [S]. Rev. 3, August, 2004.

[120] B. Vignes, B. S. Aadnoy. Well-Integrity Issues Offshore Norway [C]. IADC/SPE Drilling Conference, 4-6 March, Orlando, Florida, USA, SPE 112535, 2008.

[121] B. Vignes, S. A. Tonning, B. Aadnoy. Integrity Issues in Norwegian Injection Wells [C]. Abu Dhabi International Petroleum Exhibition and Conference, 3-6 November, Abu Dhabi, UAE, SPE 118101, 2008.

[122] 钱锋．深水套管柱载荷分析与设计方法研究 [D]. 北京：中国石油大学（北京），2012.

[123] F. Qian, D. L. Gao. A Mechanical Model for Predicting Casing Creep Load in High Temperature Wells [J]. Journal of Natural Gas Science and Engineering, 2011, 7 (3): 530- 535.

[124] D. L. Gao, F. Qian, H. K. Zheng. On a Method of Prediction of the Annular Pressure Buildup in Deep-water Wells for Oil & Gas [J]. CMES: Computer Modeling in Engineering & Sciences, 2012, 89 (1): 1-16.

[125] 钱锋，高德利．厚壁套管等效外挤载荷计算 [J]. 石油机械，2011，39 (12)：38-40，44.

[126] 钱锋，高德利，蒋世全．深水工况下套管柱载荷分析 [J]. 石油钻采工艺，2011，39 (2)：56-59.

[127] G. P. Willhite. Over-all Heat Transfer Coefficients in Steam and Hot Water Injection Wells [J]. SPE Journal of Petroleum Technology, 1967, 19 (5): 607-615.

[128] Y. S. Wu, K. Pruess, L. B. Laboratory. An Analytical Solution for Wellbore Heat Transmission in Layered Formations [J]. SPE Reservoir Engineering, 1990, 5 (4): 531-538.

[129] J. Hagoort. Ramey's Wellbore Heat Transmission Revisited [J]. SPE Journal, 2004, 9 (4): 465-474.

[130] R. F. Farris. A Practical Evaluation of Cements for Oil Wells [J]. Drilling and Production Practice. 1941, 41-283.

[131] A. F. Tragesser, P. B. Crawford, H. R. Crawford. A Method for Calculating Circulating Temperatures [J]. Journal of Petroleum Technology, 1967, 19 (11): 1507-1512.

[132] A. R. Hasan, C. S, Kabir, M. M. Ameen. A Fluid Circulating Temperature Model for Workover Operations [J]. SPE Journal, 1996, 1 (2): 133-144.

[133] A. R. Hasan, C. S. Kabir. A Simple Model for Annular Two-Phase Flow in Wellbores [J]. SPE Production & Operations, 2007, 22 (2): 168-175.

[134] A. R. Hasan, C. S. Kabir. Modeling Two-Phase Fluid and Heat Flows in Geothermal Wells [C]. SPE Western Regional Meeting, 24-26 March, San Jose, California, SPE 121351, 2009.

[135] 李嗣贵. 高温高压井井壁稳定性研究 [D]. 北京: 中国石油大学 (北京), 2004.

[136] 高永海, 孙宝江, 王志远, 等. 深水钻探井筒温度场的计算与分析 [J]. 中国石油大学学报 (自然科学版), 2008, 32 (2): 58-62.

[137] A. S. Halal, R. F. Mitchell. Casing Design for Trapped Annulus Pressure Buildup [J]. SPE Drilling & Completion, 1994, 9 (2): 107-114.

[138] G. R. Samuel, A. Gonzales. Optimization of Multistring Casing Design with Wellhead Growth [C]. SPE Annual Technical Conference and Exhibition, 3-6 October, Houston, Texas, SPE 5676, 1999.

[139] A. S. Halal, R. F. Mitchell, R. R. Wagner. Multi-String Casing Design with Wellhead Movement [C]. SPE Production Operations Symposium, 9-11 March, Oklahoma City, Oklahoma, SPE 37443, 1997.

[140] K Bybee. Transient Behavior of Annular Pressure Buildup in HP/HT Wells [J]. Journal of Petroleum Technology, 2015, 57 (3): 58-60.

[141] C. P. Leach, A. J. Adams. A New Method for the Relief of Annular Heat-Up Pressures [C]. SPE Production Operations Symposium, 21-23 March, Oklahoma City, Oklahoma, SPE 25497, 1993.

[142] D. L. Gao, F. Qian, H. K. Zheng. On a Method of Prediction of the Annular Pressure Buildup in Deep-water Wells for Oil &Gas [J]. Computer Modeling in Engineering & Sciences, 2012, 89 (1): 1-16.

[143] K. Maruyama, E. Tsuru, M. Ogasawara. An Experimental Study of Casing Performance Under Thermal Cycling Conditions [J]. SPE Drilling Engineering, 1990, 5 (2): 156-164.

[144] S. S. Rahman, G. V. Chilingarian. Casing Design Theory and Practice [M]. Elsevier, 1995.

[145] 李志明, 殷有泉. 油水井套管外挤力计算及其力学基础 [M]. 北京: 石油工业出版, 2006.

[146] 张永贵. 注蒸汽热采井套管强度理论与试验研究 [D]. 秦皇岛: 燕山大学, 2008.

[147] 李静, 林承焰, 杨少春, 等. 套管—水泥环—地层耦合系统热应力理论解 [J]. 中国石油大学学报 (自然科学版), 2009, 33 (2): 63-69.

[148] 高学仕, 张立新, 何牛仔. 热采井筒应力的数值模拟分析 [J]. 石油大学学报 (自然科学版), 2001, 25 (02): 65-69.

[149] 高亮. 稠油热采井套管应力分析 [D]. 东营: 中国石油大学 (华东), 2009.

[150] J. C. R. Placido, I. P. Pasqualino, C. E. Fonseca. Strength Analyses of Liners for Horizontal Wells [C]. SPE Annual Technical Conference and Exhibition, 9-12 October, Dallas, Texas, SPE 96870, 2005.

[151] 陈勇, 练章华, 乐彬, 等. 考虑地应力耦合的热采井套管损坏分析 [J]. 钻采工艺, 2007, 30 (5): 13-16.

[152] 王宴滨, 高德利, 房军. 深水钻采管柱力学行为模拟试验系统研制 [J]. 石油矿场机械, 2014, 43 (4): 26-29.

[153] 房军, 王宴滨, 高德利. 深水隔水管受力变形模拟试验方法研究 [J]. 石油机械, 2013, 41 (12): 53-57.

[154] 杨德兴, 姜亚军, 廖威, 等. 基于光纤布拉格光栅的深水钻井隔水管应变监测 [C] //中国光学学会. 中国光学学会 2011 年学术大会摘要集, 2011.

[155] 周守为, 刘清友, 姜伟, 等. 深水钻井隔水管“三分之一效应”的发现——基于海流作用下深水钻

井隔水管变形特性理论及实验的研究［J］. 中国海上油气, 2013, 25（6）: 1-6.

［156］刘清友, 毛良杰, 周守为, 等. 一种深水钻井工况下隔水管振动特性模拟试验装置及试验方法: 201310169000X［P］. 2015-05-13.

［157］赵焕宝, 侯晓东, 雷广进, 等. 深水钻井隔水管疲劳试验载荷分析［J］. 石油矿场机械. 2013, 42（2）: 32-35.

［158］刘秀全, 陈国明, 畅元江, 等. 海洋油气立管疲劳试验方法［C］. //中国造船工程学会, 中国海洋石油总公司. 海洋工程装备发展论坛论文集, 2011.

Experimental Investigation of the Failure Mechanism of P110SS Casing under Opposed Line Load

Deng Kuanhai[1] Lin Yuanhua[1,3,*] Liu Wanying[3] Li Hu[3]
Zeng Dezhi[2] Sun Yongxing[2]

(1. State Key Laboratory of Oil and Gas Reservoir Geology and Exploitation (Southwest Petroleum University); 2. CNPC Key Lab for Tubular Goods Engineering (Southwest Petroleum University); 3. Xinjing Oil Field Branch Company)

Abstract: Many studies focused on casing collapse strength under uniform load have been done, and the API 5C3 and ISO standards have been formed. However, only a limited number of researches on collapse failure mechanism of casing have been done under non-uniform load, especially experimental study, although the non-uniform load has a great impact on casing collapse strength. Hence, the collapse experiment is conducted for P110SS casing under opposed line load by using electro hydraulic servo pressure testing machine. The displacement variation rules of P110SS casing have been obtained under opposed line load. The strain of casing is measured in the process of collapse testing by the method of gluing strain-gauge on the outside surface of P110SS casing. The initial yield load of casing, instability load of casing and the yield load, stress-hardening rate, strain-hardening rate of casing after hardening are obtained. The stress-strain rules of casing after hardening are analyzed. The hardening characteristic and collapse failure mechanism of P110SS casing have been clarified under non-uniform load (under opposed line load). Experimental results can provide important references for the theoretical study on failure mechanism and collapse strength of casing under non-uniform load.

Keywords: Opposed line load; Experiment; Failure mechanism; Hardening characteristic; P110SS casing

1 Introduction

Casing plays an important role in protecting and reinforcing borehole wall during drilling and production. In recent years, petroleum industries encounter an increasing number of complex

* Corresponding author

E-mail: yhlin28@163.com, dengkuanhai@163.com.

formations, such as plastic creep of rock salt stratum and formation with large inclination angle[1,2], and service environment of casing is worse and worse, especially for the non-uniform load caused by mudstone creep[3], salt stratum plastic flow and stratum compaction[4,5]. The main reason is that the collapse strength of casing decreases significantly under non-uniform load[6], resulting in casing collapse failure easily. Failure data shows that the casings in the well Long-gang-001-1, Long-gang-001-2, Puguang-204-2H in Sichuan and Chongqing gas fields and the well TK1127 in Tahe oilfield in Xinjiang were collapsed due to non-uniform load caused by creep of salt rock[7].

Hence, many scholars have done many studies about casing collapse strength in the world, especially for the collapse strength under uniform load, and some important achievements[8-13] (some classical mechanical models have been built under uniform load, and the API 5C3 and ISO 10400 standards have been formed and adopted by the world oil industry[14,15]) have been obtained. In addition, some scholars have done system studies about the collapse strength of worn casing, defective casing and high collapse casing under uniform load based on the new ISO 10400 standard[16-22]. However, only a few scholars have carried out researches on collapse strength of casing under non-uniform load, and most of their studies are only in theory[6, 23-31]. For instance, Pattillo and Berger et al[28,29] have done some researches on the collapse strength of idea casing under non-uniform load based on finite element method. Ei-Sayed and Zheng Junde[6, 32,33] have built mechanic model for calculating collapse strength of idea casing under non-uniform load based on inverse method and analyzed the stress distribution rule.

It can be known that the researches on casing collapse strength under non-uniform load are mainly stay in theory at present, and all those studies are based on the same assumption that the distribution of non-uniform load is elliptical[6] (it is defined as elliptical load applied on half the outer surface of casing). In fact, there is another non-uniform load. It is opposed line load that is applied on opposite ends of a diameter on the outer surface of casing and has a greater impact on the collapse strength of casing than the elliptical load[34]. In a word, it is urgent that the experimental study on the casing failure mechanism is performed under non-uniform load, especially for the opposed line load, for improving the study on the casing collapse strength under non-uniform load. Hence, this paper aims to conduct the experimental study on the collapse failure mechanism of P110SS casing under opposed line load.

2 Collapse Testing Program of P110SS Casing under Opposed Line Load

2.1 Main Experimental Equipment and Method

In order to study the failure mechanism of casing under non-uniform load (opposed line load), the collapse test of P110SS casing is performed under opposed line load. The geometric and mechanical parameters of P110SS casing are shown in Table 1. The OD stands for outer diameter, and ID stands for inner diameter. The collapse test system mainly consists of YAW-200 pressure testing machine, YE-2533 static strain indicator, strain gauge with high precision and some special wires, as shown in Fig. 1. The YAW-200 pressure testing machine is used to record external load (opposed line load) exerted on the casing and the radial displacement of casing in the testing process. The static strain indicator is used to measure the stress and strain on the casing during the testing process.

Table 1 Geometric and mechanical parameters of P110SS casing

Grade	Length (mm)	ID (mm)	OD (mm)	Yield stress (MPa)	Tensile strength (MPa)
P110SS	1000	154.84	177.92	860	935

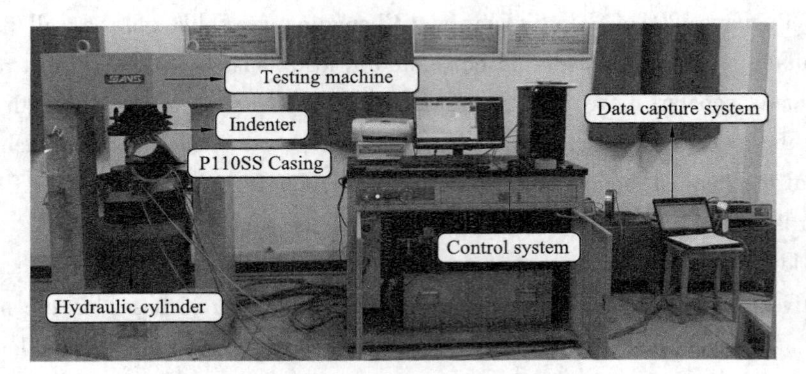

Fig. 1 Collapse test system of P110SS casing under opposed line load

2.2 Location of strain gauges

The strain gauge (TY120-3CA-10%-X) with three directions (0°, 45°, 90°) that can measure stress and strain of the three directions is used in the experimental testing. The location of every strain gauge on the P110SS casingis shown in Fig. 2. In Fig. 2 (a), the longitudinal interval between point A1 and point A3 is 150 mm, and the longitudinal interval between point A2 and point A3 is 150 mm too. The circumferential interval between point A1 and point A3 is 45°, and the circumferential interval between point A2 and point A3 is 45° too. The circumferential interval between point A1 and point B1 is 180°. The specific location of strain gauges on casing is shown in Fig. 2 (b).

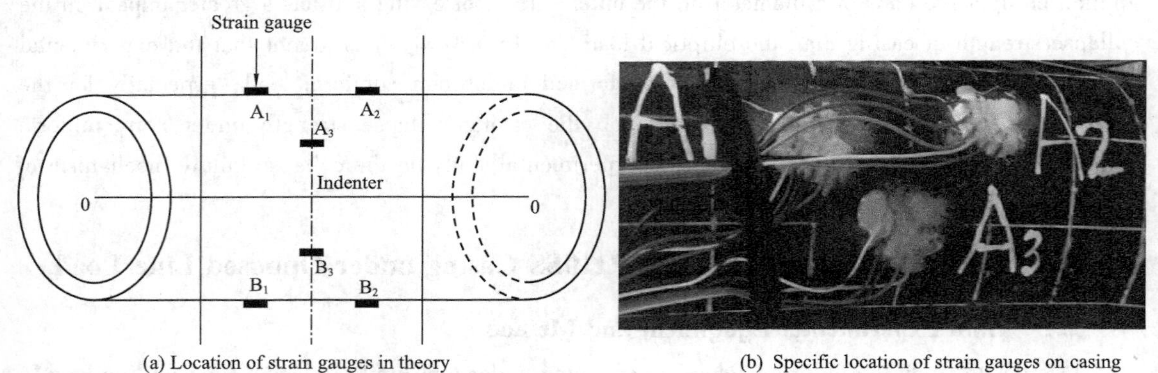

(a) Location of strain gauges in theory (b) Specific location of strain gauges on casing

Fig. 2 Location of strain gauges in theory and specific location of strain gauges on casing

3 Experimental results and analysis under opposed line load

3.1 Analysis of radial displacement

In order to obtain the deformation rules and mechanical properties of P110SS casing accurately, the loading rate of external load is slowly until the external load is equal to preset value firstly; secondly, the external load (preset value) remains unchanged for two minutes; finally, the external load is unloaded slowly until the external load is equal to zero. The other external load is loaded/ unloaded by using the same method. The loading sequence of the P110SS casing is 20 t→30 t→40 t→45 t→50 t→55 t→65 t→70 t→75 t→80 t→90 t→100 t→102 t. The radial displacement and plastic deformation of P110SS casing are shown in Table 2 and Fig. 3.

Table 2 Loading / unloading process of P110SS casing

Loading rate (t/s)	Unloading rate (t/s)	External load (t)	PD (mm)	RD (mm)	Yield	Loading rate (t/s)	Unloading rate (t/s)	External load (t)	PD (mm)	RD (mm)	Yield
0.1	0.2	20	0	0.5	No	0.1	0.2	70	2.1	6.7	Yes
0.1	0.2	30	0	1.3	No	0.15	0.3	75	3.3	8.5	Yes
0.1	0.2	40	0	2.1	No	0.15	0.3	80	5.3	11.3	Yes
0.1	0.2	45	0	2.5	No	0.2	0.3	90	7.9	21.1	Yes
0.1	0.2	50	0	2.9	No	0.3	0.4	100	31.9	42.5	Yes
0.1	0.2	55	0	3.3	No	0.3	0.4	102	36.1	46.5	Yes
0.1	0.2	65	1.3	5.3	Yes	—					

Where PD stands for the plastic deformation; RD stands for the radial displacement.

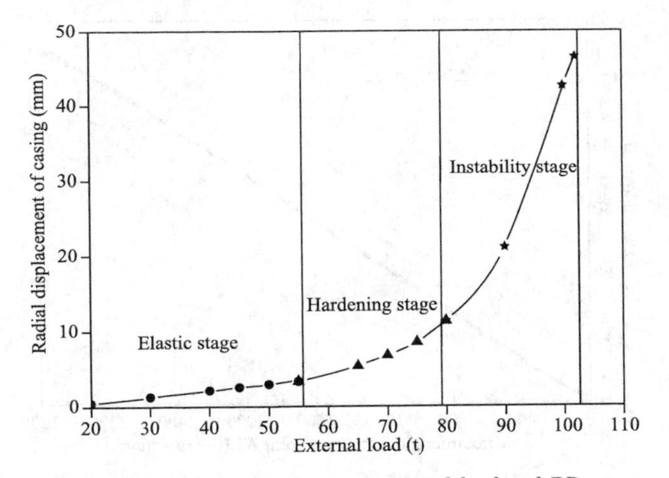

Fig. 3 Relationship between external load and RD

Fig. 3 shows that the loading process of external load can be divided into elastic stage (the black line), hardening stage (the red line) and instability stage (the blue line). P110SS casing is in the elastic stage when the external load is less than 55 t. It can be observed from Fig. 3 and Table 2 that P110SS casing begins to yield when the external load is larger than 55 t, and the plastic deformation increases with the increase of external load. In order to determine yield load accurately, the stress - strain rule of P110SS casing needs to be analyzed further. The radial displacement increases slowly with the increase of external load when the external load is smaller than 80 t, but the radial displacement increases rapidly with the increase of external load when the external load is larger than 80 t. Hence, it can be concluded that P110 casing is in hardening stage when the external load is less than 80 t, and the structural instability of P110SS casing might happen or has happened when the external load is larger than 80 t. It is necessary to analyze the stress-strain of P110SS casing in detail, in order to determine the instability load.

Finally, the maximum radial displacement and plastic deformation of P110SS casing reach up to 46.5 mm and 36.1 mm, respectively, when the external load is equal to 102 t.

3.2 Analysis of stress and strain in elastic stage

It can be observed that the testing data of point A1, A2, B1 and B2 is the same basically and much larger than point A3 and B3 in collapse testing process. In addition, the circumferential strain (strain at 90° direction) of casing is the maximum under non-uniform load (opposed line load). Hence, in the loading of 55 t process, only the circumferential strain of point A1 is analyzed in detail, as shown in Table 3 and Fig. 4.

Table 3 Strain at point A1 of P110SS casing

Strain Location	Initial strain (micro strain)	Maximum strain (micro strain)	Residual strain (micro strain)	Maximum equivalent stress (MPa)
0° direction at point A1	21	1055	32	668.6
45° direction at point A1	29	2177	40	
90° direction at point A1	40	3142	53	

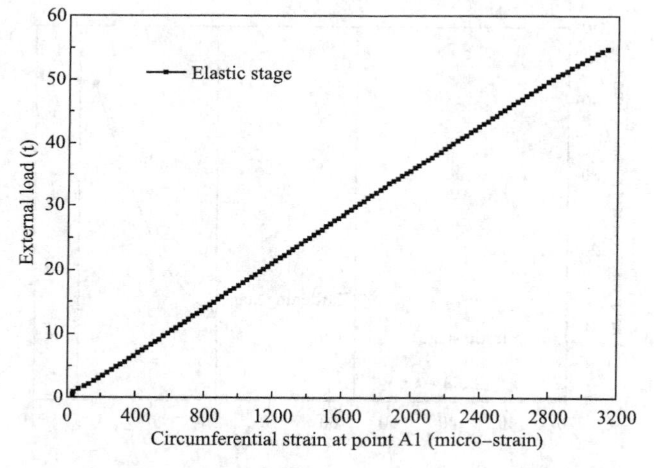

Fig. 4 Relationship between external load and CS (55 t)

Fig. 4 shows that the initial strain of loading process and residual strain after unloading are very small and equal basically. Table 3 shows that circumferential strain is linear to external load too. Hence, it can be concluded that the deformation of casing is entirely elastic deformation when the external load is not more than 55 t, which is in agreement with analysis results of radial displacement.

3.3 Analysis of stress and strain in yield stage

The P110SS casing yields for the first time in the loading of 65 t process, which is so called initial yield. Similarly, only the circumferential strain and equivalent stress (Von Mises stress) of point A1 and A3 are analyzed, as shown in Fig. 6 and Fig. 7. The CS stands for circumferential strain, and the ES stands for equivalent stress in this paper. The EL stands for external load.

Fig. 5 shows that the point A1 on the P110SS casing undergoes obviously elastic stage, yield stage, hardening stage and springback stage in the loading/unloading of 65 t process. However, it can be observed from Fig. 6 that the circumferential strain and equivalent stress of point A3 on the P110SS casing are proportional to external load, and the maximum (628.4 MPa) of equivalent stress is less than yield stress (860 MPa) of P110SS casing, which indicates that the casing is still in elastic stage. Hence, it can be concluded that the point A1 is the most dangerous under opposed line load, and the stress-strain properties and deformation rules of deferent test points (A1, A3) on P110SS casing are different each other. In addition, Fig. 5 shows that the equivalent stress, circumferential strain and external load of yield point (A1) are 861.6 MPa, 3904 (micro strain) and 59.5 t, respectively. The equivalent stress (861.6 MPa) of yield point (A1) is slightly larger than the yield stress (860.0 MPa) of P110SS casing, which further indicates that the P110SS casing yields indeed in the loading of 65 t process. Hence, the initial yield load (59.5 t) of P110SS casing can be obtained.

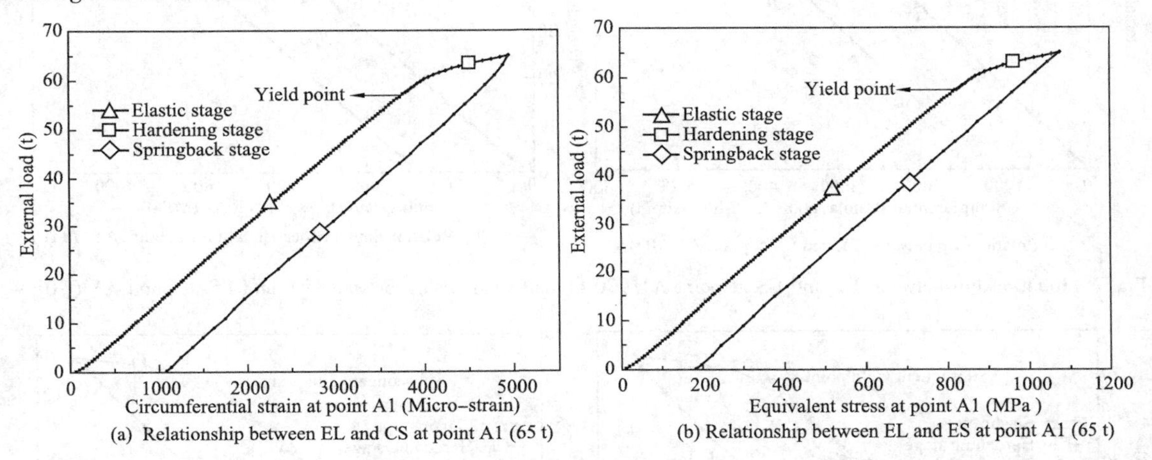

(a) Relationship between EL and CS at point A1 (65 t)

(b) Relationship between EL and ES at point A1 (65 t)

Fig. 5 Relationship between EL and CS at point A1 (65 t) and relationship between EL and ES at point A1 (65 t)

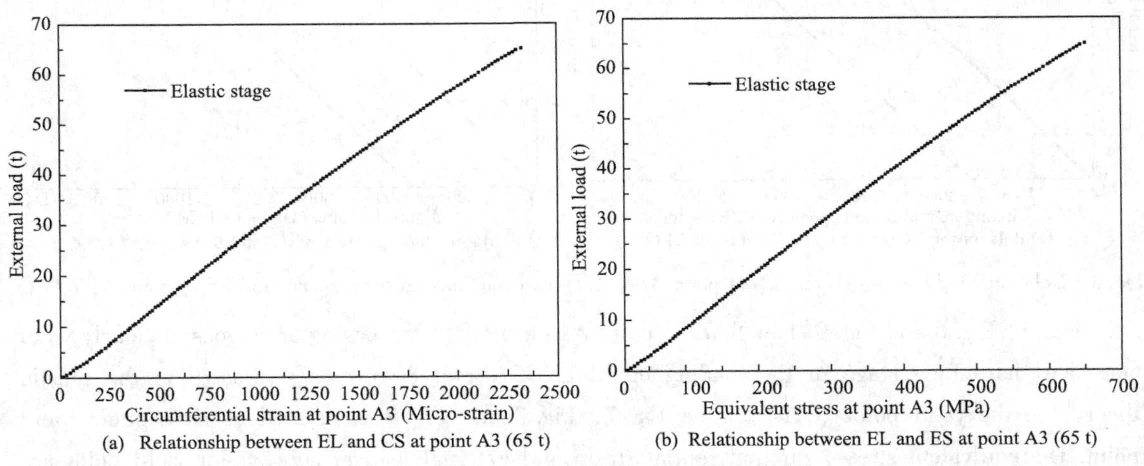

(a) Relationship between EL and CS at point A3 (65 t)

(b) Relationship between EL and ES at point A3 (65 t)

Fig. 6 Relationship between EL and CS at point A3 (65 t) and relationship between EL and ES at point A3 (65 t)

3.4 Analysis of stress and strain in hardening stage

Based on the material hardening theory[35], it can be known that the P110SS casing will be hardened under larger external load after yielding, and the bearing capacity of casing gradually increases in the hardening process until the structural instability of casing occurs. The P110SS casing has undergone different level of hardening in the loading of 70 t, 75 t and 80 t process. Hence, it is of interest to ascertain the hardening characteristic of P110SS casing under opposed line load, and analyze the hardening rules under different external load, and determine the instability load. For this purpose, the relationships between the circumferential strain of point A1 and external load as well as equivalent stress of point A1 and external load are plotted, as shown in Fig. 7, Fig. 8 and Fig. 9. The EL stands for external load.

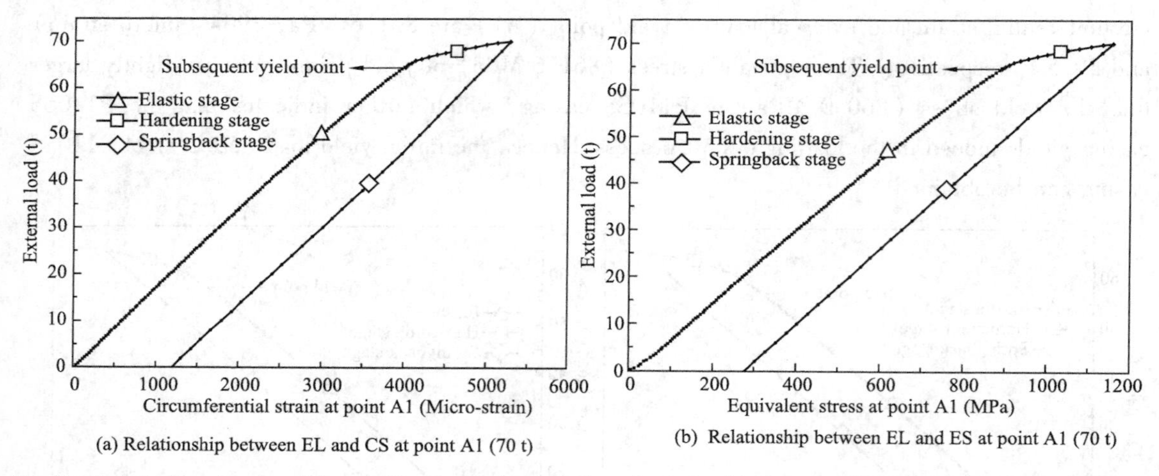

(a) Relationship between EL and CS at point A1 (70 t)

(b) Relationship between EL and ES at point A1 (70 t)

Fig. 7 Relationship between EL and CS at point A1 (70 t) and relationship between EL and ES at point A1 (70 t)

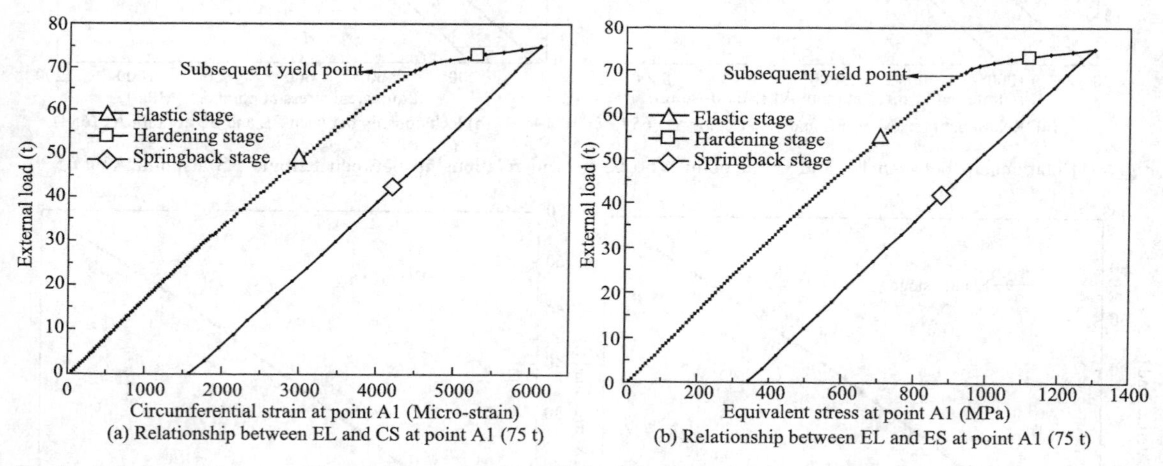

(a) Relationship between EL and CS at point A1 (75 t)

(b) Relationship between EL and ES at point A1 (75 t)

Fig. 8 Relationship between EL and CS at point A1 (75 t) and relationship between EL and ES at point A1 (75 t)

Fig. 7, Fig. 8 and Fig. 9 show that the point A1 on the P110SS casing undergoes obviously yield stage and hardening stage in the loading of 70 t, 75 t and 80 t process. Based on the plastic theory[36], the yield point (A1) in the Fig. 7, Fig. 8 and Fig. 9 is defined as subsequent yield point. The equivalent stress, circumferential strain and external load of subsequent yield point are 872.8 MPa, 3984 (micro strain) and 63.8 t, respectively, in the loading of 70 t process; the equivalent stress, circumferential strain and external load of subsequent yield point are 884.6 MPa, 4112 (micro strain) and 66.4 t, respectively, in the loading of 75 t process; the equivalent stress, circumferential strain and external load of subsequent yield point are 904.3 MPa, 4327 (micro strain) and 68.7 t, respectively, in the loading of 80 t process, as shown in Table 4.

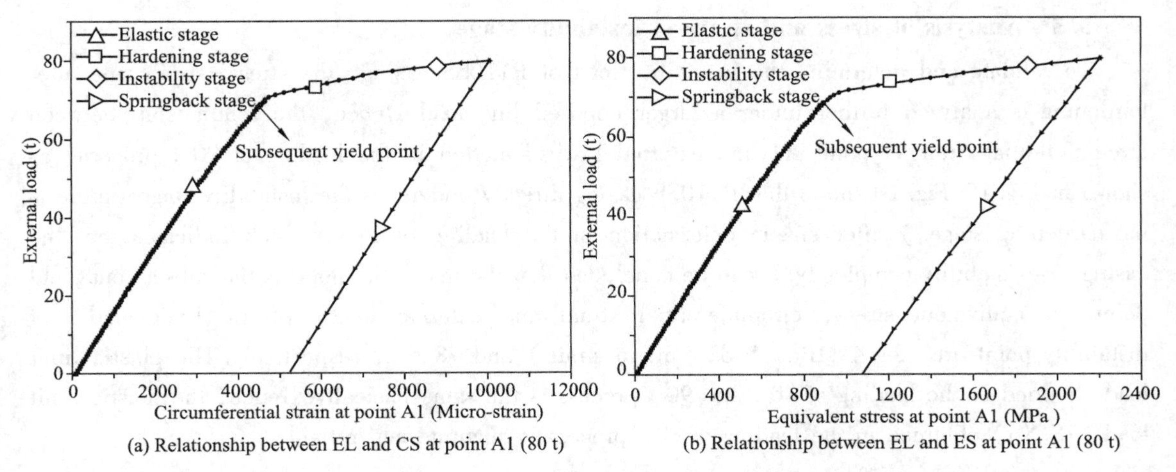

(a) Relationship between EL and CS at point A1 (80 t) (b) Relationship between EL and ES at point A1 (80 t)

Fig. 9 Relationship between EL and CS at point A1 (80 t) and relationship between EL and ES at point A1 (80 t)

From comparison and analysis of Fig. 5, Fig. 7, Fig. 8, Fig. 9 and Table 4, the following may be noted.

(1) The equivalent stress, circumferential strain and yield load of subsequent yield point are larger than the equivalent stress, circumferential strain and yield load of initial yield point in the loading of 65 t process, and increase rapidly with increase of the external load, which indicates that the level of hardening and bearing capacity of P110SS casing increase with the increase of external load.

(2) The strain rate increases obviously with the increase of external load in hardening stage, and the hardening stage is weakening and disappearing gradually, which indicates that the toughness and plastic deformation capacity of P110SS casing decrease with the increase of external load.

(3) P110SS casing only undergoes the elastic stage, hardening stage and springback stage in the loading of 70 t and 75 t process. However, the casing also undergoes the instability stage (at the beginning of the loss of stability, slightly increase of external loading leads to large plastic deformation.), and the residual strain increases significantly in the loading of 80 t process, which indicates that the structural instability occurs. The instability load (plastic limit load) is 78.8 t.

Finally, it is important to note that the above analysis and conclusions are only applicable for P110SS casing under conventional environment without the effect of acid and temperature because the P110SS casing can be used in sour environment. The analysis and conclusions may be not applicable because the environment fracture is the main failure mode of P110SS casing under sour environment. Hence, for the sour environment, the casing strength not only should meet the requirement of yield design criterion, but also the requirement of fracture mechanics design.

Table 4 Equivalent stress, circumferential strain and yield load of yield point after hardening

External load (t)	Yield point	ES (MPa)	CS (Micro strain)	Yield load (t)	Hardening
65	A1	861.6	3904	59.5	No
70	A1	872.8	3984	63.8	Yes
75	A1	884.6	4110	66.4	Yes
80	A1	904.3	4327	68.7	Yes

3.5 Analysis of stress and strain in instability stage

To validate and determine the instability load of P110SS casing, the stress-strain rule after hardening is analyzed further under a larger opposed line load. Hence, the relationship between circumferential strain of point A1 and external load is plotted in the loading of 90 t process, as shown in Fig. 10. Fig. 10 shows that P110SS casing directly undergoes the instability stage (there is not hardening stage.) after elastic deformation in the loading process, which indicates that the casing loses stability completely. It can be concluded that the instability point is the subsequent yield point. The equivalent stress, circumferential strain and external load (plastic limit load) of instability point are 924.4 MPa, 5062 (micro strain) and 78.6 t, respectively. The plastic limit load obtained in the loading of 80 t and 90 t process is the same basically. Hence, the plastic limit load (78.8 t) obtained in the loading of 80 t process is accurate and reliable.

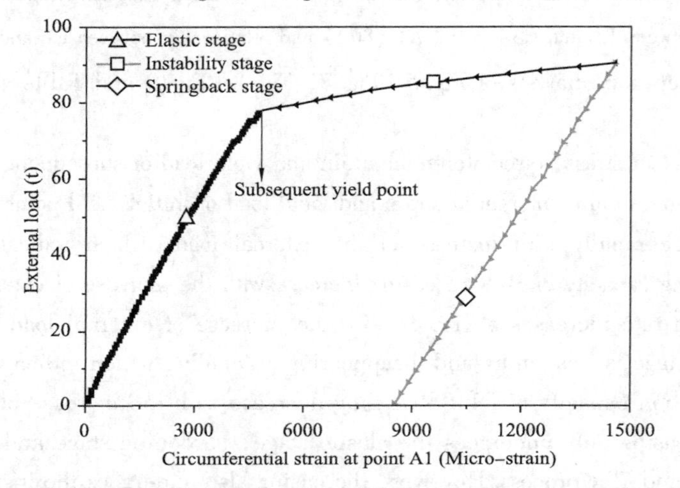

Fig. 10 Relationship between external load and CS at point A1 (90 t)

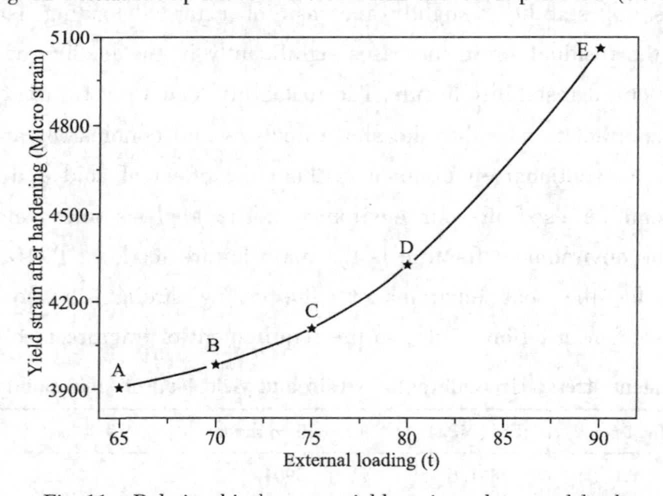

Fig. 11 Relationship between yield strain and external load

Based on the test data, the relationship between yield strain (circumferential strain) and external load can be obtained, as shown in Fig. 11. From comparison and analysis of the slopes in Fig. 11, the yield strain after hardening increases significantly when the external load is larger than

80 t, and the slope of line DE is much larger than the slope of line AB, line BC and line CD, which demonstrates that P110SS casing has lost stability completely when the external load is close to 80 t. Hence, the accuracy and reliability of plastic limit load (78. 8 t) obtained in the loading of 80 t process is validated further.

3. 6 Brief summary of collapse testing

The collapse testing results of P110SS casing under opposed line load are shown in Table 5. The increase coefficient (α) is the ratio of the yield load after hardening (subsequent yield load) to initial yield load. Table 5 shows that the initial yield load/pressure (it can be obtained based on the yield/instability load and contact area between the tested casing and indenter of YAW-200 pressure testing machine.) is 59. 5 t/38. 48 MPa, and the instability load/pressure is 78. 8 t/46. 40 MPa. The yield load, yield strain, yield stress and increase coefficient of P110SS casing increase after hardening with the increase of external load until the structural instability of P110SS casing occurs, which indicates that the P110SS casing has been hardened remarkably in the loading process, and the bearing capacity increases significantly. The maximum of yield load, strain-hardening rate, stress-hardening rate and increase coefficient are 78. 8 t, 29. 66%, 7. 29% and 1. 32, respectively. It can be concluded that it does not mean the loss of bearing capacity or service ability when P110SS casing yields initially under opposed line load. On the contrary, the bearing capacity of P110SS casing is improved to some extent when the casing slightly yields.

Table 5 Collapse testing results of P110SS casing under uniaxial compressive load

External load (t)	Yield load/ pressure (t/MPa)	Yield strain (Micro strain)	ES (MPa)	Strain-hardening rate (%)	Stress-hardening rate (%)	Increase coefficient (α)
65	59. 5 / 38. 18	3904	861. 6	0	0	1. 00
70	63. 8/ 40. 96	3984	872. 8	2. 05	1. 3	1. 07
75	66. 4 / 42. 28	4110	884. 6	5. 28	2. 67	1. 12
80	68. 7 / 43. 32	4327	904. 3	10. 83	4. 96	1. 15
90	78. 8 / 46. 40	5062	924. 4	29. 66	7. 29	1. 32

4 Conclusions

The collapse testing of P110SS casing under opposed line load is conducted for the first time. The stress-strain rules in elastic stage, yield stage, hardening stage and instability stage, and the relationship between radial displacement and external load are obtained so that the hardening characteristic and collapse failure mechanism of P110SS casing are clarified under opposed line load.

The initial yield load and instability load are 59. 5 t and 78. 6 t, respectively, and the subsequent yield loads are larger than initial yield load are 63. 8 t, 66. 4 t and 68. 7 t, respectively, in the loading of 70 t, 75 t and 80 t process, which indicates that it does not mean the loss of bearing capacity or service ability when P110SS casing yields initially under opposed line load, but on the contrary, the bearing capacity of P110SS casing is improved to some extent when the casing slightly yields. Note that the toughness and plastic deformation capacity of casing after hardening will decrease, resulting in casing failure caused by brittle fracture easily.

The research results are only applicable for P110SS casing under conventional environment without the effect of acid and temperature.

References

[1] Deyu Chang, Gensheng Li, Zhonghou Shen, Zhongwei Huang, Shouceng Tian, Huaizhong Shi, et al. The stress field of bottom hole in deep and ultra-deep wells [J]. Acta Petroleum Sinica, 2011, 32 (7): 697-703.

[2] Henglin Yang, Chen Mian, Jin Yan, Zhang Guangqing. Analysis of casing equivalent collapse resistance in creep formations [J]. Journal of China University of Petroleum, 2006, 30 (4): 94-97.

[3] Deliang Wang. Reasons and Analysis of Casing Damage in Zhongyuan Oilfield [J]. Petroleum Drilling Techniques, 2003, 31 (2): 36-38.

[4] Suping Peng, Jitong Fu, Jincai Zhang. Borehole casing failure analysis in unconsolidated formations: A case study [J]. Journal of Petroleum Science and Engineering, 2007, 59 (3-4): 226-238.

[5] Fredrich J T, Arguello J G, Deitrick G L, Rouffignac E P. Geomechanical Modeling of Reservoir Compaction, Surface Subsidence, and Casing Damage at the Belridge Diatomite Field [J]. SPE Reservoir Evaluation & Engineering, 2000, 3 (4): 348-359.

[6] EI-Sayed A. A. H, Fouad Khalaf. Resistance of Cemented Concentric Casing Strings Under Nonuniform Loading [J]. SPE Drilling Engineering, 1992, 7 (1): 59-64.

[7] Lin Yuanhua Deng Kuanhai; Zeng Dezhi, Hongjun Zhu, Dajiang Zhu, Qi Xing, et al. Theoretical and experimental analyses of casing collapsing strength under non-uniform loading [J]. J. Cent. South Univ, 2004, 9 (21): 3470-3478.

[8] Yuanhua Lin, Sun Yongxing, Shi Taihe, Kuanhai Deng, Han Liexiang, Sun Haifang, et al. Equations to calculate collapse strength for high collapse casing [J]. Journal of Pressure Vessel Technology, 2013, 135 (4): 041202.

[9] Klever F J, Tamano T. A new OCTG strength equation for collapse under combined loads [J]. SPE Drilling & Completion, 2006, 21 (3): 164-179.

[10] Sun Yongxing, Yuanhua Lin, Zhongsheng Wang, Shi Taihe, Liu Hongbin, Ping Liao, et al. A new OCTG strength equation for collapse under external load only [J]. Journal of Pressure Vessel Technology, 2011, 133, 011702: 1-5.

[11] F. Klever A design strength equation for collapse of expanded OCTG [J]. SPE Drilling & Completion, 2010, 25 (3): 391-408.

[12] Chengjing Tan, Deli Gao. Theoretic problems about calculation of casing strength [J]. Acta Petroleum Sinica, 2005, 26 (3): 123-126.

[13] Zhao Junhai, Li Yan, Zhang Changguang, Jianfeng Xu, Peng Wu. Collapsing strength for petroleum casing string based on unified strength theory [J]. Acta Petroleum Sinica, 2013, 34 (5): 969-976.

[14] America Petroleum Institute. Bulletin on Formulars and Calculations for Casing, Tubing, Drill Pipe and Line Properties [S]. API Bulletin5C3, Six Edition, 1994.

[15] International Organization for Standardization, Petroleum and Natural Gas Industries-Formulae and Calculation for Casing, Tubing, Drill Pipe and Line Pipe Properties [S]. ISO 10400, Six Edition, 2007.

[16] Liu Shaohu, Hualin Zheng, Zhu Xiaohua, Tong Hua. Equations to calculate collapse strength of defective casing for steam injection wells [J]. Engineering Failure Analysis, 2014, 42: 240-251.

[17] Zeng Dezhi, Yuanhua Lin, Shi Taihe, et al. Mew algorithm of collapsing strength of wearing casing. Nat Gas Ind 2005; 25 (2): 78-80.

[18] Hualin Liao, Guan Zhichuann, et al. Remaining strength calculation of internal wall worn casing in deep and ultra-deep wells. Eng Mech 2010; 27 (2): 250-256.

[19] N. M. M Junior, A. A Carrasquila , A. Figueiredo, et al. Worn pipes collapse strength: experimental and numerical study [J]. Journal of Petroleum Science and Engineering, 2015, 133: 328-334.

[20] Y. Lin, K. Deng, X. Qi, et al. A new crescent-shaped wear equation for calculating collapse strength of worn casing under uniform loading [J]. Journal of Pressure Vessel Technology, 2015, 137 (3): 031201.

[21] Zhonghong Yu, Liyang Wang, Liu Dawei. Investigation on production on casing in steam-injection wells and the application oilfield. J Hydrodyn 2009; 21 (1): 77-83.

[22] K. Deng, Y. Lin, H Qiang, et al. New high collapse model to calculate collapse strength for casing [J]. Engineering Failure Analysis, 2015, 58 (1): 295-306.

[23] Cai Zhengmin, Zhang Shujia, Chen Xiangkai, Chen Jun. Research on collapse strength of oil casing under non-uniform loads [J], Oil Field Equipment, 2009, 38 (12): 31-34.

[24] Han Jianzeng, Zhang Xianpu. Discussion of casing collapse strength under non-uniform loading [J]. Drilling and Production Technology, 2001, 24 (3): 48-50.

[25] Cai Zhengmin, Zhang Jun, Sheng Chaoting, Jin Meiting, Zhang Shujia. Effect of ovality on collapse strength of casing pipe under non-uniform loading [J]. Oil Field Equipment, 2010, 39 (5): 20-22.

[26] Zou Lingzhan, Jingen Deng, Zeng Yijing, Li Jing, Weizhe Dong. Investigation of casing load calculation and casing design for deep salt formation [J]. Petroleum Drilling Techniques, 2008, 36 (1): 23-27.

[27] W Huang, D. Gao, A theoretical study of the critical external pressure for casing collapse [J]. Journal of Natural Gas Science and Engineering, 2015, Accepted Paper.

[28] A. Berger, W. W. Fleckenstein, A. W. Eustes, G. Thonhauser, Effect of Eccentricity, Voids, Cement Channels, and Pore Pressure Decline on Collpase Resistance of Casing [C] // SPE Annual Technical Conference and Exhibition, 26-29 Sep 2004, Houston, Texas, Society of Petroleum Engineers, http: //dx. doi. org/10. 2118/90045-MS.

[29] P. D. Pattillo, N. C. Last, W. T. Asbill, Effect of Non-Uniform Loading on Conventional Casing Collpase Resistance [C] // SPE/IADC Drilling Conference, 19-21 Feb 2003, Amsterdam, Netherlands, Society of Petroleum Engineers, http: //dx. doi. org/10. 2118/79871-MS.

[30] Jincai Zhang, William Standifird, Chris Lensmond. Casing Ultradeep, Ultralong Salt Sections in Deep Water: A Case Study for Failure Diagnosis and Risk Mitigation in Record-Depth Well [C] // SPE Annual Technical Conference and Exhibition, 21-24 Sep 2008, Denver, Colorado, USA, Society of Petroleum Engineers, http: //dx. doi. org/10. 2118/114273-MS.

[31] Xinpu Shen. Numerical Analysis of Casing Failure Under Non-Uniform Loading In Subsalt Wells In Paradox Basin [C] // 45th U. S. Rock Mechanics / Geomechanics Symposium, 26-29 Jun 2011, San Francisco, California, American Rock Mechanics Association, ARMA-11-176.

[32] Junde Zheng, Zhang Yanqiu, Wenjun Wang, Wang Jian. Calculation of casing strength under non-uniform loading [J]. Acta Petroleum Sinica, 1998, 19 (1): 119-123.

[33] Yin Youquan, Li Ping'en. Computation of casing strength under non-uniform load [J]. Acta Petroleum Sinica, 2007, 28 (6): 138-141.

[34] J. H. Nester, Jenkins D R. Resistances to failure of oil-well casing subjected to non-uniform transverse loading [C] //Drilling and Production Practice. American Petroleum Institute, 1955.

[35] Danian Wang. Principles of Metal Plastic Forming [M]. Beijing, China Machine Press, 1986.

[36] Bingye Xu. Plasticity Mechanics [M]. Beijing, Higher Education Press, 1988.

Theoretical Study on Working Mechanics of Smith Expansion Cone

Deng Kuanhai[1] Lin Yuanhua[1] Zeng Dezhi[1] Sun Yongxing[2] Liu Wanying[2]

(1. State Key Laboratory of Oil and Gas Reservoir Geology and Exploitation, Southwest Petroleum University; 2. CNPC Key Lab for Tubular Goods Engineering, Southwest Petroleum University)

Abstract: The deformation failures of many casings are in urgent need of one reliable repair technology recently. Smith expansion cone (SEC) repair technology is high efficient to repair deformed casing. However, the reshaping force is a very important parameter for designing and optimizing the SEC and construction parameters. Hence, the mechanical mechanism of SEC repair technology is studied, and one mechanical model of SEC used to repair deformed casing is proposed based on twin-shear unified strength theory in this paper. In this model, the effects of material hardening and the ratio of yield strength to tensile strength on casing repair were taken into full account. The mechanical model can calculate reshaping force that is used to repair the deformed casing under any confining pressure, and there is a good agreement between calculated results and experimental data. Based on this model, the effects of expansion amount every time, friction efficient, cone angle and length of equal diameter section on the reshaping force were analyzed in detail, by which the correlations between the reshaping force and the expansion amount every time, friction efficient, cone angle and length of equal diameter section were obtained. Research results can provide theoretical guidance for design and optimization of the structure and construction parameters of SEC.

Keywords: Smith expansion cone (SEC); Reshaping force; Mechanical model; Twinshear unified strength theory; Repairing; Deformed casing

1 Introduction

Many casing in oil and gas wells have been deformed[1, 2] due to the geologic, engineering and corrosion factors after the production for a period of time in China. For example, the casings in well Long-gang-001-1, Long-gang-001-2, Long-gang-39, Long-gang-13, Pu-guang-204-2H in Sichuan and Chongqing gas fields and the well Yingshen-1 in Tarim oilfield as well as TK1127 in

Lin Yuanhua. PhD. E-mail: yhlin28@163.com. Tel: 13908085550. Address: Southwest Petroleum University, Chengdu, Sichuan, 610500, P. R. China.

Tahe oilfield in Xinjiang were deformed by creep and plastic flow of salt rock. The casing deformation has posed a serious threat to the safety and benefits of oil and gas field. So, to eliminate the serious threat, one reliable repair technology is needed urgently to repair the deformed casings. Of course, many repair technologies[3] have been proposed at home and aboard. Among of them, the smith expansion cone (SEC) repair technology is often used to repair the deformed casing of oil and water wells[4], especially for casing with small deformation. The advantage of SEC repair technology is that it can realize the mechanics/kinetics conversion depend on structural features of SEC.

At present, in order to restore the normal production of oil and gas field, some scholars have done many studies in the field of casing failure. It mainly includes the studies about the mechanism of casing failure[5, 6], repair technologies[7-11] (such as grind stress plastic technology, smith expansion cone technology, casing patch technology and solid expandable tubular technology) and preventive measures[12]. Only a few theory studies on grind stress plastic technology have been done by some scholars[13], all of them, Jiang's[1] work is the most detailed one. In his study, the mechanical model of casing damage well repaired by grind stress plastic technology was established according to the Hertz elastic contiguity theory. Besides, the researches about the solid expandable tubular technology for casing repair have been done in detail by many scholars[14-18], and a large number of fruitful results about those researches have been achieved.

However, the current SEC repair technology often leads to many problems[19-20]. For example, the reshaping force that is used to repair deformed casing can be determined only by the experience in a practical application due to the lack of the theoretical and experimental study on the reshaping force of SEC so that the repairing force has some randomness, which can easily lead to the secondary damage of casing and sticking accidents and makes the repair technology can't be used widely in oil and gas field.

According to the literatures surveyed, it is known that many researches about mechanism of casing failure and solid expandable tubular technologies have been done by a great number of scholars. However, theoretical studies about SEC repair technology are not enough or can't solve problems completely. In particular for the study about reshaping force of SEC, although the research is extremely important, to date, it is not enough obviously. Hence, this study aimed to provide one method to calculate reshaping force of the SEC needed for repairing deformed casing. Hence, the mechanical model of SEC and deformed casing is presented by this paper based on twin shear unified strength theory. The mechanical model can calculate reshaping force that is used to repair the deformed casing under any confining pressure, and there is a good agreement between results obtained from this model and experiment. Based on this model, the influence rules of main parameters of SEC on the reshaping force were obtained.

2 Working Principle of Smith Expansion Cone

In the repairing process of deformed casing, once the axial load (F) is applied to SEC by drill string, the contact pressure between expansion section of SEC and the deformation area of casing will be formed immediately, as shown in Fig. 1. It is the contact pressure that enables the casing to be in the plastic state and produce circumferential deformation. So, the goal to repair deformed casing can be attained by the repair technology.

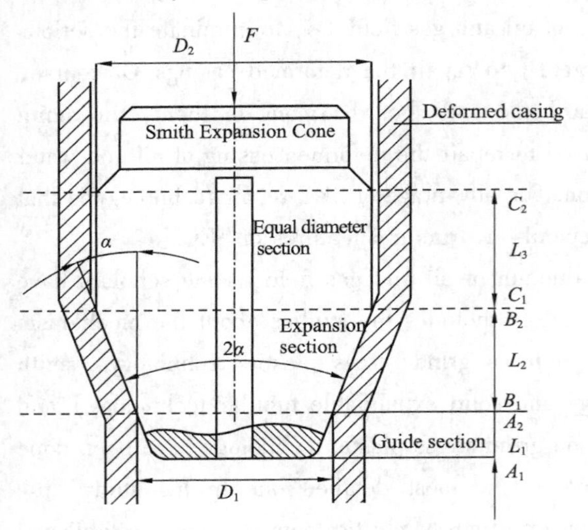

Fig. 1 Structure of smith expansion cone

Firstly, the SEC needs to be placed to the deformed area of casing when the SEC is used to repair the deformed casing. Next, the expansion section of SEC is forced into the deformation area of casing relying on the axial load applied on the SEC. Finally, the purpose of repair is achieved by the reciprocating motion of SEC in the center of deformed casing. The advantage of SEC repair technology is that it can realize the mechanics/kinetics conversion of axial loading and circumferential deformation of deformed casing in the repairing process, and its operation is simple and convenient.

In Fig. 1, α is the half-cone angle of SEC, degree; D_1 is the inner diameter of deformation area, mm; D_2 is the diameter of equal diameter section of SEC, mm; L_1 (A_1A_2) is the length of guide section, mm; L_2 (B_1B_2) is the length of expansion section, mm; L_3 (C_1C_2) is the length of equal diameter section.

3 Study of mechanical model

According to the literature surveyed, the crosssection of deformed casing is oval[21-22] in shape under non-uniform load. The ovality e is used to denote the deformational degree of casing[23], and the larger the ovality, the more serious the deformation of casing is.

In order to analyze the mechanical problem conveniently, the following hypotheses are made in this paper:

(1) The deformed casing is one idea circular tube.

(2) The outer wall of SEC is contacted with inner wall of casing completely in the repairing process for the convenience of modeling.

Based on the above hypotheses, the contact pressure between the SEC and casing can be equivalent to the uniform internal pressure (P) applied to the inner wall of casing, as shown in Fig. 2.

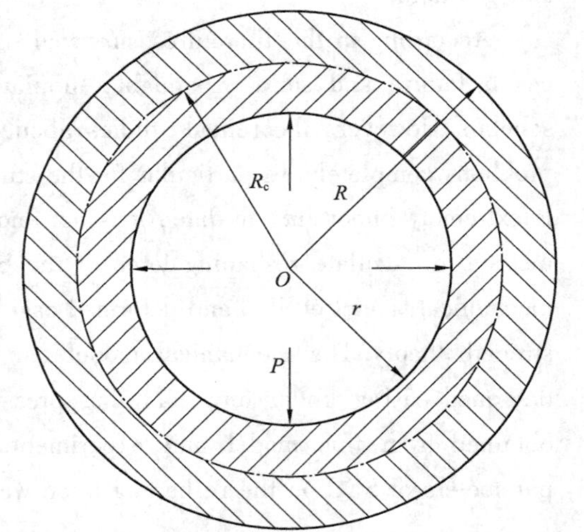

Fig. 2 Mechanical model of casing under contact pressure (P)

3.1 Plastic Limit Analysis of Casing under Internal Pressure

Based on the elastic-plastic theory, it can be known from the elastic-plastic theory that the casing will be undergone three phases in turn: elastic stage, elastic-plastic stage and plastic flow stage. Before the casing expansion, the inner wall of casing yields initially, and the contact pressure

(P) between the SEC and casing is considered to be the elastic limit load (P_e). The twin-shear unified strength theory[24-25] has been adopted to analyze this mechanical problem, as follows:

$$f=\begin{cases}\sigma_1-\dfrac{\beta}{1+b}(b\sigma_2+\sigma_3)=\sigma_t,\ \sigma_2\leqslant\dfrac{\sigma_1+\beta\sigma_3}{1+\beta}\\ \dfrac{1}{1+b}(\sigma_1+b\sigma_2)-\beta\sigma_3=\sigma_t,\ \sigma_2\geqslant\dfrac{\sigma_1+\beta\sigma_3}{1+\beta}\\ \beta=\dfrac{\sigma_t}{\sigma_c}(0\leqslant\beta\leqslant1),\ b=\dfrac{(1+\alpha)\tau_s-\sigma_t}{\sigma_t-\tau_s}(0\leqslant b\leqslant1)\end{cases}\tag{1}$$

Where σ_t is the tensile strength, MPa; σ_c is the compressive strength, MPa; τ_s is the shear strength, MPa; β is the ratio of tensile strength to compressive strength; b is the influence coefficient which reflects the effect of second/intermediate principal stress on the material failure.

Based on the Eq. (1) and complicated numerical derivations, the elastic limit load of casing under internal pressure can be obtained:

$$p_e=\frac{(1+b)(R^2-r^2)\sigma_s}{(1+b)(R^2+r^2)+[\beta(R^2-r^2-br^2)]}\tag{2}$$

Where σ_s is the yield strength of casing, MPa; R is the outer radius of casing, mm; r is the inner radius of casing, mm.

The plastic region will be formed near the inner wall of casing, when internal pressure (P) is larger than elastic limit load (P_e). Assuming that R_c is the radius of interface between elastic region and plastic region, mm, as shown in Fig. 2. The plastic region will extend from inner surface to outer surface with the increase of internal pressure (P), which will result in the range of $r\leqslant\rho\leqslant R_c$ becoming plastic region and the range of $R_c\leqslant\rho\leqslant R$ becoming elastic region, as shown in Fig. 2. The interface between the elastic region and plastic region is cylinder surface due to the axial symmetry. The whole wall of casing is in plastic state, when the radius of interface (R_c) is equal to the outer radius (R) of casing. Based on my work group research results[26], the plastic limit load (P_p) of casing can be obtained:

$$p_p=\frac{\sigma_s}{1-\beta}\left[\left(\frac{R}{r}\right)^{\frac{-2(1+b)(1-\beta)}{2+2b-b\beta}}-1\right]\tag{3}$$

Based on the twin shear yield criterion, by the Eq.(3), the calculation equation of plastic limit load can be obtained:

$$P_p=\lim_{\beta\to0,\ b=1}\frac{\sigma_s}{1-\alpha}\left(\frac{2b+2-2\beta-2b\beta}{2+2b-b\beta}\right)$$
$$\ln\left(\frac{R}{r}\right)=\frac{4}{3}\sigma_s\ln\left(\frac{R}{r}\right)\tag{4}$$

However, the Eq. (4) used to calculate the plastic limit load of real casing based on the yield strength (σ_s) of casing is not reasonable and needs to be further improved because the real casing will undergo the material hardening and large plastic deformation from inner wall yield to whole wall yield. Hence, the yield strength (σ_s) of tubing and casing has been replaced by flow stress (σ_f) to calculate the plastic limit load accurately and reasonably in this paper.

The Klever's research results[27] also demonstrated that the flow stress (σ_f) which can deal with the effect of material hardening and larger plastic deformation on the plastic limit load of casing ranged between the yield strength and tensile strength. In addition, the influence rules of the ratio of yield strength (σ_s) to tensile strength (σ_t) on the plastic limit load of casing have been obtained[28-29] by my work group. So, on the base of results of my work group, a new equation (Eq. 5) for calculating the plastic limit load of casing under internal pressure is presented:

$$P_p = \begin{cases} \frac{4}{3}\sigma_f \ln\left(\frac{D/t}{D/t-2}\right), & \sigma_f = \sigma_s,\ 0.8 \leqslant \frac{\sigma_s}{\sigma_t} \leqslant 1 \\ \frac{4}{3}\sigma_f \ln\left(\frac{D/t}{D/t-2}\right), & \sigma_f = \sigma_s + 10(\text{kpsi}),\ 0 < \frac{\sigma_s}{\sigma_t} \leqslant 0.8 \end{cases} \tag{5}$$

If the outer wall of casing is subjected to the external pressure (P_o) at the same time, the plastic limit load (P_p) should be replaced by $\frac{4}{3}\sigma_f \ln\left(\frac{R}{r}\right) + P_o$.

Where D is the diameter of idea casing, mm; t is the wall thickness of idea casing, mm; σ_t is the tensile strength of idea casing, MPa. The casing just reaches the plastic stage, when the internal pressure P (contact pressure) is equal to plastic limit load P_p, and it is the minimum contact pressure P needed for making casing plastic deformation.

3.2 Determination of Reshaping Force

The SEC consists of the guide section, expansion section and equal diameter section, as shown in Fig. 1. By the Eq. (6), the length of expansion section can be obtained:

$$L_2 = \frac{D_2 - D_1}{2\tan\alpha} \tag{6}$$

Based on the Eq. (6), the length of expansion section of SEC can be designed accurately according to the real deformed casing under the actual working condition.

In the repairing process, the expansion section and equal diameter section of SEC are the main working portions. The mechanical analysis of expansion section and equal diameter section is presented in turn as follows, as shown in Fig. 3 and Fig. 4.

Fig. 3 Mechanical analysis of the expansion section Fig. 4 Mechanical analysis of equal diameter section

3.2.1 Mechanical analysis of the expansion section

According to the Fig. 3, the equilibrium equation can be obtained:

$$F_1 = f_1\cos\alpha + N_1\sin\alpha \tag{7}$$

Where f_1 is the sliding friction force between the SEC and casing, and $f_1 = \mu N_1$ can be obtained based on the coulomb friction model, N; F_1 is the axial force applied to expansion section of SEC, N; N_1 is the reaction force on the surface of inner wall of casing, N; μ is the friction coefficient.

In order to repair the deformed casing, the reaction force (N_1) must provide the minimum contact pressure P_p [it can be obtained by the Eq. (5)] that makes the deformed casing be into plastic stage completely. So, $N_1 = P_p S_1$, S_1 is the superficial area of expansion section, mm^2. According to Fig. 1, the superficial area can be obtained:

$$S_1 = \frac{\pi}{4\sin\alpha}(D_2^2 - D_1^2) \tag{8}$$

3.2.2 Mechanical analysis of the equal diameter section

After the expansion section of SEC gets through the deformation area of casing completely, the equal diameter section of SEC will continue to expand and repair the deformed casing, and there is still contact pressure between the expansion section and inner wall of casing, as shown in Fig. 4. By the Eq. (9), the superficial area (S_2) of expansion section of SEC can be obtained:

$$S_2 = \pi D_2 L_3 \tag{9}$$

Where L_3 is the contact length between equal diameter section of SEC and inner wall of casing, mm. Similarly, in order to repair the deformed casing, the reaction force (N_2) must provide the minimum contact pressure P_p [It can be obtained by the Eq. (5)] that makes the deformed casing be into plastic stage completely, so $N_2 = P_p S_2$.

In Fig. 4, according to the equilibrium conditions of forces, the axial force F_2 can be obtained:

$$F_2 = f_2 = \mu N_2 = \mu p S_2 \tag{10}$$

By the Eq. (5), Eq. (7) and Eq. (10), the reshaping force required to repair deformed casing can be obtained:

$$F = F_1 + F_2 = \begin{cases} P_p\left[\frac{1}{4}(D_2^2 - D_1^2)(\mu\cot\alpha + 1) + \mu D_2 L_3\right], & \sigma_f = \sigma_s,\ 0.8 \leqslant \frac{\sigma_s}{\sigma_t} \leqslant 1 \\ P_p\left[\frac{1}{4}(D_2^2 - D_1^2)(\mu\cot\alpha + 1) + \mu D_2 L_3\right], & \sigma_f = \sigma_s + 10(\text{kpsi}),\ 0 < \frac{\sigma_s}{\sigma_t} \leqslant 0.8 \end{cases} \tag{11}$$

Where

$$P_p = \begin{cases} \frac{4}{3}\sigma_f \ln\left(\frac{D/t}{D/t - 2}\right), & \sigma_f = \sigma_s,\ 0.8 \leqslant \frac{\sigma_s}{\sigma_t} \leqslant 1 \\ \frac{4}{3}\sigma_f \ln\left(\frac{D/t}{D/t - 2}\right), & \sigma_f = \sigma_s + 10(\text{kpsi}),\ 0 < \frac{\sigma_s}{\sigma_t} \leqslant 0.8 \end{cases}$$

3.3 Analysis of the Effects of Parameters on the Reshaping Force

It can be observed from the Eq. (11) that the reshaping force (F) required to repair deformed casing is related to six parameters: expansion amount every time (D_2-D_1), friction coefficient (μ), yield strength (σ_s), tensile strength (σ_t), half-cone angle (α) and the length of equal diameter section (L_3). From comparison and analysis of parameters in the Eq. (11), the yield strength and tensile strength are found to be much less effects on the reshaping force than other parameters and can't be changed and improved in the repairing process based on the detailed analysis. However, the expansion amount every time, friction coefficient, half-cone angle and equal diameter section have direct impact on the reshaping force and can be changed and improved in the repairing process. Hence, in order to analyze their effects on the reshaping force conveniently, the relationships between reshaping force (F) and expansion amount every time ($D_2 - D_1$), friction coefficient (μ), length of equal diameter section (L_3) and half-cone angle (α) have been plotted, as shown in Fig. 5, Fig. 6, Fig. 7 and Fig. 8.

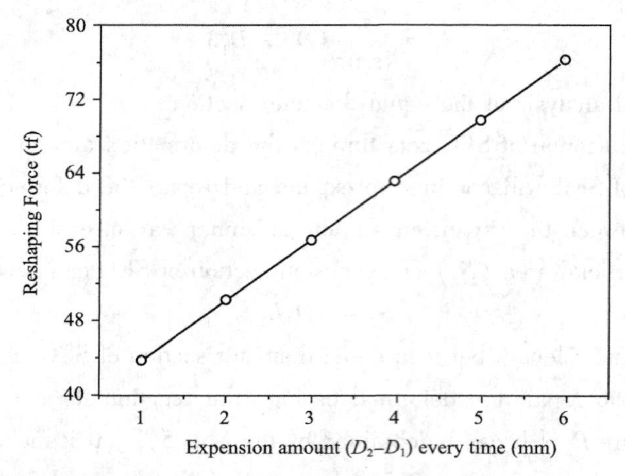

Fig. 5 Relationship between expansion amount ($D_2 - D_1$) every time and reshaping force

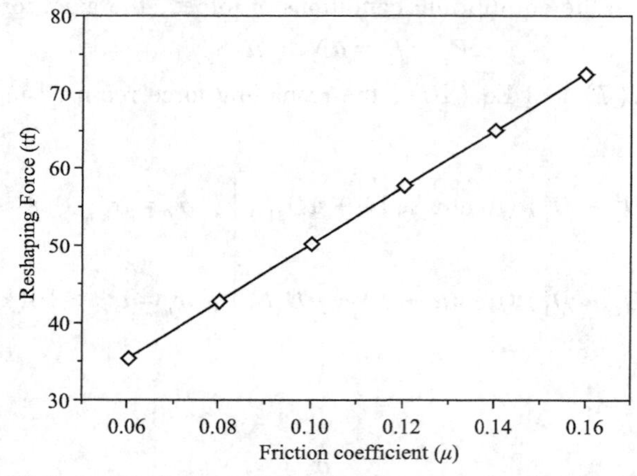

Fig. 6 Relationship between friction coefficient (μ) and reshaping force

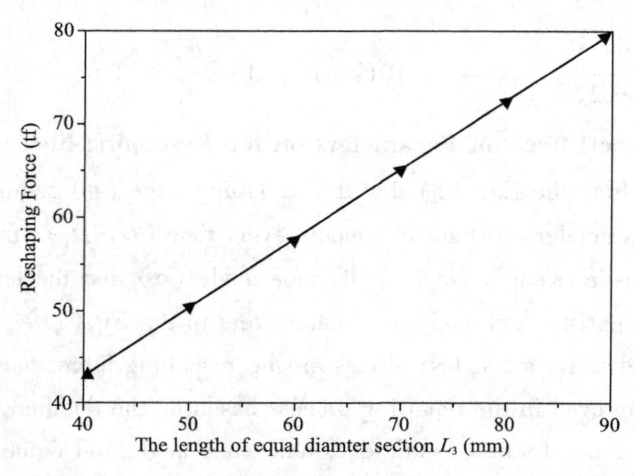

Fig. 7 Relationship between the length of length of equal diameter section (L_3) and reshaping force

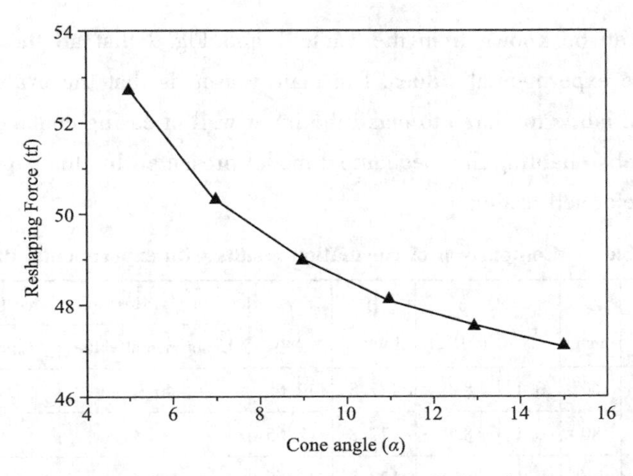

Fig. 8 Relationship between half-cone angle (α) and reshaping force

It can be observed from Fig. 5, Fig. 6 and Fig. 7 that the relationships between reshaping force and expansion amount, friction coefficient and length of equal diameter section are linear. The reshaping force increases with the increase of expansion amount, friction coefficient and length of equal diameter section. However, it can be known from Fig. 8 that the reshaping force decreases with the increase of cone angle. By comparison, the effects of expansion amount, friction coefficient and length of equal diameter section on reshaping force are larger than half-cone angle. In addition, the larger cone angle of SEC is harmful to lubrication between the SEC and inner wall of deformed casing[30] and the adhesion between SEC and inner wall of casing occurs easily, which leads to casing damage. Hence, through the reshaping force decreases with increase of cone angle, the larger cone angle should not be adopted in the process of designing the SEC. The research results can provide theoretical guidance for the design of SEC.

4 Comparison of Calculation Results with Experimental Data

The reshaping forces required to repair C110 deformed casing have been calculated by Eq. (10) and the corresponding parameters of this deformed casing and the calculated results are shown in Table 1. It is clearly showed in Fig. 9 that the calculation results obtained from the Eq. (10) are quite closer to experimental results[31], and its error is about 10 percent, which can be accepted by engineering.

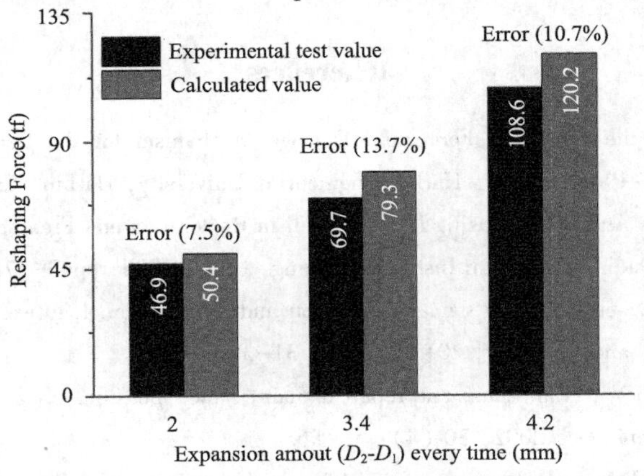

Fig. 9 Comparison of experimental data and calculation results

In addition, it can be known from the Table 1 and Fig. 9 that all the calculated results are slightly larger than the experimental values. The main reason is that the ovality of deformed casing repaired by SEC in the lab is too large to make the inner wall of casing contact with the outer wall of SEC completely, thereby enabling the mechanical model presented by this paper to deviate from true repairing process of deformed casing.

Table 1 Comparison of calculation results with experimental data

No.	D_2 (mm)	D_1 (mm)	α (°)	L_3 (mm)	μ	σ_s (MPa)	D (mm)	t (mm)	Reshaping force (tf)		Error (%)
									Experimental value	Calculated value	
1	122	120	7	60	0.1	830	177.8	12.65	46.9	50.4	7.5
2	125	121.6	7	80	0.1	830	177.8	12.65	69.7	79.3	13.7
3	128	123.8	7	140	0.1	830	177.8	12.65	108.6	120.2	10.7

5 Conclusions

Based on twin shear unified strength theory, the mechanical model of deformed casing repaired by smith expansion cone was presented by this paper; the effects of material hardening and the ratio of yield strength to tensile strength on casing repair were taken into account in this model. It can be used to calculate the reshaping force required to repair deformed casing under any confining pressure, and the reliability and accuracy of this model was validated by experimental data. According to this model, the influence rules of expansion amount every time, friction coefficient, half-cone angle and the length of equal diameter section on reshaping force were obtained by detailed analysis. The method of mechanics and kinetics analysis about axial loading and circumferential deformation of deformed casing was formed. The research results can provide theoretical guidance for design and optimization of structure and smith expansion cone repair technology.

Acknowledgements

Authors are thankful to the financial assistance provided by the National Natural Science Foundation of China (No. 51274170).

References

[1] Minzheng Jiang. Study the Repairing Force and Damage Mechanism for the Cement Loop During Repairing Destroyed Casing [D]. Ph. D. Thesis, Harbin Engineering University, Harbin, China. 2003, pp1-10.

[2] Xiuqing He. Mechanical Analysis of Casing Damage Well in Restoration and Fracture Process and Its Application [D]. Ph. D. Thesis, Daqing Petroleum Institute, Daqing, China. 2006, pp43-57.

[3] Wei Liu, Li Li, Hui L, et al. Casing damage detection and repair technology western sichuan deep well [J]. Natural Gas Technology and Economy, 2011, 5 (1): 31-33.

[4] Lixia Fu, Lijun M, Wei W, et al. Cause and repair & maintenance measures of casing in oil and water well [J]. Petroleum Drilling Techniques, 2002, 30 (4): 53-56.

[5] Jicheng Z, Kaoping S, Ruoxia D. Dongguo casing failures derive from many causes [J]. Oil and Gas Journal,

2008, 106 (3): 51-54.

[6] Li J. Casing damage mechanism and numerical simulation induced by sand producing [J]. Advances in Petroleum Exploration and Development, 2011, 2 (2): 58-62.

[7] Perveza T., Qamara S. Z., Seibib A. C., et al. Use of SET in cased and open holes: comparison between aluminum and steel [J]. Materials & Design, 2008, 29 (4): 811-817.

[8] Blane D. A., Crossland A. J.. Design and development of a tubing repair solution using an expandable metal patch [C]. SPE130718, 2010.

[9] Storaune A, Winters W. J. Versatile expandables technology for casing repair [C]. SPE/IADC 92330, 2005.

[10] Innes G, Craig J, Lavan S. The use of expandable tubular technology to enhance reservoir management and maintain integrity [C]. OTC 15148, 2003.

[11] Junsheng S, Xiaohong G, Yingchun S, et al. Rolling reshaping technology used to repair deformed casings downhole in Puguang gas field [J]. Natural Gas Industry 2009, 29 (6): 52-54.

[12] Tao W, Shenglai Y, Weihong Z, et al. Law and countermeasures for the casing damage of oil production wells and water injection wells in Tarim Oilfield [J]. Petroleum Exploration and Development, 2011, 38 (3): 352-361.

[13] Lin Y, Deng K, Zeng D, et al. Numerical and experimental study on working mechanics of rolling reshaper [J], Arabian Journal for Science and Engineering, 2014, 39 (11): 8099-8110.

[14] Stewart R. B., Marketz F., Lohbeck W. C. M., et al. Expandable Wellbore Tubulars [C]. SPE 60766, 1999.

[15] Binggui X, Yanping Z, Hui W, et al. Application of numerical simulation in the solid expandable tubular repair for casing damaged wells [J]. Petroleum Exploration and Development, 2009, 36 (5): 651-657.

[16] Mack R. D., Shell Intl. The effect of tubular expansion on the mechanical properties and performance of selected OCTG-results of laboratory studies [C]. SPE 17622-MS, 2005.

[17] Al-Abri., Omar S. Analytical and numerical solution for large plastic deformation of solid expandable tubular [C]. SPE 152370-STU, 2011.

[18] Perveza T., Seibia A. C., Karrecha A. Simulation of solid tubular expansion in well drilling using finite element method [J]. Petroleum Science and Technology, 2007, 23 (7-8): 775-794.

[19] Jingtai N, Youwen P, Congcong W, et al. Cement loop damage-based fracture mechanism during repair of casing failure well [J]. Procedia Earth and Planetary Science, 2011, 5: 322-325.

[20] Xiaopeng Z, Yishan L, mule W, et al. Research on the critical indexes for casing dressing by Smith expansion cone [J]. China Petroleum Machinery, 2012, 40 (2): 1-3.

[21] Han J Z, Shi T H. Equations calculate collapse pressures for casing strings [J]. Oil & Gas Journal, 2001, 99 (4): 44-47.

[22] Lin Y, Deng K, Zeng D, et al. Theoretical and experimental analyses of casing collapsing strength under non-uniform loading [J]. Journal of Central South University, 2014, 21 (9): 3470-3478.

[23] Petroleum and natural gas industries-Formulae and calculation for casing, tubing, drill pipe and line pipe properties [S]. ISO 10400: 2007.

[24] Yu Maohong. Unified Strength Theory and Its Applications [M]. Berlin: Springer, 2004.

[25] Junhai Z, Yan L, Changguang Z. Collapsing strength for petroleum casing string based on unified strength theory [J]. Acta Petroleum Sinica, 2013, 34 (5): 969-976.

[26] Lin Y, Deng K, Sun Y, et al. Burst strength of tubing and casing based on twin shear unified strength theory [J], PLOS ONE (Forthcoming)

[27] Lee Y K, Ghosh J. The significance of J3 to the prediction of shear bands [J]. International Journal of Plasticity, 1996, 12 (9): 1179-1197.

[28] Sun Y. Study on collapse strength and internal pressure strength of casing and tubing [D] Ph. D. Thesis, Southwest Petroleum University, Sichuan, China, 2008.

[29] Wei L. Study on the internal pressure strength of casing [D]. M. Sc. Thesis, Southwest Petroleum University, Sichuan, China, 2010.

[30] Longxiang G, Jianhong F, Yuanhua L, et al. The theoretical calculation of expansion force of expandable tubular [J]. Drilling & Production Technology, 2006, 29 (4): 76-77.

[31] Yuanhua L, Kuanhai D, Dezhi Z, et al. An experimental investigation into parameters optimization of reshaping casing by pear-shaped casing swage [J]. Natural Gas Industry, 2014, 34 (2): 71-75.

基于统一强度理论的套管全管壁屈服挤毁压力*

林元华[1]　邓宽海[1]　孙永兴[2]　曾德智[2]　仝闽闽[3]

（1. 西南石油大学油气藏地质及开发工程国家重点实验室；
2. 西南石油大学石油管工程重点实验室；
3. 西南石油大学材料科学与工程学院）

摘　要：基于对 API 5C3 屈服挤毁公式和 ISO 全管壁屈服挤毁公式的研究，发现：API 5C3 屈服挤毁设计的基本原理是油井管内壁发生屈服即认为油井管失效。实际上，内壁刚开始屈服时套管还有较大的抗挤承载能力，尤其对于厚壁和高抗挤套管，而 ISO 全管壁屈服挤毁公式并非是真正的全管壁屈服挤毁公式，导致 ISO 挤毁模型不适合于所有径厚比套管强度的计算。为此，根据统一强度理论推导出了套管全壁屈服挤毁压力的统一算法，该算法充分考虑了拉压强度不等特性（SD 效应）、中间主应力、材料硬化、屈强比对套管全壁屈服挤毁压力的影响，适合于计算具有 SD 效应和中间主应力效应的套管全壁屈服的挤毁强度；建立了 R. Von Mises、Tresca、GM、双剪屈服四种典型屈服准则下的全壁屈服挤毁公式，分析了内壁/全壁屈服、屈服准则、SD 效应和中间主应力对屈服挤毁压力的影响。新算法解决了 ISO 抗挤模型不能准确计算套管全壁屈服挤毁压力的问题，弥补了该模型不适合于预测高抗挤套管挤毁强度的不足。

关键词：套管；统一强度理论；全壁屈服；挤毁压力；SD 效应；高抗挤；中间主应力

在油气钻井和生产过程中，石油套管起着保护井眼、加固井壁、隔绝井中的油、气、水层及封固各种复杂地层的作用。近年来，随着石油和天然气勘探开发的不断深入，深井和超深井及膏岩、盐岩、页岩、软泥岩等塑性流动地层的数量也不断增加[1-3]，使得石油套管的服役环境变得非常恶劣，尤其深井和超深井中的石油套管将受到更大的外挤压力，导致许多常规的 API 套管不能完全满足深井和超深井的强度要求，严重制约着深井和超深井钻井技术的发展及钻进深度的延伸。

为满足深井和超深井套管的强度要求，国内外也研制出了许多厚壁、非 API 及高抗挤套管[4-6]，与常规 API 套管相比，厚壁、非 API 及高抗挤套管具有更高的挤毁性能[7,8]，尽管如此，近几年仍有关于高强度套管在深井和超深井中挤毁失效的相关报道[9]。因此，正确合理地预测和设计套管（尤其高抗挤套管）挤毁强度是确保油气安全开采的关键，对深井和超深井钻井技术的发展及钻进深度的延伸具有重要意义。

---

* 本文受国家自然科学基金资助（批准号：51274170）。

作者简介：林元华，男，1971 年生，博士，教授，博士生导师（西南石油大学），主要从事油井管、油气钻井工艺方面的科研和教学工作。地址：四川省成都市新都区，西南石油大学，邮政编码：610500。E-mail：yhlin28@163. com。

当然，近年来国内外也有大量学者对均匀载荷下套管的屈服挤毁强度[10-16]进行研究，并取得了一些重要的成果（建立了一些用于计算套管屈服挤毁压力的经典力学模型，形成了相应的标准，如 API 5C3 和 ISO 10400 标准[14, 16]），但大部分力学模型的设计原理是套管内壁屈服及失效，即认为使套管内壁上某点的应力达到屈服极限时的外压就是套管的屈服挤毁压力，事实上，内壁开始屈服时套管并未完全破坏，且还有很大的抗挤余量[17,18]，若按此原理设计，会造成套管材料浪费或套管选择难的问题，其次，除赵均海等人[15]基于套管内壁屈服及失效的设计原理建立了能考虑中间主应力效应和 SD 效应（尤其高强度钢具有显著的 SD 效应，且其对套管强度设计具有较大影响）的套管屈服挤毁压力计算方程之外，大部分力学模型[10-14]主要是基于 R. Von Mises 和 Tresca 屈服准则建立的，而没有充分考虑中间主应力效应和 SD 效应对套管屈服挤毁压力的影响，仅适合于抗拉和抗压强度相等的材料，此外，研究还发现 ISO 10400 挤毁模型[16]中的全壁屈服挤毁压力公式并非真正的全壁屈服挤毁压力公式，且该挤毁模型也没有考虑 SD 效应的影响，导致其可能并不适合所有径厚比套管挤毁强度的计算，尤其是高抗挤套管[9, 18]。

为此，本文基于工程研究领域广泛应用的统一强度理论[19-22]，推导出了套管全壁屈服挤毁压力计算公式，该公式充分考虑了 SD 效应、中间主应力效应、材料硬化、屈强比对套管全壁屈服挤毁压力的影响，能够准确合理地计算所有径厚比套管（如厚壁、非 API 及高抗挤套管）的全壁屈服挤毁压力，为当前深井和超深井套管强度设计提供了重要参考和新思路，更重要的是不仅解决了 ISO 10400 挤毁模型不能准确计算套管全壁屈服挤毁压力，而且弥补了 ISO 10400 挤毁模型不适合于预测所有径厚比套管挤毁强度的不足。

1　套管全壁屈服挤毁压力公式

1.1　统一强度理论

统一强度理论[19,20]可以退化为现有的多种屈服准则和一系列新的屈服准则，它考虑了中间主应力和拉压比（即 SD 效应）对材料强度的影响，可以灵活用于分析各种材料的塑性极限，其数学表达式为：

$$\begin{cases} f = \sigma_1 - \dfrac{\alpha}{1+b}(b\sigma_2 + \sigma_3) = \sigma_s, \ \sigma_2 \leqslant \dfrac{\sigma_1 + \alpha\sigma_3}{1+\alpha} \\ f' = \dfrac{1}{1+b}(\sigma_1 + b\sigma_2) - \alpha\sigma_3 = \sigma_s, \ \sigma_2 \geqslant \dfrac{\sigma_1 + \alpha\sigma_3}{1+\alpha} \\ \alpha = \dfrac{\sigma_t}{\sigma_c}, \ 0 \leqslant \alpha \leqslant 1 \\ b = \dfrac{(1+\alpha)\tau_s - \sigma_t}{\sigma_t - \tau_s}, \ 0 \leqslant b \leqslant 1 \end{cases} \tag{1}$$

式中　σ_1，σ_2，σ_3——第一、第二和第三主应力，MPa；

α——材料的拉压比；

σ_s，σ_c，τ_s——材料的抗拉屈服强度、抗压强度和剪切强度，MPa；

b——中间主应力对材料破坏的影响系数。

1.2　均匀外压作用下套管的弹性极限分析

套管在外压 P_o 的作用下，随着外压 P_o 的增加，应力分量不断增加，屈服首先在套管内壁发生，使得套管处于弹性极限状态。假设套管为一长厚壁圆筒，其内外径分别为 R_i 和 R_o，外表面受均布压力 P_o 作用，当外压 P_o 较小时，厚壁圆筒处于弹性状态，根据拉梅公式[23]可以得到其应力分量：

$$\begin{cases} \sigma_\rho = -\dfrac{R_o^2 P_o}{R_o^2 - R_i^2}\left(1 - \dfrac{R_i^2}{\rho^2}\right) \\ \sigma_\theta = -\dfrac{R_o^2 P_o}{R_o^2 - R_i^2}\left(1 + \dfrac{R_i^2}{\rho^2}\right) \\ \sigma_z = \dfrac{m}{2}(\sigma_\rho + \sigma_\theta) \end{cases} \tag{2}$$

由于长厚壁圆筒属于轴对称平面应变问题，基于文献[24，25]对于平面应变弹塑性问题的研究，中间主应力 σ_z 在弹性区，$m=2\mu$，在塑性区，$m=1$。当套管处于弹性极限状态时(即内壁屈服)，由式（2）可得：

$$\sigma_z = \frac{m}{2}(\sigma_\rho + \sigma_\theta) = -\frac{R_o^2 P_o}{(R_o^2 - R_i^2)} \tag{3}$$

基于厚壁圆筒的应力状态，可知 $\sigma_\rho \geqslant \sigma_z \geqslant \sigma_\theta$，则有 $\sigma_1 = \sigma_\rho$，$\sigma_2 = \sigma_z$，$\sigma_3 = \sigma_\theta$。由于拉压比 $\alpha<1$，可得 $\sigma_2 \leqslant \dfrac{(\sigma_1 + \alpha\sigma_3)}{(1 + \alpha)}$。因此，将式（2）和式（3）代入统一强度理论表达式（1）中可得：

$$-\frac{R_o^2 P_o}{R_o^2 - R_i^2}\left[\left(1 - \frac{R_i^2}{\rho^2}\right) - \frac{b\alpha}{2(1+b)}\left(1 - \frac{R_i^2}{\rho^2}\right) - \frac{\alpha}{2}\left(1 + \frac{R_i^2}{\rho^2}\right) - \frac{\alpha}{2(1+b)}\left(1 + \frac{R_i^2}{\rho^2}\right)\right] = \sigma_s \tag{4}$$

随着外压 P_o 的增大，厚壁圆筒内壁首先发生屈服，且 $\rho = R_i$，$P = P_y$，由式（4）可得其弹性极限压力 P_y，也称为内壁屈服挤毁压力：

$$p_y = \frac{(1+b)(R_o^2 - R_i^2)\sigma_s}{[\alpha(2+b)]R_o^2} \tag{5}$$

1.3 外压作用下套管的塑性极限分析

当外压 $P_o>P_y$ 时，在厚壁圆筒内壁附近出现塑性区，假设 R_c 为弹塑性分界面的半径，并且随着外压 P_o 的继续增加，塑性区逐渐向外扩展，使得 $R_i \leqslant \rho \leqslant R_c$ 成为塑性区（ρ 任意位置处的半径），而 $R_c \leqslant \rho \leqslant R_o$ 仍处于弹性区，如图 1 所示。由于应力分量 σ_θ、σ_r 的轴对称性，塑性区与弹性区的分界面为圆柱面，因此，分别将塑性区（内筒）和弹性区（外筒）按厚壁圆筒进行力学分析，如图 2 所示。在外筒（弹性区）的内壁和外壁分别作用有径向应力 P_i 和外压 P_o，如图 2 所示，而内筒外壁作用有径向压力 P_i，如图 3 所示。

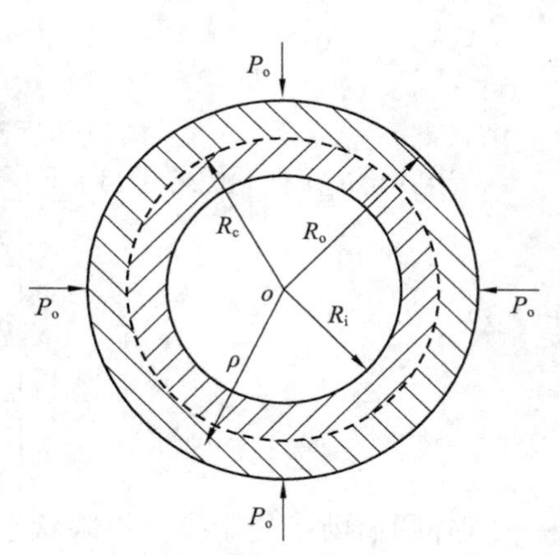

图 1　套管弹塑性力学模型

在 $R_i \leqslant \rho \leqslant R_c$ 的塑性区，由塑性力学平衡方程可得：

$$\frac{d\sigma_\rho}{d\rho} + \frac{\sigma_\rho - \sigma_\theta}{\rho} = 0 \tag{6}$$

由式（2）、式（4）可得：

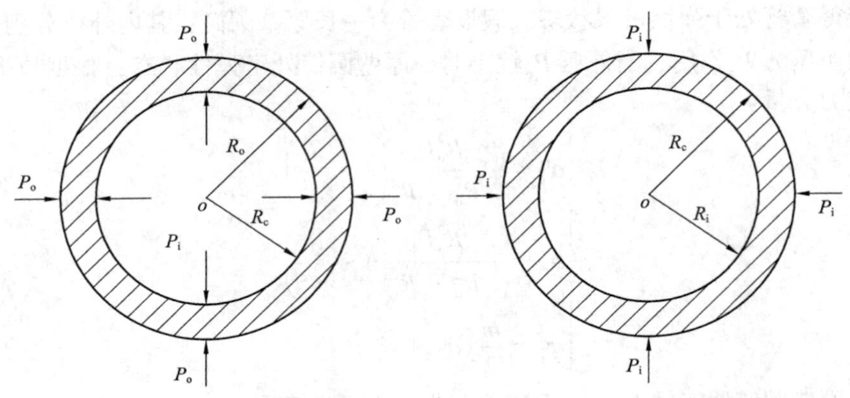

图 2 弹性区（外筒）　　　　图 3 塑性区（内筒）

$$\sigma_{\rho} - \sigma_{\theta} = \frac{2(1+b)(\alpha-1)}{\alpha(2+b)}\sigma_{\rho} + \frac{2(1+b)}{\alpha(2+b)}\sigma_{s} \tag{7}$$

将式（7）代入式（6）并化解可得：

$$\frac{\mathrm{d}\sigma_{\rho}}{\mathrm{d}\rho} + \frac{A\sigma_{\rho} + B\sigma_{s}}{\rho} = 0 \tag{8}$$

将边界条件（$\rho = R_{i}$，$\sigma_{\rho} = 0$）代入式（8）求解可得：

$$\sigma_{r} = \left(\frac{\sigma_{s}}{\alpha-1}\right)\left(\frac{\rho}{R_{i}}\right)^{\frac{2(1-\alpha)(1+b)}{\alpha(2+b)}} - \frac{\sigma_{s}}{\alpha-1} \tag{9}$$

$$\sigma_{\theta} = \left[\frac{2+2b-b\alpha}{\alpha(2+b)}\right]\left[\left(\frac{\sigma_{s}}{\alpha-1}\right)\left(\frac{\rho}{R_{i}}\right)^{\frac{2(1-\alpha)(1+b)}{\alpha(2+b)}} - \frac{\sigma_{s}}{\alpha-1}\right] - \left[\frac{2(1+b)}{\alpha(2+b)}\right]\sigma_{s} \tag{10}$$

在弹性区，即 $R_{c} \leqslant \rho \leqslant R_{o}$ 之间的外圆筒，根据拉梅公式可得外圆筒的应力分量：

$$\begin{cases} \sigma_{\rho} = \dfrac{R_{c}^{2}P_{i} - R_{o}^{2}P_{o}}{R_{o}^{2} - R_{c}^{2}} - \dfrac{(P_{i} - P_{o})}{R_{o}^{2} - R_{c}^{2}}\dfrac{R_{c}^{2}R_{o}^{2}}{\rho^{2}} \\ \sigma_{\theta} = \dfrac{R_{c}^{2}P_{i} - R_{o}^{2}P_{o}}{R_{o}^{2} - R_{c}^{2}} + \dfrac{(P_{i} - P_{o})}{R_{o}^{2} - R_{c}^{2}}\dfrac{R_{c}^{2}R_{o}^{2}}{\rho^{2}} \\ \sigma_{z} = m(\sigma_{\rho} + \sigma_{\theta})/2 \end{cases} \tag{11}$$

当 $\rho = R_{c}$ 时，由式（11）可得外圆筒内壁的应力分量：

$$\begin{cases} \sigma_{\rho} = -P_{i} \\ \sigma_{\theta} = \dfrac{(R_{c}^{2} + R_{o}^{2})P_{i} - 2R_{o}^{2}P_{o}}{R_{o}^{2} - R_{c}^{2}} \\ \sigma_{z} = m\dfrac{R_{c}^{2}P_{i} - R_{o}^{2}P_{o}}{R_{o}^{2} - R_{c}^{2}} \end{cases} \tag{12}$$

外圆筒内壁处于弹性极限状态，将 m 等于 1 代入式（12）中，比较式（12）中的应力分量可得：

$$\sigma_{\rho} \geqslant \sigma_{z} \geqslant \sigma_{\theta} \tag{13}$$

基于式（13）的等量关系，将式（12）代入统一强度理论表达式（1）中求解可得径向应力 P_{i}：

$$P_{i}=\frac{\alpha(b+2)R_{o}^{2}P_{o}}{R_{o}^{2}(1+b+\alpha)-R_{c}^{2}(1+b-b\alpha-\alpha)}-\frac{(1+b)(R_{o}^{2}-R_{c}^{2})\sigma_{s}}{R_{o}^{2}(1+b+\alpha)-R_{c}^{2}(1+b-b\alpha-\alpha)} \tag{14}$$

根据应力分量的连续性，在弹性区与塑性区的分界面上 ($\rho=R_{c}$)，弹性区与塑性区的径向应力 σ_{ρ} 相等，根据式（9）、式（11）和式（14）可得：

$$P_{o}=-\frac{(C+E)}{F} \tag{15}$$

其中

$$C=\sigma_{r}=\left(\frac{\sigma_{s}}{\alpha-1}\right)\left(\frac{R_{c}}{R_{i}}\right)^{\frac{2(1-\alpha)(1+b)}{\alpha(2+b)}}-\frac{\sigma_{s}}{\alpha-1}$$

$$E=\frac{(1+b)(R_{o}^{2}-R_{c}^{2})\sigma_{s}}{R_{o}^{2}(1+b+\alpha)-R_{c}^{2}(1+b-b\alpha-\alpha)}$$

$$F=\frac{\alpha(b+2)R_{o}^{2}}{R_{o}^{2}(1+b+\alpha)-R_{c}^{2}(1+b-b\alpha-\alpha)}$$

随着压力的增加，塑性区不断的扩大，当 $R_{c}=R_{o}$时，套管整个截面进入塑性状态，根据式（15）可得套管全壁屈服挤毁压力 P_{ty}：

$$P_{ty}=\frac{-\sigma_{s}}{\alpha-1}\left[\left(\frac{D/t}{D/t-2}\right)^{\frac{2(1+b)(1-\alpha)}{\alpha(2+b)}}-1\right] \tag{16}$$

式（16）可以充分考虑材料的拉压比和中间主应力对套管全壁屈服挤毁压力的影响，因此，通过改变式（16）中 α 和 b 的值可以计算不同径厚比、钢级套管（如 API、高抗挤、非 API 套管）的全壁屈服挤毁压力，解决了 ISO 10400 挤毁模型中全壁屈服挤毁压力的计算问题及其未考虑 SD 效应和中间主应力效应对套管全壁屈服挤毁压力影响的问题，进一步改善了不同规格套管的强度设计，也为 ISO 10400 挤毁模型的进一步发展和完善提供了理论基础。

2 全壁屈服挤毁压力的分析与讨论

2.1 全壁屈服挤毁压力的分析

2.1.1 中间主应力效应对全壁屈服挤毁压力影响的分析

为了分析中间主应力效应（b）对套管全壁屈服挤毁压力的影响规律，假设套管材料的抗拉屈服极限与抗压屈服极限相等，即 $\alpha=1$，对式（16）求极限可得：

$$P_{ty}=\sigma_{s}\frac{2(1+b)}{(2+b)}\ln\left(\frac{D/t}{D/t-2}\right) \tag{17}$$

基于式（17）可以分析中间主应力对套管全壁屈服挤毁压力的影响规律，下面分别取 $b=0.0$，0.25，0.5，0.75 这四种情况，以 P110 套管为例，分别计算该套管在径厚比 $5\leqslant D/t\leqslant30$ 的全壁屈服挤毁压力与屈服强度之比（后面简称挤毁比：P_{ty}/σ_{s}），如图 4 所示。

由图 4 可知，挤毁比（P_{ty}/σ_{s}）随着 b 的增加而增加，与文献得到的趋势基本相似[15]，且在所有的径厚比范围内，中间主应力（b）对挤毁比的影响规律基本一致，其最大影响比率为 20%（即 $b=0.75$ 的全壁屈服挤毁压力与 $b=0$ 相比，其全壁屈服挤毁压力可提高 20%），故中间主应力对全壁屈服挤毁压力的影响不容忽视，在套管强度设计时，合理考虑中间主应力的影响，可充分发挥套管材料的强度潜能。

2.1.2 SD 效应对全壁屈服挤毁压力影响的分析

为了分析 SD 效应（α）对套管全壁屈服挤毁压力的影响规律，假设 $b=1$，将 $b=1$ 代入式（16）可得：

$$P_{ty}=\frac{-\sigma_s}{\alpha-1}\left[\left(\frac{D/t}{D/t-2}\right)^{\frac{4(1-\alpha)}{3\alpha}}-1\right] \tag{18}$$

基于式（18）可以分析 SD 效应对套管全壁屈服挤毁压力的影响规律，由于套管材料的抗拉强度与抗压强度之比一般在 0.8 左右[26]，故分别取 $\alpha=0.7$，0.8，0.9，1.0（表示拉压强度相等）这四种情况，以 P110 套管为例，分别计算该套管在径厚比 $5\leqslant D/t\leqslant 30$ 的挤毁比（P_{ty}/σ_s），如图 5 所示。

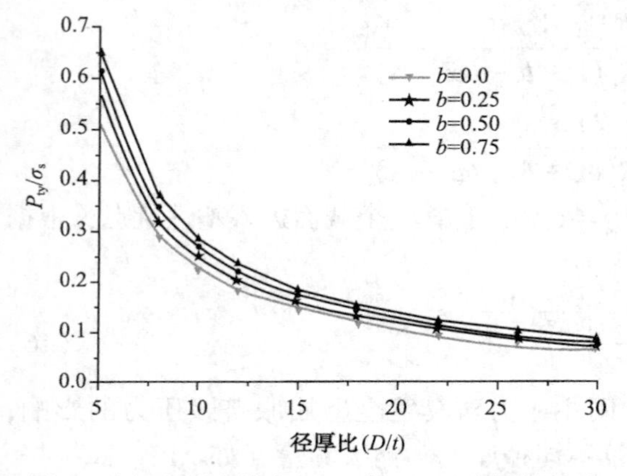

图 4 中间主应力效应（b）与 P_{ty}/σ_s 之间的关系

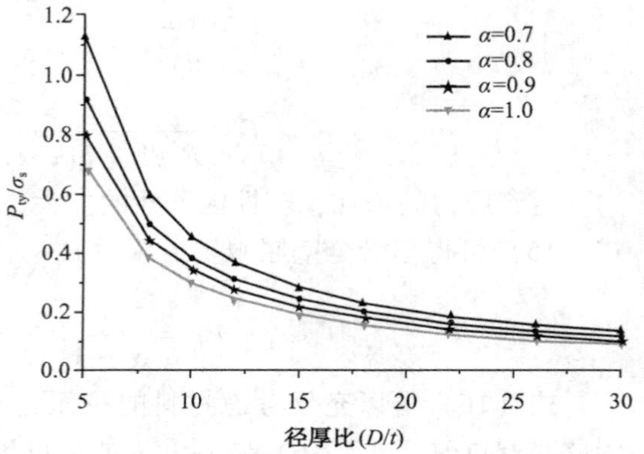

图 5 SD 效应（α）与 P_{ty}/σ_s 之间的关系

由图 5 可知，挤毁比（P_{ty}/σ_s）随着 α 的增加而降低，即抗拉强度与抗压强度的差异越大，其挤毁比越大，且随着径厚比的增加，SD 效应对挤毁比的影响程度逐渐降低，其最大影响比例为 39.5%（$D/t=5$），最小影响比例为 31.1%（$D/t=30$），故 SD 效应对全壁屈服挤毁压力的影响较大。因此，在套管强度设计时，合理考虑 SD 效应的影响，可以充分发挥套管材料的强度潜能，提高套管的挤毁性能。

2.1.3 屈服准则对全壁屈服挤毁压力影响的分析

由于统一强度理论可退化为多种重要的屈服或屈服准则，如 R. Von Mises、Tresca、GM[27,28] 和双剪屈服等屈服准则，因此通过改变式（16）中 α 和 b 的值可以获得多种计算套管全壁屈服挤毁压力的公式，下面依次给出上述屈服准则所对应的计算公式并分析屈服准则对全壁屈服挤毁压力的影响。

2.1.3.1 R. VonMises 屈服准则

当 $\alpha=1$，$b=1/\left(1+\sqrt{3}\right)$，统一强度理论退化为 R. Von Mises 屈服准则，由式（16）求极限可得 R. Von Mises 屈服准则对应的套管全壁屈服挤毁压力计算公式：

$$\lim_{\alpha\to 1}P_{y-new}=\lim_{\alpha\to 1}\frac{-\sigma_s}{\alpha-1}\left[\left(\frac{R_o}{R_i}\right)^{\frac{2(1+b)(1-\alpha)}{\alpha(2+b)}}-1\right]=1.155\sigma_s\ln\left[\frac{(D/t)}{(D/t-2)}\right] \tag{19}$$

2.1.3.2 Tresca 屈服准则

当 $\alpha=1$，$b=0$，统一强度理论退化为 Tresca 屈服准则，由式（16）求极限可得 Tresca 屈服准则对应的套管全壁屈服挤毁压力计算公式：

$$\lim_{\alpha \to 1} P_{y-new} = \lim_{\alpha \to 1} \frac{-\sigma_s}{\alpha - 1}\left[\left(\frac{R_o}{R_i}\right)^{\frac{2(1+b)(1-\alpha)}{\alpha(2+b)}} - 1\right] = \sigma_s \ln\left[\frac{(D/t)}{(D/t-2)}\right] \quad (20)$$

2.1.3.3 GM 屈服准则

当 $\alpha = 1$, $b = 2/5$，统一强度理论退化为 GM 屈服准则，由式（16）求极限可得 GM 屈服准则对应的套管全壁屈服挤毁压力计算公式：

$$\lim_{\alpha \to 1} P_{y-new} = \lim_{\alpha \to 1} \frac{-\sigma_s}{\alpha - 1}\left[\left(\frac{R_o}{R_i}\right)^{\frac{2(1+b)(1-\alpha)}{\alpha(2+b)}} - 1\right] = 1.167\sigma_s \ln\left[\frac{(D/t)}{(D/t-1)}\right] \quad (21)$$

2.1.3.4 双剪屈服准则

当 $\alpha = 1$, $b = 1$，统一强度理论退化为双剪屈服准则，由式（16）求极限可得该屈服准则对应的套管全壁屈服挤毁压力计算公式：

$$\lim_{\alpha \to 1} P_{y-new} = \lim_{\alpha \to 1} \frac{-\sigma_s}{\alpha - 1}\left[\left(\frac{R_o}{R_i}\right)^{\frac{2(1+b)(1-\alpha)}{\alpha(2+b)}} - 1\right] = 1.333\sigma_s \ln\left[\frac{R_o}{R_i}\right] \quad (22)$$

为了分析上述屈服准则对套管全壁屈服挤毁压力的影响，以 N80 和 J55 两种钢级的套管为例，分别采用不同屈服准则下的全壁屈服挤毁公式［即式（19）、式（20）、式（21）和式（22）］计算 N80 和 J55 两种套管的全壁屈服挤毁压力，计算结果体详见图 6 和图 7。

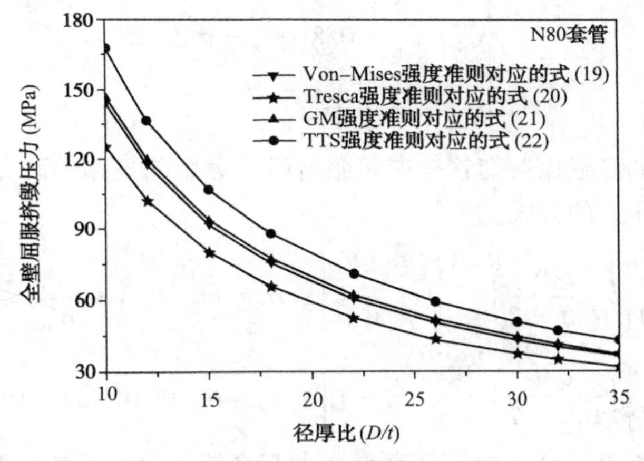

图 6　N80 套管全壁屈服挤毁压力

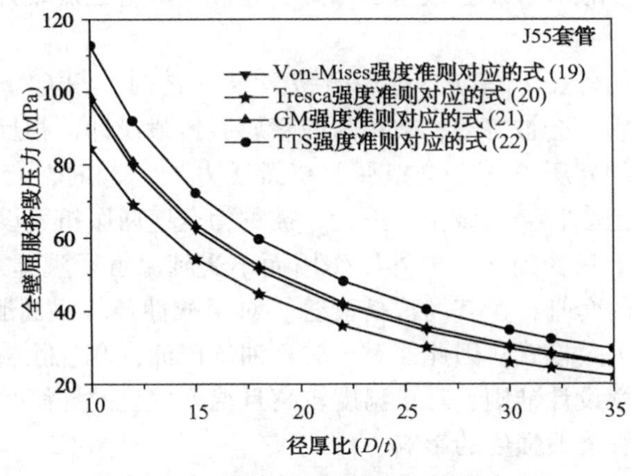

图 7　J55 套管全壁屈服挤毁压力

由图 6 和图 7 可知，套管全壁屈服挤毁压力随着径厚比（D/t）的增加而急剧降低；双剪屈服准则对应的全壁屈服挤毁公式计算值最大，Tresca 屈服准则对应的全壁屈服挤毁公式计算值最小，R. Von Mises 屈服准则和 GM 屈服准则对应的全壁屈服挤毁公式计算值次之且基本相同，其中最大计算值［式（22）的计算值］与最小计算值［式（20）的计算值］的差异高达 33.3%，由此说明屈服准则对套管全壁屈服挤毁压力具有显著的影响。因此，双剪屈服准则对应公式的计算值可作为全壁屈服挤毁压力设计的上限，Tresca 屈服准则对应公式的计算值可作为全壁屈服挤毁压力设计的下限，而 R. Von Mises 屈服准则和 GM 屈服准则对应公式的计算值可作为全壁屈服挤毁压力的平均值。

2.2 全壁屈服挤毁压力公式的讨论

由于套管从内壁屈服到外壁屈服要经历硬化和塑性流动两个阶段，采用屈服强度 σ_s 计算套管的全壁屈服挤毁压力是不合理的，因此，本文引入参数 σ_f，即“流动应力”代替屈服强度 σ_s 来计算套管的全壁屈服挤毁压力。Klever[29] 的研究结果表明：流动应力是介于屈服强度与抗拉强度之间的一个值，它考虑了材料硬化对套管全壁屈服挤毁压力的影响。大量学者针对不同金属管线，给出了计算流动应力经验公式[30-32]：

$$\begin{cases} \sigma_f = 1.1\sigma_s \\ \sigma_f = \sigma_s + 10(\mathrm{ksi}) \\ \sigma_f = \sigma_s + 0.8(\sigma_{uts} - \sigma_s) \\ \sigma_f = \dfrac{\sigma_{uts} + \sigma_s}{2} \end{cases} \tag{23}$$

课题组根据材料的屈强比对套管全壁屈服挤毁压力影响规律的研究[33]，笔者给出了计算套管全壁屈服挤毁压力的公式：

$$P_{iR-New} = \begin{cases} \dfrac{-\sigma_f}{\alpha - 1}\left[\left(\dfrac{D/t}{D/t-2}\right)^{\frac{2(1+b)(1-\alpha)}{\alpha(2+b)}} - 1\right], & \sigma_f = \sigma_s,\ 0.8 \leqslant \dfrac{\sigma_s}{\sigma_{uts}} \leqslant 1 \\ \dfrac{-\sigma_f}{\alpha - 1}\left[\left(\dfrac{D/t}{D/t-2}\right)^{\frac{2(1+b)(1-\alpha)}{\alpha(2+b)}} - 1\right], & \sigma_f = \sigma_s + 10(\mathrm{ksi}),\ 0 < \dfrac{\sigma_s}{\sigma_{uts}} \leqslant 0.8 \end{cases} \tag{24}$$

采用公式（24）能准确合理的计算套管（尤其高抗挤套管）的全壁屈服挤毁压力，在保证材料安全的基础上，很大程度上改善了深井、超深井套管强度设计，为套管强度设计提供了重要参考和新思路。

因此，当套管受到的外压 P_o 等于内壁屈服挤毁压力 P_y 时，只有套管内部开始屈服而发生局部塑性变形，但套管不会被挤毁，被称为内壁屈服挤毁设计，相反，套管还能承受更高的外压，直到套管受到的外压 P_o 等于全壁屈服挤毁压力 P_{iR-NEW} 时，套管内壁到外壁均屈服而发生塑性变形，理论上套管会被该压力挤毁，被称为全壁屈服挤毁设计。套管内壁屈服和全壁屈服挤毁设计适用于只受到均匀内外压作用而不受到温度及温差（酸化压裂引起）影响的常规非酸性环境下的套管，尤其高抗挤套管。对于酸性环境，屈服失效准则已不适用，环境断裂是套管的主要失效模式，因此，对于酸性油气田油，套管除满足屈服失效设计准则之外，还需满足断裂力学设计准则；对于温度较高且温差较大的井筒，需要考虑温度及温差产生的附加热应力对套管挤毁强度的影响。

最后，需要说明的是，本文推导的全壁屈服挤毁压力公式仅适用于均匀外压的情况，而针对非均匀外压的情况，需要先对套管受到的非均匀载荷进行简化并做相关假设，然后采用统一强度理论及弹塑性极限分析方法对非均匀外压下套管全壁屈服挤毁压力公式进行推导和求解。由于非均匀载荷对套管挤毁强度具有显著的影响，因此，有必要基于统一强度理论及相关分析方法开展非均匀载荷下套管全壁屈服挤毁压力的研究。

3 计算结果与试验数据的对比

为验证全壁屈服挤毁压力公式［式（24）］正确性和可靠性，用本文提出的双剪屈服准则所对应的全壁屈服挤毁公式［式（24），其中 $\alpha=1$，$b=1$］代替 ISO 10400 挤毁模型中所谓的“全壁屈服挤毁公式”，形成新的挤毁强度计算模型，并将新模型与 API 模型[14]、ISO 10400 模型[16]的计算结果与实验值进行对比，图 8 和表 1 是用上述模型对 DEA-130[34]提供的 9 组典型测试数据计算的结果与真实挤毁值 P_{test} 的对比情况，图 8 中横坐标为径厚比（D/t），纵坐标为挤毁压力（MPa），P_{ISO} 表示用 ISO 模型计算的挤毁强度，P_{New} 为新模型计算的挤毁强度，P_{API} 表示用 API 模型计算的挤毁强度。

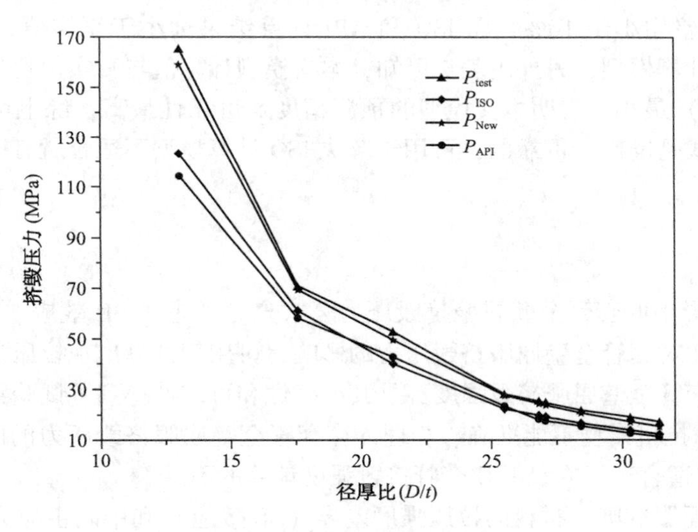

图 8 套管挤毁强度计算值与试验值的对比

表 1 套管实测爆裂值与模型计算值对比

DEA-130 中的编号	D/t	钢级	P_{test}（MPa）	P_{ISO}（MPa）	P_{new}（MPa）	P_{API}（MPa）	P_{test}/P_{ISO}	P_{test}/P_{new}	P_{test}/P_{API}
77	13. 03	Q125	165. 18	123. 57	159. 23	114. 29	1. 337	1. 037	1. 445
62	17. 57	A110	70. 40	61. 31	69. 34	58. 15	1. 148	1. 015	1. 211
144	21. 21	Q125	52. 54	40. 28	49. 67	42. 79	1. 304	1. 058	1. 228
139	25. 49	A95	27. 65	22. 07	27. 37	23. 22	1. 253	1. 010	1. 191
140	26. 81	P110	25. 08	19. 77	24. 25	18. 12	1. 269	1. 034	1. 384
115	27. 08	N80	24. 42	19. 29	23. 67	17. 59	1. 266	1. 032	1. 389
136	28. 41	P110	21. 24	16. 59	20. 33	15. 16	1. 280	1. 045	1. 401

续表

DEA-130 中的编号	D/t	钢级	P_{test}（MPa）	P_{ISO}（MPa）	P_{new}（MPa）	P_{API}（MPa）	P_{test}/P_{ISO}	P_{test}/P_{new}	P_{test}/P_{API}
146	30. 30	N80	18. 40	13. 88	16. 94	12. 44	1. 325	1. 087	1. 479
148	31. 41	N80	16. 64	12. 42	15. 15	11. 15	1. 340	1. 098	1. 492

表 2　套管挤毁强度的预测精度

DEA-130 数据表	P_{test}/P_{API}			P_{test}/P_{ISO}			P_{test}/P_{new}		
	Mean	CV	SD	Mean	CV	SD	Mean	CV	SD
9 根不同径厚比套管	1. 36	0. 086	0. 117	1. 28	0. 059	0. 046	1. 05	0. 030	0. 029

由图 8 和表 1 可知，本文模型的计算结果与套管挤毁强度的真实值最接近，且计算值与真实值之间的误差均小于 10%，而 ISO 和 API 计算结果远小于真实值，且其误差均大于 10%，尤其 API 计算模型；另外由表 2 可知，本文模型计算值与真实值比值的标准差（SD）和变异系数（CV）最小，表明本文模型的预测精度及可靠性最高，综上可知，本文提出的全壁屈服挤毁公式是准确、可靠的，可用于解决 ISO 计算模型不适合所有壁厚段套管挤毁强度计算的问题。

4　结论

（1）本文推导出的套管全壁屈服挤毁压力公式充分考虑了 SD 效应、中间主应力效应、材料硬化、屈强比对套管全壁屈服挤毁压力的影响，可用于计算出现拉压强度差异和存在中间主应力效应的套管全壁屈服挤毁强度，且其准确性和可靠性得到了试验数据的验证，从而解决了 ISO 10400 抗挤模型不能准确、合理计算套管全壁屈服挤毁压力的问题，弥补了 ISO 10400 挤毁模型不适合于所有径厚比套管挤毁强度计算的不足。

（2）分析了屈服准则、材料的拉压强度差异（SD 效应）和中间主应力效应均对套管屈服挤毁压力的影响规律，给出了四种典型屈服准则下的套管全壁屈服挤毁压力计算公式，其中双剪屈服准则下的计算公式的计算值最大，可作为全壁屈服挤毁压力设计的上限，Tresca 屈服准则下的计算公式的计算值最小，可作为全壁屈服挤毁压力设计的下限，而 R. Von Mises 和 GM 屈服准则下的计算公式的计算值居中且十分接近，可作为全壁屈服挤毁压力设计的平均值。

参 考 文 献

[1] 常德玉，李根生，沈忠厚，等．深井超深井井底应力场［J］．石油学报，2011，32（7）：697-703.

[2] 杨恒林，陈勉，金衍，等．蠕变地层套管等效破坏载荷分析［J］．中国石油大学学报，2006，30（4）：94-97.

[3] 高德利，郑传奎，覃成锦．蠕变地层中含缺陷套管外挤压力分布的数值模拟［J］．中国石油大学学报，2007，31（1）：56-61.

[4] 王军，毕宗岳，韦奉，等．国内 SEW 油套管开发现状［J］．钢管，2014，43（4）：7-10.

[5] 王少华，王军，张峰．BSG-80TT 高抗挤套管的研制［J］．焊管，2014，37（5）：35-39.

[6] 严泽生，高德利，张传友，等．一种新型高抗挤套管的研制［J］．钢铁，2004，39（4）：35-38.
[7] 王军，田晓龙，樊振兴，等．SEW 高抗挤套管抗外压挤毁性能研究［J］．钢管，2014，43（2）：16-21.
[8] 张月敏，王俊芳，栗广科．盐膏层段高抗挤厚壁套管的开发与应用［J］．西南石油学院学报，2004，26（3）：79-81.
[9] Lin Y H, Sun Y X, Shi T H, et al. Equations to Calculate Collapse Strength for High Collapse Casing [J]. Journal of Pressure Vessel Technology, 2013, 135 (4): 041202.
[10] Klever F J, Tamano T. A new OCTG strength equation for collapse under combined loads [J]. SPE Drilling & Completion, 2006, 21 (03): 164-179.
[11] Sun Y X, Lin Y H, Wang Z S et al. A New OCTG Strength Equation for Collapse Under External Load Only [J]. Journal of Pressure Vessel Technology, 2011, 133, 011702: 1-5.
[12] Han, J Z, Shi T H. Equations Calculate Collapse Pressures for Casing Strings [J]. Oil Gas Journal, 2001, 99 (4): 44-47.
[13] 覃成锦，高德利．套管强度计算的理论问题［J］．石油学报，2005，26（3）：123-126.
[14] America Petroleum Institute. Bulletin on formulars and calculations for casing, tubing, drill pipe and line properties [S]. API Bulletin 5C3, Six Edition, 1994.
[15] 赵均海，李艳，张常光，等．基于统一强度理论的石油套管柱抗挤强度［J］．石油学报，2013，34（5）：969-976.
[16] International Organization for Standardization. Petroleum and natural gas industries-Formulae and calculation for casing, tubing, drill pipe and line pipe properties [S]. ISO 10400, Six Edition, 2007.
[17] 韩建增，李中华，张毅，等．特厚壁套管抗挤强度计算及现场应用［J］．天然气工业，2003，23（6）：77-79.
[18] 孙永兴，林元华，施太和，等．套管全管壁屈服挤毁压力计算［J］．石油钻探技术，2011，39（1）：48-51.
[19] Yu M H. Unified strength theory and its applications [M]. Berlin: Springer, 2004.
[20] Yu M H. Advances in strength theories for materials under complex stress state in the 20th century [J]. Applied Mechanics Reviews, 2002, 55 (3): 169-218.
[21] 赵均海，梁文彪，张常光，等．非饱和土库仑主动土压力统一解［J］．岩土力学，2013，34（3）：609-614.
[22] 金乘武，王立忠，张永强．薄壁管道爆破压力的强度差异效应与强度准则影响［J］．应用数学和力学，2012，33（11）：1266-1274.
[23] 徐芝纶．弹性力学（第四版）［M］．北京：高等教育出版社，2006.
[24] Yu M H, Yang S Y, Liu C Y, et al. Unified plane-strain slip line field theory system [J]. China civil engineering Journal, 1997, 30 (2): 182-185.
[25] Lee Y K, Ghosh J. The significance of J_3 to the prediction of shear bands [J]. International Journal of Plasticity, 1996, 12 (9): 1179-1197.
[26] Law M, Bowie G. Prediction of failure strain and burst pressure in high yield-to-tensile strength ratio linepipe [J]. International journal of pressure vessels and piping, 2007, 84 (8): 487-492.
[27] 赵德文，张雷，章顺虎，等．用 GM 屈服准则解析薄壁筒和球壳的极限载荷［J］．东北大学学报（自然科学版），2012，33（4）：521-523，532.
[28] 章顺虎，高彩茹，赵德文，等．GM 准则解析受线性荷载简支圆板的极限载荷［J］．计算力学学报，2013（2）：292-295.
[29] Klever F J. Burst strength of corroded pipe: "Flow stress" revisited [R] //Offshore Technology Conference. Offshore Technology Conference, 1992. Available: https://www.onepetro.org/conference -

paper/OTC-7029-MS.

[30] Zhu X K, Brian N L Influence of yield to tensile strength ratio on failure assessment of corroded pipeline [J]. Journal of pressure vessel technology, 2005, 127: 436-442.

[31] American National Standards Institute. Manual for Determining the Remaining Strength of Corroded Pipelines, ASME Guide for Gas Transmission and Distribution Piping Systems B31G [S]. American Society of Mechanical Engineers, 1984.

[32] Kiefner JF, Vieth PH. A Modified Criterion for Evaluating the Remaining Strength of Corroded Pipe, Pipeine Research Committee Report PR3-805 [R]. Available: http://www.osti.gov/scitech/biblio/7181509. Accessed 22 Dec 1998.

[33] 孙永兴．油套管抗内压抗挤强度研究［D］．四川成都：西南石油大学石油与天然气工程学院，2008.

[34] Asbill W T, Crabtree S, Payne M L. DEA-130 modernization of tubular collapse performance properties [R]. API/HSE/MMS Participant Report, 2002.

三、腐蚀与防腐

Influence of Stray Alternating Current on Corrosion Behavior of Pipeline Steel in Near-neutral pH Carbonate/Bicarbonate Solution

Fu Anqing Yuan Juantao Li Lei Long Yan Song Chengli
Bai Zhenquan Lin Kai

(CNPC Tubular Goods Research Institute, CNPC Key Laboratory for Petroleum Tubular Goods Engineering)

Abstract: The AC induced corrosion of pipeline steel was investigated in near-neutral pH carbonate/bicarbonate solution by potentiodynamic technique and weight loss test. It is found that superimposed AC is able to cause the negative shift of corrosion potential of pipeline steel, the severity of the polarization curve oscillation is proportional to the ratio of AC current density to DC current density, the most severe oscillation can be observed in the region near corrosion potential. The results of weight loss test indicated that the greater AC current density resulted in the higher AC-induced corrosion rate. The corrosion morphology viewed by optical camera and SEM showed that the form of corrosion is closely related to the magnitude of AC current density.

Keywords: AC; Pipeline Steel; Corrosion; Polarization curve

1 Introduction

It is acknowledged[1-5] that most metallic materials or structures corroded at an accelerated rate due to the presence of stray alternating current (AC). The corrosion problems caused by stray AC on pipeline have received increasing attention from pipeline corrosion scientists during the last 30 years[4]. Generally, there are three AC sources[6] causing pipeline corrosion. Inductive coupling arises where pipeline has same route with powerline in length. Resistive coupling arises if pipeline is located at AC potential gradient caused by discharged current from grounding systems into earth. Capacitive coupling arises during pipeline construction when long pipeline is exposed in the air before burial. AC interference from high voltage transmission lines has been reported as one of the main factors to cause corrosion problems on buried pipelines by several distribution companies[7-10]. A harmful situation is when the high voltage lines and transmission pipelines parallel to each other for certain distance, particularly the pipeline steel coated with high dielectric coatings, like extruded polyethylene, polypropylene, or composite coating, if coating defects exist, these areas will subject to a severe AC induced corrosion, even under cathodic protection (CP) conditions[9,11,12]. However, CP criterion takes AC induced corrosion into consideration has not been established[12].

There has been a number of studies performed to investigate the corrosion effects of AC on steel[13-15], copper[2,4] nickel[3], etc. in acidic, sulfate, and seawater solutions. Most of the above research demonstrated that AC behaves as a depolarizer with the ability to reduce the polarization of both anode and cathode, as well as the electrochemical passivity. The reduced cell polarization or passivity increases the corrosion rate. Even though the AC induced corrosion of common metallic materials in different solutions have been well investigated, little work has been performed on pipeline steel in simulated solution extracted from disbonded coating. Song and coworkers[6] reported that AC corrosion rate was affected not by AC voltage but by both of frequency and AC current density, AC corrosion rate increased linearly with effective AC current density. Literature review of AC induced corrosion on pipeline steel made by Wakelin and coworkers[9] revealed that corrosion did not occur at AC current densities of less than 20 A/m^2, while the corrosion is unpredictable at AC current densities of 20 ~ 100 A/m^2, the corrosion rate can be expected if the AC current densities are greater than 100 A/m^2. Linhardt and Ball[16] found that AC could induce pitting corrosion of pipeline steel in soil, interestingly, the pH of adhering soil collected from corroded pipeline surface were measured in the range 9 to 11, which is higher than the pH of soil (pH=8). It suggests that it is more practical to do the AC induced corrosion in simulated solution extract from disbonded coating, instead of soil.

In the present work, the AC induced corrosion of pipeline steel in near-neutral pH carbonate/bicarbonate solution is studied by potentiodynamic technique and weight loss test. It is anticipated that this research would advance the understanding of AC induced corrosion of pipeline steel in near-neutral pH solution.

2 Experimental

2.1 Electrode and solution

All test specimens were fabricated from a sheet of X65 pipeline steel. The chemical composition (wt %) of X65 steel is: C 0.04, Si 0.2, Mn 1.5, P 0.011, S 0.003, Mo 0.02 and Fe balance. The specimens were precisely machined with a dimension of 10mm×10mm×9mm, three equivalent samples were prepared for each test. All the samples were subsequently polished with 600 grit, 800grit, and 1200 grit emery papers followed by cleaning with distilled water and acetone.

NS4 solution had been widely used to simulate the dilute electrolyte trapped between coating and the pipeline steel, with the chemical composition: 0.483 g/L $NaHCO_3$, 0.122 g/L KCl, 0.181 g/L $CaCl_2 \cdot 2H_2O$, and 0.131 g/L $MgSO_4 \cdot 7H_2O$. Prior to test, NS_4 solution was purged with 5% CO_2/N_2 gas for 1 h to achieve an anaerobic and near-neutral pH condition (pH 6.8). The gas flow was maintained throughout the test. The solution was made from analytic grade reagents (Fisher Scientific) and ultra-pure water (18 MΩ cm in resistivity). All tests were conducted at room temperature (~22°C).

2.2 Electrochemical measurements

Electrochemical measurements were performed using a Solartron 1280C electrochemical test

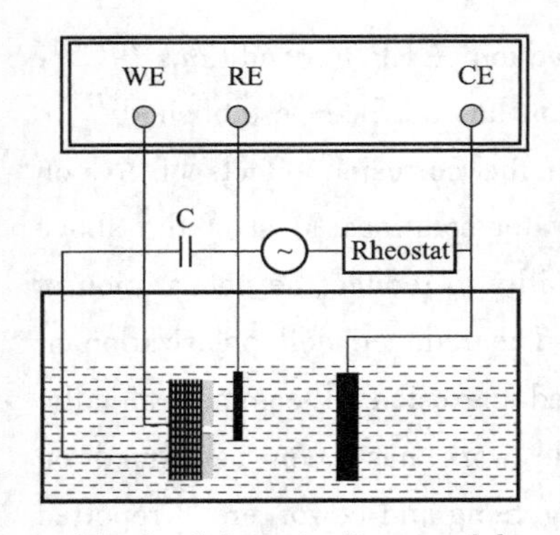

Fig. 1 Schematic diagram of the apparatus for investigating AC induced corrosion

system. A three-electrode test cell was used with the test specimen as working electrode, a platinum foil as counter electrode, and a saturated calomel electrode (SCE) as the reference electrode. The AC signal (f= 60Hz) was applied between the working electrode and counter electrode, as shown in Fig. 1, a rheostat was employed to adjust the AC current, and a capacitor was introduced to avoid the AC signal interfering with DC measurements, assuring the AC current flow only between working electrode and counter electrode instead of DC measurement circuit.

Prior to the electrochemical tests, the sample was conditioned for 30 min in the test solution at the corrosion potential to ensure that the steady state was reached. Potentiodynamic measurements were conducted at a scanning rate of 0.5mV/s under various AC current densities, six different AC current densities of 0, 20A/m^2, 50A/m^2, 100A/m^2, 200A/m^2, and 500A/m^2 were chosen in the present work. All potentials are reported relative to the SCE.

2.3 Weight loss test

Weight loss test was performed to study the effects of AC on corrosion rate and corrosion morphology in near-neutral pH solution. The AC signal was applied between the working electrode and the counter electrode, the values of AC current densities chosen for weight loss tests are the same as that of the electrochemical measurements. Prior to each test, the specimen was cleaned with distilled water and acetone, dried, and then weighed using Satorius ED124S 4-digit electronic balance for gravimetric weight loss measurements. Weight loss tests were carried out for the duration of 48 h. After the completion of each test, the specimens were cleaned, and dipped briefly into a 50% (v/v) HCl solution according to ASTM procedure G1-81 to remove the corrosion products on the surface, rinsed and dried again, finally reweighed to get the resulting weight loss due to corrosion. The corrosion rate was reported in mm/a according to the obtained weight loss. Scanning electron microscope (SEM), combined with optical photography was used to characterize the corrosion morphology of the specimens after removal of corrosion products.

3 Results and discussion

3.1 Corrosion potential measurements on pipeline steel under interference of various AC current densities in near-neutral solution

Fig. 2 shows the corrosion potentials of pipeline steel measured in near-neutral pH solution under interference of various AC current densities. The corrosion potentials shifted to negative direction with the increasing of applied AC current densities. A number of studies[4,17-19] related to the influence of AC current on corrosion of metallic materials have concluded that the applied AC current could cause the negative shift of corrosion potential. Lalvani and Lin[18] proposed a model to derive the corrosion potential shift induced by applied AC current, the explanation in terms of mathematical equations are shown as follows:

$$E_{corr,\ AC} = E_{corr,\ DC} - \alpha \tag{1}$$

$$\alpha = \left(\frac{m_a m_c}{m_a - m_c}\right) \ln \left[\frac{\sum_{k=1}^{\infty} \frac{1}{(k!)^2} \left(\frac{E_p}{2m_c}\right)^{2k} + 1}{\sum_{k=1}^{\infty} \frac{1}{(k!)^2} \left(\frac{E_p}{2m_a}\right)^{2k} + 1} \right] \tag{2}$$

where $E_{corr,AC}$ and $E_{corr,DC}$ are the corrosion potentials with and without applied AC current, m_a and m_c are the anodic and cathodic Tafel slopes, respectively, E_p is the peak voltage of applied AC signal. The variation of corrosion potential caused by AC is a function of the ratio of anodic Tafel slope to cathodic Tafel slope ($r = m_a / m_c$) and E_p, and the corrosion potential shifted to negative direction as r and E_p increased. In the present work, it is seen that the variations of m_a and m_c are not obvious to cause the significant change of their ratio as presented in Fig. 3, therefore, the increase of E_p is a main contribution to the negative shift of corrosion potential, E_p is proportional to AC current density when the sample has a constantly exposed surface area in a given solution.

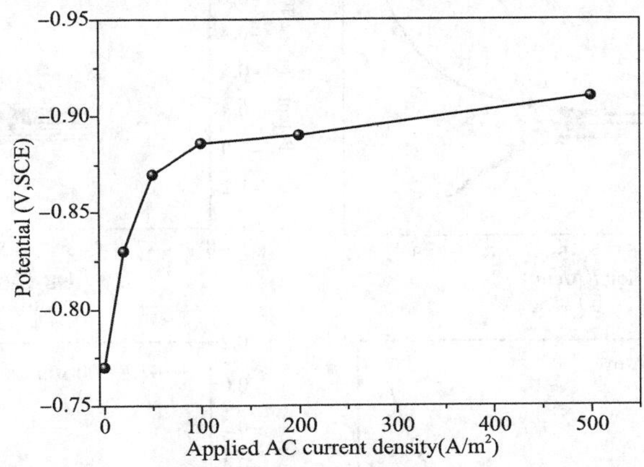

Fig. 2 Corrosion potentials of X65 pipeline steel in near-neutral pH carbonate/bicarbonate solution under interference of various AC current densities

3.2 Polarization curves of pipeline steel under interference of various AC current densities in near-neutral solution

The corresponding polarization curves are shown in Fig. 3, there is no oscillation on polarization curve if AC current was not superimposed. The oscillations of polarization curves become more and more severe as superimposed AC current increased from Fig. 3 (b) to (f), particularly in the region near corrosion potential.

A portion of the anodic polarization curves at various AC current densities are replotted in Fig. 3 (g). It is seen that the curves shifted to the positive direction with the increase of AC current density. The enhanced anodic dissolution rate could be attributed to the superimposed AC caused the negative shift of corrosion potential and the increase of exchange current density, thermodynamically, the more negative the corrosion potential, the easier the corrosion occurs, and kinetically corrosion will take place in a higher rate if exchange current density increases. Apparently, the magnitude of the polarization curve oscillation is dependent on the ratio of AC current density to DC current density, which can be expressed as:

$$A_{DC} = K \frac{i_{AC}}{i_{DC}} \tag{3}$$

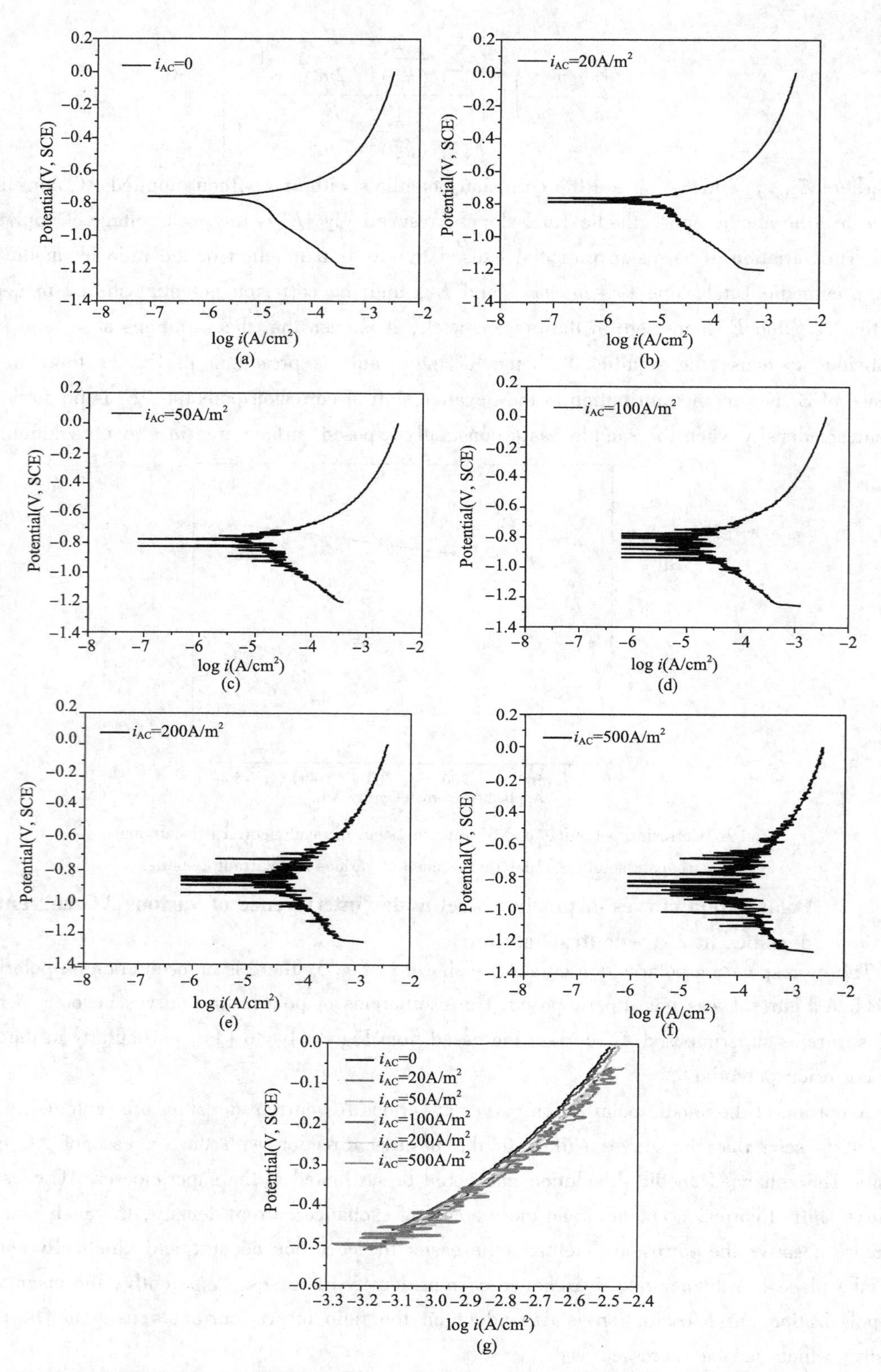

Fig. 3 Polarization curves of X65 pipeline steel in near-neutral pH carbonate/bicarbonate solution under interference of various AC current densities

where A_{DC} is the magnitude of the oscillation, K is a coefficient, i_{AC} and i_{DC} are the AC current density and DC current density, respectively.

3.3 Corrosion rate and morphology of pipeline steel under interference of various AC current densities in near-neutral pH solution

The corrosion rates of pipeline steel in near-neutral pH solution under interference of various AC current densities are plotted in Fig. 4. It is seen that the corrosion rate gradually increased with the AC current density. Fig. 5 and Fig. 6 show the corrosion morphologies of pipeline steel in near-

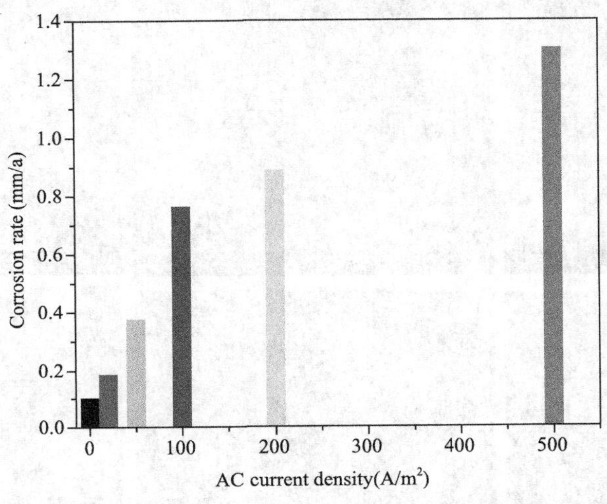

Fig. 4 Corrosion rates of X65 pipeline steel in near-neutral pH carbonate/bicarbonate solution under interference of various AC current densities

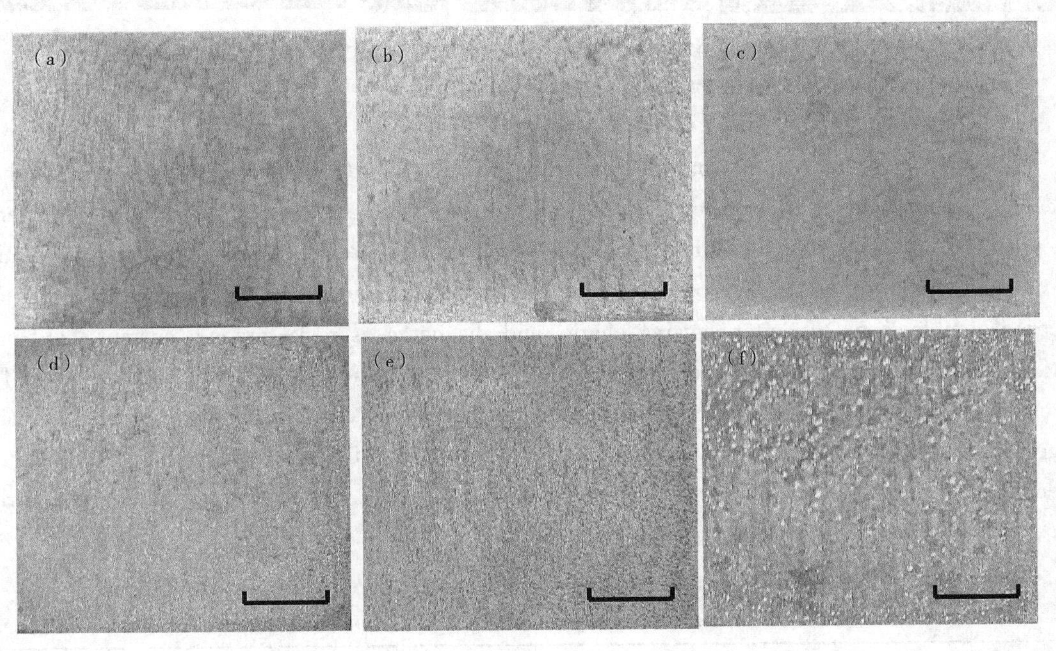

Fig. 5 Optical observation of X65 pipeline steel in near-neutral pH carbonate/bicarbonate solution under interference of various AC current densities after two-day test (a. 0; b. 20 A/m^2; c. 50A/m^2; d. 100 A/m^2; e. 200 A/m^2; f. 500A/m^2)

neutral pH solution under interference of various AC current densities viewed by optical camera and SEM, respectively. It is seen that uniform corrosion occurred when AC current density is less than 500 A/m^2, a porous structure can be observed on sample surface at AC current densities of 100 A/m^2 and 200 A/m^2. If the AC current density reached 500 A/m^2, the sample undergoes pitting corrosion, the shape of corrosion pits is like a crater. The occurrence of pitting corrosion is possible in a non-passive environment if the superimposed AC current density is sufficiently high.

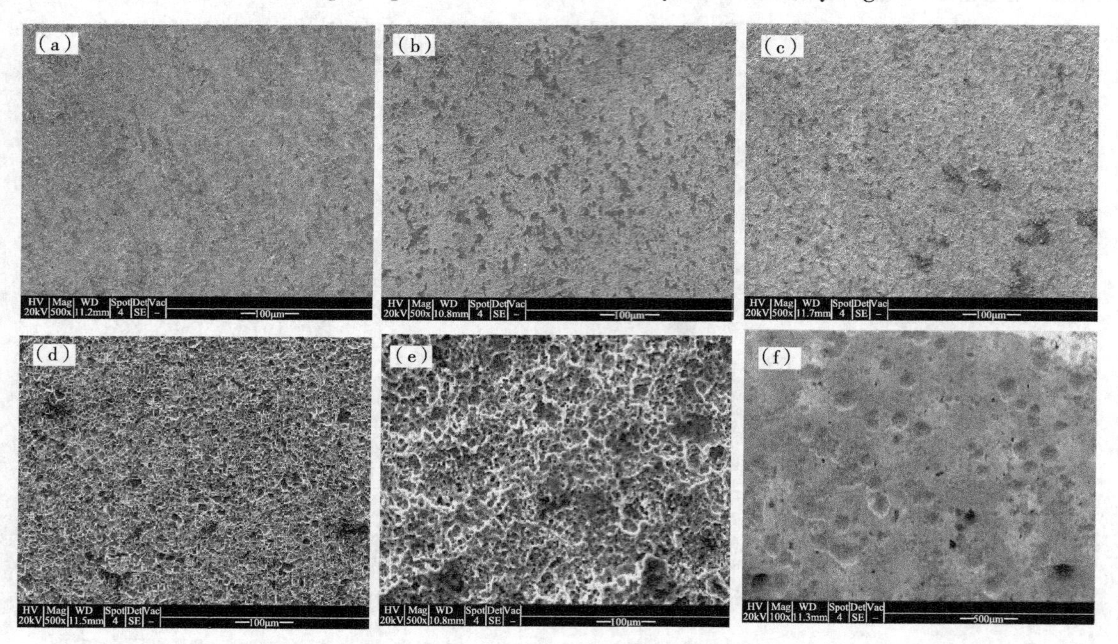

Fig. 6 SEM images of X65 pipeline steel in near-neutral pH carbonate/bicarbonate solution under interference of various AC current densities after two-day test (a. 0; b. 20 A/m^2; c. 50A/m^2; d. 100 A/m^2; e. 200 A/m^2; f. 500A/m^2)

It is reported[6] that the amount of corrosion caused by AC current could be expressed as percentage of the amount of corrosion that would be caused by an equivalent intensity of DC current, generally, AC current causes less than 1% of the corrosion of the equivalent DC current. The relative low current efficiency in AC corrosion due to the fact that majority of AC current flow through electrical double layer capacitor causing charge and discharge as non-faradaic current, and only minority of AC current results in charge transfer causing corrosion as faradaic current[6, 13]. The corrosion rates shown in Fig. 4 indicated that superimposed AC current could accelerate corrosion process compared to that without applied AC current, the AC current densities expressed as percentages of equivalent DC current densities is shown in Table 1, indicating that the corrosion rate caused by AC current was less than 1% of equivalent DC current.

Table 1 AC current densities expressed as the percentage of equivalent DC current densities in near-neutral pH carbonate/bicarbonate solution

AC current density (A/m^2)	0	20	50	100	200	500
Equivalent DC current density (%)	N/A	0.76	0.63	0.64	0.38	0.22

4 Conclusions

(1) Applied AC current is able to cause the negative shift of corrosion potential of pipeline steel in near-neutral pH carbonate/bicarbonate solution. The greater the AC current density, the more negative the corrosion potential shifted.

(2) The polarization curve oscillation increased with the superimposed AC, the severity of oscillation is proportional to the ratio of AC current density to DC current density, the most severe oscillation occurred in the region near corrosion potential.

(3) The greater the AC current density, the higher is the AC corrosion rate. The form of corrosion is closely related to the magnitude of AC current density. Localized corrosion took place when the superimposed AC current density is greater than 500 A/m^2.

References

[1] Mark A. Pagano, Shashi B. Lalvani, Corrosion of mild steel subjected to alternating voltages in seawater, Corrosion Science, 1994, 36: 127.

[2] W. W. Qiu, M. Pagano, G. Zhang, S. B. Lalvani, A periodic voltage modulation effect on the corrosion of Cu-Ni alloy, Corrosion Science 37 (1995) 97.

[3] T. C. Tan, D-T. Chin, A. C. corrosion of nickel in sulphate solutions, Journal of Applied Electrochemistry 18 (1988) 831.

[4] Sara Goidanich, Luciano Lazzari, Marco Ormellese, MariaPia Pedeferri, Influence of AC on corrosion kinetics for carbon steel, zinc and copper, Corrosion 2005, Paper No. 05189.

[5] J. L. Wendt, D-T. Chin, The AC corrosion of stainless steel-Ⅰ. The breakdown of passivity of SS304 in neutral aqueous solutions, Corrosion Science 25 (1985) 889.

[6] Hongseok Song, YoungGeun Kim, SeongMin Lee, YoungTai Kho, Competition of AC and DC current in AC corrosion under cathodic protection, Corrosion 2002, Paper No. 02117.

[7] S. L. Eliassen, S. M. Hesjevik, Corrosion management of buried pipelines under difficult operational and environmental conditions, Corrosion 2000, Paper No. 00724.

[8] Robert G. Wakelin, Christopher Sheldon, Investigation and mitigation of AC corrosion on a 300m diameter natural gas pipeline, Corrosion 2004, Paper No. 04205.

[9] R. G. Wakelin, R. A. Gummow, S. M. Segall, AC corrosion - case histories, test procedures, &mitigation, Corrosion'98, Paper No. 98565.

[10] Christopher M Movley, Pipeline corrosion from induced A. C., two UK case histories, Corrosion 2005, Paper No. 05132.

[11] I. Ragault, AC corrosion induced by V. H. V electrical lines on polyethylene coated steel gas pipelines, Corrosion 1998, Paper No. 98557.

[12] F. Kajiyama, Y. Nakamura, Effect of induced alternating current voltage on cathodically protected pipelines paralleling electric power transmission lines, Corrosion 55 (1999) 200.

[13] S. B. Lalvani, G. Zhang, The corrosion of carbon steel in a chloride environment due to periodic voltage modulation: Part Ⅰ, Corrosion Science 37 (1995) 1567.

[14] S. B. Lalvani, G. Zhang, The corrosion of carbon steel in a chloride environment due to periodic voltage modulation: Part Ⅱ, Corrosion Science 37 (1995) 1583.

[15] S. Muralidharan, et al., Influence of alternating, direct and superimposed alternating and direct current on the corrosion of mild steel in marine environments, Desalination 216 (2007) 103.

[16] P. Linhardt, G. Ball, AC corrosion: Results from laboratory investigations and from a failure analysis, Corrosion 2006, Paper No. 06160.
[17] Rong Zhang, Pon Rajesh Vairavanathan, Shashi B. Lalvani, Perturbation method analysis of AC-induced corrosion, Corrosion Science 50 (2008) 1664.
[18] S. B. Lalvani, X. A. Lin, A theoretical approach for predicting AC-induced corrosion, Corrosion Science 36 (1994) 1039.

Failure Analysis of Girth Weld Cracking of Mechanically Lined Pipe Used in Gasfield Gathering System

Anqing Fu[1,2*] Xianren Kuang[1,2] Yan Han[1,2]
Caihong Lu[1,2] Juntao Yuan[1,2] Youguo Wei[3] Qiang Tang[3] Yang Yang[3]

(1. Tubular Goods Research Institute of CNPC;
2. State Key Laboratory of Performance and Structurd Safety for Petroleum Tubular Goods and Equipment Materials;
3. Tarim Oilfield Company, PetroChina Company Limited)

Abstract: Girth weld cracking of mechanically lined pipe was occurred after 75 days operation under design parameters. Failure cause was determined based on available field documents and a series of laboratory tests including X-ray inspection, chemical element, mechanical performance, metallurgical properties and cracking behavior. Results revealed that the girth weld cracking failure of mechanically lined pipe was initiated from outer carbon steel, and propagated in intergranular mode along the weld-fusion line, crack source zone and propagation zone can be seen in the outer carbon steel layer, which was mainly caused by the martensite structure formed at sealing weld zone and transition weld zone, martensite is a typical hard and brittle structure with high hardness of HV350~450. Moreover, the failed pipe area had undergone the heavy rain for 2 days, the external stress generated by soil movement lead to the relative motion between outer L415 carbon steel and 316L liner resulted in high shear stress concentration at sealing weld zone.

Keywords: Mechanically lined pipe; Weld joint; Cracking; Martensite; Shear stress.

1 Introduction

With the continuously growing demand in oil and gas energy globally, the common oil and gas reservoir was extensively exploited and which became less and less. Therefore, high pressure high temperature (HPHT) gas reservoir was increasingly developed all around the world, the typical

* Corresponding author. Tel.: +86-29-81887902; Fax: +86-29-88223416. E-mail address: fuanqing@cnpc.com.cn (A. Q. Fu).

HPHT gas well broadly exist in Gulf of Mexico, Tarim Basin of China, North Sea, South East Asia, Africa and Middle East[1,2]. The nature of HPHT gas reservoir fluids places demands upon material selection for linepipe that can only be met by the use of corrosion resistant alloys (CRAs) as an internal clad layer combined with a carbon steel substrate[3]. Although the solid CRAs are the best choice with overall desirable properties, the biggest disadvantage is that the CRAs are too expensive for using as gathering pipes. Hence, development of mechanically lined pipe is an alternative choice with consideration of cost and corrosion resistance. Mechanically lined pipe is composed of external carbon steel pipe and a thin internal layer of CRA, in which the outer carbon steel is to provide structure strength, and the inner CRA layer is designed to resist corrosion[4,5], as shown in Fig. 1.

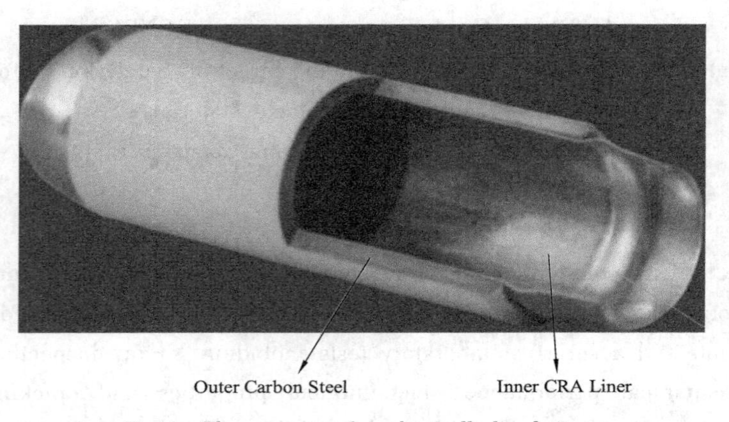

Fig. 1 The structure of mechanically lined pipe

Mechanically lined pipe combines the advantages of carbon steel (low cost and high strength) and CRA (high corrosion resistance) together, this combination enables that the mechanically lined pipe was widely used in HPHT gas field gathering systems and offshore flowlines in subsea gathering systems. Moreover, the liner of mechanically lined pipe can be customized according to the fluids corrosivity, various liner materials are available for selection, i. e, 304, 316L, 825, 625, G28. Mechanically lined pipe has been considered as a preferred pipe to replace traditional carbon steel pipe in HPHT gas field gathering systems in China. However, the dissimilar nature of the materials abutting at the weld joint presents challenges in terms of welding processes, flaw assessment and inspection methods[6], it was found that most of the failures were related to the girth weld cracking and perforation[3,7,8]. Pipe failures may cause loss of product, temporary shutdown of production, pollution, and other unpredictable losses. Therefore, it is of significance to decrease the failure risk, failure analysis is one of the best ways to provide failure reason and prevention measure.

The objective of this work is to analyze the causes about the girth weld cracking failure of mechanically lined pipe used in northwest China gasfield, although extensive laboratory studies have been done on the girth welding of mechanical lined and metallurgical clad pipes, i. e. , welding procedure, fracture assessment, fatigue behavior, full-scale mechanical test and NDT test[3,5-9]. Few works were found on the girth welding failure after period of use in oil/gas field, the weld joint of bi - metal pipe is more complicated than that of single metal pipe, i. e, welding material, welding method, welding parameters, and pre - treatment process, as shown in Fig. 2. It is expected that this work would provide the insight of mechanically lined pipe girth welding failure reasons and prevention measures.

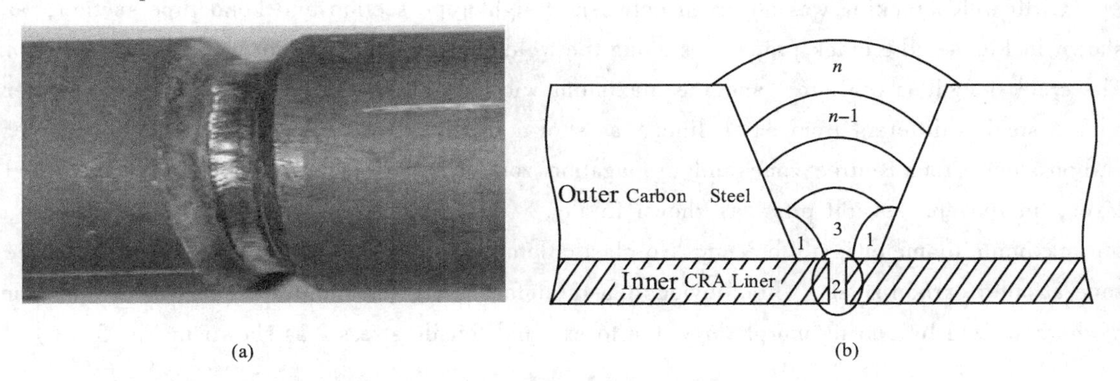

Fig. 2 Weld joint of mechanically lined pipe and weld structure

2 Background of the failure

The failed mechanically lined pipe is composed of L415 carbon steel and 316L stainless steel, the outer diameter is 508mm, the wall thickness of outer carbon steel and stainless steel liner is 14. 2 and 2. 5mm, respectively. The welding methods are Gas Tungsten Arc Weld (GTAW) and Shielded Metal Arc Welding (SMAW), and the X-ray inspection of weld joint was judged as Class-Ⅰ without defect. Prior to operation, the pipeline was conducted strength pressure test and sealing pressure test at 22MPa and 16MPa, respectively. The fluid flowing in pipeline is natural gas, the designed pressure is 16MPa, and the operation pressure is between 10. 5 and 12. 5MPa, as shown in Fig. 3. The mechanically lined pipe failure occurred after 75 days operation with operating parameters shown in Fig. 3.

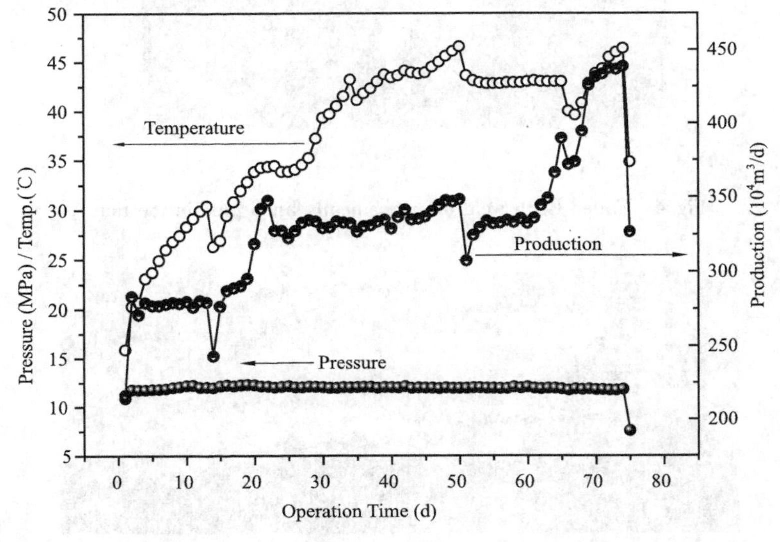

Fig. 3 Operation parameters of failed mechanically lined pipe within 75 days

3 Failure description and characterization

3.1 Generals of failed girth weld

Girth weld cracking was observed between straight pipe section and bend pipe section, as shown in Fig. 4, the crack propagates along the weld-fusion line, as shown in Fig. 5 (a, b). The crack length is 660mm, and the maximum width is 30mm. The fracture surface of outer carbon steel is different from 316L liner, as shown in Fig. 5 (a), crack initiated from outer carbon steel. Crack source zone and propagation zone can be seen in the outer carbon steel layer, moreover, several pits, as shown in Fig. 5 (b), were observed on fracture surface, the maximum diameter is up to 3 mm. No plastic deformation was observed near the crack source zone at outer carbon steel, while obvious plastic deformation was found at 316L liner, the liner is characterized by tearing morphology due to external tensile stress, as shown in Fig. 5 (c).

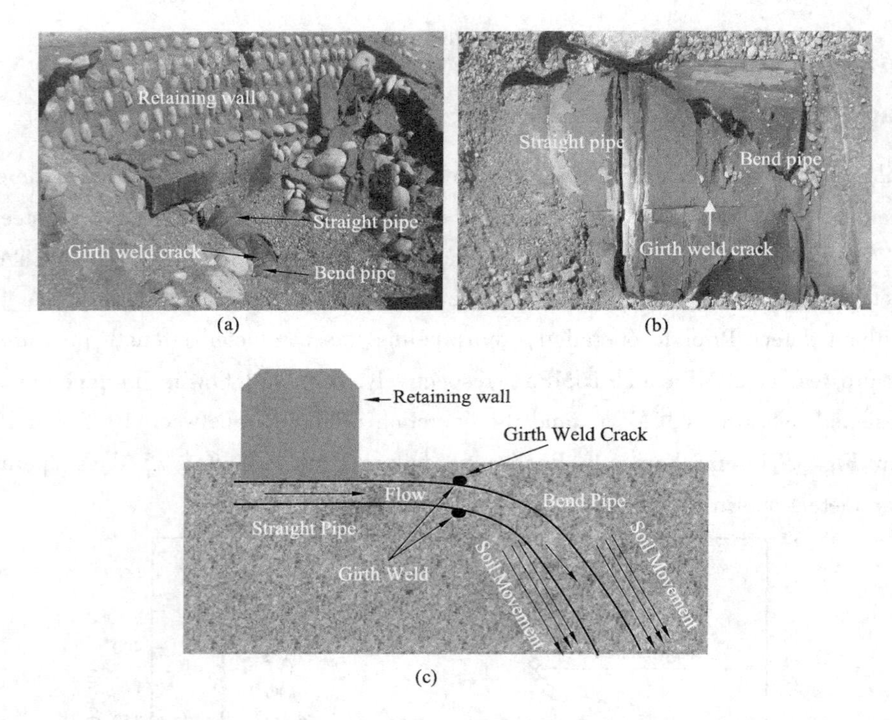

Fig. 4 Failed Girth weld of mechanically lined pipe in the field

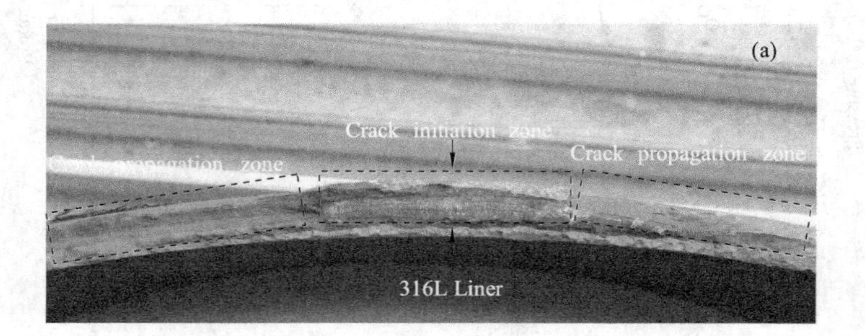

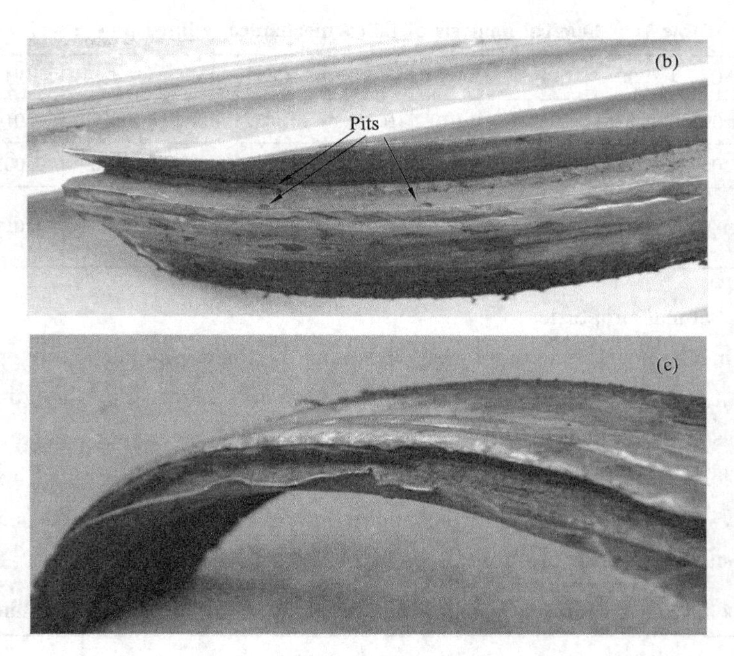

Fig. 5 Images of failed girth weld of mechanically lined pipe

3.2 Inspection and testing of failed girth weld

3.2.1 X-Ray inspection

The failed girth weld was inspected by XXH2505 X-ray detectscope according to JB/T 4730. 2-2005 Ⅱ [10]. Five cracks with lengths of 20mm were detected around the girth weld, in which one crack is connected to the main crack, and others are around the main crack, one detected crack is shown in Fig. 6.

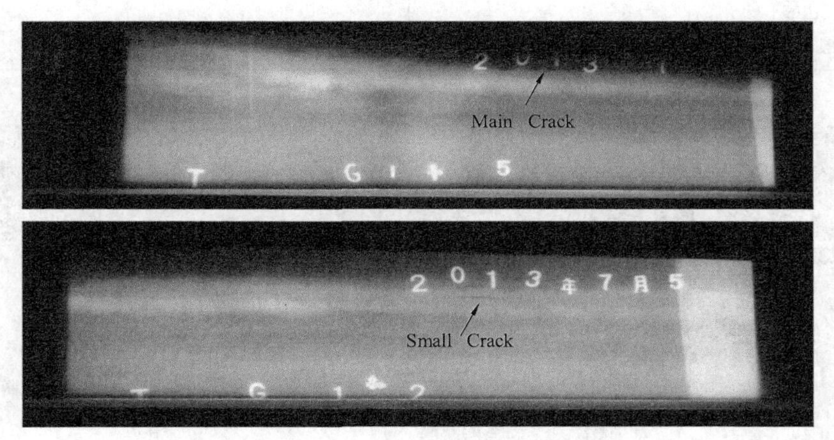

Fig. 6 Cracks detected by X-ray detectoscope near failed girth weld

3.2.2 Chemical element analysis

Chemical element of outer carbon steel pipe and weld joint were analyzed by spectrometer, the results are giving in Table 1, it is seen that all the elements contained in base metal of the outer carbon steel is in accordance with the requirement of GB 9711—2011[11], while there is no specific requirement for the weld joint.

Table 1 Chemical analysis of failed mechanically lined pipe (wt%)

Element	C	Si	Mn	P	S	Cr	Mo	Ni	V	Ti	Cu
Weld joint	0. 091	0. 336	1. 31	0. 019	0. 008	0. 885	0. 107	0. 699	0. 014	0. 008	0. 039
Base metal	0. 069	0. 273	0. 53	0. 009	0. 005	0. 025	0. 006	0. 009	0. 001	0. 015	0. 010
GB 9711—2011 Requirements	≤0. 24	≤0. 45	≤1. 4	≤0. 025	≤0. 015	≤0. 3	≤0. 15	≤0. 3	≤0. 1	≤0. 04	≤0. 5

3. 2. 3 Mechanical properties

Due to the mechanical strength of the mechanically lined pipe is mainly provided by outer carbon steel pipe, so mechanical properties of the outer carbon steel pipe including ultimate tensile strength, hardness, and impact energy were tested. All the specimens for mechanical behavior tests include the weld joint in the middle, the sample for the ultimate tensile strength test is rod-shaped, as shown in Fig. 7 (a), the sample for the impact energy test is cuboid-like, as shown in Fig. 7 (b), and the results are given in Table 2.

Table 2 Mechanical properties and impact energy of failed mechanically lined pipe

	Mechanical properties		Impact Energy
	UTS (MPa)	Fracture Area	-20℃ (J)
1	519	Base metal	46
2	530	Base metal	25
3	519	Base metal	44
Average value	523	/	38
Tarim Oilfield Technical Requirement	≥520	/	≥30

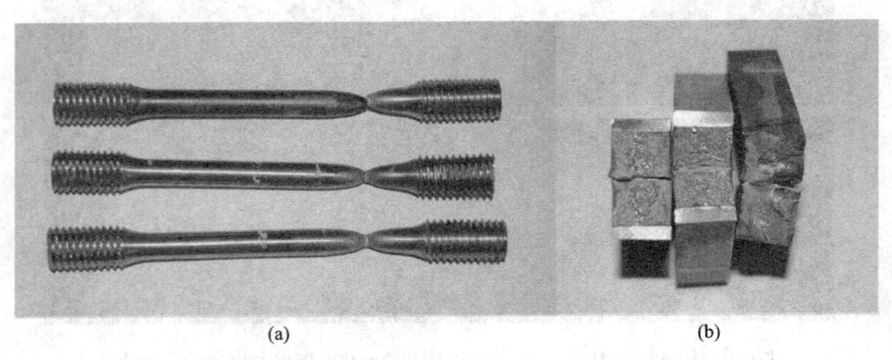

(a) (b)

Fig. 7 Specimens for tensile strength (a) and Charpy energy (b) test

The sample for the hardness test is trapezoid-like, as shown in Fig. 8, in which Line AB and Line GH with 50 points are for hardness test of base metal (outer carbon steel pipe), Line CD with 9 points and Line EF with 25 points are for hardness test of weld joint including sealing weld zone, root weld zone, transition weld zone and filling weld zone, the results are shown in Fig. 9.

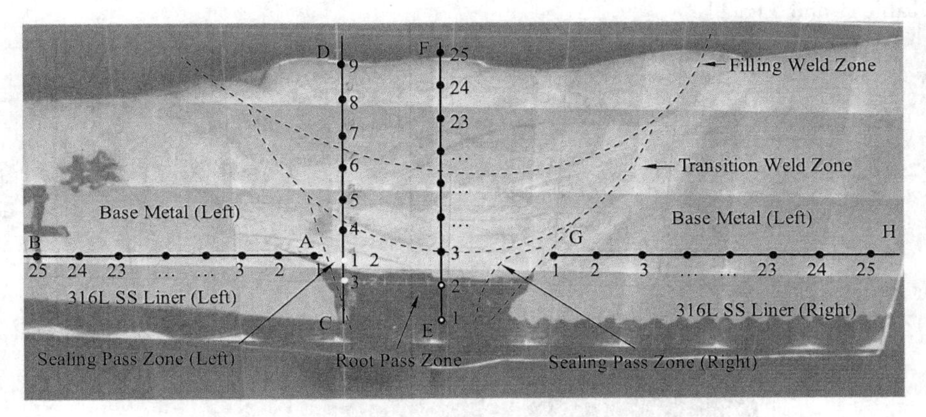

Fig. 8 Weld joint specimen for hardness test

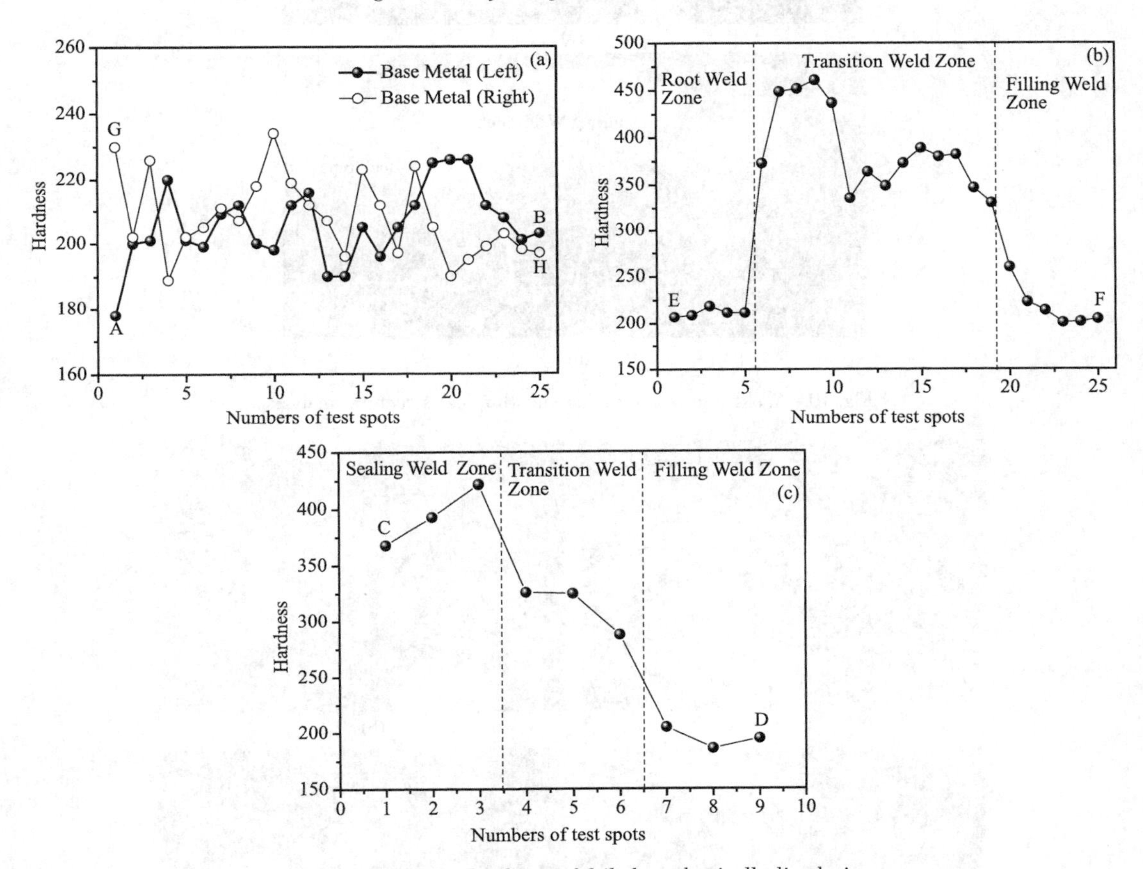

Fig. 9 Weld joint hardness of failed mechanically lined pipe

3.2.4 Metallurgical properties

Three areas were selected to take weld joint samples for metallurgical structure analysis, three areas are located in main crack (#1), crack detected by X-ray inspection (#2), and uncracked area (#3), as illustrated in Fig. 10. With regard to three weld joint samples, the metallurgical structure of root pass zone, sealing pass zone, filling pass zone and cover pass zone are characterized by austenite, martensite, martensite and few bainite, bainite and few ferrite, as given in Table 3 and Fig. 11.

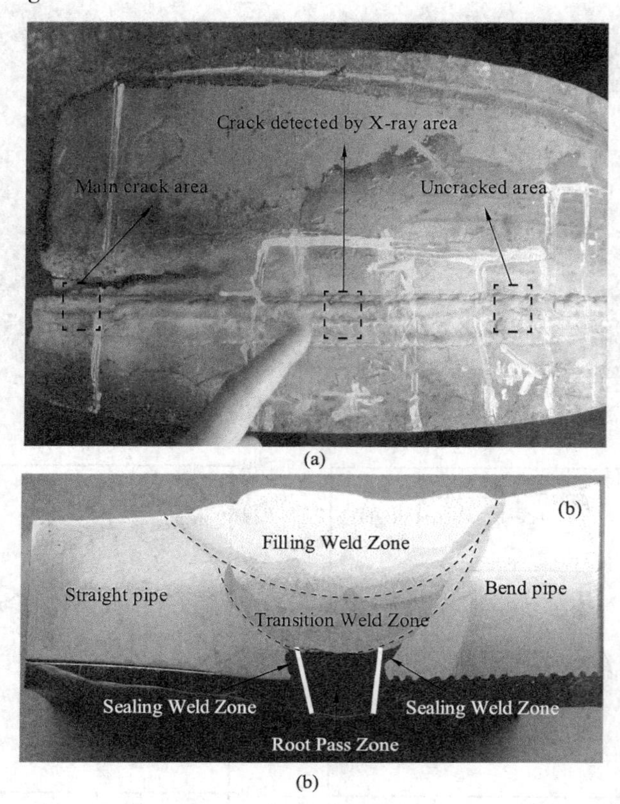

Fig. 10 Weld joint samples for metallurgical structure analysis

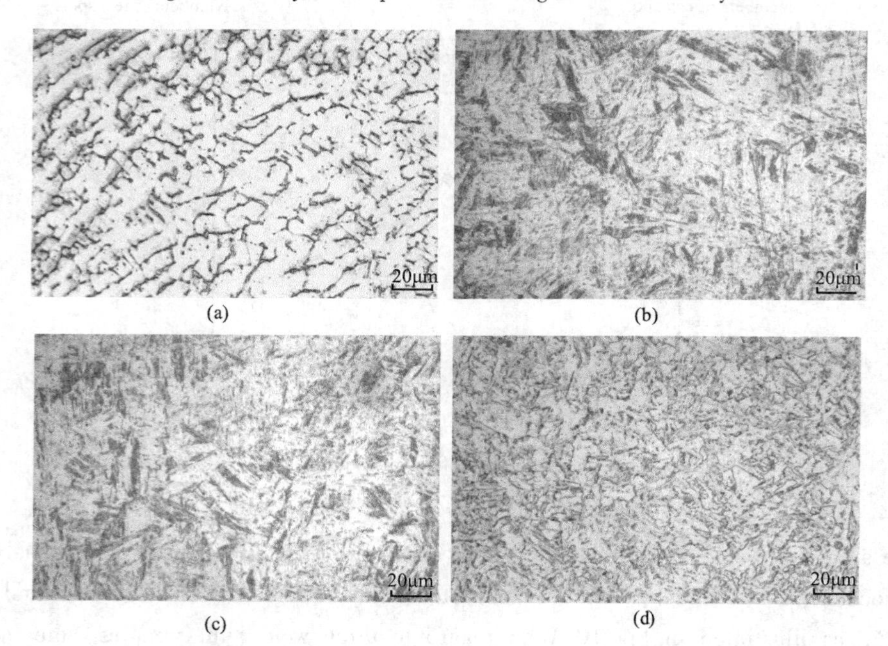

Fig. 11 Metallurgical structure of weld joint

(a. Root weld zone, b. Sealing weld zone, c. Transition weld zone, d. Filling weld zone)

Table 3 Metallurgical structure of weld joint samples taking from four zones

Sample No.	Root Weld Zone	Sealing Weld Zone	Transition Weld Zone	Filling Weld Zone
1	Austenite	Martensite	Martensite+Bainite (few)	Bainite+Ferrite (few)
2	Austenite	Martensite	Martensite+Bainite (few)	Bainite+Ferrite (few)
3	Austenite	Martensite	Martensite+Bainite (few)	Bainite+Ferrite (few)

3.2.5 Cracking behavior characterization

3.2.5.1 Micro-morphology of the main crack

The main crack, as illustrated in Fig. 10 (a), was analyzed by scanning electron microscope. The crack source zone (the area close to 316L liner) is characterized by typical tearing morphology, and the tearing ridge can be obviously observed as shown in Fig. 12 (a), incomplete fusion was found in the crack source zone, as shown in Fig. 12 (b). The crack propagated in intergranular mode, as shown in Fig. 12 (c), secondary cracks were observed in the crack propagation zone, moreover, small amount of corrosion product was found in the crack propagation zone.

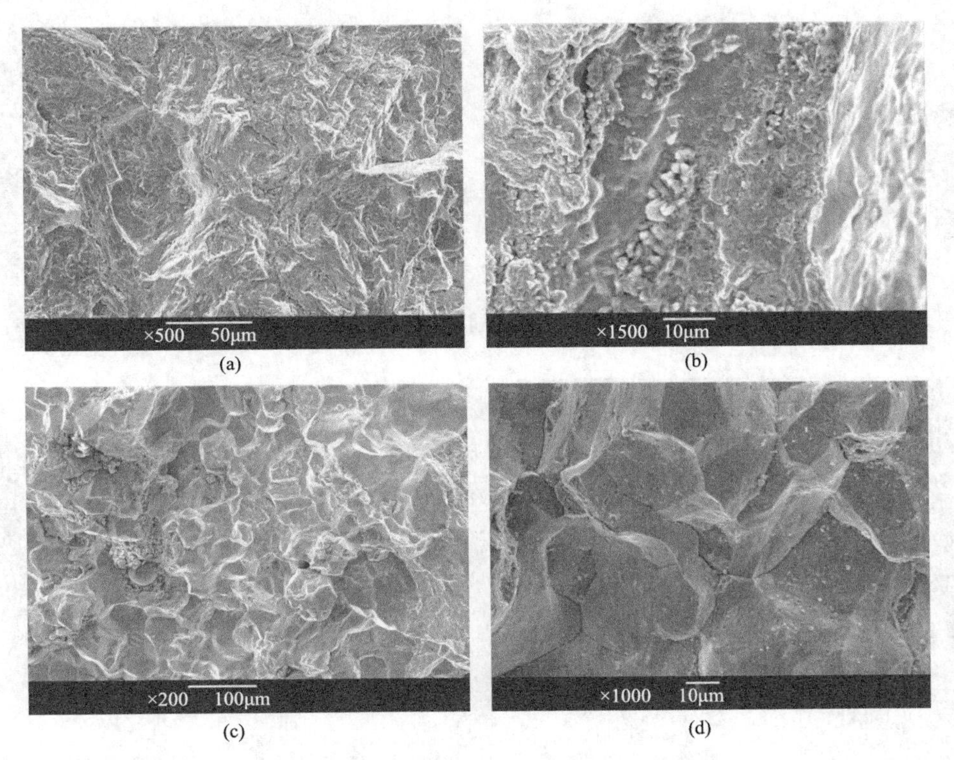

Fig. 12 Morphologies of main crack

(a. Crack source zone, b. Incomplete fusion in crack source zone, c. Crack propagation zone, d. Intergranular cracking in crack propagation zone)

3.2.5.2 Cracking behavior of sealing weld zone

The weld joint sample was taken from crack detected by X-ray, as illustrated in Fig. 10 (a), the general image of weld joint and sealing weld area is given in Fig. 13. In order to obtain the microstructure of the sealing weld area, the sample was analyzed by SEM in different magnifications, as shown in Fig. 14. The dendritic crack was observed and black material filled in crack, it is found that the black material is corrosion product. The element of the cracking zone was analyzed, it is seen that there was no element difference on left and right sides of the crack, as shown in Fig. 15.

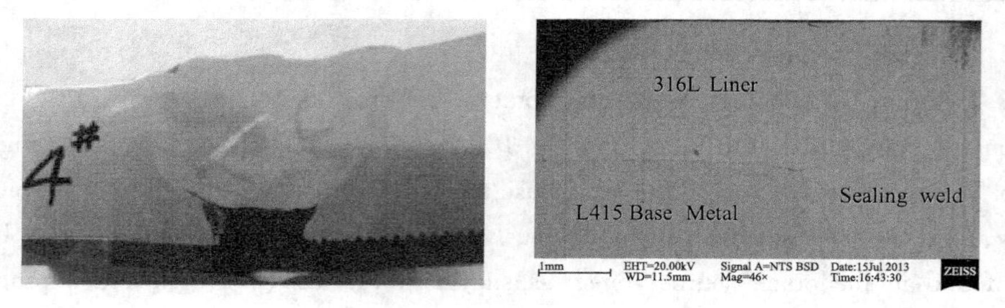

Fig. 13 Images of weld joint and sealing pass zone

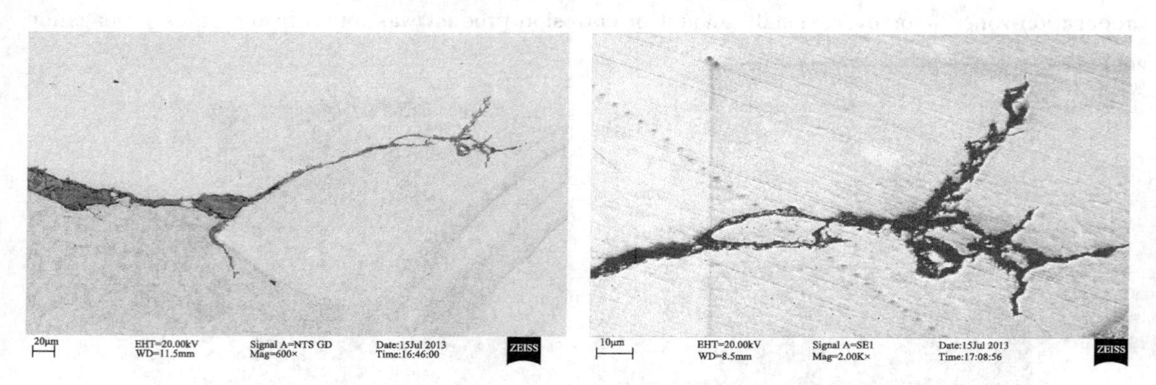

Fig. 14 Cracking morphology of sealing pass zone

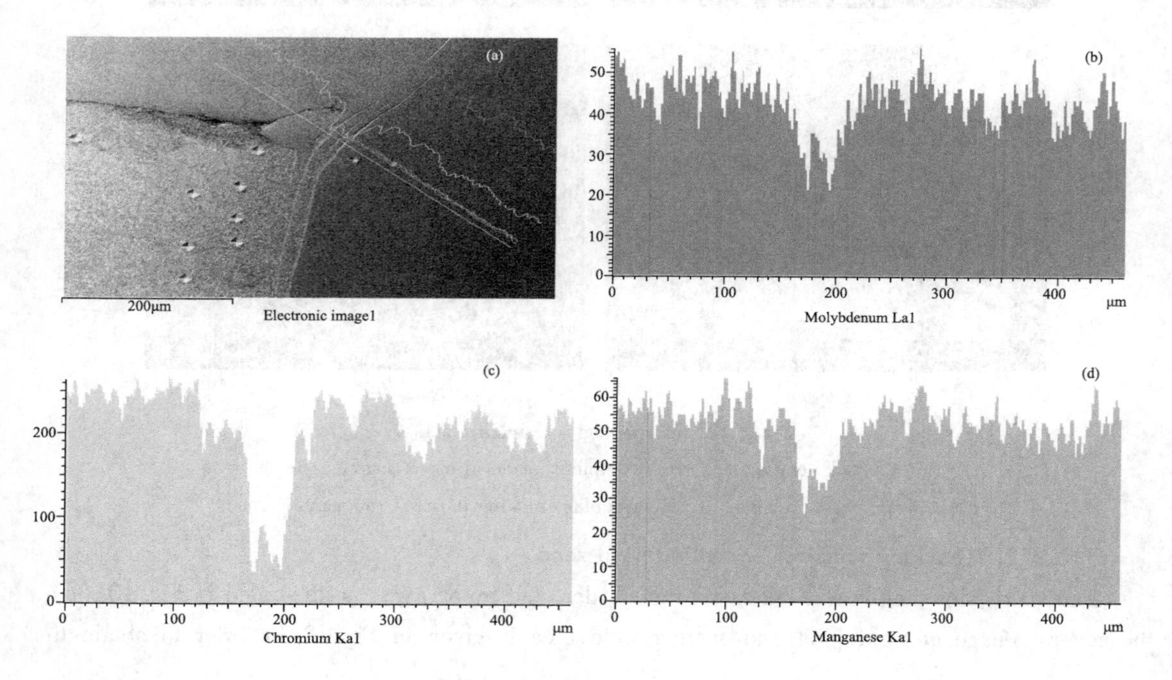

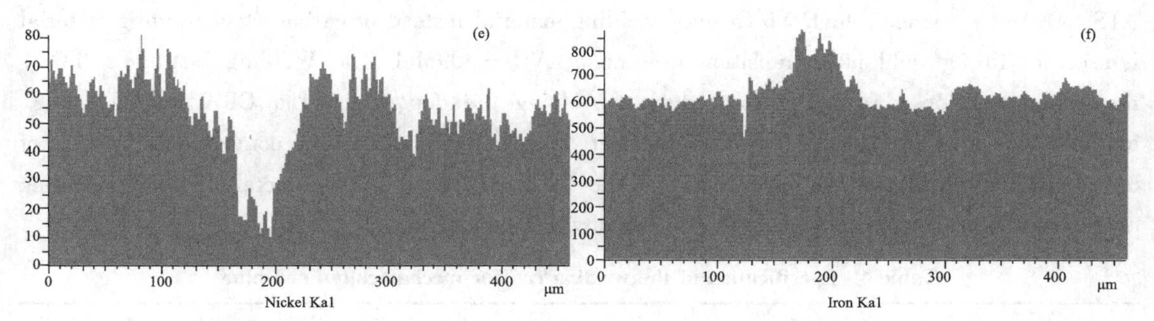

Fig. 15 Element analysis of sealing weld zone with crack (b. Mo, c. Cr, d. Mg, e. Ni, f. Fe)

4 Failure analysis

4.1 Material mechanical performance

The chemical composition (Table 1) of the base metal meets the GB 9711—2011 Requirements[11]. The average ultimate tensile strength of failed mechanically lined pipe are 523MPa, as given in Table 2, which meets the value of 520 ~ 760MPa as required in "Mechanically Lined Pipe Material Performance Standard of KLS Gasfield Gathering Pipeline Project". Moreover, the average impact energy of failed mechanically lined pipe is 38J at −20℃, which also meets the requirement of no less than 30J (−20℃) as mentioned in "Mechanically Lined Pipe Material Performance Standard of KLS Gasfield Gathering Pipeline Project".

4.2 Weld joint and welding process

"API SPEC 5LD-2009: Specification for CRA clad or lined steel pipe" specified that the maximum hardness of weld joint is HV 248[4]. Fig. 9 shows that the hardness of sealing weld zone and transition weld zone is higher than HV 248, generally, the weld with high hardness is very susceptible to crack initiation and with poor ability for crack arrest. The metallurgical structure given in Fig. 11 demonstrated that high hardness zone is martensite, which is a typical hard and brittle structure. Therefore, crack initiated from the boundary between L415 base metal, 316L liner and sealing weld zone, as shown in Fig. 13, then the crack propagated across transition weld zone and filling weld zone successively.

According to the "Mechanically Lined Pipe Welding Procedure Specification of KLS Gasfield Gathering Pipeline Project", four kinds of welding rod was used for welding of mechanically lined pipe, ATS-F309L (SS) was used for sealing weld, ATS-F316L (SS) was used for root weld, ATS-309MoL (SS) was used for transition weld, and CHE507 (CS) was used for filling weld, the detail information of the welding rod is given in Table 4. It is generally acknowledged that carbon steel or low alloy steel welding rod is not allowed to use for welding on stainless steel layer, for the failed mechanically lined pipe, CHE507 welding rod was used for welding on top of the transition welding layer (ATS-309MoL welding layer), it is equivalent to carbon steel and stainless steel mixed together and formed a medium alloy steel weld joint, this medium alloy layer are highly inclined to form martensite structure during cooling process, this kind of metallurgical structure is the root cause of crack. It is suggested that two possible ways to avoid the formation of martensite structure, first, a pure iron layer is introduced by building-up welding after transition welding of

ATS-309MoL; second, high Ni/Cr alloy welding material instead of carbon steel welding material is used for filling weld after transition welding of ATS-309MoL. The Welding Institute (TWI) researchers[6] proposed a hybrid procedure, involving the deposition of a CRA root/hot pass, followed by a high strength C-Mn fill, moreover, it is pointed out that the conventional practice of use a pure iron intermediate layer between the CRA and C-Mn weld metal had poor welding characteristics and could be susceptible to porosity and other defects.

Table 4 Specification of the welding rod for mechanically lined pipe

Type	Grade	Material	Inspection Standard	Layer
Sealing Weld	ATS-F309L	Stainless Steel	AWS A5. 22-95	1
Root Weld	ATS-F309L	Stainless Steel	AWS A5. 9-06	1
Transition Weld	ATS-309MoL	Stainless Steel	AWS A5. 22-95	2
Filling Weld	CHE507	Carbon Steel	AWS A5. 9-06	3~4
Shielding Gas	99. 99% Ar			

4. 3 Stress analysis

4. 3. 1 Operation stress analysis

Usually, it is necessary to consider the circumferential and axial stress for the pipeline during operation. As mentioned above, the mechanical strength of the mechanically lined pipe is mainly provided by outer carbon steel pipe, the outer carbon steel pipe was considered for stress analysis accordingly. The wall thickness (δ) is 14. 2mm, outer diameter (D) is 508mm, and the testing pressure (P) is 22MPa. Due to the ratio of $\delta/D=0.028$ is less than 0. 05, the axial stress (σ_a) and circumferential stress (σ_h) can be expressed as:

$$\sigma_a = PD/(4\delta) = 22\text{MPa}\times508/(4\times14.2) = 196.76\text{MPa} \quad (1)$$

$$\sigma_h = PD/(2\delta) = 22\text{MPa}\times508/(2\times14.2) = 98.38\text{MPa} \quad (2)$$

While the outer carbon steel pipe of the failed mechanically lined pipe grade is L415, the tensile strength is ≥520MPa, and the tensile strength of CHE507 welding rod is ≥490MPa, both of them are much bigger than axial stress and circumferential stress. Therefore, it is concluded that the strength of L415 carbon steel base metal and CHE507 weld joint are satisfied with the strength design requirement.

4. 3. 2 Environment stress analysis

The field documents show that the failed pipe area had undergone the heavy rain for 2 days before the failure occurred, which caused the ground settlement as result of part pipe section in strained condition. As schematically shown in Fig. 4 (c), the straight pipe section is fixed by the retaining wall, while the bend pipe section is pulled by soil movement, therefore, the stress concentration occurred in weld joint due to pulling force and pipe self-weight, the top half weld joint is under the tensile stress condition, and the bottom half weld joint is under the compressed stress condition. The external stress would cause the relative motion between outer L415 carbon steel and 316L liner, however, the outer carbon steel and liner was fixed by sealing weld, and the relative motion is impeded by the sealing weld, consequently, the sealing weld zone, as the boundary of outer L415 carbon steel and 316L liner, is a high shear stress concentration area. As mentioned above, the sealing weld zone is characterized by martensite structure with high hardness, which is a typical hard and brittle structure and very susceptible to cracking. T. Tkaczyk et al. [5] also reported that the liner to clad transition region is the potential critical location for fracture failure during plastic straining, in which the liner to clad transition region is equivalent to the sealing weld zone as shown in Fig. 8.

5 Conclusions and mitigation measures

5.1 Conclusions

(1) The chemical composition, tensile strength and impact energy of outer carbon steel pipe is in accordance with relevant technical requirements of GB 9711-2011 and Tarim Oilfield technical requirement for mechanically lined pipe.

(2) The sealing weld zone and transition weld zone is with high hardness of HV350~450, while the hardness of root weld zone and filling weld zone is around HV200, which is close to carbon steel base metal.

(3) The metallurgical structure of root weld zone, sealing weld zone, transition weld zone and filling weld zone are characterized by austenite, martensite, martensite and few bainite, bainite and few ferrite, respectively.

(4) The crack initiated from outer carbon steel, and propagated in intergranular mode along the weld-fusion line, crack source zone and propagation zone can be seen in the outer carbon steel layer.

(5) The internal factor for the mechanically lined pipe cracking failure is martensite structure formed at sealing weld zone and transition weld zone, which is a typical hard and brittle structure with high hardness.

(6) The external factor for the mechanically lined pipe cracking failure is external stress generated by soil movement, the relative motion between outer L415 carbon steel and 316L liner resulted in high shear stress concentration at sealing weld zone.

5.2 Mitigation Measures

(1) It is necessary to optimize the welding procedure, such as the weld joint structure design, welding rod material selection, and heat treatment temperature control.

(2) It is compulsory to conduct hardness, metallurgical structure, and corrosion resistance test of the weld joint.

(3) It is an alternative option to improve the mechanically lined pipe end manufacturing process, i. e. , metallurgical bonding or build-up welding.

(4) It is indispensable to install thrust blocks during pipe construction, which can effectively prevent the external stress induced by soil movement.

Acknowledgements

This work was supported by the Key Laboratory for Mechanical & Environment Behavior of Tubular Goods, China National Petroleum Corporation.

References

[1] A. Shadravan, M. Amani, What Every Engineer or Geoscientist Should Know about High Pressure HighTemperature Wells. 2012 SPE Kuwait International Petroleum Conference and Exhibition, 2012, Kuwait City, Kuwait.

[2] M. Ueda, T. Omura, S. Nakamura, T. Abe, K. Nakamura, P. I. Nice, J W. Martin, Development of 125ksi grade HSLA steel OCTG for mildly sour environments. Corrosion'2005, Paper No. 05089, NACE International, Houston, 2005.

[3] K. A. Macdonald and M. Cheaitani, Engineering Critical Assessment in the Complex Girth Welds of Clad and Lines Pipe Materials, IPC2010-31627, Proceedings of the 8th International Pipeline Conference, Calgary, Alberta, Canada, 2010.

[4] API Specification 5LD: Specification for CRA Clad or Lined Steel Pipe.

[5] T. Tkaczyk, A. Pépin, S. Denniel, Fatigue and Fracture Performance of Reeled Mechanically Lined Pipes. Proceedings of the Twenty-second International Offshore and Polar Engineering Conference, Rhodes, Greece, 2012.

[6] D. Howse, H. Pisarski, C. Nageswaran, M. Hoekstra, A. Bourgeon, P. Sinker, Improved Welding, Inspection and Integrity of Clad Pipeline Girth Welds, TWI Report 18807/13/11, 2011.

[7] D. Gentile, A. Carlucci, N. Bonora, G. Lannitti, Crack Initiation and Growth in Bimetallic Girth Welds. Proceedings of the ASME 2014 33rd International Conference on Ocean, Offshore and Arctic Engineering, San Francisco, California, 2014.

[8] A. Carlucci, N. Bonora, A. Ruggiero, G. Iannitti, G. Testa, Integrity Assessment of Clad Pipe Girth Welds. ASME 2014 33rd International Conference on Ocean, Offshore and Arctic Engineering, San Francisco, California, 2014.

[9] S. Denniel, T. Tkaczyk, A. Pepin, Reeled Mechanically Lined Pipe: Cost Efficient Solution for Static and Dynamic Applications in Corrosive Environment. Deep Offshore Technology Conference 2012, Perth, Australia, 2012.

[10] JB/T 4730. 2—2005, Non-destructive testing of pressure equipment-Part 2: Radiographic testing.

[11] GB 9711—2011, Petroleum and Natural Gas Industries-Steel Pipe for Pipeline Transportation systems.

Downhole Corrosion Behavior of Ni-W Coated Carbon Steel in Spent Acid & Formation Water and Its Application in Full-scale Tubing

Fu Anqing Feng Yaorong Cai Rui Yuan Juntao

(1. Tubular Goods Research Institute of CNPC;
2. State Key Laboratory of Performance and Structural Safety for Petroleum Tubular Goods and Equipment Materials;
3. Northwest Oilfield Company, China Petroleum & Chemical Corporation)

Abstract: Downhole corrosion behavior of Ni - W coated carbon steel tubing was investigated in spent acid (dilute HCl) solution and formation water by using autoclave and electrochemical techniques, in which dilute HCl with different pH is used to simulate the spent acid during flowback in acidizing process before well production, and the formation water is used to simulate the fluid during well production. Weight loss test in autoclave and electrochemical measurement indicated that Ni - W coated carbon steel exhibited higher corrosion resistance than carbon steel in spent acid and formation water, especially in spent acid with low pH and formation water at high temperature. According to weight loss test, corrosion rate ratio of carbon steel and Ni-W coated carbon steel in spent acid with pH = 1, pH = 2, and pH = 4 is 13.7, 14.5, and 6.9, respectively, while corrosion rate ratio of carbon steel and Ni - W coated carbon steel in formation water at 30℃, 60℃, and 90℃ is 85.5, 73.5, and 125.9, respectively. Moreover, the sealing performance of full-scale Ni-W coated carbon steel tubing was evaluated by using make-and-break test, hydraulic bursting test, and extreme downhole condition corrosion test. No sticky thread was found after 4 times of make up and three times of break out by using maximum recommended makeup torque of 4258 lbf · ft, no leakage was detected after hydraulic bursting test at 95.0 MPa for 30 mins, and it is observed very slight corrosion on the tubing shoulder experienced make - and - break test under extreme downhole condition.

Keywords: Ni-W coated steel tubing; Downhole corrosion; Sealing; Spent acid; Formation water; Make-and-break test

1 Introduction

With the continuously growing demand in oil and gas energy globally, the search for new sources of oil and gas makes the operation condition became more and more severe. High temperature high pressure (HTHP) gas well was developed in northwest China increasingly with years. The average reservoir temperature is 90℃ and the initial reservoir pressure is nearly 100MPa, the downhole partial pressure of CO_2 and H_2S is up to 4MPa and 2MPa, respectively, and chloride concentration is as high as 150000 mg/L. Acidizing is employed to enhance the productivity of HTHP gas well, the acid system is HCl-based including 15% HCl, 1.5% HF, 3% HAc, and inhibitor. Super 13Cr martensitic stainless steels tubing was used for well completion and production, field statistics shows several gas well failure after 1~3 years of production due to pitting corrosion of 13 Cr tubing, moreover, over 70% failed gas well was acidized before production. Besides the corrosion attack by acid during acidizing process, in the long-term production process, the presence of CO_2, H_2S, and high chloride concentration in formation water plays a key role in the corrosion failure of downhole tubing. The corrosivity of downhole environment is further complicated and enhanced by the high temperature and high pressure, it is well acknowledged that temperature accelerated the kinetics of corrosion reactions, as well as pressure contributes to internal corrosion in terms of higher downhole pressure increasing the partial pressure or solubility of naturally occurring corrosive acid gases, such as CO_2 and H_2S[1,2].

Both field data and lab research indicated that 13Cr martensitic stainless steel tubing is no longer the best choice for HTHP well, especially in spent HCl solution during flowback and high chloride concentration at high temperature[3-5]. The correct strategy in the choice of tubing material is becoming increasingly significant, safety and cost have to be preferentially considered for tubing material selection, in the petroleum exploration and production industry, the major corrosion resistant alloys used fall into three categories: martensitic stainless steels, duplex stainless steels, and Ni-based alloys[6]. It is well-known that the Ni-based alloys are the best choice with overall desirable properties, but the biggest disadvantage is that the Ni-based alloys are prohibitively expensive used as downhole tubing. However, development of Ni-based alloy coating on carbon steel is an economical alternative choice, providing excellent corrosion resistance and wear resistance. Extensive studies have been carried out to investigate the mechanical and corrosion performance of Ni-based alloy coating, such as Ni-P[7], Ni-P-W[8], Ni-SiC[9], Ni-TiN[10], under various environmental conditions.

In this work, downhole corrosion behavior of Ni-W coated carbon steel tubing was studied in HCl solution with different pH and formation water by using autoclave, HCl solution with different pH is used to simulate the spent acid during flowback in acidizing process before well production, and the formation water is used to simulate the fluid during well production. Meanwhile, corrosion mechanism in terms of electrochemical corrosion behavior was characterized by potentiodynamic polarization technique, electrochemical impedance spectroscopy (EIS). For all the tests, the carbon steel sample without coating was used for comparison. Moreover, Ni-W coating was applied to full-scale carbon steel tubing, the full-scale tubing sealing performance was evaluated by using make-and-break test, hydraulic bursting test, and extreme downhole condition corrosion test. It is expected that this work would provide technical support for potential application of Ni-W coated carbon steel tubing in HTHP gas well.

2 Experimental

2.1 Electrode and solution

Carbon steel tubing and Ni-W coated carbon steel tubing (Hunan Nanofilm New Material Technology Co., Ltd.) were used in this work, in which Ni-W coated carbon steel is being planned to use for gas well tubing in western China. The chemical composition is given in Table 1, and microstructure is shown in Fig. 1. The specimens for weight loss test in autoclave were machined into pieces with a dimension of 40mm×10mm×3mm. The specimens for electrochemical measurement were machined into pieces with a dimension of 10mm×10mm×2mm, and then embedded in epoxy resin with an exposed working area of 1 cm^2. For Ni-W coated carbon steel sample, part of the machined coupons, including autoclave test and electrochemical measurement sample, were electrodeposited with Ni-W coating. Prior to experiment, the working surface of carbon steel specimen was sequentially grounded with 320 grit, 600 grit, 800 grit, 1000 grit and 1200 grit SiC papers, polished with 0.1μm alumina polishing powder. Then the surface-treated carbon steel sample and as-prepared Ni-W coated carbon steel sample degreased with alcohol, cleaned in water, and finally dried in air.

Table 1 Chemical composition analysis of Ni-W coated carbon steel and carbon steel (wt%)

Element	C	Fe	Si	P	Ni	W	Cr	Mn
Ni-W coated carbon steel	/	2.02	/	3.29	73.37	21.32	/	/
Carbon steel	0.62	97.78	0.21	/	/	/	0.88	0.51

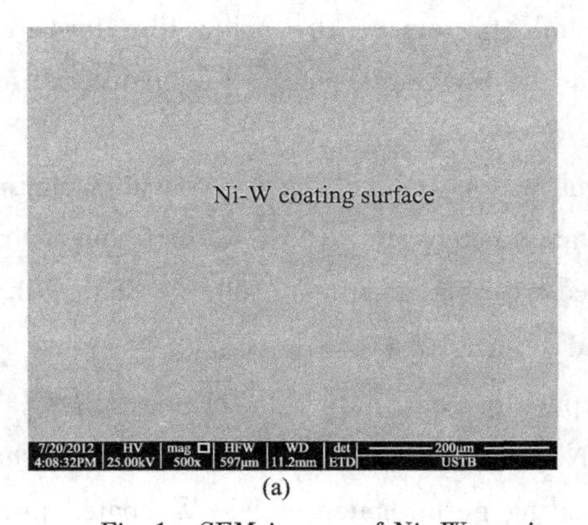

(a)

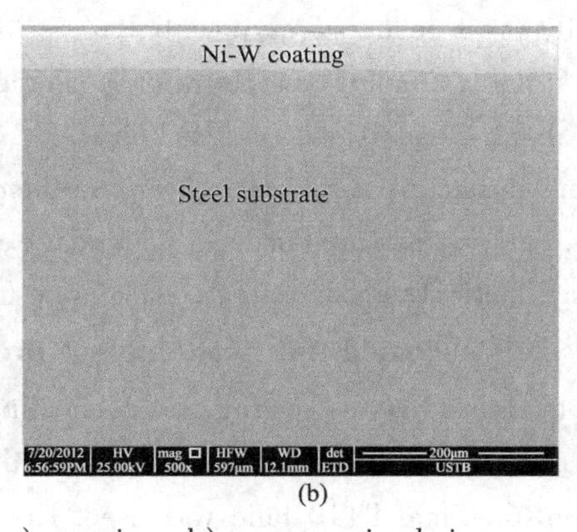

(b)

Fig. 1 SEM images of Ni-W coating: (a) top view; b) cross-sectional view

Two types of solution including spent acidizing fluid (dilute HCl) and formation water were used to study the corrosion performance during acidizing process and after acidizing in formation water, respectively. Detailed composition of formation water is given in Table 2. The solution was made from analytic grade reagents and ultra-pure water (18 MΩ cm in resistivity).

Table 2 Chemical composition of oilfield formation water (mg/L)

Na^+	Mg^{2+}	Ca^{2+}	Ba^{2+}	Cl^-	HCO_3^-	SO_4^{2-}	pH	Total Mineralization
71560	912.6	8037	18.48	127000	466.6	537.6	5.86	209000

2.2 Weight loss test in autoclave

In order to simulate the downhole corrosion of high temperature high pressure gas well, weight loss test was performed in autoclave to characterize the corrosion behavior of carbon steel and Ni-W coated carbon steel tubing. With consideration of reproducibility, three equivalent coupons were used for each test condition. Before the weight loss test, the specimens were cleaned with distilled water and acetone, dried, and then weighed using METTLER TOLEDO 4-digit electronic balance with a precision of 0. 1mg, and the weight before test was recorded as the original weight (W_0). After completion of test, the corroded samples were rinsed with distilled water, and then cleaned in corrosion film removing solution to remove the corrosion product formed on the sample surface, and then rinsed and dried again, finally reweighed to obtain the final weight (W_f). The corrosion rate (CR) was reported in mm/y according to the obtained weight loss via Eq. (1). The average corrosion rate of three specimens for each test condition was used.

$$CR=\frac{(W_0-W_f)\times1000\times365\times24}{t\times\rho\times S}$$

where W_0 and W_f are the original weight and final weight of specimen, g, respectively; S is the exposed surface area of specimen, mm^2; t represents the immersion time, h; and ρ is the density.

2. 3 Electrochemical measurements

Electrochemical measurements were performed using a Princeton Applied Research 273 electrochemical workstation on a three-electrode cell, where the X60 pipeline steel specimen was used as working electrode, a platinum foil as counter electrode, and a saturated calomel electrode (SCE) as reference electrode. Prior to electrochemical measurement, the working electrode was cathodically pre-polarized at potential of -800 mV to remove the oxide film formed on steel electrode surface. Finally, the corrosion potential of the working electrode was monitored for 30 min to ensure that a steady state was reached.

Potentiodynamic polarization curve measurement was conducted at a potential scanning rate of 0. 5 mV/s. EIS was measured under a sinusoidal excitation potential of 10 mV in the frequency range from 100 kHz to 10 mHz. The obtained EIS results were fitted by using commercial software ZSimpWin3. 2.

2. 4 Ni-W coated full-scale tubing thread sealing test

Due to thread acts as sealing and connecting part for each individual tubing string, therefore, the surface of pin and box have to be coated with Ni-W coating as well. For all the full-scale tests, ϕ88. 90mm×6. 45mm P110 tube was used. The sealing performance of Ni-W coated pin and box coupling was evaluated by using make-and-break test system, hydraulic bursting test system, and complex loading system. (1) Make-and-break test: the maximum recommended makeup torque of 4258 lbf · ft was used, the speed for makeup torque is 25 rpm, for each thread, 4 times of make up and 3 times of break out were performed, the high pressure API thread sealant SHELL TYPE3 was used. (2) Hydraulic bursting test: the test medium for hydraulic bursting test is water, which is conducted at 95. 0 MPa for 30 mins.

3 Results and discussion

3. 1 Corrosion behavior of Ni-W coated carbon steel in spent acid solution

Fig. 2 shows the corrosion rate of Ni-W coated carbon steel and carbon steel in spent acid with different pH at 90℃. The corrosion rate ratio of carbon steel and Ni-W coated carbon steel is 13.7, 14.5, and 6.9 in spent acid with pH=1, pH=2, and pH=4, respectively. Compared to carbon steel, Ni-W coating exhibits higher corrosion resistance in acid solution, especially in lower pH solution. Fig. 3 shows the micro-morphology of corroded Ni-W coated carbon steel and composition of corroded Ni-W coating surface. It is seen that uniform corrosion occurred in spent acid solution, according to the comparison of EDS analysis results of corroded Ni-W coated carbon steel surface and chemical composition of Ni-W coating in Table 1, few Fe originally contained in coating was totally dissolved in spent acid solution, and relative content of Ni decreases from 73.37% to 54.21%, while the relative content of W increases from 21.32% to 35.57%. It is indicated that Ni was partially dissolved in spent acid solution, and W was oxidized resulting in formation of WO_3, several previous papers have reported the influence of tungsten in enhancing corrosion resistance of Ni-P based coating, it is reported that tungsten preferentially migrated to the coating surface and formed W-riched oxide film during corrosion process, which inhibited the further corrosion[11-13]. Ni-W coating was reported more corrosion resistant than stainless steel 304 in acidic medium[14].

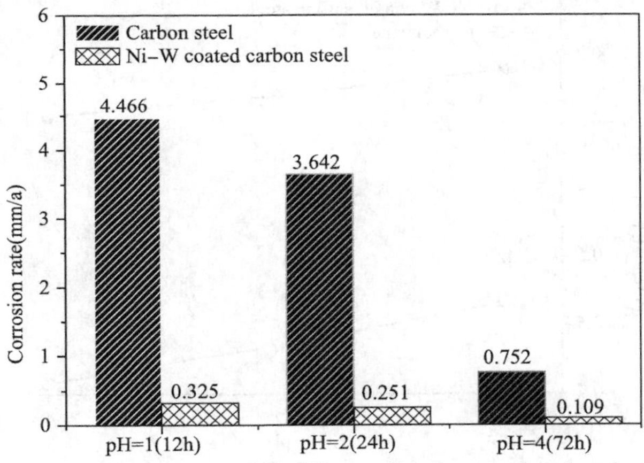

Fig. 2 Corrosion rate of Ni-W coated carbon steel and carbon steel in spent acid solution with different pH at 90℃

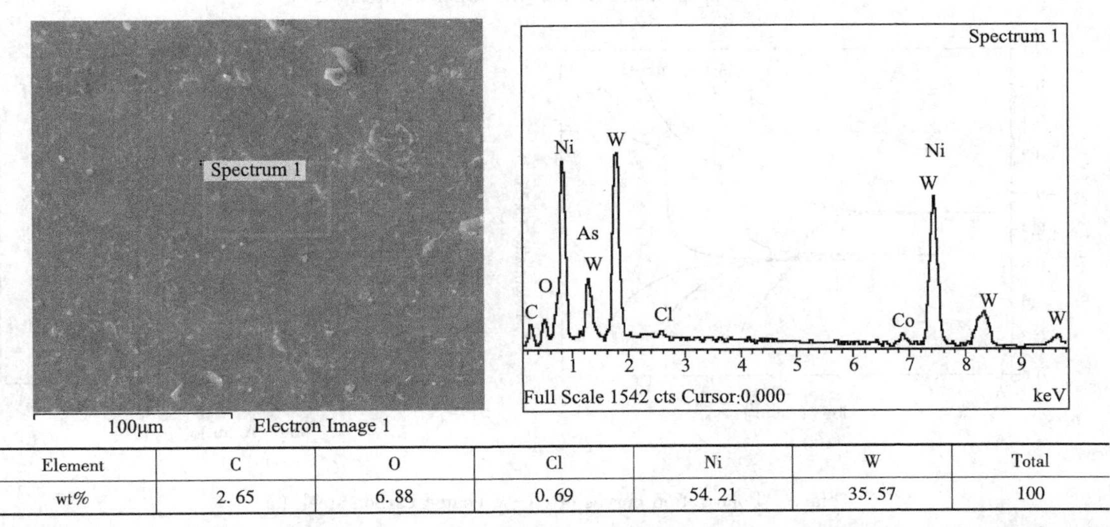

Element	C	O	Cl	Ni	W	Total
wt%	2.65	6.88	0.69	54.21	35.57	100

Fig. 3 SEM images and EDS results of corroded Ni-W coated carbon steel in spent acid solution at 90℃

Electrochemical measurements were carried out to investigate corrosion mechanism of Ni-W coated carbon steel. Fig. 4 shows the corrosion potential of Ni-W coated carbon steel and carbon steel in spent acid solution with different pH at 90℃, corrosion potential shifted negatively with the increase of solution pH. Correspondingly, Fig. 5 shows polarization curves of Ni-W coated carbon steel and carbon steel in spent acid solution with different pH at 90℃. It is seen that the polarization curves of Ni-W coated carbon steel are characterized by passivation when the solution pH is equal to 1 and 2, the passive current density decreased with the increasing of pH, while no passive region was observed when the pH increased up to 4, which indicated that corrosion mechanism of Ni-W coating varied with acid pH, passivity dominated the anodic reaction if spent acid solution pH is less than 2, while active dissolution dominated the anodic reaction if spent acid solution pH is greater than 2. In contrast, all the polarization curves of carbon steel measured in different pH spent acid solutions have similar characteristics, active dissolution instead of passivation dominated the anodic reaction.

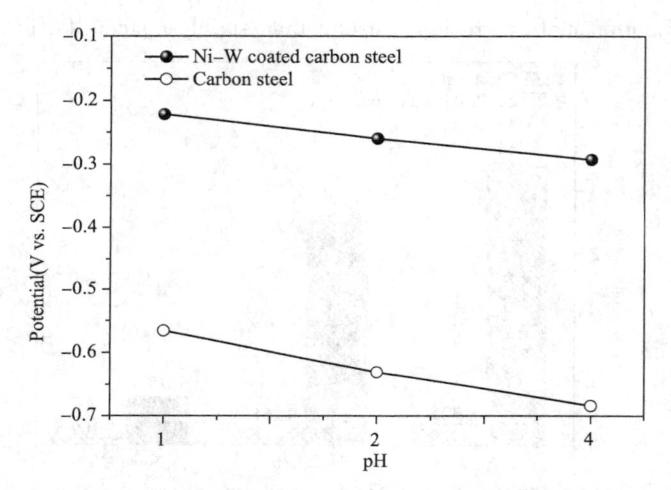

Fig. 4 Corrosion potential of Ni-W coated carbon steel and carbon steel in spent acid solution with different pH at 90℃

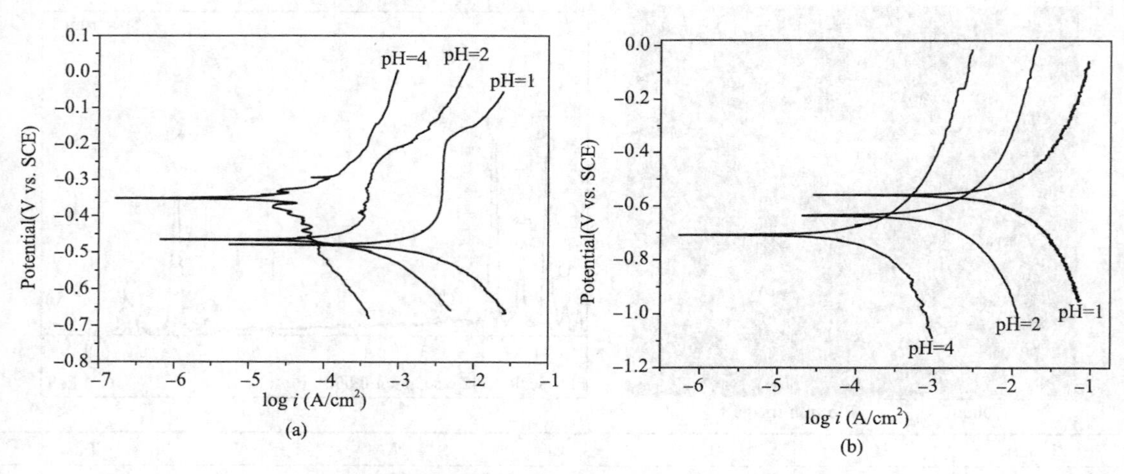

Fig. 5 Polarization curves of Ni-W coated carbon steel (a) and carbon steel (b) in spent acid solution with different pH at 90℃

EIS diagrams of Ni-W coated carbon steel and carbon steel measured in different pH spent acid solutions are shown in Fig. 6, EIS diagrams of Ni-W coated carbon steel and carbon steel are characterized by one semicircle. Electrochemical equivalent circuits shown in Fig. 7 (a) are proposed to fit the EIS data, where R_s is solution resistance, R_{ct} is charge-transfer resistance, Q_{dl} is double-charge layer capacitance. The fitted electrochemical impedance parameters are given in Table 3. The charge-transfer resistance and solution resistance of Ni-W coated carbon steel increases with solution pH, similar trends observed on carbon steel. Since corrosion rate is inversely proportional to the charge-transfer resistance, consequently, the corrosion rate ratio of carbon steel and Ni-W coated carbon steel calculated based on charge-transfer resistance is 97, 62, and 12 in spent acid with pH = 1, pH = 2, and pH = 4, respectively. The corrosion rate ratio obtained both from autoclave weight loss test and electrochemical test indicated that Ni-W coated carbon steel exhibited higher corrosion resistance than carbon steel in spent acid, especially in low pH condition. The reason why Ni-W coated carbon steel exhibits high corrosion resistance than carbon steel in low pH solution (pH = 1, 2) than that in high pH solution, which is mainly due to passive nature of Ni-W coating in low pH solution, as shown in Fig. 5.

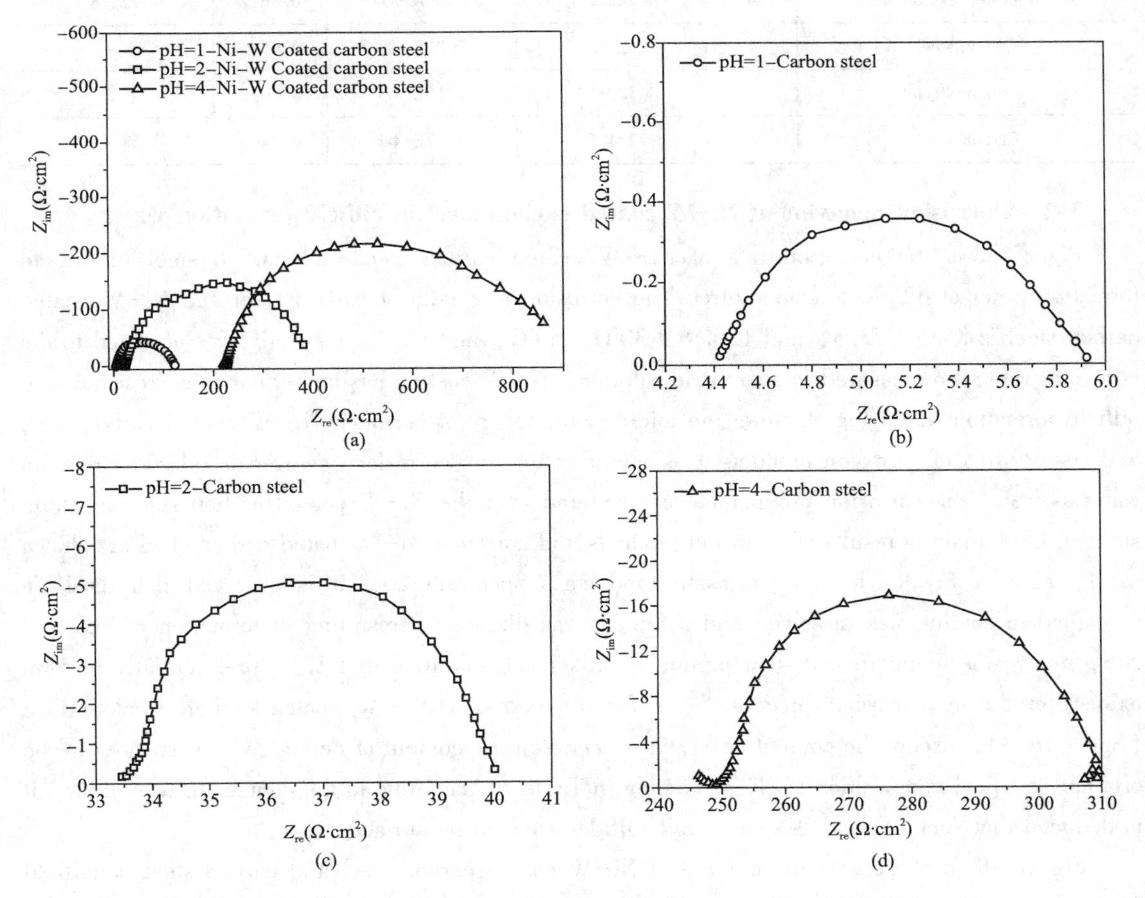

Fig. 6 EIS diagrams of Ni-W coated carbon steel (a) and carbon steel (b, c, d) in spent acid solution with different pH at 90℃

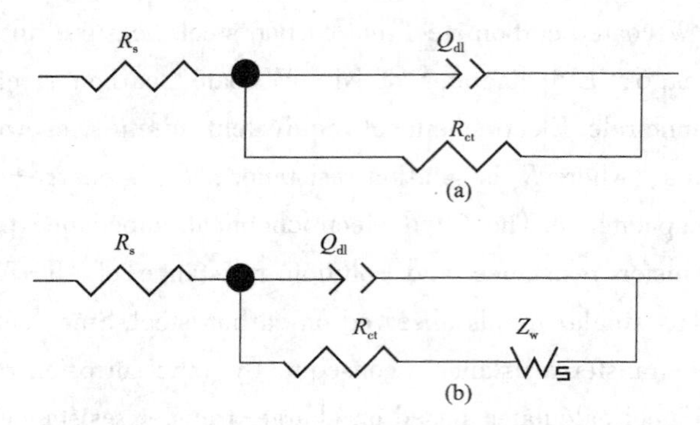

Fig. 7 Electrochemical equivalent circuit (EEC) for EIS data fitting

Table 3 EIS fitted data of Ni-W coated carbon steel and carbon steel in spent acid solution

Steel	pH	R_s (Ω · cm²)	Q_{dl} (F/cm²)	n	R_{ct} (Ω · cm²)
Ni-W coated carbon steel	1	4.535	7.32 E-04	0.8079	138.5
Ni-W coated carbon steel	2	23.71	3.125E-04	0.8338	396.5
Ni-W coated carbon steel	4	242.2	2.688E-04	0.8612	722.8
Carbon steel	1	4.42	3.20E-03	0.6589	1.43
Carbon steel	2	33.86	1.15E-03	0.8077	6.37
Carbon steel	4	254.9	5.87E-04	0.7469	58.45

3.2 Corrosion behavior of Ni-W coated carbon steel in oilfield formation water

Fig. 8 shows the corrosion rate of Ni-W coated carbon steel and carbon steel in oilfield formation water at different temperatures. The corrosion rate ratio of carbon steel and Ni-W coated carbon steel is 85.5, 73.5, and 125.9 at 30℃, 60℃, and 90℃, respectively. Compared to the corrosion rate ratio measured in spent acid solution, Ni-W coating has higher corrosion resistance in oilfield formation water. Fig. 9 shows the micro-morphology of corroded Ni-W coated carbon steel and composition of corrosion products. It is seen that uniform corrosion occurred in oilfield formation water as well, the corrosion product pieces scattered over the Ni-W coated carbon steel substrate surface. EDS analysis results of corrosion products and corroded Ni-W coated carbon steel are shown in the table of Fig. 9, for the corrosion products (Spectrum 2), it is observed that most Fe contained in coating was dissolved and a few Ni was dissolved, resulting in formation of FeS and NiS, no W was found in corrosion products, this result confirms that W is preferentially to form oxide film during corrosion process[11-13]; for the corroded Ni-W coated carbon steel surface (Spectrum 3), except the content of Ni, it is seen that the content of Fe and W is very close to the original chemical composition of Ni-W coating in Table 1. According to the content in two tables, it is deduced that some Ni and W oxides and sulfides covered on surface.

Fig. 10 shows the corrosion potential of Ni-W coated carbon steel and carbon steel in oilfield formation water at different temperatures, corrosion potential shifted negatively with temperature. Correspondingly, Fig. 11 shows the polarization curves of Ni-W coated carbon steel and carbon steel in oilfield formation water at different temperatures, it is seen that all the polarization

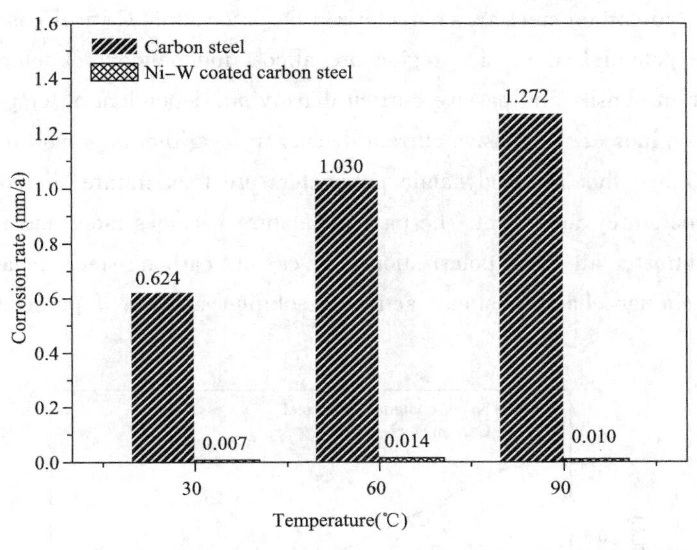

Fig. 8 Corrosion rate of Ni-W coated carbon steel and carbon steel in oilfield formation water with CO_2 (5MPa), H_2S (2MPa), and Cl^- (100000mg/L)

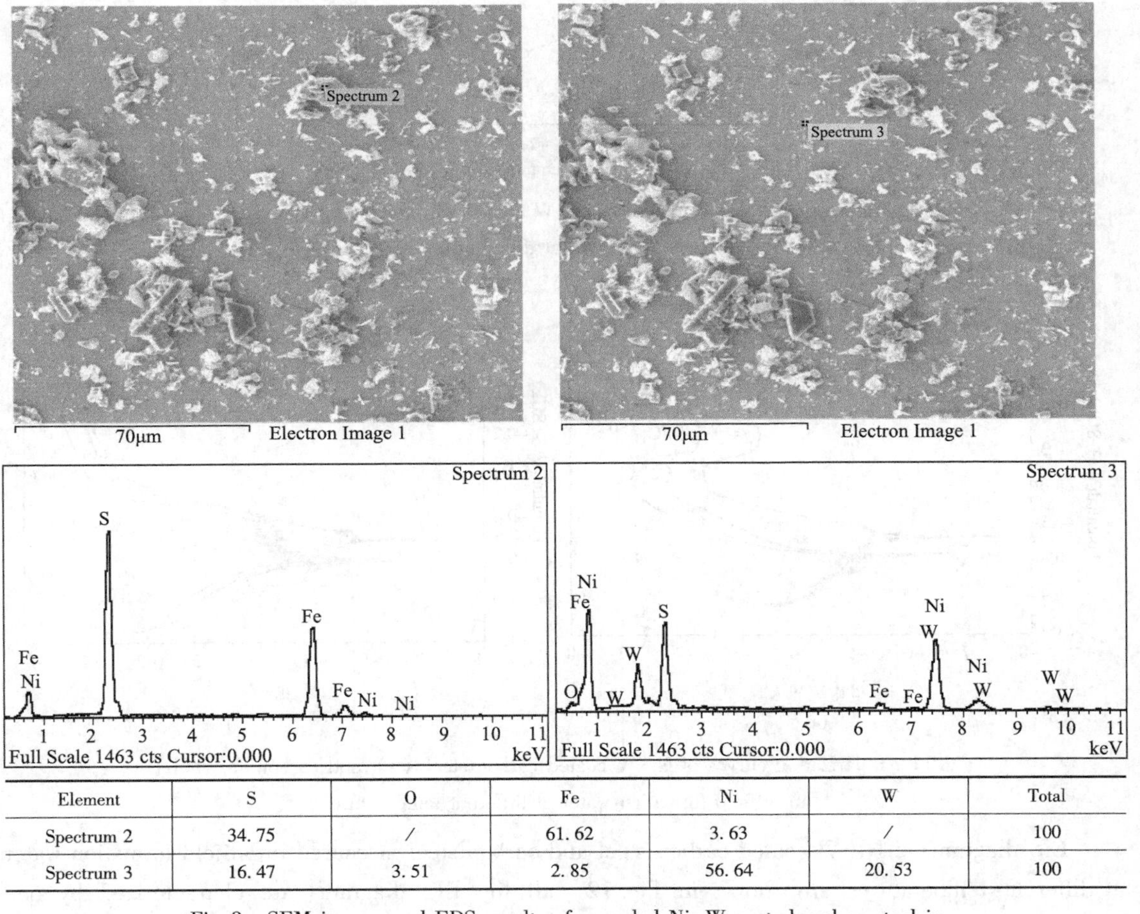

Element	S	O	Fe	Ni	W	Total
Spectrum 2	34. 75	/	61. 62	3. 63	/	100
Spectrum 3	16. 47	3. 51	2. 85	56. 64	20. 53	100

Fig. 9 SEM images and EDS results of corroded Ni-W coated carbon steel in oilfield formation water with CO_2 (5MPa), H_2S (2MPa), and Cl^- (100000mg/L)

curves of Ni-W coated carbon steel are characterized by passivation. Critical passive potential, start and end of passive potential in passive region are almost independent of temperature, while the critical passive current density and passive current density are dependent of temperature, moreover, it is observed that the increase of passive current density in logarithm is proportionally to temperature increase. It is concluded that thermodynamic parameters are temperature-independent and kinetic parameters are temperature-dependent. The passivity nature becomes more and more indistinct with temperature. In contrast, all the polarization curves of carbon steel measured at different temperatures have similar characteristics, active dissolution instead of passivation dominated the anodic reaction.

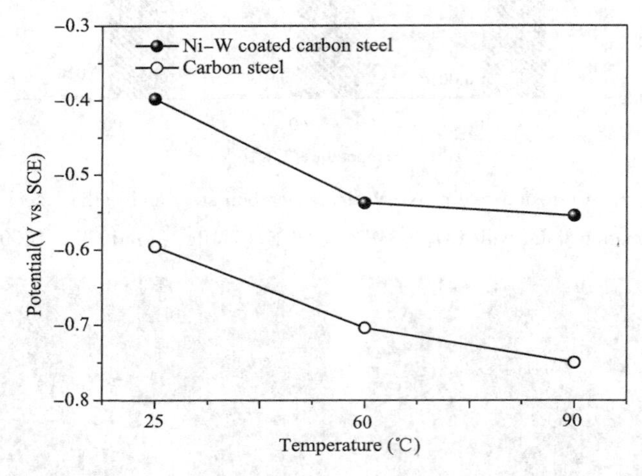

Fig. 10 Corrosion potential of Ni-W coated carbon steel and carbon steel in formation water at different temperatures

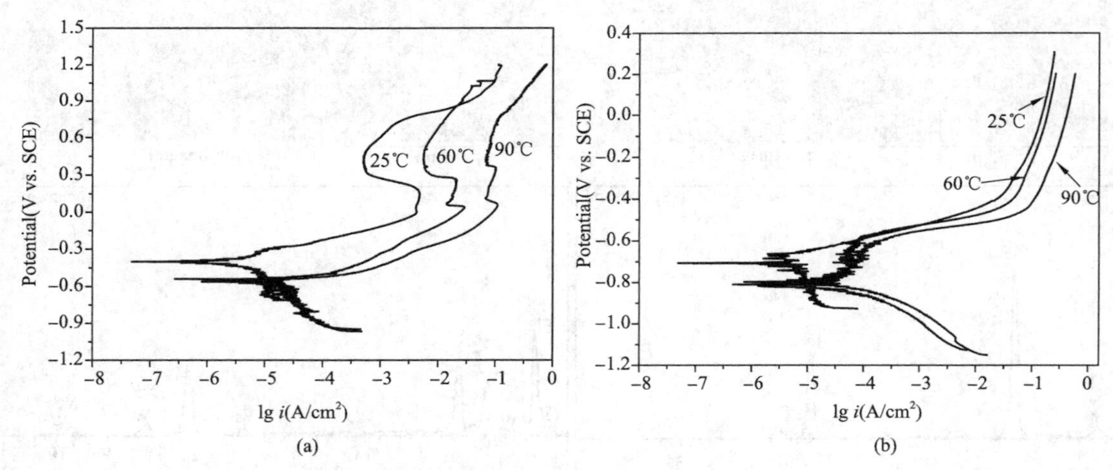

Fig. 11 Polarization curves of Ni-W coated carbon steel (a) and carbon steel (b) in oilfield formation water at different temperatures

EIS diagrams of Ni-W coated carbon steel and carbon steel measured in oilfield formation water at different temperatures are shown in Fig. 12, all the EIS diagrams are characterized by one semicircle, while diffusion control is observed on carbon steel at 60℃ and 90℃. EIS diagrams in Fig. 12 (a, c, d) are fitted by electrochemical equivalent circuits shown in Fig. 7 (a), and EIS diagrams in Fig. 12 (b) with diffusion are fitted by electrochemical equivalent circuits shown in Fig. 7 (b). The fitted electrochemical impedance parameters are given in Table 4. The charge-transfer resistance of Ni-W coated carbon steel and carbon steel decreases significantly with temperature increasing, the corrosion rate ratio of carbon steel and Ni-W coated carbon steel calculated based on charge-transfer resistance is 3.2, 5.0, and 29.2 in oilfield formation water at 30℃, 60℃, and 90℃, respectively. The corrosion rate ratio obtained both from autoclave weight loss test and electrochemical test indicated that Ni-W coated carbon steel exhibited higher corrosion resistance than carbon steel, especially in high temperature.

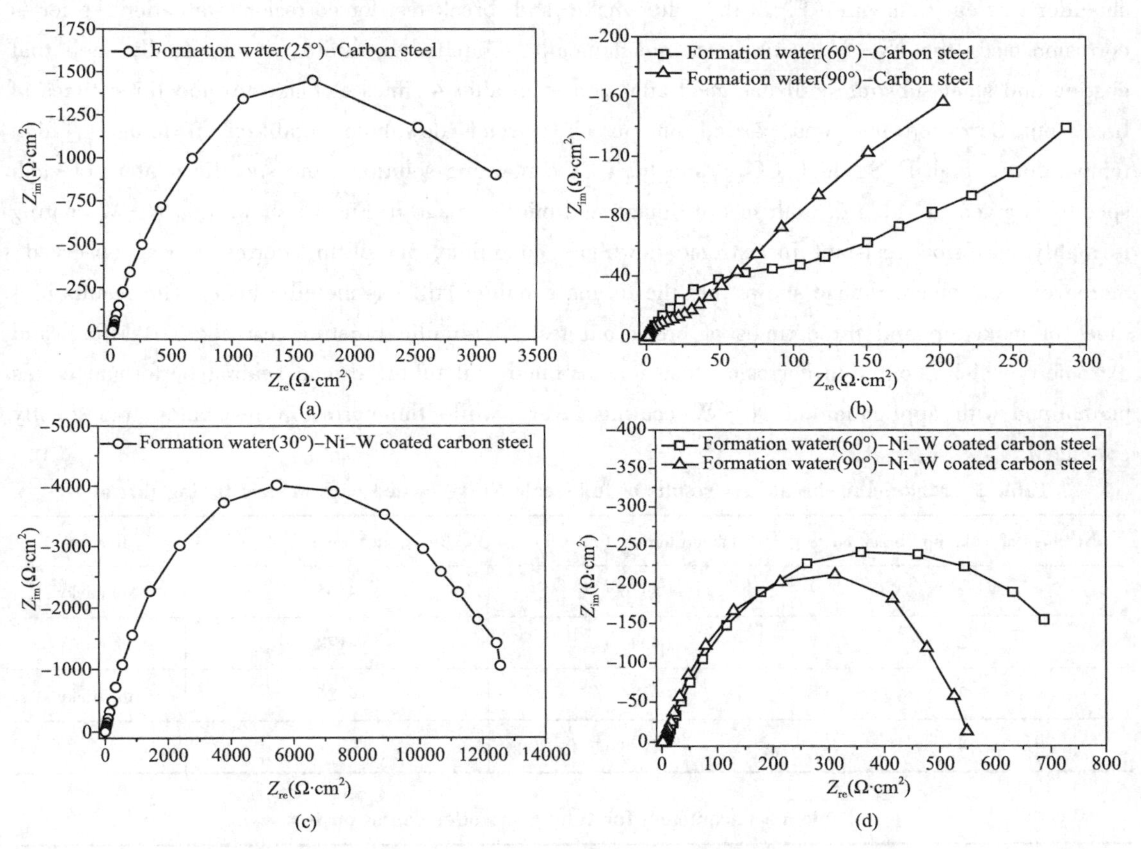

Fig. 12 EIS diagrams of Ni-W coated carbon steel (a, b) and carbon steel (c, d) in oilfield formation water at different temperatures

Table 4 EIS fitted data of Ni-W coated carbon steel and carbon steel in oilfield formation water

Steel	T (℃)	R_s ($\Omega \cdot cm^2$)	Q_{dl} (F/cm^2)	n	R_{ct} ($\Omega \cdot cm^2$)	Y_w ($\Omega^{-1} \cdot cm^{-2} s^{1/2}$)
Ni-W coated carbon steel	30	2.887	2.622E-05	0.7616	12320	/
Ni-W coated carbon steel	60	8.069	4.505 E-04	0.7193	800.8	/
Ni-W coated carbon steel	90	1.532	5.17E-04	0.7203	555.6	/
Carbon steel	30	21.91	1.077E-03	0.8025	3886	/
Carbon steel	60	1.764	1.827E-03	0.6928	160	0.01825
Carbon steel	90	0.7646	5.59E-04	0.8532	19.01	0.01605

3.3 Thread sealing test of full-scale Ni-W coated carbon steel tubing

Table 5 shows the make-and-break test results of full-scale Ni-W coated carbon steel tubing thread, Fig. 13 shows that the largest makeup torque (4220 lbf · ft) was reached after 2.75 turns of screw. It is seen that after 4 times of make up and three times of break out, the coating on pin and box still exhibits good appearance and adhesion, no sticky thread and damage was observed, as shown in Fig. 14. There is no leakage was found after hydraulic bursting test at 95.0 MPa for 30 mins. Since the shoulder plays a key role in tubing sealing performance, therefore, the tubing shoulder was cut from pin (Fig. 14b) after make-and-break test for corrosion evaluation, prior to corrosion test, the Ni-W coating was mechanically scratched by knife, it is visually seen that coating and steel substrate still has good adhesion even after 4 times of make up and three times of break out. Corrosion test was carried out in an extreme downhole condition in terms of high temperature, high H_2S, high CO_2, and high Cl^- containing solution, the specific content of each species is given in Table 6. Both macro-image and micro-image in Fig. 15 shows that Ni-W coating is highly corrosion resistant in extreme downhole condition, no pitting corrosion was observed, moreover, the macro-image shows that the tubing shoulder still has metallic luster. The results of 4 times of make up and three times of break out test, hydraulic bursting test at 95.0 MPa, and extreme downhole condition corrosion test demonstrated that tubing thread sealing performance was maintained with application of Ni-W coating layer, while the corrosion resistance was greatly enhanced.

Table 5 Make-and-break test results of full-scale Ni-W coated carbon steel tubing thread

Numbers of make up/ break out	Makeup torque (lbf · ft)	Break out torque (lbf · ft)	Results
1	4476	4435	Not sticky
2	4347	4178	Not sticky
3	4384	4126	Not sticky
4	4578	/	/

Table 6 Parameters for tubing shoulder corrosion test

T (℃)	P_{H_2S} (MPa)	P_{CO_2} (MPa)	Cl^- (mg/L)	Flow rate (m/s)	Duration (d)
120	2.07	5.96	100000	5	7

4 Conclusions

(1) Ni-W coated carbon steel exhibited higher corrosion resistance than carbon steel in spent acid, especially in low pH spent acid. The corrosion rate ratio of carbon steel and Ni-W coated carbon steel in spent acid with pH = 1, pH = 2, and pH = 4 is 13.7, 14.5, and 6.9, respectively.

(2) Ni-W coated carbon steel exhibited higher corrosion resistance than carbon steel in formation water, especially at high temperature. The corrosion rate ratio of carbon steel and Ni-W

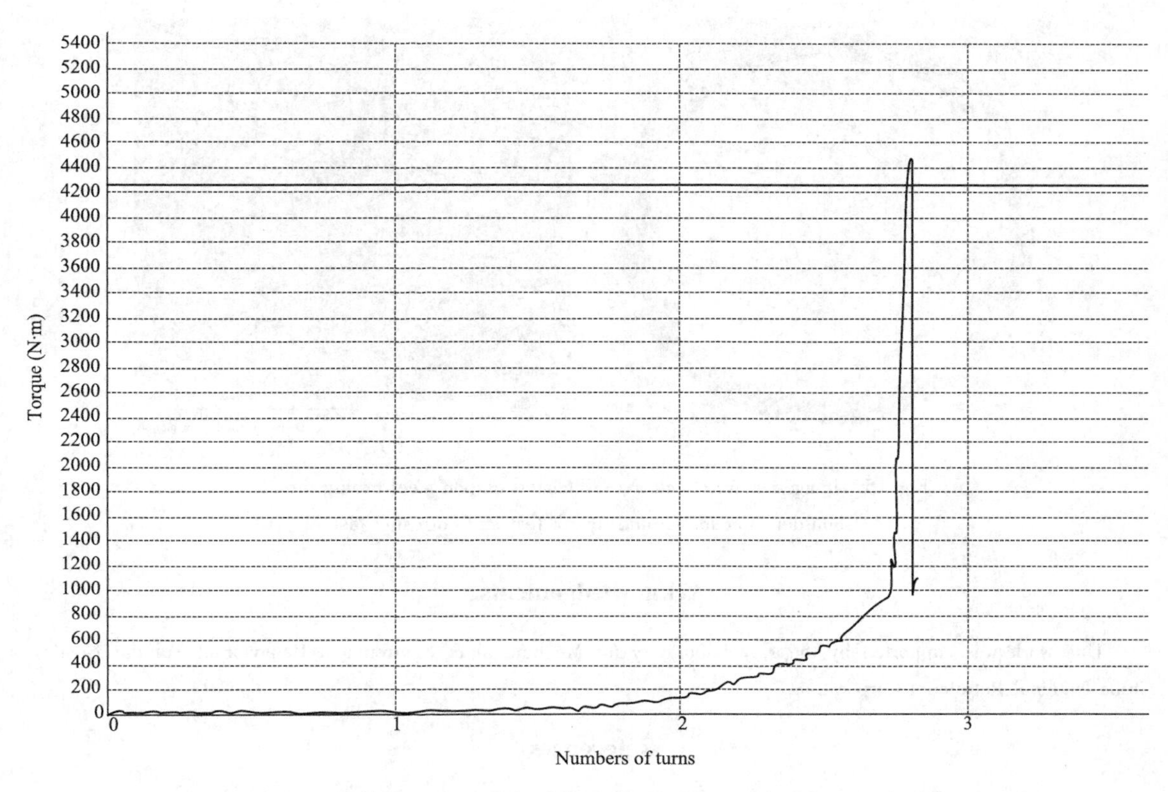

Fig. 13 Torque vs. numbers of turns for full-scale Ni-W coated carbon steel tubing thread

(a) (b)

Fig. 14 Images of full-scale Ni-W coated carbon steel tubing thread after make-and-break test: (a) box; (b) pin

coated carbon steel in formation water at 30℃, 60℃, and 90℃ is 85.5, 73.5, and 125.9, respectively.

(3) No sticky thread was found after 4 times of make up and three times of break out by using maximum recommended makeup torque of 4258 lbf · ft, no leakage was detected after hydraulic bursting test at 95.0 MPa for 30 mins, and it is observed very slight corrosion on the tubing shoulder experienced make-and-break test under extreme downhole condition.

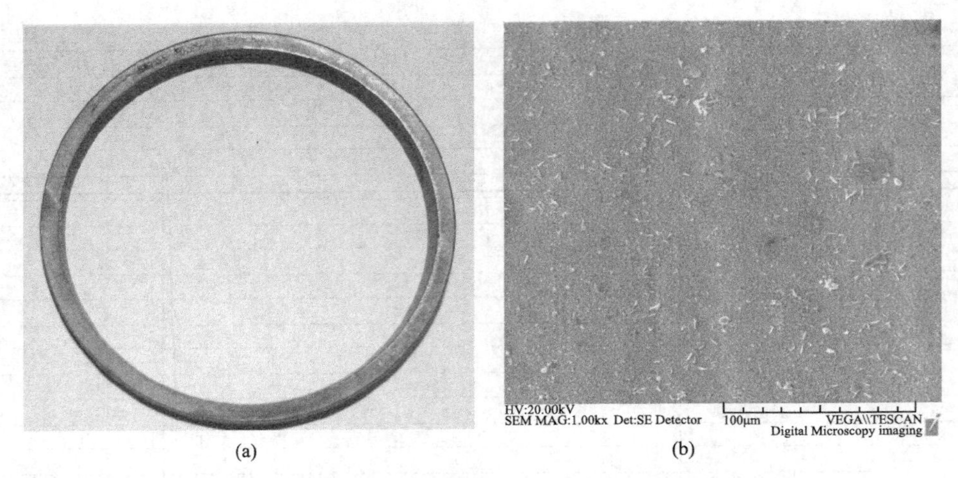

(a) (b)

Fig. 15 Images of full-scale Ni-W coated carbon steel tubing thread shoulder after make-and-break test and corrosion test

Acknowledgements

This work was supported by the Key Laboratory for Mechanical & Environment Behavior of Tubular Goods, China National Petroleum Corporation.

References

[1] G. M. Abriam, Controlling corrosion of carbon steel in sweet high temperature and pressure downhole environments with the use of corrosion inhibitors. NACE Northern Area Western Conference, Calgary, Alberta, 2010.

[2] J. A. Carew, A. Al-Sayegh, A. Al-Hashem, The effect of water-cut on the corrosion behaviour L80 carbon steel under downhole conditions. Corrosion'2000, Paper No. 00061, NACE International, Houston, 2000.

[3] Q. J. Meng, B. Chambers, R. Kane, J. Skogsberg, M. Kimura, K. Shimamoto, Evaluation of localized corrosion resistance of high strength 15Cr steel in sour well environments. Corrosion'2010, Paper No. 10320, NACE International, Houston, 2010.

[4] H. A. Nasr-EI-Din, S. M. Driweesh, G. A. Muntasherr, Field application of HCl-formic acid system to acid fracture deep gas wells completed with super Cr-13 Tubing in Saudi Arabia, SPE international, Paper No. 84925, Kuala Lumpur, Malaysia, 2003.

[5] H. Marchbois, H. EL Alami, J. Leyer, A. Gateaud, Sour service limits of 13% Cr and super 13% Cr stainless steels for OCTG: effect of environmental factors. Corrosion'2009, Paper No. 09084, NACE International, Houston, 2009.

[6] T. Henke, J. Carpenter, Cracking tendencies of two martensitic stainless alloys in common heavy compeltion brine systems at down-hole conditions: A laboratory investigation. Corrosion'2004, Paper No. 04128, NACE International, Houston, 2004.

[7] H. Liu, R. X. Guo, Z. Yun, B. Q. He, Z. Liu, Comparative study of microstructure and corrosion resistance of electroless Ni-W-P coatings treated by laser and furnace-annealing. Transactions of Nonferrous Metals Society of China 20 (2010) 1024-1031.

[8] Y. W. Yao, S. W. Yao, L. Zhang, H. Z. Wang, Electrodeposition and mechanical and corrosion resistance properties of Ni-W/SiC nanocomposite coatings. Materials Letters 61 (2007) 67-70.

[9] M. R. Vaezi, S. K. Sadrnezhaad, L. Nikzad, Electrodeposition of Ni - SiC nano - composite coatings and evaluation of wear and corrosion resistance and electroplating characteristics. Colloids and Surfaces A: Physicochem. Eng. Aspects 315 (2008) 176-182.
[10] F. F. Xia, C. Liu, F. Wang, M. H. Wu, J. D. Wang, H. L. Fu, J. X. Wang, Preparation and characterization of Nano Ni - TiN coatings deposited by ultrasonic electrodeposition. Journal of Alloys and Compounds 490 (2010) 431-435.
[11] W. H. Hui, J. J. Liu, Y. S. Chaug, Surf. Coat. Technol. 68/69 (1994) 546-551.
[12] M. Obradovic, J. Stevanovic, A. Despic, R. Stevanovic, J. Stocii, Journal of Serbian Chemical Society 66 (2001) 899-912.
[13] R. Z. Valiev, R. K. Islamgaliev, I. V. Alexandrov, Bulk nanostructured materials from severe plastic deformation. Progress in Material Science 45 (2000) 103-89.
[14] S. Yao, S. Zhao, H. Guo, M. Kowaka, A new amorphous alloy deposit with high corrosion resistance. Corrosion 52 (1996) 183-186.

Effects of Cl^- concentration on corrosion behavior of carbon steel and Super13Cr steel in simulated oilfield environments

Zhao Xuehui[1,2] Yin Cenxian[1] Li Fagen[1] Han Yan[1]

(1. Tubular Goods Research Institute of CNPC, State Key Laboratory of performance and Structural Safety for Petroleum Tubular Goods and Equipment Materials;
2. School of Materials Science and Engineering, xi'an Jiao tong University)

Abstract: In order to understand the changing tendency of the corrosion behavior of carbon steel and stainless steel with different chloride ion concentration, Electrochemical method and high temperature and high pressure immersion method combined with scanning electron microscopy were used in this paper. The results show that the corrosion potential of carbon steel is almost the same at 60℃. While for super13Cr martensitic stainless steel, its critical pitting potential sharply decreases from −48 mV to −213 mV (vs Ag/AgCl). However, with the increase of Cl^- content at 90℃, the corrosion potential of carbon steel is gradually reduced. While for the super13Cr, the corrosion potential reduces sharply, and when the Cl^- concentration is greater than or equal to 40g/L, the corrosion potential basically remained constant, the critical pitting potential also is not obviously changed. Carbon steel corrosion is mainly controlled by the diffusion effect. Super13Cr martensitic stainless steel displays the single capacitive reactance arc.

Keywords: Corrosion resistant alloy; Electrochemistry; Corrosion behavior; Impedance spectroscopy

1 Introduction

The corrosion problem of casing and tubing have attracted more attention of more oil and gas fields. Especially along with the increasingly extended length of oil and gas drilling time, many oil and gas wells are working in the later period, and fluid moisture content and aggressive ions increased and complicated, corrosion failures of casing and tubing are becoming more and more serious[1-3]. Cl^- is the most common and corrosive ions in the oil field environment, the corrosion failure and perforation leakage events of tubings happened repeatedly due to the presence of Cl^-. Especially in acid environment and in the presence of Cl^- with higher concentration, corrosion failure and serious damage of stainless steel tubings can be seen everywhere[4,5]. Many research about stainless steel have been done, however, most focus on the temperature, partial pressure of

CO_2/ H_2S and corrosion product film, and the results consider that with the increase of chloride concentration, the corrosion rates are becoming higher[6]. But combined with the results of corrosion online monitoring, the higher of the Cl^- concentration (above 10^5 mg/L) in service medium conditions, the corrosion rates are relatively not very higher. On the contrary, in the solution environment with relatively lower Cl^- concentration, the corrosion rates are relatively more higher. So the applicability between the tubing materials and service environment is considered a key question. Many reference show that Cl^- has a strong corrosive to oil pipe, but without clearly defined as the concentration range of the applicability[7-9], and little attention was paid on the corrosion resistance and variation trend of carbon steel and stainless steel material at the different Cl^- concentration. Carbon steel and stainless steel material have different chemical composition and organizational structure, so the corrosion mechanism of two kinds of materials are not the same in the solution medium containing Cl^- [10-12]. The present work aims at investigating the electro-chemical corrosion behavior of carbon steel and super13Cr stainless steel in acidic environment containing Cl^-, and comparative analysis the corrosion property differences and change trend of two kinds of materials with the different Cl^- concentrations. The results of this study have an important guiding significance to oil fields for rational selection materials and hope to serve as a lead to the choosing of tubings for the oil fields.

2 Experimental

2.1 Immersion tests

Corrosion behavior were carried out on two commercially available tubings, one was P110 and the other was super13Cr, which were denoted as 1# and 2#, respectively. The compositions of the samples are given in Table 1. It can be seen from Table 1 that Ni, Cr, Mo contents were the main differences between the two materials. The specimens for immersion test were machined as rectangle coupons with dimension of 40mm×15mm×3mm. A 6 mm diameter hole at one end serves to hang the specimens from a specimen stand with a non-metallic wire inside the autoclave. The specimens were first ground successively with 200, 400, 600 and 800 grit sandpaper, cleaned with alcohol and acetone, and then dried immediately. The specimens were weighed with a precision of 0.1 mg.

The simulated solution was made of Na^+ and Cl^-. N_2 gas was bubbled for 3 h to remove oxygen before CO_2 was introduced in the autoclave. Field environmental conditions were simulated at a different Cl-concentration. The test duration was 168 h. The corrosion rates were calculated from the data obtained by the weight-loss method. After 168 h of exposure, the specimens were rinsed with distilled water and then dried. Each specimen was weighed with a precision of 0.1 mg. However, the corrosion rate is the average value for 168 h. Through weight-loss method, corrosion rates were calculated as follows:

$$V=\frac{\Delta m}{\rho st}\times 24\times 365 \tag{1}$$

where V is corrosion rate in mm/a, Δm is the weight loss in g, ρ is the material density in g/mm^3, s is the corrosion area in mm^2, and t is the test duration in h.

Table 1 Chemical composition of two kinds of materials used in this study (wt%)

Number	C	Si	Mn	P	S	Cr	Mo	Ni	Ti	Cu
1#	0. 26	0. 25	1. 02	0. 012	0. 0026	0. 62	0. 017	0. 032	0. 008	/
2#	0. 016	0. 45	0. 46	0. 013	0. 0047	12. 98	2. 25	5. 04	/	1. 37

2. 2 Electrochemical measurements

The potentiodynamic polarization measurements were carried out using a 273A Electrochemical Measurement System manufactured by EG&G. The tests were carried out in a three – electrode system. The samples for electrochemical test were employed as the working electrode, the size of samples was ϕ10mm × 5mm. The reference electrodes are silver, silver chloride electrode (Ag/AgCl) with saturated potassium chloride (KCl) solution. The auxiliary electrode was a pair of graphite poles. Working electrodes were polished with silicon carbide papers from 400–grit to 1000–grit, degreased with acetone and rinsed with deionized water, dewatered with ethanol prior to the experiment. The test solution was deaerated by purging CO_2 (99. 95%) for 2 h. Gas exit was sealed with water.

The potentiodynamic polarization curves were recorded at a constant sweep rate of 0. 3 mV/s from–100 mV ~ 800mV with respect to the open circuit potential (E_{corr} vs Ag/AgCl). According to the oil field environment, the electrochemical experiments were performed at 60℃ and 90℃ and the Cl^- concentration are 20 g/L, 40 g/L and 120g/L, respectively. The choice of Cl^- concentration in the test was based on the environment of the Tarim Oilfield, which can show the trend of corrosion of materials.

Theelectrochemical impendence spectroscopy (EIS) measurements were carried out at open circuit potential using an alternating current voltage amplitude of 5 mV. The frequency varied from 10^5 Hz to 10 mHz. Data were presented as Nyquist plots.

3 Results and Discussion

3. 1 Effect of Cl^- on corrosion rate and corrosion product films

Table 2 shows the corrosion rates of samples at the high temperature and high pressure with different Cl^- concentrations. The test temperature was 110℃ and the CO_2 partial pressure was 4 MPa. We can see that the corrosion rates of 1# were higher than that of 2#, the corrosion rates increases firstly and then decreases with increasing Cl^- concentration from 20g/L to 120g/L. When the Cl^- concentration was 40g/L, the result appears that the corrosion rates have a peak, and reaches the maximum value. This means that certain amounts of Cl–concentration can accelerate the anodic reaction[13] and increase corrosion rates. The anodic reaction as shows:

$$Fe \rightarrow Fe^{2+} + 2e$$

$$Fe^{2+} + CO_3^{2-} \rightarrow FeCO_3$$

$$Fe + HCO_3^{2-} \rightarrow FeCO_3 + 2e + H^+$$

$$Fe + CO_3^{2-} \rightarrow FeCO_3 + 2e$$

Table 2 Corrosion rates of samples in solution with different Cl^- concentration

Materials	Corrosion rates (mm/a)		
	20g/L	40g/L	120g/L
1#	2.56	3.92	3.68
2#	0.0041	0.0069	0.0092

And for the 2# sample, the corrosion rates increases with increasing Cl^- concentrations. Fig. 1 shows the SEM profiles of the corrosion product films on 2# samples, and obvious pitting corrosion phenomena were observed on the surface of 2# with increasing Cl^- concentrations. At a content of 20 g/L [Fig. 3 (a)], a thin layer of corrosion product film was observed and the tiny pitting also can be found on the surface of the sample. At a content of 40 g/L [Fig. 3 (b)], the pitting is small but more obvious, at a content of 120 g/L [Fig. 3 (c)], the pitting is relatively bigger. As we all know that the radius of Cl^- is small, Cl^- can penetrate through the corrosion product film and aggregate on metal surface. In addition, when Cl-and other anions coexist, particularly when the content of Cl^- is higher, Cl^- is preferentially absorbed on metal surface[14]. So a high amount of Cl^- can aggregate in the link boundaries between the metal surface and corrosion product films. Therefore, the adhesion of the link boundaries decreases, making the corrosion product films easily remove from the metal surface automatically. Thus, pitting appear on the surface of samples. At this test conditions, only the Cl - concentration is a variable, and the higher Cl^- concentration can significantly reduce the CO_2 solubility and decrease the opportunities of H^+, H_2O, H_2CO_3 and HCO_3^- to participate in the reaction, so we can conclude that Cl^- can accelerate corrosion rates within certain concentration. The main corrosion type of 1# samples was uniform corrosion.

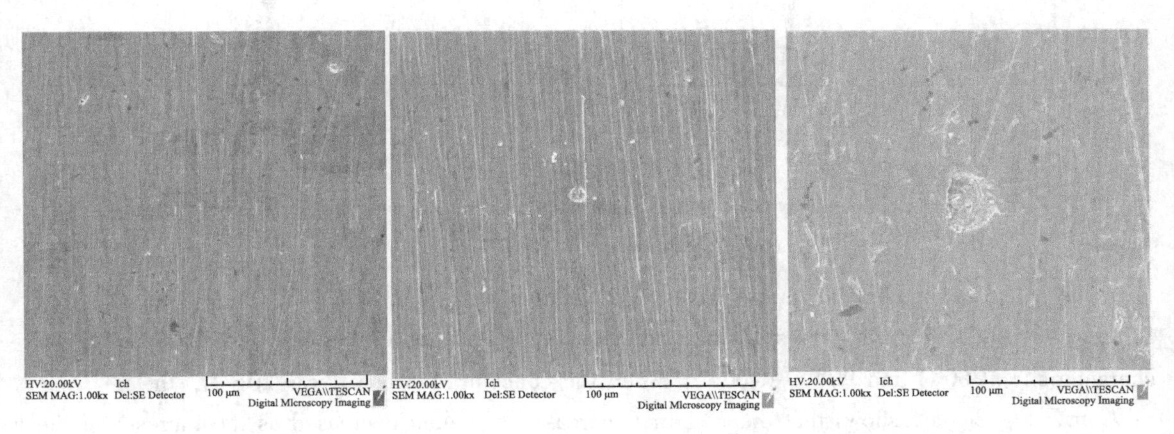

Figure 1 SEM morphology results of 2# for immersion containing different Cl^- concentration at 110℃. (a) 20 g/L, (b) 40 g/L, (c) 120 g/L

3.2 Effect of Cl^- concentration on Electrochemical performance

Fig. 2 shows the polarization curves of 1# sample at the different temperatures with various Cl^- concentrations. It can be seen that the corrosion potential (E_{corr}) and corrosion current density (I_{corr}) keep a slightly change as Cl^- concentration increasing from 20 to 120 g/L at 60℃ [Fig. 2

(a)], and E_{corr} was almost−700 mV. At 90℃, the E_{corr} moves first to a more negative direction, and then to positive direction as Cl^- concentration increasing from 20 to 120 g/L [Fig. 2 (b)]. It is obvious that increasing Cl^- concentration can increase the rate of the anodic reaction. In this presence of Cl^- in corrosive medium, Cl^- could promote the iron dissolution through a catalytic mechanism and formed intermediate corrosion species and accelerated corrosion reaction[15-18]. So Cl^- is the main reason in increasing the corrosion rate of samples. The anodic reaction process is as follows[19]:

$$Fe+Cl^-+H_2O=[FeCl(OH)]^-_{ad}+H^++e$$

$$[FeCl(OH)]^-_{ad}\rightarrow FeClOH+e$$

$$FeClOH+H^+=Fe^{2+}+Cl^-+H_2O$$

The Fig. 2 also indicated that the corrosion properties of 1# sample was mainly affected by the test temperature. At higher temperature (90℃), the activity of material surface was strengthen from the thermodynamic point of material, and electron transferring between the solution medium and material surface were intensified, the reaction speed was improved. Meanwhile the activity of Cl^- penetrate through the corrosion product film also be strengthened, therefore, the effect of Cl^- on corrosion properties of materials is also affected by the test temperature.

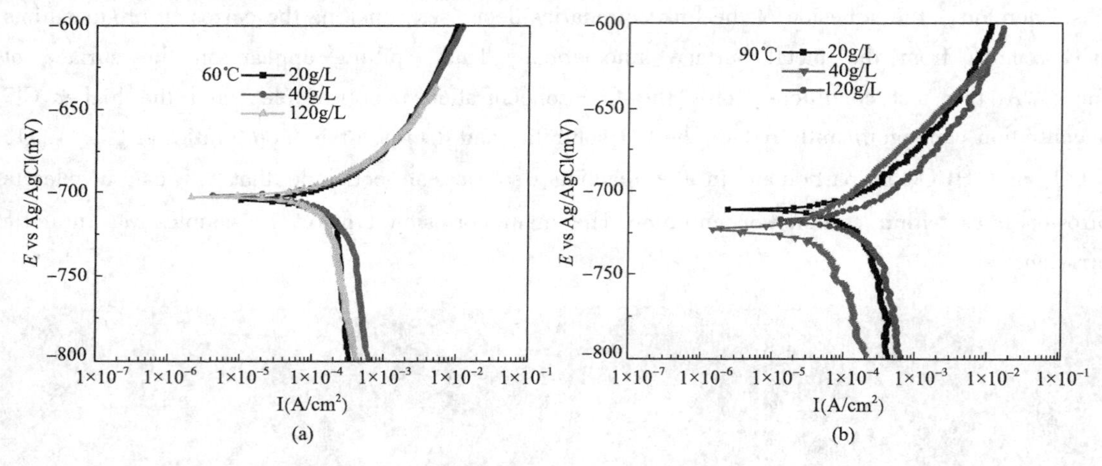

Figure 2 Potentiodynamic polarization curves of 1# tubing materials in different Cl^- containing solution at (a) 60℃ and (b) 90℃

The potentiodynamicpolarization curves of 2# sample were shown in Fig. 3, and the E_{corr} and I_{corr} were obtained from the curves. It was clear that the polarization curves show a better passivation characteristic at 60℃ and 90℃. At 60℃, when the concentration of Cl^- was 20g/L, the E_{corr} was −499 mV. Fig. 3 (a) shows that the E_{corr} first increased and then decreased as the increase of the Cl^- concentration, and the critical pitting potential (E_{pit}) of 2# appeared an obvious change. When the test temperature was 90℃, we can see from the Fig. 3 (b) that the polarization curves also have an obvious passivation region, but the change trend of E_{corr} and I_{corr} are different from that of the low temperature conditions. The Fig. 3 (b) shows the E_{corr} decreased with the concentration of Cl^- increased from 20g/L to 120g/L, meanwhile when the concentration of Cl^- increased from 40g/L to 120g/L, the E_{corr} changed almost no longer. This indicated that the corrosion resistance of 2#

material was not affected obviously by the concentration of Cl^- when the Cl^- content increased to a value. As a result, the E_{pit} showed a similar values with the different concentration of Cl^- at the higher test temperature of 90℃. The results proved that when the concentration of Cl^- over a certain range of values, the pitting corrosion sensitivity tended to a steady value, and the pitting corrosion sensitivity of super13Cr was controlled by the test temperature. Passivation film was affected by a combined action of test temperature and Cl^-, and broken suddenly at some moment, the pitting corrosion happened.

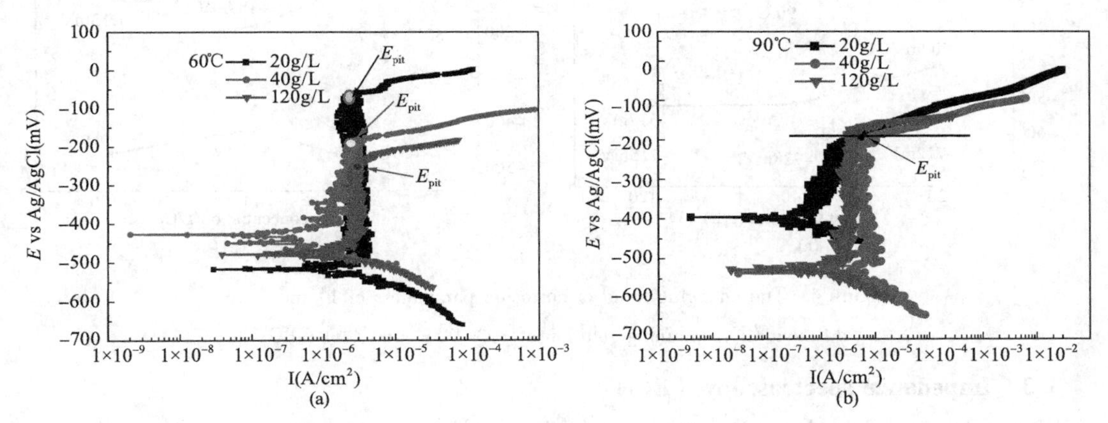

Figure 3 Potentiodynamic polarization curves of 2# tubing materials in different Cl^- containing solution at (a) 60℃, (b) 90℃

Comparative analysis the material composition and reaction mechanism, it was clear that the chemical composition of super13Cr contain relatively higher alloying elements, especially Ni, Cr, Mo elements, which have better corrosion resistance properties at high temperature and formed passivation film easily. So the compactness of corrosion scale and corrosion resistance have a close relationship with reaction temperature. Combined with the test results of the polarization curves in Figs. 2 and 3, the change trend of corrosion properties of carbon steel and super 13Cr material were shown in Fig. 4.

It can be easily seen from Fig. 4 that when the concentration of Cl^- was 20g/L, the Epit has a big change with the temperature increased from 60℃ to 90℃, and the potential difference (ΔE_{pit}) equal to 161 mV. The result indicated that when the temperature increased, the reaction rate of the solution medium and the sample surface were accelerated, and strengthened the activity of this corrosion system. Meanwhile, the penetrating ability of erosion ion into the corrosion film also was strengthened along with the rise of temperature, and the film was breakdown at relatively low corrosion potential, the pitting failure happened suddenly.

From the Fig. 4 (a), when the concentration of Cl^- increased from 20g/L to 120g/L, the E_{pit} reduced from −48mV to −213 mV, and the Icorr have not obviously changed, and kept the same order of magnitude. Comparative analysis the results of the different test temperatures, it was clear that when test temperature was higher as 90℃ [as Fig. 4 (b)], the E_{pit} of 2# sample almost maintained around −200 mV with the increase of the Cl^- concentration, but the E_{pit} gradually reduced at 60℃. This indicated that the E_{pit} was mainly controlled by the breakdown potential of passivation film at high temperature and was mainly controlled by penetration of Cl^- at low temperature. However, the combined effect of the test temperature and chloride ions increased the corrosion susceptibility leading to a lower corrosion resistance. The E_{corr} of 1# carbon steel showed a slightly tended toward more negative values as the Cl^- concentration increased [as Fig. 4 (b)].

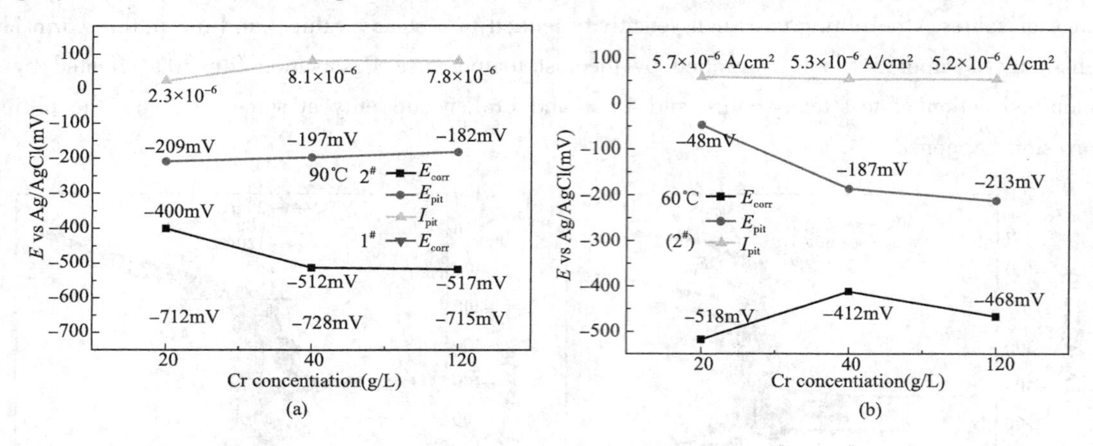

Figure 4 The changing trend of corrosion parameters of 1# and 2# in different Cl^- containing solution at (a) 90℃ and (b) 60℃

3.3 Impedance spectroscopy (EIS)

Electrochemical Impedance Spectroscopy (EIS) has been used as an interesting technique to characterize the electrochemical behavior of metal materials. However, In these cases, the most common equivalent circuit also used to describe the carbon steel tubing electrochemical behavior. Fig. 5 (a) shows the EIS of 1# sample at 90℃ in different Cl^- concentration. In order to keeping the samples in the same conditions before testing, the test samples were immersed in the test solution and keep 60 mins, then testing and analyzing the impedance. It was clear that the shape and size of these impedance spectra strongly depended on the Cl^- content. When the concentration of Cl^- was 20g/L, it shows a single impedance arc characteristics, and this indicated that the corrosion scale was dense and stable, and the Cl-adsorbing on the surface have a smaller influence on the corrosion scale. The reaction process is mainly controlled by electrochemical process. With the increase of Cl-concentration and greater than or equal to 40g/L, the impedance espectra consisted of two capacitive loops. But the high frequency capacitive loop was very small, which indicated the very severe local corrosion occurred. However, Warburg impedance was found in low frequency region as the Cl^- content increased to 40g/L. It revealed that with the increase of corrosive ion concentration, the corrosion product film forming on the sample surface were not enough dense, and existed the ion diffusion channels, corrosion product film could not fully cover the surface of the sample, the electrode system was mainly controlled by the diffusion. When the Cl^- concentration increased to 120g/L, the impedance spectrum characteristics changed not obvious, but the capacitive reactance arc radius in high frequency significantly less than that of 40g/L values, it shows that under the high Cl^- concentration the polarization resistance of the sample reduced, the corrosion resistance decreased. Whereas, obvious departure from the beeline the Warburg impedance was. It has been explored that a capacitive loop would add on the Warburg impedance when slight local corrosion bourgeoned[20].

Fig. 5 (b) shows the impedance spectra of corrosion resistant alloy 2# sample under the different Cl^- concentrations. It indicated that the impedance spectra have the similar characteristics, and all of them show the single capacitive reactance arc with different radius, this illustrated that the corrosion film has a good capacitive resistance and has good protection for the materials. And when the Cl^- concentration was greater than 20g/L, the capacitive reactance arc radius reduced obviously, this showed that when corrosive ions concentration exceeds a certain range, the corrosion resistance of 2# material greatly reduced.

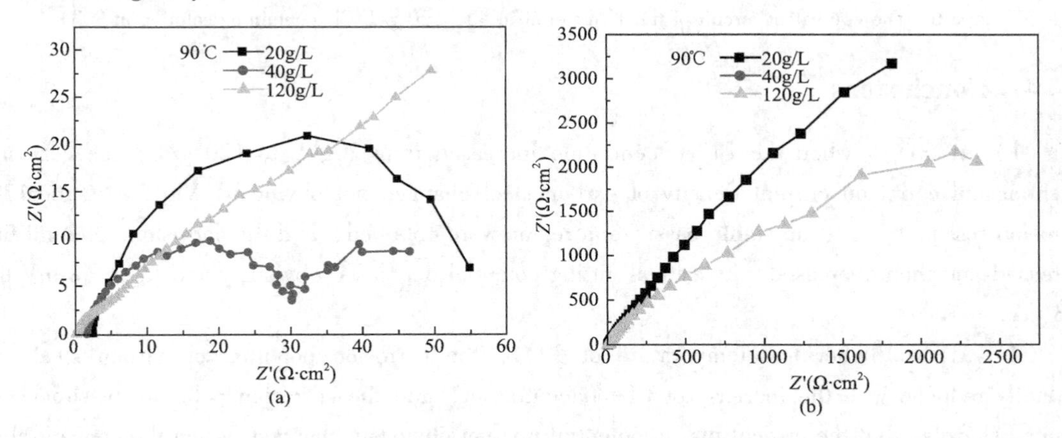

Fig. 5 EIS plots of 1#, 2# materials in different Cl^- containing solution at 90℃ (a) 1# and (b) 2#

So fitted the experimental results of 1# with ZSimpWin software, and got the equivalent circuit and can be expressed as the R_s (RPW) (Q). Due to the electrode frequency response characteristics of the electric double layer capacitance existed deviation with the capacitance, and phase Angle component Q instead of pure capacitance, the impedance expressions as denote equation 2[21], and the Warburg diffusion impedance as denote equation 3[21]. The electrode system existed dispersion effect.

$$Z_Q = \frac{1}{Y_0}(jw) - n \tag{2}$$

$$Z_W = \frac{1}{Y_0}(2w) - \frac{1}{2}(1-j) \tag{3}$$

The fitting results as shown in Table 3, according to the equations 2 and 3, it was clear that with the increase of Cl^- concentration, impedance values are reduced. This implied that increasing Cl^- concentration can speeded up the damaging process of corrosion product film, corrosion were more likely to happen. Fitting error of the EIS parameters were less than 10%.

Table 3 The fitting results of impedance spectrum under different Cl^- concentration

Concentration	R_s (Ω · cm²)	Q (CPE)		R_p (Ω · cm²)	Warburg
		Y_0 (S · sec^n/cm²)	n		Y_0 (S · sec^n/cm²)
40g/L	1.39	0.0024	0.76	27.3	0.1119
%Error	1.56	6.5	1.5	2.4	8.1
120g/L	1.30	0.0048	0.8	10.62	0.2398
%Error	3.08	7.2	3.1	6.5	6.1

Fitting results agreed well with the results of the measured impedance, as shown in Fig. 6. This explained that the equivalent circuit can reflects the actual corrosion process of 1# carbon steel in test solution which Cl^- concentration greater than 20g/L.

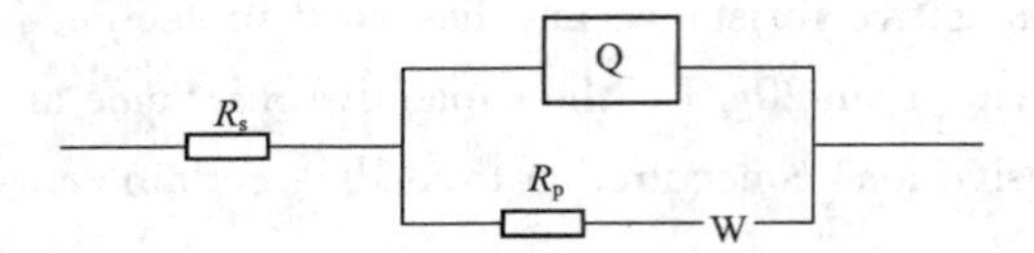

Figure 6 The equivalent circuit of the 1# material in 40, 120 g/L Cl^- containing solution at 90℃

4 Conclusions

(1) At 60℃, when the Cl^- concentration increased from 20g/L to 120 g/L, the corrosion potential and corrosion current density of carbon steel changed not obviously. While for super13Cr corrosion resistant alloy, an stable passivation region were obtained, and the corrosion potential first increased and then decreased, its critical pitting potential decreased obviously and from−48 mV to−213 mV.

(2) At the higher test temperature of 90℃, the corrosion potential of carbon steel was gradually reduced with the increase of Cl−concentration, and the corrosion potential of super13Cr reduced sharply, but the critical pitting potential was not obviously changed. When the concentration of Cl^- was higher than 20g/L, the corrosion potential of super13Cr almost remain unchanged.

(3) When the concentration of Cl^- was greater than 20g/L, the impedance spectroscopy of carbon steel showed a Warburg impedance characteristics. Carbon steel corrosion was mainly controlled by the diffusion effect. Super13Cr corrosion resistant alloy displayed the single capacitive reactance arc.

(4) when the Cl^- concentration is higher than 40g/L, The pitting sensitivity of Super 13Cr was higher at 90℃, an obvious pitting phenomenon was observed.

Acknowledgements

The authors are grateful for the financial supports fromthe China National Petroleum Corporation for the technology project (2011D−4603−0102), and the Shaan xi Province Nature Science Foundation of China under Contracts of 2012JQ6014。

Conflicts of Interest

The authors declare no conflict of interest.

References

[1] Zhu,S. D., Fu, A. Q., Yin, Z. F., *Corrosion Science.*, 2011, vol. 53, p. 3156.
[2] Ren,C. Q., Liu, D. X., Bai, Z. Q., *Materials Chemistry and Physics.*, 2005, vol. 93, p. 305.
[3] Li,D. G., Feng, Y. R., Bai, Z. Q., *Applied Surface Science.*, 2007, vol. 253, p. 8371.
[4] Machuca,L. L., Bailey, S. I., Gubner, R., *Corrosion Science.*, 2012, vol. 64, p. 8.
[5] Nesic,S., John, P., Vrhova, M., *Corrosion Reviews.*, 1997, vol. 15, p. 112.
[6] Gray,L. G., *Corrosion/*98. Houston: NACE, 1998, 40.

[7] Banas, J., Lelek-Borkowska, U., Mazurkiewicz, B., Solarski, W., *Electrochim Acta.*, 2007, vol. 52, p. 5704.
[8] Chen, C. F., Lu, M. X., Zhao, G. X., *Journal of Chinese Society for Corrosion and Pro.*, 2003, vol. 23, p. 21.
[9] Kermani, M. B., Morshed, A., *Corrosion.*, 2003, vol. 8, p. 659.
[10] Moreira, R. M., Franco, C. V., *Corrosion Science.*, 2004, vol. 46, p. 2987.
[11] Stephen, N. S., Michael, W. J., *Corrosion* 2006. No. 06115, San Diego. 2006, 1-26.
[12] MA, L. P., WANG, Y. Q., Zhao, S. H., *West-China Exploration Engineering.*, 2006, vol. 11, p. 50.
[13] Liu, Q. Y., Mao, L. J., Zhou, S. W., *Corrosion science.*, 2014, vol. 84, p. 165.
[14] Jiang, X., Nesic, S., Kinsella, B., Brown, B., Young, D., Corrosion., 2013, vol. 69, p. 15.
[15] Zhang, G. A., Cheng, Y. F., *Electrochimica Acta.*, 2011, vol. 56, p. 1676.
[16] Marcelina, S., Pébèrea, N., Régnierb, S., *Electrochimica Acta.*, 2013, vol. 87, p. 32.
[17] Bai, Z. Q., Chen, C. F., Lu, M. X., Li, J. B., *Applied Surface Science.*, 2006, vol. 252, p. 7578.
[18] Lin, C., Li, X. G., Dong, C. F., Materials., 2007, vol. 14, p. 5.
[19] Nesic, S., Thevenot, N., Crolet, J. L., Corrosion/1996, NACE, Houston, 1996, paper No. 3.
[20] Ren, C. Q., Liu, D. X., Bai, Z. Q., Materials chemistry and physics., 2005, vol. 93, p. 305.
[21] Cao, C. N., Zhang, J. Q., An Introduction to Electrochemical Impedance Spectroscopy, 2002, Science Press: Beijing, p. 231.

CCUS 腐蚀控制技术研究现状及建议

赵雪会[1]　何治武[2]　刘进文[3]　何淼[2]

（1. 中国石油集团石油管工程技术研究院国家重点实验室；
2. 长庆油田分公司油气工艺研究院；3. 长庆油田机械制造总厂建安公司）

摘　要：本文主要阐述了 CCUS 技术的发展需求和国内外应用现状，探讨分析了 CCUS 技术发展过程中 CO_2 驱油中存在的腐蚀问题和防腐措施的研究现状，结合油田 CO_2 驱注采现场需求，提出了着重开展腐蚀研究的热点问题及建议，进一步为 CCUS 技术的利用及深入发展提供理论指导和技术支撑。

关键词：CCUS；CO_2 驱油；腐蚀；研究现状

CCUS 是指碳的捕获（Carbon Capture）、利用（Utilization）和封存（Storage）的简称。CCUS 技术是指将 CO_2 从电厂、煤化等工业或其他排放源分离出来，经富集、压缩并运输到特定地点，注入储层封存以实现被捕集的 CO_2 与大气长期分离，或注入储层驱油进行合理利用的一项技术[1]。CCUS 作为一项新兴的、具有大规模二氧化碳减排潜力的技术，它可以实现石化能源的低碳利用，被国际能源组织（IEA）认为有望减少全球碳排放的 20% 以上[2]。因此发展 CCUS 技术是我国煤化工、钢铁、水泥等高排放行业温室气体减排的迫切需求，同时 CCUS 技术作为削减温室气体排放以减缓气候变化的新兴技术，对我国中长期应对气候变化、推进低碳发展具有重要意义。

CCUS 技术的发展也受到国家领导人以及各界专家的高度重视和关注，在 2015 年 9 月习近平与美国总统奥巴马共同发布了《中美元首气候变化联合声明》[3]，声明明确提出 CCUS 项目，两国已选定项目开展实施的场址，因此环保工程利国利民，治理环境，刻不容缓。目前随着 CCUS 技术的不断进步与发展，国内外实践经验证明，CCUS 中利用环节对于油田来讲，利用 CO_2 驱油技术提高油田采收率、提高经济效益具有广泛应用前景。

但是 CO_2 遇水形成的碳酸对管柱及管道管网的腐蚀问题如影随形，无处不在，在利用 CO_2 驱油的注采过程中，腐蚀问题层出不穷，对 CCUS 技术的发展带来严峻的考验。因此解决 CO_2 驱油注采过程中的腐蚀问题是目前关注的重点和首要工作。目前石油行业市场低迷，油价持续降低现象，对油田来说，降低成本、提高油田采收率是应对目前市场的一种有效措施。而 CCUS 技术中利用环节的 CO_2 驱注采技术是既可减少 CO_2 的排放，又能有效提高石油采收率，实现了油田节约开支，提高产量的需求。因此积极发展 CO_2 驱注采技术、减缓腐蚀隐患具有重要的应用价值。

1　CCUS 应用技术国内外现状

CCUS 应用技术发展主要侧重于利用 CO_2 驱油来提高产量的技术的发展与应用。美国自 1952 年 Whorton 等发明第一个利用 CO_2 采油的专利以来，利用 CO_2 驱油技术始终是石油开采领域的研究重点[4]。在美国、加拿大、英国等国家 CO_2 采油技术应用较多。近 20 多年来

CO_2驱油技术已经成为美国提高使用采收率的主导技术。美国是世界上利用 CO_2驱油技术最多的国家。应用 CO_2驱油技术增产量，约占世界总 EOR 产量的93%[5]。

国内是在2006年开始开展CCUS项目的研究，设立了“973”项目——“温室气体的资源化利用及地下埋存”；2007年中石油重大科技专项“温室气体 CO_2资源化利用及地下埋存”；2008年，大庆油田建设了全国首家 CO_2注气站；2010年吉林油田开展 CO_2驱油先导试验，目前已取得重要的应用研究成果；延长油田利用煤化工生产成本低廉 CO_2（120元/t），已建成第一期 50×10^4t/a CCUS项目，该工程作为中美联合项目，由美国能源部出资。胜利油田在近年也建立了CCUS全流程工程实验基地。

在“十二”五至“十三”五期间，随着国家在环保减排立项上的重视和科技人员的关注和技术攻关，CO_2驱油示范基地逐渐在国内较多的油田得到了较大的发展[2]，其中大庆油田、吉林油田以及胜利油田等建立的 CO_2驱油示范基地的实践应用成果不断创新，技术的推广应用为各油田提高采收率提供了积极的技术支撑。

2　CO_2驱注、采环节腐蚀情况

CO_2驱油注采过程中由于温度、压力的不同，CO_2将处于不同状态，图1为 CO_2所处的温度和压力下的气液平衡图[6]。可以看出，在临界温度31.06℃以下根据压力分为固相、液相和气相，而在临界温度31.06℃以上以及压力超过7.39MPa时为超临界状态，气液相达到平衡的混相状态。此时如果 CO_2气体含有一定水分，则会均匀的分部，因此湿 CO_2在超临界状态下对管柱的运行存在明显的腐蚀隐患。

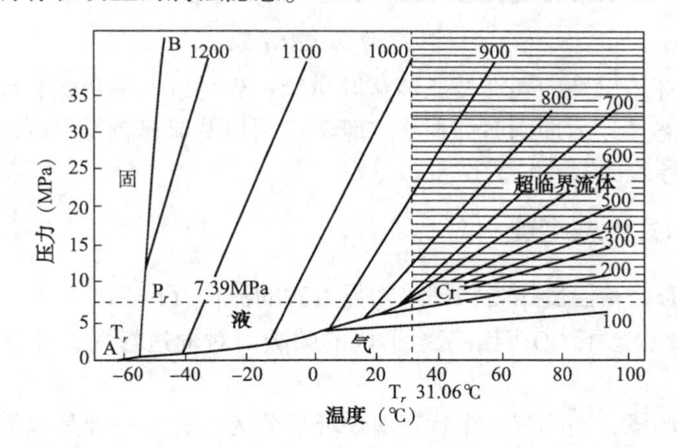

图1　为 CO_2所处的温度和压力下的气液平衡图

2.1　CO_2驱注入

CO_2驱注入的腐蚀问题主要包括管道输送过程中服役介质对管道的腐蚀损伤和井口注入过程中低温的 CO_2对油管柱的损伤隐患问题。主要腐蚀种类有：气源的纯净度（SO_2/NO_2/H_2S等）、超临界 CO_2含水率（湿度）导致的腐蚀问题、-20℃以下低温对管柱性能的影响以及超临界 CO_2状态管柱应力敏感性的变化问题。从典型管柱失效案例分析，注入 CO_2的纯净度对管柱的运行安全存在较大的安全隐患，2015年某油田在注入阶段发生油套管柱断裂落井，管体裂纹较明显，失效导致停产作业，造成严重的经济损失。失效结果分析显示，管柱属于硫化物应力腐蚀开裂，究其原因，一方面气源含有一定量的 H_2S，油管柱用普通圆螺

纹连接，管柱密封性存在隐患，另一方面油套环空是否存在 SRB 生成 H_2S 而导致硫化物应力腐蚀开裂。目前正进一步现场分析导致管柱失效的 H_2S 的具体来源。但此案例说明气源的纯净度必须避免有害气体的介入，避免形成失效隐患。

2.2 CO_2驱采出环境

油田开采环境较为苛刻，一般伴生气 CO_2 含量较高，有的含有 H_2S 等腐蚀气体，这些腐蚀气体溶于地层水成为酸性介质，对高温高压井底油管柱造成较严重的腐蚀和损害，管柱穿孔、刺漏等现象层出不穷。结合现场调研，长庆由于井深较浅，大约 3000m，井底温度一般最高 80℃，为腐蚀最敏感温度。地层压力 20MPa，因此在高压 CO_2条件下腐蚀问题较为严重。主要发生腐蚀种类有：高浓度 CO_2电化学腐蚀、管柱应力及腐蚀介质协同作用、管柱气密封性、超临界状态 CO_2腐蚀以及由于地层水质矿化度高而造成的油管柱结垢问题。图 2 所示为起出的油管柱腐蚀失效形貌，腐蚀穿孔、螺纹处点蚀现象较为普遍。

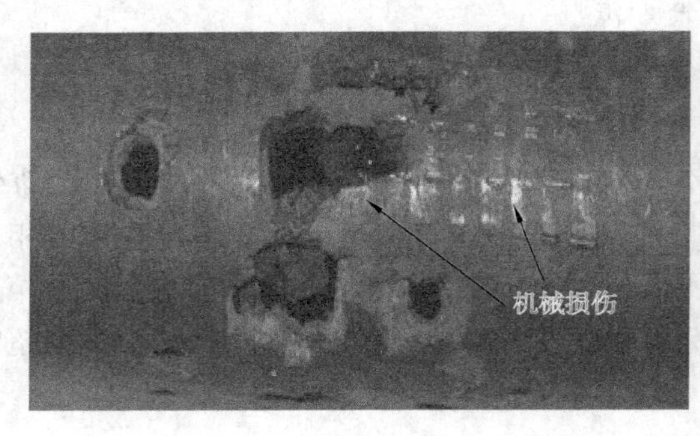

图 2　腐蚀失效形貌

另一方面开采环境腐蚀问题涉及原油从井口输送到处理站的集输管线腐蚀情况，比如流体的不同流速、流态导致的冲刷腐蚀损伤，管道弯头以及流速较低的区域，由于冲蚀和高浓度 CO_2电化学腐蚀导致的穿孔较为严重。

3　超临界 CO_2状态腐蚀

超临界 CO_2状态是指温度压力均超过临界点时的流体（$t=31℃$，$P=7.39MPa$），超临界 CO_2状态腐蚀指在此状态下 CO_2因溶于水形成的碳酸而对输送管网、井下管柱造成不同的腐蚀损伤。

结合油田工况环境，超临界 CO_2状态腐蚀环境分为两种，一种是以 CO_2为主体含少量水和杂质的超临界状态；另一种是 CO_2饱和的水体系的超临界状态。由于 CO_2存在体系环境的不同，超临界 CO_2状态引起的腐蚀问题就存在差别。

3.1　超临界以 CO_2为主体的腐蚀

众所周知，CO_2在捕集、分离后高压压缩富集、运输以及井口注入过程中，CO_2一般处于高压、低温的超临界状态，而且是以 CO_2为主体的流体。在此状态下，主要发生的腐蚀损坏是流体引起的冲蚀、应力腐蚀、螺纹连接及密封问题以及对管柱完整性的影响，尤其当 CO_2驱气源有一定含水率、含有杂质气体更加剧了服役介质对管材的腐蚀，给油田带来了较大的经济损失。2015 年 6 月某油田 CO_2驱油注入过程中多口井油套管均出现断裂失效、接头螺纹断裂，调研发现气源含 30~40ppm 微量 H_2S，管柱服役在 50℃且压力 20MPa 的超临界状态，管柱在运行期间失效断裂，导致多口井停井、关井，损失严重，主要失效形貌如图 3 所示。

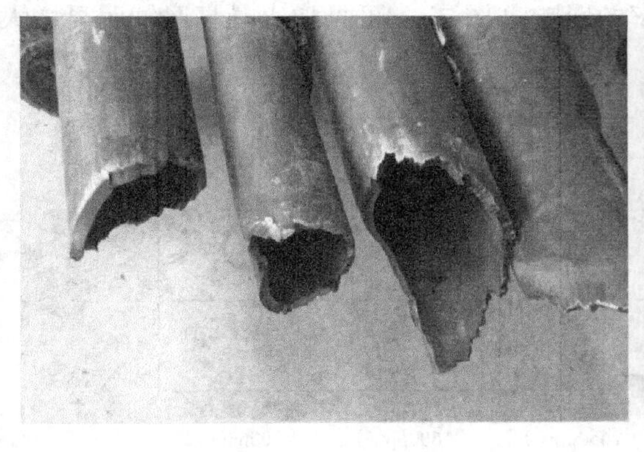
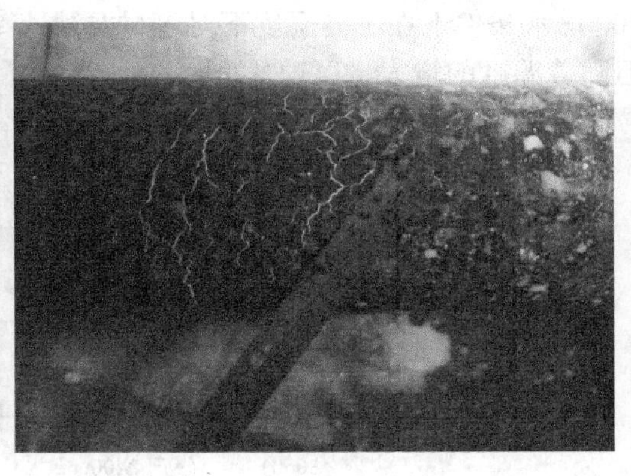

图 3　注入井油管柱失效形貌

针对超临界 CO_2 的腐蚀，国内外学者研究热点主要有几点：

（1）气体杂质含量。清华大学向勇等对比分析了湿气 CO_2 中不同的 SO_2 含量对管线钢 X70 腐蚀行为的影响，表明随着杂质气体 SO_2 含量的增大，腐蚀先增大再减小，并且当 CO_2 中混合 SO_2 时，腐蚀性相对更加剧[7]。Choi 等研究了不同 CO_2 水饱和状态下，SO_2、O_2 含量对金属钢管 CO_2 腐蚀的影响（图 4），可见 O_2、SO_2 的加入均有促进材料腐蚀的加重，比较可见 SO_2 加入后材料的平均腐蚀速率大幅度增大，对材料的腐蚀损伤相对 O_2 明显增大。两种杂质气体的混入使得材料腐蚀速率由 0.5mm/a 增加到 7mm/a。Hua[9] 等研究成果也得到相似的结论，SO_2 的浓度增大促进腐蚀加剧（图 5）。

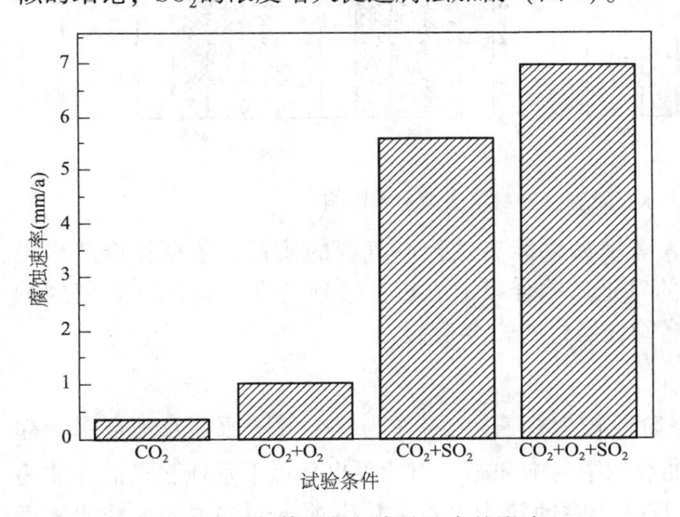

图 4　杂质气体对材料腐蚀速率的影响

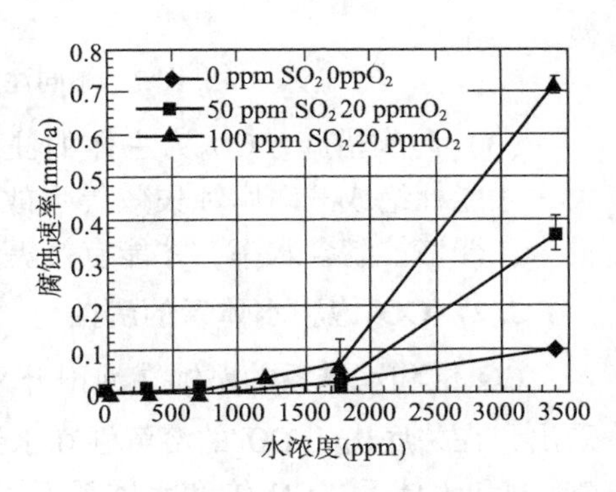

图 5　SO_2 浓度对材料腐蚀程度的影响

（2）含水率的影响。含水率对管道的腐蚀影响一直是争论较多的问题，尤其在油田处于中后期含水率较高的情况下，但是在以 CO_2 为主体的腐蚀体系中，含水率控制范围一般较低，尤其在超临界 CO_2 注入环境下，更是必须严格控制的一个参数。Hua[8] 等人研究了 CO_2 水—饱和、不饱和以及在 CO_2 中水饱和、不饱和条件下的材料的腐蚀状况（图 6），可见 CO_2 水饱和条件腐蚀最严重，在 CO_2 环境下水饱和和不饱和条件下腐蚀稍有减缓，并且随着 CO_2 条件下水含量的降低而腐蚀速率降低，进一步说明严格控制 CO_2 注入环境下水含量的比率，能有效缓解腐蚀的发生。Zhang Yucheng[9]等人分别研究了600gCO_2+100g水和450gCO_2+1000g水条件下在不同温度下对管材腐蚀速率的影响（图7），可见加水量高腐蚀明显增大，但随着温度的增大，腐蚀速率存在一个最值。

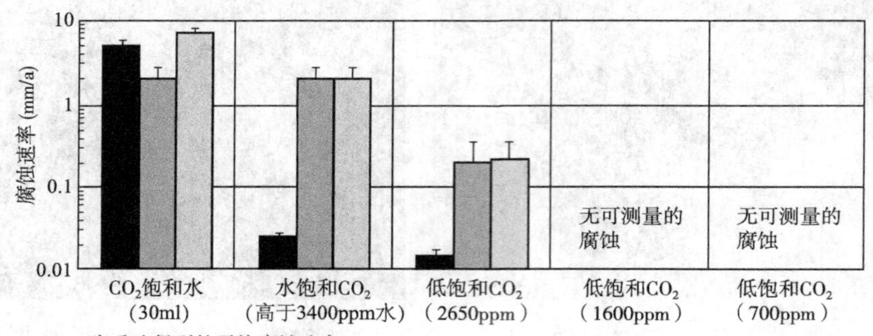

图6　不同含水条件下腐蚀速率比较

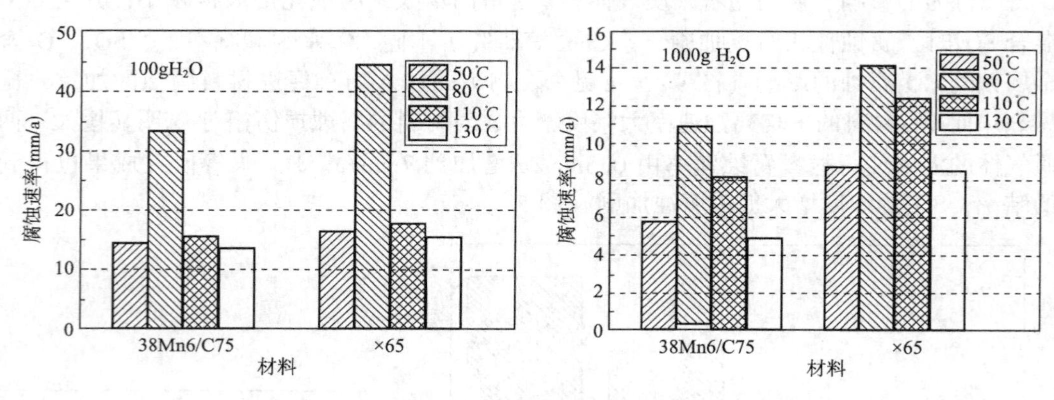

图7　不同注入水量及温度对材料腐蚀速率影响

（3）防腐措施。CO_2为主体的注入环境如果能有效控制气源的质量，依据标准严格把关，才能杜绝或减缓腐蚀失效等事故的发生。结合以上分析，做到有害（腐蚀性）气体的混入，严格控制含水率，才能有效提高管柱运行的安全系数。

3.2　CO_2饱和水体系的腐蚀

CO_2饱和水体系的腐蚀在油田开采阶段较为普遍，CO_2作为伴生气随原油及地层水一起采出，在井底由于CO_2的溶解性在水油混相中为饱和态，因此采出环境下管柱服役的介质为CO_2饱和水体系。CO_2饱和水体系下对管柱的腐蚀较为严重，从腐蚀机理讲主要为铁的阳极氧化过程和表面阴极氢离子催化还原反应。CO_2腐蚀主要影响因素有：CO_2的分压，介质温度，流速流态，pH值以及矿化度等。

（1）CO_2分压。国内外学者在CO_2的分压方面研究成果较多，有的认为高CO_2分压会增加腐蚀速率。理论依据是高CO_2分压下，H_2CO_3增加，促进阴极反应，最终增加腐蚀速率。另一种认为高CO_2分压会降低腐蚀速率，高CO_2分压会导致碳酸氢盐和碳酸氢根浓度增加，加速沉淀，促进保护层的形成。管研院大量的CO_2腐蚀工作表明[10]，CO_2分压的影响呈现一个抛物线状的趋势，当分压增加到一定值时，腐蚀产物的致密程度反而保护了腐蚀的进一步发生，腐蚀速率降低（图8）。

（2）流速。George 等研究了流速的影响，表明流速在保护层形成前、后对腐蚀影响不同；Mohammed Nor 等发现，在低的 CO_2 分压下随着流速增加，碳钢腐蚀速率增加，在高的 CO_2 分压下随着流速增加，碳钢腐蚀速率降低，且腐蚀速率对流速变化敏感性降低（图9）。

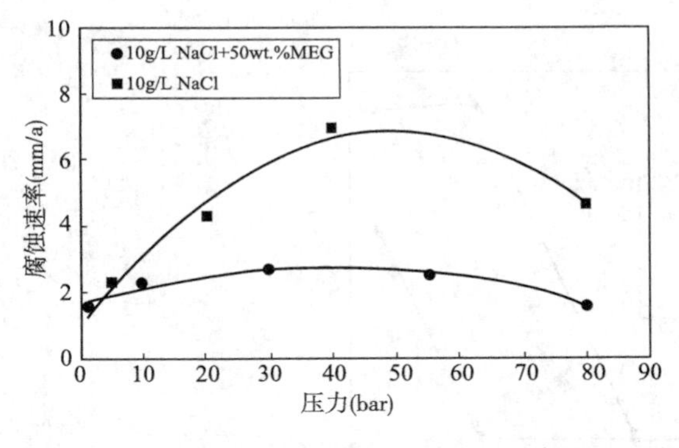

图8　CO_2 分压对腐蚀速率的影响关系

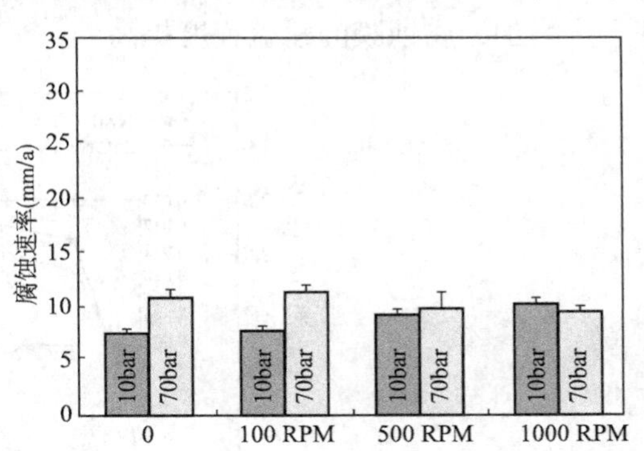

图9　不同流速、CO_2 分压对腐蚀速率的影响

（3）原油含水率。在油田开采过程中，石油管材面临着不同比例的油水混相流体的腐蚀，原油含水率是影响 CO_2 腐蚀的一个重要因素。一般来说原油含水率较低时，可以形成油包水乳状液，水相对钢铁表面的浸湿会受到抑制，发生 CO_2 腐蚀倾向较小；原油含水率较高时，可以形成水包油乳状液，水相对钢铁表面发生浸湿而引发严重的 CO_2 腐蚀[11]。相关的研究结果表明[12]，在30%~80%含水率区间内，可能会出现一个腐蚀速率剧变的临界含水率，低于临界含水率，腐蚀速率非常低且增长缓慢，高于临界含水率，腐蚀速率急增。Z. D. Cui[13] 等人研究了超临界 CO_2 状态下含水率对三种碳钢油管腐蚀状况的影响（图10），随含水率的增大腐蚀加剧，三种材料变化趋势相似。同时研究表明在一定流速下随着含水率的增大，油、水、气混相结构也发生变化，图11表明含水率低时呈现油包水形态，当含水率增大到90%时，流体结构为水包油状态，因此流体结构的不同对材料的腐蚀影响存在明显的不同，图9的腐蚀结果证实了这一结论。

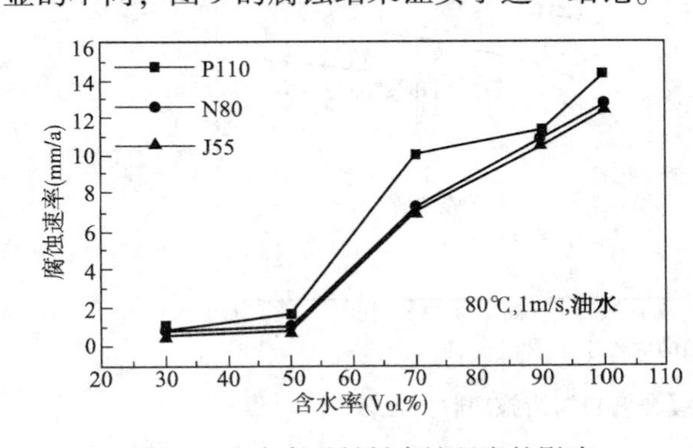

图10　含水率对材料腐蚀程度的影响

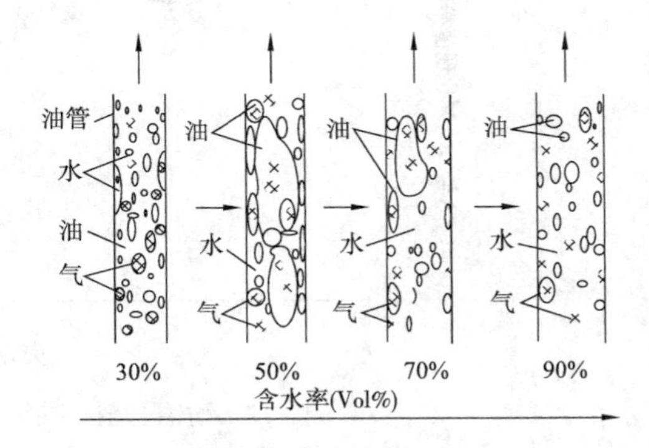

图11　含水率与油水气混相结构的关系

王世杰[14] 等人研究了含水率变化对材料在超临界状态下的腐蚀影响，结果如图12所示。研究表明当原油含水率低于50%时，油水混相流体能够形成稳定的油包水型乳状液，水相对钢表面的润湿作用受到抑制。原油在钢整个表面吸附阻碍水与钢表面的接触，减少表面腐蚀反应的活性点，抑制腐蚀反应过程，因此，钢的腐蚀速率低于0.1mm/a。当原油含

水率在50%~75%之间时，由于水含量增加，油水混相流体处于由“油包水”向“水包油”转变的过渡状态。原油的保护作用减弱，水与钢表面的接触机会增加，表面的活性点增加，腐蚀速率明显增加。当原油含水率高于75%以后，油水混相流体形成稳定的水包油型乳状液。原油对钢表面的润湿作用受到抑制，水能够润湿整个钢表面，加剧金属的腐蚀，腐蚀速率大大增加，腐蚀膜的覆盖程度提高。

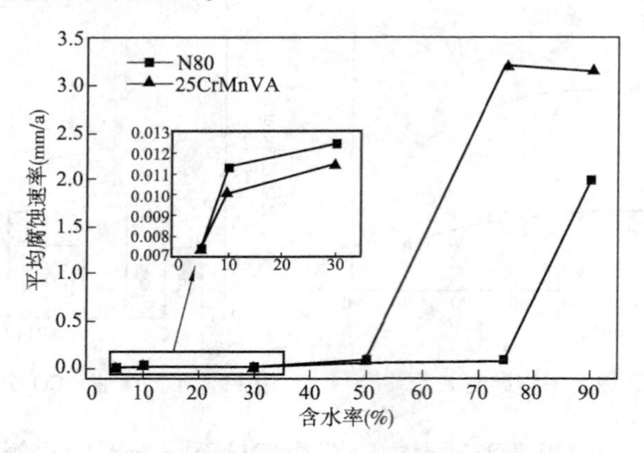

图12　含水率变化对材料在超临界状态下的腐蚀影响

（4）防腐措施。油田环境防腐措施种类较多，主要有材料升级防腐、缓蚀剂加注、涂镀层等。随着油田开采的不断加深以及大多油田处于中后期，开采环境日益复杂以及苛刻化，因此油套管柱的安全运行面临着严峻的挑战。选材方面由于长远的投资成本考虑，选择耐蚀合金也是一种有效的防腐手段之一（图13），相对净收益处于一直平稳水平。

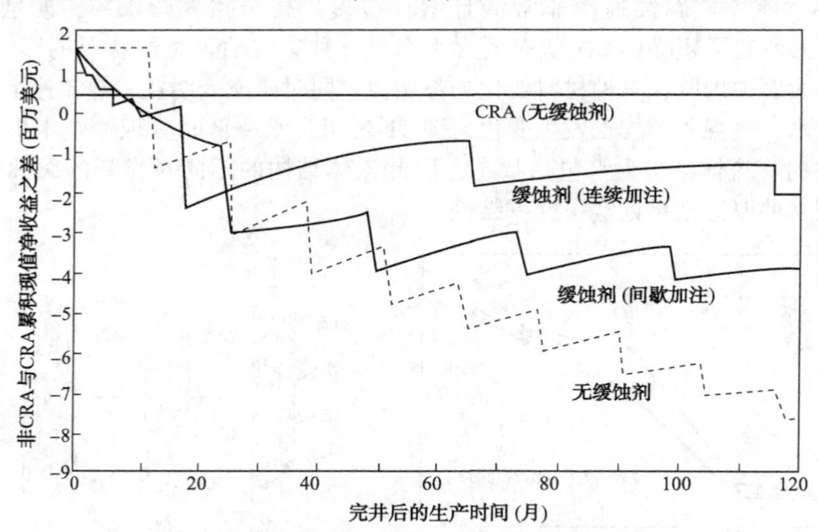

图13　耐蚀合金与碳钢+缓蚀剂防腐措施净收益比较

4　存在问题及建议

（1）CO_2腐蚀机理及规律成果较多，借鉴性不足。建议管材腐蚀特性及规律研究针对性要强，结合管材的经济性和现场环境的适应性，掌握管材在服役工况下的腐蚀规律，设计科学合理的配套的防腐措施。

（2）低温环境管材的性能变化目前研究较少。针对 CO_2驱注入环境特点，加强对管柱低温条件以及超临界 CO_2环境下应力疲劳的敏感性进行深入规律，开展低温高压条件受力状态与环境介质耦合作用对管柱失效机制的影响规律。

（3）气源的纯净度对管材存在一定的腐蚀隐患。有必要开展腐蚀性杂质气体含量对 CO_2体系腐蚀行为的影响，界定临界范围，严格控制注入气体的纯净度。

（4）全尺寸实物模拟试验能更好地接近油田环境，反映管子腐蚀失效状况。开展全尺寸管柱苛刻工况环境腐蚀演化规律研究，为油田选材提供重要的技术支撑。

参 考 文 献

[1] 郭敏晓，蔡闻佳. 全球碳捕捉、利用和封存技术的发展现状及相关政策 [J]. 研究与探索，2015，35（3）：39-42.

[2] 韩桂芬，张敏，包立. CCUS 技术路线与发展前景探讨 [J]. 电力科技与环保，2012，28（4）：8-10.

[3] 中国气候变化信息网. http：//www. ccchina. gov. cn/Detail. aspx.

[4] 刘卓亚. 大庆油田 CO_2驱油经济效益评价方法研究 [D]. 北京：中国地质大学（北京），2014.

[5] Rosa M. Cuellar-Franca，Adisa Azapagic. Carbon capture，storage and utilization technologies：A critical analysis and comparison of their life cycle environmental impacts [J]. Journal of CO_2 Utilization. 2015，9：82-102.

[6] 张颖，李春福，王斌. 超临界 CO_2对刚才的腐蚀试验研究 [J]. 西南石油学院学报，2006，28（2）：17-23.

[7] Yong Xiang，Zhe Wang，Chao Xu.，et al. Impact of SO_2 concentration on the corrosion rate of X70 steel and iron in water-saturated supercritical CO_2 mixed with SO_2 [J]. The Journal of Supercritical Fluids. 2011，58：286-294.

[8] Yong Hua，Richard Barker，Anne Neville. Comparison of corrosion behavior for X-65 carbon steel in supercritical CO_2-saturated water and water-saturated / unsaturated supercritical CO_2 [J]. The Journal of Supercritical Fluids. 2015，97：224 -237.

[9] Yucheng Zhang，Kewei Gao，Guenter Schmitt. Water effect on steel under supercritical CO_2 Condition [J]. Corrosion 2011，No：11378.

[10] 林冠发，白真权，赵国仙，等. CO_2压力对金属腐蚀产物膜形貌结构的影响 [J]. 中国腐蚀与防护学报，2004，24（5）：284-288.

[11] 张学元，邸超，雷良才. 二氧化碳腐蚀与控制 [M]. 北京：化学工业出版社，2000.

[12] 李建平，赵国仙，郝士明. 几种因素对油套管钢 CO_2腐蚀行为影响 [J]. 中国腐蚀与防护学报，2005，25（4）：241.

[13] Z. D. Cui，S. L. Wu，C. F. Li. et al. Corrosion behavior of oil tube steels under conditions of multiphase flow saturated with super-critical carbon dioxide [J]. Materials Letters. 2004，58：1035-1040.

[14] 王世杰. 原油含水率对油气管材超临界 CO_2腐蚀行为的影响 [J]. 腐蚀科学与防护技术，2015，1（27）：73-77.

300M钢在油积水环境中的腐蚀行为研究

徐秀清

（中国石油集团石油管工程技术研究院）

摘　要：通过分析探讨腐蚀面积、模拟油积水中的溶解氧含量和pH值等变化，研究了300M高强钢在模拟油积水环境中的腐蚀行为。研究发现：随着腐蚀的进行，300M钢在模拟油积水中的溶解氧含量逐渐降低，而pH值迅速增大，24h后氧含量和pH值均趋于稳定。电化学测试结果表明，300M钢在模拟油积水中发生了明显的钝化现象，腐蚀速率的大小关系为$v_5<v_6<v_7<v_4$。

关键词：300M高强钢；模拟油积水；电化学测试；腐蚀

腐蚀和疲劳是结构材料的主要损伤形式，它们的共同作用影响使用寿命，严重威胁材料的结构安全可靠性，甚至成为发生事故的重要原因。研究表明，高强钢因其具有高强度、高韧性等优异的力学性能，以及良好的加工性能而广泛用于储油罐结构的连接中，尤其是高强钢自攻螺钉、固定螺钉、连接螺栓等部位。这些部位是储油罐腐蚀高发部位，这是由于外部雨水（或雪、雾、霜、露等）和内部形成的冷凝水在油箱内积存，这些积水中含有较多的氯离子和微量硫酸根离子，都是强腐蚀介质，是构成结构材料发生腐蚀的重要条件之一。一般高强钢螺栓表面均有电镀层，但是装配过程可能导致表面镀层损伤而出现基体的裸露，高强钢基体对腐蚀环境相当敏感，易发生腐蚀。因此，研究模拟油积水环境下高强钢材料的腐蚀行为，不仅为以后此类研究提供了参考依据，而且具有重要的理论意义和工程实践意义。

本文以300M高强钢为对象，通过腐蚀损伤面积、模拟油积水中的溶解氧含量、pH值等参数的变化，探讨其在模拟油积水环境中的腐蚀特征和腐蚀规律，为300M高强钢构件在油积水腐蚀环境中的安全使用提供依据。

1　实验方法

实验材料为300M高强钢（化学成分见表1），试样形状为长方形片状，尺寸为50mm×25mm×2mm。对油积水水样进行成分分析，发现积水中含有较多数量的Cl^-、微量SO_4^{2-}，以及多种金属离子（主要有Cd^{2+}、Na^+、Ca^{2+}、Mg^{2+}等）。实验选择的模拟油积水成分如表2所示，溶液初始pH值为4.2±0.2。

表1　300M高强钢成分

元素	C	Si	Mn	S	P	Cr	Ni	Mo	Cu	Fe
含量（wt %）	0.38~0.43	1.45~1.80	0.60~0.90	≤0.020	≤0.020	0.70~0.95	1.65~2.00	0.30~0.50	0.08	余量

表 2　模拟油积水溶液成分

成分	含量（mg/L）
$CaCl_2$	50
$CdCl_2$	1000
$MgCl_2$	50
NaCl	100
$ZnCl_2$	10
$PbCl_2$	1
$CrCl_3 \cdot 6H_2O$	1
$CuCl_2 \cdot 2H_2O$	1
$FeCl_3$	5
$MnCl_2 \cdot 4H_2O$	5
$NiCl_2 \cdot 6H_2O$	1
去离子水	余量

将 300M 钢试样置入玻璃容器中。在试样腐蚀过程中，每 24h 测量模拟油箱积水中的溶氧含量和 pH 值（初始 12h 内，每 2h 测量一次），其测量仪器分别为 YSI DO 200 溶解氧测定仪，PHS-25 数字酸度计。

观察不同腐蚀时间材料的宏观腐蚀形貌，用 Matlab 软件对腐蚀形貌图像进行二值化处理，得出灰度图像并统计腐蚀损伤面积。二值化是数字图像处理中一项最基本的变换方法，通过非零取一、固定阈值、双固定阈值等不同的阈值化变换方法，使一幅灰度图变成黑白二值图像，将所需的目标部分从复杂的图像背景中脱离出来，以便于统计分析。高强钢材料腐蚀形貌图像中腐蚀部分的灰度值较大，而其他部分灰度值较小，因此本文采用固定阈值法处理使腐蚀损伤部分分离出来，二值图像中黑色区域为腐蚀部分，白色区域为未腐蚀的材料表面。

电化学测试在三电极体系中进行，电解液为不同 pH 值的模拟油积水溶液，参比电极为饱和甘汞电极（SCE），辅助电极为铂片。电化学测试使用的仪器为 A 370 电化学工作站。交流阻抗测试的激励信号幅值为 5mV，频率范围为 0.001～100kHz。Tafel 极化曲线测试扫描速率为 20mV/min，扫描电位区间为-0.7～-0.3V。使用 ZSimpWin 软件对交流阻抗谱进行拟合，得到腐蚀过程的等效电路。

2　结果与讨论

2.1　腐蚀面积

图 1 为 300M 高强钢在模拟油积水环境中腐蚀不同时间的宏观腐蚀形貌及二值化图像。由图 1a 可以看出，腐蚀 24h 时表面出现红棕色的腐蚀斑，随着腐蚀时间的延长腐蚀斑面积扩大并在该层下面生成黑色的腐蚀产物。用 Matlab 软件对图 1（a）进行二值化处理，得出灰度图像见图 1（b），并统计出腐蚀损伤面积，结果见表 3。可知，随着腐蚀时间延长，腐蚀速率逐渐增大，腐蚀面积由 22%增大到 89%左右。

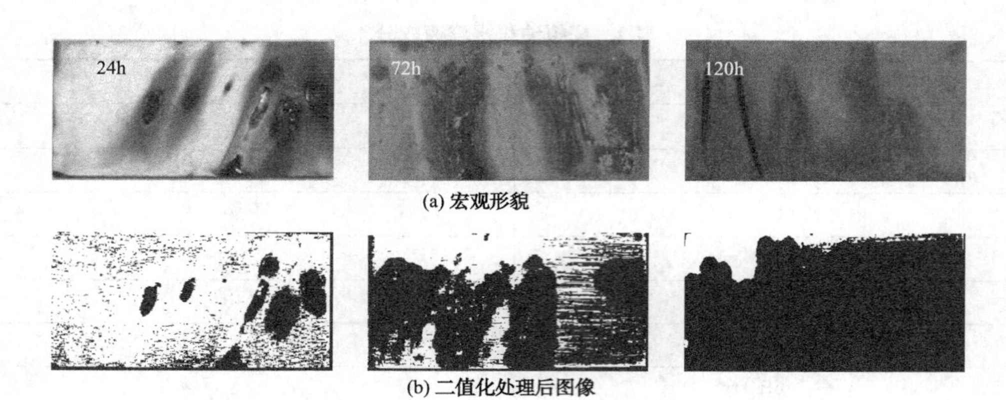

(a) 宏观形貌

(b) 二值化处理后图像

图 1　300M 钢在不同腐蚀时间下的宏观腐蚀形貌及二值化图像

表 3　300M 钢不同腐蚀时间的表面腐蚀损伤度

腐蚀时间（h）	24	72	120
腐蚀损伤度（%）	22.57	69.28	89.48

图 2 为 300M 钢腐蚀 120h 后的 SEM 微观形貌及 EDS 结果。由图可知，试样在模拟油箱积水中生成的腐蚀产物呈疏松的小颗粒团聚状态，腐蚀产物为 Fe 的氧化物，另外还存在少量 Cr 和 Cd 的氧化物。

图 2　300M 钢在腐蚀 120h 后的 SEM 形貌及 EDS 结果

2.2　腐蚀过程中溶解氧含量及 pH 值的变化

图 3 是腐蚀过程中模拟油积水中的溶解氧含量随时间的变化曲线。可以看出，模拟油箱积水中的初始溶解氧含量为 8.0mg/L，在腐蚀初期，溶解氧含量急剧下降，24h 后氧含量的变化趋于平稳，平稳值为 4.3mg/L。这是由于开始时腐蚀过程的快速进行消耗了溶液中较多的氧气，生成了大量腐蚀产物，覆盖在试样表面，阻碍氧的扩散，使其反应速率减慢，耗氧量逐渐减少直至基本不变。

图 4 为腐蚀过程中模拟油积水 pH 值随时间的变化曲线。可知模拟油箱积水的初始 pH 值为 4.2，在腐蚀初期，随着溶液中氧的消耗，阴极反应使得 OH^- 增多，致使溶液 pH 值增

大；当24h后，腐蚀产物膜的生成使得溶液趋于稳态平衡，溶液pH值稳定在5.3左右。

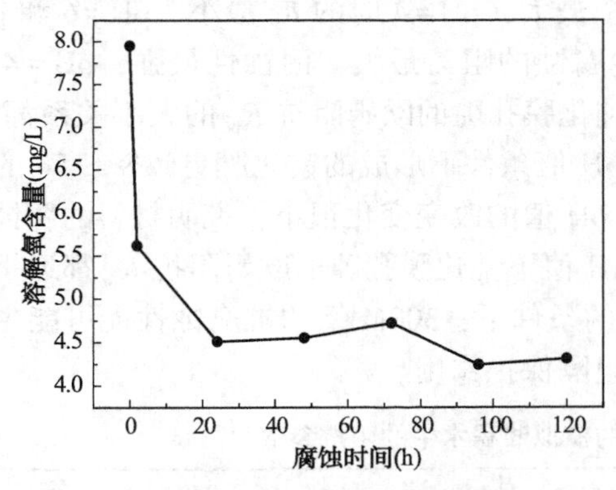

图3　模拟油积水中的溶解氧含量变化曲线

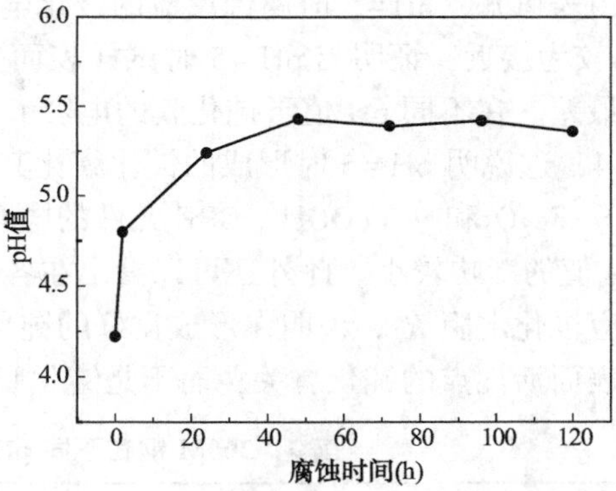

图4　模拟油积水中的pH值变化曲线

2.3　电化学腐蚀行为

图5为300M钢在不同pH值（pH=4、5、6、7）的模拟油积水中的交流阻抗谱。每个阻抗谱中均出现了两个容抗弧，第一个小容抗弧为对电极表面的容抗，第二个大容抗弧为工作电极的容抗弧。从图中看出，试样的阻抗值随着pH值的变化而改变，并且十分明显。对于不同的pH值，容抗弧的半径大小为$R_5>R_6>R_7>R_4$。对试验结果用ZSimpWin软件进行拟合，得到的等效电路图见图6。电路中C_c为300M钢钝化膜的电容，R_e为模拟油箱积水溶液的电阻，R_{po}为300M钢钝化膜孔隙的欧姆阻抗，R_p为300M钢腐蚀的极化电阻，Q为CPE等效电容元件，Z_w表示Warburg阻抗，对应于半无限扩散过程的阻力。

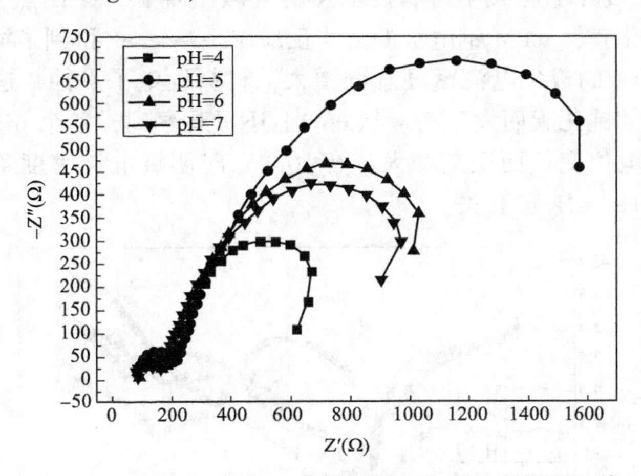

图5　300M钢在不同pH值模拟油积水中的交流阻抗谱

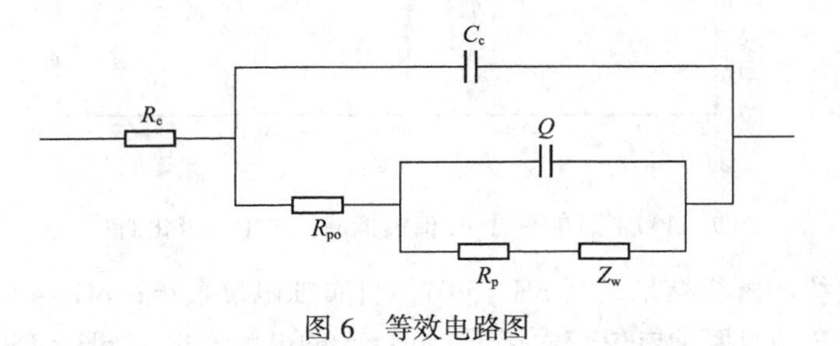

图6　等效电路图

表 4 为运用该等效电路对试样在不同 pH 值的模拟油积水中的阻抗参数进行分析的结果。由表可知，pH=5 时腐蚀反应的极化电阻 R_p 最大，pH=4 时的 R_p 最小，pH=6 和 pH=7 的 R_P 较为接近，证明当 pH=5 时试样表面发生腐蚀的阻力最大，耐蚀性最强，pH=4 时耐蚀性最差。在不同 pH 值下钝化膜的电容 C_c 和钝化膜孔隙的欧姆阻抗 R_{po} 的大小关系都为 5>6>7>4，这说明 pH=5 时形成的钝化膜比其他 pH 值条件下形成的钝化膜更致密，Fe 的钝化膜是 $\gamma-Fe_2O_3$ 和 $\gamma-FeOOH$。CPE 元件的电容随 pH 值的改变变化很小，说明材料的腐蚀界面受 pH 值的影响较小。此外还可以看出在各个 pH 值下钝化膜孔隙的欧姆阻抗 R_{po} 都远小于腐蚀反应极化电阻 R_p，说明在形成良好的钝化膜的条件下，300M 钢的抗腐蚀性能可能主要与金属表面活性点的钝化有关，而不是钝化膜的阻隔保护作用。

表 4　300M 钢在不同 pH 值的模拟油积水中的阻抗参数

pH 值	4	5	6	7
R_e（$\Omega\cdot cm^2$）	85.9	84.12	83.95	83.87
C_c（$\mu F\cdot cm^{-2}$）	1.406×10^{-6}	2.196×10^{-6}	1.8×10^{-6}	1.466×10^{-6}
R_{po}（$\Omega\cdot cm^2$）	91.85	122.7	117.1	96.82
Q（$\mu F\cdot cm^{-2}$）	3.687×10^{-4}	2.344×10^{-4}	2.699×10^{-4}	3.154×10^{-4}
R_p（$\Omega\cdot cm^2$）	735.4	1886	1277	1162
Z_w/Y_O（$S\cdot s^{0.5}\cdot cm^{-2}$）	2.461×10^5	1.629×10^{13}	8.059×10^4	0.5574

图 7 是 300M 钢在不同 pH 值的模拟油积水中的 Tafel 极化曲线。从图中可以看出，不同 pH 值的阳极区极化曲线在电位大于自腐蚀电位的区域出现了一个凹点，说明 300M 钢在此电位下表面形成了钝化膜，300M 钢电极的溶出反应 $Fe-2e=Fe^{2+}$ 得到了抑制。当电极电位升高到-0.45V 时，pH=4 的极化电流密度急剧增大，材料发生了点蚀，这是由于钝化膜表面吸附的阴离子穿透局部钝化膜而发生的。从曲线上还可观察到，各个 pH 值下的极化曲线阴极区的电流密度随着电位变负而逐渐增大，在-0.9V 时阴极电流密度急剧增大，这可能与 H^+ 的还原析出（$2H^++2e=H_2$）有关。

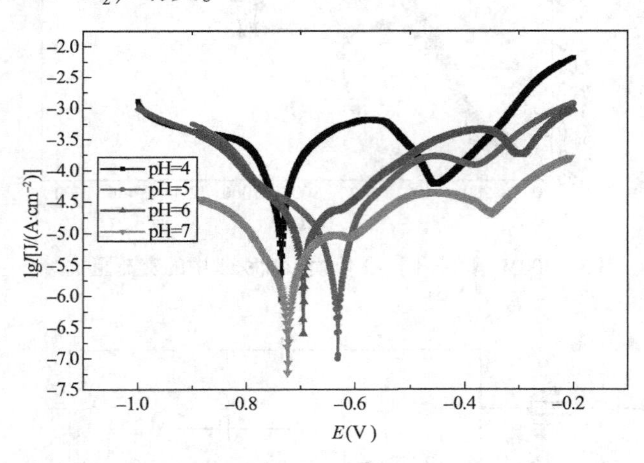

图 7　300M 钢在不同 pH 值模拟油积水中的极化曲线

表 5 是极化曲线的解析结果。当 pH=5 时，自腐蚀电位最正；pH=4 时，自腐蚀电位最负；pH=6 和 pH=7 的自腐蚀电位较为接近。pH=5 的电流密度比 pH=4 的电流密度下降了

两个数量级，比 pH=6 和 pH=7 的电流密度下降了一个数量级，在不同 pH 值下电流密度的大小为 $I_5>I_6>I_7>I_4$。由此可知，300M 钢在模拟油积水中腐蚀时，腐蚀速率的大小关系为 $v_5<v_6<v_7<v_4$，这与交流阻抗谱的测试结果一致。

表 5　300M 钢在不同 pH 值的模拟油积水中极化曲线的特征值

pH 值	腐蚀电位（V）	腐蚀电流（$A\cdot cm^{-2}$）
4	-0.735	2.157×10^{-4}
5	-0.622	6.502×10^{-6}
6	-0.695	1.782×10^{-5}
7	-0.719	6.425×10^{-5}

腐蚀初期，活性阴离子 Cl^- 在材料表面的不均匀吸附诱发了材料局部腐蚀，阳极发生溶解，消耗了大量的氧气；阴极受氧去极化控制，发生吸氧反应，生成了大量的 OH^-，使得 pH 值增大，腐蚀速度加快，同时腐蚀电位以较快速度降低。随着腐蚀时间的延长，腐蚀产物 $Fe(OH)_2$ 表面发生电子转移催化溶解使其快速转化为 γ-FeO（OH），并且分布在外层，逐渐覆盖在试样表面阻碍氧的扩散，使耗氧速度减慢。最终腐蚀电池反应逐渐停止，溶解氧含量、pH 值等都趋于稳定，腐蚀速率缓慢。

3　结论

（1）300M 钢在油积水环境中 24h 即发生腐蚀，120h 后试样表面腐蚀损伤达 89.48%。

（2）腐蚀过程中，油积水溶液中的溶解氧含量和 pH 在 24h 后趋于稳定。

（3）电化学交流阻抗谱和 Tafel 极化曲线测试结果表明，不同 pH 值的阳极区极化曲线在电位大于自腐蚀电位的区域都出现了一个凹点，材料表面发生了明显的钝化现象，腐蚀速率的大小关系为 $v_5<v_6<v_7<v_4$。

某 Q345R 焊接接头应力腐蚀开裂分析 *

韩燕

（石油管材及装备材料服役行为与结构安全国家重点实验室，
中国石油集团石油管工程技术研究院）

摘　要：通过宏观分析、化学分析、金相分析，以及扫描电镜和能谱分析试验，对断裂样品进行了分析。结果表明：该 Q345R 样品发生硫化氢应力开裂的主要原因是母材组织中带状偏析严重，建议从生产工艺查找原因，减轻或避免带状偏析，方可避免再次失效。

关键词：Q345R 焊接接头；应力腐蚀开裂；带状偏析；分析

在对某 Q345R 钢板焊接接头进行抗硫化氢应力开裂评价试验时，发现该样品多次发生断裂，且断口出现分层现象，遂对断裂样品进行分析，以明确其断裂原因。

抗硫化氢应力开裂评价试验依据 NACE TM0177—2005 A 法恒拉伸载荷法进行，试验溶液采用 A 溶液（5%NaCl+0.5%冰乙酸溶解在去离子水中），加载应力 247MPa，试验温度 24℃，试验时间 720 小时[1]。

1　分析方法及分析结果

1.1　宏观分析

该焊接接头样品分别在硫化氢饱和溶液中浸泡 46h、242h、286h 时发生断裂，宏观照片见图 1，断裂样品断口有分层特征，典型断口形貌见图 2。

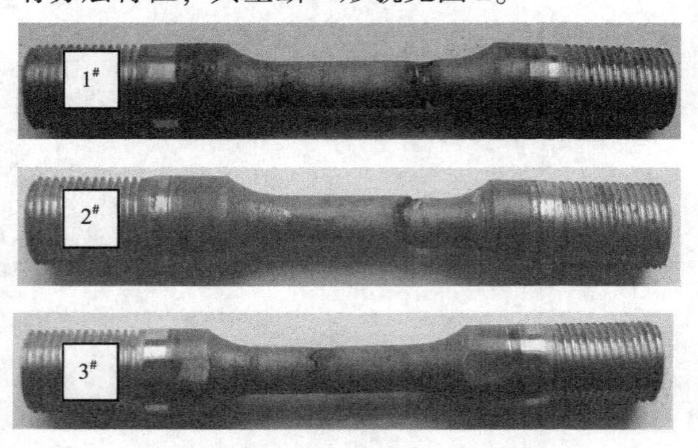

图 1　Q345R 试样 SSC 试验后宏观照片

* 基金项目：中国石油天然气集团公司应用基础课题，编号：2014A-4214，课题名称：复杂工况气井油套管柱失效控制与完整性技术研究。

作者简介：韩燕（1981—），性别：女，籍贯（陕西西安），职称：工程师，学位：硕士研究生，主要从事石油管材腐蚀与防护研究，联系电话：029-81887912；E-mail：hanyan003@cnpc.com.cn。

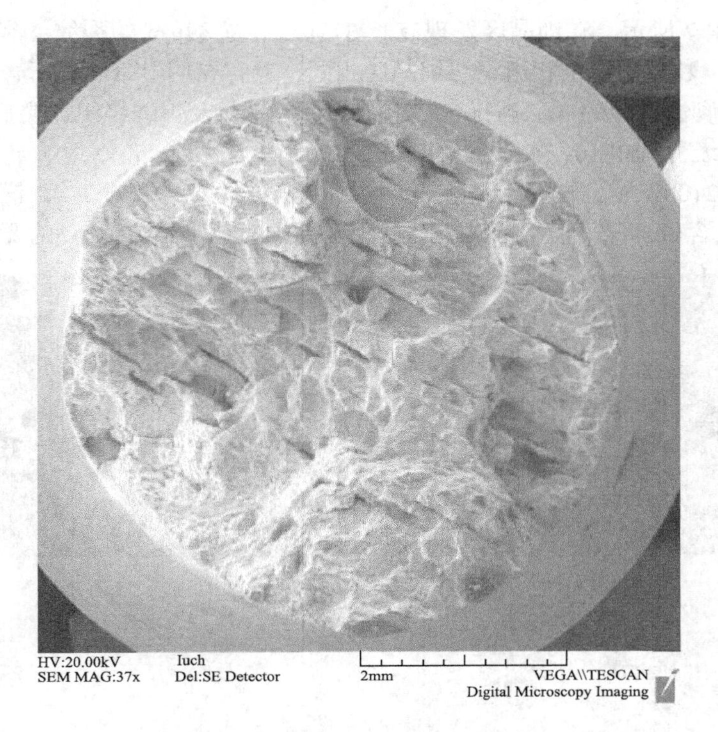

图 2　Q345R 试样 SSC 试验断口照片（取自 $3^{\#}$样品）

样品断口无明显塑性变形，断裂位于拉伸试样工作段，与拉应力方向呈一定倾斜角度，断口表面粗糙，颜色发黑，局部有分层现象，由于浸泡于腐蚀溶液中使得断口产生了腐蚀。

1.2　理化性能检测

1.2.1　化学成分

从同批样品母材上取样，用 ARL 4460 直读光谱仪，按照 GB/T 4336—2002 标准，进行化学成分分析，结果见表 1。由表可知，母材的化学成分符合 GB 713—2014 锅炉和压力容器用钢板标准要求[2]。

表 1　化学成分分析结果（wt %）

元素	C	Si	Mn	P	S	Cr	Mo	Ni
Q345R	0.18	0.24	1.25	0.0060	<0.002	0.13	<0.005	0.11
GB 713—2014	≤0.20	≤0.55	1.20~1.70	≤0.025	≤0.010	≤0.30	≤0.08	≤0.30
元素	Nb	V	Ti	Cu	B	Al	其他	
Q345R	0.015	<0.005	0.0019	0.0054	<0.0005	0.027	—	
GB 713—2014	<0.050	<0.050	≤0.30	≤0.30	—	—	Cu+ Ni+ Cr+ Mo≤0.70	

1.2.2　金相分析

从 $3^{\#}$样一侧断口处取样，采用 MEF4M 金相显微镜及图像分析系统，进行了金相组织观察和分析。断裂位置位于母材近焊缝处的细晶区，见图 3。母材组织为 $S_{回}$+B，焊缝组织为 PF+IAF+$B_{粒}$+P，熔合区组织为 $B_{粒}$+ PF+P+WF，细晶区组织为 PF+P，细晶区存在带状偏析现象，如图4~图7所示。在细晶区发现有长377μm、宽89μm的裂纹，裂纹沿带状偏析走向分布，见图8。断口附近金相组织与细晶区相同，且存在明显带状偏析，见图9。

取同批未做腐蚀试验的样品进行金相组织分析，母材及焊缝组织无明显差异，母材带状组织评级为3.0级，见图10，夹杂物等级A 0.5，B 0.5，D 0.5。

采用Tukon 2100B显微硬度计，进行了硬度测试，测试结果见表2。满足ISO 15156-2：2015规定的，在含硫工况下使用的碳钢和低合金钢硬度值小于22HRC的要求[3]。

图3 断裂位置宏观照片

图4 母材组织

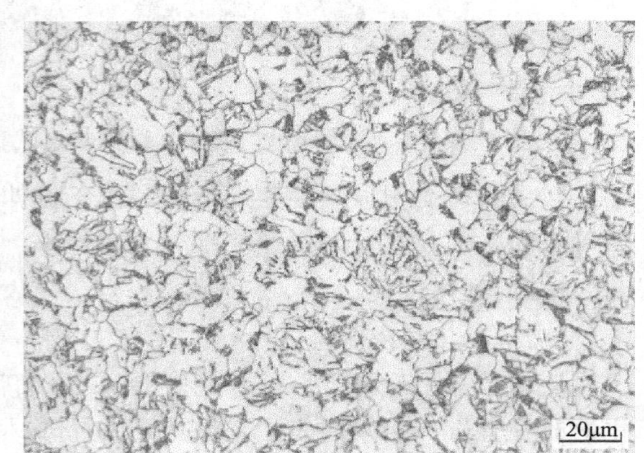

图5 焊缝组织

图6 熔合区组织

图7 细晶区组织

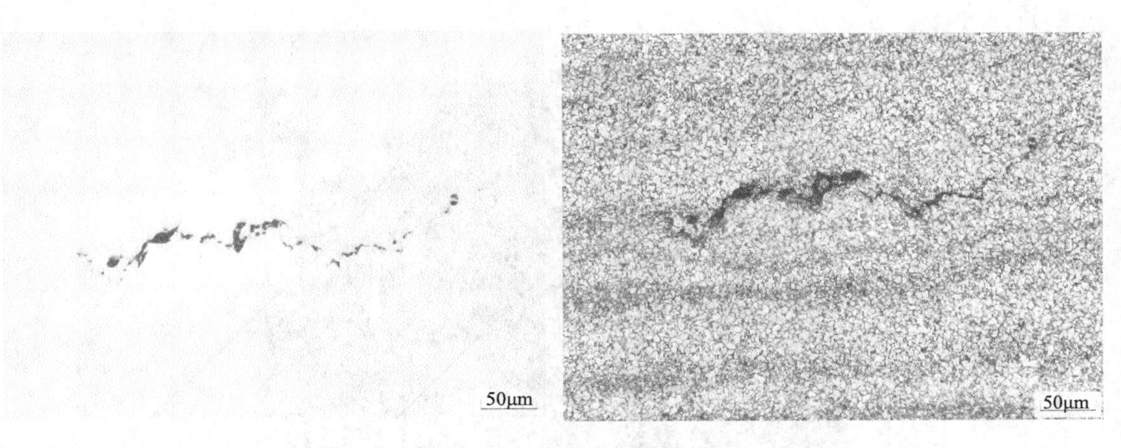

图 8　细晶区裂纹形貌及裂纹走向

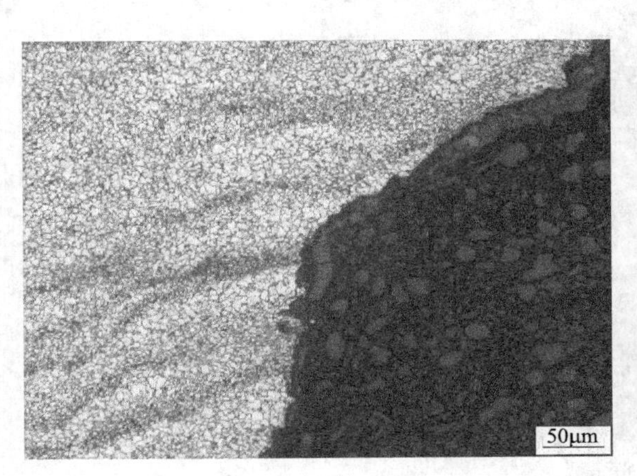

图 9　断口周围组织

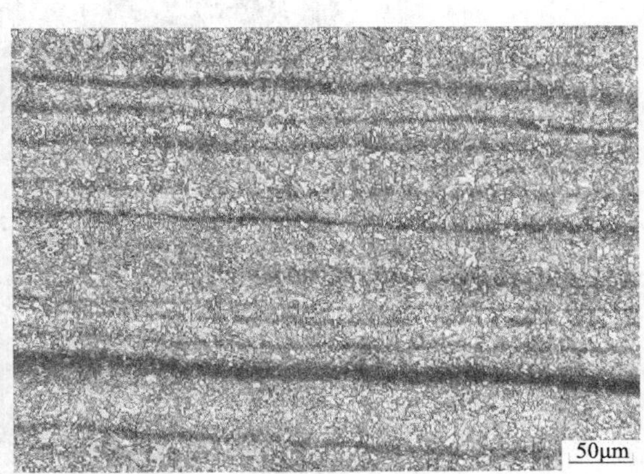

图 10　母材带状组织

表 2　显微维氏硬度检测结果

试样	显微维氏硬度值（$HV_{0.5}$）		
	靠近焊缝母材	细晶区	另一侧细晶区
Q345R 焊接接头	190，189，195	162，156，155，161，163	165，166，150，156，158

1.3　微观形貌及能谱分析

取 $3^{\#}$样断口进行微观形貌及能谱分析。断口表面凹凸不平，断面上存在多处微裂纹和扇形形貌，见图 11～图 13。由图可见，裂纹起源于试样内部，呈层状分布，多个裂纹连接在一起导致了最终的断裂。裂纹有典型的氢致开裂（HIC）裂纹形貌，呈阶梯状、沿壁厚方向分布。每一个扇形形貌为一个小的裂纹扩展面，扇形扩展面上放射棱收缩方向为裂纹源位置，多数扇形扩展面起源于内部裂纹处。

能谱分析显示：断口腐蚀产物中主要元素为 Fe、O 和 S 元素，分析结果见图 14。

2　综合分析

由试验结果可知，该样品的化学成分符合 GB 713—2014 标准要求，样品母材及细晶区金相组织中存在明显带状偏析。带状偏析是钢材内部缺陷之一，出现在热轧低碳结构钢显微组织中，沿轧制方向平行排列、成层状分布、形同条带的铁素体晶粒与珠光体晶粒。带状偏析是由于钢材在热轧后的冷却过程中发生相变时铁素体优先在由枝晶偏析和非金属夹杂延伸而成的条带中形成，导致铁素体形成条带，铁素体条带之间为珠光体，两者相间成层分布。

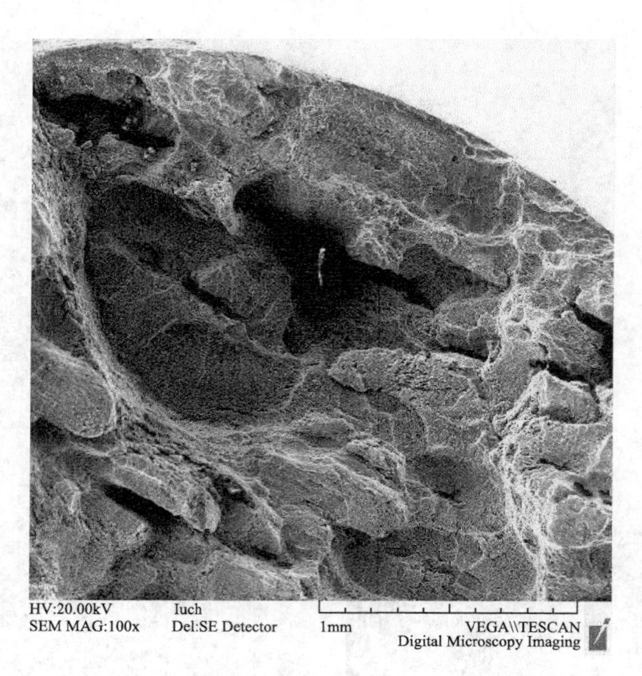

图 11　断口微观形貌 1

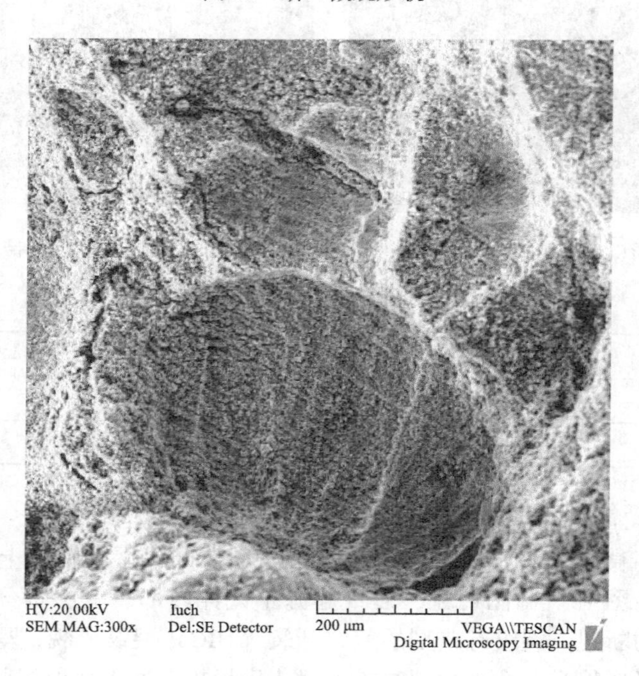

图 12　断口微观形貌 2

带状组织的存在使钢的组织不均匀，并影响钢材性能，形成各向异性，降低钢的塑性、冲击韧性和断面收缩率。带状组织对材料的抗氢致开裂性能也有显著的影响，随着带状组织增

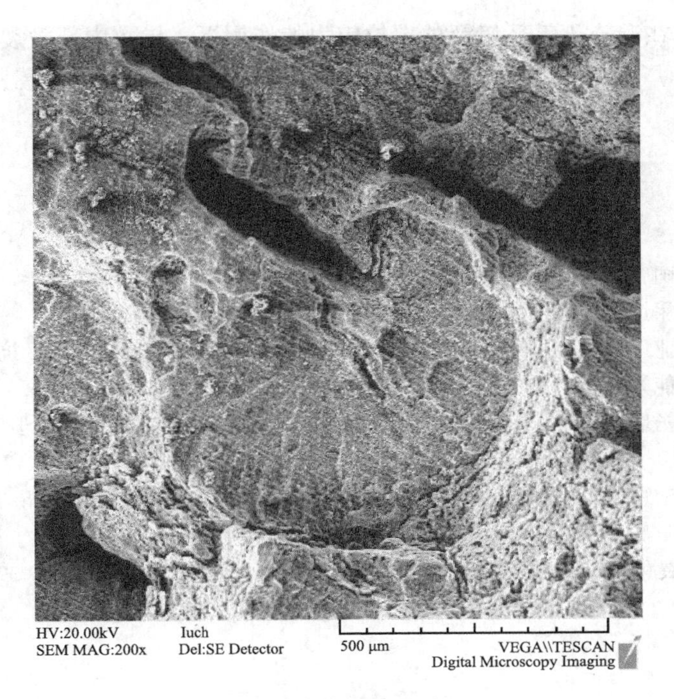

图 13 断口微观形貌 3

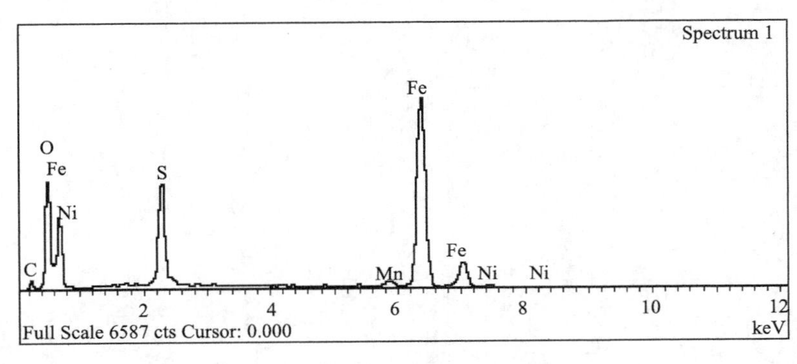

图 14 断口腐蚀产物能谱图

多，HIC 敏感性提高[4]。同时，该样品组织中还有贝氏体存在，由于贝氏体组织中的位错密度相对铁素体大，因此 HIC 裂纹也较易在贝氏体周围形成[5]。

当金属材料接触湿硫化氢环境时，溶液中大量离子态的氢会通过电化学反应进入到材料的内部，过饱和的氢在钢中缺陷部位如位错、夹杂物等界面处聚集，形成分子氢，从而产生巨大的内压，它能单独或协同外应力使氢致裂纹形核和扩展[6]。本研究中的断口形貌上出现大量层状分布的裂纹与带状组织分布相吻合，金相分析也表明裂纹沿带状组织分布，而显微分析中发现的扇形形貌即是微裂纹面的扩展阶段，这也表明这些微裂纹的形成经历了一定的时间，符合氢致滞后开裂的原理。内部微裂纹的形成，加之外部拉伸载荷的作用，使得材料在远低于其屈服强度的外应力作用下发生了过早的失效。

3 结论及建议

该 Q345R 样品发生断裂的主要原因是母材组织中带状偏析严重，为氢原子在钢材内部聚集提供了条件，在湿硫化氢环境及外部拉应力的作用下，导致内部微裂纹的产生，最终发生断裂。建议生产厂家进一步查找母材带状偏析原因，改进生产工艺，避免此问题再次发生。

参 考 文 献

[1] 金属材料在含 H_2S 环境下耐硫化物应力开裂和应力腐蚀开裂的实验室试验：NACE TM0177-2005 [S]. 美国休斯顿：美国腐蚀工程师协会，2005.

[2] 锅炉和压力容器用钢板：GB 713—2014 [S]. 北京：中国标准出版社，2014.

[3] 石油和天然气工业　石油和天然气生产中用于硫化氢环境的材料　第 2 部分：抗应力碳钢、低合金钢及铸铁的抗开裂能力：ISO 15156-2-2015 [S]. 瑞士：国际标准化组织，2015.

[4] 周琦，季根顺，杨瑞成，等. 管线钢中带状组织与氢致开裂 [J]. 甘肃工业大学学报，2002，28 (2)：30-33.

[5] 仝珂，韩新利，宋寰，等. L245NS 抗硫弯管力学性能及 HIC 和 SSC 试验分析 [J]. 焊管，2012，35 (12)：45-49.

[6] 褚武扬. 氢致开裂和应力腐蚀机理新进展 [J]. 自然科学进展，1991，1 (5)：393-399.

双金属复合管失效原因分析及对策

李发根

（中国石油集团石油管工程技术研究院，
石油管材及装备材料服役行为与结构安全国家重点实验室）

摘　要：双金属复合管成为油田地面集输管网防腐的重要解决方案之一，现已广泛应用于油田油气集输及注水等工程中。不过由于产品结构特殊性，在应用过程中存在较多的质量问题和使用问题，影响了双金属复合管的进一步推广应用。文章围绕典型失效事故原因分析，提出了相应措施建议，为规范双金属复合管的使用提供借鉴。

关键词：双金属复合管；衬层塌陷；焊接工艺；失效分析

双金属复合管以其低廉的价格、较高承压能力和优异耐腐蚀性能，被石油工业领域看作是解决高含 H_2S/CO_2 油气田管材腐蚀问题的一种安全和经济的办法。2005 年开始，国内便将 316L 内衬复合管应用于含 CO_2 腐蚀环境天然气集输管网。然而由于双金属复合管结构的特殊性，对产品的生产及焊接施工过程提出了更高的要求，应用中也出现了一些问题。本文主要围绕油田典型复合管失效事故，在分析事故原因的基础上，提出了相应的解决措施建议。

1　典型失效问题

1.1　产品质量问题

双金属复合管产品不但要兼顾基管强度和衬管耐蚀性，还要保证复合管材的整体使用性能，对工艺制造要求极高。产品在国内应用早期，出现过使用后衬层焊缝开裂失效（图 1）。随着工艺改进，衬管焊缝开裂问题已得到解决，但当前复合管产品还依旧存在质量不稳定的情况。某工程应用的双金属复合管产品，部分规格在做管道外防腐时，衬层发生了严重的鼓包、塌陷现象（图 2），比例甚至达到 28%。

图 1　衬管直焊缝开裂

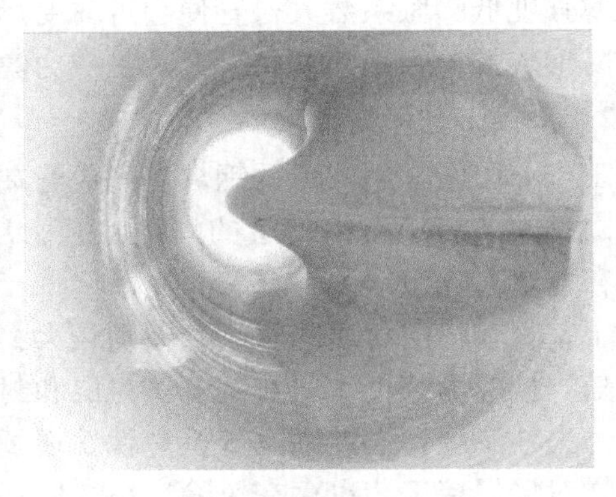

图 2　衬管鼓包、坍陷

1.2 焊接工艺问题

在对待316L内衬复合管的焊接问题上，国内通常采用不锈钢焊丝完成管端封焊、不锈钢层和过渡层焊接，选用强度性能匹配的碳钢焊条焊接碳钢基层。不过该类焊接工艺在焊接过程中也存在较多问题[1]，焊接成功率较低，有时一次合格率甚至只有65%。而且焊接质量不易保证，封焊后的复合管现场焊接难度仍然较大，对焊工焊接手法要求较高，封焊层容易产生孔洞和裂纹等缺陷，焊接接头在使用过程中陆续出现了基管焊接接头开裂及衬管焊接接头腐蚀等失效事故（图3和图4）。

图3 基管焊接接头开裂

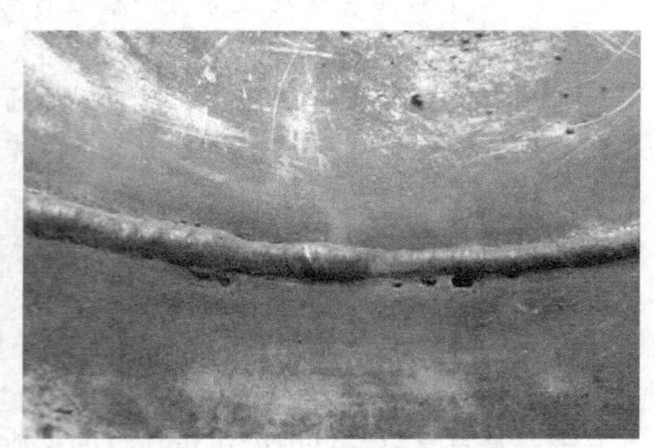

图4 衬管焊接接头腐蚀

2 失效因素分析

2.1 产品自身局限性

由于管材结构和工艺的限制，衬层塌陷问题目前还未能完全突破。首先双金属复合管由碳钢基管和不锈钢衬管组成，衬管材质热膨胀系数远大于基管材质，一旦管材受热膨胀衬管变形幅度会明显超过基管；其次在将两种材质复合的过程中，目前还不能完全清除基管与衬层之间水与空气等杂质，而成品管在出厂前一般又做了端部封焊处理，这样水与空气等杂质将会一直残留在基衬之间；最后，衬管一般都带有直焊缝，焊缝两侧易留有凹陷空间，形成形状突变区，容易成为应力集中区。

在以上三因素的共同作用下，双金属复合管在做外防腐过程中，管体承受近200℃高温后受热膨胀，而残留的水与空气受热后还会产生蒸气压力作用在管材上，在双重应力的作用下，壁薄强度低膨胀系数大衬管便会首先变形，但受制于端部封焊束缚又不能自由伸展，因此便会易形成鼓包、塌陷现象，尤其是应力集中的焊缝地带更是主要发生区域。

2.2 焊接工艺不成熟

对于当前国内常用的不锈钢焊丝完成管端封焊、不锈钢层和过渡层焊接，而采用匹配相应强度碳钢焊条焊接碳钢基层的焊接工艺，从理论上看，无论是端部封焊还是对接焊接工艺，都存在难以克服缺陷。

由于基管和衬层材质膨胀系数差别较大，管材运输存放过程中存在温差应力和焊接过程中也伴随焊接应力都会作用在基管、衬层和封焊交界位置的薄弱位置，容易导致产生裂纹缺陷，成为失效源区，失效案例分析中就发现了这种情况。

同样对接焊接工艺也存在风险。在焊接基层时，采用碳钢成分焊接材料在不锈钢成分过渡层焊道上焊接，碳钢金属被不锈钢母材稀释后形成中合金钢焊缝金属。在焊接快速冷却条件下容易形成高硬度马氏体组织，进而影响焊接接头的塑性、韧性，严重时还可能产生冷裂纹，给管道安全运行埋下了隐患。失效案例分析时，就曾发现了微裂纹，并且填充及盖面层有马氏体组织，HV10 硬度高达 400。另外焊接工艺中根焊使用气保护药芯焊丝焊接，且根焊和过渡焊背面不加氩气保护，但从实际效果来看，衬管氧化严重，焊缝热影响区域耐蚀性变差，易形成局部点蚀坑，进而影响复合管的安全使用[2]。

2.3 焊接施工缺乏规范

双金属复合管产品质量控制，目前国际上已经形成了 API 5LD 规范。但对于双金属复合管对接焊接工艺评定，API 5LD 规范没有明确指导方法。传统管道焊接工艺评定标准 SY/T 4103，评价对象也只是针对单一碳钢管或不锈钢管，而复合板焊接工艺评定标准 GB/T 13148 也只能评价复合板产品，并不适用于双金属复合管焊接工艺评定，而且还缺少焊接接头耐蚀性能评价要求[2]。

另外，对于复合管这种特殊管材的现场焊接施工目前也同样缺乏相关标准做出针对性规定，双金属复合管现场焊接施工也得不到有效规范，焊接质量难以保证。

3 应用建议

鉴于双金属复合管在油田现场应用过程中的存在上述问题，建议从生产厂家、检测机构、施工单位到油田用户从以下几个方面着手对其进行有效控制：

（1）产品工艺质量改进。

机械复合管产品有其自身的固有问题，但也存在改善空间。生产厂家改进生产工艺入手，比如文献[3]就提出了一种复合管制造方法，借助管端开孔后水压排气的方法可有助于解决衬层塌陷问题。同样还可从着眼于产品结构优化，比如提高产品椭圆度等规格要求，减小衬层受力后局部应力集中，降低塌陷概率。另外生产厂家还可加大研发力度，开发廉价冶金复合管，彻底解决衬层塌陷问题。

（2）产品检验项目完整性。

API 5LD 标准对于复合管产品性能要求相对宽泛，比如对产品尺寸规定主要基于碳钢管原材料规格要求制定，对于复合管的应用也考虑不够，无形中增加了应用时管端堆焊及对接焊接施工难度。另外 API 5LD 只是产品质量标准，对于管材适用性却不涉及。复合管材在实际应用中，往往会面临衬层腐蚀环境适应性，衬层的塌陷敏感性等问题。建议油田用户制定高于标准要求的产品技术规格书，补充适宜的管材适用性评价项目。

（3）焊接工艺的研究。

焊接一直是双金属复合管应用的棘手问题。当前国外普遍全程采用耐蚀合金焊丝完成管端堆焊和对接焊缝焊接工艺，焊接操作简单，焊缝质量容易保证，在国内也有成功应用案例[4-6]。当然国外焊接工艺也存在经济性不足问题，文献[7]提出了一种相对经济性的焊接方法，通过在不锈钢过渡层之后基层焊接之前，增加一层微碳纯铁焊材焊接的焊缝金属，将不锈钢与碳钢隔离开，随后基层焊接采用与基层母材强度匹配的碳钢焊接材料，广大工程建设单位可考虑尝试应用。

（4）编制焊接施工与验收标准。

双金属复合管焊接工艺评定和施工验收标准的缺乏，关乎复合管焊接质量，直接影响复合管的安全应用，编制相关标准要求非常迫切。当前工作主要是结合双金属复合管特性，围绕焊接接头力学性能和耐蚀性能等要求，尽快形成有效评定方法和焊接施工验收规定，打破无标准可依的局面。

参 考 文 献

[1] 许爱华，等. 双金属复合管的施工焊接技术 [J]. 天然气与石油，2010，28 (06)：22-28.
[2] 李发根，等. 双金属复合管焊接技术分析 [J]. 焊管，2012，37 (6)，40-43.
[3] 李为卫，等. 新型双金属复合管及制造方法：ZL200710118131 [P]. 2009-10-14.
[4] 王晓燕，等. 复合管管端堆焊 Inconel625 合金工艺及性能研究 [J]. 热加工工艺，2011，
[5] 杨刚. X65/316L 复合管的焊接工艺及焊接质量控制 [J]. 焊接技术，2012，41 (12)，56-57.
[6] 汪建明. Inconel 625/X65 复合管焊接工艺及接头性能研究 [J]. 焊接，2012 (8)，42-46.
[7] 李为卫，等. 一种双金属复合管环焊缝焊接方法：201310202717. X [P]. 2013-10-02.

天然气管路球阀失效分析

来维亚　尹成先　李金凤　路彩虹

（石油管材及装备材料服役行为与结构安全国家重点实验室）

摘　要：失效不锈钢球阀裂纹起源于阀体内表面，裂纹沿晶界扩展，裂纹线扫描分析表明，晶界边沿并未发现贫铬现象，裂纹内部也未发现铬元素富集，不存在晶间腐蚀失效。失效球阀阀体材料存在沿晶铸造缺陷和枝晶露头，水淬激冷过程中形成沿晶微裂纹，承压状态下微裂纹沿晶扩展，因内螺纹根部承压厚度最薄且应力集中，裂纹由内向外扩展，最终沿螺纹根部环向断裂，断口形貌为沿晶断裂。

关键词：球阀；失效；分析

2015年6月2日，中国石油长庆油田第六采气厂送检球阀2件，规格材质均为DN15-6.4MPa-316。其中一件断裂失效，另一件是该球阀的新产品。失效球阀是第六采气厂西二干线天然气管道编号为2#的不锈钢导气球阀，工作温度是20~50℃，工作压力5.7MPa。中国石油天然气集团公司管材研究所对送检断裂的球阀进行了失效分析。图1是失效阀门与新阀门的宏观照片；图2与图3显示失效阀门断裂面沿环向分布；图4显示阀体鼓包并有径向裂纹分布。

图1　失效与新球阀宏观照片

图2　失效球阀环向断裂

图3　失效球阀环向断口

图4　球阀鼓包及微裂纹

1　球阀宏观尺寸检测

图 5 是失效球阀宏观尺寸；图 6 是新阀门宏观尺寸；图 7 是球阀关闭后阀体形成压力腔实际承压厚度。失效阀门全长 52. 00mm，螺纹端外径 24. 30mm，内径 18. 84mm，压力腔内径 27. 50mm，压力腔外径 29. 64mm，承压厚度 1. 07mm。新阀门全长 62. 78mm，螺纹端外径 24. 30mm，内径 18. 84mm，压力腔内径 27. 02mm，压力腔外径 32. 08mm，承压厚度 2. 53mm。图 5 失效球阀与图 6 新阀门宏观尺寸有差异，具体到压力腔承压厚度，失效球阀承压厚度为只有 1. 07mm，而新球阀承压厚度为 2. 53mm，二者相差 1. 46mm。图 7 显示内螺纹加工过程已延伸至压力腔，压力腔实际承压厚度为 b_1，通过断口壁厚测量，断口最小厚度为 0. 86mm，比实际厚度 b（1. 07mm）还要小。再之，螺纹加工形成局部应力集中。

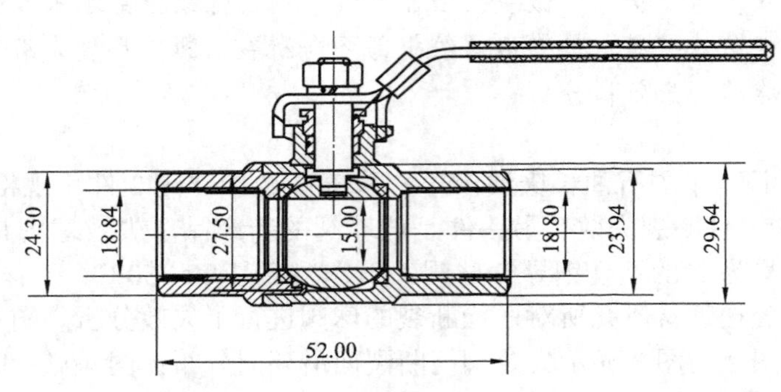

图 5　失效球阀宏观尺寸

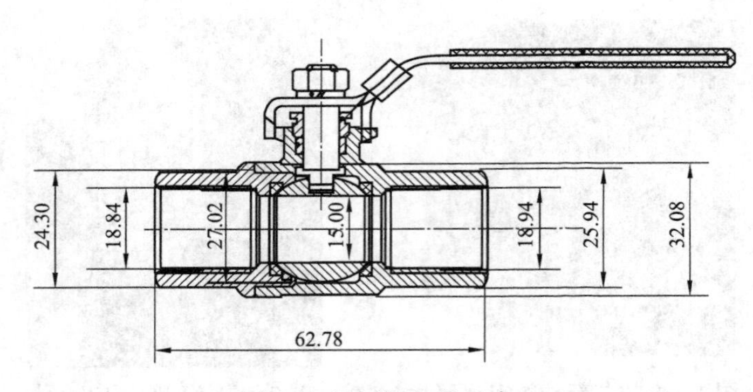

图 6　新阀门宏观尺寸

2　化学成分分析

按照长庆石油第六采气厂的要求，管研院对失效球阀与新球阀材质进行了分析，取样位置均为非螺纹连接端面，分析结果见表 1。

表 1　球阀材质化学成分分析结果（wt %）

球阀 / 元素	失效球阀	新球阀
碳（C）	0. 12	0. 066
硅（Si）	1. 21	0. 65

续表

元素 \ 球阀	失效球阀	新球阀
锰（Mn）	8.96	0.94
磷（P）	0.030	0.033
硫（S）	0.0058	0.0045
铬（Cr）	13.18	17.64
钼（Mo）	0.026	0.086
镍（Ni）	1.40	8.00
铌（Nb）	/	/
钒（V）	/	/
钛（Ti）	0.0016	0.0015
铜（Cu）	/	/
硼（B）	/	/
铝（Al）	/	/
氮（N）	/	/

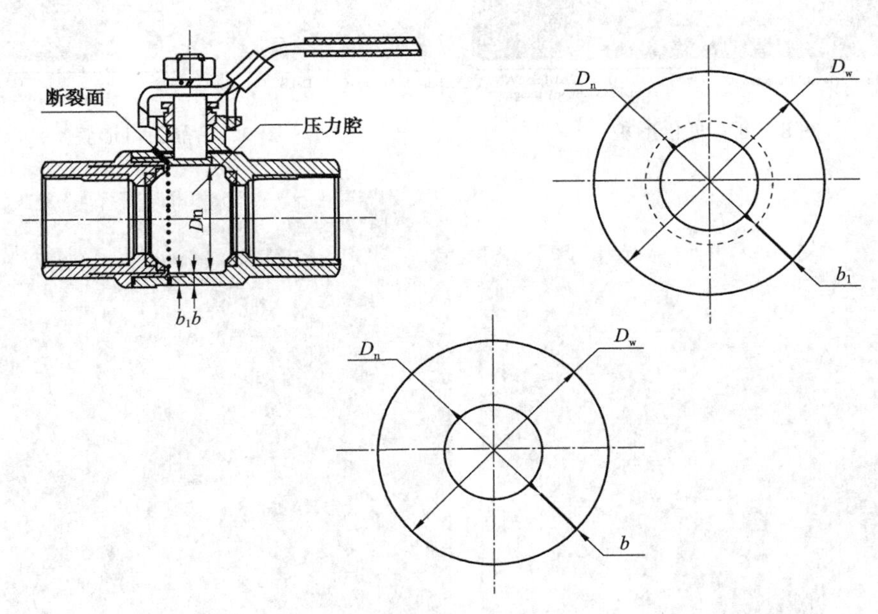

图 7　球阀关闭后阀体形成压力腔

b_1—因螺纹过深实际承压厚度；b—承压厚度

由表 1 中检测结果可以看出：失效球阀镍含量不到 8.00，铬含量不到 18.00，钼含量不到 2.00~3.00，所以失效球阀选材并非 316 不锈钢。送检的新球阀也非 316 不锈钢，分析结果为 304 不锈钢。

3　电镜与能谱分析

对失效球阀断口进行扫描电镜观察，其断口形貌为沿晶断裂，裂纹从阀体内表面起裂并从里向外扩展，沿内螺纹根部断裂。断裂面呈多元台阶，自由表面较多，断面可见材料内部沿晶裂纹和枝晶露头现象，阀体内外表面微裂纹沿晶网状分布，内表面微裂纹数量多于外边面，如图 8~21 所示。失效球阀断口附近区域观察到纵向裂纹与沿晶裂纹。非失效区内壁扫描电镜观察到细微沿晶裂纹。断口能谱分析结果表明，断口铬元素，锰元素含量高，而镍元素含量低，能谱分析结果与化学分析结果相吻合。

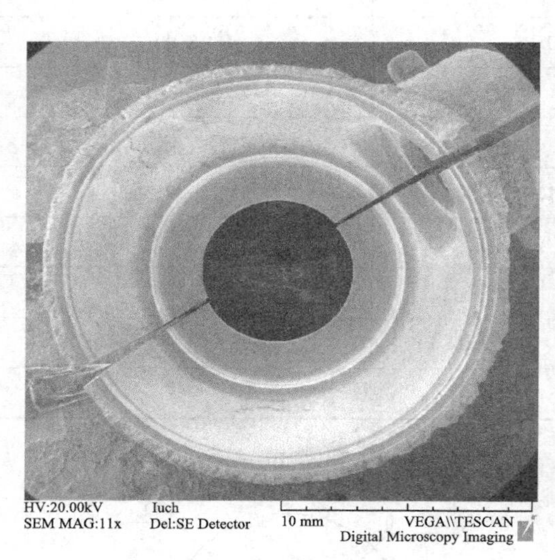

图 8　断口低倍形貌

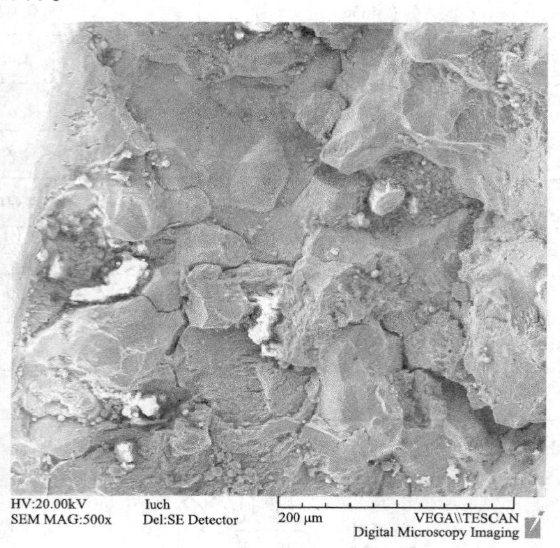

图 9　沿晶断口形貌

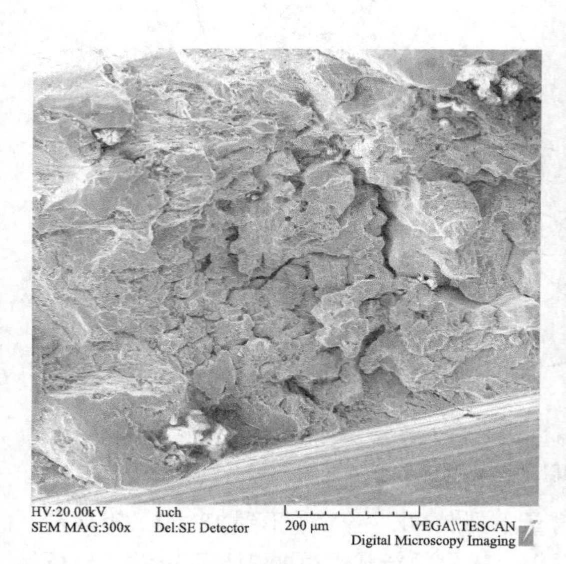

图 10　断口含缺陷形貌 1

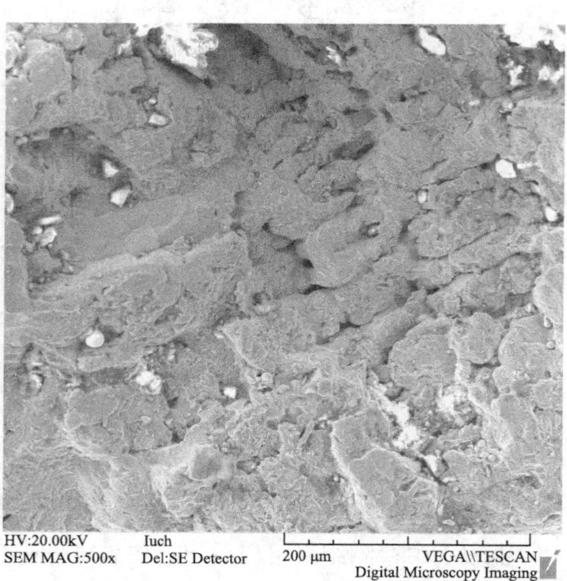

图 11　断口含缺陷形貌 2（枝晶露头）

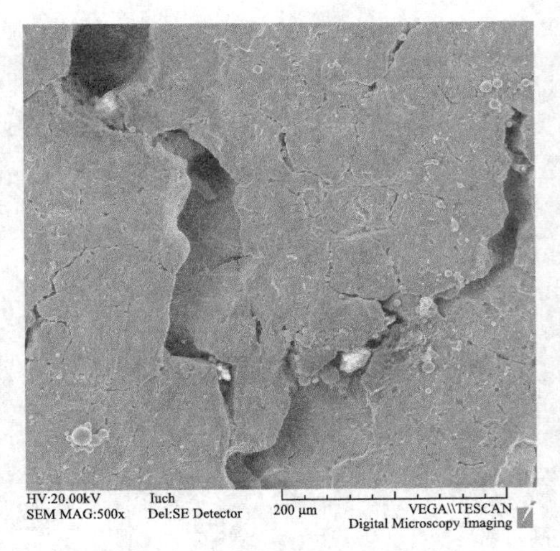

图 12　断口附近内壁纵向裂纹 1

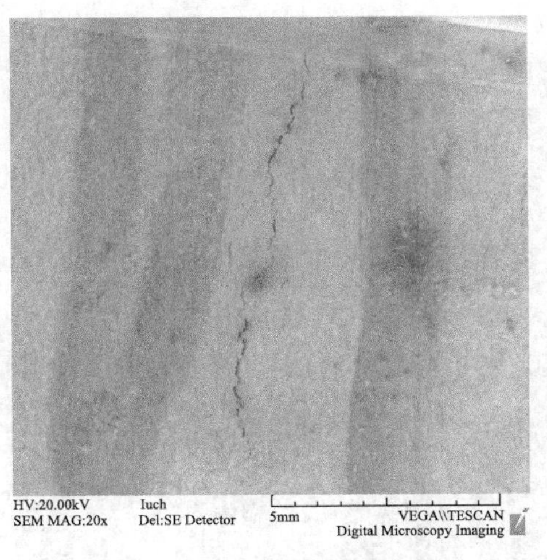

图 13　断口附近内壁纵向裂纹 2

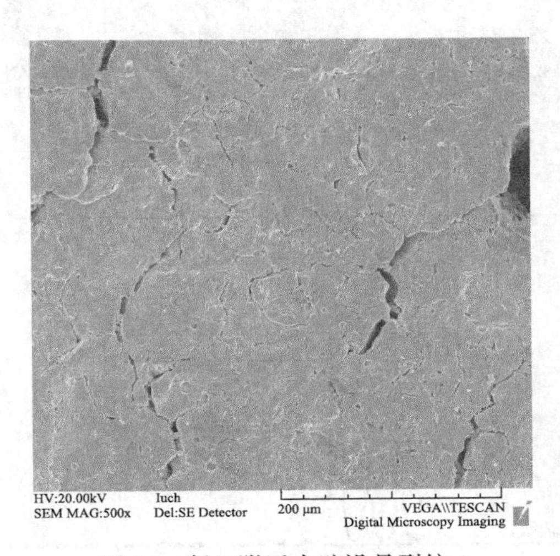

图 14　断口附近内壁沿晶裂纹 1

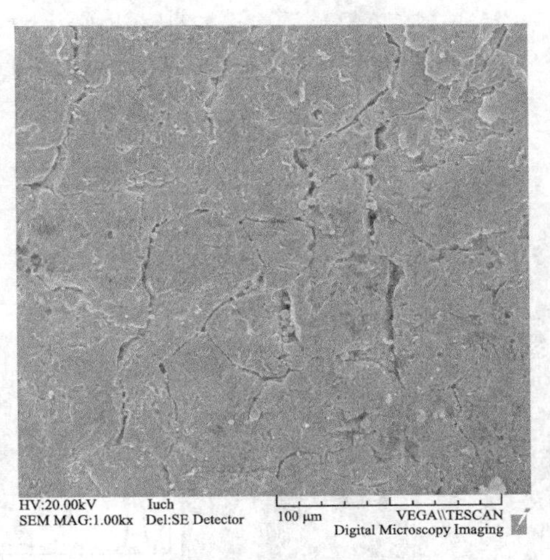

图 15　断口附近内壁沿晶裂纹 2

图 16　阀体外壁鼓包区低倍形貌

图 17　阀体外壁非失效区低倍形貌

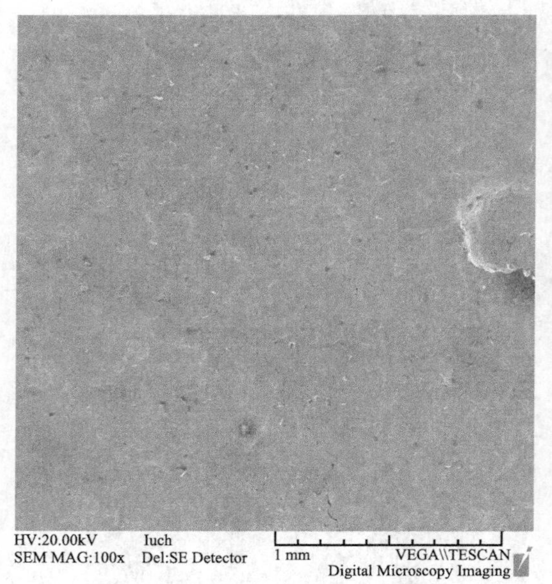

图 18　阀体内壁非失效区微裂纹 1

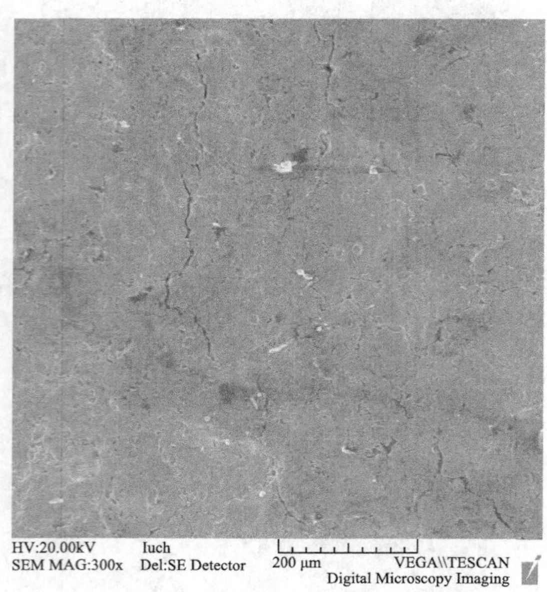

图 19　阀体内壁非失效区微裂纹 2

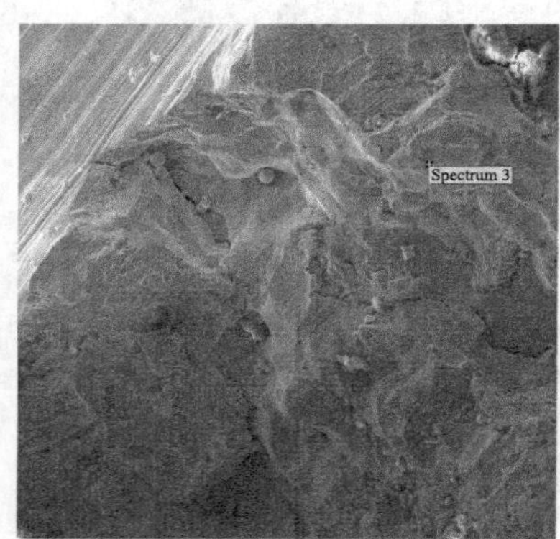

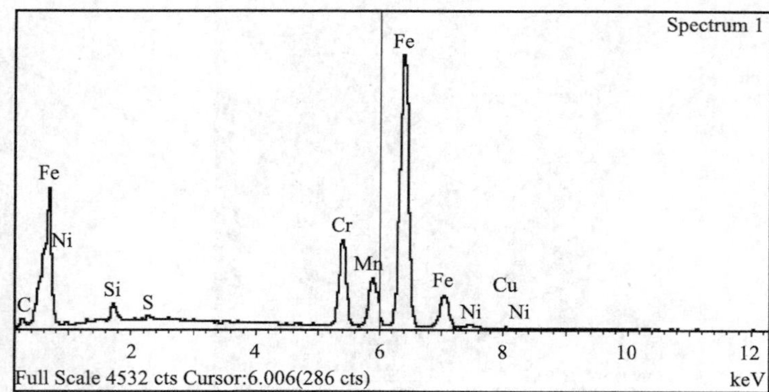

图 20　断口能谱分析位置及图谱之一

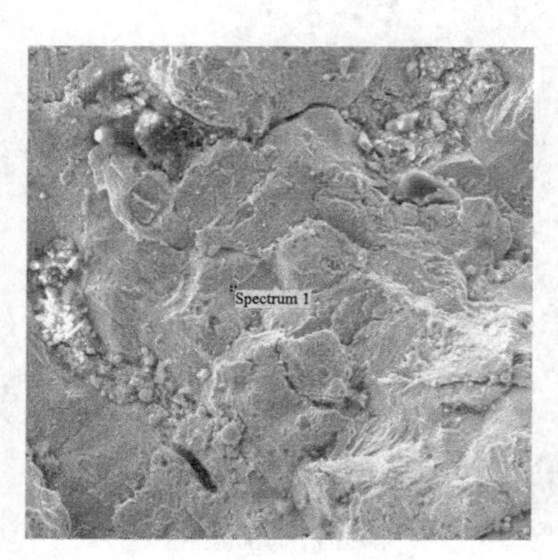

Spectrum 1

Fe

Fe

Ni

Cr

Mn

Fe

C

Si

S

Cu

Ni

Ni

2

4

6

8

10

12

Full Scale 4532 cts Cursor:6.006(286 cts)

keV

图 21　断口能谱分析位置及图谱之二

4 金相分析

依据 GB/T 13298—1991 标准，利用 MEF4M 金相显微镜及图像分析系统对失效球阀不同部位试样进行金相分析。失效球阀内表面有裂纹，内表面有明显凹陷，该区较光滑，没有加工痕迹，低倍见图 22。图 22 凹陷处样品金相分析结果，该区样品内外表面均有裂纹，裂纹沿晶扩展，有的裂纹几乎穿透阀体，且裂纹开口相对较大，特征见图 23 和图 24。失效区附近样品与失效区样品相比，该区内外表面也均有裂纹，内表面裂纹数量多于外表面，裂纹横向与纵向基本特征相同，均沿晶扩展。特征见图 25 和图 26。非失效区样品（失效区对面）样品与失效区样品比较，该区样品裂纹数量相对较少，开口较小，但裂纹基本特征相同，即内外表面均有裂纹，但内表面裂纹数量多于外表面，裂纹沿晶扩展，特征见图 27。球阀所有样品，金相分析发现材料内有铸造疏松孔洞，见图 28。经对标准球阀金相分析，其组织为奥氏体+铁素体，未发现沿晶裂纹与铸造缩孔，见图 29。

图 22 内表面裂纹和凹陷

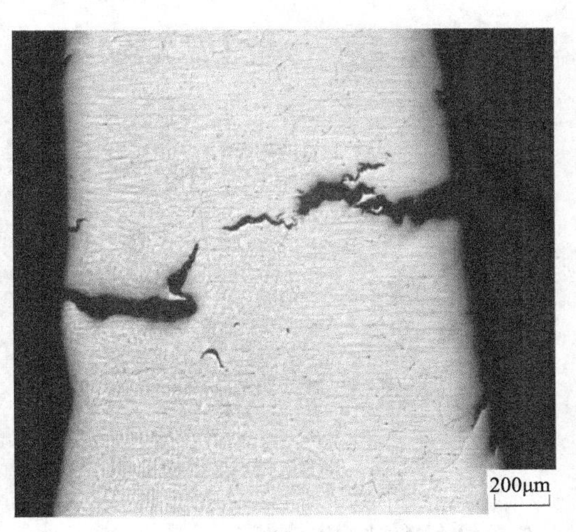

图 23 样品内外表面裂纹

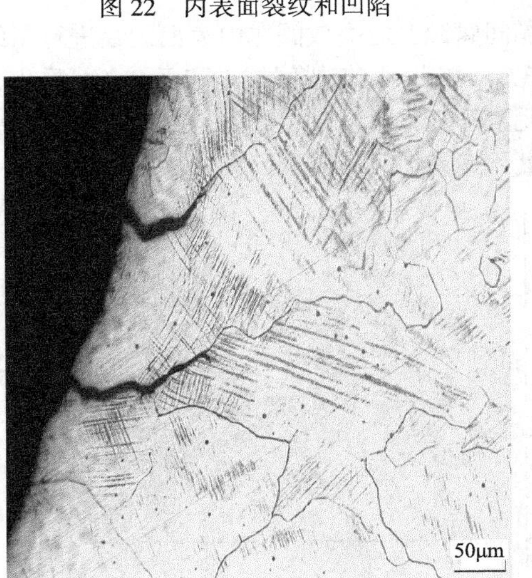

图 24 失效区内表面裂纹图

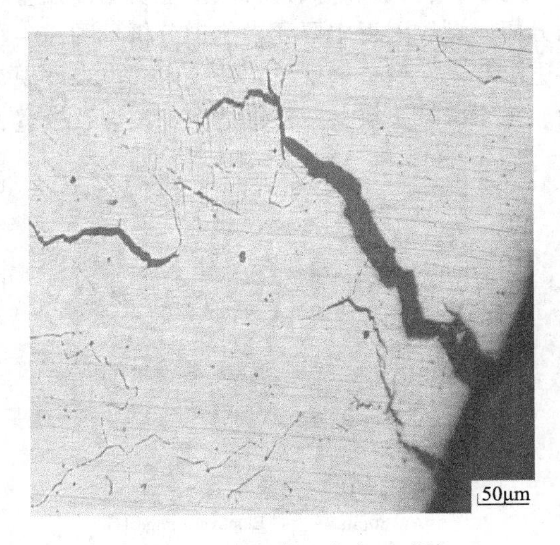

图 25 失效区附近内表面裂纹

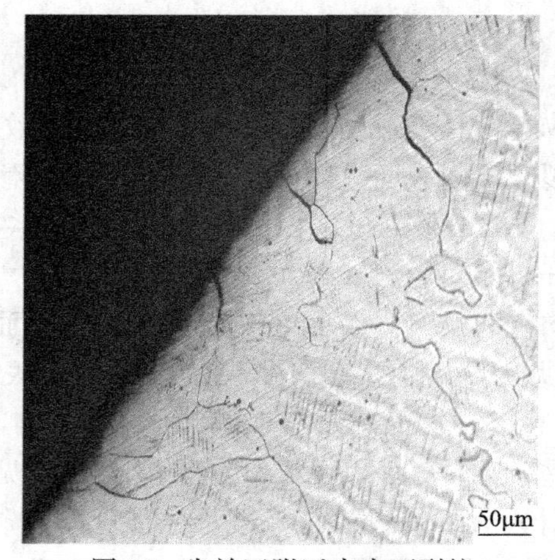

图 26　失效区附近内表面裂纹

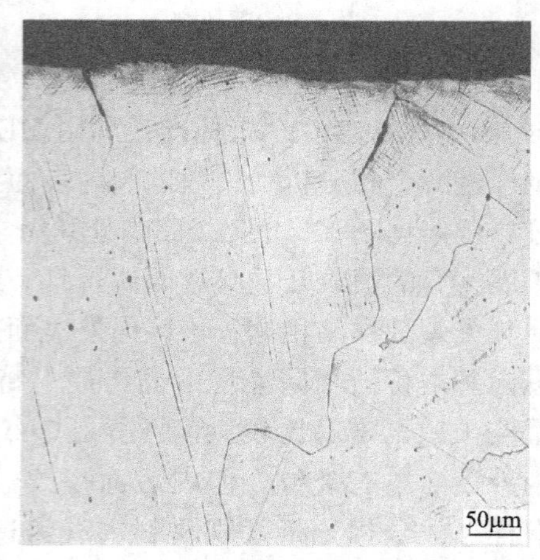

图 27　非失效区内表面裂纹

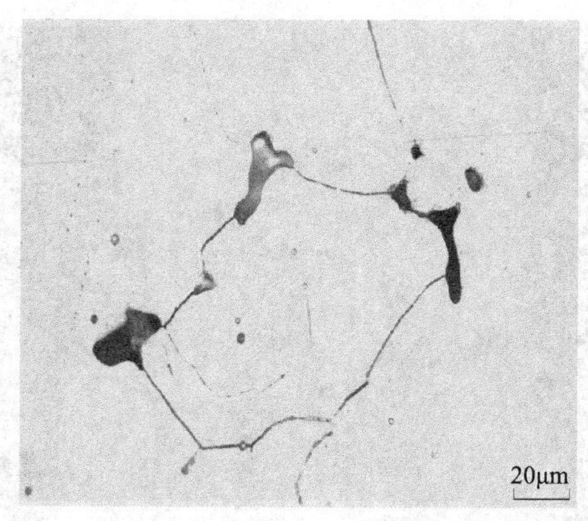

图 28　球阀材料内疏松孔洞

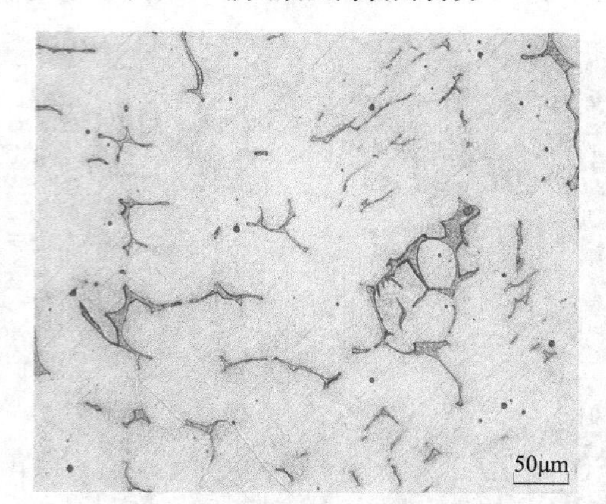

图 29　新球阀奥氏体+铁素体组织

5　裂纹线扫描分析

为了检验失效球阀是否因沿晶贫铬而发生了晶间腐蚀，对失效阀体内表面裂纹进行了线扫描分析，裂纹附近主要元素能谱分析见图 30~图 37。从裂纹线扫描区域 1 和区域 2 可看出，晶界边沿并未发现贫铬现象，裂纹内部也未发现铬元素富集，所以微裂纹不是由于晶间腐蚀引起的。至于区域 2 氧元素富集是由于该裂纹较宽，氧元素易扩散形成金属氧化物所致。

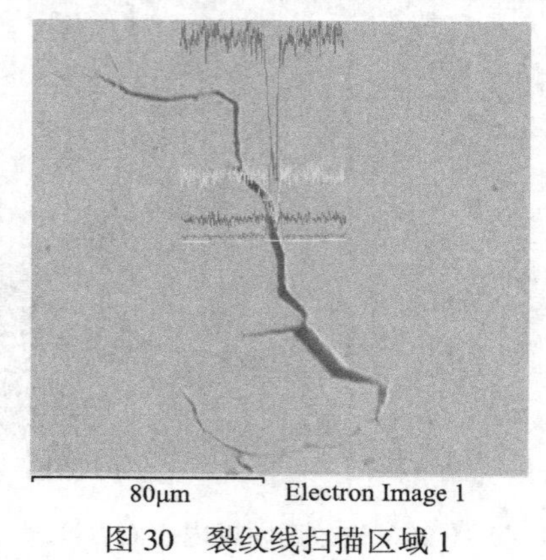

图 30　裂纹线扫描区域 1

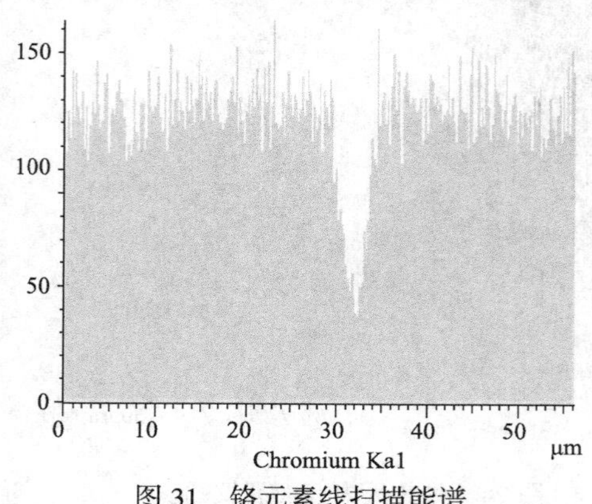

图 31　铬元素线扫描能谱

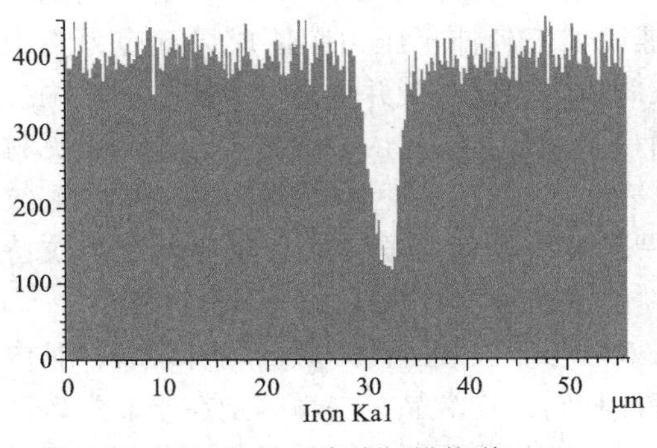

图 32　铁元素线扫描能谱

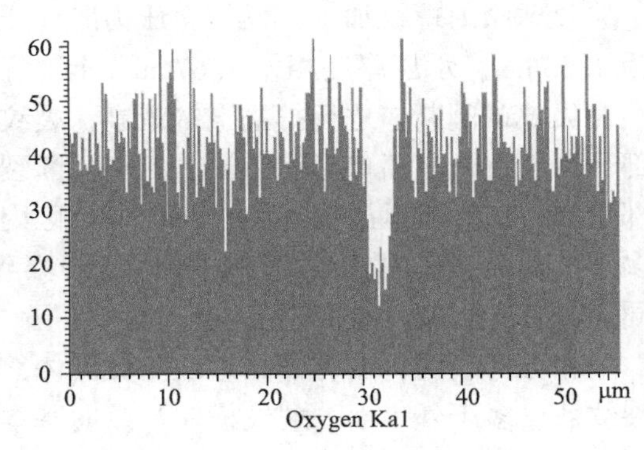

图 33　氧元素线扫描能谱

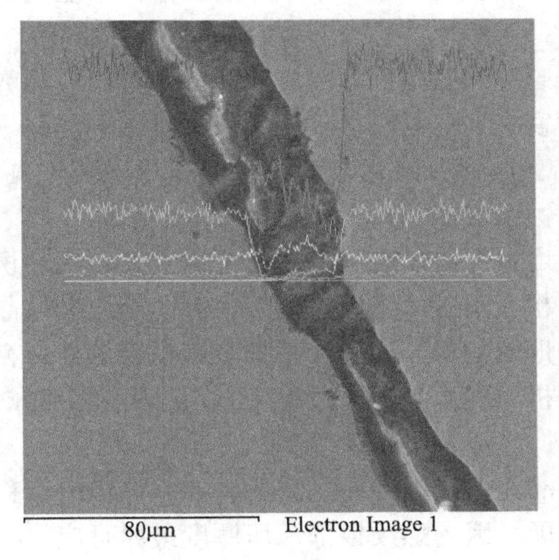

图 34　裂纹线扫描区域 2

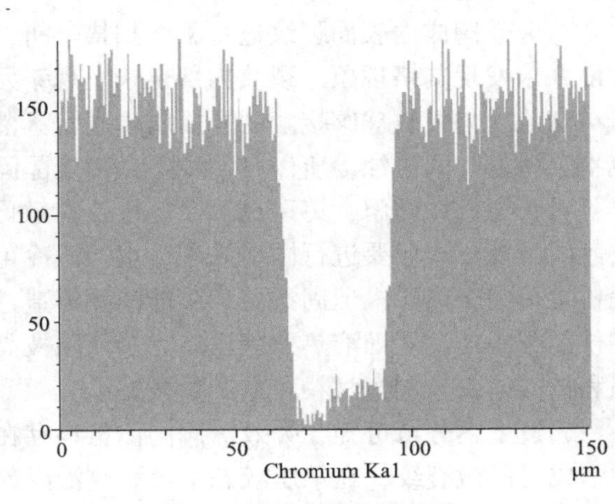

图 35　铬元素线扫描能谱

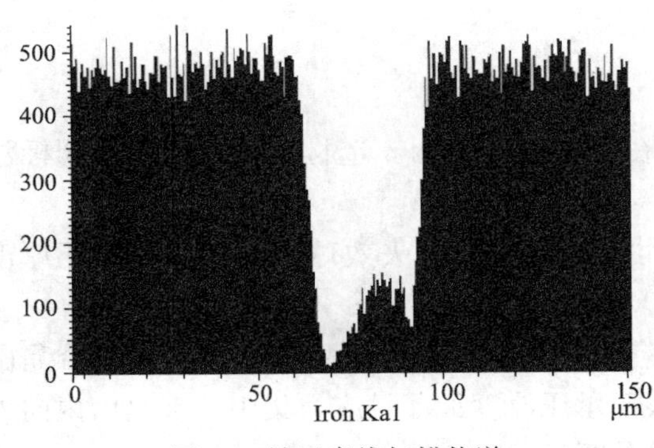

图 36　铁元素线扫描能谱

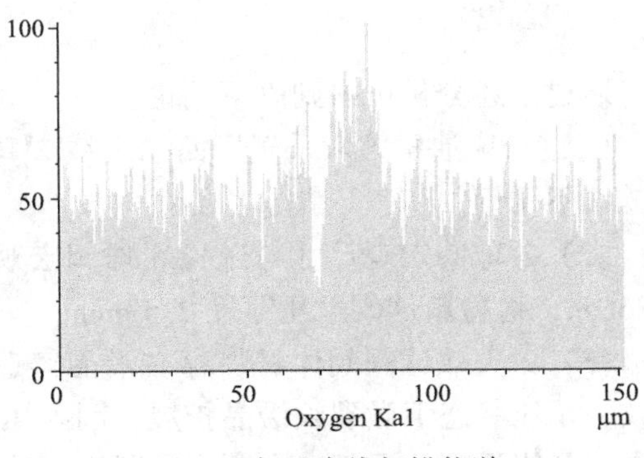

图 37　氧元素线扫描能谱

6　分析与讨论[1-7]

经对送检两样品材质分析，新球阀虽然宏观尺寸满足制造要求，但其材质是 304 不锈钢，并非制造要求的 316 不锈钢；失效球阀材质铬元素含量不足 18.00，镍元素含量不足 8.00，不含有钼元素，而锰元素含量高，并非制造要求的 316 不锈钢。

失效球阀内螺纹加工已延伸至压力腔，压力腔实际承压厚度为0.86mm，新球阀承压厚度为2.53mm，承压厚度相差1.67mm。同时，螺纹加工延伸至压力腔造成局部应力集中现象。对失效球阀断口进行扫描电镜观察，裂纹从阀体内表面起裂并从里向外扩展，沿内螺纹根部环向断裂，其断口形貌为沿晶断裂。断裂面呈多元台阶，自由表面较多，断面可见材料内部沿晶裂纹和枝晶露头现象，阀体失效区内外表面微裂纹沿晶网状分布，内表面微裂纹数量多于外边面。非失效区内壁扫描电镜观察到沿晶裂纹，其基本特征与断口附近裂纹特征相同。

金相分析表明，所有样品组织为奥氏体，失效区与非失效区内外表面均有裂纹，且内表面裂纹数量多于外表面，裂纹横向与纵向基本特征相同，均沿晶扩展，区别在于非失效区裂纹数量相对较少，开口较小，但裂纹基本特征相同。失效球阀所有样品，金相分析发现材料内均有铸造疏松孔洞沿晶分布。经对新球阀金相分析，其组织为奥氏体+铁素体，未发现沿晶裂纹与铸造缺陷。

对失效阀体内表面裂纹进行了线扫描分析，从裂纹线扫描区域1和区域2可看出，晶界边沿并未发现贫铬现象，裂纹内部也未发现铬元素富集。经与长庆油田第六采气厂现场技术人员交流，与失效球阀接触的介质属于脱水天然气，虽含有硫化氢气体，但属于干硫化氢，也不会诱发晶间腐蚀，所以微裂纹不是由于晶间腐蚀引起的。

对于奥氏体阀门，铸造过程冷却慢，在敏化温度范围停留时间较长，不进行热处理易诱发晶间腐蚀，一般要进行固溶处理，水淬激冷得到粗大奥氏体组织。固溶处理时间合理枝晶偏析也可完全消除，但固溶处理不能消除铸造过程形成的沿晶缺陷。扫描电镜观察到断口存在枝晶露头，说明固溶热处理没有完全消除枝晶偏析，在水淬激冷过程中，由于存在沿晶缺陷和枝晶露头，阀体材料沿晶产生微裂纹。

通过上述分析可知，失效球阀阀体材料存在沿晶铸造缺陷和枝晶露头，在水淬激冷过程中形成沿晶微裂纹，在承压状态下微裂纹沿晶界扩展，因内螺纹根部承压厚度最薄且应力集中，裂纹由内向外扩展，最终沿螺纹根部环向断裂。

7 结论

通过上述分析可得到以下结论：

（1）长庆油田第六采气厂西二干线2#球阀在铸造过程中阀体选材错误，并非球阀标注的316不锈钢。

（2）2#球阀宏观尺寸与送检新球阀宏观尺寸有较大差异，失效球阀阀体实际承压厚度0.86mm，送检新球阀承压厚度2.53mm。

（3）2#失效球阀阀体材料存在沿晶铸造缺陷和枝晶露头，水淬激冷过程中形成沿晶微裂纹，承压状态下微裂纹沿晶扩展，因内螺纹根部承压厚度最薄且应力集中，裂纹由内向外扩展，最终沿螺纹根部环向断裂，断口形貌为沿晶断裂。

参 考 文 献

[1] 王凤平，李晓刚. 316L不锈钢法兰腐蚀失效分析与对策［J］. 腐蚀科学与防护技术，2003，15（3）：180-183.

[2] 张振杰. 奥氏体不锈钢应力腐蚀破裂探讨［J］. 石油化工腐蚀与防护，2006，23（2）：48-50.

[3] 邱宏斌. 奥氏体不锈钢输油管道焊缝的应力腐蚀失效分析［J］. 化工设备与管道，2011，48（4）：

68-72.

[4] 张国华. 奥氏体不锈钢应力腐蚀分析研究［J］. 焊接技术，2002，31（6）：54-58.

[5] 雷阿利. 奥氏体不锈钢焊接接头在含硫介质中的腐蚀性［J］. 焊接学报，2006，27（1）：89-92.

[6] 廖景娱. 金属构件失效分析［M］. 北京：化学工业出版社，2003.

[7] 熊金平，左禹. 波纹不锈钢换热板腐蚀开裂失效分析［J］. 腐蚀科学与防护技术，2005，17（6）：435-438.

Insights into the Corrosion Perforation of UNS S32205 Duplex Stainless Steel Weld in Gas Transportation Pipelines

Yuan Juntao[1] Zhang Huihui[2] Fu Anqing[1] Yin chengxian[1]
Zhu Ming[2] Lv Naixin[1] Xu xiuqing[1] Miao Jian[1]

(1. State Key Laboratory of Performance and Structural Safety for Petroleum Tubular Goods and Equipment Materials, Tubular Goods Research Institute of CNPC;
2. College of Materials Science and Engineering, Xi'an University of Science and Technology)

Abstract: Microstructure and pitting corrosion resistance of UNS S32205 DSS weld were systematically studied, and its failure causes were analyzed carefully. The results showed the uneven metallographic microstructures with majority of austenite in the HAZ and then the significantly reduced pitting corrosion resistance of the weld. Furthermore, large potential difference between the both sides of the insulation joint and high salinity of the formation water in the pipeline make the leakage current preferably flow out from the weld and enter the transmission fluid, so localized corrosion happens at the inner surface of the weld and leads to the corrosion perforation eventually.

Keywords: Duplex stainless steel; Polarization curves; Pitting corrosion; Welding; Cathodic protection

1 Introduction

The oil and gas industry has received much attention from researchers worldwide. During oil exploration and transportation, corrosive media such as carbon dioxide, chloride ion and hydrogen sulfide have caused severe corrosion problems[1-3]. In the most corrosive regions, corrosion resistant alloys (CRAs) are applied to reduce safety accidents. In recent years, considerable amount of duplex stainless steels (DSSs) have been utilized as oil-gas-gathering pipelines in China to insure the safety.

DSSs possess beneficial combinations of austenitic (fcc) and ferritic (bcc), and exhibit good mechanical properties and corrosion resistance in acidic, caustic and marine environments[4-7]. DSSs with a ferrite/austenite phase fraction near to 50 : 50 usually exhibit the highest resistance to pitting corrosion[8]. The resistance to pitting corrosion generally decreases with a rise in the phase imbalance in the DSSs[9]. In practice, welding would cause disturbance of this phase balance, because rapid cooling is involved in most weld thermal cycles[10-12].

For applications of DSSs in many industries, welding is an inevitable fabrication process, because microstructures in the weld metal (WM) and heat affected zone (HAZ) undergo fast heating and cooling cycles which would cause excessive ferritization. Furthermore, several harmful phases like sigma phase, carbide, chi phase, secondary austenite and chromium nitride are also prone to form[13-15], which is detrimental to the pitting corrosion resistance and toughness[16]. For the welding process, heat input is the most important parameter[17]. Low heat input leads to a relatively fast cooling rate, and then results in an extremely unbalance microstructure with excess of ferrite phase. In addition, plenty of chromium nitrides would also precipitate in the interior of the ferrite grains or the interface of austenite and ferrite grains. High heat input leads to a relatively slow cooling rate, and then results in microstructure with more austenite reformation during cooling stage. However, intermetallic phases would be prone to form. In recent years, many efforts have been put forward to optimize the welding technology (e. g. thermal cycles and chemical compositions) and made remarkable advances in laboratories and factories. However, some irresistible factors in remote oil filed such as poor working environment and inadequate working equipment limit the fully successful application of novel welding technologies in the field, so that the welding of DSS pipelines is still a hidden danger to the safe production.

On the other hand, cathodic protection is usually employed to slow down or prevent the external corrosion of underground metallic pipelines in oil fields. In order to prevent the loss of protection current, electrical insulation is required between the protected pipeline and non protected one. In recent years, serious internal corrosion perforation of insulation joints happened[18,19] possibly due to the poor insulation effectiveness. Such issues are extremely catastrophic due to the wide utilization of cathodic protection and insulation joints in underground metallic pipelines. However, few insights into the causes of such issues have been reported until now.

Most recently, serious corrosion perforation happened in a UNS S32205 DSS gas-gathering pipeline which was welded with an insulation joint manufactured with the same DSS material. Fig. 1 illustrates the failed UNS S32205 DSS pipeline schematically. The welding process was argon welding union in which argon tungsten arc welding was for the bottom area and manual arc welding was for the filling and cover area. On the inner surface of the insulation joint, organic silicon high temperature resistant paint with thickness of ~100μm was coated to insure good insulation. Serious corrosion and perforation happened at the weld in the non protected side, as shown in Fig. 2.

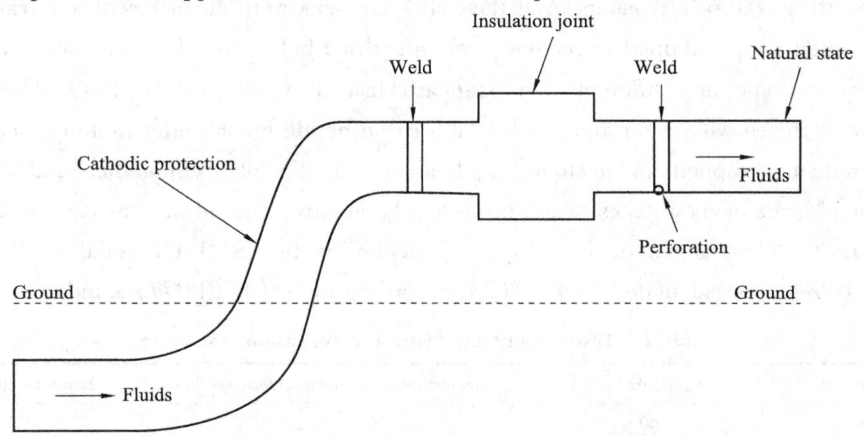

Fig. 1 Schematic illustration of the failed UNS S32205 DSS pipeline

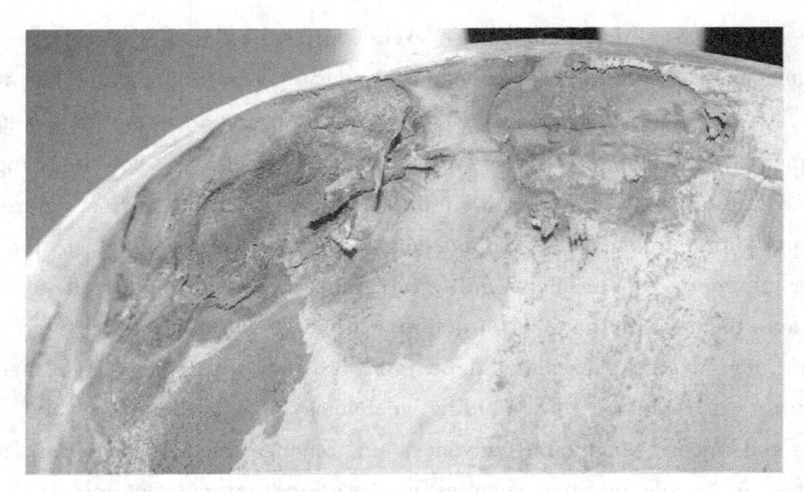

Fig. 2 Macro corrosion morphology of the failed UNS S32205 DSS weld

The present work was aimed to find out the causes of corrosion perforation for the UNS S32205 DSS weld by systematic investigation on the microstructure, chemical compositions, pitting resistance, and polarization behavior. At last, a possible corrosion mechanism was proposed.

2 Experimental

2.1 Chemical and metallographic characterization

The chemical compositions of perforated DSS pipe were analyzed by a direct reading spectrometer (ARL 4460) and a nitrogen oxygen analyzer (LECO TC600) on the basis of ASTM A751. The metallographic structure and inclusions were analyzed by optical microscopy (OM, MEF4M) based on ASTM E381, ASTM E112 and ASTM E562, after etching in Murakami reagent solution (30 g potassium ferricyanide + 30 g potassium hydroxide + 60 ml deionized water) for 3 min at 90℃.

2.2 Test of pitting corrosion resistance

Pitting corrosion resistance was evaluated by immersing in ferrite chloride solutions according to the procedures in GB/T 19897-1999. Test specimens were cut from the base metal (BM) and weld of the failed pipeline, with dimensions of 40mm×15mm×3mm. All surfaces of the specimens were polished by 120-grit abrasive papers. And then all specimens were cleaned with equivalent, rinsed well with distilled water, dipped in acetone, and air-dried before use. Test parameters are shown in Table 1, where some are different from standard test in GB/T 19897-1999. After tests, all specimens were rinsed with water and scrub with a nylon bristle brush under running water to remove corrosion products, dipped in acetone, and air-dried. Visual examination and photographic reproduction of specimen surfaces were used to characterize the pitting resistance. The average corrosion rate (*CR*) was obtained as Eq. (1) according to ASTM G1 standard. The maximum pitting rate (*PR*) was calculated by Eq. (2) according to NACE RP0775 standard.

Table 1 Test parameters of pitting corrosion resistance

Test parameter	Temperature (℃)	Concentration of ferric chloride (%)	Exposure duration (h)
Values	40	6, 14, 20	24

$$CR\ (\mathrm{mm/y}) = (8.76\times10^4\times W) / (A\times T\times D) \quad (1)$$

where, W is the mass loss (g), A is the area (cm^2), T is the time of exposure (h), D is the density (g/cm^3).

$$PR\ (\mathrm{mm/y}) = (d_m\times365) / t \quad (2)$$

where d_m is the depth of deepest pit (mm), t is the exposure time (d).

2.3 Electrochemical characterization

The resistance to pitting corrosion was evaluated by electrochemical tests. Specimens were embedded in epoxy resin to be used as working electrodes with a copper wire attached to the rear part for electric contact. Prior to use, specimens were ground with 1200 grit SiC paper, rinsed with distilled water and degreased with acetone. A conventional three electrode electrochemical cell was used with the duplex stainless steel samples as working electrodes, a saturated silver chloride electrode as the reference electrode and a graphite bar as the counter electrode.

Test solutions were configured with analytical reagents according to the ion content of the field water analysis report (Table 2) where heavy metal ions were not detected. The pH value was adjusted to 5.3 with dilute hydrochloric acid. Prior to immersion of the test specimen, high purity carbon dioxide gas was bubbled at a rate of 150 cm^3/min for 1h to reduce oxygen levels in solution. During the whole procedure, the high purity carbon dioxide was continuously passed into the solution.

Table 2 Chemical compositions of the field water

Component	HCO_3^-	Cl^-	SO_4^{2-}	Ca^{2+}	K^++Na^+
Content (mg/L)	99.65	178426.30	123.33	14726.39	96885.98

Potentiodynamic polarization tests were conducted in a CS370 electrochemical workstation at 40℃. After 55 min immersion, the open-circuit specimen potential was recorded. The potential scan was started 1 h after specimen immersion, beginning at the open circuit potential (OCP), and stopped at the potential value at which the anodic current density exceeded 300 μA/cm^2. The potentiodynamic potential sweep rate was set as 10 mV/min. Corrosion potential (E_{corr}), breakdown potential (E_b), passivation current density (i_p) were obtained from the polarization curves. Corrosion current density (i_{corr}) was obtained by the extrapolation of the cathodic and anodic slopes between 50 and 100 mV away from E_{corr}. The E_b marks the end of the passive potential region and the transition from passive to transpassive behavior and it was determined as the current density reached a value of 100 μA/cm^2. In order to characterize the passive layer integrity and pitting resistance, potentiostatic measurements were also performed in the anodic potential of 0.46 V versus OCP.

2.4 Characterization of morphology

A smart digital microscope (ZEISS Smartzoom 5) was employed to investigate the 3D pitting morphology and measure the pit depth. Corrosion morphology was analyzed using a Scanning Electron Microscopy (SEM, Philips XL-30), and chemical composition was analyzed by Energy Dispersive Spectrum (EDS, INCA - 350). Phases of the corrosion products and scaling deposits were identified by an X-ray Diffractometer (XRD, D8 ADVANCE) with filtered Cu kα radiation.

3 Results

3.1 Chemical compositions and metallographic structure

The chemical compositions of the failed UNS S32205 DSS pipe are shown in Table 3. Compared to API Spec 5LC standard, the chemical contents of the base metal (BM) meet the technical requirements. For the weld, the chemical contents meet the requirements of E2209 welding wire in YB/T 5092-2005 and E2209-16 welding electrode in GB/T 983-2012. It is evident that the contents of Cr and Ni in the weld are slightly higher than those in the BM. The chromium equivalents (Cr_{eq}) and nickel equivalents (Ni_{eq}), which represents the ability of stabilizing the ferrite and austenite structure from the perspective of alloying elements, were calculated by Eqs. (3) - (4)[20]. The ratio Cr_{eq}/Ni_{eq} for the BM and weld is 2.63 and 1.93 respectively. The decrease of Cr_{eq}/Ni_{eq} value would reduce the decrease of ferrite proportion in the microstructures[17].

Table 3 Chemical compositions of the UNS S32205 DSS pipeline and its weld (wt. %)

Element	C	Si	Mn	P	S	Cr	Mo	Ni	Nb	V	Ti	Cu	B	Al	N
BM	0.024	0.50	1.16	0.026	0.0012	22.9	3.3	5.7	0.012	0.11	0.0021	0.032	0.0032	0.0036	0.17
Weld	0.036	0.67	1.45	0.024	0.0030	23.4	3.2	9.1	0.028	0.089	0.014	0.029	0.0041	0.0048	ND

$$Cr_{eq} = \%Cr + \%Mo + 0.7\%Nb \tag{3}$$

$$Ni_{eq} = \%Ni + 35\%C + 20\%N + 0.25\%Cu \tag{4}$$

Fig. 3a presents the macro morphology of the weld, where three layers representing backing welding (region 1), filling welding (region 2), and capping welding (region 3) can be seen evidently. Fig. 3b shows the OM microstructures of the BM in which the α/γ phase ratio quantification indicates 55.6% α phase (dark) and 44.4% γ phase (white). The microstructures of the WM and the HAZ of the weld are shown in Fig. 3c-3h. For the WM (in Fig. 3c, Fig. 3e, and Fig. 3g), the microstructure is featured with irregular strip shape, where the strip, block and feather austenite phases are placed on the ferrite matrix. In some regions, segregation is significant. For the HAZ (in Fig. 3d, Fig. 3f, and Fig. 3h), the microstructure is featured with needle like ferrites embedded in large austenite matrix, and the content of ferrite phase is no more than 30%, far from equilibrium phase content (~50%). Generally, the time for passing through the two phases during the welding cooling process is very short, so that the austenite is too late to form, which makes it difficult to obtain the balance ratio of austenite in the weld. DSSs at high temperature are composed of 100% ferrite phase. If the line energy is too small, the heat affected zone cooling rate is fast, so that the austenite is too late to precipitation, excess ferrite will keep down at room temperature. If the line energy is too large, the cooling speed is too slow so that ferrite can fully transform to austenite, together with the formation of large-size ferrite grains in HAZ and the precipitation of intermetallic compound phase like σ phase. When the ferrite content in the BM, WM, and HAZ of the DSS weld is out of control in the range of 30% ~ 65%, the corrosion resistance will be greatly reduced.

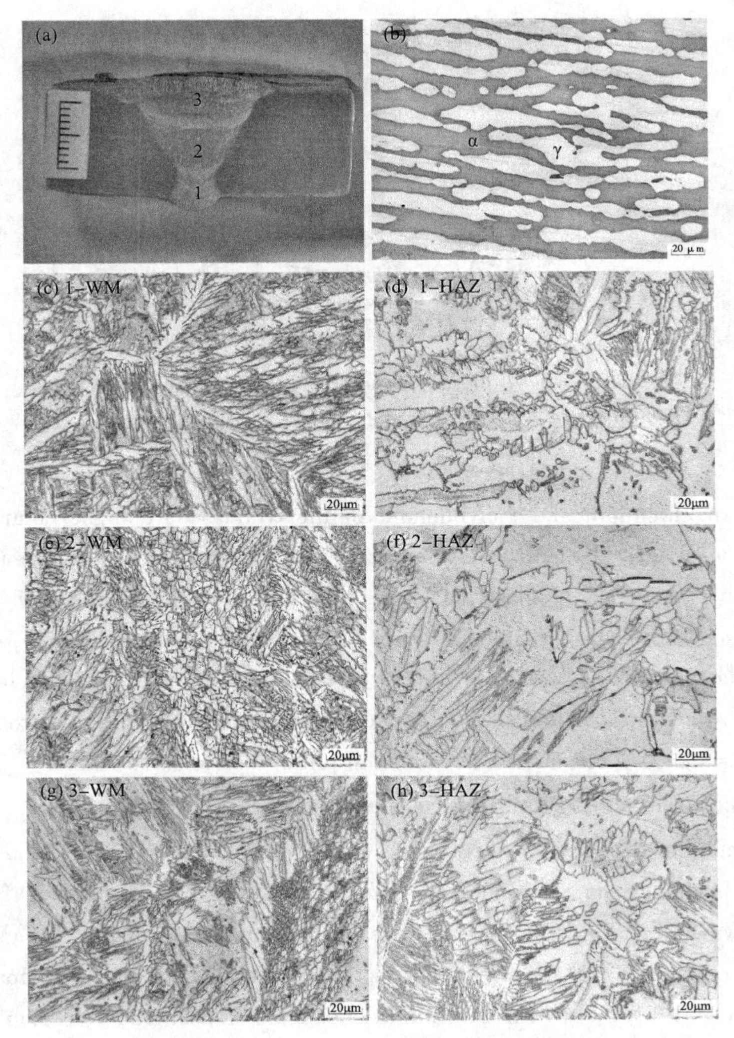

Fig. 3 (a) Macrostructure of UNS S32205 DSS welded joints and microstructures of (b) base alloy, (c)-(d) backing welding, (e)-(f) filling welding, and (g)-(h) capping welding

Fig. 4 presents the content of main alloying elements in the different phases in the BM, WM, and HAZ determined by EDS, where the pitting resistance equivalent numbers (PRENs) of the α and γ phases ($PREN_{\alpha}$ and $PREN_{\gamma}$, respectively) are also shown. The PREN values were calculated based on the chemical analysis and following formula (Eq. (5))[21,22], where percentage mass fraction of the total composition for each element was used for calculation. The nitrogen content of the phases could not be obtained due to the detection limit of EDS equipment, thus, the data of nitrogen content for the calculation of PREN was taken from published literatures[23]. Considering the fact that the nitrogen concentration in the γ phase increases with increase in the α phase fraction, 0.364% nitrogen was considered in the γ phase in the BM and WM, and 0.261% nitrogen was considered in the γ phase in the HAZ. For the α phase, the nitrogen content was taken as 0.06%.

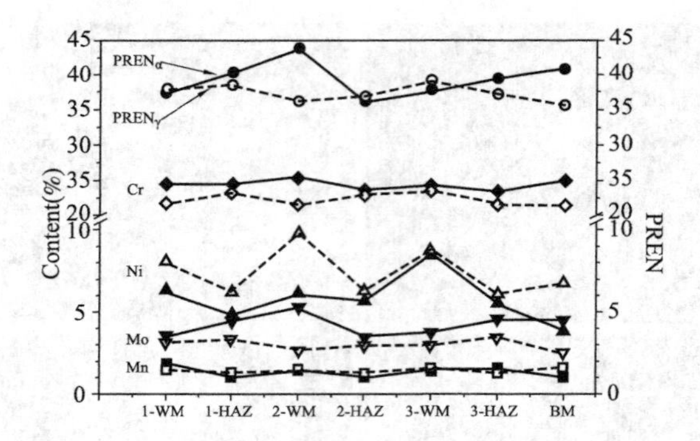

Fig. 4 Contents of alloying elements and PREN values for the α phase (solid line) and γ phase (dash line)

$$PREN = \%Cr + 3.3\%Mo + 16\%N \tag{5}$$

The contents of chromium and molybdenum in the α phase are higher than those in the γ phase, because these two elements are stabilizers for the α phase. The contents of nickel, manganese, and nitrogen in the γ phase are higher than those in the α phase, since these elements are the stabilizers for the γ phase. It is found from Fig. 4 that the value of PREN for the α phase is generally higher than that of the γ phase, indicating the higher pitting resistance for the former[24]. For the weld, both nitrogen and nickel could increase austenite content, but nitrogen increases the pitting corrosion resistance while nickel degrades the pitting corrosion resistance[25].

3.2 Corrosion morphology and phase identification

Fig. 5a presents the cross section view of the corrosion pit in the UNS S32205 DSS pipeline at low magnification. In the Region 1 as indicated in Fig. 5a, there is a corrosion film with uneven thickness, as shown in Fig. 5b. EDS analysis reveals that this film is mainly composed of carbon, oxygen, iron and chromium. On the inner wall of the pipeline, there is a corrosion film with a thickness of ~ 25 μm (Fig. 5c), with the similar chemical compositions to Region 1. In the Region 2 as indicated in Fig. 5a, lamellar structure composed of corrosion products (dark phase) and metal (white phase) can be seen in Fig. 5d. For the dark phase, 42.78% carbon, 28.63% oxygen, 5.62% chromium, 19.20% iron, and 1.31% nickel (in weight percent) was detected by EDS. For the white phase, 19.53% chromium, 6.02% nickel, 2.31% molybdenum, 1.74% manganese, and iron as the balance (in weight percent) was detected.

On the inner wall of the UNS S32205 DSS pipeline, three kinds of corrosion products and scaling deposits were collected and identified by XRD and EDS. Fig. 6 shows the XRD spectra for the different collected solids. For the dark brown solids in the corrosion pit, they were identified as ferrous silicate (Fe_2SiO_4), sodium chloride and iron-nickel. These corrosion products are not mainly the direct corrosion products but the deposits of the surrounding soil during the later treatment process. However, the presence of iron-nickel may be related to the exfoliation of the lamellar structure as observed in Fig. 5d. For the soil red scaling deposits on the edge of the corrosion pit and the gray scaling deposits far away from the corrosion pit, they were identified as strontium/barium sulfate, calcium carbonate, and sodium chloride. The formation of strontium/barium sulfate and

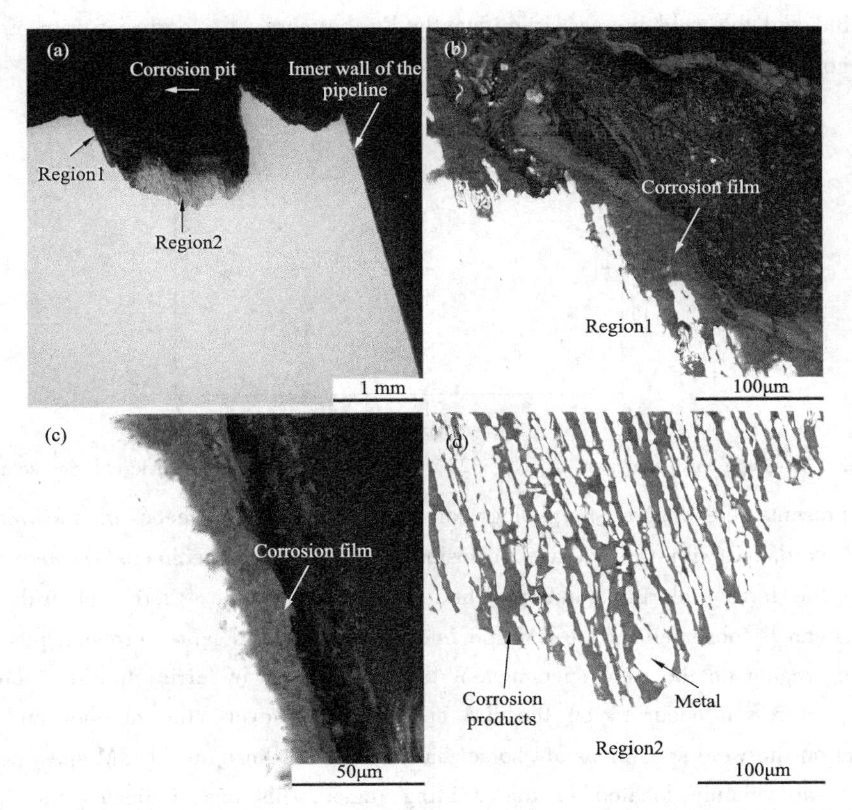

Fig. 5 Cross-section morphologies of the corrosion pit in the UNS S32205 DSS weld: (a) the low magnification view, (b) the microstructure of Region 1, (c) the microstructure of the inner wall, and (d) the microstructure of Region 2

calcium carbonate is ascribed to the high scaling sensitivity and high content of salts of the formation water.

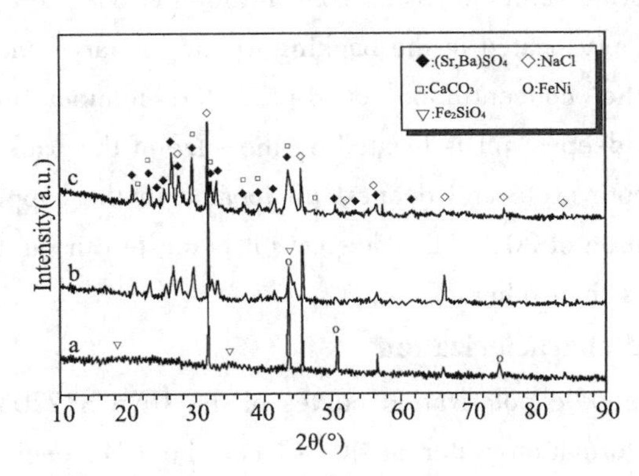

Fig. 6 XRD spectra for the different solids collected from the inner wall of the UNS S32205 DSS pipeline: (a) the dark brown solids in the corrosion pit, (b) the soil red scaling deposits on the edge of the corrosion pit, (c) the grey scaling deposits far away from the corrosion pit

3.3 Pitting corrosion performance

Fig. 7 shows the average corrosion rates of UNS S32205 DSS BM and its weld in the ferric chloride solution at 40 °C. With concentration of ferric chloride increasing from 6% to 20%, the

CRs of the BM and the weld increase significantly. Furthermore, the corrosion rate of the BM at a given concentration of ferric chloride is much lower than that of the weld.

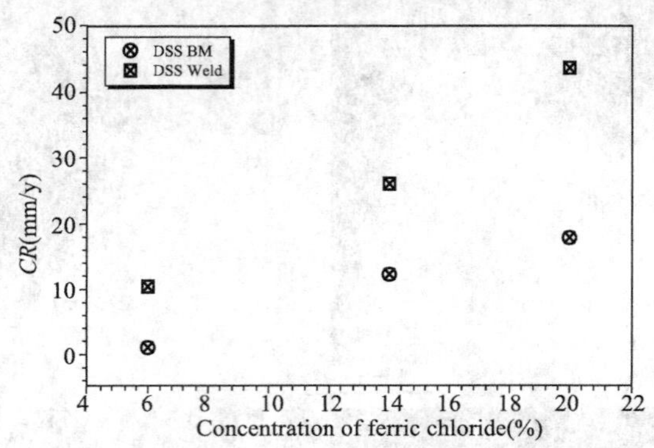

Fig. 7　Average corrosion rates of UNS S32205 DSS and its weld in the ferric chloride solution

Fig. 8 presents the macro morphology of the corroded specimens in the ferric chloride solution. In accordance with the average corrosion rate, the BM specimens showed better pitting resistance to the ferric chloride solution. When the concentration of ferric chloride is 6%, no corrosion pits can be observed on the BM specimen, while several large corrosion pits can be seen in the welding region on the weld specimen. With concentration of ferric chloride increasing up to 20%, pitting corrosion occurred on the BM specimens. However, the number and size of the corrosion pits on the weld specimens are somewhat larger than those on the BM specimens, and the corrosion pits are mainly located in the welding region. This also indicates the better pitting resistance of BM region than the welding region. At high concentration of ferric chloride, severe damage can be seen on the edges of the weld specimens.

The depth of corrosion pits on the UNS S32205 DSS weld specimens were measured by a smart digital microscope and then the pitting rates were calculated. Fig. 9 shows the pitting rates of the weld specimens in the ferric chloride solution. At the concentration of 6%, the number of corrosion pits is small, but the depth of the pit located in the backing welding is large enough so that the pitting rate reaches 314.3mm/y. At the concentration of 14%, the number of corrosion pits increase significantly. However, the deepest pit is located on the edge of the backing welding region, and is difficult to measure. The depth of second deepest pit located in the capping welding region reaches 373.7μm. At the concentration of 20%, the deepest pit is located in the filling welding region, and it pitting rate is much larger than others.

3.4 Electrochemical characterization

Fig. 10 illustrates the anodic polarization curves of the UNS S32205 DSS BM and its weld in carbon-dioxide-saturated formation water at 40 °C. For the BM specimen, a typical passivation phenomenon with a wide potential range can be seen. The passive potential is −0.427 V, the passive current density is 2.1 $\mu A/cm^2$, the breakdown potential is 1.07 V, and the corrosion current density is 2.4 $\mu A/cm^2$. It is evident that the anodic current density started to increase sharply after the potential of 1.07 V, where pitting corrosion occurred. For the weld specimen, no obvious passivation can be seen. The potential corresponding to the anodic current density 100 $\mu A/cm^2$ is 0.22 V. There is a crossing point between the anodic polarization curves of the UNS S32205

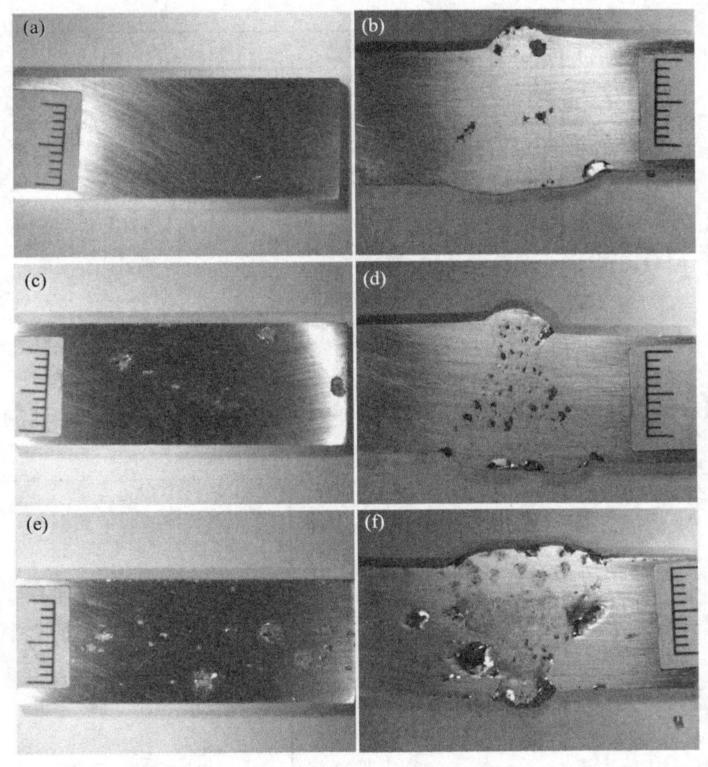

Fig. 8 Macro morphology of the corroded UNS S32205 DSS specimens in solutions with different concentrations of ferric chloride: (a and b) 6%, (c and d) 14%, and (e and f) 20%

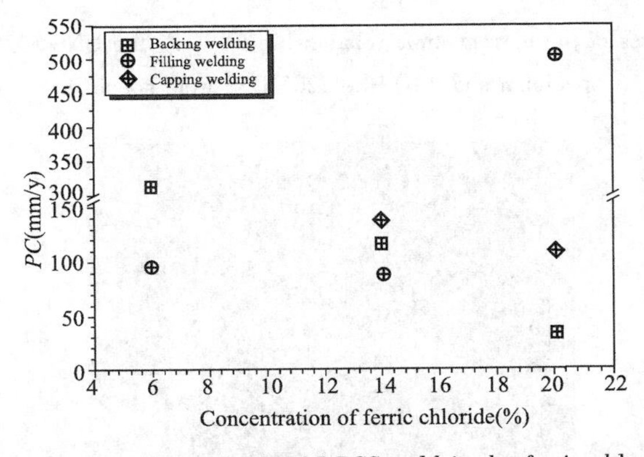

Fig. 9 Pitting rates of the UNS S32205 DSS weld in the ferric chloride solution

DSS BM and its weld. When the potential exceeds the value at the crossing point −68.8 mV, the anodic current density of the weld specimen exceeds that of the BM specimen, and increases sharply.

Fig. 11 presents the polarization curves of the UNS S32205 DSS BM and its weld at a constant potential 0.46 V (versus OCP) in carbon-dioxide-saturated formation water. For the BM specimen, the current density maintains in a narrow range of 1.347~1.349 mA/cm^2. For the weld specimen, the current density increases significantly with time, and increases sharply after 2800 s, suggesting the occurrence of pitting corrosion (Fig. 12).

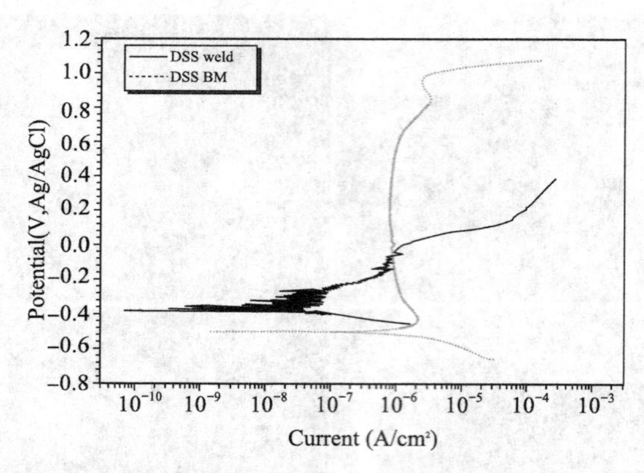

Fig. 10 Polarization curves of the UNS S32205 DSS and its weld in CO_2-saturated formation water

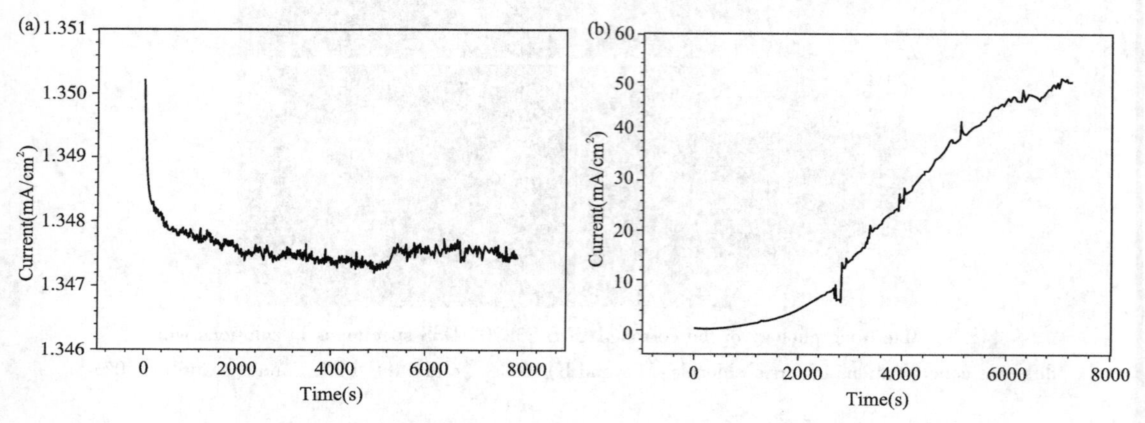

Fig. 11 Curves of the current-time relationship for (a) the UNS S32205 DSS BM specimen and (b) the 2205 DSS weld specimen

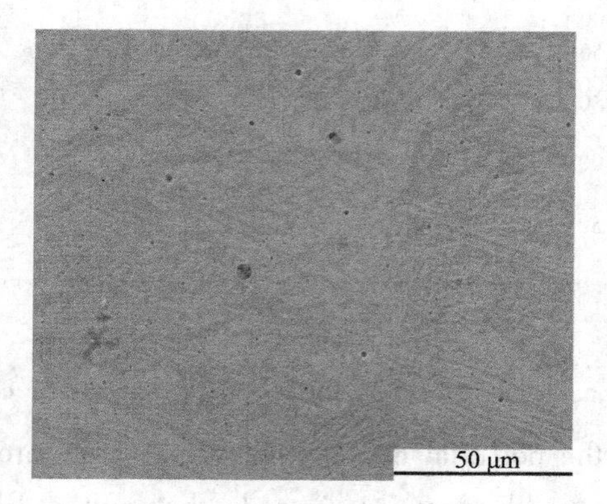

Fig. 12 Corrosion pits on the UNS S32205 DSS weld specimen after polarization at 0.4 V (vs OCP) for 600 s

The corrosive attack starts in the weakest phase of the DSS and preferential corrosion of one of the phases may happen. In order to identify the preferential initiation sites of pitting, potentiostatic measurements were performed in the test solutions under a potential above the breakdown potential. It was found that pitting was initiated at the austenite side of the ferrite-austenite boundary as shown in Fig. 13, in accordance to the observation in Ref. [24].

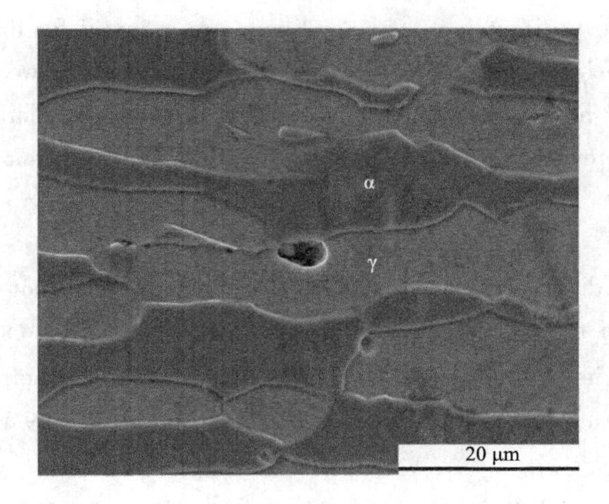

Fig. 13 Metastable pit image of UNS S32205 DSS after potentiostatic measurements

4 Discussion

Fig. 14 illustrates the schematic diagram for the corrosion-perforated UNS S32205 DSS pipeline. Considering the insulation joint as a dividing point, the left section is under cathodic protection with a potential of −1.1 V, while the right section is under the natural state with a potential of −0.5 V. During the service period, the output voltage value of the constant potential instrument increased with time, indicating the leakage of the protection current due to the weak insulation effectiveness. From this perspective, the partial current leakage of the insulation joint may be the direct cause of the serious inner corrosion of the pipeline in the non-protected end.

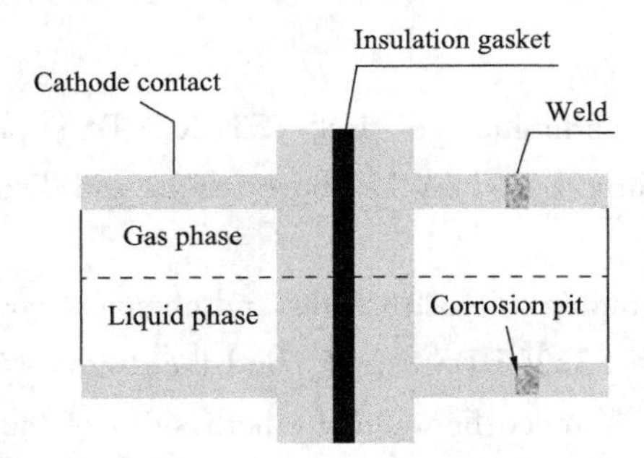

Fig. 14 Schematic diagram for the corrosion-perforated 2205 DSS pipeline

The corrosion mechanism can be described as follows. A large potential difference between the both sides of the insulation joint would produce a localized electric field with a direction from the unprotected side to the cathodically protected side. If the internal transportation medium in the pipeline possessing high conductivity is accumulated at the bottom of the insulating joint, local anodic current, which represents the rate at which the metal loses its electrons and becomes a cation into the water, would be generated. FEM analysis by Chen et. al.[26] indicated that the leakage current density at the edge of the internal coating of the insulation joint increases with the potential difference between the both sides of the insulation joint, the conductivity of the transmission fluid and the internal coating defects.

According to the above discussion, two conditions are required for the inner corrosion of the non protected side of the insulating joint: large potential difference between the both sides of the insulation joint, and the high salinity water in the bottom of the insulation joint. In the present study, the potential difference between the both sides of the insulation joint was calculated by Eq. (6),

$$E_d = E_N - E_P \tag{6}$$

where E_d is the potential difference in volts, E_N is the potential of the non protected side in volts, and E_P is the potential of the cathodically protected side in volts. The conductivity of transmission fluid is determined by its salinity, and increases with salinity generally[25]. Concerning the perforated pipe in the present study, E_d varies in the range of 0.46~0.59 V, which is high enough according to the literature[26].

Under the lowest potential difference, potentiostatic results (Fig. 11) reveal the poorer pitting corrosion resistance of the DSS weld compared to that of the DSS BM, which is also indicated by pitting corrosion resistance measurements in ferric chloride solution (Fig. 7-9). Furthermore, corrosion pits are mainly distributed at inclusions or at the ferrite-austenite boundary, because of the higher energy, the more active property on these sites.

Based on above discussion, once current leakage takes place, it would be preferred to flow out through the HAZ in the weld, enters the transmission fluid. Under the high potential difference and high salinity water of the pipeline, leakage current of the insulation joint would be preferred to flow out through the weld and enter the transmission fluid, so that localized corrosion happens at the inner surface of the weld and leads to the corrosion perforation eventually.

5 Conclusions

The causes of corrosion perforation of UNS S32205 DSS pipeline in gas transportation applications was analyzed carefully. Based on the above results and discussion, following conclusions can be made.

(1) The uneven metallographic microstructures and chemical compositions in the weld weaken the passivation behavior of UNS S32205 DSS weld, and then reduce its pitting resistance.

(2) The large potential difference between the both sides of the insulation joint and the high salinity of the transmission fluid in the pipeline make the leakage current preferably flow out from the weld and enter the transmission fluid, so that localized corrosion happens at the inner surface of the weld and leads to the corrosion perforation eventually.

Acknowledgement

The authors are grateful for the financial supports from the National Nature Science Foundation of China under Contracts of 21506256, 51301202 and 51201131, and the Special Research Project sponsored by the Education Department of Shaanxi Provincial Government under grant No. 15JK1489.

References

[1] P. Bai, H. Zhao, S. Zheng, C. Chen, *Corros. Sci.* 2015, 93, 109.

[2] L. Sow, J. Idrac, P. Mora, E. Font, *Mater. Corros.* 2015, 66, 1245.
[3] W. Liu, S. -L. Lu, Y. Zhang, Z. -C. Fang, X. -M. Wang, M. -X. Lu, *Mater. Corros.* 2015, 66, 1232.
[4] K. W. Chan, S. C. Tjong, *Materials* 2014, 7, 5268.
[5] K. H. Lo, C. H. Shek, J. K. L. Lai, *Mater. Sci. Eng. R.* 2009, 65, 39.
[6] Z. Zhang, Z. Wang, Y. Jiang, H. Tan, D. Han, Y. Guo, J. Li, *Corros. Sci.* 2012, 62, 42.
[7] C. Ornek, D. L. Engelberg, *Corros. Sci.* 2015, 99, 164.
[8] S. T. Kim, S. H. Jang, I. S. Lee, Y. S. Park, *Corros. Sci.* 2011, 53, 1939.
[9] H, Tan, Y. Jiang, B. Deng, J, Xu, J, Li, *Mater. Charact.* 2009, 60, 1049.
[10] D. H. Kang, H. W. Lee, *Corros. Sci.* 2013, 74, 396.
[11] J. Xiong, M. Y. Tan, M. Forsyth, *Desalination* 2013, 327, 39.
[12] G. R. M. Arturo, L. M. V. Hugo, G. H. Rafael, B. B. Egberto, G. S. J. Antonio, *Procedia Mater. Sci.* 2015, 8, 950.
[13] D. Wang, D. R. Ni, B. L. Xiao, Z. Y. Ma, W. Wang, K. Yang, *Mater. Design* 2014, 64, 355.
[14] H. Sieurin, R. Sandstraum, *Mater. Sci. Eng. A.* 2007, 444, 271.
[15] P. G. Normando, E. P. Moura, J. A. Souza, S. S. Tavares, L. R. Padovese, *Mater. Sci. Eng. A.* 2010, 527, 2886.
[16] B. Deng, Z. Wang, Y. Jiang, T. Sun, J. Xu, J. Li, *Corros. Sci.* 2009, 51, 2969.
[17] Y. Jiang, H. Tan, Z. Wang, J. Hong, L. Jiang, J. Li, *Corros. Sci.* 2013, 70, 252.
[18] B. Peng, B. Chen, J. Chen, P. Yao, J. Du, *Corros. Prot.* 2015, 36, 1202.
[19] S. Tang, Q. Cai, X. He, *Chem. Eng. Oil Gas* 2008, 37, 156.
[20] H. Tan, Z. Wang, Y. Jiang, Y. Yang, B. Deng, H. Song, J. Li, *Corros. Sci.* 2012, 55, 368.
[21] L. F. Carfias-Mesias, J. M. Sykes, C. D. S. Tuck, *Corros. Sci.* 1996, 38, 1319.
[22] M. Gholami, M. Hoseinpoor, M. H. Moayed, *Corros. Sci.* 2015, 94, 156.
[23] H-Y Ha, M-Ho Jang, T-H Lee, J Moon, *Corros. Sci.* 2014, 89, 154.
[24] H. Luo, C. F. Dong, X. G. Li, K. Xiao, *Electrochim. Acta* 2012, 64, 211.
[25] T. Ogawa, T. Koseki, *Weld. J.* 1989, 68, 181.
[26] L. Chen, H. Dong, C. Chen, S. Zheng, *Corros. Sci. Prot. Technol.* 2010, 22, 452.

Analysis of Corrosion Perforation of a Gas Gathering Carbon Steel Pipeline in a Western Oilfield in China

Yuan Juntao[1,a] Zhang Huihui[2] Tong Ke[1] Li Fagen[1] Long Yan[1] Li Lei[1]

(1. CNPC Tubular Goods Research Institute; 2. College of Materials Science and Engineering, Xi'an University of Science and Technology)

Abstract: In the west of China, corrosion perforation and scaling of pipelines occur frequently in the ground gathering system in oil and gas field, and cause serious safety and environmental problems. In the present work, the corrosion perforation of a carbon steel pipeline with a grade of 20 in a western oil field in China was investigated. Optical metallographic microscopy, scanning electron microscopy (SEM), energy dispersive spectroscopy (EDS), and X-ray diffraction were used to determine the most probable causes of the corrosion perforation. The results showed that the internal corrosion of the studied carbon steel pipe was ascribed to the corrosive CO_2 gas and the high content of formation water in the internal transporting medium. The formation of loose and porous iron carbonate layer and the scaling layer with poor adhesion could not protect the carbon steel effectively, so that the high concentration of chloride ions in the formation water caused the initiation of localized corrosion possibly by changing the solubility of iron carbonate, and then the localized corrosion pit propagated and eventually caused the perforation.

Keywords: Corrosion perforation; Scaling; Failure analysis; Pipeline; Carbon dioxide

1 Introduction

With the continuously growing demand in oil and gas energy globally, the search for new sources of oil and gas makes the operation condition became more and more severe[1]. In west China, with the increase of dosage of chemical in the crude oil exploitation process, pH value of the produced oil decreases and salinity increases, therefore, the corrosion rate of oil pipeline accelerates sharply, and the corrosion mechanism also becomes much more complicated[2]. Corrosion perforation and scaling occur frequently in surface gathering pipeline of western oil field in China in recent years, and this problem seriously affects the safe production of oil field.

Recently, corrosion perforation occurred in a carbon steel pipeline with a grade of 20 which

[a] Corresponding author: yuanjuntaolly@163.com

was utilized in the ground gathering system in a western oil field in China, and caused safety accident. In the present work, we analyzed this perforated carbon steel pipe carefully by macro/micro investigation, chemical analysis, and phase identification; and then proposed the failure causes for the corrosion perforation.

2 Experimental

The failed carbon steel pipe under investigation was subjected to the following procedures to assess corrosion attack.

2.1 Visual examination of inner and outer surfaces

The inner and outer surface of the perforated pipe was examined visually, the amount of the obvious corrosion pits was investigated, and the typical macro morphology was taken using the camera (NIKON L100). Then the sample pieces comprising the perforated carbon steel pipe were cut from the corresponding parts based on the requirements of the different tests.

2.2 Composition, metallographic structure tests

The chemical compositions of perforated carbon steel pipe were analyzed by a direct reading spectrometer (ARL 4460) on the basis of GB/T 4336—2002. The metallographic structure and inclusions were analyzed by optical metallographic microscopy (MEF4M) based on GB/T 13298—1991, GB/T 10561—2005 and GB/T 6394—2002.

2.3 SEM examination of the corroded surfaces and scaling

Morphologies of the corrosion films were investigated using a Scanning Electron Microscopy (SEM, Philips XL-30), and chemical compositions of the surface films were analyzed by Energy Dispersion Spectrum (EDS, INCA). Phases of the corrosion products were identified using an X-ray Diffractometer (XRD, D8 ADVANCE) with filtered Cu kα radiation.

3 Results and discussion

3.1 Visual inspection

Fig. 1 shows the outer appearance of the perforated carbon steel pipe. On the outside surface of the pipe, the red anticorrosive paint and white insulation layer can be observed. In addition, a corrosion pore with a diameter of ~8mm can be seen in Fig. 1. Near the corrosion pore, the red anticorrosive paint dropped locally and slight corrosion can be seen.

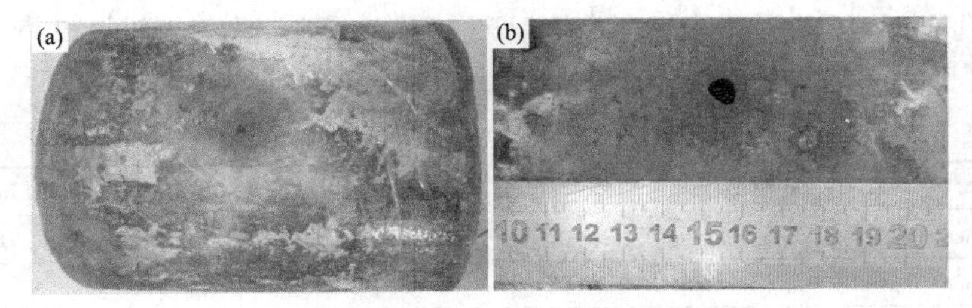

Fig. 1 The outside appearance of the corroded carbon steel pipe

Fig. 2 presents the internal appearance of the perforated carbon steel pipe. There are several characteristic features. First, a wide-range area of corrosion thinning can be seen around the corrosion pore. The largest size of the corrosion pits is 120mm×40mm. Second, significant exfoliation of corrosion products can be seen. Fig. 3 shows the appearance of the spalled corrosion layer.

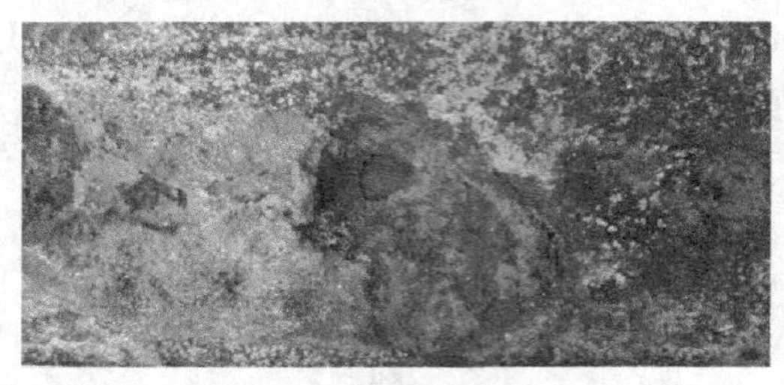

Fig. 2 The internal appearance of the corroded carbon steel pipe

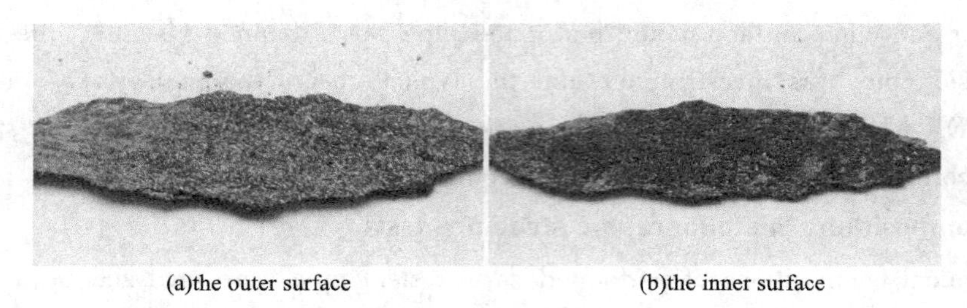

(a)the outer surface (b)the inner surface

Fig. 3 Macro appearances of the corrosion layer

3.2 Composition test of carbon steel pipe

The chemical compositions of the carbon steel pipe are shown in Table 1, which is compared with the parameter requirements of GB/T 699—1999 standard. The results showed that the contents of C, Si, Mn, P, S, Cr, Ni, and Cu all meet the technical requirements of GB/T 699—1999.

Table 1 Chemical compositions of the corroded carbon steel pipe (wt %)

Element	C	Si	Mn	P	S	Cr	Ni	Cu	Fe
Carbon steel pipe	0.20	0.27	0.49	0.0088	0.0081	0.020	0.018	0.043	Bal.
GB/T 699—1999	0.17~0.23	0.17~0.37	0.35~0.65	≤0.035	≤0.035	≤0.25	≤0.30	≤0.25	Bal.

3.3 Light optical microscopy

Table 2 lists the results of metallographic microstructure, grain size and non metal inclusions of the perforated carbon steel pipe. It indicates that the metallographic structure of the carbon steel pipe is ferrite (shown in Fig. 4a), the grain grade is 6.0, and the inclusions are A0.5, B0.5, D1.5, and D0.5e. For the specimen containing the corrosion perforation, its metallographic structure is ferrite with no significant deformation, while some corrosion products is visible in the corrosion pore as shown in Fig. 4b.

Table 2 Results of metallographic analysis

Specimen	Metallographic structure	Grain grade	Non metal inclusions
NF[a] specimen	Ferrite	6.0	A0.5, B0.5, D1.5, D0.5e
CP[b] specimen	Ferrite	ND[c]	ND[c]

a. NF represents the Non Failed specimen.

b. CF represents the specimen containing corrosion pore.

c. ND represents the issue are not detected in the present work.

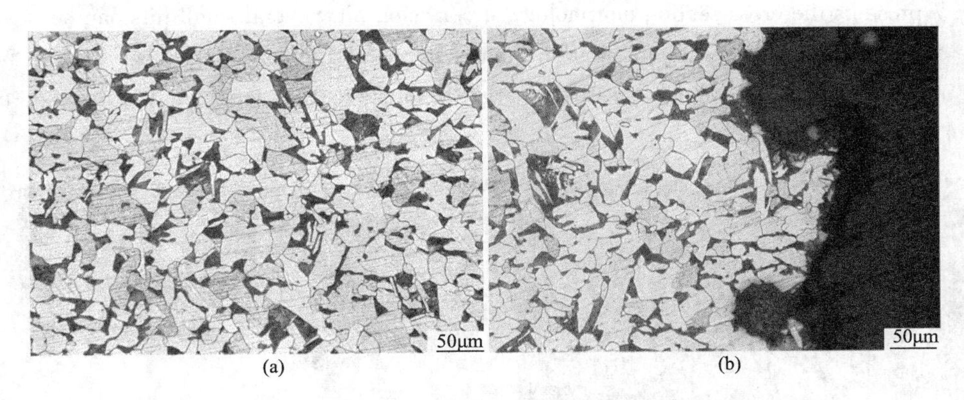

Fig. 4 Microstructures of the (a) NF specimen and (b) CP specimen

3.4 Scanning electron microscopy

Three coupons were cut from the perforated carbon steel pipe and were used to investigate the microstructure by SEM and analyze the chemical compositions by EDS. No. 1 specimen was cut from the region near to the corrosion pore. No. 2 specimen was cut from the region far away from the corrosion pore. No. 3 specimen was cut from the region nearby the corrosion pore and prepared as cross-section specimen.

Fig. 5 presents the surface morphologies of the No. 1 specimen. On the bottom of the corrosion pit, several small pits make the surface rough enough (Fig. 5a), and EDS analysis indicates the high content of C, O, Fe, as well as some content of S and Cl. In the region outside of the corrosion pit, evident exfoliation can be seen in Fig. 5b. EDS analysis indicates that the chemical compositions are similar to those in the bottom of the corrosion pit.

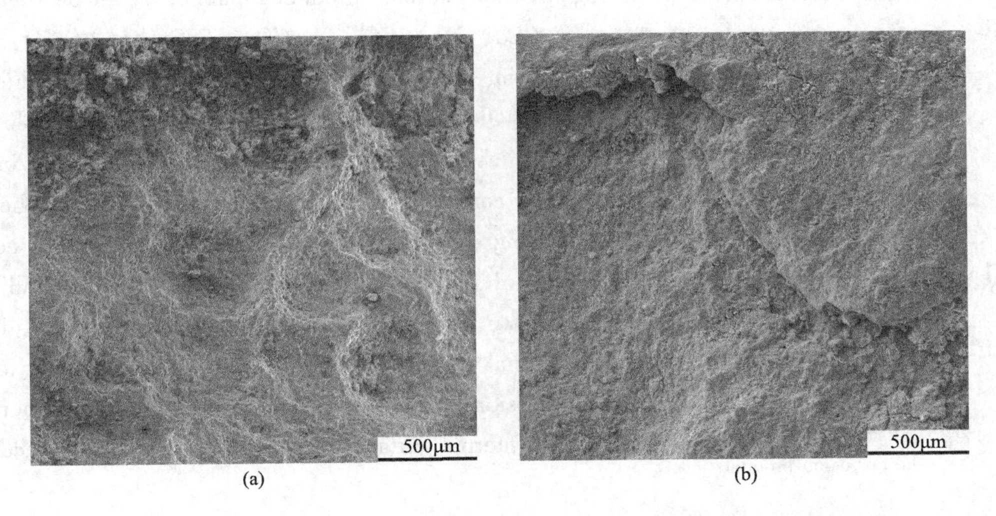

Fig. 5 Surface morphologies of the No. 1 specimen in the region of
(a) the bottom of the corrosion pit and (b) the plat area near to the corrosion pit

Fig. 6 shows the surface morphologies of the No. 2 specimen. Far away from the corrosion pit, the corrosion products seem dense, however, exfoliation is still evident. EDS analysis indicates that the similar chemical compositions as above.

Fig. 7 presents the cross section morphology of corrosion pit. Several small pits can be observed, which is in accordance with the observation from surface morphology as shown in Fig. 5a. EDS analysis implies that the chemical compositions are similar to the corrosion products in others, except for the small content of Ca.

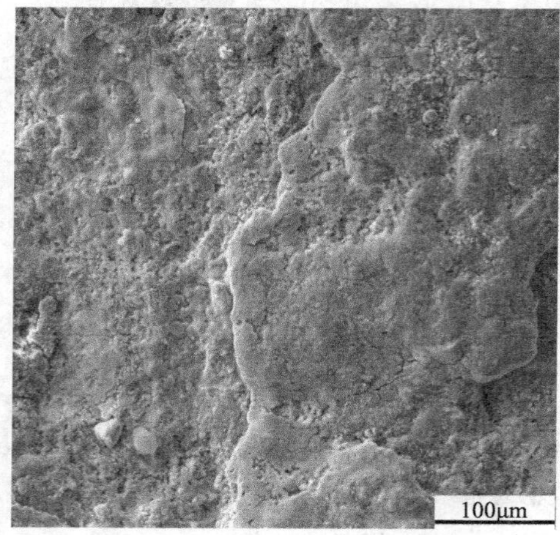

Fig. 6 Surface morphology of the No. 2 specimen

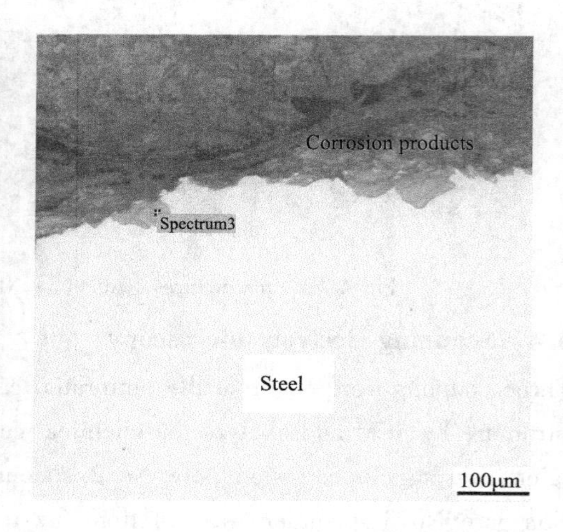

Fig. 7 Cross section morphology of the corrosion pit (No. 3 specimen)

3.5 Phase identification of corrosion products

Fig. 8 shows the XRD spectrum diagrams of the corrosion products, where the outer corrosion products were made from the exfoliated corrosion layers as shown in Fig. 3 and the internal corrosion products were taken from the region where surface products spalled. It can be seen that the outer corrosion products are composed of (Ca, Mg) CO_3, $BaCO_3$, $FeCO_3$ and $SrSO_4$, and the internal corrosion products consists of $FeCO_3$, FeO(OH) and Fe_2O_3. The formation of insoluble salts such as (Ca, Mg) CO_3, $BaCO_3$ and $SrSO_4$ in the outer corrosion products is generally ascribed to the high scaling sensitivity of the formation water which containing high concentrations of Ca^{2+}, Mg^{2+} and Sr^{2+} (shown in Table 3). The formation of iron oxides in the internal corrosion products is usually related to the corrosion during the later transporting process where the wet internal surface was exposed to air and formed iron oxides.

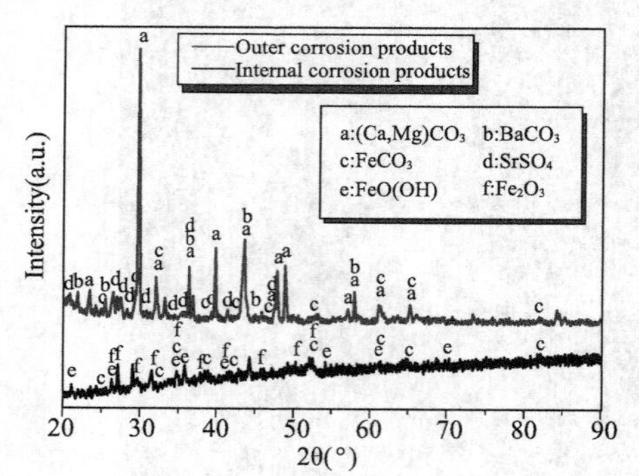

Fig. 8 XRD spectrum diagrams of the corrosion products

Table 3 Chemical compositions of the formation water

Component	Cl^-	HCO_3^-	SO_4^{2-}	K^+	Na^+	Ca^{2+}	Mg^{2+}	Sr^{2+}	Ba^{2+}
Content (g/L)	111	0.03	0.27	1.84	56.71	11.31	0.94	0.46	0.007

4 Failure mechanism analysis

During the service period, this pipeline was utilized to transport wet gas. According to the service conditions listed in Table 4, the internal medium possesses high content of water (84.23%) and corrosive acid gas (0.906% CO_2 and 0.013% H_2S). As mentioned above, high contents of Ca^{2+}, Mg^{2+} and Sr^{2+} in the formation water show high trend to form scaling deposit (i.e. $CaCO_3$ and $SrSO_4$).

Table 4 Service conditions of the studied carbon steel pipeline

Parameters	Value	Parameters	Value
Temperature (℃)	49	Flow rate (m/s)	1.4
Pressure (MPa)	1.5	H_2S content in gas (%)	0.013
Water content (%)	84.23	CO_2 content in gas (%)	0.906

From the above discussions, it is possible to infer that the reason of the studied carbon steel pipe suffered the severe local corrosion was mainly the interaction of the associated gas CO_2 and formation water. $FeCO_3$ in the corrosion deposits were formed from CO_2, (Ca, Mg) CO_3 and $SrSO_4$ in the scaling deposit were formed from the formation water.

Research indicates that at temperatures below 60℃, the iron carbonate ($FeCO_3$) product film formed by corrosion is thick, but loose, and not dense, and does not prevent further corrosion. In addition, CO_2 corrosion increases with velocity of flow. The rate of corrosion increases ~68% as the velocity of flow increases from 0.1 to 1.0m/s [3]. From this perspective, the low service temperature and high flow rate in the present work would lead to a rapid formation of loose iron carbonate layer.

In the present work, the service temperature of the studied carbon steel pipeline was 49℃, and the flow velocity was 1.4m/s. During the initial service period, iron carbonate film formed as the CO_2 corrosion products. Because the iron carbonate film formed at temperatures below 60℃ is loose and rough, so that the formation water could go through the loose and porous scale and arrived at the surface of the matrix, and the electrochemical reactions took place subsequently. At the same time, scaling deposits such as (Ca, Mg) CO_3 and $SrSO_4$ deposited on the outer rough surface.

The possible mechanism of the corrosive perforation is as follows.

First, the associated gas CO_2 dissolved in formation water and carbonic acid was formed (Eq. (1)), which corroded the carbon steel. As a weak acid, carbonic acid partially dissociates into hydrogen ions and bicarbonate ions (Eq. (2)), which is followed by the dissociation of bicarbonate ions that form additional hydrogen ions and carbonate ions (Eq. (3)). Hydrogen ions are reduced at the steel surface to form hydrogen gas by Eq. (4). In addition to the reduction of hydrogen ions, it is often assumed that carbonic acid is directly reduced at the metal surface by Eq. (5). Surface iron atoms, give up electrons and dissolve into the aqueous solution to produce ferrous ions by Eq. (6), resulting steel corrosion [4]. When steel corrodes, it releases Fe^{2+} ions which accumulate in the solution. When their concentration reaches and exceeds the saturation level, iron carbonate will precipitate according to Eqs. (7) - (9) [5-6].

$$CO_2+H_2O \rightarrow H_2CO_3 \quad (1)$$

$$2H_2CO_3+2e \rightarrow H_2+2HCO_3^- \quad (2)$$

$$2HCO_3^-+2e \rightarrow H_2+2CO_3^{2-} \quad (3)$$

$$2H^++2e \rightarrow H_2 \quad (4)$$

$$2H_2CO_3+2e \rightarrow H_2+2HCO_3^- \quad (5)$$

$$Fe \rightarrow Fe^{2+}+2e \quad (6)$$

$$Fe^{2+}+CO_3^{2-} \rightarrow FeCO_3 \quad (7)$$

$$Fe^{2+}+2HCO_3^- \rightarrow Fe(HCO_3)_2 \quad (8)$$

$$Fe(HCO_3)_2 \rightarrow FeCO_3+CO_2+H_2O \quad (9)$$

Second, the loose and porous iron carbonate layer made the deposition of scaling deposits on the outer rough surface possible. Scaling deposits may from in following ways (Eqs. (10) - (11)).

$$Ca^{2+}+CO_3^{2-} \rightarrow CaCO_3+CO_2+H_2O \quad (10)$$

$$Sr^{2+}+SO_4^{2-} \rightarrow SrSO_4 \quad (11)$$

Third, formation water could go through the loose and porous iron carbonate layer, arrive at the surface of matrix, and cause further corrosion.

Fourth, the localized corrosion may be related to the failure of iron carbonate layer and/or scaling deposit layer. The scale deposit layer seems be detached from the outer surface layer by layer as mentioned in Fig. 3. In the formation water, the high concentration of chloride ions (~ 111g/L) could cause the initiation of localized corrosion possibly by changing the solubility of iron carbonate[7]. When an initiation process occurred that would partially damage the iron carbonate layer and/or scaling deposits layer to leave a small bare steel area coupled with the larger iron carbonate covered surface area, the localized corrosion would propagate and grow[8]. With the pitting deepening and widening, carbon steel pipe would undergo corrosion perforation eventually.

5 Conclusions

Based on the above results and discussions, the internal corrosion of the studied carbon steel pipe is ascribed to the corrosive CO_2 gas and the high content of formation water in the internal transporting medium. At the relatively low service temperature, loose and porous iron carbonate layer initially formed and the rough surface made the deposition of scale deposits [i. e. $(Ca, Mg)CO_3$ and $SrSO_4$] possible. Corrosive medium went through the loose and porous iron carbonate layer and/or localized scale deposits, arrived at the surface of the metal matrix, and then induced further corrosion. With time elongated, the thick scaling deposit layer was pat to drop locally. The high concentration of chloride ions in the formation water caused the initiation of localized corrosion possibly by changing the solubility of iron carbonate, and then the localized corrosion pit propagated and eventually caused the perforation.

References

[1] A. Q. Fu, Y. R. Feng, R. Cai, J. T. Yuan, C. X. Yin, D. M. Yang, Y. Long, Z. Q. Bai. Eng. Fail. Anal. 66, 566 (2016).

[2] Y. Liu, Y. Zhang, J. Yuan, M. Ye, J. Xu. Eng. Fail. Anal. 34, 25 (2013).

[3] S. Lu, J. Xiang, Y. Kang, Z. Chang, X. Dong, T. Zhai. Mater. Performance 47, 66 (2008).
[4] T. Tran, B. Brown, S. Nesic. *NACE CORROSION CONFERENCE*, paper No. 5671 (2015).
[5] T. Tanupabrungsun, B. Brown, S. Nesic. *NACE CORROSION CONFERENCE*, paper No. 2348 (2013).
[6] J. K. Heuer, J. F. Stubbings. Corros. Sci. 41, 1231 (1999).
[7] X. Gao, B. Brown, S. Nesic. *NACE CORROSION CONFERENCE*, paper No. 3880 (2014).
[8] J. Han, B. N. Brown, S. Nesic. Corrosion 66, 095003-1 (2010).

含硫气田用国产与进口 UNS N08825 合金耐腐蚀性能对比研究

李　科[1]　曹晓燕[1]　李　星[2]　陈勇彬[1]　崔　磊[1]　郑　初[1]

（1. 中国石油集团工程设计有限责任公司西南分公司；2. 中国石油工程建设公司）

摘　要：为确保国产 UNS N08825 合金材料的成功应用，本文采用国际通用的镍基合金耐腐蚀性能评定方法，包括晶间腐蚀试验、点腐蚀试验、模拟环境均匀腐蚀试验及模拟环境应力腐蚀开裂试验等，进一步验证了国产 UNS N08825 合金材料的耐腐蚀性能，并对国产与进口材料进行了综合评比及分析。

关键词：UNS N08825；耐腐蚀；晶间腐蚀；点腐蚀；均匀腐蚀；应力腐蚀开裂

对于“六高”气田（高产量、高温、高压、高含 H_2S、高含 CO_2、高含氯离子）的开发，由于介质的腐蚀性较强，运行风险较高，与高腐蚀性原料气接触的集输管道材质耐腐蚀性能要求较高。目前，对于集输用原料气管道的内腐蚀控制难题，国内外工程已逐步采用耐腐蚀合金材料替代碳钢加注缓蚀剂方案进行内腐蚀控制，从材质本身解决管道的内腐蚀问题，同时也便于后期管理与维护。UNS N08825 合金作为一种具有较强耐腐蚀性能的材料，已广泛用于含硫气田，但国产的 UNS N08825 合金材料应用较少，前期市场基本为国际知名公司所垄断。目前，国内公司也开发了 UNS N08825 合金材料。为确保国产材料的成功应用，本文采用国际通用的镍基合金耐腐蚀性能评定方法，进一步验证了国产 UNS N08825 合金材料的耐腐蚀性能，并对国产与进口材料进行了综合评比及分析。

1　试验材料

试验用材料分别为国产和进口的 ASTM B424 N08825 合金薄板，热处理状态均为固溶退火，其化学成分及力学性能详见表 1 和表 2。两种材料化学成分以及力学性能总体差异较小。

表 1　试验用材料化学成分　　［单位：%（质量分数）］

材料	C	Si	Mn	P	S	Cr	Ni	Cu	Mo	Ti	Al	Fe
ASTM B424 N08825	≤0.05	≤0.5	≤1.0	≤0.02	≤0.02	19.5~23.5	38.0~46.0	38.0~46.0	38.0~46.0	0.6~1.2	≤0.20	余量
国产 UNS N08825	0.01	0.32	0.62	0.015	0.002	22.30	39.20	1.98	2.87	0.70	0.11	30.10
进口 UNS N08825	0.01	0.30	0.40	—	0.001	22.50	39.20	2.0	3.20	0.70	0.10	31.60

作者简介：李科，1984 年出生，硕士，工程师，主要从事酸性油气田材料焊接、腐蚀与防护工作。

表 2　试验用材料力学性能

材料	屈服强度（MPa）	抗拉强度（MPa）	延伸率（%）
ASTM B424 N08825	≥241	≥586	≥30
国产 UNS N08825	312	650	42
进口 UNS N08825	316	620	43

试验用材料的金相组织形貌如图 1 所示。两种材料金相组织均为奥氏体，且未见明显析出相。其中进口 UNS N08825 材料较国产的晶粒度粗。

(a)国产UNS N08825合金，100倍

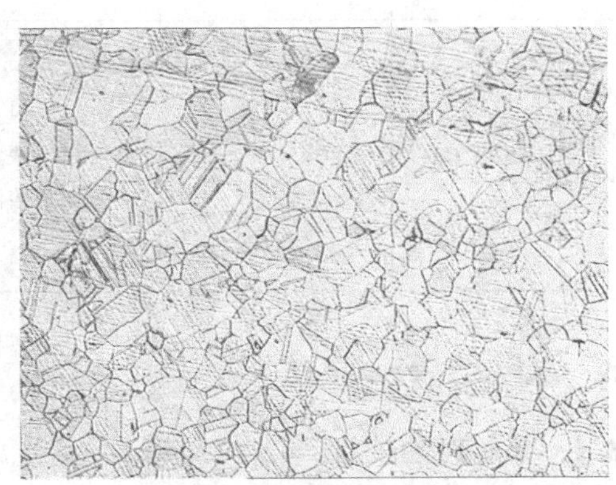

(b)进口UNS N08825合金，100倍

图 1　不同材料金相组织形貌

2　试验方法

结合 UNS N08825 合金材料耐蚀性能特点，采用国际通用的试验方法，分别对两种材料进行晶间腐蚀试验、点腐蚀试验、模拟环境均匀腐蚀试验以及模拟环境应力腐蚀开裂试验，以便验证材料制造工艺可靠性以及材料在模拟苛刻环境中的耐蚀性能差异性。不同的试验方法及要求见表 3。

表 3　UNS N08825 合金不同腐蚀试验方法及要求

试验项目	试验方法或标准	推荐试样尺寸（mm×mm×mm）	试验时间（h）	试验溶液	备注
晶间腐蚀	ASTM G28 方法 A	25×25×3	120	400mL 蒸馏水 + 236mLH_2SO_4(浓度为 95%~98%)+25g$Fe_2(SO_4)_3$(浓度约 75%)	试验温度：沸腾
	ASTM A262 方法 C	30×20×3	240，每个周期 48	65%HNO_3 溶液，共 600mL	试验温度：沸腾
点腐蚀	ASTM G48 方法 A	50×25×3	72	溶液配置：100g$FeCl_3\cdot 6H_2O$+900mL 蒸馏水，溶解后，取 600mL 用于试验	试验温度，22±2℃

续表

试验项目	试验方法或标准	推荐试样尺寸（mm×mm×mm）	试验时间（h）	试验溶液	备注
模拟环境应力腐蚀	NACE TM 0177 及 ISO 7539. 2	80×5×3	720	模拟环境试验溶液（含氯离子浓度 120000mg/L 的脱氧蒸馏水溶液）	试验载荷：100%实测屈服强度； 试验温度：120℃； 试验周期：720h； $p_{总}$：1. 44MPa； p_{H_2S}：0. 6MPa； p_{CO_2}：0. 8MPa
模拟环境均匀腐蚀	ASTM G31	80×5×3	720	模拟环境试验溶液（含氯离子浓度 120000mg/L 的脱氧蒸馏水溶液）	试验温度：120℃； 试验周期：720h； $p_{总}$：1. 44MPa； p_{H_2S}：0. 6MPa； p_{CO_2}：0. 8MPa

3　试验结果

3. 1　晶间腐蚀试验

3. 1. 1　ASTM G28 方法 A

按照 ASTM G28 A 法要求进行晶间腐蚀试验，不同材料试验结果如表 4 所示。

表 4　ASTM G28A 法晶间腐蚀试验结果

材质	试样编号	均匀腐蚀速率（mm/a）	平均值（mm/a）
国产 UNS N08825	1	0. 1925	0. 1887
	2	0. 1844	
	3	0. 1892	
进口 UNS N08825	1-1	0. 0960	0. 1023
	1-2	0. 1032	
	1-3	0. 1076	

由表 4 可见，国产和进口 UNS N08825 合金材质同一材料不同试样之间的均匀腐蚀速率值波动较小，且腐蚀速率值均较低，两种材料试验结果相差不大。

3. 1. 2　ASTM A262 方法 C

根据 ASTM A262 C 法要求，试验分五个分阶段，每个周期 48h，每个试验周期前后对试样进行称重。同时，试验前，试样进行敏化处理，加热至温度 650℃并保温 1h。

不同试验阶段，两种材料的试验结果如表 5 所示。试验完成后试样表面形貌如图 2 所示。

不同材料在不同试验阶段的平均腐蚀速率变化如图 3 所示，两种材料在不同的试验阶段，腐蚀速率变化趋势较为接近，在第二个和第五个试验段腐蚀速率均较高，但进口 UNS N08825 合金材料的腐蚀速率值较国产的低。

表 5　ASTM A262 方法 C 晶间腐蚀试验结果

试验材料	试样编号	第一周期腐蚀速率（mm/a）	第二周期腐蚀速率（mm/a）	第三周期腐蚀速率（mm/a）	第四周期腐蚀速率（mm/a）	第五周期腐蚀速率（mm/a）	平均腐蚀速率（mm/a）
国产 UNS N08825	1	0. 0875	0. 2202	0. 1026	0. 0905	0. 1284	0. 1258
	2	0. 0799	0. 3541	0. 1628	0. 1522	0. 2626	0. 2023
	3	0. 1046	0. 2702	0. 2509	0. 5871	0. 5670	0. 3560
进口 UNS N08825	4	0. 0691	0. 1262	0. 0721	0. 0781	0. 0903	0. 0872
	5	0. 0720	0. 0915	0. 0555	0. 0630	0. 0916	0. 0747
	6	0. 0660	0. 1229	0. 0690	0. 0765	0. 0871	0. 0843

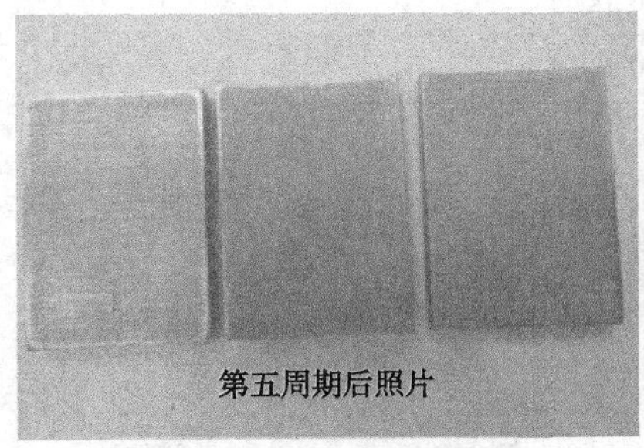

(a)国产UNS N08825试样正面

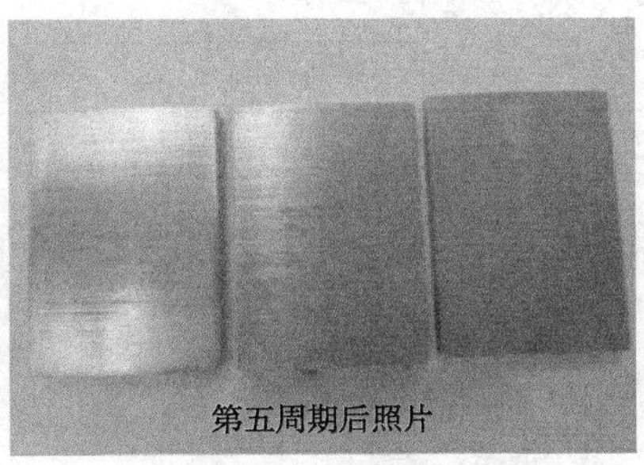

(b)国产UNS N08825试样背面

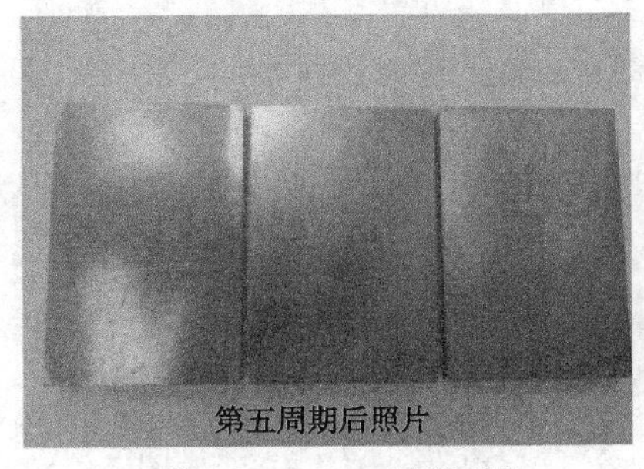

(c)进口UNS N08825试样正面

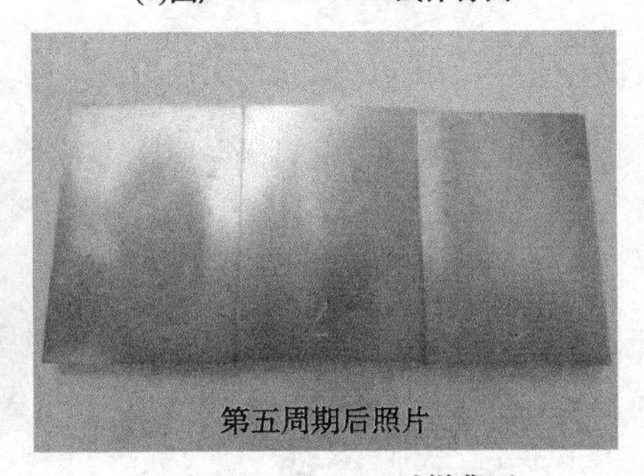

(d)进口UNS N08825试样背面

图 2　ASTM A262 C 法试验后试样宏观形貌

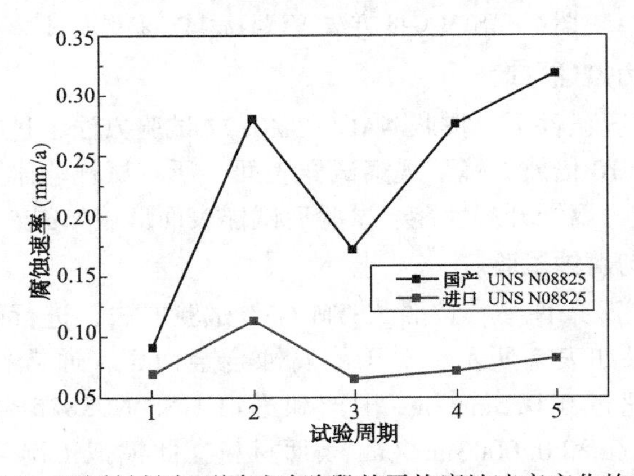

图 3　不同材料在不同试验阶段的平均腐蚀速率变化趋势

3.2 点腐蚀试验

按照 ASTM G48 方法 A 的要求，分别对国产和进口 UNS N08825 合金进行点腐蚀试验，试验完成后试样表面均无点腐蚀痕迹。试样腐蚀速率如表 6 所示，两者平均腐蚀速率值均较低，国产 UNS N08825 合金材料的略高。试验后试样表面形貌如图 4 所示。

表 6　ASTM G48 方法 A 点腐蚀试验结果

试验材料	试样编号	腐蚀速率［$g/(m^2 \cdot h)$］	平均腐蚀速率［$g/(m^2 \cdot h)$］
国产 UNS N08825	1	0.0056	0.0056
	2	0.0084	
	3	0.0028	
进口 UNS N08825	D	0.0004	0.0007
	E	0.0004	
	F	0.0014	

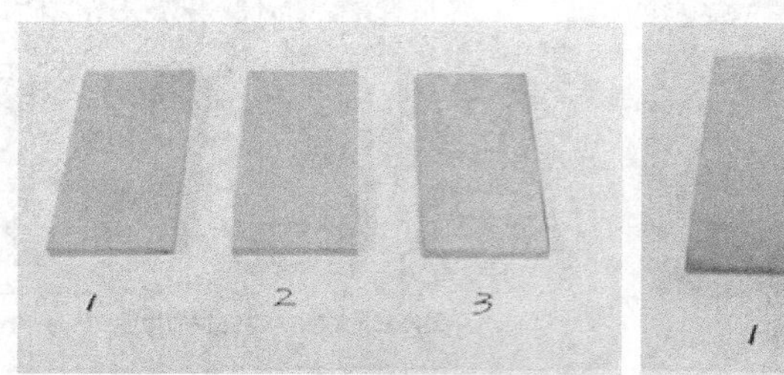

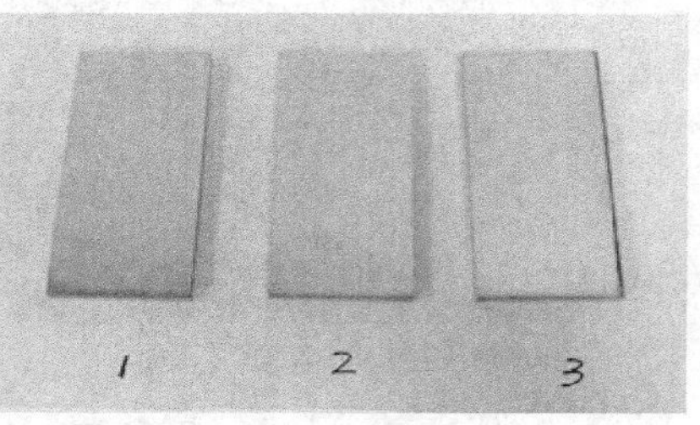

(a)国产UNS N08825合金材料试验后试样正、反面宏观形貌

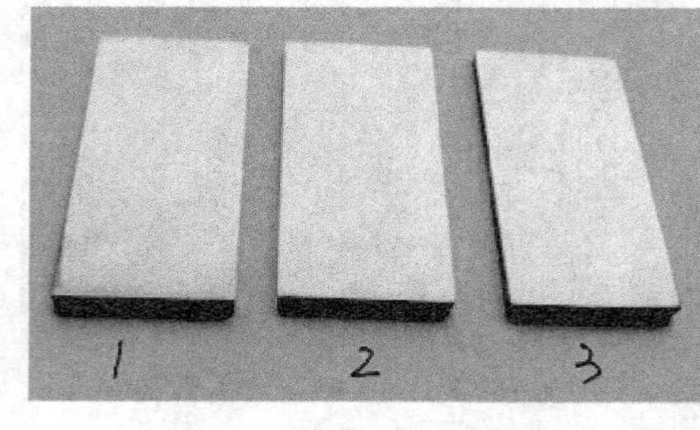

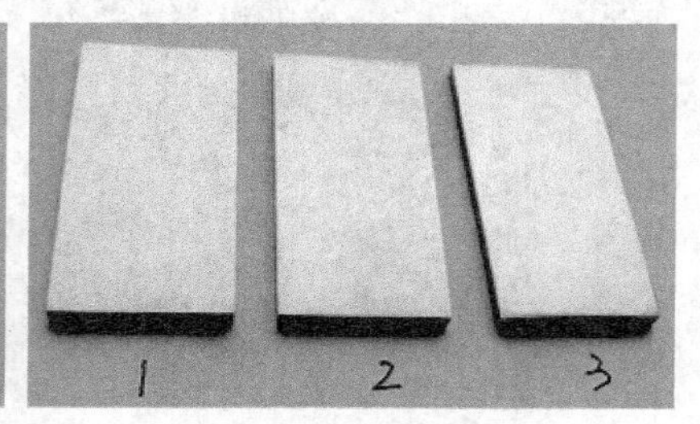

(b)进口UNS N08825合金材料试验后试样正、反面宏观形貌

图 4　ASTM G48 方法 A 试验后试样宏观形貌

3.3 模拟环境应力腐蚀试验

在模拟含硫气田工况条件下，按照 NACE TM0177 试验方法，进行两种材料的应力腐蚀试验。试验结束后，在 10 倍放大镜下观察试样表面，所有试样均未开裂，表明材料在此试验条件下有较好的耐应力腐蚀开裂性能。试验后试样表面形貌如图 5 所示。

3.4 模拟环境均匀腐蚀试验

在模拟含硫气田工况条件下，按照 ASTM G31 试验方法，进行两种材料的模拟环境均匀腐蚀试验。试验结果如表 7 所示，对于耐腐蚀合金而言，通常材料在模拟环境下的耐均匀腐蚀速率要求不超过 0.025mm/a，国产和进口 UNS N08825 合金材料腐蚀速率均较低，分别为 0.0002mm/a 和 0.0003mm/a，表面材料在此模拟工况条件下具有较好的耐均匀腐蚀性能。

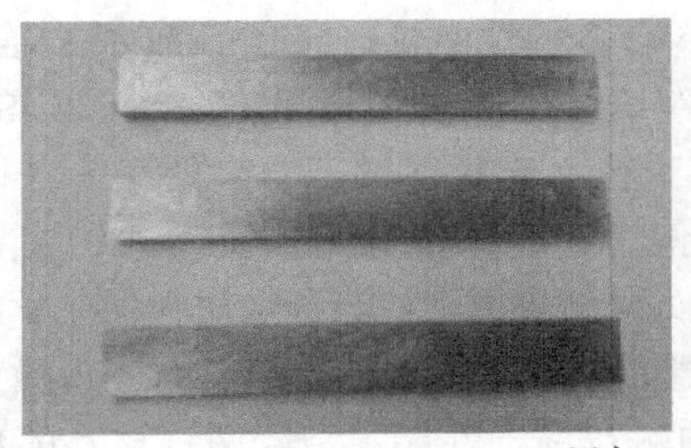

(a)国产UNS N08825合金

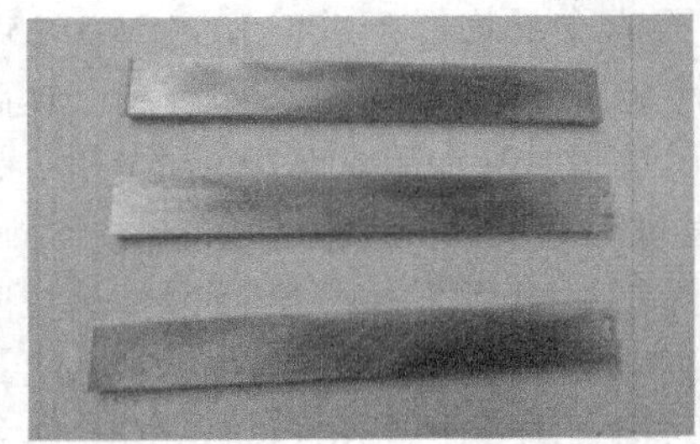
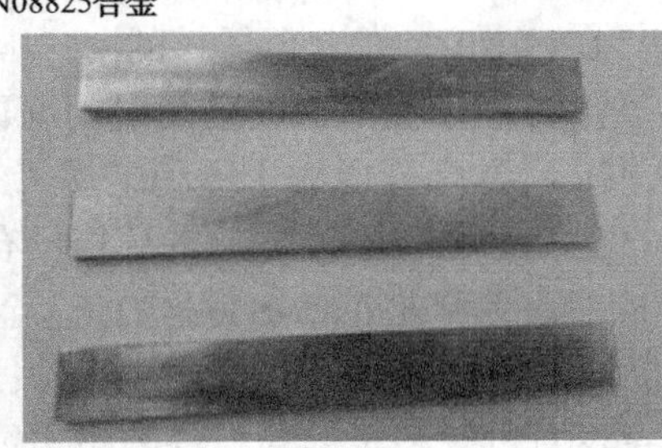

(b)进口UNS N08825合金

图 5　模拟环境应力腐蚀试验后试样宏观形貌

表 7　模拟环境均匀腐蚀试验结果

试验材料	试样编号	试样尺寸（mm×mm×mm）	腐蚀速率（mm/a）	平均值（mm/a）
国产 UNS N08825	4	80×10×3	0.0004	0.0002
	5		0.0001	
	6		0.0002	
进口 UNS N08825	1	80×10×3	0.0006	0.0003
	2		0.0001	
	3		0.0001	

4　试验结果分析

国产和进口 UNS N08825 合金材料的晶间腐蚀试验、点腐蚀试验、模拟环境的均匀腐蚀试验和应力腐蚀开裂试验结果对比如表 8 所示。

表 8　国产和进口 UNS N08825 合金材料耐腐蚀性能对比

试验类型	晶间腐蚀		点腐蚀，平均腐蚀速率［g/(m^2·h)］	模拟环境抗应力腐蚀开裂	模拟环境均匀腐蚀，平均腐蚀速率（mm/a）
	ASTM G28 方法 A，平均腐蚀速率（mm/a）	ASTM A262 方法 C，五个周期平均腐蚀速率（mm/a）			
国产 UNS N08825 合金	0.1023	0.2280	0.0056	未开裂	0.0002
进口 UNS N08825 合金	0.1887	0.0821	0.0007	未开裂	0.0003
相对综合评级	国产 UNS N08825 合金和进口 UNS N08825 合金耐腐蚀性能均较好，但进口 UNS N08825 合金耐腐蚀性能略好				

由表 8 试验结果可知，国产 UNS N08825 合金和进口 UNS N08825 合金在模拟环境下的抗应力腐蚀开裂性能和耐均匀腐蚀性能试验结果基本一致，且性能较好。两种材料耐晶间腐

蚀和耐点腐蚀性能略有差异，进口 UNS N08825 合金的性能较国产 UNS N08825 合金均较好。经分析，这与两种材料的化学成分构成和微观组织结构有较大的关系。对于 UNS N08825 合金而言，晶间腐蚀试验用于检测材料随着处理工艺（包括热加工、热处理等）以及化学成分变化时耐晶间腐蚀敏感性，其中材料的晶界偏析状态和晶间析出相，以及 Cr、Mo、Fe 等元素含量直接影响材料的耐晶间腐蚀性能。点腐蚀试验用于检测合金材料的化学成分、热处理状态和表面光洁度对材料的耐点腐蚀性能的影响，其中材料的热处理状态、微观组织结构（析出相）以及 Mo 元素的含量直接影响材料的耐点腐蚀性能。结合两种材料的化学成分含量分析，国产 UNS N08825 合金材料的 Cr、Mo 元素含量均低于进口 UNS N08825 合金材料，Cr 元素含量增加有利于提高材料的耐晶间腐蚀性能，Mo 元素含量增加有利于提高材料的耐点腐蚀性能，Fe 元素含量增加均不利于材料的耐晶间腐蚀和耐点腐蚀性能[1-3]。另外通过观察两种材料的金相组织结构，材料本身均没有发现明显的析出相，但国产 UNS N08825 合金材料的晶粒度明显细于进口 UNS N08825 合金材料。对于耐腐蚀合金而言，晶粒度大小与材料的耐蚀性能以及力学性能的综合性能有关，细化晶粒有利于材料的力学性能，降低晶粒度、减少晶界表面积有利于提高材料的耐腐蚀性能。国产 UNS N08825 合金材料的晶粒度偏细，一定程度上影响了材料的耐腐蚀性能，晶粒度的大小可通过热处理行为，如加热最高温度、保温时间、冷却速度等参数进行控制[4]。

结合以上因素分析，国产 UNS N08825 合金材料虽然具有较好的耐蚀性能，但其化学元素成分控制、材料微观结构、热处理等质量控制方面还可进行进一步优化，以保证材料本身质量的可靠性，从而提高国产 UNS N08825 合金的市场竞争能力。

5 结论及建议

5.1 结论

（1）国产 UNS N08825 合金材料具有较好的耐晶间腐蚀、耐点腐蚀、耐模拟环境均匀腐蚀以及耐模拟环境应力腐蚀开裂性能。

（2）相比进口 UNS N08825 合金材料，国产 UNS N08825 合金材料的耐点腐蚀性能和耐晶间腐蚀性能略差，这与材料的 Ni、Cr、Mo、Fe 等化学成分含量以及微观组织结构有关。

5.2 建议

为进一步提高国产 UNS N08825 合金材料的质量可靠性，建议材料在制造生产过程中，进一步优化 Cr、Ni、Mo、Fe 等主要元素的成分含量，同时不断调整材料的热处理工艺，确保国产 UNS N08825 合金有较好的综合耐腐蚀性能和力学性能，并逐步提高材料的市场竞争能力。

参 考 文 献

[1] 杨俊峰，范芳雄，等 . Incoloy825 合金晶间腐蚀原因分析［J］. 材料开发与应用，2009，24（4）：26-29.

[2] Luo H，Xiang D，Guo X F. The relation between austenitic stainless stel crystal grain size and speed of intercrystalline corrosion［J］. J. Shangdong Jianzhu Univ.，2008，23（5）：406-409.

[3] Nate C. Eisinger，Lew E. Shoemaker. An explantion of corrosion acceptance tests and their applicability to field use［C］//NACE CORROSIN CONFERENCE AND EXPO，2007.

[4] 俞树荣，何燕妮，等 . 晶粒尺寸对奥氏体不锈钢晶间腐蚀敏感性的影响［J］. 中国腐蚀与防护学报，2013，33（1）：70-74.

基于动电位极化测试研究缓蚀剂对焊缝作用的不确定度分析研究

张金钟[1]　骆　俊[2]　张　米[3]　许　帅[3]　崔　磊[1]　施岱艳[1]　孙亚军[4]　吉　灵[5]

（1. 中国石油集团工程设计有限责任公司西南分公司；2. 西南石油大学石油与天然气工程学院；3. 西南石油大学化学化工学院；4. 中国石油工程建设公司土库曼分公司；5. 公安边防部队士官学校）

摘　要： 本文以油气田中常见的集输管线焊缝腐蚀作为研究对象，使用动电位极化的电化学方法研究了焊接接头各部位的腐蚀电位，同时对加有缓蚀剂时的焊接接头腐蚀进行了研究。通过试验结果可以发现，使用腐蚀电位的方法可以对焊接接头的腐蚀电位进行腐蚀预测，同时，缓蚀剂很明显可以降低焊接接头的腐蚀，对焊接接头的局部也具有一定的抑制作用。经过不确定度分析，该方法可靠，具有很高的置信度。

关键词： 不确定度；焊缝接头腐蚀；腐蚀电位；缓蚀率

在石油天然气行业中，腐蚀是一种最为常见危害之一，是困扰油气行业的一大难题，其存在不仅仅会影响油气资源生产，而且其潜在威胁巨大，是造成环境污染、安全事故等问题的重要原因。尤其是在油气田的开发中后期，腐蚀现象愈加严重，造成了巨大经济损失。在各式腐蚀现象中，油气集输管线焊缝处的腐蚀最为突出，一直以来无论是科研研究，还是工程技术，都是热点对象[1-3]。

基于成本的考虑，集输管线材料往往都为碳钢，在考虑防腐方案时，国内一直推荐使用碳钢加上缓蚀剂的防腐方案，这一方案已经在国内，尤其是四川地区，甚至国外的油气建设项目得到了成功的运用[4-6]。前人已经对管线材料的缓蚀剂腐蚀防护作用进行了大量的研究，但是对于在缓蚀剂作用下焊缝的腐蚀研究还比较少见。

研究表明[7-9]，焊接接头的腐蚀原因主要在于在焊接接头处材料的腐蚀电位不同，在焊接过程中，由于焊接工艺影响，在焊接接头焊缝处，更多的在临近焊缝处的热影响区的腐蚀电位高于其他处材料的腐蚀电位，容易导致该处的局部腐蚀。因此，焊缝接头处的腐蚀动力来源于腐蚀各处材料的腐蚀电位差。当加有缓蚀剂时，这一差值会发生变化，使得缓蚀剂对焊缝具有一定的保护作用。由于这一差值的数值很小，只能使用精度很高的测试仪器，并且需要控制好测试影响因素，才能准确测得这一数值，因此使用这一方法来测定缓蚀剂的作用，容易遭到质疑。本文针对这一议题，经过多组试验，并对这些数据进行了不确定度分析，得到了研究结论。

作者简介：张金钟（1984—），男（汉族），四川平昌人，工程师，硕士研究生，研究方向：油气田腐蚀与防护，电话：02886014991，手机：15882067315。

项目名称：高温高含 H_2S、CO_2、Cl^-气田地面集输防腐技术研究。

项目编号：2011GJTC-09-02。

1 焊缝腐蚀试验

1.1 试验材料

参比电极：饱和甘汞电极。

辅助电极：石墨电极。

工作电极：自制工作电极，电极材料为从管线焊接接头处使用冷切割加工制成，切割时对焊接接头的各个部位进行划分，分别划分成焊缝焊肉部分、热影响区部分、母材部分。加工切割工作电极材料示意图及实物图分别如图 1 和图 2 所示。

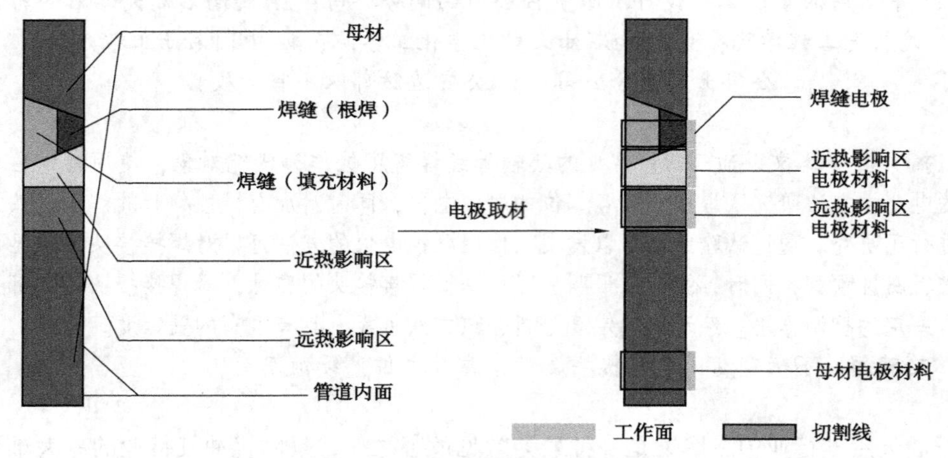

图 1 加工切割工作电极示意图

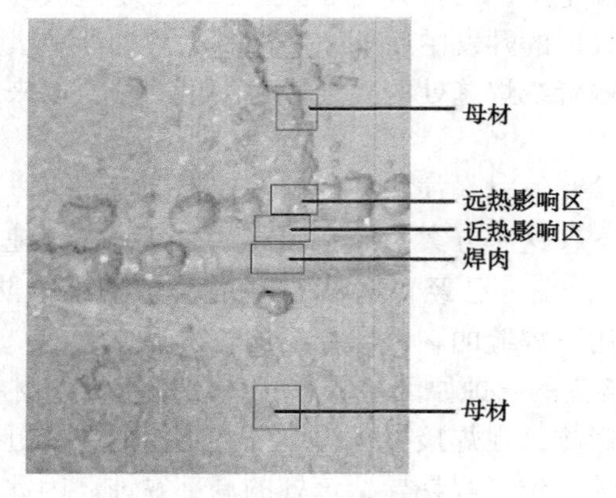

图 2 焊接接头实物图

缓蚀剂：国内某厂家提供。

实验溶液：Cl^- 浓度为 1.2×10^5 mg/L 的 NaCl 溶液，溶液先经过 N_2 赶氧，以 100mL/(min·L) 速度，持续通入 24 h；然后通入 CO_2 至饱和，通入 CO_2 时间为 6 h/L。

实验仪器：美国 Gamry 公司生产的 Reference350 电化学工作站，电化学工作站电位分辨率为 0.0001 V，电位示值误差不超过 0.01%。

水浴锅：温度控制精度为 0.1℃。

1.2 试验方法

试验时，先逐个测定开路电位确定电路连接效果，当体系稳定后马上测定极化曲线，各个电极在测定极化曲线前浸泡在被测溶液时间要一致。在恒温水浴锅内加热到 50℃，先进行开路电位扫描，再进行动电位极化测试。试验参数设置：开路电位测试时间 600s；极化范围（相对开路电位）：阳极+200mV，阴极-150mV，扫描速度：0.6mV/s。

使用动电位极化扫描的方法，在不加缓蚀剂的情况下，测试 10 组焊接接头的母材、焊肉、热影响区的腐蚀电位。然后加入缓蚀剂，从新测试加入缓蚀剂后的电位变化情况。

2 试验结果

取同一区段焊接接头的母材和热影响区，进行 10 次重复性试验，并分别测试加入缓蚀剂前后的腐蚀电位，测试结果见表 1。

表 1 腐蚀电位

序号	加入缓蚀剂前		加入缓蚀剂后	
	母材电位（V）	热影响区电位（V）	母材电位（V）	热影响区电位（V）
1	0.7234	0.7244	0.7152	0.7153
2	0.7235	0.7246	0.7153	0.7154
3	0.7237	0.7247	0.7151	0.7155
4	0.7239	0.7245	0.7150	0.7151
5	0.7235	0.7245	0.7151	0.7152
6	0.7236	0.7246	0.7151	0.7151
7	0.7235	0.7246	0.7152	0.7155
8	0.7257	0.7247	0.7150	0.7154
9	0.7235	0.7246	0.7149	0.7154
10	0.7236	0.7245	0.7147	0.7153

3 不确定度

3.1 数学模型

$$X=1-\frac{\overline{b'}-\overline{a'}}{\bar{b}-\bar{a}}$$

式中 $\bar{a}$——加入缓蚀剂前母材电位的平均值；

$\bar{b}$——加入缓蚀剂前热影响区电位的平均值；

$\overline{a'}$——加入缓蚀剂后母材电位的平均值；

$\overline{b'}$——加入缓蚀剂后热影响区电位的平均值。

X 为缓蚀剂的缓蚀效率，根据焊缝腐蚀的原理，焊缝的腐蚀动力来源于腐蚀电位差，腐蚀电位差和腐蚀电流成正比，腐蚀电流和腐蚀速度成正比，同时结合缓蚀效率计算公式：

$$X=\frac{\Delta\overline{W_1}-\Delta\overline{W_2}}{\Delta\overline{W_1}}$$

$\Delta\overline{W_1}$ 为未加入缓蚀剂腐蚀速度，$\Delta\overline{W_2}$ 为加入缓蚀剂后的腐蚀速度。因此可以计算得出使用电位差计算缓蚀剂效率的计算公式。

3.2 A 类不确定度的评定

实验过程中具体操作和人员影响因素之间无具体的函数关系，所以评定过程中的可采用

A类方法进行[10]。

3.2.1　a 的最佳估值

$$\bar{a}=\frac{1}{n}\sum_{i=1}^{n}a_i=0.7238$$

同理：

$$\bar{b}=0.7246$$

$$\overline{a'}=0.7151$$

$$\overline{b'}=0.7153$$

3.2.2　标准差 S 按贝塞尔公式

$$S(\bar{a})=\sqrt{\frac{1}{n-1}\sum_{i=1}^{n}(\bar{a}-a_i)^2}=6.85$$

同理：

$$S(\bar{b})=9.49$$

$$S(\overline{a'})=1.71$$

$$S(\overline{b'})=1.48$$

3.2.3　平均值标准差

$$S_1=\frac{S(\bar{a})}{\sqrt{n}}=2.17$$

$$S_2=\frac{S(\bar{b})}{\sqrt{n}}=3$$

$$S_3=\frac{S(\overline{a'})}{\sqrt{n}}=0.54$$

$$S_4=\frac{S(\overline{b'})}{\sqrt{n}}=0.47$$

3.2.4　分量不确定度

$$U(\bar{a})=\frac{\partial X}{\partial a}\times S_1=-0.068$$

同理：

$$U(\bar{b})=0.009$$

$$U(\overline{a'})=0.068$$

$$U(\overline{b'})=0.059$$

3.3　B类标准不确定度的评定[10]

3.3.1　电化学工作站仪器测量示值误差引入的不确定度分量

$U_1=\frac{0.01\%}{\sqrt{3}}=5.8\times10^{-6}$工作站的示值误差不超过0.01%。

3.3.2　电化学工作站仪器电位分辨率引入的不确定度分量

$U_2=0.29\delta_x=2.9\times10^{-6}$工作站的电位分辨率为 0.0001V。

3.4　合成标准不确定度的评定

$$U_x=\sqrt{U_1^2+U_2^2+U(\bar{a})^2+U(\bar{b})^2+U(\overline{a'})^2+U(\overline{b'})^2}=0.059$$

3.5　扩展不确定度 *U* 的评定

$$U=K\cdot U_x=2\times0.059=0.118$$

其中，*K* 为包含因子，一般取 2。

3.6　不确定度报告

$$X=0.75\pm0.118$$

4　结论

通过使用电化学工作站，采用测试动电位极化曲线的方法测试了 10 组焊缝处母材及热影响区的腐蚀电位，得到缓蚀剂的缓蚀效率为 0.75。经过对测试结果的不确定度计算分析，将 A 类不确定度与 B 类不确定度合成后的不确定度为 0.059，扩展不确定度为 0.118。得到的缓蚀剂对焊缝的缓蚀率为 0.75±0.118。从结果上来看，使用电化学动电位极化曲线测量腐蚀电位，计算缓蚀剂对焊缝的缓蚀率方法测试基本可行，但是由于焊缝区域的母材和热影响区电位相差不大、焊缝取材区分不明显以及缓蚀剂影响效果不够明显等原因，从不确定分析结果上来看，存在不确定性，但是从整体上来看，此方法仍然具有相当的可行性，可为研究缓蚀剂对焊缝的作用效果提供理论支持。

参　考　文　献

[1] 叶凡，杨伟．塔河油田集输管道腐蚀与防腐技术［J］．油气储运，2010，29（5）：354-360.

[2] 甘振维．凝析气田集输管道内腐蚀分析［J］．油气储运，2010，29（1）：41-47.

[3] 叶凡，高秋英．凝析气田单井集输管道内腐蚀特征及防腐技术［J］．天然气工业，2010，30（4）：96-101.

[4] 刘有超，苏万军，牟建，等．含 H_2S 气田地面工程中碳钢材料/设备的质量控制［J］．天然气与石油，2014，32（6）：92-95.

[5] 李克敏，施岱艳，林普，等．广元、射洪和宝鸡试验站的碳钢土壤腐蚀研究［J］．天然气与石油，2009，27（3）：19-23.

[6] 鲜宁，汤晓勇，施岱艳，等．CO_2 凝析气田碳钢管道穿孔失效分析［J］．天然气与石油，2012，30（3）：64-70.

[7] 赖春晓．焊缝腐蚀的原因和解决方法［J］．全面腐蚀控制，2004，18（6）：10-12.

[8] 殷名学，姜放，张维臣，等．K 气田集输管道焊缝腐蚀失效研究［J］．天然气与石油，2011，29（4）：73-78.

[9] 鲁强，顾宝珊，杨培燕，等．X65 输油管及焊缝的腐蚀失效原因［J］．腐蚀与防护，2015，36（10）：1004-1008.

[10] 刘军，温亚雄，雷晓红，等．缓蚀剂常压静态缓蚀率检测不确定度的评定［J］．计量与测试技术，2012，39（2）：58-59.

埋地集输管线的细菌腐蚀研究

张仁勇[1,2]　施岱艳[1,2]　姜　放[1,2]　李林辉[1,2]　曹晓燕[1,2]　廖　芸[1,2]

（1. 中国石油集团工程设计有限责任公司西南分公司；2. 中国石油天然气集团公司石油管工程重点实验室酸性气田管材腐蚀与防护研究室）

摘　要：通过对A气田集气管线内壁的腐蚀坑形貌、腐蚀坑垢样成分分析以及用垢样培养细菌的结果等方面综合分析了管线的内腐蚀原因。结果表明，该L360QS管线内壁的腐蚀坑为典型的细菌腐蚀造成的半球状。通过腐蚀坑垢样的细菌培养，培养出了硫酸盐还原菌。垢样的EDS分析结果显示其主要为Fe、O和C元素，还有少量S元素。而S元素主要来源于硫酸盐还原菌的腐蚀产物。

关键词：细菌腐蚀；输气管线 ；细菌培养；形貌；成分

有资料显示，输油气管道中有50%的泄漏是由于腐蚀引起的，而这些腐蚀案例中有近27 % 是由于微生物腐蚀（MIC，Microbiologically Induced Corrosion）引起的[1]。微生物腐蚀会造成输油气管线的腐蚀，在油田注水系统中，厌氧菌硫酸盐还原菌（SRB）成团簇附着在管壁上，会对金属表面产生的去极化作用使管道和设备的腐蚀速率大增。当输油气管线流速较慢或管道低点位置形成积液时，细菌腐蚀或沉积物下的腐蚀会更明显。国内长庆油田、塔里木油田等一些集输管线均发生过细菌腐蚀[2,3]。本文结合川内A气田输气管线的腐蚀现象，从管道内壁腐蚀坑形貌、腐蚀坑垢样的成分分析以及通过垢样培养的细菌情况等方面研究了该集输管线钢管内腐蚀的原因。

1　分析背景

A气田集输管道采用的是L360Q PSL2无缝钢管，该管道在进行水压试验后约16个月，在未投运的前提下，低洼地段的埋地钢管内壁发生了明显的腐蚀，内壁6点钟位置及附近区域存在分布不均、大小不一、深浅不同的腐蚀坑，如图1所示。最深的腐蚀坑深度为2. 2mm，已达到整个管道壁厚的15. 5%。

图1　钢管内壁的腐蚀形貌

2　实验

从钢管内壁选取腐蚀坑取样，去除表层物

作者简介：张仁勇（1984. 12— ），男，工程师，2011年毕业于西南石油大学获硕士学位。现就职于中国石油集团工程设计有限责任公司西南分公司，主要从事油气田材料选择和腐蚀控制研究方面的工作。联系地址：四川省成都市高新区天府三街升华路6号CPE大厦（邮编：610041）。电话：028-86978190。E-mail:zhangreny84@ 126. com。

质后，将坑下层的垢样取出放入已经过杀菌的容器内，并按照 SY/T 0532—2012《油田注入水细菌分析方法 绝迹稀释法》制备相应培养基。在管道上选取两个腐蚀坑分别编号 A、B，并对腐蚀坑 B 使用 ZEISS AXIO 金相显微镜进行三维扫描。腐蚀坑 A 的宏观腐蚀形貌和腐蚀坑 B 的三维形貌如图 2 所示，形状均呈半球状，与 API RP571[4] 中碳钢的细菌腐蚀破坏形貌特征的描述一致。

(a)腐蚀坑A宏观形貌

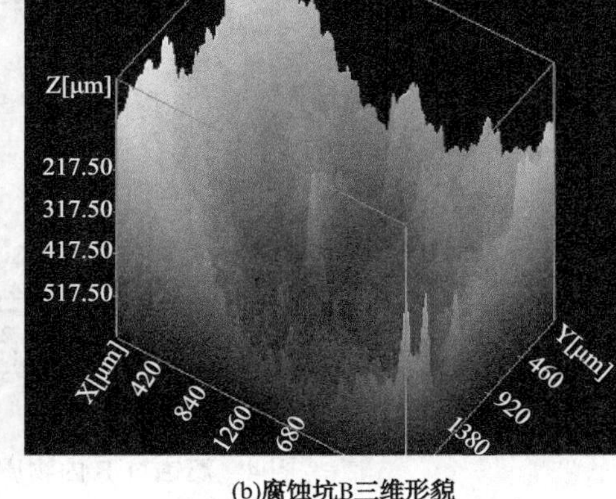

(b)腐蚀坑B三维形貌

图 2　腐蚀坑宏观形貌和三维形貌

3　结果与讨论

3.1　细菌检测结果

将取出的样品按照 SY/T 0532—2012《油田注入水细菌分析方法 绝迹稀释法》制备相应培养基，将固体样在无菌状态下加入测试瓶中，混匀后用无菌注射器逐级稀释，送培养箱内培养，根据细菌瓶阳性反应和稀释的倍数。其中，硫酸盐还原菌培养基的成分包含 KH_2PO_4，NH_4Cl，Na_2SO_4，$CaCl_2 \cdot 2H_2O$，$MgSO_4 \cdot 7H_2O$，60%乳酸钠，酵母粉以及 1%的硫酸亚铁氨。最终依据绝迹稀释法进行计数，测得硫酸盐还原菌 4.5×10^3 个/g、腐生菌 1.5×10^3 个/g、铁细菌 1.1×10^3 个/g。细菌培养结果表明钢管内壁的确存在硫酸盐还原菌（SRB）。

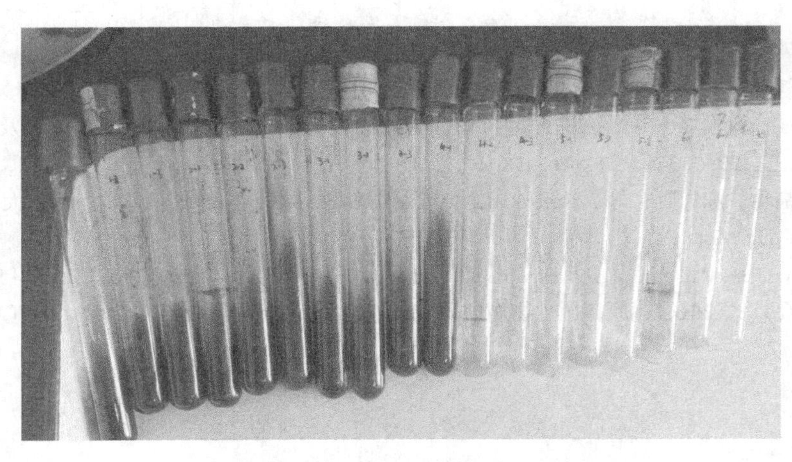

图 3　样品不同稀释梯度 SRB 的生长情况

3.2 腐蚀坑垢样成分分析

对选取的腐蚀坑 B 在去除腐蚀坑的表层物质后，对坑内的垢样进行 EDS 分析。测试结果显示，腐蚀坑中的物质主要为 Fe、O 和 C 元素，还含有少量 S 元素。由于样品暴露在空气中，腐蚀产物中的 FeS 可能发生氧化，导致 EDS 分析结果中的 O 元素含量较高。S 元素主要来源于 SRB 的腐蚀产物，在厌氧条件下，硫酸盐还原菌能够利用附着于金属表面的有机物作为碳源，并利用细菌生物膜内产生的氢，将硫酸盐还原成硫化氢[5]。

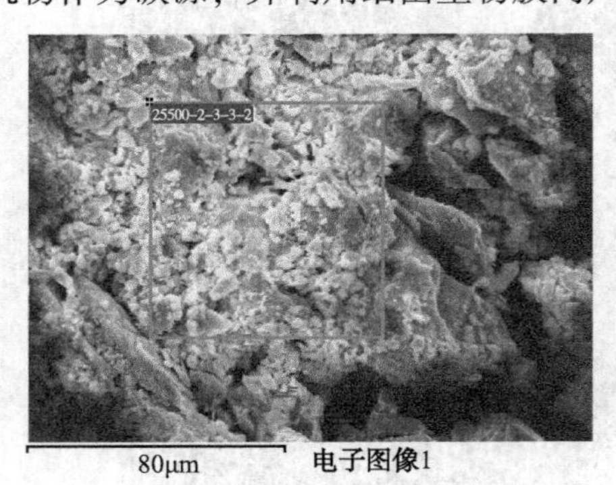

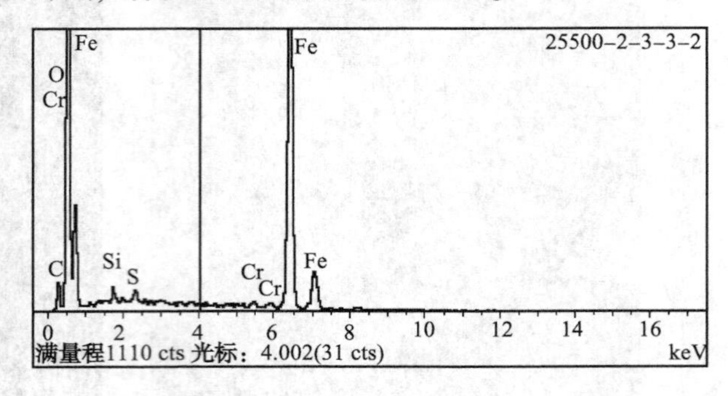

图 4　腐蚀坑 B 内物质的 EDS 分析谱图

表 1　腐蚀坑 B 内物质的 EDS 分析结果

元素	C	O	Si	S	Fe
含量［%（质量分数）］	8.90	45.28	0.68	0.49	43.93

4　结论

（1）从腐蚀坑中掘取垢样进行细菌培养检测，依据绝迹稀释法计数得到硫酸盐还原菌 4.5×10^3 个/g，腐生菌 1.5×10^3 个/g，铁细菌 1.1×10^3 个/g。

（2）在管道内表面 6 点及其附件区域，存在分布不均、大小不一、深浅不同的腐蚀坑，较深的腐蚀坑形状呈半球状，与 API RP571 中碳钢的细菌腐蚀破坏形貌特征一致。

（3）EDS 分析显示腐蚀坑中的物质主要为 Fe、O 和 C 元素，还有少量 S 元素。S 元素主要来源于 SRB 的腐蚀产物。综上所述，该集输管道的内壁发生了细菌腐蚀。

参　考　文　献

［1］杨筱蘅．油气管道安全工程［M］．北京，中国石化出版社，2005.

［2］刘黎，敬加强，谢俊峰，等．一株分离自输油管线中的硫酸盐还原菌生理生化特性及腐蚀行为研究［J］．当代化工，2016，45（2）：263-264.

［3］黄建新，马艳玲，陈志昕，等．长庆油田金属管材的腐蚀性细菌类群研究［J］．石油大学学报，2002，26（2）：66-69.

［4］Damage Mechanisms Affecting Fixed Equipment in the Refining Industry：API RP571 2011［S］.

［5］李迎霞，弓爱君．硫酸盐还原菌微生物腐蚀研究进展［J］．全面腐蚀控制，2005，19（1），30-33.

Corrosion Behaviors of High Strengthen Steel in Simulated Deep Sea Environment of High Hydrostatic Pressure and Low Dissolved Oxygen

Han Wenli[1] Tong Hui[1, 2] Wei Shicheng[2] Xu Binshi[2] Zhang Yanjun[1] Lin Zhu[1]

(1. CNPC Research Institute of Engineering Technology; 2. National Key Laboratory for Remanufacturing, Academy of Armored Forces Engineering)

Abstract: This paper investigated high strength steel in a simulated deep sea environment of the high hydrostatic pressure (HP) and low dissolved oxygen (DO) using microstructures, rust compositions, and electrochemical measurements. The result showed that the primary form of steel corrosion in this simulated high HP and low DO environment was pitting. After 48 h of immersion, we performed a Warburg impedance test of the material surface on the Nyquist plot of electrochemical impedance spectroscopy (EIS). This paper further analyzed the cyclic voltammetry (CV) and laser Raman spectrum, and found that the HP and the DO all accelerated the corrosion rates; though they affected the corrosion processes in different ways. The HP simultaneously accelerated the anodic and cathodic reaction rates, whereas the DO mainly accelerated the cathodic reaction. In addition, in the high HP and low DO environment, oxygen concentration cell corrosion had a tendency to form on the electrode surface, which could accelerate pitting.

Key words: deep sea corrosion; electrochemical analysis; laser Raman spectrum; electrode reaction rate; oxygen concentration cell corrosion

1 Introduction

With the rapid development of deep ocean engineering, many investigators have started to investigate ways to protect deep-sea engineering materials from corrosion. A deep ocean environment has the characteristics of high hydrostatic pressure (HP) and low dissolved oxygen (DO)[1,2], so the corrosion mechanism in the deep ocean is different from that in found in a shallow sea.

The deep ocean is very complicated, because there are various environmental characteristics in different parts of the deep sea. The primary difference is the level of the DO. For insistence, in tropic seas, the DO is lowest from 300 m to 1000 m depth, because many decaying microorganisms have depleted the DO. But in the seas of the north and south poles, the DO is relatively higher, because the melted ice contains plenty of DO[3].

Some investigators exposed a few materials to explore the corrosion laws. For instance,

Schumacher[4] conducted corrosion tests on steel at 2060 m, indicating that the corrosion rates of steel varied with the level of the DO at some depths[5]. Sawant et al. [6] investigated the corrosion of mild steel in the Arabian Sea and Bay of Bengal at 1000 m and 2900 m for one year. They reported that the corrosion rate at 2900m was less than that at 1000 m, and concluded that deep water corrosion of mild steel was less rapid than that of shallow water. Venkatesan[7] also agreed that oxygen is an effective cathodic depolariser and the cathodic reaction in sea water is generally oxygen reduction. However, when considering the DO, sea-water contains many mineral substances that reduce bacteria, which can also have an effect on the corrosion behaviors of different materials. Therefore, the influences of the HP and the DO are difficult to ascertain exactly. In other words, although investigators can obtain some inferences using practical simulations, they cannot entirely grasp how the environment influences all corrosion processes.

Other researchers have studied corrosion rates using simulating lab equipment to investigate how certain factors influence corrosion and to measure corrosion signals using an electrochemical workstation. For example, Yang's[8] investigation of the corrosion laws of Ni-Cr-Mo-V high strength steel under high HP showed that the HP enhanced pitting sensitivity. However, the researchers did not study the effects of the DO, so we cannot obtain the corrosion laws in the high HP and low DO environment in this study. Sun[9] investigated the corrosion behaviors of low alloy steel under a low DO and high HP environment, but their equipment could not measure the exact amount of DO; thus, the simulation experiment does not represent the real conditions.

Concerned with above problems, we designed and manufactured equipment that could control and adjust the HP, the DO, and the temperature. Fig. 1 illustrates the schematic of the designed equipment. The equipment can achieve measurement accuracy of up to 0.01 mg/L. High strength steel is widely used to build deep ocean engineering structures. Therefore, we investigated 10CrNiCu, which is a common high strength steel in marine environments, in the simulation high HP and low DO environment.

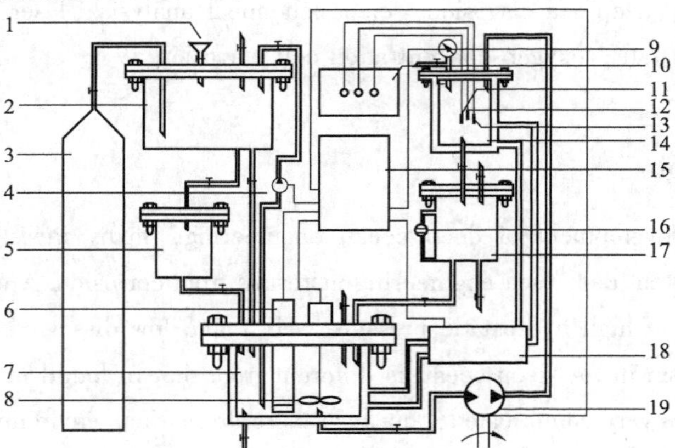

Fig. 1 Structural drawing of deep sea environment simulator

1—filler opening; 2—liquid adding reservoir; 3—air collector; 4—circulating pump; 5—disoxidation reservoir; 6—oxygen-temperature detector; 7—oxygen-controlling reservoir; 8—stirrer paddle; 9—HP gauge; 10—electrochemical workstation; 11—working electrode; 12—reference electrode; 13—counter electrode; 14—pressure reactor; 15—central controller; 16—liquidometer; 17—fluid reservoir; 18—cooler; 19—booster pump

2 Experimental procedures

2.1 Specimen preparation

We investigated 10CrNiCu high strength steel with the following composition: (wt%): C≤ 0.11, Si 0.35~0.80, Mn 0.60~1.20, S≤0.015, P≤0.025, Cr 0.60~0.80, Ni 0.50~ 0.80, Cu 0.40~0.60, and Fe balance. The selected specimens had dimensions of 10mm×10mm× 4mm, and were prepared for electrochemical tests, polished with grit paper (200#, 400#, 600#, 800#, and 1000#), dusted with acetone, and rinsed with distilled water after having been sealed by epoxy resin.

2.2 Electrochemical test

In this study, we carried out high HP electrochemical measurements using the self-made equipment with a test HP value of 30 atm (equal to the HP at 300 m depth of the sea) and a DO content of 3±0.2 mg/L (the DO content of 300 m), which is compared with other test conditions in Table 1. All tests were executed for 48 h at 283 K in 3.5% NaCl solution. We measured the electrochemical impedance spectroscopy (EIS) and the cyclic voltammograms (CV) using an IM6ex electrochemical workstation, with the test sample as working electrodes, Pt electrodes, and Ag/AgCl electrodes as the counter and reference electrodes, respectively. All impedance measurements were performed at open circuit potential and an applied AC amplitude of 5mV, with applied frequencies ranging from 10^5 Hz to 10^{-2} Hz. We analyzed the EIS data using Zsimpwin 3.10 software. All CV measurements were performed with the sweep rate 10mV/s and sweep range from −1.5 V to 0 V; the tests started at the open circuit potentials and ended with the initial sweep potentials.

Table 1 Test conditions

Specimens	1#	2#	3#	4#
HP (atm)	30	1	30	1
DO (mg/L)	3.0±0.2	3.0±0.2	7.8	7.8

2.3 Rust detection and analysis

We observed the rust morphologies in different conditions using an optical microscope (OM, BX51M) and a scanning electron microscope (SEM, Quanta 200). Simultaneously, we detected and analyzed the iron oxides using a laser Raman spectrometer (JY-T64000) with 532nm wave length.

3 Results and discussion

3.1 Morphology analysis

Fig. 2 illustrates the four morphologies that were immersed for 48 h. The Figure indicates that 1# specimen was not corroded heavily, except for some local areas, but 2#, 3# and 4# were largely corroded, 3# and 4# especially. This result occurred because 3# and 4# were immersed in a higher DO NaCl solution and the DO, as a depolarizer, reacted with more Fe atoms, as shown in Fig. 2 (c) and (d), so the rust layers were thicker. After the oxides were cleared away by diluted HCl, the steel substrates were revealed, as shown in Fig. 3. Fig. 3 (a) and (b) showed that the microstructure of 1# contained some deep pitting holes, which occurred without oxides. In other words, corrosion found in 1# mainly developed from pitting. At the same time, pitting holes occurred on the surface of 3#, as shown in Fig. 3 (d). However, 2# and 4# demonstrated mainly uniform corrosion, as shown in Fig. 3 (c) and (e).

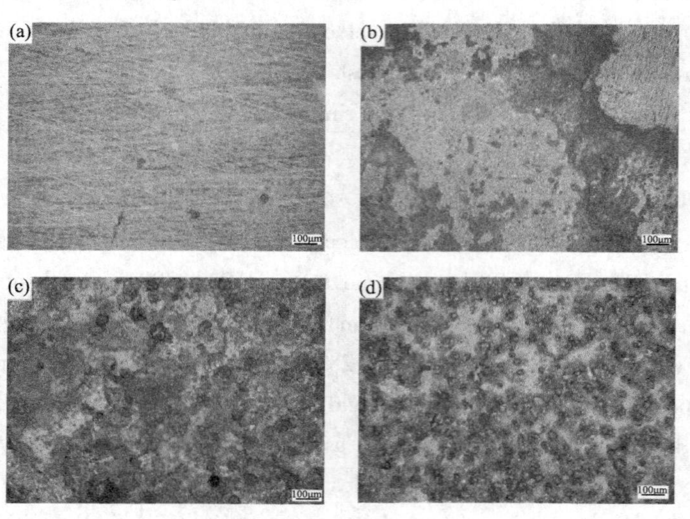

Fig. 2 OM morphologies in different test conditions

(a) 1#, (b) 2#, (c) 3#, (d) 4#

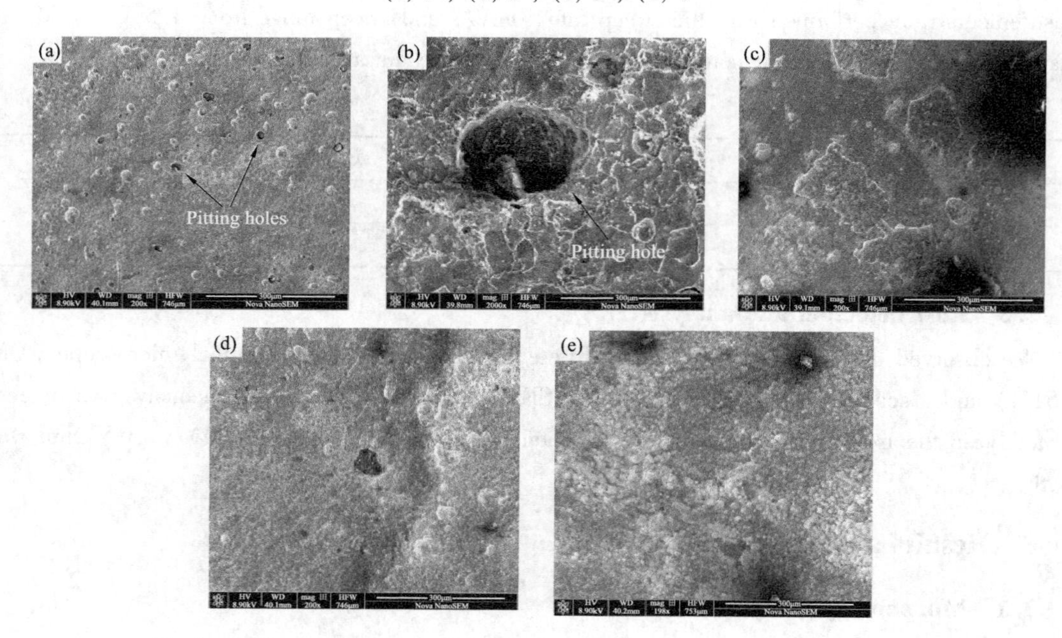

Fig. 3 SEM morphologies in different test conditions without corrosion products

(a) and (b) 1#, (c) 2#, (d) 3#, (e) 4#

3.2 Component analysis

Fig. 4 shows the Raman peaks of the rust layers and Table 2 lists the rust components. The Raman peaks were all broad, except for some peaks of 2#, which indicated that most of the rusts were mixed[10]. The strong peaks that represented the oxides uniformly contained low DO and ordinary HP, as shown in Fig. 4 (b), which may have occurred because of a phenomenon during slow reactions. The corrosion products were mainly Fe_2O_3 and FeOOH, but Fe_3O_4 only appeared on the surface of 1#. Fe_3O_4 crystal was composed of an inverse spinel structure, including Fe^{2+}, which could be oxidized to Fe^{3+}; likewise, Fe_3O_4 could be oxidized to Fe_2O_3 under certain conditions.

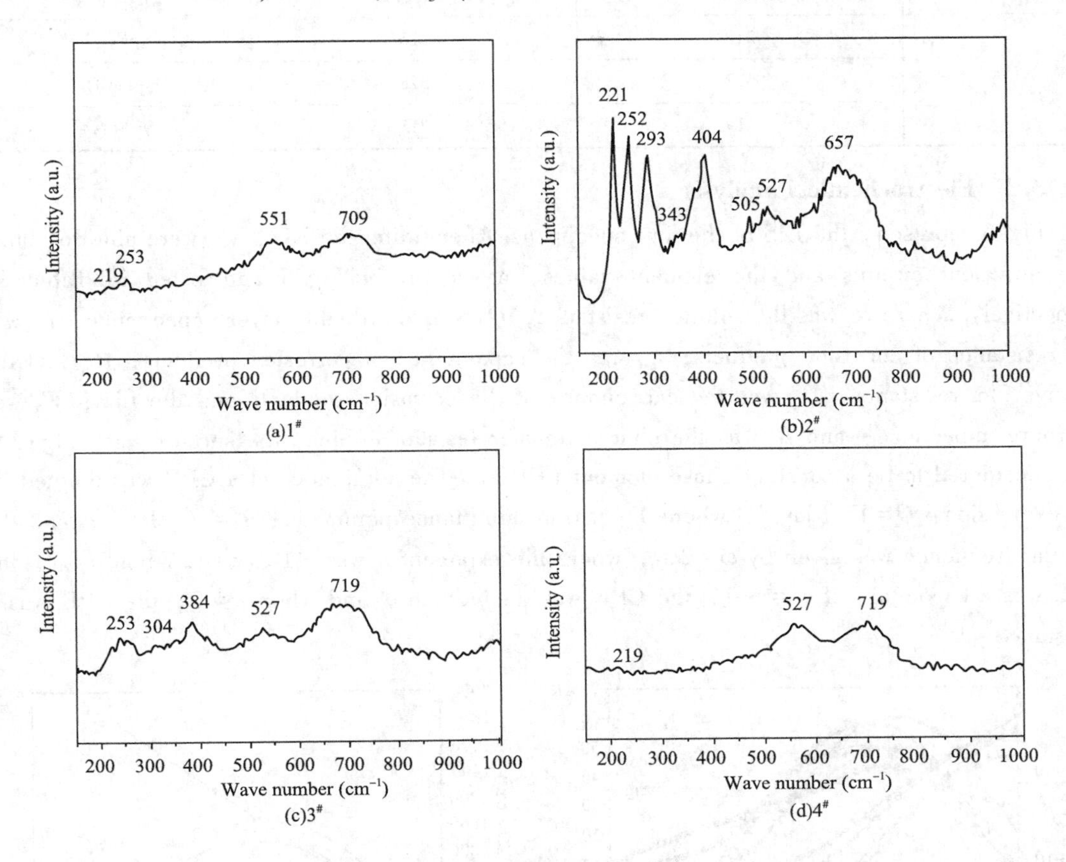

Fig. 4 Laser Raman spectrum of the rust

Table 2 Phase component and its Raman peaks of rust in different test conditions

Specimens	Locations of Raman peaks of rust (cm^{-1})	Locations of standard Raman peaks (cm^{-1})	Components of rust
1#	219	225	α-Fe_2O_3
	253	255	γ-FeOOH
	551	560	Fe_3O_4
	709	715	γ-Fe_2O_3
2#	221, 293, 505	225, 295, 500	α-Fe_2O_3
	252, 527, 657	255, 528, 654	γ-FeOOH
	343	345	γ-Fe_2O_3
	404	413	δ-Fe_2O_3

Continued

Specimens	Locations of Raman peaks of rust (cm^{-1})	Locations of standard Raman peaks (cm^{-1})	Components of rust
3#	253, 384, 527	255, 380, 528	γ-FeOOH
	304, 719	300, 715	$\gamma-Fe_2O_3$
4#	219	225	$\alpha-Fe_2O_3$
	527	528	γ-FeOOH
	719	715	$\gamma-Fe_2O_3$

3.3 Electrochemical analysis

Fig. 5 represents the EIS of the four specimens. After fitting the EIS, we were able to obtain the equivalent circuits and the element values, as shown in Fig. 6 and listed in Table 3, respectively. Where R_s was the solution resistance, Q_{dl} was the double layer capacitance; R_f was the resistance of corrosion products; Q_f was the capacitance of corrosion products; R_m was the passive film resistance; Q_m was the capacitance of the corrosion products and the film; W_s was Warburg impedance; and R_t was the charge transfer resistance. Since the surface was rough, it was considered to be a constant phase element (CPE). The admittance of a CPE was defined by the expression: $Q = Y_o (j\omega)^n$, where Y_o was an admittance parameter; $j = (-1)^{1/2}$; and the angular frequency was given by $\omega = 2\pi f$, where the exponent n was $-1 < n < +1$. When $n = 1$, the CPE was a capacitor; if $n = -1$, the CPE was an inductor; and when $n = 0$, the CPE was a resistance[11].

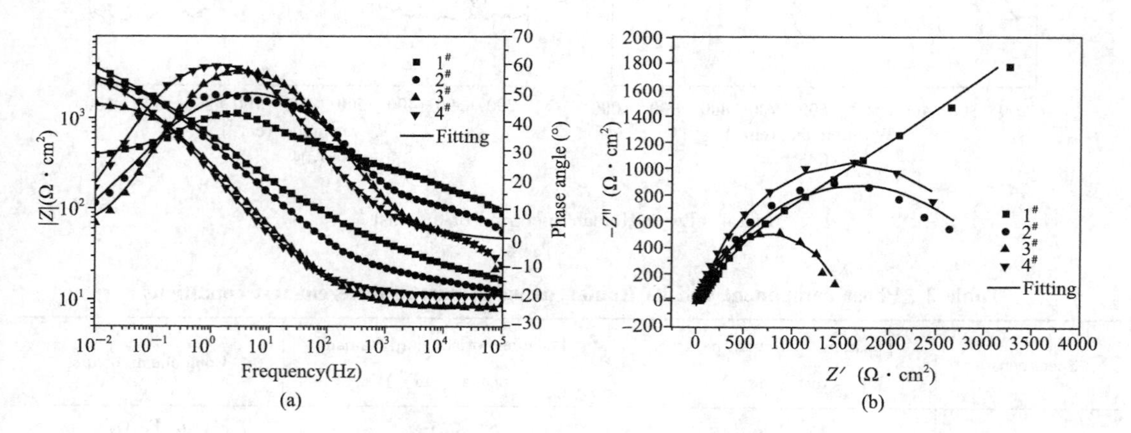

Fig. 5 EIS in different test conditions:
(a) Bode plots, (b) Nyquist plots

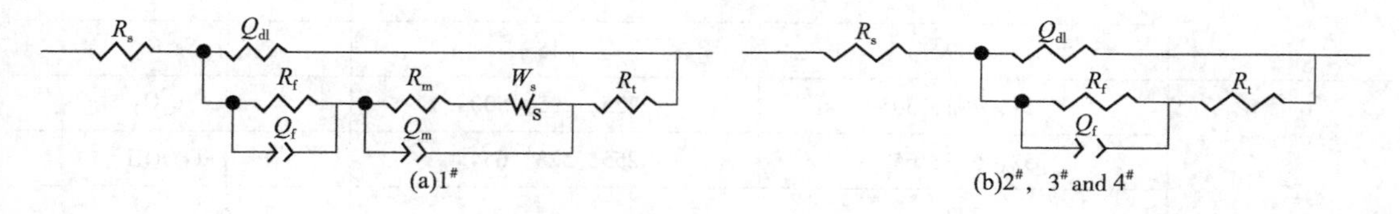

Fig. 6 Equivalent circuits of EIS data

Table 3 Element values of equivalent circuit to fit the impedance data in Fig. 6

Specimens	1#	2#	3#	4#
R_s ($\Omega \cdot cm^2$)	12.44	12.76	8.31	11.13
Q_f-Y_o ($\Omega^{-1} \cdot cm^{-2} \cdot s^n$)	5.404×10^{-4}	4.324×10^{-4}	4.477×10^{-4}	4.090×10^{-4}
Q_f-n	0.432	0.800	0.789	0.792
Q_f-Y_o ($\Omega^{-1} \cdot cm^{-2} \cdot s^n$)	9.203×10^{-5}	5.516×10^{-4}	3.739×10^{-4}	6.467×10^{-4}
Q_f-n	0.800	0.800	0.698	0.699
R_f ($\Omega \cdot cm^2$)	4246	3332	1467	3215
Q_m-Y_o ($\Omega^{-1} \cdot cm^{-2} \cdot s^n$)	6.023×10^{-4}	—	—	—
Q_m-n	0.349	—	—	—
R_m ($\Omega \cdot cm^2$)	1.783×10^{6}	—	—	—
W_s-Y_o ($\Omega^{-1} \cdot cm^{-2} \cdot s^{0.5}$)	2.924×10^{-9}	—	—	—
R_t ($\Omega \cdot cm^2$)	35.43	5.36	52.90	11.85

From Fig. 5 (a) Bode plots, the electrode surface of 1# was obviously different than that of the other three, which demonstrated that the steel with the test conditions of high HP and low DO showed different corrosion mechanisms. The same conclusion could be drawn from the Fig. 5 (b) Nyquist plots. After 48 h of immersion, the Nyquist plots of 2#, 3#, and 4# all represented a capacitive loop, because a large number of oxides formed on the surface. But the Nyquist plot of the 1# specimen indicated Warburg impedance in the low frequency ranges, which suggested that the diffusion of the DO was the rate-controlling step. In other words, the DO could not diffuse throughout the electrode surface quickly[12] because the electrode reaction rate was too high, so the DO did not have enough time to diffuse throughout the electrode surface.

In order to clarify the influence of the electrode reaction rates, we carried out further CV studies and analysis. In the anodic ranges, the retrace lines all represented different deviations, as shown in Fig. 7, which indicated that the HP and the DO affected the anodic dissolving process. Clearly, the retrace line of the 3# specimen had the largest deviation, because the steel surface's passive films bonded with the Cl^- under high HP[13], which dissolved quickly. In that case, the metal surface would reveal the fresh substrate, so the anodic corrosion density increased. The rank of the deviations was 3#>1#>4#>2#, from which we gathered that the high HP and DO could all accelerate the anodic dissolving, but the influence of the HP played a more important role than that of the DO in the anodic

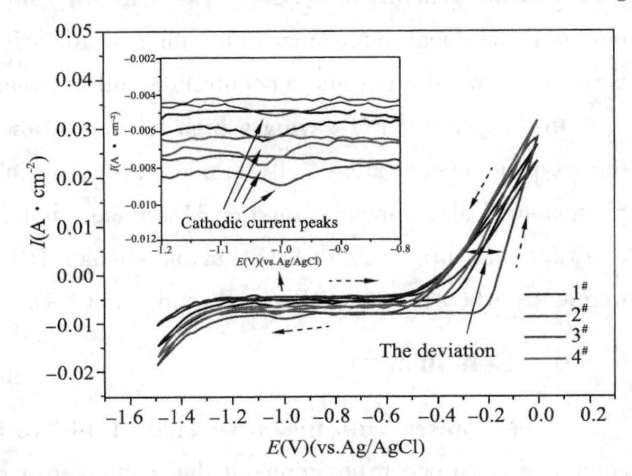

Fig. 7 CV curves in different test conditions

reaction.

At the same time, in the cathodic ranges, the current peaks appeared, whose rank was 3#>4#>1#>2# (absolute value). The cathode reaction was:

$$O_2+2H_2O+4e \rightarrow 4OH^-$$

It could be seen that the DO was more significant than the HP in the cathodic reaction. In a low DO environment, the cathodic current of 1# was larger than that of 2#, which indicated that the HP also accelerated the cathode reaction, because H_2O and the DO molecules acutely bumped the electrode surface under the high HP, so the reaction (1) is very fast. At the same time, the reaction could decrease the DO density and increase the OH^- density, which contributed to the acceleration of an anodic reaction. Given the above information, we can gather that the high HP simultaneously accelerated the anodic and cathodic reactions and the DO mainly accelerated the cathodic reaction.

Based on the results from the OM and SEM, all of the specimens pitting appeared under high HP, because the dissolving rates of the surface metals were different. The steel surface was non-uniform, so some Fe atoms preferentially reacted with the DO. The DO under high HP reacted with the Fe atom quickly; as a result, oxygen-rich areas and oxygen-poor areas appeared on the electrode surface. In the solution of oxygen-rich area, the (1) reaction predominantly occurred, so the current density of the cathode was larger than that of the anode, and the steel passive films were stable. This could explain that an impedance of film was fitted within the equivalent circuit, as shown in Fig. 6 (a). But in the solution of the oxygen-poor area, the density of the metal ions increased as the metal dissolved, so the current density of the anode was larger than that of cathode. In that case, the current density of the anode and cathode on the metal surface was not balanced, so another current streamed from interface of oxygen-rich area to the oxygen-poor one, which led to the quick formation of Fe^{2+}. At the same time, the solution ions (mainly Cl^- and Na^+) distributed on the surface again. Due to electric field action, Cl^- transferred to oxygen-poor interfaces, which were predominantly adsorbed by the Fe passive films, thus accelerating the dissolving of the films. In addition, because the DO transferred slowly, the local shortage of the DO promoted the generation of Fe_3O_4, as listed in Table 2. Since Fe_3O_4 as a good conductor, it could decrease the local impedance; in other words, it could accelerate local corrosion, i. e. pitting corrosion. Thus the oxygen concentration cell corrosion formed[14].

In the condition of having a high HP and a low DO, the electrode reaction rate increased and the oxygen concentration cell formed easily, which accelerated the non-uniform corrosion. This phenomenon also appeared on the 3# surface, but it was not so aggravated, because the uniform corrosion was primarily due to having enough DO. In other terms, pitting holes appeared more frequently when there is a high HP and a low DO.

4 Conclusions

In our present research, we studied the corrosion of 10CrNiCu high strength steel in a simulated deep ocean environment that contained a high HP and a low DO. We compared the high HP and low DO to ordinary HP vs. low DO, high HP vs. ordinary DO, and ordinary HP vs. ordinary DO. The results revealed that the HP and the DO accelerated the corrosion rates, but they affected the corrosion processes in different ways. The HP simultaneously accelerated the anodic and cathodic reactions, whereas the DO mainly accelerated the cathodic reaction.

In a high HP and low DO environment, the electrode reaction rate was very high; as a result, the DO did not have enough time to diffuse throughout the surface of the steel, so the oxygen concentration cell corrosion formed easily on the electrode surface, which accelerated pitting. Therefore, pitting occurred more frequently with a high HP and a low DO.

Acknowledgements

The authors are grateful to the National Science Foundation of China (No. 51222510), the National Basic Research Program of China (No. 2011CB013403), and the High Technology Research and Development Program of China (No. 2013AA040203).

References

[1] J. Liu, X. B. Li, J. Wang, Acta Met. Sin. 2011, 47, 697.

[2] R. Venkatesan, E. S. Dwarakadasa, M. Ravindran, Corros. Prev. Control 2004, 51, 98.

[3] R. Venkatesan, M. A. Venkatasamy, T. A. Bhaskaran, E. S. Dwarakadasa, M. Ravindran, Br. Corros. J. 2002, 37, 257.

[4] M. Schumacher, Sea water corrosion handbook, Park Ridge, New Jersey, 1979.

[5] N. Morgan, Marine technology reference book, Butterworths, London, 1990, 8.

[6] S. S. Sawant, K. Venkat, A. B. Wagh, Indian J. Technol. 1993, 1, 862.

[7] R. Venkatesan, Br. Corros. J. 2002, 37, 257.

[8] Y. G. Yang, T. Zhang, Y. W. Shao, G. Z. Meng, F. H. Wang, Corros. Sci. 2010, 52, 2697.

[9] H. J. Sun, L. Liu, Y. Li, F. H. Wang, J. Electrochem. Soc. 2013, 160, 89.

[10] S. M. Cao, L. J. Qi, Q. H. Guo, Z. Q. Zhong, Z. L. Qiu, Z. G. Li, Spectrosc. Spect. Anal. 2008, 28, 847.

[11] J. M. Bastidas, J. L. Polo, C. L. Torres, E. Cano, Corros. Sci. 2001, 43, 269.

[12] H. L. Hu, Electrochemical Measurement, National Defence Industry Press, Beijing, 2011, 235.

[13] H. J. Sun, L. Liu, Y. Li, L. Ma, Y. G. Yan, Corros. Sci. 2013, 77, 77.

[14] C. N. Cao, Principle of Corrosion Electrochemistry, Chemistry Industry Press, Beijing, 2004, 281.

Effect of Hydrostatic Pressure on The Corrosion Behaviors of High Velocity Arc Sprayed Al Coating

Tong Hui[1,2] Han Wenli[1] Wei Shicheng[2] Xu Binshi[2] Zhang Yanjun[1] Lin Zhu[1]

(1. CNPC Research Institute of Engineering Technology; 2. National Key Laboratory for Remanufacturing, Academy of Armored Forces Engineering)

Abstract: In this paper, the effect of hydrostatic pressure on the corrosion behaviors of high velocity arc sprayed Al coating was investigated using microstructures, rust compositions and electrochemical measurement. Microstructure observation showed that the Al coating surface appeared corrosion pits with diameter about 180μm after 72h immersion under high pressure, and the corrosion products was very loose. By X - ray diffraction (XRD) and Fourier transform infrared (FTIR) spectra analysis, the hydrostatic pressure accelerated the corrosion products hydroxyl hydrate $Al_5Cl_3(OH)_{12} \cdot 4H_2O$ etc to generate and make the corrosion products loose like powders. Electrochemical impedance spectroscopy (EIS) indicated that the corrosion process under high pressure was different from that under atmospheric pressure, and potentiodynamic polarization curves represented that hydrostatic pressure accelerated the anodic dissolution. Analyzed the corrosion process using cyclic voltammetry (CV), hydrostatic pressure simultaneously accelerated the anodic and cathodic reactions. On the one hand, hydrostatic pressure accelerated Al coating film dissolving, that was, anodic reaction rate. On the other hand, hydrostatic pressure increased the cathode electrode reaction rate the because of H_2O molecules acute bumping.

Key words: Deep sea corrosion; Rust compositions; FTIR; Hydroxyl hydrate; EIS; Electrode reaction rate

1 Introduction

Al coating has excellent corrosion resistance, which is used to protect steel substrate from corrosion. Meanwhile, high velocity arc sprayed Al coating is very cost - effective and high - efficiency, especially applied to large steel structure such as bridge, derrick and ship. Al coating is anodic sacrifice protection to steel with the potential approximate −0.85 V to −0.95 V in 3% NaCl solution, and in the open air, the surface is inclined to generate a layer of passive film Al_2O_3, which can protect the coating from O_2 invading. However, in the ocean Al coating tends to grow pit corrosion due to Cl^-[1].

A lot of researchers focused on Al coating performance and corrosion procedures in the shallow ocean[2,3], but only a few references referred to the corrosion behaviors of the materials in deep ocean. In the 20th century 60 s, some investigators from the USA, Russia, Japan and India had a great deal of deep ocean tests on a certain materials[4-8], however, the tests had uncertainties due to changeable deep ocean environment, so many researchers carried out a series of experiments in the laboratory in order to simplify the experimental conditions[9,10]. Venkatesan[11] found that the corrosion rates of Al alloys 1060 and 2000 increased with increasing of seawater depth, but it could not be declared that all materials became more serious corrosion with increasing of the depth because of different laws of deep sea corrosion. Reinhart et al[12,13]. investigated the corrosion behaviors of Ni in deep sea of 760 m and 2100 m, and the results revealed that the corrosion rates decreased with increasing of the depth and decreasing of the oxygen content, at the same time, Delluccia et al[14]. also draw a similar conclusion. However, other materials could perform different or opposite corrosion laws. For instance, Zhang et al[15]. investigated the corrosion law of Ni-Cr-Mo-V high strength steel under high pressure, which showed that the hydrostatic pressure enhanced the pitting sensitivity, and Yang et al[16]. found that the hydrostatic pressure accelerated the pitting of Fe-20Cr, but inclined the material to uniform corrosion at a lower rate.

To sum up, the corrosion laws of different materials in deep sea are dissimilar, so the deep sea corrosion theories should be studied deeply. High velocity arc sprayed Al coating as a frequently-used anti-corrosion coating for ocean, should be studied on the resistance in deep sea. Therefore, in this study, the corrosion behaviors of high velocity arc sprayed Al coating under high pressure was investigated by microstructures, rust compositions and electrochemical measurement.

2 Experimental

2.1 Sample material and preparation

The aluminum arc sprayed wire (2.0mm) with 99.7% purity was used for A3 steel substrate, and the spraying technological parameters were optimized with spraying voltage at 34 V to 36 V, spraying current at 120A to 140A, spraying distance 200mm, atomization air pressure 0.7MPa and the coating thickness was about 200μm. The specimens with dimension of 10mm × 10mm × 4mm, prepared for electrochemical tests, were polished with grit paper (200#, 400#, 600#, 800#, and 1000#), dusted with acetone, then rinsed with distilled water after sealed by epoxy resin.

2.2 Rust detection

The rust morphologies for immersion different time under different pressures were observed using scanning electron microscope (SEM, Quanta 200), and the compositions were detected using XRD (D08) and FTIR (S-V70).

2.3 Electrochemical measurement

High pressure electrochemical measurements were carried out in the electrochemical pressure-bearing kettle system, with the test pressure value 30 atm (equal to the pressure at 300 m depth of the sea), as compared with 1 atm (atmospheric pressure tests). All tests were executed in 3.5% NaCl solution at 298 K for different time. The EIS, potentiodynamic polarization curves and cyclic voltammograms (CV) were measured using ZAHNER IM6ex electrochemical workstation, with the test sample as working electrode, Pt electrode and Ag/AgCl electrode as the counter and the reference electrode respectively. All impedance measurements were performed at open circuit potential and applied AC amplitude of 5mV, applied frequencies ranging from 10^5Hz to 10^{-2} Hz, and the EIS data were analyzed using Zsimpwin 3.21 software. The potentiodynamic polarization curves of Al coating were measured at a constant scan rate of 0.333mV/s and analyzed using C-View 2 software. The potential ranges of CV measurement were -1.5V to 1V and the sweep rate was controlled at 10mV/s.

3 Results and discussion

3.1 Coating surface morphology analysis

As shown in Fig. 1, the microstructure of high velocity arc sprayed Al coating was not uniform. A lot of large volume particles were on the coating surface. Fig. 2 showed the element contents of the Al coating by EDS, which demonstrated that the particle was aluminum oxide.

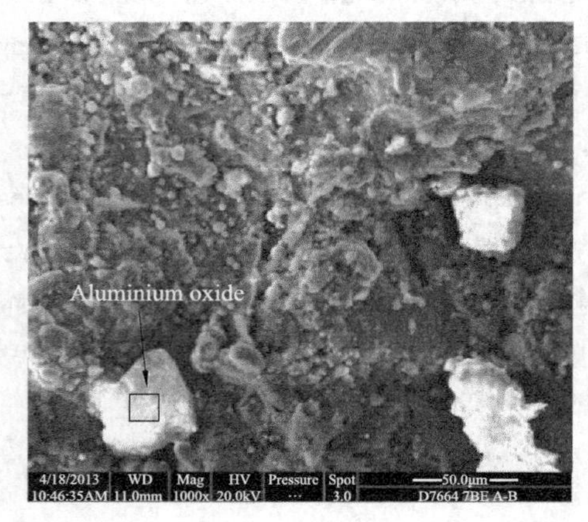

Fig. 1 Surface morphology of high velocity arc sprayed Al coating

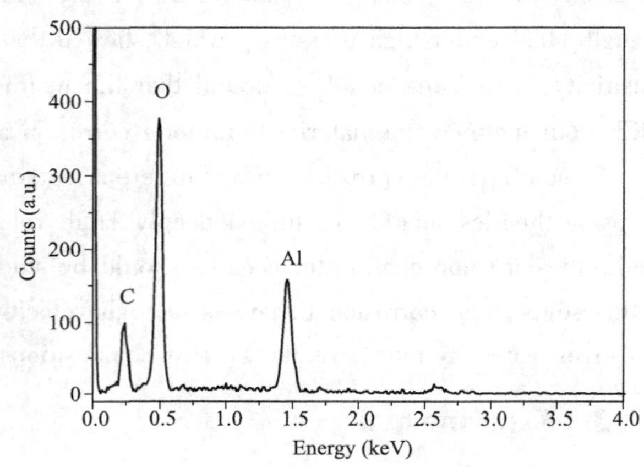

Fig. 2 EDS results of high velocity arc sprayed Al coating surface

3.2 Corrosion morphology analysis

Corrosion morphologies of Al coating with different immersion time under different pressures were discrepant, as shown in Fig. 3.

After 72 h of immersion at 1 atm, the surface still retained some scratches when the sample was prepared, which could be explained that the Al coating was corroded lightly, as shown in Fig. 3 (a). Only a little white corrosion products and cracks appeared on the surface, which meant that the Al coating was corroded partly, as shown in Fig. 3 (b). As the immersion time increased, after 720 h, the corrosion morphology changed a lot, as shown in Fig. 3 (c) and 3 (d). The surface was covered by thick and large volume corrosion products. The massive corrosion products were poor electrical conductivity, which blocked the corrosion channels, so the corrosion resistance increased[17]. However, the corrosion morphology under high pressure was very different from that at atmospheric pressure. After 72 h of high pressure immersion, large pitting holes with diameter about 180μm appeared on the surface, as shown in Fig. 3 (e). Moreover, the corrosion products were very loose like powders, as shown in Fig. 3 (f). The powder corrosion products could not stop the solution passing through the coating under high pressure, so the Al coating corrosion was accelerated.

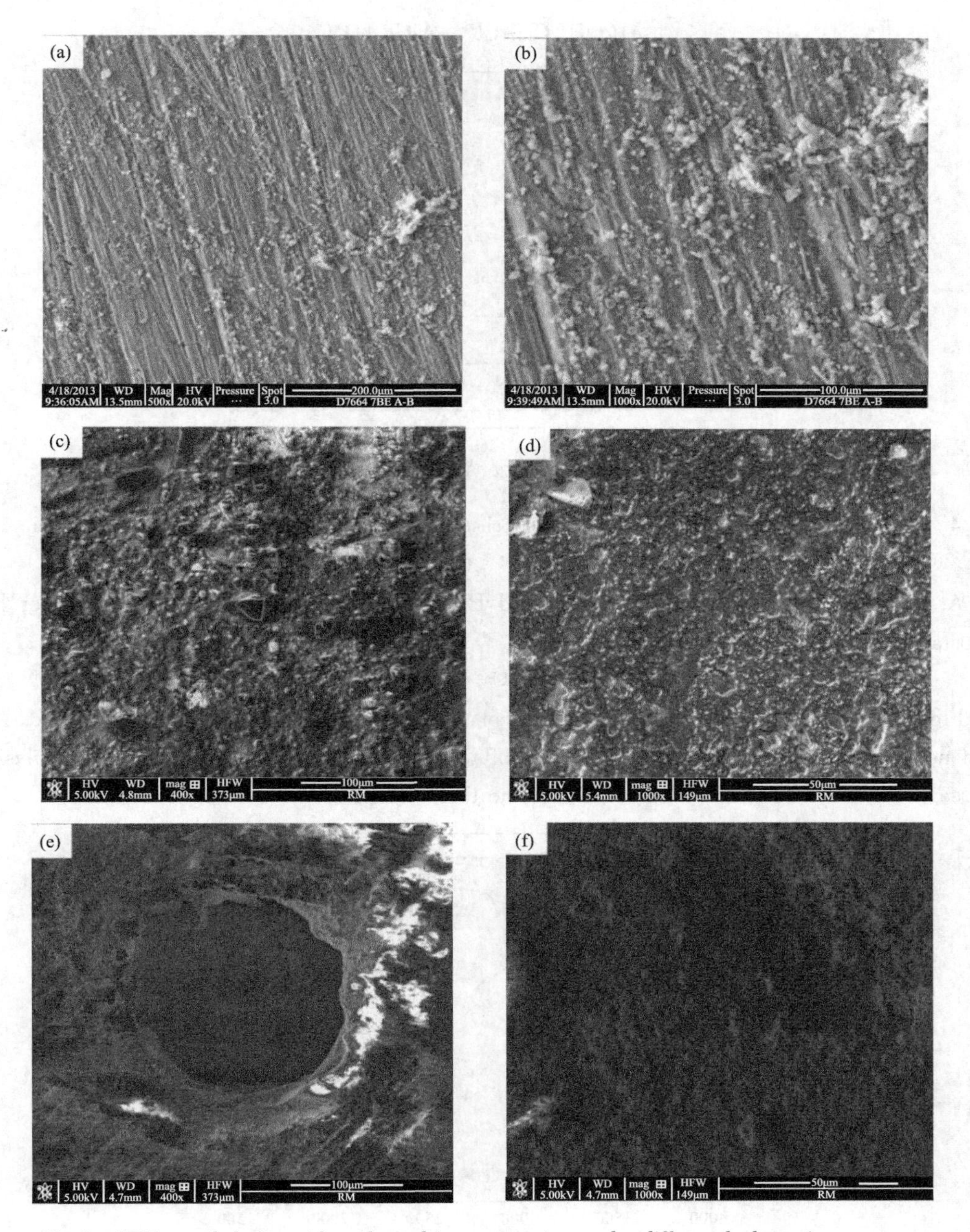

Fig. 3 SEM morphologies with prolonged immersion time under different hydrostatic pressures: (a) and (b) 72 h of immersion at 1 atm, (c) and (d) 720 h of immersion at 1 atm, (e) and (f) 72 h of immersion at 30 atm

3.3 Rust composition analysis

Fig. 4 showed the XRD patterns of the three samples. It could be seen that the corroded Al coatings showed the similar XRD graphs, which could be inferred that the corrosion products were all homologous. $Al(OH)_3$ dissolving procedure was shown as follows[18]:

$$Al(OH)_3+Cl^- \rightarrow Al(OH)_2Cl+OH^- \quad (1)$$

$$Al(OH)_2Cl+Cl^- \rightarrow Al(OH)Cl_2+OH^- \quad (2)$$

$$Al(OH)Cl_2+Cl^-\rightarrow AlCl_3+OH^- \quad (3)$$

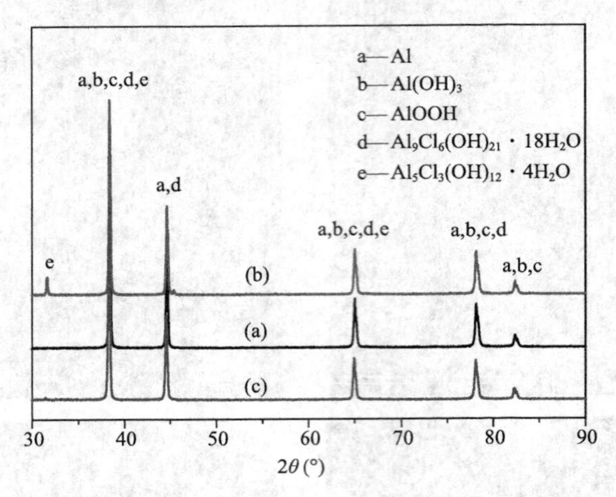

Fig. 4 XRD of corrosion products of Al coating with prolonged immersion time under different hydrostatic pressures: (a) 72 h of immersion at 1 atm, (b) 720 h of immersion at 1 atm, (c) 72 h of immersion at 30 atm

At the same time, $Al(OH)_3$, $AlCl_3$ and H_2O were inclined to form hydrate under higher hydrostatic pressure. The reaction was:

$$xAl(OH)_3+yAlCl_3+zH_2O\rightarrow Al_{x+y}Cl_{3y}(OH)_{3x}\cdot zH_2O \quad (4)$$

From the XRD patterns, the compositions of the corrosion products could be revealed, but it could not explain how the hydrostatic pressure influenced the compositions. Thus, FTIR analysis was necessary. Fig. 5 showed the FTIR patterns of the three samples.

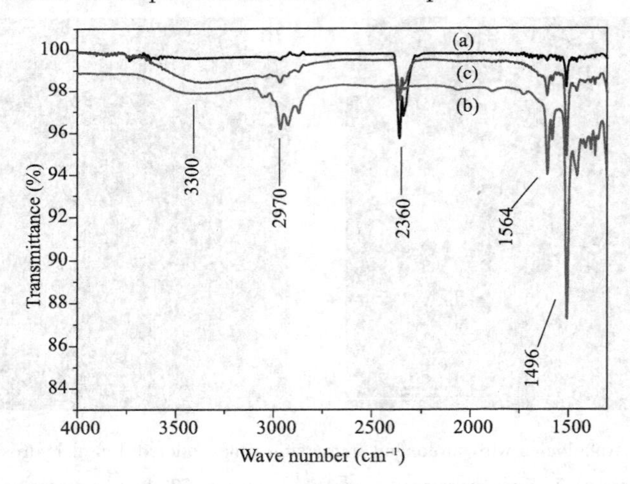

Fig. 5 FTIR of Al coating with prolonged immersion time under different hydrostatic pressures: (a) 72 h of immersion at 1 atm, (b) 720 h of immersion at 1 atm, (c) 72 h of immersion at 30 atm

From the FTIR graph (a) of the sample which immersed for 72 h under atmospheric pressure, only a sharp peak appeared near $2360cm^{-1}$, which was the characteristic peak of CO_2 stretching vibration[19], and other peaks were very weak, which meant that only a few functional group molecules reacted with corrosion products. That CO_2 was from air in an experiment, and other two graphs had the same peaks, therefore, the CO_2 peaks could be ignored here. Graphs (b) and (c) all appeared board absorption band (the shift zone from $3600cm^{-1}$ to $2750cm^{-1}$ is stretching vibration peak of O-H, and central peak is about $3300cm^{-1}$) and two sharp peeks (the shifts near $1564cm^{-1}$ and $1496cm^{-1}$ were bending vibration peak of O-H), which could be explained that a mass of O-H bond on the corrosion products and form hydrogen bonds with compound molecule[20]. According to the solution system, in functional group area ($4000cm^{-1}$ to $1300cm^{-1}$), Graph (c) showed a characteristic broad band of O-H, because corrosion products had bond abundant O-H after 720 h of immersion under atmospheric pressure. However, graph (b) also appeared O-H characteristic broad band when the sample had immersed after 72 h at 30 atm, which indicated that the hydrostatic pressure accelerated the formation of the hydrates $Al_9Cl_6(OH)_{21} \cdot 18H_2O$ and $Al_5Cl_3(OH)_{12} \cdot 4H_2O$.

3.4 EIS analysis

Fig. 6 showed the EIS of the Al coating after different immersion time under atmospheric and high pressures. As shown in Fig. 6 (a) Bode plot, from the impedance module value curves, the Al coating impedance under atmospheric pressure was obvious higher than that under high pressure, at the same time, Nyquist plot came to the same conclusion, as shown in Fig. 6 (b). From the phase curves, the shapes of curves after 72 h and 720 h of atmospheric pressure immersion were similar, which could demonstrate that the two corrosion procedures were the same, but the two atmospheric pressure curves were different from the high pressure curve, which represented that the two corrosion mechanisms were nonuniform.

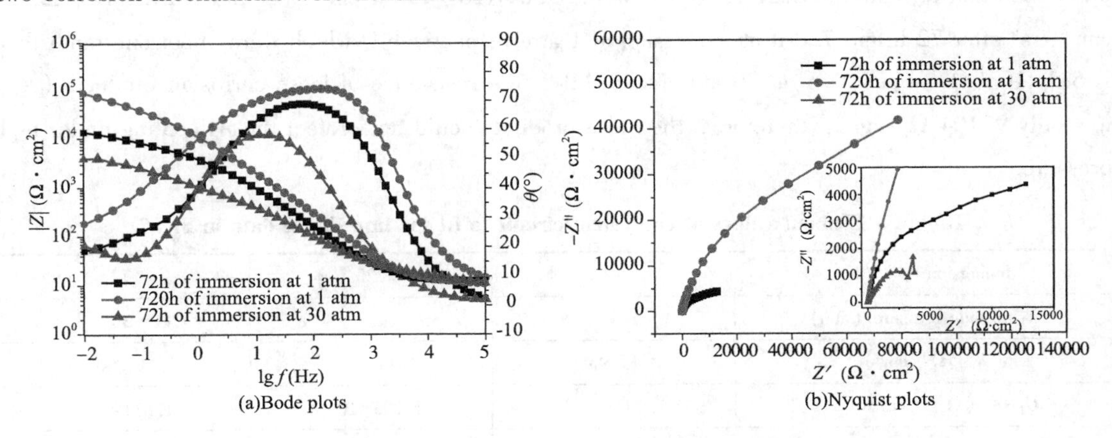

Fig. 6 EIS of Al coating with prolonged immersion time under different hydrostatic pressures

The two different corrosion procedures of Al coating were researched further. According to the fitting curves and SEM micrographs, the corrosion models and equivalent circuits at 1 atm and 30 atm pressures were proposed, as shown in Fig. 7.

Where R_s was the solution resistance, Q_f was the capacitance of the solution and the surface of density corrosion products, R_f was the resistance of density corrosion products film, Q_{dl} was the double layer capacitance, R_L was the resistance of loose corrosion products, Q_m was the capacitance of the solution and the surface of loose corrosion products, R_t was the charge transfer resistance and Z_w was Warburg impedance. Due to the surface was rough, it was considered as a constant phase element (CPE). The admittance of a CPE was defined by the expression: $Q = Y_o (j\omega)^n$, where

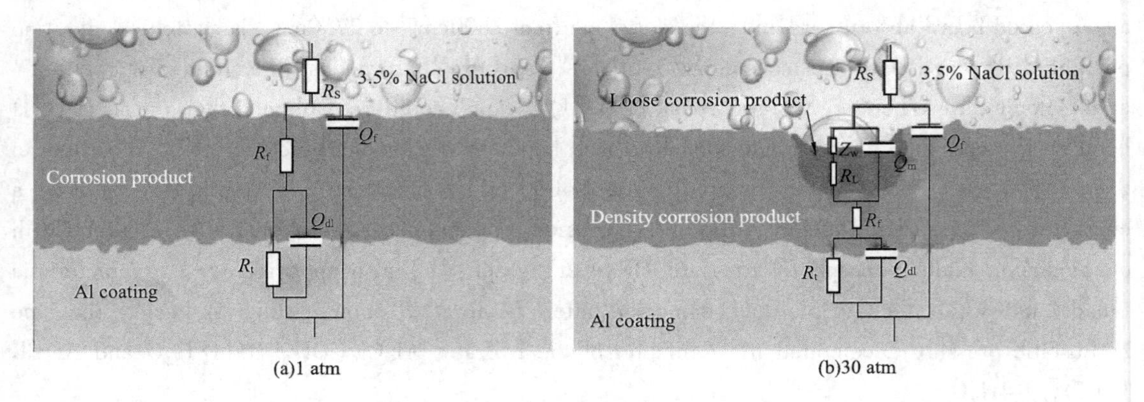

Fig. 7 Equivalent circuits of EIS data

Y_o was a admittance parameter, $j = (-1)^{1/2}$, the angular frequency was given by $\omega = 2\pi f$, and the exponent n was $-1<n<+1$. When $n = 1$, the CPE was a capacitor, if $n = -1$, the CPE was an inductor, and when $n = 0$, the CPE was a resistance[21].

The fitted results were listed in Table 1. By the data analyzing, Q_f and R_f changed a lot at 30 atm, as compared with that at 1 atm. When hydrostatic pressure was 1 atm, R_f were 3951 $\Omega \cdot cm^2$ and 7800 $\Omega \cdot cm^2$ after 72 h and 720 h of immersion respectively, which were much higher than 894.9 $\Omega \cdot cm^2$ at 30 atm for 72 h, i. e., the resistance of corrosion products at 30 atm was much lower than that at 1 atm. Moreover, Q_f-Y_o were 3.004×10^{-5} $\Omega^{-1} \cdot cm^{-2} \cdot s^n$ and 1.295×10^{-5} $\Omega^{-1} \cdot cm^{-2} \cdot s^n$ after 72 h and 720 h of immersion at 1 atm respectively, which were about one tenth than 1.631×10^{-4} $\Omega^{-1} \cdot cm^{-2} \cdot s^n$ at 30 atm. Meanwhile, the resistance of loose corrosion products (R_L) was only 9.104 $\Omega \cdot cm^2$, therefore, the loose products could not protect the Al coating under high pressure.

Table 1 Element values of equivalent circuit to fit the impedance data in Fig. 7

Immersion time (h)	72	720	72
Hydrostatic pressure (atm)	1	1	30
R_s ($\Omega \cdot cm^2$)	11.93	11.22	11.00
Q_f-Y_o ($\Omega^{-1} \cdot cm^{-2} \cdot s^n$)	3.004×10^{-5}	1.295×10^{-5}	1.631×10^{-4}
Q_f-n	0.8268	0.8594	0.6122
R_f ($\Omega \cdot cm^2$)	3951	7800	894.9
Q_{dl}-Y_o ($\Omega^{-1} \cdot cm^{-2} \cdot s^n$)	1.848×10^{-5}	2.409×10^{-5}	2.702×10^{-5}
Q_{dl}-n	0.4114	0.4118	0.7748
R_L ($\Omega \cdot cm^2$)	—	—	9.104
Q_m-Y_o ($\Omega^{-1} \cdot cm^{-2} \cdot s^n$)	—	—	6.088×10^{-5}
Q_m-n	—	—	0.7979
R_t ($\Omega \cdot cm^2$)	2.474×10^4	2.099×10^5	5.312×10^2
Z_w-Y_o ($\Omega^{-1} \cdot cm^{-2} \cdot s^{0.5}$)	—	—	3.244×10^{-4}

3.5 Potentiodynamic polarization curves analysis

The cathodic and anodic polarization curves of Al coating under different hydrostatic pressures were shown in Fig. 8, and the fitted results were listed in Table 2.

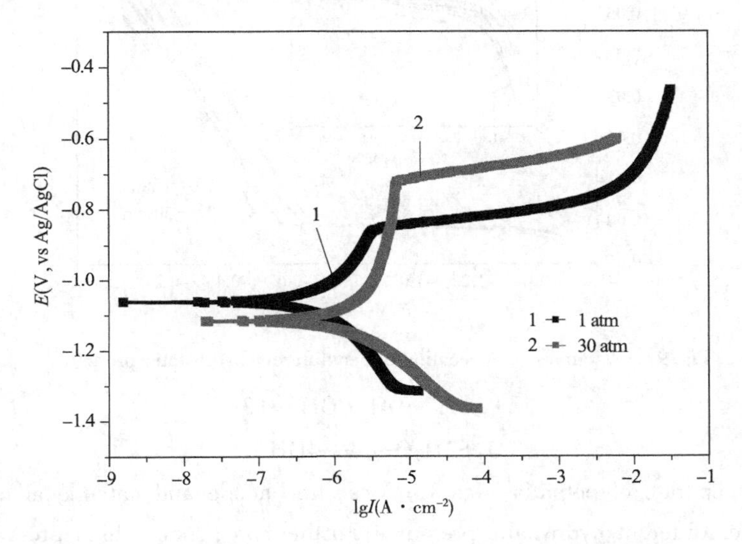

Fig. 8 Potentiodynamic polarization curves of Al coating under different hydrostatic pressures

Table 2 Comparison of corrosion behaviors of Al coating after different immersion time under different pressures

Hydrostatic pressure (atm)	E_{corr} (V)	I_{corr} ($A \cdot cm^{-2}$)	R_p ($\Omega \cdot cm^2$)	β_a (V/dec)	β_c (V/dec)
1	-1.0620	4.4947×10^{-7}	5.8040×10^4	0.63695	-0.42511
30	-1.1165	3.0936×10^{-6}	2.4435×10^4	0.79892	-0.22209

The self-corrosion current density I_{corr} were $4.4947 \times 10^{-7}\ A \cdot cm^{-2}$ at 1 atm, which was approximate one tenth than $3.0936 \times 10^{-6}\ A \cdot cm^{-2}$ at 30 atm, meanwhile, the polarization resistance R_p was $5.8040 \times 10^4\ \Omega \cdot cm^2$ and $2.4435 \times 10^4\ \Omega \cdot cm^2$ at 1 atm and 30 atm respectively. From the data of potentiodynamic polarization, the Al coating resistance under high pressure was much lower than that under atmosphere pressure.

Anodic polarization curves indicated a weaker passivating range, which could be explained that Al could form a little passive film, but the film was not continuous on the thermal spraying coating. Furthermore, anodic current density at 30 atm was much higher than that at 1 atm, which indicated that the hydrostatic pressure accelerated the anodic dissolution. The cathodic polarization curves under different hydrostatic pressures showed the similar behaviors, which indicated that the cathodic process was nearly unchanged. These conclusions were consistent with the corrosion behavior of Ni-Cr-Mo-V under hydrostatic pressure[22].

3.6 CV analysis

In order to research the effect on the electrode process, CV curves were measured under different hydrostatic pressures, as shown in Fig. 9.

Anodic and cathodic reactions were respectively:

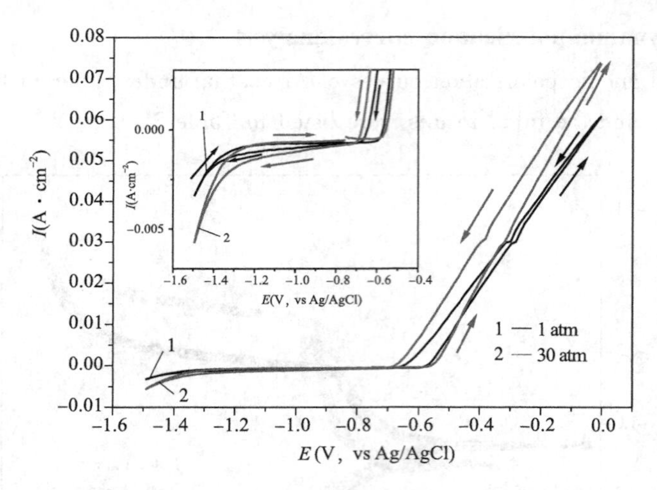

Fig. 9 CV curves of Al coating under different hydrostatic pressures

$$Al+3OH^- \rightarrow Al(OH)_3+3e \tag{5}$$

$$O_2+2H_2O+4e \rightarrow 4OH^- \tag{6}$$

When the sweep rate of potential was 10mV/s, the anode and cathode all showed different reaction rates under different hydrostatic pressures. Furthermore, under high pressure, the current density of anode and cathode all got larger. Analyzed the Fig. 9, in the anodic range, the retrace lines showed that the retrace current density got larger than that in the first scanning. This represented that the Al coating film dissolved in the range from the open potential to 1.5V[23]. Compared in the atmosphere pressure, the retrace line at 30 atm diverged more heavily. This further demonstrated that hydrostatic pressure accelerated the anodic reaction (5), which was consistent with the former conclusions. Meanwhile, in the cathodic range, cathodic current density was lower in the retrace process (cathodic current density absolute value was lower), because dissolved oxygen could not diffuse to the electrode surface in a short time, therefore, the diffusion of oxygen became the rate-controlling step of the cathode. But the cathodic reaction rate increased. Hydrostatic pressure accelerated the cathodic reaction (6), because the H_2O molecules acutely bumped the electrode surface under high pressure, the reaction (6) reacted to the right quickly. That was why the cathodic current density under hydrostatic pressure was higher than that under atmosphere pressure. At the same time, the reaction could cause dissolved oxygen density decreasing, OH^- density increasing, which contributed to anodic reaction (5) to the right. Consequently, hydrostatic pressure simultaneously accelerated the anodic and cathodic reactions.

4 Conclusions

The corrosion behaviors under high hydrostatic pressure were very different from that under atmospheric pressure. The corrosion resistance of high velocity arc sprayed Al coating under high pressure was lower than that under atmospheric pressure, and the corrosion products changed a lot. On one hand, hydrostatic pressure accelerated the formation of the hydrates and decreased the binding force of the Al coating molecules, moreover, the high speed H_2O molecules destroyed the Al coating surface with the loose corrosion products, where the large volume pits appeared, meanwhile, the loose corrosion products could not stop the NaCl solution into the coating substrate, so the Al coating inclined to be failure faster. On the other hand, under high pressure H_2O molecules acutely bumped the electrode surface to increase the cathodic reaction rate, and the anodic reaction was accelerated at the same time, so under high pressure the whole corrosion rate increased.

Acknowledgements

The authors are grateful to the National Science Foundation of China (No. 51222510), the National Basic Research Program of China (No. 2011CB013403), and the High Technology Research and Development Program of China (No. 2013AA040203).

References

[1] Trueman, A. R. Corros. Sci. 2005, 47, 2240.

[2] Wang, F., Zhang, J., Zou, J., Fan, Z. K., Zhang, F. S., Liu, X. H. Rare Met. Mater. Eng. 2010, 11, 1934.

[3] Zhao, J. J., Wang, W. X., Cai, Z. H., Zhang, P. Trans. Nonferr. Met. Soc. China 2006, S3, 1524-1525.

[4] Dexter, S. Corros. 1980, 36, 423.

[5] Ulanovskii, I., Egorova, V. Prot. Met. 1978, 14, 137.

[6] Sawant, S., Wagh, A. Corros. Prev. Contr. 1990, 37, 154.

[7] Venkatesan, R., Venkatasamy, M., Bhaskaran, T., Dwarakadasa, E. Br. Corros. J. 2002, 37, 257.

[8] Chen, S., Hartt, W., Wolfson, S. Corros. 2003, 59, 721.

[9] Beccaria, A. M., Poggi, G., Arfelli, M., Mattongno, D. Corros. Sci. 1993, 34, 989.

[10] Chen, S., Hartt, W. Corros. 2002, 58, 38.

[11] Venkatesan, R. Studies on Corrosion of Some Structural Materials in Deep Sea Environment. Ph. D. Dissertation, India Institute of Science, Bengaluru, 2000.

[12] Reinhart, F. M. Corrosion of Materials in Hydrospace, Naval Facilities Engineering Command, 1st ed., Springer. Alexandria Virginia, 1964, 297-312.

[13] Reinhart, F. M. Nickel and Nickel Alloys, Port Hueneme, 1st ed., Springer. Alexandria Virginia, 1971, 121-125.

[14] Deluccia, J. J. Mater. Prot. 1966, 6, 26.

[15] Zhang, T., Yang, Y. G., Shao, Y. W., Meng, G. Z., Wang, F. H. Corros. Sci. 2009, 54, 3915.

[16] Yang, Y. G., Zhang, T., Shao, Y. W., Meng, G. Z., Wang, F. H. Corros. Sci. 2010, 52, 2706.

[17] Dong, C. F., Xiao, K., Xu, L., Sheng, H., An, Y. H., Li, X. G. Rare Metal Mater. Eng. 2011, 40, 277.

[18] Graedel, T. E. J. Electrochim. Soc. 1989, 136, 204.

[19] Weng, S. F. Fourier Transform Infrared Spectroscopy Analysis, 2nd ed., Chemical Industry Press: Beijing, 2009, 13-14.

[20] Weng, S. F. Fourier Transform Infrared Spectroscopy Analysis, 2nd ed., Chemical Industry Press: Beijing, 2009, 325-327.

[21] Bastidas, J. M., Polo, J. L., Torres, C. L., Cano, E. Corros. Sci. 2001, 43, 272.

[22] Liu, B., Zhang, T., Shao, Y., Meng, G., Wang, F. Mater. Corros. 2012, 63, 270.

[23] Zhao, Y., Lin, C. J., Li, Y., Du, R. G., Wang, J. R. Acta Phys. -Chim. Sin. 2007, 23, 1344.

咪唑啉季铵盐缓蚀剂对N80钢在盐酸中的腐蚀行为影响研究

杨耀辉[1,2]　韩文礼[1,2]　张彦军[1,2]　林　竹[1,2]　李玲杰[1,2]

（1. 中国石油集团工程技术研究院；2. CNPC石油管工程重点实验室涂层材料与保温结构研究室）

摘　要：采用油酸和二甲苯、三乙烯四胺合成一种咪唑啉季铵盐缓蚀剂。通过腐蚀失重、电化学测试对缓蚀剂的性能进行了研究，结果表明，咪唑啉季铵盐对N80钢在1.0mol/L盐酸中的腐蚀有良好的抑制作用，在8mmol/L时缓蚀效率达97%以上。热力学参数分析表明，咪唑啉季铵盐在N80表明的吸附符合Langmuir等温吸附方程，吸附是自发的过程，且物理吸附和化学吸附同时存在。

关键词：咪唑啉；季铵盐；缓蚀剂；腐蚀

大多数缓蚀剂是含有极性基团（氨基-NH_2、醛基-CHO、羧基-COOH、羟基-OH）[1,2]，或者杂环原子（O、N、S、P）[3,4]的有机化合物。极性基团和杂环分子含孤对电子和π电子，能在金属表面产生强烈的吸附作用，从而起到缓蚀作用。尤其是含N杂环化合物，通过N杂环原子吸附在金属表面，是防止金属在酸性环境腐蚀的优良缓蚀剂。

咪唑啉型化合物不仅存在N原子上孤对电子对金属表面的吸附作用，而且烷基支链也能在金属表面产生吸附，因此，对钢铁等金属具有优良的缓蚀性能[5,6]。咪唑啉型化合物作为缓蚀剂，还具有无特殊的刺激气味、毒性低等特点在国内外的油田中大量使用[7,8]。

本文研究了咪唑啉季铵盐在HCl介质中对N80钢缓蚀作用。采用失重法、动电位极化曲线研究了缓蚀剂在酸液中的缓蚀性能，同时应用吸附理论和腐蚀动力学公式求出了相应的吸附热力学参数，详细讨论了缓蚀剂在HCl中对N80钢的吸附及缓蚀作用机理。

1　试验

1.1　原料和试验溶液

测试用的N80试片化学组成如表1，测试用1.0mol/L HCl溶液用37%的浓盐酸和蒸馏水配制。

表1　N80试片化学组成

元素	C	Si	Mn	P	S	Cr	Mo	Ni	Fe
含量（wt%）	0.24	0.22	1.19	0.013	0.004	0.036	0.021	0.028	余量

1.2　咪唑啉季铵盐的合成

（1）在三颈瓶中加入按1：1：1的油酸和二甲苯、二乙烯三胺，程序升温至130~160℃进行酰胺化3~5h，程序升温至230℃，反应5~8h，至不再有H_2O生成时结束反应，减压蒸馏二甲苯，得到红棕色黏稠液体即为咪唑啉化合物。

(2) 用适量去离子水与氯乙酸钠配成氯乙酸钠水溶液，加入咪唑啉化合物，升温至80~90℃，反应2 h，反应结束后蒸馏出H_2O，得到黏稠状咪唑啉季铵盐化合物。

1.3 失重测量

在不同咪唑啉季铵盐缓蚀剂浓度、温度下测试N80钢在1.0mol/L的盐酸溶液中的腐蚀速率。用350#~800#耐水砂纸逐级打磨至表面呈镜面光泽；用丙酮除油，去离子水冲洗、干燥、称量。在设定的实验条件下浸泡4h后取出，用软毛刷轻轻擦除其表面的腐蚀产物，再以去离子水和丙酮清洗，干燥至恒质量。由浸泡前后试片的质量变化（精确至0.1mg）。缓蚀率IE%、缓蚀剂覆盖度θ和腐蚀速率V_{corr}（单位是mm/a）按如下公式计算：

$$IE\% = \frac{W_0 - W}{W_0} \times 100$$

$$\theta = \frac{W_0 - W}{W_0}$$

$$V_{corr} = \frac{W_0 - W}{st}$$

式中 W_0，W——分别为无缓蚀剂和添加缓蚀剂时的腐蚀失重；

s——试片的表面积；

t——挂片试验时间。

1.4 电化学测试

本文中动电位扫描测试试验使用普林斯顿2273电化学工作站进行，采用三电极体系：工作电极材质为N80钢材圆柱电极，辅助电极为铂电极，参比电极为饱和甘汞电极，动电位扫描范围为：-0.25~+0.25V，扫描速度为0.5mV/s。试验介质为含不同浓度缓蚀剂的盐酸溶液。测试体系在恒温水浴中保持温度恒定。缓蚀效率按下式计算：

$$IE\% = \frac{I_{corr} - I_{corr(inh)}}{I_{corr}} \times 100$$

式中 I_{corr}，$I_{corr(inh)}$——分别为空白和添加缓蚀剂时的腐蚀电流密度。

1.5 表面分析

利用JSM-5600LV型扫描电镜观察试验后N80的表面形貌。

2 结果和讨论

2.1 腐蚀失重

表2为N80钢在60℃和90℃、1.0mol/L盐酸中添加不同浓度缓蚀剂时的试验结果。可以发现，在盐酸溶液中加入缓蚀剂后腐蚀速率大幅度降低，并且随着咪唑啉季铵盐浓度的增加，腐蚀速率持续降低。由表2可以看出，在浓度为8mmol/L时，N80钢在60℃的腐蚀速率分别从15.77mm/a降到0.17mm/a，在90℃时的腐蚀速率从34.54mm/a降到0.83mm/a，从结果还可以发现，随着温度升高，缓蚀效率下降，可能是由于温度升高对咪唑啉季铵盐在N80钢表面的吸附能力造成了影响。

表 2　N80 钢在 60℃和 90℃、1.0mol/L 盐酸中添加不同浓度缓蚀剂时的试验结果

试验温度（℃）	缓蚀剂浓度（mmol/L）	腐蚀速率（mm/a）	IE（%）	θ
60	Blank	15.77	—	—
	0.06	2.99	81.03	0.8103
	0.4	0.98	93.78	0.9378
	0.9	0.57	96.38	0.9638
	4.0	0.25	98.41	0.9841
	8.0	0.17	98.92	0.9892
90	Blank	34.54	—	—
	0.06	8.19	76.89	0.7689
	0.4	4.13	88.35	0.8835
	0.9	2.33	93.42	0.9342
	4.0	1.15	96.75	0.9675
	8.0	0.83	97.65	0.9765

2.2　表面形貌分析

为了考察缓蚀剂的添加对 N80 钢在 1.0mol/L 盐酸中腐蚀形貌的影响，对未添加和添加缓蚀剂的 N80 钢片在 60℃、1.0mol/L 盐酸中浸泡 4h 后进行了 SEM 形貌分析，见图 1。由图可以看出，在未添加缓蚀剂时，试验后的试片表面遭到了严重的腐蚀。而添加了 4mmol/L 的缓蚀剂时，试验后试片腐蚀轻微，咪唑啉季铵盐对腐蚀有明显的抑制作用。这也间接说明了咪唑啉季铵盐在 N80 表面形成了保护膜，从而减缓了腐蚀。

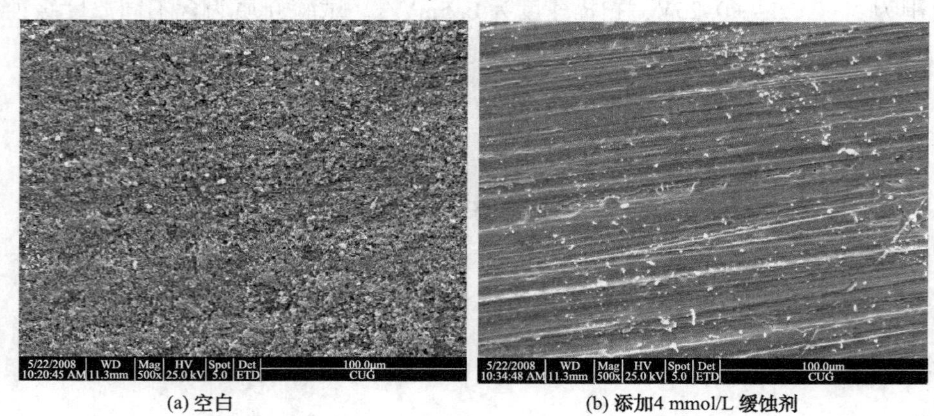

(a) 空白　　(b) 添加4 mmol/L 缓蚀剂

图 1　N80 钢在 60℃、1mol/L 盐酸中浸泡 4h 后的 SEM 照片

2.3　动电位测试

图 2 为 N80 钢在 60℃、添加不同浓度缓蚀的 1.0mol/L HCl 中的极化曲线。由图 2 可以看出，随着咪唑啉季铵盐浓度的增加，阴极和阳极腐蚀电流密度减小，电流的减小可能是由于缓蚀剂占据了部分金属表面活性点。极化曲线的阴极和阳极部分随着缓蚀剂浓度的增加没有发生明显变化，说明阴极的氢还原和阳极的金属溶解的机理没有发生变化。

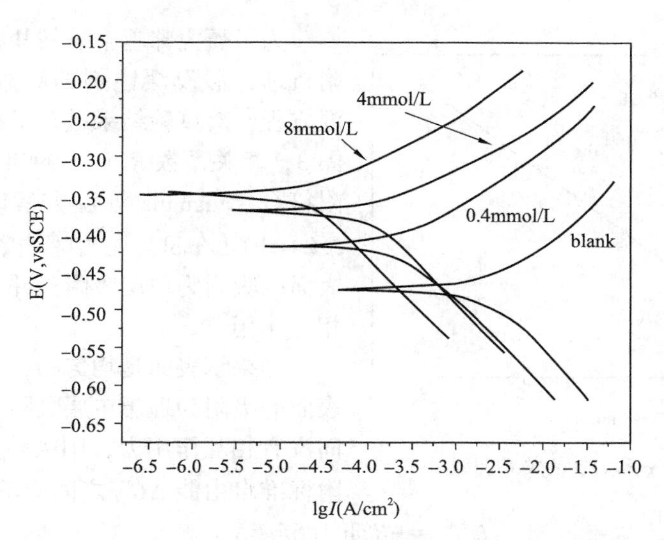

图 2　N80 钢在 60℃、添加不同浓度缓蚀的 1.0mol/L HCl 中的极化曲线

由极化曲线通过计算得到腐蚀电流密度、腐蚀电位、阴极和阳极塔菲尔常数、缓蚀效率等参数，见表 3。由表 2 看出，阳极塔菲尔常数在咪唑啉季铵盐浓度增加时基本保持不变，而阴极塔菲尔常数在添加缓蚀剂后变大。随着缓蚀剂浓度的增加，极化曲线向阳极方向移动，在缓蚀剂浓度从 0 增加到 8mmol/L 时，腐蚀电位从-473.2mV 上移至-349.4mV，上移了大约 124mV，加入缓蚀剂后电位正向移动并且 b_c 变大，说明缓蚀剂时混合型的缓蚀剂，通过同时抑制阳极和阴极反应达到减缓腐蚀。

在加入 8mmol/L 的缓蚀剂后，腐蚀电流密度从 532.0μA · cm^{-2} 降到了 7.0μA · cm^{-2}，缓蚀效率高达 98.7%，缓蚀剂有较好的缓蚀性能。

表 3　N80 钢在 60℃、添加不同浓度缓蚀的 5.0M HCl 中的参数

Concentration (mmol/L)	E_{corr} (mV)	b_c (mV · dec^{-1})	b_a (mV · dec^{-1})	I_{corr} (μA · cm^{-2})	IE (%)
Blank	-473.2	75.4	54.8	532.0	—
0.4	-414.5	90.7	55.2	52.2	90.2
4	-371.2	121.2	47.6	30.4	94.3
8	-349.4	113.5	45.5	7.0	98.7

2.4　吸附方程和热力学参数

表面活性剂在金属/溶液界面的吸附能改变金属的抗腐蚀性。缓蚀剂的缓蚀效率主要取决于其在金属表面的吸附能力，因此有必要研究缓蚀剂的吸附、脱附等温方程。缓蚀剂的表面覆盖率 θ 由腐蚀失重数据得到，θ 与缓蚀剂浓度之间的关系可用 Langmuir 等温方程来描述：

$$\frac{C_{inh}}{\theta}=C_{inh}+\frac{1}{K_{ads}}$$

式中　C_{inh}——缓蚀剂浓度；

θ——缓蚀剂表面覆盖率；

K_{ads}——吸附脱附过程的平衡常数。

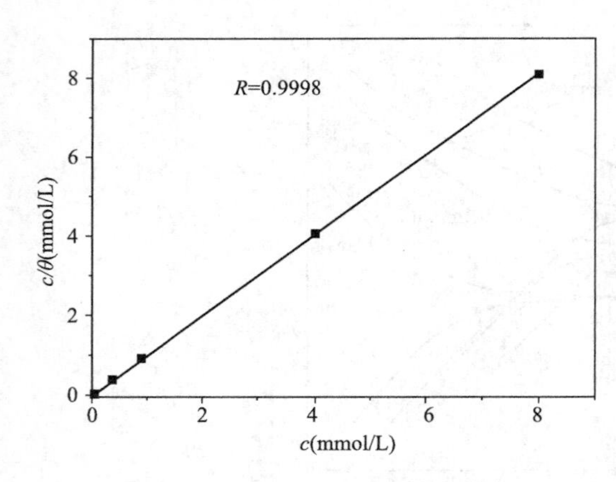

图 3　60℃时 c 和 c/θ 的关系曲线

为了研究咪唑啉季铵盐在 N80 钢表面的吸附机理，假设缓蚀剂的吸附符合 Langmuir 等温方程，对试验结果进行了拟合，拟合结果见图 3，相关系数 R 为 0. 9998，说明缓蚀剂的吸附符合 Langmuir 等温方程。K_{ads} 的平均值为 3.24×10^{4} L/mol，说明咪唑啉季铵盐在 N80 钢表面的吸附为单层吸附，并且吸附分子之间无相互作用[9]。

如果钢表面是均匀的，且缓蚀剂分子在其表面的吸附为理想单层吸附，即相邻的分子之间没有相互作用力，则吸附平衡常数 K_{ads} 与吸附标准自由能 ΔG^{0}_{ads} 之间应满足下面的关系式：

$$\Delta G^{0}_{ads}=-RT\ln(55.5K_{ads})$$

一般来说，吸附标准自由能 ΔG^{0}_{ads} 数值为负说明缓蚀剂在金属表面的吸附是自发的。ΔG^{0}_{ads} 的数值在 -20kJ mol^{-1} 以上时，说明缓蚀剂的吸附主要为物理吸附，ΔG^{0}_{ads} 的数值在 -40kJ mol^{-1} 或更负时，表面缓蚀剂的吸附主要为化学吸附[10,11]，由计算得到标准吸附自由能 ΔG^{0}_{ads} 的数值为-30. 46kJ/mol，在 20~40kJ/mol 范围内，表明咪唑啉季铵盐在 N80 钢表面的吸附属于物理吸附和化学吸附的混合吸附[12]。

3　结论

（1）失重试验和电化学测试结果表明，咪唑啉季铵盐对 N80 钢在 1. 0mol/L 盐酸中的腐蚀有良好的抑制作用，缓蚀效率随浓度的增加而提高，在浓度为 8mmol/L 时缓蚀效率高达 97%以上。

（2）数据拟合结果表明，咪唑啉季铵盐在 N80 表明的吸附符合 Langmuir 等温吸附方程，吸附是自发的过程。咪唑啉季铵盐在 N80 钢上的吸附不是单一的物理吸附或者化学吸附，而是属于物理吸附和化学吸附的混合吸附。

参　考　文　献

[1] S. M. A. Hosseini，A. Azimi，The inhibition of mild steel corrosion in acidicmedium by 1-methyl-3-pyridin-2-yl-thiourea，Corros. Sci. 51（2009）728-732.

[2] M. Behpour，S. M. Ghoreishi，N. Mohammadi，N. Soltani，M. Salavati-Niasari. Investigation of some Schiff base compounds containing disulfide bond as HCl corrosion inhibitors for mild steel Corros. Sci. 52（2010）4046-4057.

[3] M. Lagrenee，B. Mernari，M. Bouanis，M. Traisnel，F. Bentiss，Corros. Sci. 44（2002）573-588.

[4] M. A. Quraishi，I. Ahamad，A. K. Singh，S. K. Shukla，B. Lal，V. Singh，Matar. Chem. Phys. 112（2008）1035-1039.

[5] 汪的华，甘复兴，姚禄安．缓蚀剂吸附行为研究进展与展望［J］. 材料保护，2000，33（1）：29-32.

[6] 朱丽琴，刘瑞泉，王吉德，等. 席夫碱基咪唑啉化合物对 Q235 钢在盐酸介质中缓蚀性能研究［J］. 中国腐蚀与防护学报，2006，26（6）：336-341.

[7] 张静，杜敏，于会华，等. 分子结构对咪唑啉缓蚀剂膜在 Q35 钢表面生长和衰减规律的影响［J］. 物理化学学报，2009，25（3）：231-525.

[8] 胡建春，胡松青，石鑫，等. CO_2 分压对碳钢腐蚀的影响及缓蚀性能研究 [J]. 青岛大学学报（工程技术版）2009，24（2）90-94.

[9] E. E. Ebenso，I. B. Obot，L. C. Murulana，Quinoline and its derivatives as effective corrosion inhibitors for mild steel in acidic medium [J]. Electrochem. Sci. 5（2010）1574-1586.

[10] F. Bentiss，M. Lebrini，M. Lagrenée，Thermodynamic characterization of metal dissolution and inhibitor adsorption processes in mild steel/2，5-bis(n-thienyl)-1，3，4-thiadiazoles/hydrochloric acid system [J]. Corros. Sci. 47（2005）2915-2931.

[11] X. Li，S. Deng，H. Fu，Triazolyl blue tetrazolium bromide as a novel corrosion inhibitor for steel in HCl and H_2SO_4 solutions [J]. Corros. Sci. 53（2011）302-309.

[12] Ashassi-Sorkhabi H，Majidi M R，Seyyedi K. Investigation of inhibition effect of some amino acids against steel corrosion in HCl solution [J]. Appl. Surf. Sci.，2004，225（1-4）：176-185.

四、非金属与复合材料

CO_2 在高密度聚乙烯中的渗透特性及机理研究

李厚补[1]　羊东明[2]　张冬娜[1]　朱原原[2]　葛鹏莉[2]　戚东涛[1]　张志宏[2]

(1. 石油管材及装备材料服役行为与结构安全国家重点实验室，中国石油集团石油管工程技术研究院；2. 中国石化西北油田分公司工程技术研究院)

摘　要：耐腐蚀性能优良的热塑性塑料内衬复合管在含 H_2S/CO_2 油气集输领域得到广泛应用。但在使用过程中，气体组分在热塑性塑料中的渗透导致部分内衬起泡失效，给管线运行和生态环境带来安全隐患。为了了解气体介质在热塑性塑料中的渗透特性并明确其渗透机理，从根本上提出控制气体渗透的有效措施，本文研究了典型酸性气体 CO_2 在广泛采用的高密度聚乙烯（HDPE）内衬材料中的渗透行为，并检测分析了渗透样品的微观形貌、结构成分和耐热性能，探讨了 CO_2 在 HDPE 中的渗透机理。结果表明：CO_2 在 HDPE 中渗透系数随温度升高而提高，渗透样品的微观形貌、结构成分和耐热性能未发生明显变化，气体渗透主要模式为物理渗透过程，不存在化学侵蚀破坏。利用分子动力学模拟得到的扩散系数与实验结果吻合，表明 CO_2 在 HDPE 中的扩散属于正常扩散（空穴间跃迁式的跳跃）。

关键词：高密度聚乙烯；渗透；油田环境；扩散；集输

含 H_2S 天然气已成为我国天然气资源的重要组成部分[1]。对于集输用管线，普通碳钢管在这类酸性环境中（含 H_2S/CO_2）腐蚀非常严重。近年来，施工方便且具有优异的耐腐蚀、柔韧性、抗疲劳性能等特点的增强热塑性塑料复合管（也称 RTP 管、柔性复合管等）正成为含硫油气输送的一个重要选择，已在塔里木油田、塔河油田、新疆油田等得到成功应用[2]。增强热塑性塑料复合管是以热塑性塑料管为内衬层，以金属（钢丝、钢带等）或非金属（芳纶纤维、聚酯纤维等）作为增强层，外敷热塑性塑料保护层复合而成。当前，该产品的热塑性塑料内衬层广泛采用高密度聚乙烯（HDPE）[1]。但鉴于热塑性塑料的本质特性，气体分子的自由运动不可避免地会在 HDPE 中发生渗透现象。气体渗透不仅会导致输送气体的浪费，还会造成 HDPE 起泡、坍塌等失效现象[3,4]。而吸附在 HDPE 内衬表面的气体介质会沿内衬厚度方向发生扩散，进而增加了增强层（尤其是金属增强层）的腐蚀失效风险，复合管承压能力及使用寿命由此降低。

为了有效预防气体渗透损伤破坏 HDPE 内衬，降低增强热塑性塑料复合管在油气输送过程中的失效风险，必须首先了解气体介质在 HDPE 中的渗透特性并明确其渗透机理，才能从根本上提出控制气体渗透的有效措施，也能为改善 HDPE 的气体阻隔性能提供直接依据。本文参考 GB/T 1038—2000，采用压差法气体渗透仪，研究了不同温度下典型酸性气体

基金项目：国家自然科学基金（51304236），陕西省自然科学基金（2014JQ6227）。

作者简介：李厚补（1981—），男，博士，高级工程师，主要研究方向为油气田用非金属与复合材料管材。E-mail：lihoubu@ cnpc. com. cn。

CO_2 在 HDPE 中的渗透特性，结合渗透样品微观形貌、结构成分、耐热性能分析，探讨了 CO_2 在 HDPE 中的渗透机理，为油气介质在热塑性塑料中的渗透性评价及控制提供支撑。

1 实验材料与方法

试验用 HDPE 样品为挤塑制备的 ϕ60mm 圆状薄膜，厚度为 300μm。采用 VAC-V2 型压差法气体渗透仪测试 CO_2 气体在 HDPE 中的渗透系数。试验温度为 30℃、40℃、60℃和 80℃，每种条件下测试 3 个样品。

采用 Hitachi S-4800 型冷场发射扫描电镜观察分析 HDPE 的形貌；采用 VERTEX 70 型傅里叶红外光谱仪测试 HDPE 结构成分变化；利用 DTG204F1 热分析系统测试 HDPE 的 TG-DSC 曲线，升温速率 10℃/min，流动 N_2 保护；采用 Material Studio 6.0（MS）并辅助 Visual Studio 分析 CO_2 在 HDPE 中的渗透机理。

2 结果与讨论

2.1 CO_2 在 HDPE 中的渗透系数

不同温度下 CO_2 在 HDPE 中的渗透系数见图 1。随着温度的升高，CO_2 气体在 HDPE 中的渗透系数不断增加。在恒定压力下，温度对气体在聚合物中扩散系数的影响服从阿累尼乌斯（Arrhenius）方程[5,6]。根据自由体积模型，低温时（如靠近或低于聚合物材料的玻璃化转变温度时），材料自由体积减少，激活能上升。随温度升高，总的自由体积增加，渗透气体分子可在无定型材料中更加自由的运动，此时介质分子遵循“似液体”渗透机理，激活能较低，气体渗透系数随之增加[5]。

2.2 渗透样品性能分析

2.2.1 微观形貌分析

PE 原始样品的微观 SEM 形貌（5k 倍）见图 2。可以看出，样品表面平整，不存在明显的裂纹、孔洞、起泡等缺陷，但具有挤塑拉挤痕迹。不同温度下，CO_2 气体在 HDPE 样品渗透后的形貌见图 3。与 HDPE 的原始 SEM 形貌（图 2）相比，不同温度下 CO_2 气体渗透后的样品微观形貌并未发生明显变化，样品挤塑拉挤痕迹仍较清晰，不存在裂纹、孔洞、起泡等缺陷。

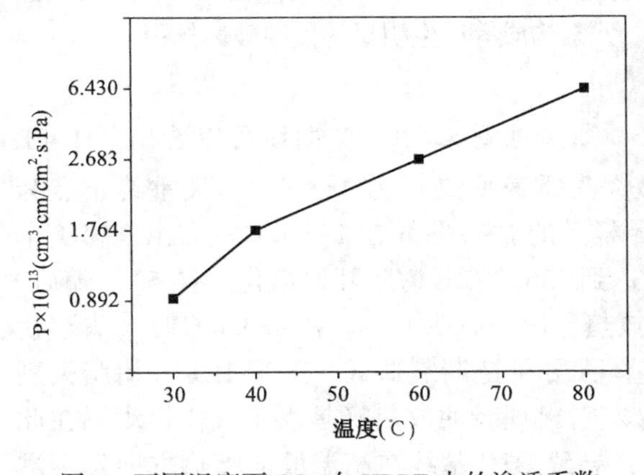

图 1 不同温度下 CO_2 在 HDPE 中的渗透系数

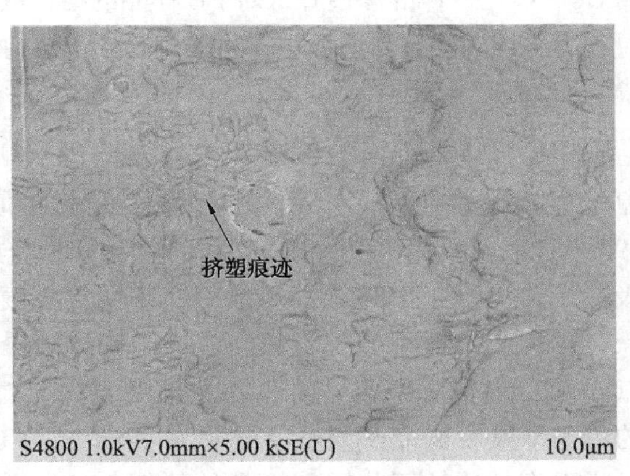

图 2 HDPE 样品原始 SEM 形貌（5k）

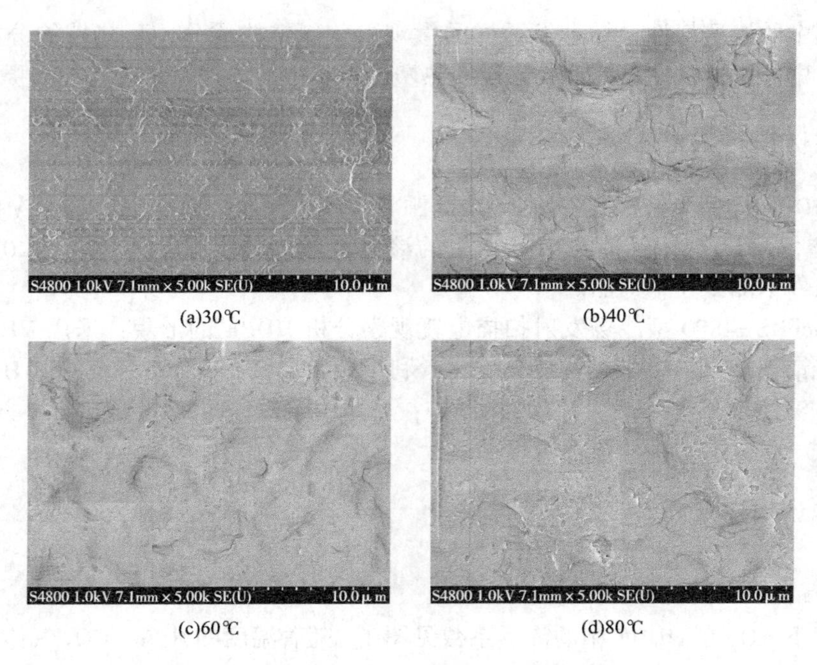

图 3 CO_2 气体渗透后 HDPE 样品 SEM 形貌（5k）

以上结果表明，CO_2 气体的渗透并未对 HDPE 的微观形貌造成明显影响。温度的升高虽然加快了 CO_2 气体在 HDPE 中的渗透速率，但气体渗透进程的加快并未破坏 HDPE 样品的微观形貌。由此可以推断 CO_2 气体渗透过程并未对 HDPE 产生明显的物理或化学破坏作用。

2.2.2 结构成分分析

CO_2 气体在 HDPE 样品渗透前后的 FTIR 图谱见图 4。HDPE 分子式为：—［CH_2—CH_2］$_n$—。由图 4 可以看出，HDPE 原始样品的红外光谱图的特征吸收峰分别在 2912cm^{-1}、2854cm^{-1}、1465cm^{-1}、721cm^{-1}等处。其中 2912cm^{-1}、2854cm^{-1}为亚甲基—CH_2 的非对称和对称伸缩振动吸收峰，1465cm^{-1}为亚甲基—CH_2 的对称弯曲振动吸收峰，721cm^{-1}为亚甲基—CH_2 的面内摇摆振动吸收峰[7]。与 HDPE 原始样品的 FTIR 图谱相比，不同温度下 CO_2 气体渗透后样品 FTIR 图谱中的各处特征吸收峰位置和峰强度均未发生明显变化（图 4），表明不同温度下（30℃、40℃、60℃、80℃），CO_2 气体的渗透并未对 PE 样品的结构成分造成破坏。可以推断，气体在 HDPE 中的渗透过程为单纯的物理渗透过程，并未对其分子结构、元素构成等产生明显的侵蚀破坏作用。

2.2.3 耐热性能分析

CO_2 气体在 HDPE 样品渗透前后的 TG-DSC 图谱见图 5。由 TG 曲线可以看出，HDPE 原始样品起始失重温度为 464.9℃；拐点（最大失重速率点）为 483.9℃，失重终止点为 494.7℃；质量损失 96.1%（图 5）。HDPE 原始样品的 DSC 曲线在 135.0℃左右出现第 1 个吸热峰，该峰是 HDPE 的熔融峰，表明当温度达到 135.0℃时树脂开始熔化（图 5）。随后，聚合物树脂侧链小分子受热断裂逸出导致缓慢放热，使得曲线上升。在 483.0℃时（失重拐点附近）出现一个明显放热峰，是由于树脂基体大分子链断裂造成。随着 HDPE 树脂失重的趋于平缓，树脂热量变化趋近平稳。717.3℃左右出现新的放热峰是由于氧化铝坩埚在此温度下变的通透不能屏蔽样品的热辐射和吸收，导致 DSC 基线在高温段向吸热方向发生漂移，干扰了样品 DSC 信号[7]。

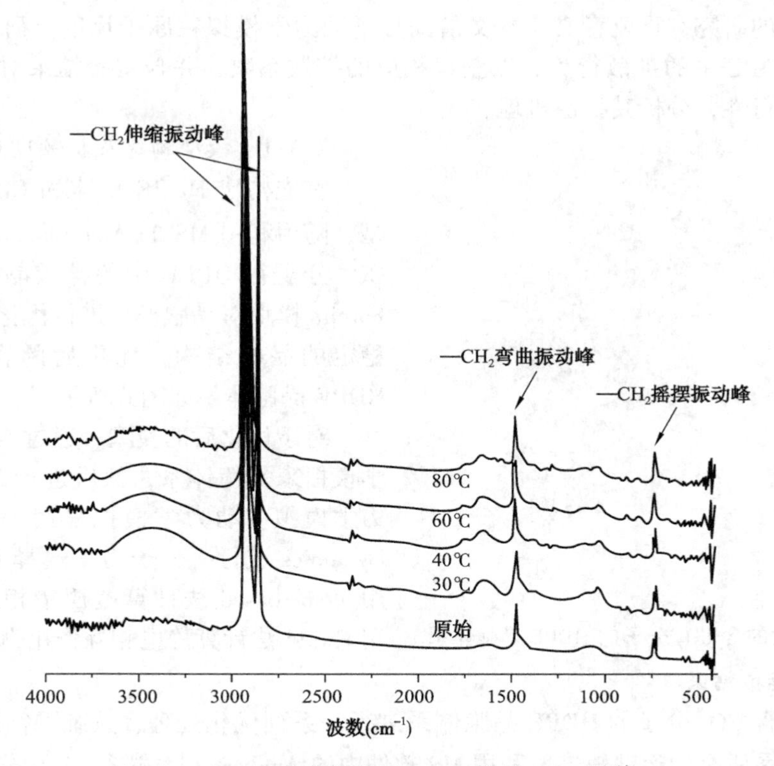

图 4　CO_2 气体渗透前后 HDPE 样品 FTIR 图谱

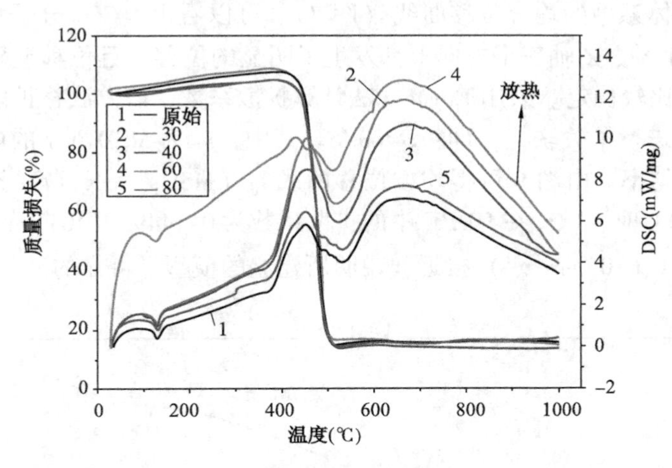

图 5　CO_2 气体渗透前后 HDPE 样品 TG-DSC 曲线

与 HDPE 原始样品的 TG-DSC 曲线相比，不同温度下（30℃、40℃、60℃、80℃）CO_2 气体在 HDPE 样品渗透后的 TG-DSC 曲线变化趋势类似（图 5），吸热峰和放热峰的位置及强度基本保持一致，且并未有其他吸热峰或放热峰出现，表明不同温度下 CO_2 气体的渗透未对 HDPE 样品的耐热性能和热解行为造成影响。

2.3　CO_2 在 HDPE 中渗透的分子动力学模拟

气体在聚合物中的渗透过程通常可以用溶解-扩散机理来描述，它是过去几十年来最为广泛接受的机理模型[8]。即高压气体吸附（溶解）进入材料的高压侧表面；渗透的气体分子在压差的作用下向材料内部进行扩散，最后是气体从低压侧表面脱附。因此，研究扩散过程对渗透过程的研究有重要意义。本文借助分子动力学模拟在原子尺度上研究30℃时CO_2气体分子在HDPE中的扩散行为，以获得相应的扩散系数，并与实验结果对比，验证分子模拟手段的可行性，分析其扩散机理。

2.3.1 模型构建及扩散过程模拟

首先构建HDPE单体和CO_2扩散分子构型，随后利用MS的Amorphous Cell模块构造CO_2分子在HDPE中的扩散晶胞模型。利用Forcite模块对晶胞模型进行优化，以获得相对稳定的晶格结构。优化后的含CO_2分子的HDPE晶胞体系如图6所示。

对于优化后的模型，通过NVT+NVE动力学模拟来平衡体系，然后进行NVE全轨迹动力学模拟。动力学模拟采用Discover模块中Dynamics进行。分子力学选择COMPASS，采用atom-based法计算范德华相互作用力，采用Ewald法计算静电相互作用力。

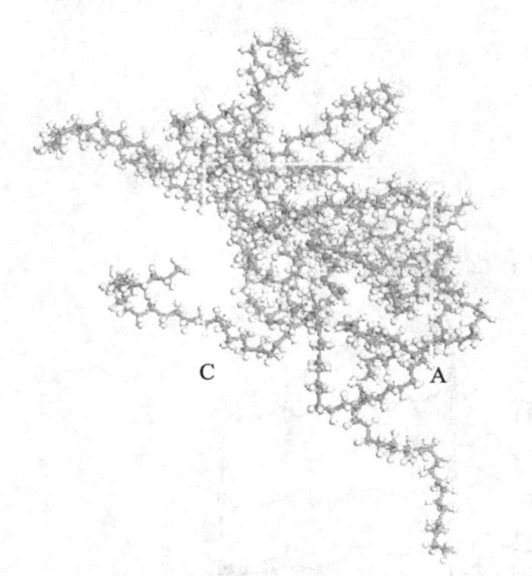

图6 优化后的含CO_2分子的HDPE晶胞体系

2.3.2 模拟结果

对建立的含CO_2分子的HDPE晶胞体系进行一系列优化处理、系统平衡后，进行NVE模拟，保存体系所有的运动轨迹。利用MS软件中的Analysis模块进行MSD分析，得到CO_2分子在HDPE晶胞体系中的均方位移曲线（图7）。可以看出，CO_2分子均方位移曲线线性关系良好，X轴、Y轴、Z轴三个方向上都发生了明显的位移，且位移差别不大，即CO_2分子在体系中的扩散比较稳定。采用Einstein法计算扩散系数，该方法将扩散系数D与分子的均方位移（MSD）进行了关联[9]，即：$D=m/6$。其中，m为MSD对t的曲线通过最小二乘法拟合得到的直线斜率。由图8可得均方位移曲线lg（MSD）—lg（t）拟合后的直线斜率m为0.9694，则30℃时，CO_2在HDPE中的扩散系数为0.14493（$10^{-6}cm^2s^{-1}$），与该条件下实验测试值0.16341（$10^{-6}cm^2s^{-1}$）接近，说明所建立的模型是可靠的。

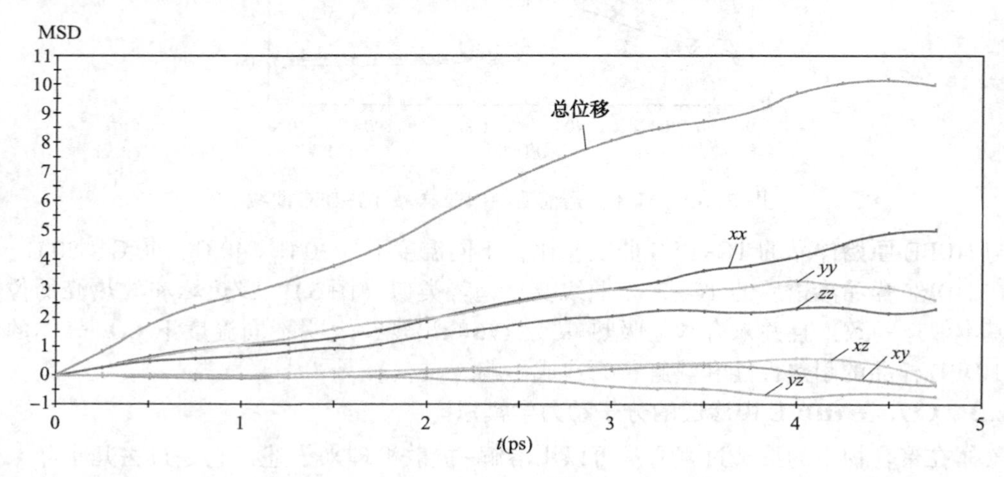

图7 CO_2分子在HDPE中扩散的均方位移图

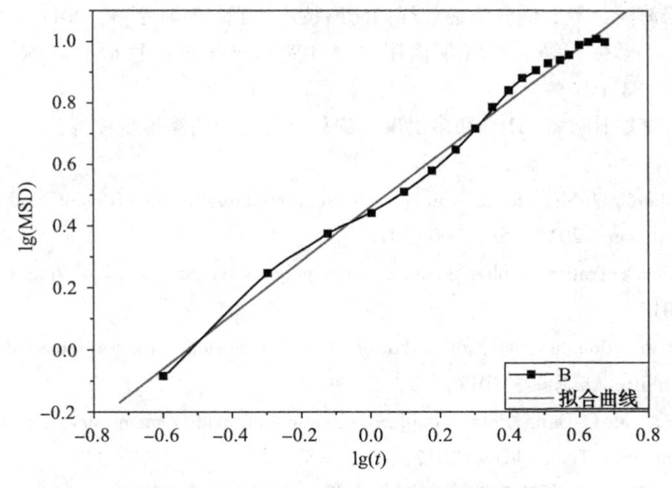

图 8 lg（MSD）对 lg（t）的拟合曲线

2.4 扩散机理初探

聚合物体积由高分子占有体积和未被占有的以空穴形式存在的自由体积组成[10]。自由体积的大小和形态对渗透分子在聚合物中的扩散行为起着重要作用，它为链运动提供了必要的活动空间和小分子的扩散空间。小分子在聚合物内是不断运动的，大部分情况下运动幅度较小，可视为振动。偶尔出现一次大的跃迁，跳跃到另一个区域继续进行小幅振动，这样一个区域就是一个空穴，即自由体积。当气体分子在空穴内振动到某一位置，此时这个聚合物链段发生扭转或相对运动，为小分子发生大的跳跃提供了通道，这样小分子就跳到另一个空穴内继续振动。

以上通过分子动力学模拟，获得了 CO_2 分子在 HDPE 中的均方位移曲线（MSD），利用 lg（MSD）对 lg（t）关系的曲线斜率 m 可判断小分子的扩散行为，若 $m<1$ 属于非正常扩散（空穴内的运动）；若 m 接近 1 时，表明扩散属于正常扩散（空穴间跃迁式的跳跃）[9]。本文通过模拟获得的 m 为 0.9694，接近 1，说明 CO_2 在 HDPE 中的扩散属于正常扩散，其扩散运动轨迹可表示为[9]：（1）在一定时间内，气体分子在一个以某一点为中心的较小空间内跳跃，但不会超过这个“准稳定”区域；（2）“准稳定”区域被打破，气体分子从一个“准稳定”区域快速跳跃进入另一个“准稳定”区域。这样一个“准稳定”区域其实就是一个空穴，气体分子从一个空穴运动到另一个空穴前会在原空穴附近做空穴内的运动。

3 结论

（1）CO_2 气体在 HDPE 中的渗透系数随着温度的升高不断增加，服从阿累尼乌斯方程。

（2）渗透样品的微观形貌、结构成分和耐热性能未发生明显变化，表明 CO_2 在 HDPE 中渗透的主要模式为物理渗透过程，不存在化学侵蚀破坏。

（3）利用分子动力学模拟得到的扩散系数与实验结果基本吻合，扩散系数计算结果表明 CO_2 在 PE 中的扩散属于正常扩散（空穴间跃迁式的跳跃）。

参 考 文 献

[1] 朱光有，张水昌，梁英波，等．四川盆地高含 H_2S 天然气的分布与 TSR 成因证据［J］．地质学报，2006，80（8）：1208-1218.

[2] 李林辉，李浩，屠海波，等．油气集输管线内防腐技术［J］．上海涂料，2011，49（5）：31-33.
[3] 李厚补，张学敏，毛学强，等．油气集输用热塑性塑料气体渗透性研究现状［J］．天然气与石油，2016，34（1）：84-88.
[4] 戚东涛，任建红，李厚补，等. H_2S 在多层聚合物复合管材中的渗透规律［J］．天然气工业，2014，34（5）：126-130.
[5] F. Mozaffari，H. Eslami，J. Moghadasi. Molecular dynamics simulation of diffusion and permeation of gases in polystyrene［J］. Polymer，2010，51：300-307.
[6] C. G. Soney，S. Thomas. Transport phenomena through polymeric systems［J］. Progress in Polymer Science，2001，26：985-1017.
[7] Houbu Li，Milin Yan，Dongtao，Qi，et al. Failure analysis of steel wire reinforced thermoplastics composite pipe. Engineering Failure Analysis，2012，20，88-96.
[8] X. L. Ren，J. Z. Ren，M. C. Deng. Poly（amide-6-b-ethylene oxide）membranes for sour gas separation［J］. Separation and Purification Technology，2012，89：1-8.
[9] 钟颖．分子模拟研究气体在高渗透性膜中扩散溶解行为［D］．重庆：西南大学，2012.
[10] R. Scheichl，M. H. Klopffer，Z. Benjelloun - Dabaghi，B. Flaconnèche. Permeation of gases in polymers：parameter identification and nonlinear regression analysis［J］. Journal of Membrane Science. 2005，254：275-293.

自蔓延高温合成陶瓷内衬油管的性能研究

李厚朴[1]　王守泽[2]　杨永利[2]　戚东涛[1]　丁　楠[1]　段国栋[3]

（1. 石油管材及装备材料服役行为与结构安全国家重点实验室；2. 松原大多油田配套产业有限公司；3. 长庆油田分公司第五采油厂）

摘　要：采用离心自蔓延高温合成法将氧化铝（$\alpha-Al_2O_3$）陶瓷衬入J55油管（ϕ73mm×5.5mm）内表面，制备出陶瓷内衬复合油管。系统评价了陶瓷内衬油管的力学、耐腐蚀、抗结垢结蜡、结合、抗冲击、弯曲等各方面的性能，并与J55油管的力学、耐腐蚀、抗结垢结蜡性能进行了对比。结果表明：陶瓷内衬油管的力学性能与J55油管基本相近；相同腐蚀条件下，J55油管的失重率达到陶瓷内衬油管失重率的600倍以上；相同结垢结蜡条件下，J55油管的增重率达到陶瓷内衬油管增重率的70倍以上。陶瓷内衬油管（ϕ73mm×5.5mm，J55基管）的界面结合强度为36.68MPa，最大抗冲击能量为73.5J，最小弯曲半径为11.47m。

关键词：陶瓷内衬；油管；结垢结蜡；连接；结合；冲击；弯曲

油管腐蚀与磨损一直是困扰采油发展的严重问题，它不仅给采油带来大量的人力、物力浪费，而且直接影响原油开采效率和经济效率的提高[1]。随着金属表面处理工艺的不断改进，各种防腐、耐磨油管不断投入应用[2,3]。但不论是涂料涂层，还是镀渗金属衬层，或是玻璃钢内衬、高密度聚乙烯内衬，一般都只有某项性能指标突出，综合性能差：防腐的不耐磨，耐磨的不防腐；既防腐又耐磨的又怕冲击或怕高温、气浸，不能很好地满足恶劣油井环境的要求[2-5]。近年发展起来的离心自蔓延高温合成法（SHS）把氧化铝陶瓷（$\alpha-Al_2O_3$）衬在油管内壁，制备出的陶瓷内衬复合油管具有优良的抗蚀性、耐磨性、隔热性等特点，为解决以上问题提供了新途径[6]。由于制作工艺简单灵活，设备要求不高，生产率高，成本低，陶瓷内衬复合油管产品在油田开采领域具有极大的应用潜力和前景[6,7]。

当前，针对陶瓷内衬油管的研究报道相对较少，研究热点主要集中在陶瓷内衬油管产品的质量控制领域，如陶瓷衬层的损伤控制及质量优化[8]，陶瓷内衬油管管端强韧化[9,10]，陶瓷内衬油管的力学性能数值模拟[11,12]等。然而，广大油田用户普遍较为关注产品如下实际应用问题：内衬陶瓷后的油管力学性能；陶瓷内衬复合管连接处的耐蚀性；陶瓷内衬油管的抗结垢结蜡能力；陶瓷内衬层与金属油管基体的结合；陶瓷内衬油管的抗外力冲击性能及抗弯曲能力等。由于当前没有相应的产品标准或成熟的试验方法来评价及验证陶瓷内衬油管的以上性能，油田用户对具有广阔应用前景的陶瓷内衬油管大多还持有怀疑态度，严重阻碍了该产品的进一步推广应用。基于此，本文首先对比测试了陶瓷内衬油管（J55油管作为基管）和J55普通油管的拉伸性能和承压性能，随后模拟陶瓷内衬油管实际应用情况，开发设

基金项目：国家自然科学基金（51304236，51504037）。

作者简介：李厚朴（1981—），男，江苏徐州人，博士。

计了相应的试验装置和测试系统，评价了陶瓷内衬油管的耐蚀性（包括连接部位）、抗结垢结蜡性能、内衬结合性能、抗冲击性能及弯曲性能等，旨在解答用户共同关注的产品实际应用问题，为今后陶瓷内衬油管的质量控制和推广应用提供借鉴。

1　试样制备与试验方法

松原大多油田配套产业有限公司制备陶瓷内衬油管（ϕ73mm×5.5mm，J55 基管）试验样品。首先将氧化铁粉和铝粉按照重量 3∶1 的比例，添加少量硅粉后混合均匀（形成铝热剂体系），放入 200℃烘箱中烘干后将其置入经热解除油、喷砂矫直处理的 J55 油管。铝热剂用量控制在 1.1g/cm^2 范围内。随后将装好铝热剂的 J55 油管放入离心机并调整油管中心，启动离心机待其转速达到 2400r/min 时点燃铝热剂体系。SHS 反应完成后平稳降低离心机转速至停止，自然冷却后即完成陶瓷内衬油管的制备。另外，选择该公司提供的 J55 普通油管（ϕ73mm×5.5mm）做为对比测试分析样品。

依据 API RP 5C5—2003，分别采用复合加载试验系统和 HY-MLF-50K-WI 静水压爆破试验机检测陶瓷内衬油管的室温拉伸性能及承压性能，试验样品长度均为 3m，数量均为 3 根。依据 GB/T 13298—1991，采用 MEF4M 金相显微镜及图像分析系统检测陶瓷内衬油管 J55 基管的金相组织。

采用自制耐蚀性评价试验装置测试样品的耐腐蚀性能。采用高温陶瓷胶将两个陶瓷内衬油管管段（长度 50mm）的陶瓷内衬层粘接密封后，再对两管段外层钢管进行焊接。随后将 J55 普通油管管段（长度 50mm）、陶瓷胶粘接陶瓷层管样用聚四氟乙烯垫圈隔离并密封，然后用法兰盘螺杆拉紧固定。循环泵驱动硫酸池内的硫酸溶液流经试验管段后返回硫酸池，形成循环回路。到达设定试验时间后，取出试样，清洗、烘干、称重，计算试验结果并观察试样腐蚀形貌。试验条件为：室温、600h、10wt%的硫酸溶液，循环泵排量为 19L/min。

采用自制结垢结蜡试验装置测试样品的抗结垢结蜡性能。将陶瓷内衬油管管段（长度 200mm）和 J55 油管管段（长度 200mm）同时置于试验装置上，用聚四氟乙烯垫圈隔离后采用法兰盘螺杆拉紧固定。循环泵驱动储液池内的原油流经试验管段后返回储液池，形成循环回路。到达设定试验时间后，取出试样称重，并观察试样内表面形貌。试验条件为：50℃、168h、油田现场输送原油介质，循环泵排量为 6.3m^3/h。

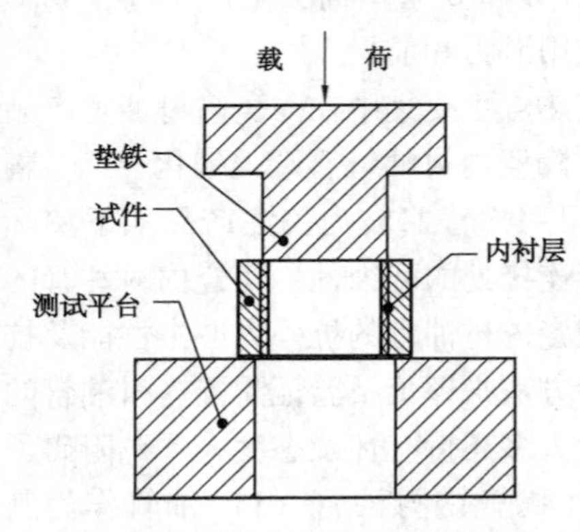

图 1　内衬结合性能测试示意图

参考 CJ/T 136—2007，采用图 1 所示方法，利用 UH-F500KNI 万能试验机测试陶瓷内衬层与基管的结合强度。试验管环样品宽度为 15mm，数量为 3 个。

参考 GB/T 6112—1985，采用 XJL-300C 落锤冲击试验机测试复合管的抗冲击性能。选择半径为 5mm 的冲击锤（R5），落锤总质量设置为 10kg。对不同试验管段（长度 200mm）分别冲击，冲击高度从 0.2m 以上递增，确定复合管陶瓷内衬出现破坏时的最终高度，并计算此时的冲击能量。试样经冲击作用后陶瓷内衬层出现在自然光线下肉眼可见的裂纹、龟裂和破碎的现象均视为破坏。

采用自制装置测试陶瓷内衬油管的抗弯曲性能。将陶瓷内衬油管（长度 10m）两端固定，并将内窥镜探头置入管材内部中间点位置，该位置处内衬层形貌实时传输到计算机影像系统。随后，采用手动链式滑轮对复合管中间点施加拉力，逐步将中间点提升至设定的系列位移增量。当计算机影像系统中观测到内衬层出现裂纹时终止实验，记录此时复合管中间点的累积位移，并结合复合管两固定点长度计算管材的最小弯曲半径。

2 试验结果与讨论

2.1 力学性能

以 J55 油管为基管的陶瓷内衬油管拉伸性能检测结果见表 1。3 根陶瓷内衬油管的平均拉伸失效载荷为 925kN，满足 API TR 5C3 标准要求的≥443.5kN。陶瓷内衬油管轴向抗拉强度平均值为 699.2MPa，满足 API SPEC 5CT 标准要求，且与 J55 普通油管的抗拉强度（约为 705.6MPa）相当[13]，表明内衬陶瓷后陶瓷内衬油管的拉伸性能未发生明显变化。3 根陶瓷内衬油管静水压试验后管体均未发生泄漏（表 2），水压爆破强度分别为 106.2MPa、108.7MPa 和 105.9MPa，均满足 API TR 5C3 标准要求的≥50MPa。

表 1　陶瓷内衬油管拉伸性能检测结果

基管钢级	试样编号	失效载荷（kN）	抗拉强度（MPa）	失效位置和失效形貌
J55	1Y	899	677.9	管体断裂失效
	2Y	966	720.5	管体断裂失效
	3Y	911	705.4	管体断裂失效

表 2　陶瓷内衬复合油管静水压及水压爆破检测结果

基管钢级	试样编号	测试项目	试验压力（MPa）	保载时间（s）	失效位置和失效形貌
J55	1Z	静水压	46.0	600	未发生泄漏
		水压爆破	106.2	/	管体爆破失效
	2Z	静水压	46.0	600	未发生泄漏
		水压爆破	108.7	/	管体爆破失效
	3Z	静水压	46.0	600	未发生泄漏
		水压爆破	105.9	/	管体爆破失效

经检测，内衬陶瓷后的 J55 基管金相组织（图 2）与 J55 普通油管的金相组织[13]基本一致，包括珠光体（P）和铁素体（F）。内衬陶瓷后的 J55 基管非金属夹杂物 A 类、B 类、C 类和 D 类的最高等级为 1.5 级，均满足 API SPEC 5CT 标准≤2.5 级要求。内衬陶瓷后的 J55 基管晶粒度检测值为 10.0 级，高于 J55 普通油管的晶粒度等级（约为 7.0 级）[13]。离心 SHS 制造陶瓷内衬油管过程中放热反应的燃烧温度高达 2450K[14]，虽然整根复合油管 SHS 反应时间仅为几秒钟，但整个 SHS 制备陶瓷内衬的过程相当于对 J55 油管进行了正火处理，这是其晶粒出现细化并保持其力学性能基本不变的主要原因。

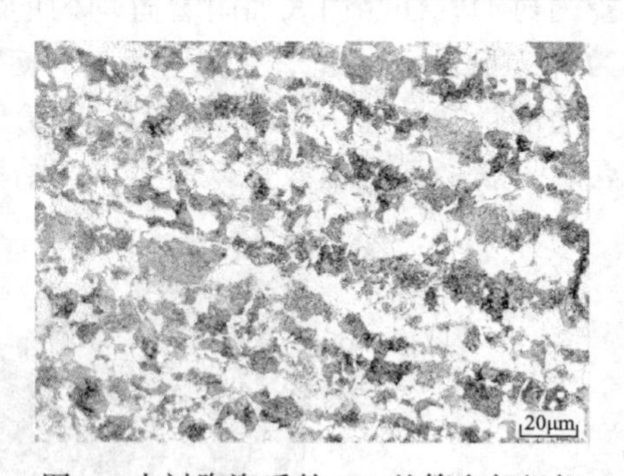

图 2　内衬陶瓷后的 J55 基管金相组织

2.2 耐腐蚀性能

耐腐蚀性能试验前的陶瓷内衬油管内表面平整、均匀、连续（图 3a），得益于陶瓷层优异的耐蚀性能，试验后的复合管样品内表面形貌基本无变化，未发现陶瓷层开裂、脱落、孔洞等缺陷（图 3b）。试验前后陶瓷胶粘接样品内径保持一致，经 600h 腐蚀试验后样品失重率仅为 0.09%，且陶瓷层粘接缝隙形貌完整（图 3b），表明陶瓷胶粘接陶瓷层的连接部位具有良好的耐蚀完整性。与此相比，J55 普通油管样品在相同条件下试验后的失重率高达 51.11%，与试验前平整光滑的内表面（图 4a）相比，腐蚀试验后样品的内表面出现深度不一的腐蚀沟壑（图 4b），导致试样内径无法测量。

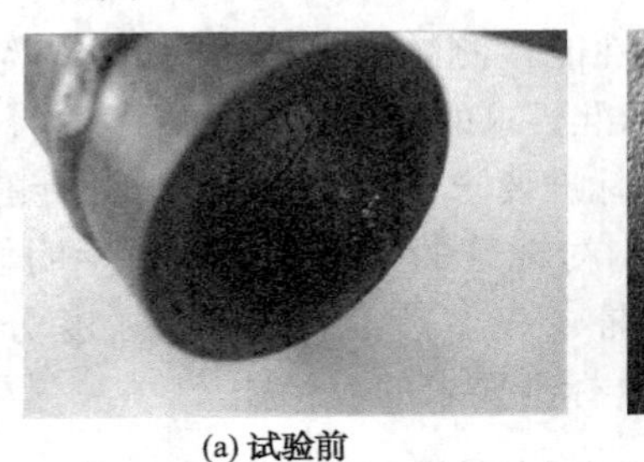

(a) 试验前　　(b) 试验后

图 3　陶瓷胶粘接复合管样品腐蚀试验前后内表面形貌

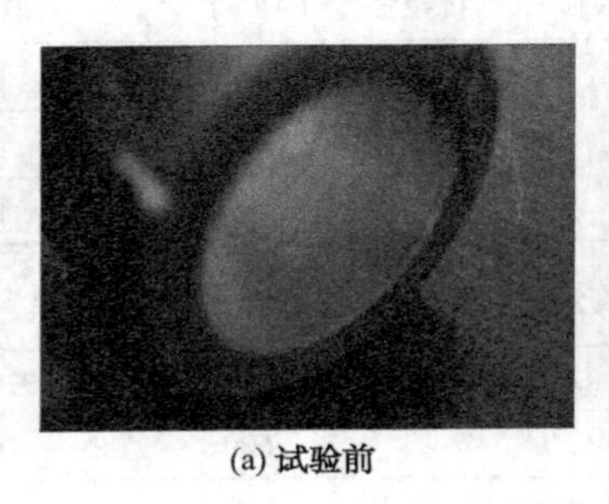

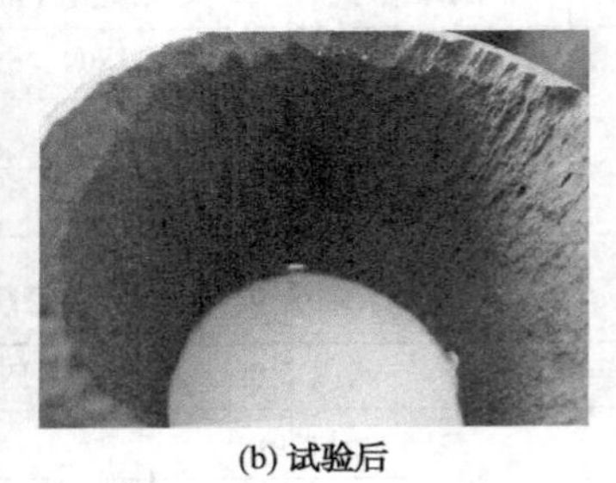

(a) 试验前　　(b) 试验后

图 4　J55 普通油管样品腐蚀试验前后内表面形貌

2.3　抗结垢结蜡性能

陶瓷内衬油管和 J55 普通油管样品经结垢结蜡试验后均出现增重现象。陶瓷内衬油管样品的增重率仅为 0.45%，而 J55 普通油管样品的增重率则高达 31.5%。结垢结蜡试验后的陶瓷内衬油管样品内表面附着有均匀的沉淀物（图 5a），复合管内径由原来的 57.94mm 略减为 57.76mm。从管壁上取下的该沉淀物可熔融、可燃烧，应为以蜡为主的结垢产物。而 J55 普通油管样品经结垢结蜡试验后，管体内部被沉积物完全填充（图 5b）。该填充物同样可熔融、可燃烧，也应该是以蜡为主的结垢产物。

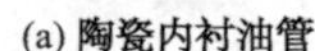

(a) 陶瓷内衬油管　　(b) J55普通油管

图 5　抗结垢结蜡试验后管样内表面形貌

管道结垢原因包括腐蚀结垢、静电吸附结垢、结晶结垢、沉淀结垢等几种原因[15]。陶瓷内衬层具有优异的耐蚀性且无电荷存在，大大降低了腐蚀结垢和静电吸附结垢发生的概率。另外，与金属管道相比，陶瓷内衬层具有较低的表面能，与油气介质的润湿性差[16]，使得蜡和垢无法在陶瓷表面生根集结，难以出现结晶结垢和沉淀结垢，因此表现出良好的抗结垢结蜡性能。

2.4 结合性能

图 6 为结合性能测试之后的样品形貌。可以看出，样品的陶瓷内衬层整体推出约 5mm，陶瓷层结构保持完整（图 6），依据 CJ/T 136—2007 标准计算其界面结合强度为 36.68MPa。SHS 过程中的放热反应使生成物 Al_2O_3 和 Fe 瞬时熔化，由于 Fe 的密度比 Al_2O_3 密度大，因而在离心力的作用下液态反应产物相互分离，Fe 层紧靠 J55 油管内表面成为钢管与陶瓷之间的过渡层，进而实现了陶瓷内衬层与钢管基体的冶金结合[4]，因此其界面结合强度远大于超高分子量聚乙烯内衬油管等机械结合的界面结合强度要求（SY/T 6947—2013 要求超高分子量聚乙烯内衬油管的界面结合强度不低于 0.15MPa）。

图 6　结合性能测试后样品形貌

2.5 抗冲击性能

当落锤冲击高度分别为 0.2m、0.5m 和 0.7m 时，陶瓷内衬油管外表面有明显的冲击坑，但陶瓷内衬层形貌完好，未观察到明显的破坏现象。当落锤冲击高度增加至 0.75m 时，不仅管材外表面存在明显的冲击坑（图 7a），陶瓷内衬层也发现明显的冲击破坏剥落现象（图 7b），依据 GB/T 6112—1985 计算此时管材的冲击能量为 73.5J，该冲击破坏能量能够确保陶瓷内衬复合管满足现场施工过程中的磕碰与应用过程中的振动要求。SHS 过程中陶瓷层与钢管胀缩不同步引起的拘束力作用使得外层油管紧压内衬陶瓷层[17]，这种压应力状态使陶瓷层具有较好的耐外机械冲击及耐热冲击能力。

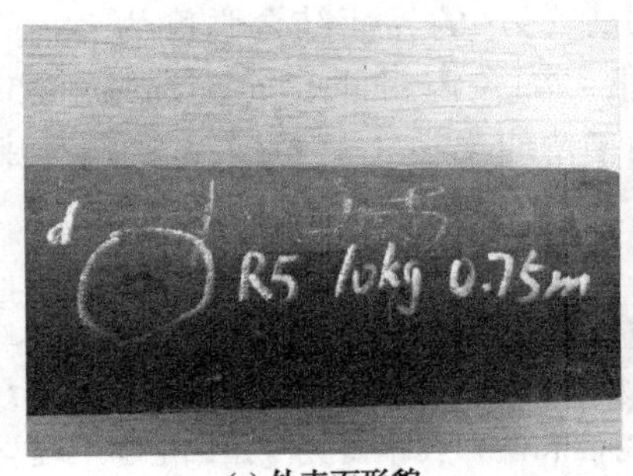

(a) 外表面形貌

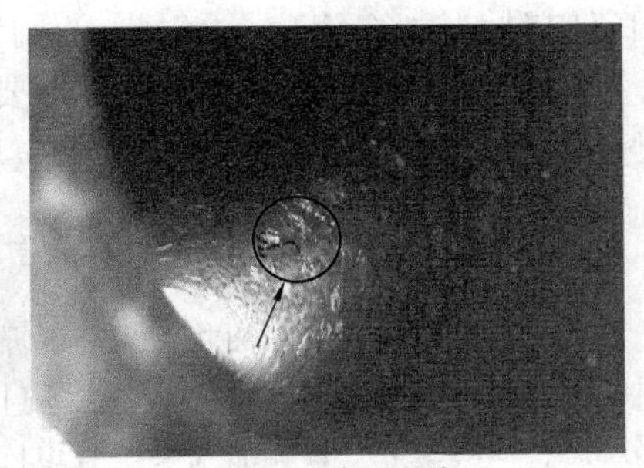
(b) 内表面形貌

图 7　冲击高度 0.75m 时的样品形貌

2.6 弯曲性能

陶瓷内衬油管在中心累积位移达到 870mm 时，内衬层仍保持完好（图 8a）。当累积位移增加至 890mm 时，陶瓷内衬层开始出现裂纹（图 8b），此时陶瓷内衬油管呈现明显弯曲，计算其弯曲半径为 11.47m。陶瓷（Al_2O_3）材料的塑性形变量有限，尤其在离心 SHS 工艺过程中，虽然陶瓷材料一度处于极高的温度，但由于冷却速率太高，亦即应变速率很大，所以产生的塑性形变量也就有限[17]。这是陶瓷层相对容易产生开裂的本质原因。虽然可以通过改善铝热剂配方（如添加提高陶瓷韧性的添加剂）、选择合理的工艺控制技术（如控制涂料厚度、预热温度和离心力）等改善陶瓷层的抗开裂能力[17]，但在斜井结构中应用陶瓷内衬油管时，仍需注意相应规格管材的最小弯曲半径要求，防止弯曲角度过大导致陶瓷内衬层开裂。

(a) 累积位移870mm

(b) 累积位移890mm

图 8　弯曲试验中陶瓷内衬表面形貌（20 倍）

3　结论

（1）陶瓷内衬油管（ϕ73mm×5. 5mm，J55 基管）的拉伸性能和承压性能与 J55 普通油管基本一致，均满足 API 标准要求。内衬陶瓷后的基管金相组织与 J55 普通油管类似，晶粒度等级得到提高。

（2）相同腐蚀试验条件下，钢管的失重率达到陶瓷内衬油管失重率的 600 倍以上。与钢管相比，陶瓷内衬油管（含连接部位）腐蚀试验前后的质量、尺寸和形貌均未发生明显变化，表现出良好的抗腐蚀性能。

（3）相同结垢结蜡试验条件下，钢管的增重率达到陶瓷内衬油管增重率的 70 倍以上。与钢管相比，陶瓷内衬油管试验前后的质量、尺寸和形貌均未发生明显变化，表现出良好的抗结垢结蜡性能。

（4）陶瓷内衬油管（ϕ73mm×5. 5mm，J55 基管）的界面结合强度为 36. 68MPa，最大抗冲击能量为 73. 5J，满足现场施工及应用要求。陶瓷内衬油管（ϕ73mm×5. 5mm，J55 基管）内衬层开裂时弯曲半径为 11. 47m，可应用于弯曲半径大于 11. 47m 的斜井中。

参　考　文　献

[1] 霍光春，石茂才，杜素珍，等．油井腐蚀最新研究进展［J］．辽宁化工，2015，44（5）：591-594.

[2] 杨蕾，侯明明．国内油井防腐油管技术综述［J］．全面腐蚀控制，2014，28（4）：26-27.

[3] 段旋，郭悠悠，徐峰，等．含硫油气集输管道内腐蚀和防护技术研究［J］．化学工程与装备，2015，（1）：45-48.

[4] 李厚补，严密林，戚东涛，等．离心 SHS 陶瓷内衬复合钢管存在问题评述［J］．热加工工艺，2010，39（24）：130-134.

[5] 李厚补，严密林，戚东涛，等．重力分离 SHS 陶瓷内衬复合钢管存在问题评述［J］．热加工工艺，2011，40（2）：100-104.

[6] 何鹏，刘寒梅．陶瓷内衬复合油管在靖边油田的应用［J］．延安职业技术学院学报，2011，25（3）：128-130.

[7] 田加明．离心-SHS 法制备 Al_2O_3 复合陶瓷及耐磨性研究 [D]. 哈尔滨：哈尔滨理工大学，2005.6.
[8] 张洪霖．油管内衬 SHS 陶瓷涂层的变形损伤研究 [D]. 大庆：东北石油大学，2013.6.
[9] 沈于森．陶瓷内衬油管端部强韧化及抽油泵摩擦副磨损性能研究 [D]. 长春：吉林大学，2010.5.
[10] 吴耀达，沈于森，安健，等．P110 陶瓷内衬复合油管管端感应加热强化处理 [J]. 石油矿场机械，2010，39（7）：60-63.
[11] 孙立强，解明，朱红波．内衬 SHS 陶瓷复合油管抗压性能有限元分析与试验研究 [J]. 广州化工，2014，42（12）：105-107.
[12] 刘赛寅．离心自蔓延陶瓷复合油管的力学性能及热应力数值模拟 [D]. 长春：吉林大学，2010.5.
[13] 张明信，杜义．J55 油管管体材料的组织和力学性能 [J]. 科技情报开发与经济，2001，11（3）：37-38.
[14] 符寒光，邢建东．自蔓延高温合成技术在石油工业中的应用展望 [J]. 石油机械，2001，29（7）：48-51.
[15] 王兵，李长俊，廖柯熹，等．管道结垢原因分析及常用除垢方法 [J]. 油气储运，2008，27（2）：59-62.
[16] 石成刚，王国庆，刘锋．陶瓷功能梯度涂层的防垢机理研究 [J]. 石油机械，2006，34（9）：10-13.
[17] 张曙光，张宝平，李俊．离心 SHS 陶瓷复合钢管裂纹的控制 [J]. 稀有金属，2002，26（3）：225-230.

增强热塑性塑料连续管标准现状及发展建议

李厚补[1]　羊东明[2]　戚东涛[1]　朱原原[2]　葛鹏莉[2]　张志宏[2]

（1. 石油管材及装备材料服役行为与结构安全国家重点实验室，中国石油集团石油管工程技术研究院；2. 中国石化西北油田分公司工程技术研究院）

摘　要：增强热塑性塑料连续管（RTP 管）因其具有柔性好、接头少、单根长、重量轻、易安装等一系列优点，现已广泛应用于我国油田的地面油气集输、高压注醇、油田注水、污水处理等，在井下管、海洋管等领域也得到成功试用。但与之相配套的 RTP 管标准化工作还相对滞后。本文总结了我国 RTP 管产品标准现状，分析了现有各标准存在的主要技术问题。在总结国内 RTP 管标准化工作存在问题的基础上，提出了今后 RTP 管的标准化工作建议，力求助推国内 RTP 管新产品和新技术的快速有形化，促进石油工业质量效益不断提高。

关键词：增强热塑性塑料管；RTP 管；集输；标准化；石油天然气

增强热塑性塑料连续管（也称 RTP 管、柔性复合管、盘卷管等）因其连续成型，单根可达数百米、接头少，柔性好、抗冲击性能优良，重量轻、运输成本低，安装快速简单等一系列优点，在国内外油气田得到广泛推广应用，成为最具有发展潜力的非金属管材之一[1-3]。相比较而言，国外 RTP 管产品已相当成熟，且已开发了海洋用柔性高压输送管。目前相关标准也已实现系列化，包括 API RP 15S—2006《柔性管验收规范》、API SPEC 17K—2010《黏结柔性管规范》、API SPEC 17J—2014《非黏结柔性管规范》、API 17B—2014《柔性管规程》等[4,5]。截至目前，国内 RTP 管产品主要应用于地面油气集输、高压注醇、油田注水、污水处理等[6-10]。2011 年以来，在输气（油气混输）、井下注水等领域已开始拓展试验，同时也已开启了海洋柔性管产品的研发和试验[11-13]。可以看出，国内 RTP 管的产品及应用技术向着系列化、多样化方向发展，进步显著。但与之相配套的 RTP 管标准化工作还存在相对滞后的问题[14,15]。本文分析了国内 RTP 管标准化现状，对比说明了各标准存在的主要问题，围绕现有标准亟须规划的领域，提出了今后 RTP 管标准化的发展方向和建议。

1　国内 RTP 管标准化现状

我国生产 RTP 管的历史并不长，目前国内产品尚无统一的正式名称，有的使用“RTP 管”，有的则根据产品特点自行命名如“柔性复合高压输送管”“连续增强塑料复合管”等。国内生产厂家的情况也是各不相同，有的引进国外成熟的生产技术，有的则是自行设计研制生产工艺；有的选用高性能材料生产高档产品，有的则选用相对低档的材料走经济型路线。

基金项目：国家自然科学基金“油气耦合介质在热塑性塑料中的渗透特性及机理研究”（项目编号：51304236）。

作者简介：李厚补，男，博士，高级工程师，从事油气田用非金属与复合材料管材的研究。lihoubu@cnpc. com. cn。

目前市场上的产品可谓是品种多样、性能各异。为了规范 RTP 管产品及市场，标准化工作引起了大家的普遍重视。2008 年至今，针对不同的产品类型和应用领域，陆续制定了 SY/T 6716—2008、SY/T 6794—2010、SY/T 6662.2—2012、SY/T 6662.4—2014、SY/T 6662.5—2014、SY/T 6662.6—2014 等 6 项 RTP 管产品标准，同时配套 SY/T 6662.4—2014 的钢骨架增强塑料复合连续管产品标准，颁布了 SY/T 6770.4—2012 和 SY/T 6770.4—2012 等质量验收和设计、施工验收规范，有效保障了国内 RTP 管的规范化生产及应用。

2 国内 RTP 管产品标准简介

2.1 SY/T 6716—2008

2008 年，针对有机纤维增强的 RTP 管（产品名称：柔性复合高压输送管）首次起草了产品标准 SY/T 6716。该标准规定了 RTP 管的产品分类、技术要求、试验方法、检验规则、标志、包装、运输、贮存和安装要求。标准描述的 RTP 管为多层结构，主要由聚合物内衬层、增强层、外护套构成，典型的管层结构如图 1 所示。

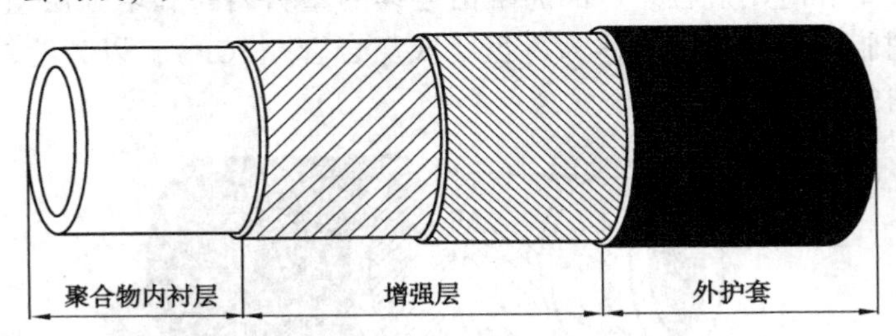

图 1 SY/T 6716 规定的 RTP 管管层结构

该标准的颁布对于 RTP 管在石油天然气行业的规范应用起到了积极作用。但在后续应用过程中发现，标准中还存在以下主要问题：

（1）理化性能检测不明确。本标准并未给出 RTP 管的静水压强度试验条件（如温度、保压压力、保压时间等），且未提及 RTP 管的水压爆破强度要求。管材的承压性能测试与评定存在较大问题。

（2）技术指标不合理。标准中提及的 RTP 管公称压力修正系数普遍较高，如在 0~70℃ 范围内修正系数仍选择为 1.0。但实际测试表明，RTP 管随温度的升高（<70℃），其水压爆破强度下降明显。

（3）标准内容涵盖的管材规格有限。随着制造技术的进步，RTP 管规格型号日益增多，压力等级逐渐系列化。但该标准涵盖的管材规格有限，不能满足现有市场需求。

2.2 SY/T 6662.2—2012

鉴于 SY/T 6716—2008 标准存在以上主要问题，严重制约了 RTP 管产品的发展，2012 年对其进行了全面修订，形成了 SY/T 6662.2—2012。该标准描述的产品类型未发生变化，主要修订了复合管的温度和压力范围，增加了 RTP 管管层结构、接头结构及其材料说明，修改了复合管规格尺寸、最小弯曲半径、公称压力修正系数等。同时重点修订了复合管理化性能检测要求，增加了爆破强度试验要求等。该标准的发布对于国内 RTP 管的制造、质量监督、施工安装等提供了完善的指导。但该标准在修订时仍未考虑以下主要问题：

（1）高温工况的选择及评价。本标准虽然明确提出了 RTP 管静水压强度、水压爆破强度等承压性能试验条件及方法，但对用户普遍关心的高温性能（如设计温度）评价还尚未提及。

（2）内衬材料适用性评价。RTP管用热塑性塑料内衬在油气田工况条件下的适用性成为油田用户普遍关注的焦点。但当前标准中尚未规定系统的适用性评价方法，无法评定材料的适用性，给材料的选择和评价带来极大困难。

（3）内衬材料气体渗透性能评价。RTP管现已试用于油气集输或天然气输送等领域，气体分子的自由运动不可避免地在热塑性塑料表面产生吸附、扩散等渗透现象，带来了起泡和坍塌失效风险。标准并未给出材料气体渗透性评价方法，制约了含气体介质（尤其是酸性气体）输送用热塑性塑料内衬的选材。

2.3 SY/T 6794—2010

为了适应国内油气田市场发展需要，国内某些厂家引进了国外整套先进设备，制造了与国外产品结构相似的高性能RTP管。在此背景下，SY/T 6794—2010由该产品对应的API RP 15S：2006等同采标制定。该标准主要规定了油田集输管线用可盘绕式增强塑料管线管的设计、制造、评定和应用规程。产品通常由连续的塑料内衬和增强层组成（图2），以玻璃纤维增强环氧树脂作为增强层的称为可盘绕式复合管（SCP），以芳纶纤维为增强层的称为增强热塑性塑料管（RTP）。

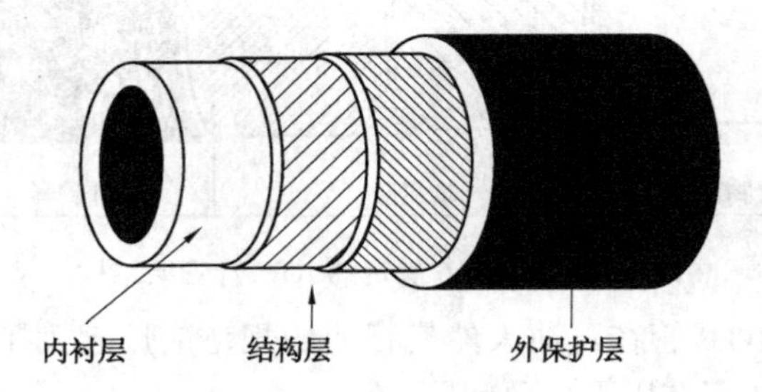

图2　SY/T 6794规定的可盘绕式增强塑料管结构

本标准主要为RTP管产品评定标准，评定内容包括内衬材料的适用性、渗透性、耐温性，还包括复合管的压力等级、提高温度、接头或套筒、最小弯曲半径等评定内容。标准理念先进，重点考核管材的安全性和可靠性，对RTP管的设计具有非常重要的指导意义。该标准提出的试验评定方法能有效评定复合管的各项性能，但试验周期明显较长，试验样本数量众多，对国内大多数RTP管制造商来讲，可操作性相对较差，应用频率不高。

2.4 SY/T 6662.4—2014

在修订了管材及接头规格，介质输送层、热熔胶、保温层材料性能指标，管材短期静水压、最小弯曲半径试验要求及压力修正系数的基础上，SY/T 6662.4—2014代替SY/T 6795—2010规定了一种钢带增强RTP管（产品名称：钢骨架增强热塑性塑料复合连续管）的产品分类、技术要求、试验方法、检验规则、标志、包装、运输和贮存。该类产品同样为多层结构的柔性管，分为传输层、钢带增强层、隔离层和保护层（图3）。产品接头分为内胀接头和内胀外扣接头，均由芯管和外套组成。

比标准SY/T 6662.2—2012进步的是，本标准对RTP管产品的高温承压性能、内衬材料适用性、气体渗透性、轴向拉伸强度、最小弯曲半径等评价内容进行了详细的规定。但与SY/T 6794—2010相比，该标准对管材的压力等级评定尚未提出相应要求。

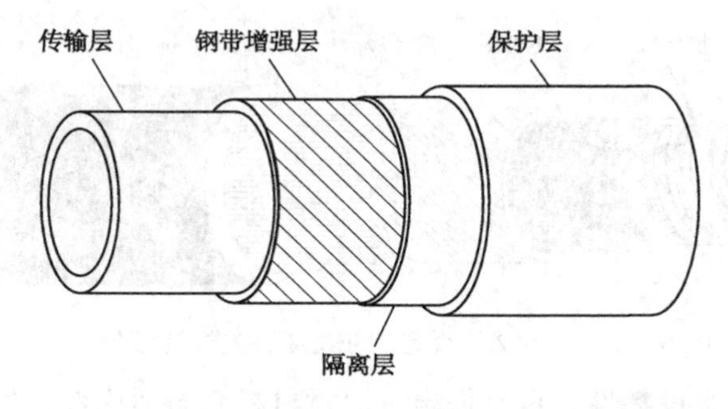

图 3　SY/T 6662.4 规定的钢带增强塑料连续管结构

2.5　SY/T 6662.5—2014

本标准规定的产品结构类型与 SY/T 6662.2—2012 相似（图 4）。但由于所用内衬层材料选用超高分子量聚乙烯或改性超高分子量聚乙烯，因此该标准对内衬材料的特殊性能进行了明确的规定。同时也提出了材料介质适用性、气体渗透性能等评价方法。但对管材的最小弯曲半径评价、轴向拉伸强度等未作要求。与 SY/T 6662.4—2014 相同的是，该标准同样未提出管材压力等级评定要求。

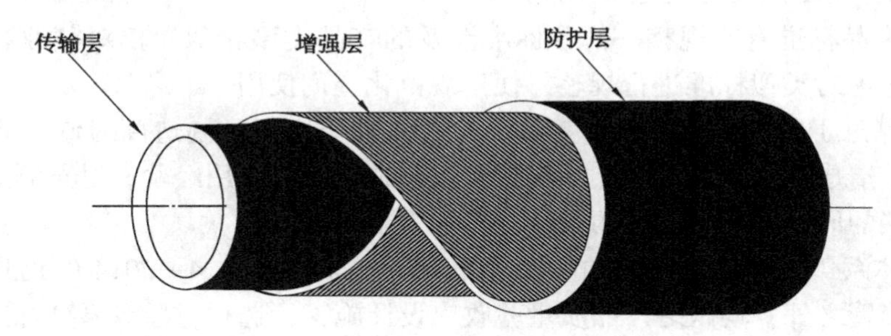

图 4　SY/T 6662.5 规定的增强超高分子量聚乙烯复合连续管结构

2.6　SY/T 6662.6—2014

2011 年，RTP 管在国内井下注水领域得到成功试用。随着产品及应用技术的日益成熟，井下用 RTP 管产品标准 SY/T 6662.6 于 2014 年制定完成。该标准涵盖合注型 RTP 管（图 5）和分注型 RTP 管（图 6）两类产品。其中合注型 RTP 管（代号：Ⅰ）主要用于井液的合注。它主要由聚合物内衬层、增强层、抗拉层、外护套构成（图 5）。分注型 RTP 管（代号：Ⅱ）主要用于井液的分注。它相当于在合注型 RTP 管内部增加了抗压层（图 6）。

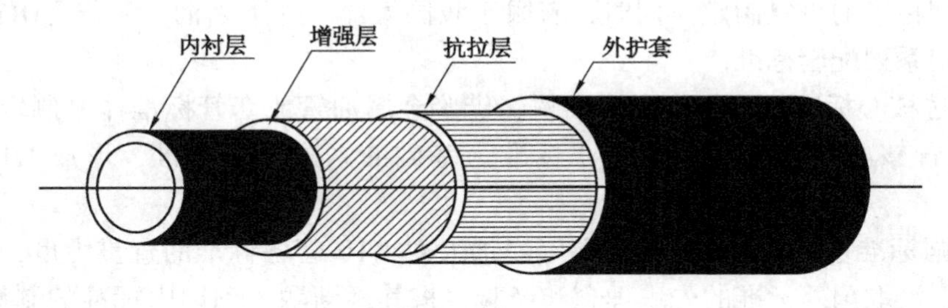

图 5　SY/T 6662.6 规定的Ⅰ型柔性复合连续管结构

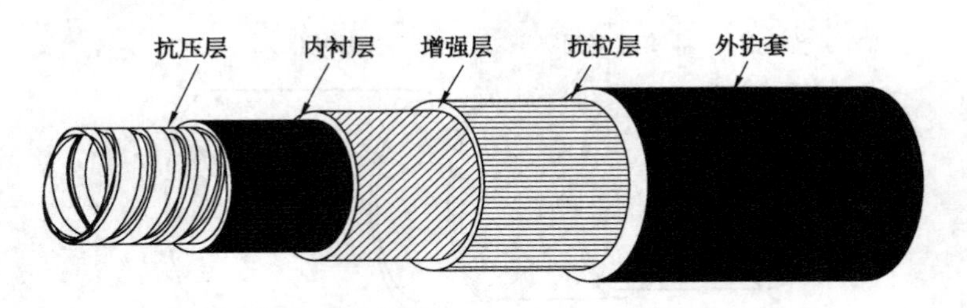

图 6 SY/T 6662.6 规定的Ⅱ型柔性复合连续管结构

本标准针对井下管服役特性，重点提出了内衬材料介质适用性、气体渗透性能等评价方法，以及复合管拉伸性能和抗压溃性能评价方法。但由于当前分注型 RTP 管尚处在试用阶段，因此对其拉伸能和抗压溃性能的具体指标要求还不是非常精确。另外，RTP 管在采油或注聚等其他井下应用领域的试验数据和标准要求还相对缺乏。

3 国内 RTP 管产品标准存在问题总结

随着我国 RTP 管产品的日益成熟及其应用技术的不断进步，相应标准制修订及标准体系建设进步明显，但仍存在以下主要问题：

（1）部分产品标准有待规整。部分标准涉及的产品类型相似，仅对其原材料要求有所不同，因此应考虑对类似标准进行规整，便于规范化查询使用。

（2）新产品急于制定行业标准。为了尽快进入油田市场，制造商对制定新产品行业标准的热情很高，造成部分标准应用范围窄，且由于缺乏实际应用经验，过早制定的标准在限制产品自身发展的同时，也给油田的安全生产埋下了隐患。

（3）标准体系仍不完善。已有的 RTP 管标准除 SY/T 6662.4—2014 的钢骨架增强热塑性塑料复合连续管之外，均缺乏产品质量验收、设计施工、选材指导等系列标准。

（4）大多数标准内容亟须补充完善。当前，多数产品标准对用户普遍关注的适用性评价、压力等级评定、服役性能评估、气体渗透性评价、结垢结蜡性能评价、各结构层壁厚限定等均未提及。

4 国内 RTP 管标准化工作建议

（1）完善标准内容，发展标准体系。继续完善各标准内容，将用户关注的焦点问题反映到标准中；制定 RTP 管质量验收、设计选材、施工指导等标准体系。

（2）规整现有标准，制定前沿标准。配合石油管材标准化体系建设要求，规整现有标准，避免相似标准的重复制定。同时，着眼于我国海洋、井下采油、采气等 RTP 管前沿应用领域，制定系列配套标准。

（3）筛选核心标准，建立配套标准树。借鉴金属油管或套管标准体系制定经验，确立 1~2 个 RTP 管核心标准，建立以核心标准为基础的系列配套标准，形成 RTP 管标准组织树。

（4）加强标准宣贯，缩小国际化差距。加大对 RTP 管各标准的宣贯力度，积极学习国外先进标准，汲取国外标准的先进理念和经验，将其逐渐融入到国内标准的制修订过程中，力争尽快缩小与国际先进标准的差距，促进我国 RTP 管产品的不断优化和应用技术的不断进步。

参 考 文 献

[1] 张玉川，王德禧，吴念．增强热塑性塑料（RTP）复合管材的发展［J］. 化学建材，2007，（1）：20-23.

[2] 王登勇，吴念．增强热塑性塑料复合管国内外发展比较［J］. 国外塑料，2011，29（5）：44-50.

[3] 郭金明，谈述战，于水，等．长纤维增强热塑性塑料制品技术及应用进展［J］. 塑料，2013，42（6）：24-27.

[4] 张玉川．高压增强热塑性塑料管［J］. 国外塑料，2008，26（10）：42-45.

[5] 孙哲，孙振国，刘惠明，等．柔性增强热塑性塑料管道的发展和市场前景［J］. 新型建筑材料，2013，12：97-101.

[6] 李凯，张贤波，李康锐，等．柔性复合管在油田集输管线的应用［J］. 石油工业技术监督，2013，13-14.

[7] 罗贞礼．柔性复合管在油气田中的开发应用探讨［J］. 化工新材料，2011，（6）：55-57.

[8] 李志，王艳令，刘持政，等．超高分子量聚乙烯成型技术研究进展［J］. 塑料，2014，43（2）：38-43.

[9] 王非，刘丽超，薛平．超高分子量聚乙烯纤维制备技术进展［J］. 塑料，2014，43（5）：31-35.

[10] 李忠，王宏军，卫杰．非金属管道在塔里木油田集油系统的应用［J］. 油气储运，2003，22（1）：37-39.

[11] 魏斌，戚东涛，李厚补，等．增强热塑性复合管在酸性环境下的耐蚀性能［J］. 天然气工业，2015，35（6）：87-92.

[12] 张建强．热塑性塑料管材在油气田腐蚀介质中性能退化规律研究［D］. 西安：西安石油大学，2015.

[13] 董云滨，张恒群，侯双英．增强热塑性塑料管现场施工质量控制［J］. 油气储运，2014，33（10）：1135-1138.

[14] 齐国权，李鹤林，李循迹，等．油田非金属管国内标准的发展与应用［J］. 油气储运，2014，33（10）：1029-1033.

[15] 李厚补，李鹤林，戚东涛，等．油田集输管网用非金属管存在问题分析及建议［J］. 石油仪器，2014，28（6）：4-8.

柔性海洋管隆起屈曲有限元分析

魏　斌[1]　李　兵[2]

（1. 石油管材及装备材料服役行为与结构安全国家重点实验室；
2. 西安交通大学机械工程学院）

摘　要：柔性复合管在海洋油气输送领域发挥着重要作用，管道隆起屈曲分析是海洋管道设计的重要技术之一。柔性复合管具有多层结构，每层材料性质差异较大，且层间应力关系复杂，使得隆起屈曲分析有别于传统的单层刚性管道。本文研究利用有限元分析软件 ANSYS，研究了管道内流体温度，压力以及埋泥深度三种因素对多层柔性管道隆起屈曲程度的影响。结果表明：对于多层柔性管道，输送流体温度越高，管道隆起高度越高，输送流体压力升高将限制管道的隆起，埋泥深度增大将限制管道的隆起。研究工作和分析结果将为海底管道铺设提供理论依据和设计参考。

关键词：海底管道；隆起屈曲；有限元分析；温度；压力；埋泥深度

柔性海洋管是连接海底井口与浮式平台之间传输油气或注水的重要海洋装备。它由聚合物阻隔层和各种金属螺旋铠甲层、非金属耐磨层等结构层复合而成。柔性管主要优点有：良好的柔性、安装速度快、安全可靠性高、可采用模块化的设计方式、抗腐蚀性好、抗压性能好、可回收重新利用等七方面。柔性管在海洋石油资源开发中具有良好的环境适应能力和经济效益，在国内外多个海洋油气田的开发过程中得到了广泛的应用[1]。

目前石油输送管线的工作温度随着海上石油工业的发展不断的提高，很多管线设计温度普遍达到或者超过 80℃，甚至出现 120℃的高温[2]。海底埋设管道在这种高温高压条件下运行时，有时会从原来位置上突然隆起，甚至拱出埋设土层，这种现象称为海底管道隆起屈曲[3]（图 1）。该过程类似于梁在轴向载荷达到临界值时发生的欧拉屈曲[4]。当管道隆起屈曲时，管会产生较大的弯矩以及较大的塑性变形，甚至可能进一步出现裂纹、疲劳、局部屈曲[5]。这些问题给管道安全运行带来巨大安全隐患，挪威船级社（DNV）规范 RP4-110 明确要求在海洋管道设计时对这种情况进行详细分析[6]。

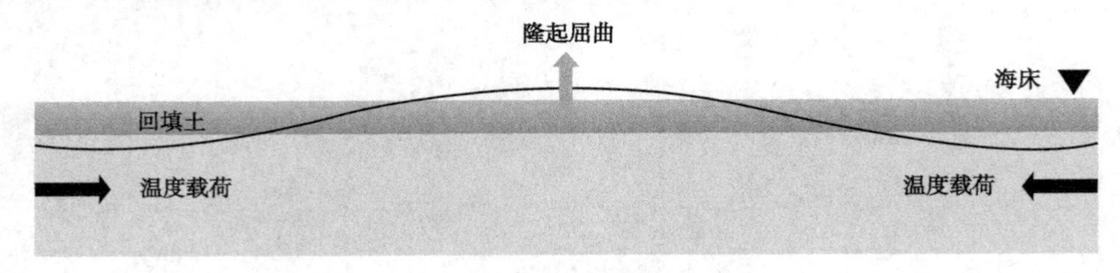

图 1　管道隆起示意图

现有的管道设计规范虽然都详细地描述了管道温度应力的计算方法，但都是以单层管为对象的[7]。并且当存在初始缺陷情况下，解析及求解非常困难，因此通常采用数值模拟分析具有初始隆缺陷的海底管道的隆起屈曲[8]。由于多层柔性管道在结构上与传统的单层刚性管道差别较大，然而传统的分析单层刚性管道方法运用在多层柔性管道中有着计算精度低，计算量大等问题。

本文针对一新型七层柔性海洋管道，采用有限元软件 ANSYS，建立了七层管道的高精度有限元模型，分析了管道温度、压力和埋泥深度等不同因素对其隆起屈曲变形的影响，研究工作为海底管道施工提供重要的理论依据和技术参考。

1 柔性海洋管有限元建模

海洋柔性管的内径为 140mm，外径为 250mm，主要结构为 7 层，端面结构如图 2 所示，有里层到外层依次为内压屏蔽层、压力铠装层、抗拉铠装层、中间屏蔽层、抗外压铠装层和外保护层。各层材料类型和功能如表 1 所示。

图 2　多层海洋柔性管道横截面图

表 1　柔性复合管主要结构层与功能

层数	名称	材料	功能
1	内压屏蔽层	聚偏氟乙烯	提供内部流体完整性通道
2	压力铠装层	Q195L 钢	承受内压荷载
3	抗拉铠装层	Q235B 钢	平衡扭转，承受轴向及部分环向荷载
4	抗拉铠装层	Q235B 钢	
5	中间屏蔽层	耐热聚乙烯	阻止内外介质二次渗透，保护金属结构层
6	抗外压铠装层	Q195 钢	承受外压荷载
7	外保护层	耐热聚乙烯	阻止外部流体进入，保护柔性管内部结构

多层柔性管道主要受轴向载荷影响，轴向载荷的大小取决于多种因素，如铺设过程中产生的张力、管道运行过程中内压、温度的变化所产生的轴向载荷以及管道与海底的摩擦等。温度载荷是管道发生隆起屈曲的关键控制因素[9-11]。假设有初始形状的管道铺设在海床上，两端无轴向位移，内管受温度载荷作用。对该问题建立温度载荷有限元分析模型，内外管采用 ANSYS 中 Elbow290 单元[12,13]。该单元是 ANSYS 公司最新开发的专门针对多层管道有限元建模的多层管单元。该单元能够适应线性大转角情况，同时也适用于非线性大应力等非线性情况。该单元可以定义每一层的材料性质进行建模分析，适用于复合材料管道分析，其复合管道计算精确度受第一剪切应力准则控制。

由于管道隆起对于管道长度有所限制，长度过短会影响隆起部分的隆起形状和应力分布，因此建模长度选取为 200m。初始几何缺陷是海底管道整体屈曲分析中非常关键的因素，基于有限元分析软件建立的海底管道整体屈曲分析的有限元模型必须引入几何缺陷才能进行整体屈曲分析[14]。管道初始形状可用式（1）表示[15]：

$$f=\omega_0\left(0.707-0.2617\pi^2x^2/L_0^2+0.293\cos\left(2.86\pi x/L_0\right)\right) \tag{1}$$

式中　ω_0——初始缺陷高度；

L_0——初始缺陷长度。

缺陷形状假设关于 $x=0$ 对称，初始缺陷高度为 0.3m，初始缺陷长度为 10m，建模后效果如图 3 和图 4 所示。

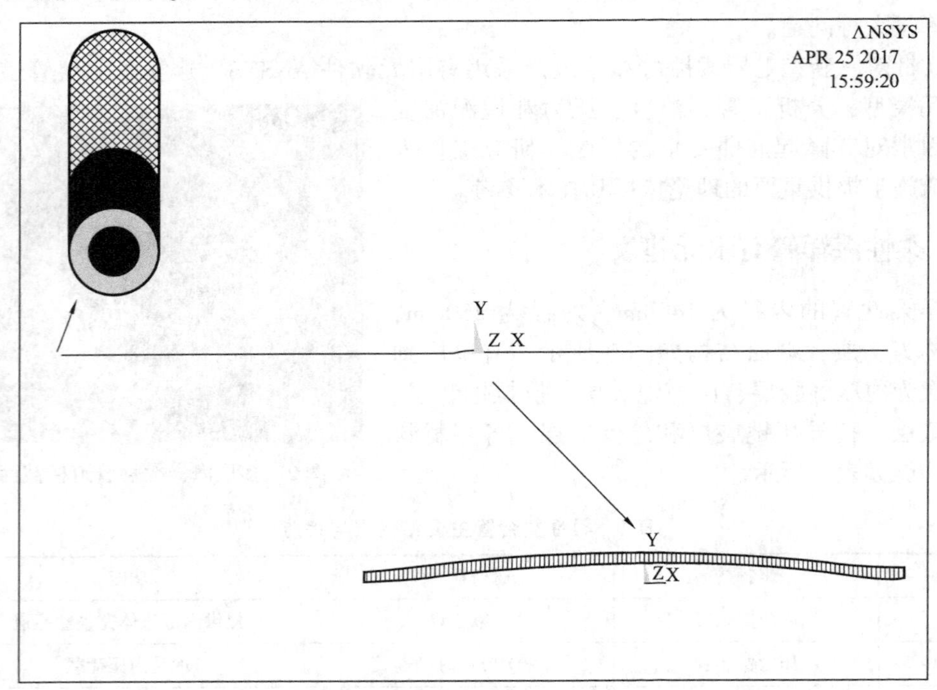

图 3　管道建模整体示意图

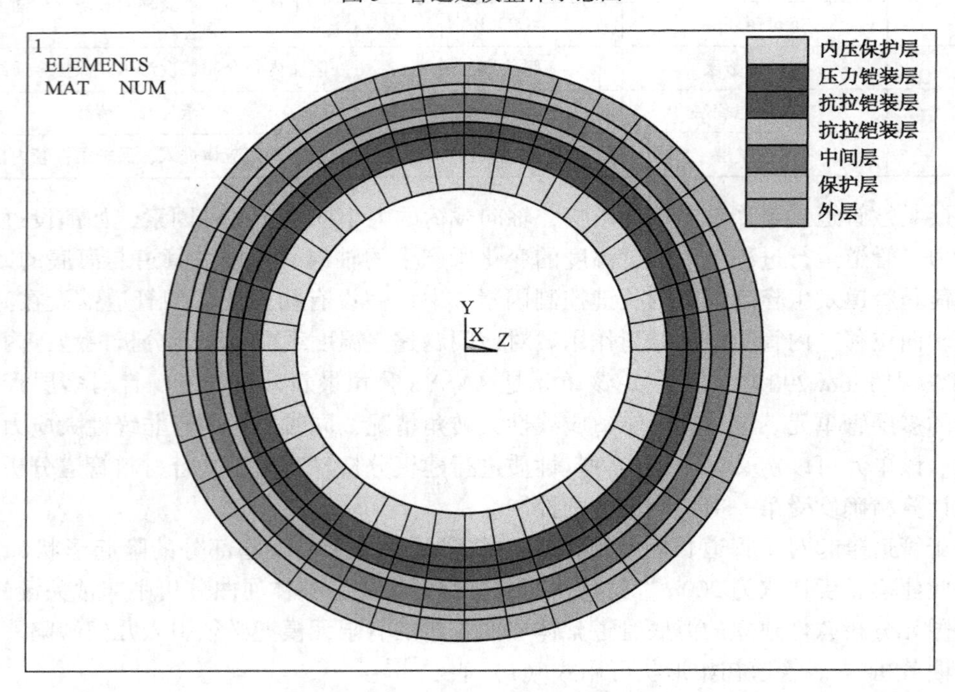

图 4　管道建模截面示意图

在隆起屈曲 −5～5m 处每 0.1m 划分一个单元，其余每 1m 划分一个单元。管道沿周向划分 40 等份，建模完成共划分 290 个单元。

2 边界条件及载荷施加

计算过程中需要考虑两种边界条件，一是管道接头两端的固定作用，二是地面的支撑作用。柔性复合管单支长度一般为 50~100m，中间靠金属接头连接，因此计算过程中可将其简化为管道两端的六自由度完全约束。其物理意义在于接头会限制管道的轴向的移动转动，垂直于地面的移动和转动还有平行于地面方向的移动和转动，恰好完全限制了六个自由度。对于地面支撑，没有隆起的管道直接限制垂直于地面的移动。根据管道三种实际工况分别施加载荷。

（1）温度载荷：对于 Elbow290 单元，温度载荷对其来说为体载荷，可直接设定每一层表面的温度进行计算。

（2）内压载荷：对于 Elbow290 单元，压力载荷对其为面载荷，可将压力直接施加在某一层的表面上。处理时将压力载荷直接施加在管道内层表面上，压力载荷正方向的定义是垂直于表面法向向外为正方向，对于内表面来说，内压其方向为垂直于表面向内，因此内压载荷的设定为负。

（3）埋泥载荷：对于 Elbow290 单元，可以将埋泥深度转化为节点力直接施加在变形单元的节点上，其线载荷可表述为式（2）[6]。

$$R_{clay}=\gamma HD+D^2\left(1/2-\pi/8\right)+2S_u\left(H+D/2\right) \tag{2}$$

式中 γ——土壤浮重，γ 取 6900N/m^3；

H——管道上部覆盖土体高度（海床至管道最上端）；

D——管道外径；

S_u——管道中心到沟槽顶部的平均不排水剪切强度。S_u取 2000N/m^2。

管道输送流体温度为 80℃，内部压力为 20MPa。

表 2 柔性复合管各层材料性质一览表

层数	名称	层厚度（mm）	弹性模量（MPa）	泊松比	热导率（W/K）	线膨胀率（K^{-1}）
1	内压屏蔽层	12	1400	0.3	0.13	1.20×10^{-4}
2	压力铠装层	8	2.10×10^5	0.33	52.3	1.20×10^{-5}
3	抗拉铠装层	9	2.10×10^5	0.33	52.3	1.20×10^{-5}
4	抗拉铠装层	8	2.10×10^5	0.33	52.3	1.20×10^{-5}
5	中间屏蔽层	8	750	0.3	0.48	2.40×10^{-4}
6	抗外压铠装层	3	2.10×10^5	0.33	52.3	1.20×10^{-5}
7	外保护层	7	750	0.3	0.48	2.40×10^{-4}

3 计算结果

3.1 热应力对管道隆起屈曲的影响

由于需要对七层管道每一层施加温度载荷，因此在管道屈曲分析之前进行传热分析，以确定每一层管道的每个面的温度情况。进行建模之后，施加边界条件，内层温度为变量从 30~80℃变化，外层温度定为 0，进行计算得到温度分布云图如图 5 所示。

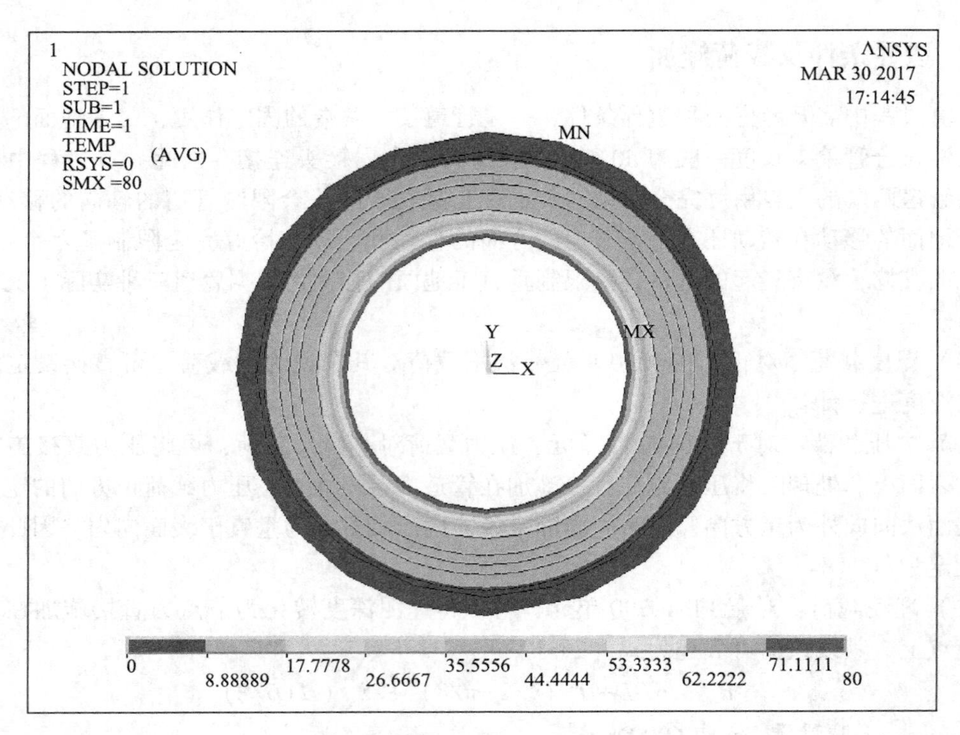

图 5　管道横截面温度分布云图

将以上热分析结果带入到 Elbow290 单元体载荷中进行及计算，打开非线性大变形开关，将管道横截面温度分布数据带入，以初始不平度高度为 0.3m，初始不平度长度为 10m，内层温度为 80℃为例，管道屈曲形状如图 6 和图 7 所示。最大应力以及截面应力分布如图 8 所示。

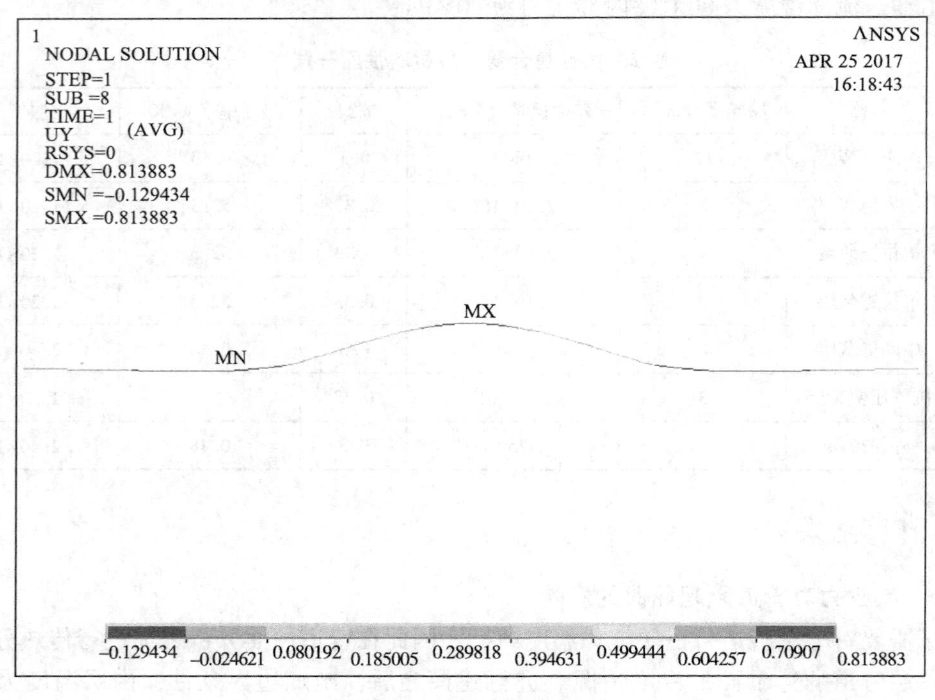

图 6　管道变形图

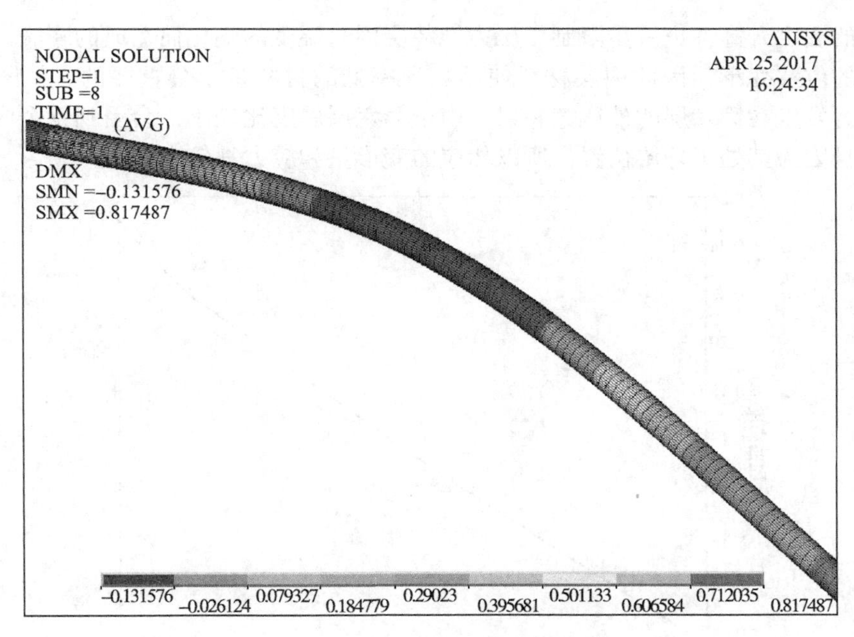

图 7　管道变形局部变形云图

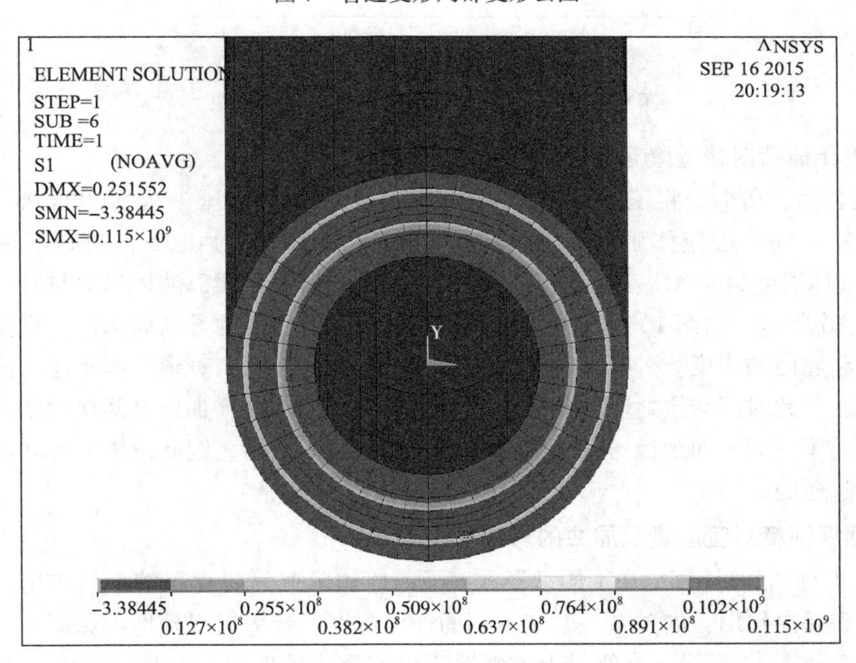

图 8　管道屈曲界面应力分布云图

改变内层温度，经计算得到温度与管道隆起屈曲垂直方向位移关系如图 9 所示。由图 9 可以看出，内层温度直接影响了管道的隆起屈曲程度。随着温度的升高，隆起屈曲的位移不断变大。产生上述结果的原因是温度升高使材料发生了膨胀，并在轴向方向产生了压应力，又由于初始不平度的影响导致的欧拉杆的失稳，最终导致管道发生隆起屈曲。另一方面是由于压力铠装层的材料弹性模量较大，在同等变形的情况下会产生更大的应力，容易造成整个结构的塑性屈服。但是热导率很低的最内层材料聚偏氟乙烯减小了向外的热量传递，限制了压力铠装层的温度升高，进一步限制了压力铠装层因为温度的变化而造成的膨胀，最终减小了整个结构的位移变形，因此内层材料使用热导率低的材料是十分重要的。从图 8 可以看出，最大应力发生在第六层的外压铠装层。由于第六层厚度比较小，会受到整个内层的挤压而变形，并且因为其处于外层位置，所以相应的变形量和应力都较大。

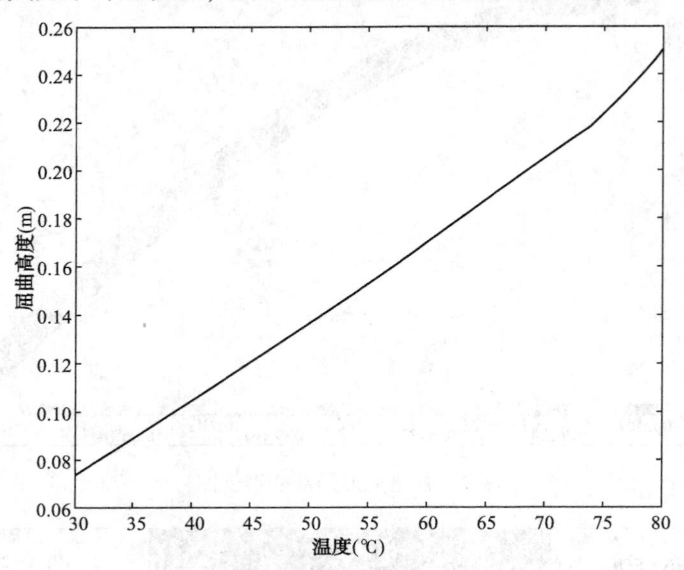

图 9　隆起屈曲高度与输送流体温度关系图

3.2　内压载荷对管道隆起屈曲的影响

使用 SFE 面载荷命令将压力加载到内层管道中，管道内部压力从 0~20MPa 每隔 2MPa 进行一次计算，以输送流体压力为 20MPa，初始不平度高度为 0.3m，初始不平度长度为 10m，内层液体温度为 80℃。改变输送流体压力，计算后得到内部液体压力与管道隆起屈曲的关系如图 10 所示。由图 10 可以看出，隆起高度随管道内压的升高而降低，提高压力能够减小管道隆起屈曲的程度。分析原因是因为管道内部压力越大，管道会因为压力的作用有一定的体积膨胀，此时层与层之间的正压力就会增加，正压力的增加就会间接地增加最大静摩擦力以及滑动摩擦力。而摩擦力的增加会阻碍管道的层间滑动。因此增大内部液体压力会限制管道的隆起屈曲。

3.3　埋泥深度对管道隆起屈曲的影响

根据式（2）能够计算出在不同埋泥深度下，管道隆起需要克服的垂向作用力。经过计算埋泥深度和垂向作用力的关系，改变不同的埋泥深度，计算后得到埋泥深度与管道隆起高度之间的关系如图 11 所示。在管道上方埋泥不仅有重力的作用，并且还有管道突破土层需要克服排水剪切压力，因此埋泥是最能够限制管道隆起屈曲的因素。但是考虑到生产成本和管道铺设难度的问题，埋泥深度不宜过大。计算结果表明埋泥深度在 0.3m 左右，管道隆起现象已经被极大地抑制。

4　结论

通过有限元计算，计算了海洋柔性管道在海床屈曲的过程，分别分析了温度，内部液体压力，埋泥深度这三个因素对管道隆起屈曲的作用影响。

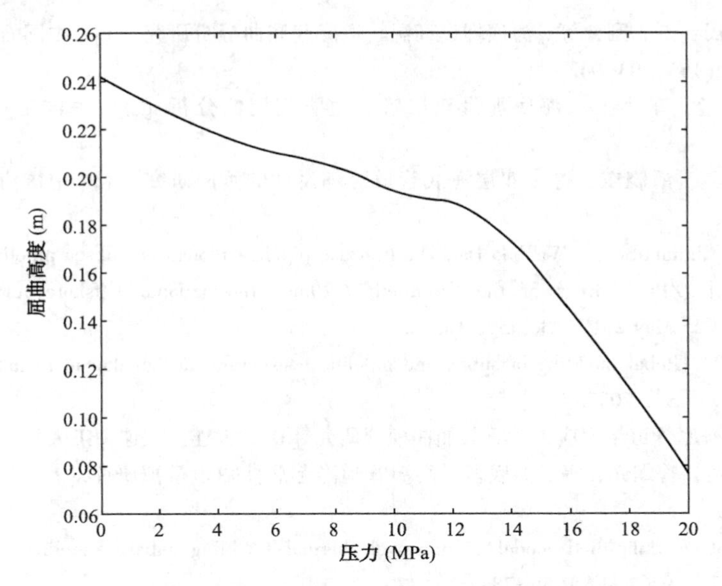

图 10　隆起屈曲高度与管道内压关系图

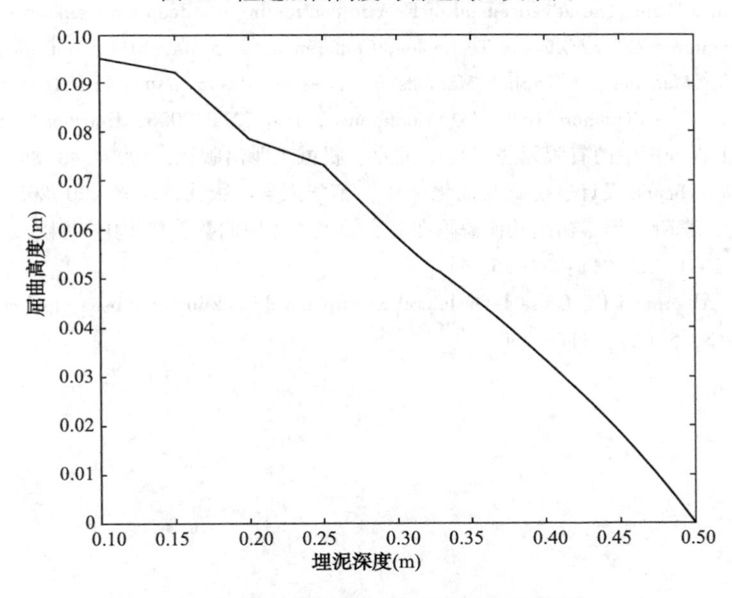

图 11　隆起屈曲高度与埋泥深度关系图

（1）随温度的升高，隆起程度不断增加。

（2）内部液体压力能够减小管道隆起屈曲的程度。当内压达到 14MPa 之后，管道隆起高度下降的斜率相比于 0~14MPa 部分有了明显减小。因此内压保持在 14~20MPa 对管道隆起屈曲的限制效果较好。

（3）埋泥也能够限制管道的隆起屈曲。计算结果表明埋泥深度在 0.3m 左右，管道隆起现象已经被极大地抑制。

参　考　文　献

[1] 余志兵，孙国民，王辉，等．海底长输油气管道的隆起屈曲分析［J］．中国海洋平台，2015，30（2）：64-69.

[2] 刘新帅，孙国民，刘志刚，等．海底埋设高温管道隆起屈曲分析研究［J］．中国石油与化工标准与质量，2016，36（15）：93-94.
[3] 曾霞光，段梦兰，车小玉．海底埋设双层管管道隆起屈曲分析［J］．海洋工程，2014，32（2）：72-77.
[4] 赵天奉，段梦兰，潘晓东，等．双层海底管道跨越设计的垂向研究［J］．中国海上油气，2010，22（3）：197-201.
[5] Huifeng Deng，Chunna Song，Wanbao Dai. The first use pipeline plough on subsea pipeline trenching in South China Sea ［C］//Proceedings of the Twentieth （20m） International Offshore and Polar Engineering Conference，20-25 May 2010，Beijing，China.
[6] DNV-RP-F110，Global buckling of submarine pipelines-structural design due to high temperature and high pressure pipelines［S］. 2007.
[7] 赵天奉．高温海底管道温度应力计算与屈曲模拟研究［D］．大连：大连理工大学，2008.
[8] 车小玉，段梦兰，曾霞光，等．海底埋设高温管道隆起屈曲数值模拟研究［J］．海洋工程，2013，31（5）：103-111.
[9] James GA Croll. A Simplified model of upheaval thermal buckling subsea pipelines ［J］. Thin - Walled Structures，1997，29（1-4）：59-78.
[10] Jason Sun，James Wang. The advancement of FEA in confronting the deepwater pipelines under high pressure and high temperature［C］//Offshore Technology Conference，2-5 May 2011，Houston，USA.
[11] NFA Parsloe，P Manchec，A Dorbec. Methods for assessing design issues with HP/HT pipelines at early stages a project［C］//Offshore Technology Conference，1-4，May 2006，Houston，USA.
[12] K. L. 巴斯．工程分析中的有限元法［M］．北京：机械工业出版社，1999：48-86.
[13] 李兵．Ansys Workbench 设计、仿真与优化［M］．清华大学出版社，2008：20-80.
[14] 刘羽霄，葛涛，李昕，等．初始几何缺陷在海底管道整体屈曲数值建模中的引入方法［J］．水资源与水工程学报，2011，22（3）：32-35.
[15] Karampour H，Albermani F，Gross J. On lateral and upheaval buckling of subsea pipelines［J］. Engineering Structures，2013，5（2）：317- 330.

Failure Analysis of the Fluorine Rubber Sealing Ring Used in Acidic Gas Fields

Qi Guoquan Qi Dongtao Wei Bin Li Houbu
Zhang Dongna Ding Nan Shao Xiaodong

(Tubular Goods Research Institute of CNPC; State Key Laboratory of Performance and Structural Safety for Petroleum Tubular Goods and Equipment Materials)

Abstract: The failure of fluorine rubber sealing ring occurs when it is used in the sulfur condition after one year. By visual observation, there are some cracks and the color changed to black from brown near the location of the contact delivery medium. In order to identify the cause of the failure, the physical properties, structure and composition were tested and compared with the same batch of the rings but without using. Firstly, the physical properties including hardness and density were measured. The results showed that the hardness and density of the material were increased because of gas permeation. It is evidenced by the microstructural observation, which shows that there are holes in the material near the transport medium. To check any significant chemical modification of the fiber, the specimens were characterized by Fourier transform infrared spectroscopy (FT-IR) and X-ray photoelectron spectroscopy (XPS), respectively. The results show that the three-dimensional network structure of the ring is destroyed and the peak strength of $-CF_2$ and $-CF$ is weakened while the CH_3, $-CH_2$ and $-C=O$ increase by the action of external stress and gas permeation. As a result of comprehensive evaluation, the reason of sealing ring failure is fatigue aging during the long-term service life. The gas permeation leads to the destruction of the network structure, and accelerating the aging process.

Keywords: Fluorine rubber; Sealing ring; Fatigue aging; H_2S

1 Introduction

The gas field contained hydrogen sulfide and carbon dioxide has become an important part of the natural gas resources for exploitation in China nowadays[1,2]. Unfortunately, ordinary carbon steel pipe corrosion is very serious in such high acidic environment. In recent years, the non-metallic pipe with good corrosion resistance is researched and applied in the oil and gas field contained H_2S or CO_2, and turned into an important direction for sulfur-containing conveying pipe, gradually[3].

During the oil and gas field exploitation, sealing ring is used in non-metallic pipe as an important accessory[4]. Sealing performance and service life of rubber seal products are closely related to oil and gas field environment, such as temperature, pressure, chemical corrosion, and et al. Furthermore, with the increase of temperature and pressure in the pipeline increase and service environment becomes worse. With the deepening of the exploitation depth and increasing severe environment, the quality of rubber seal is put forward higher requirements and face enormous challenges. In the process of transmission, due to the high temperature, high pressure and corrosive medium[5], the seal products often failure caused by performance degradation, which lead to the contact stress release between seal products and joints. In the service environment with H_2S/CO_2, accident such as perforation or fracture often happened[6,7], usually results in oil and gas leak, even cause serious accident, which may lead to huge economic losses, casualties and ecological damage.

It has a great significance that study on the corrosion damage behavior of sealing ring serviced in corrosion environment, but there is few research report about this aspect. In this study, failure analysis of fluorine rubber sealing ring used in non-metallic pipe which anti hydrogen sulfide is investigated. For reliable operation and efficient design, the security of this engineering non-metallic pipe especially, it is urgent to investigate the failure reasons of the sealing ring and to further analyze the factors which affect the pipeline's serving ability. It will be of great benefit to prevent events which could trigger disastrous incidents, thus can reduce much loses in terms of service life and economics.

2 Material and methodologies

2.1 Background of the Failure

Non-metallic pipe (DN80 PN16MPa), as is shown in Fig. 1, was used for oil gathering and transportation in sour oil and gas field. The transmission medium is mostly gas, with a small amount of oil and water. The pressure and temperature of transmission medium are 8.9MPa and 35℃, respectively. The content of H_2S is 51800mg/m³ while CO_2 of 3.80mol%. The whole construction project had been completed on April 2015, while the pipeline had been started to use on May 2015. After running for half a year, the inspection of pipeline application effect was carried out. The result shows that the fluorine rubber sealing rings have failed with cracks and color changed shown in Fig. 2. Meanwhile, new sealing rings produced in the same batch were also collected for comparison.

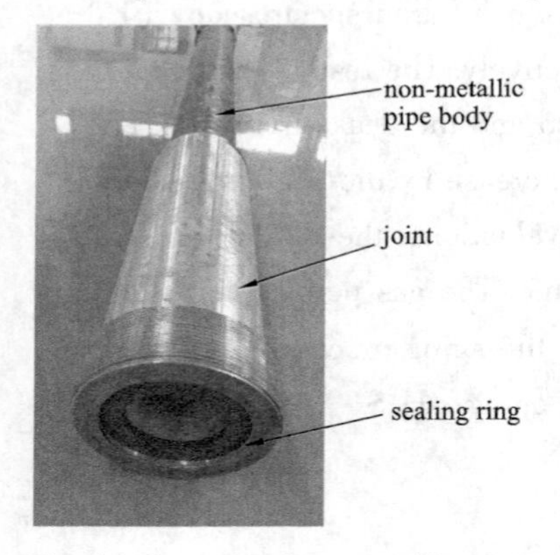

Fig. 1 Structure of non-metallic pipe

2.2 Failure description

It can be seen from Fig. 2 that the color of the used sealing inside changes from brown to dark obviously, while the outer color is not changed. Compared with the new one, the color changed border of the used sealing ring is at a third place from inside to outside. Moreover, we can easily find by visual observation that the color changed border have some cracks, also in the black area. We can also find that it has an obvious squeeze traces in the sealing ring surface, and the thickness decreases from 7.72mm to 6.54mm, approximately.

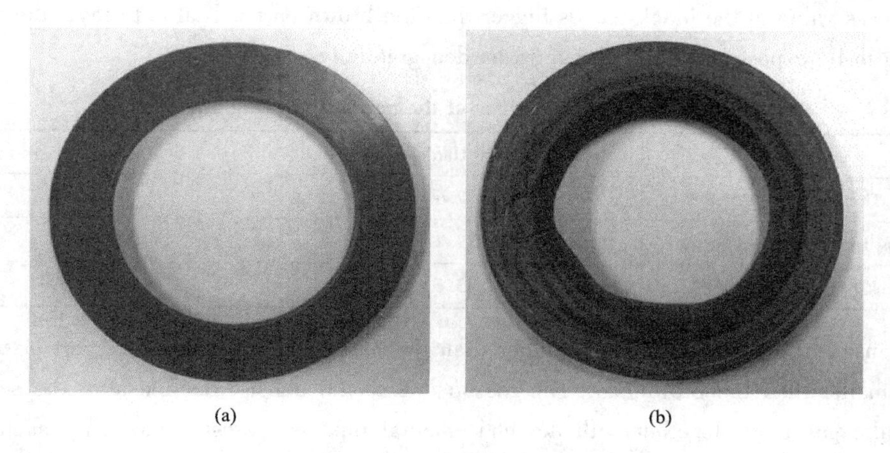

Fig. 2 Macro morphology of (a) the new sealing ring and (b) used one

2.3 Methodologies

Many factors may cause the failure of sealing ring and more studies are needed. Firstly, the background information and the operation conditions that might lead to failures of the ring were investigated in detail. Secondly, probable causes for failure of ring were systematically analyzed by various measurements, such as Shore hardness, density, morphology observations, composition analysis, etc.

The specimens of sealing ring were cut from the same failed one which was taken from the pipeline after running for half a year. By comparison, specimens for above tests were also taken from a new ring which produced from the same batch with the failed one. At least three specimens taken along the circumference of the ring were tested for each measurement to evaluate the results statistically. The "used" samples were collected from the area close to the damage region of the failed ring (as seen in Fig. 2b). The "new" samples were taken from the new ring (as seen in Fig. 2a).

Surface morphologies of rings were detected by digital microscope (KH-7700, Hirox, Japan) and scanning electron microscope (SEM, JEOL-6700F, Tokyo, Japan). To check any significant chemical modification of the fiber, the specimens were characterized by Fourier transform infrared spectroscopy (FT-IR) and X-ray photoelectron spectroscopy (XPS), respectively, which were obtained by using a Nicolet Avator 360 Spectrometer (Wisconsin, USA) and a PHI5300 X-ray photoelectron spectrometer (PE Corp., USA).

3 Results and analysis

Analysis contain morphology and structure will be conducted as follows to study the actual reason for such gradual failure of the sealing ring.

3.1 Analysis of physical properties

Firstly, the physical properties of the new sealing ring and used one were investigated, can be seen in Table 1, the Shore hardness value of the used sealing ring are increasing compared to the new one's (the value of Shore hardness is 27.6). For the old sealing ring, of particular note is

Shore hardness value of the black part is bigger than the brown part's. That is to say, the part of the sealing ring to be exposed to fluid change to harden gradually.

Table 1 The physical properties of the new sealing ring and used one

Types	Shore Hardness	Density (g/cm^3)
The new ring	27.6	2.1296
The brown part of the old one	29.2	1.8391
The black part of the old one	33.6	1.7539

The density of old sealing ring is smaller than the new one, and the black part of the sealing ring even smaller than the brown part. The reason due to the density decline after the sealing ring used in acidic gas fields together with the non-mental pipe is corrosion medium, such as H_2S, CO_2, permeate into the body of sealing ring, and then the pores become more and more bigger with the service time increasing. The pressure difference between inside and outside of the pipe lead to the occurrence of gas permeation.

3.2 Analysis of microstructure

For researching the microstructure of the invalid sealing ring, the optical microscope was used to characterize the structure. From the Fig. 3, we can find that the color of the sealing ring used for half year changed from brown to black partially and obviously.

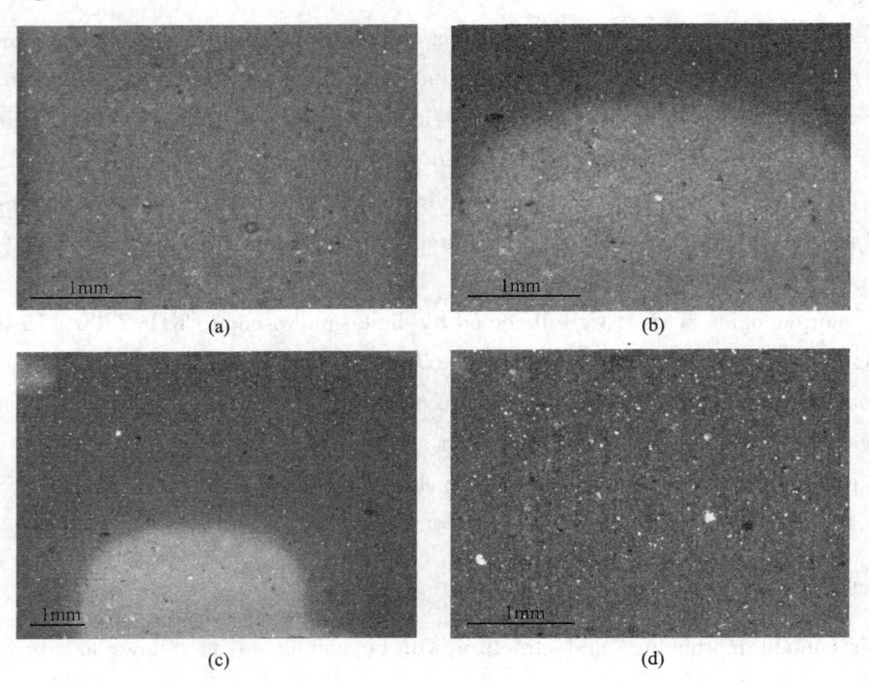

Fig. 3 Optical microscope of the sealing ring used for half year

For further research, scanning electron microscope was used to characterize the microstructure. There are some micrometer grade pore in the used sealing ring (Fig. 4 b), compared to the new one (Fig. 4 a). The arising pores are resulted from gas permeation.

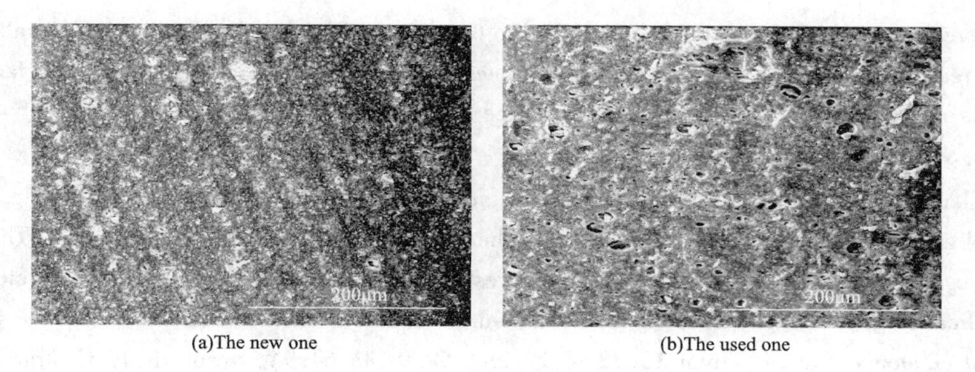

Fig. 4 SEM of the sealing ring

3.3 FT-IR analysis

Due to service in acidic gas fields, the sealing ring, along with the time increasing, with some aging, which have the characteristics of the physical properties deterioration, some cracks and color change. For understanding what have changed for compositions of the sealing ring with aging, The structures of the sealing ring were characterized by Fourier transform infrared spectroscopy. Fig. 5 compares the FT-IR spectrum of the used sealing ring with the spectra of new one. The peaks at around $884cm^{-1}$, $2852cm^{-1}$ and $2922cm^{-1}$ are assigned to $=CH$, $-CH_2$ and $-CH_3$ vibration, respectively. The absorption peaks at $1742cm^{-1}$ is due to $-C=O$ stretching. Moreover, $-CF$ and $-CF_2$ bending stretching at $1078cm^{-1}$, $1125cm^{-1}$, and $1181cm^{-1}$ respectively, reveals the existence of fluorine in the sealing ring.

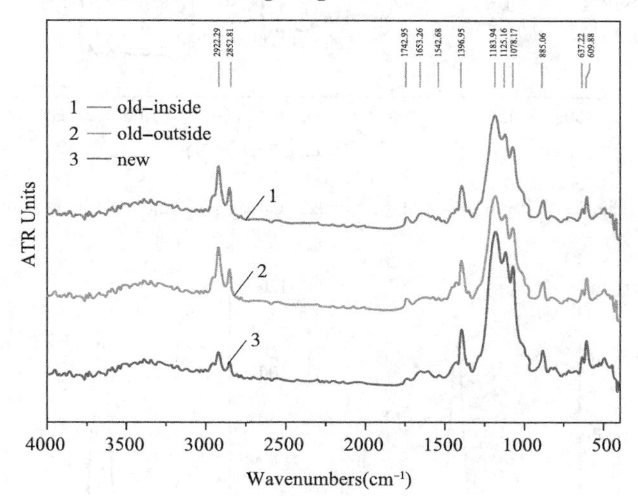

Fig. 5 FTIR of sealing ring

Through the tests (Fig. 5), the results show that the old ring absorption peak is close to the new one, but the absorption intensity is different. It can be seen from Fig. 5 that the absorption intensity of $-CF$ and $-CF_2$ decreases while the absorption intensities of CH, CH_2, CH_3 and $C=O$ increase with the service life of the fluorine rubber seal ring increasing. During the aging of the sealing ring, the position of the unsaturated bond in the rubber molecule is oxygenated to $C=O$ structure. On the contrary, $-CF$ and $-CF_2$ bond strength decreased, indicating that $-CF$ and $-CF_2$

bond breakage and recombination, which lead to the degree of branching increase, eventually. That is the reasons why the -CH group absorption intensity increase. Therefore, the cause of failure of fluorine rubber seals is due to molecular breakage and cross linking.

3.4 XPS analysis

Elemental ID and Quantification of the new and the used sealing ring are shown in Fig. 6. It is seen obviously that the element exist in the rubber mainly include Si, S, O, C and F. The trace additives such as Na, Ca, Cl and N not be investigated in this study. By comparing the elemental quantification of two rings (Table 2), the content of O and C element increase, with the percent content of atomic (at. %) from 13.82 to 19.25, 59.91 to 64.94, respectively. Contrarily, the content of F element decrease with the percent content of atomic from 14.34 to 7.11, the elemental F peak of both used and new rings counts to binding energy relation is shown in Fig. 7.

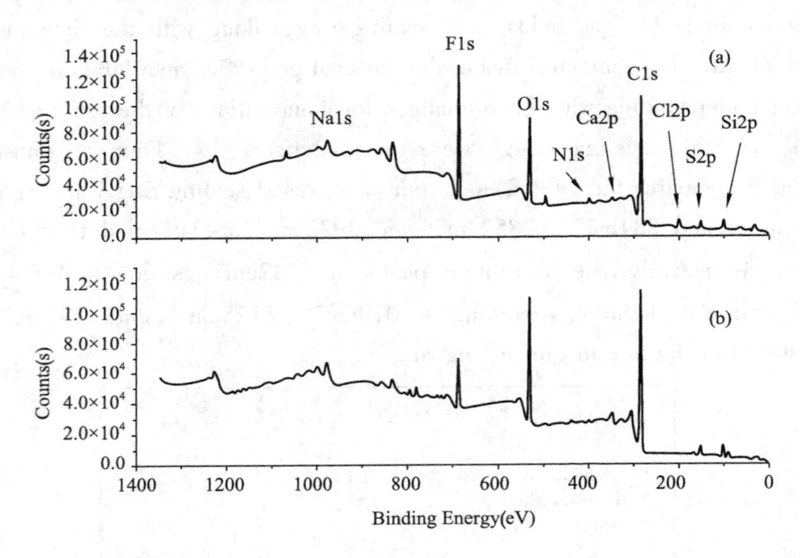

Fig. 6 XPS analysis of (a) used sealing ring and (b) new one

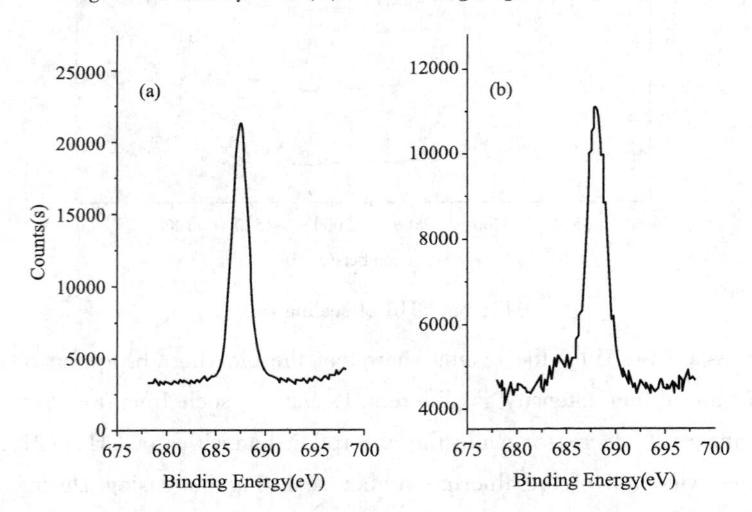

Fig. 7 F1s scan in (a) the used sealing ring and (b) the new one

Table 2 Elemental Quantification of the new sealing ring (at%)

sample	O	F	C
New sealing ring	13.82	14.34	59.91
Used sealing ring	19.25	7.11	64.94

4 Discussion

In the actual process of service, the sealing ring need to withstand a certain compressive stress, even more alternating pressure. In this case, the aging of the sealing material is directly affected by the stress. In addition, the aging of the seal ring is also affected by the results of the rubber material, components and contact with the external environment (such as temperature, H_2S and CO_2 and other transmission medium). From the pressure gauge reading for pipe connecting, the ring by the value of compressive stress at sealing ring reach 24MPa. The macroscopic analysis of the fracture morphology shows that the occurrence of the crack due to the compressive stress is too large, and then the molecular chain of rubber is cut off by stress (Fig. 5). However, the latter part of the study found that, there is not appear similar phenomenon in the air environment when applied stress at 24MPa. Through mechanical analysis, the reason is that partial molecules of the transmission medium, with the service time increases, will be permeated into the inside of the rubber and make volume expansion. Further more, rubber network structure of the molecular chain will expand to three-dimensional space. The excessive deformation will cause loss of elasticity, reduce the rubber material resilience, and result in decreased sealing performance, eventually. The applied stress and gas permeation promote each other, and accelerating the aging of the sealing ring[8-10].

For the fluorine rubber aging process, it can be divided into three stages[11,12]. Firstly, the rubber make an elastic deformation after loading, also known as softening. Secondly, as the stress or deformation is relatively slow and can not be evenly distributed, it will focus and generate rupture nucleus somewhere, in the rubber surface or internal. Thirdly, the rupture nucleus increases until the rubber is destroyed as a whole.

Therefore, according to the analysis the structure, composition and related physical properties of the failure of the seal, comparison with the same batch of non-service sealing ring, it is shown that the failure of the used sealing ring is due to fatigue aging. During the process of service life, the sealing ring structure change and properties degradation due to the applied stress, and the gas permeation accelerates the fatigue aging process.

5 Conclusions

(1) The reason of sealing ring failure is fatigue aging during the long-term service life.

(2) The gas permeation leads to the destruction of the network structure, and accelerating the aging process.

(3) This type of sealing ring is not suitable for using in acidic gas fields, high-grade sealing materials should be selected. For preventing the sealing ring failure used in the acidic gas fields, the most important strategies is evaluations and detections before using.

Acknowledgements

The project was supported by the National Natural Science Foundation of China (Grant No. 51304236).

References

[1] H. B. Li, M. L. Y, D. T. Qi, N. Ding, X. H. Cai, S. H. Zhang, Q. Li, X. M. Zhang, J. L. Deng. Failure analysis of steel wire reinforced thermoplastics composite pipe [J]. Engineering Failure Analysis, 2012, 20: 88-96.

[2] D. T. Qi, M. L. Yan, N. Ding, X. H. Cai, H. B. Li, S. H. Zhang, . The 7th international MERL Oilfield Engineering with Polymers Conference, London, UK, 2010.

[3] Yong Bai, Fan Xu, Peng Cheng. Investigation on the Mechanical Properties of the Reinforced Thermoplastic Pipe (RTP) Under Internal Pressure [C]. The Twenty-second International Offshore and Polar Engineering Conference, June 17-22, Rhodes, Greece: International Society of Offshore and Polar Engineers, 2012.

[4] K. Drake, R. Callaway, New polymeric materials development for extreme environments [C]. OTC-25304, Offshore Technology Conference, Houston, 5-8, 2014.

[5] Yamabe J, Nishimura S. Failure behavior of rubber O-ring under cyclic exposure to high-pressure hydrogen gas [J]. Engineering Failure Analysis, 2013, 35: 193-205.

[6] Stevenson A. A fracture mechanics study of the fatigue of rubber in compression. Int. J. Fracture 1983, 23: 47-59.

[7] Mars W. V., Cracking energy density as a predictor of fatigue life under multiaxial conditions [J]. Rubber Chemistry and Technology, 2002, 75: 1-17.

[8] Ayoub G., Naït-Abdelaziz M., Zaïri F., Gloaguen J.-M., Charrier P., Fatigue life prediction of rubber-like materials under multiaxial loading using a continuum damage mechanics approach: effects of two-blocks loading and R ratio [J]. Mechanics of materials, 2012, 52: 87-102.

[9] Le Saux V., Marco Y., Calloch S., Doudard S., Charrier P., An energetic criterion for the fatigue of rubbers: an approach based on a heat build-up protocol and μ-tomography measurements [J]. Procedia engineering, 2010, 2: 949-958.

[10] Poisson J.-L., Lacroix F., Méo S., Berton G., Ranganathan N., Biaxial fatigue behavior of a polychloroprene rubber, Int. Jour. Of Fat., 2011, 33: 1151-1157.

[11] Bathias C, Legorju C, Chuming LU, Menabeuf L. Fatigue crack growth damage in elastomeric materials [C]. In: Elastomeric materials ASTM. STP, 1996: 505-513.

[12] Cruanes C., Berton G., Lacroix F., Méo S., Ranganathan N., Study of the fatigue behavior of the chloroprene rubber for uniaxial tests with infrared method [J]. Elastomery, 2014, 18: 3-9.

油气田用热塑性塑料管材 ESC 研究进展

齐国权[1,2,3]　戚东涛[2,3]　魏　斌[2,3]　李厚补[2,3]

(1. 西北工业大学；2. 中国石油集团石油管工程技术研究院；3. 石油管材及装备材料服役行为与结构安全国家重点实验室)

摘　要：耐腐蚀性能优良的热塑性塑料在油气集输领域的应用受到广泛关注。但在使用过程中，由于环境应力开裂（ESC）导致管材失效事故时有发生，给管线运行和生态环境带来巨大隐患。文中首先调研了国内外关于油气田用热塑性塑料管材 ESC 的研究现状，总结了该类管材的 ESC 行为及其机理，指出了目前研究存在的不足之处并对今后的研究方向给出了建议。

关键词：环境应力开裂；热塑性塑料；油气田；集输

在管道服役过程中，直接接触于油气耦合介质的热塑性塑料管（聚乙烯 PE、聚丙烯 PP、聚氯乙烯 PVC、聚苯乙烯 PS、超高分子量聚乙烯 UHWMPE 和聚偏氟乙烯 PVDF 等）不仅需要承受高温条件下化学介质侵蚀的作用，而且要承受输送流体的内液压载荷等多场耦合作用[1-4]。高含 H_2S、CO_2 等酸性腐蚀气体的油气介质会造成塑料管材溶胀、鼓泡、软化等性能下降[5]，介质与内液压载荷相结合，将会造成管材环境应力开裂（Environmental Stress Cracking，ESC）行为发生[6,7]。

研究表明，热塑性塑料管材在大载荷、较短时间下破环形式是韧性破坏（塑形破坏），而在长时间、小载荷作用下破环形式是脆性破坏[8]。管道要求长的使用寿命，因而必须认识管材的脆性破环。韧性破环与脆性破环的机理与破环形式是不同的。脆性破环的本质在于应力开裂。根据脆性断裂的裂缝理论，所有实际存在的管材表面不可能不存在裂缝和缺陷（如表面划痕、内部夹杂、微孔等），裂缝尖端处的应力集中，达到或超过某一临界条件时，裂缝会失去稳定性而发生扩展，当裂缝贯穿管壁，管子便受到破环，此即为塑料管的环境应力开裂[6,7]。

1　油气田用热塑性塑料管材 ESC 行为研究

国内外学者从不同角度对热塑性塑料的 ESC 行为进行了研究。大多聚焦于以下四个方面：

（1）ESC 行为评价方法研究。

ESC 行为评价，即 ESCR 性能检测主要有基于恒定应变或者恒定应力试验以及替代传统

基金项目：国家自然科学基金（油气耦合介质在热塑性塑料中的渗透特性及机理研究，51304236），陕西省自然科学基金（酸性气体在热塑性塑料中的渗透特性及机理研究，2014JQ6227）。

作者简介：齐国权，1983 出生，男，工程师，硕士，主要从事油气田用非金属及其复合材料管材研究工作，地址：（710077）陕西省西安市锦业二路 89 号。电话：029-81887685，E-mail：qgqstar@163.com。

方法的 ESCR 快速分析方法。

典型的恒定应变测试方法主要有[9] BELL 弯曲法（ASTM D1693）等；恒定应力方法有切口管试验（NPT）（ISO 13479）、宾夕法尼亚州切口试验（PENT）（ASTM F1473）、全切口蠕变试验（FNCT）［ISO 16770：2004（E）］等。Wang 等[10]认为不同种类的塑料在 ESC 试验中所施加的应变值不同，他们发现对于脆性塑料说应变应该选在应力应变曲线的弹性区域，而对于韧性塑料来说塑性区域是最好的选择。

Alan Turnbull[9]评估了塑料材料环境应力开裂的一系列标准，比较了不同测试方法的主要特征，为发展统一的 ESC 测试标准给出了一个良好的框架。Michaeli[11]等设计了一种新的测试设备，通过短时实验来预测制品长期耐环境应力开裂行为，该方法利用时温等效原理来计算蠕变曲线，为预测高分子制品的性能不稳定性提出一种新的思路。近年来，许多学者着力于用拉伸试验[12,13]和结晶实验[14]来分析聚乙烯材料的耐慢速应力开裂的能力。这样的试验比起建立在断裂力学原理上的 ESCR 实验更简单快捷。

（2）不同聚合物/溶剂体系的相互作用。

不同的化学溶剂对聚合物的敏感性不同，主要体现在溶解度参数上，所以溶剂加速塑料发生环境应力开裂的程度也不同。D. C. Wright[8]根据所有的溶剂作用于某种塑料时候产生不同的加速效果，将溶剂分为强、中、弱三类。Lutfi F. Al-Saidi[15]等研究了聚碳酸酯在异丙醇、乙二醇甲醚和甲醇三种不同溶液中的 ESC 行为，结果表明，由于三种溶剂扩散作用而导致了聚碳酸酯发生了 ESC 行为，扩散速度由大到小依次为甲醇、在乙二醇甲醚、异丙醇，扩散过程符合菲克第二定律[16]。从微观角度可以认为，溶剂的渗透作用降低了分子间作用力，加速了分子链解缠结的过程，这种加速效应造成了银纹/裂纹萌生的时间明显缩短，并且增加了裂纹的扩展速度，从而缩短了失效时间。

$$\frac{\partial C}{\partial t}=\frac{\partial}{\partial x}\left(\mathrm{D}\frac{\partial C}{\partial x}\right) \tag{1}$$

式中 C——扩散物质的体积浓度，kg/m^3；

t——扩散时间，s；

x——距离，m。

（3）环境因素对 ESC 行为的影响。

外界环境因素（包括温度升高、应力集中、循环载荷等）会加速这种自然发生的脆化过程，即降低了初始裂纹形成时间。韩冬雪[17]通过对光氧老化前后 PC 注塑试样力学性能及开裂行为的研究发现，在乙醇环境下，光氧老化后 PC 应力松弛速率更高，表面裂纹增多，其原因主要是 PC 光氧老化后形成新的基团与乙醇有更高的亲和性以及出现分子断链，加剧了应力开裂效应。不同条件下的力学实验结果和裂纹形态对比可以发现应力松弛试验更适合体现光氧老化和 ESC 对 PC 试样的协同作用。J. H. Greewnood[18]通过对高聚物材料拉伸强度试验结果总结得出，如果对材料施加一个应力，只要时间足够长，即使应力很低，即使外界暴露环境为空气，受到应力作用的塑料制品最终会发生破裂而失效。

（4）材料改性对 ESCR 性能影响。

通过增加聚合物分子间链缠结，晶粒间连接分子数增多，因而提高了它的 ESCR。Joy J. Cheng 等[19]研究了熔融态分子链缠绕行为用于推测在固态聚乙烯中的分子链缠绕量，从而分析了聚合物网络结构对 HDPE 的 ESCR 性能影响。研究发现，聚合物体系中的分子链缠绕数量可以通过其间的分子量推测而得。对于相近的链长度的聚合物链，分子量越小就意味着缠绕越多。聚合物的分子链缠绕数量利用高弹模量（G）来描述，它和分子量有关并符合流变学理论[20]。随着分子链缠绕数量的增加，HDPE 的 ESCR 性能也明显提高。

$$G_N^0=\frac{\rho RT}{M_e} \tag{2}$$

式中 ρ——液体密度；

R——通用气体常数；

T——绝对温度。

李巧娟等[21]研究了低密度聚乙烯接枝聚硅氧烷后的 ESC，结果表明，LDPE 与聚硅氧烷接枝后，增加了聚乙烯分子间链缠结，晶粒间连接分子增多，因而提高了聚乙烯的耐环境应力开裂性。LDPE 与聚硅氧烷接枝以提高聚乙烯的 ESCR 时，聚硅氧烷的用量不应太多，以2%为最好。

2 油气田用热塑性塑料 ESC 机理研究

通过以上 ESC 行为研究，国内外学者分析了热塑性塑料 ESC 机理主要包括以下三个方面[22]：

（1）增塑理论。

基于 Griffith 强度理论[23]，为了使塑化引起沿银纹发生脆性破坏，一定要局部塑化，可能因为：（a）由于应力增强溶解度，使高度受力不断生长的银纹尖端或应力集中区域塑化增加；（b）由于更大的表面能-体积比，使不断增长的缺陷尖端高度空穴区域塑化增加；（c）较高的局部应力使缺陷的尖端增加扩散，产生更大的增塑作用。如果聚合物和溶剂的溶解度参数很接近时，容易发生局部塑化，由于应力增强溶解度，使高度受力不断生长的银纹尖端或应力集中区域塑化增加。较高的局部应力使缺陷的尖端增加扩散，产生更大的增塑作用。如果聚合物局部应力-应变足够大，ESC 行为就会发生。

（2）表面能的影响。

表面能对 ESC 的影响主要体现在两个方面：（a）减少流体环境的表面张力，减少产生新的表面所需的能量，从而更容易形成空穴和银纹；（b）由于表面能较低导致空隙内的静水压力使银纹和空穴通过环境的毛细管充填能力增加，从而有助于银纹形成。

（3）分子链断裂导致 ESC 发生。

在应力和活性溶剂的共同作用下，热塑性塑料产生由形变不均一性引起的微裂纹，从而形成了具有不同程度的内在缺陷，随着活性溶剂渗透到热塑性塑料的内部，降低了其表面能，导致分子链的滑脱及解缠，当滑脱和解缠的分子链达到某临界值时，便出现了宏观的脆性开裂。

3 油气田用热塑性塑料管材 ESC 研究存在问题

虽然不同介质在聚合物材料中的 ESC 行为已有研究基础，但通过以上分析可以看出，当前对多场耦合下酸性油气田用热塑性塑料管材的 ESC 行为研究仍存在以下明显不足。

（1）油气田用热塑性塑料管材 ESC 行为快速与分析方法尚未建立。

材料的破坏时间是慢速裂纹增长的抵抗能力的直接度量，但通常热塑性塑料管材正常的慢速裂纹增长时间远超过一般试验所能容许的范围，因此需要开发加速试验方法。目前国际上有多种环境应力开裂试验方法，譬如 BELL 弯曲法、切口管试验（NPT）、全切口蠕变试验（FNCT）方法等。然而，这些方法在评价酸性油气田用热塑性塑料管材 ESC 行为仍存在以下缺点：(a) 实验所用介质未能模拟真实工况。例如在 BELL 弯曲法中采用的试剂为壬基酚聚氧乙醚烯（TX-10），而在 NPT 和 FNCT 实验中，所用的介质为水。(b) 实验耗时较长。例如，标准的 NPT 实验，破坏时间大于 2200h 中。通过以上分析可知，由于合适的评价与分析方法尚未建立，严重阻碍了快速评价 ESC 性能、指导生产实践和进行质量控制，建立适应于酸性油气田用热塑性塑料管材评价的实验方法是本课题目标之一。

(2) 热塑性塑料结构形态对 ESC 的影响未予充分考虑。

相关报道虽然研究了不同聚合物材料结构形态对其 ESC 的影响，但并未将结构形态参数引入到相关数学或物理 ESC 模型中，不能为酸性油气集输用热塑性塑料的选择和 ESC 机理的分析提供直接依据。另外，热塑性塑料以聚合物树脂为主要成分，同时添加了各种助剂如填料、增塑剂、抗氧化剂等。这使得介质在热塑性塑料中的 ESC 过程更加复杂，不能完全沿用单纯聚合物的 ESC 研究结果，也就无法利用现有方法评价常用热塑性塑料的 ESCR 性能，仍需花费大量时间和精力，采用试错法测试材料的 ESCR 性能。

(3) 尚无多场耦合下酸性油气混输用热塑性塑料管材的 ESC 机理研究。

油气集输运行工况复杂，温度变化范围较大（室温～100℃），压力达到了 10MPa 以上。这种高温、高压和流动场环境对热塑性塑料 ESC 行为的影响规律尚不明确。温度的升高会影响聚合物材料的结晶率，而压力的提升则会增加 ESC 驱动力。因此，常规状态下材料的 ESC 评价方法，将不适于指导油气集输用热塑性塑料的正确选材，粗略套用将会给复合管的使用埋下质量隐患。另外，在油气混输过程中，输送介质中的气体（H_2S、CO_2）及液体（含 Cl^- 溶液）会发生吸附、扩散、渗透、溶胀等系列物理反应过程。复杂的混合气体组分（CH_4、H_2S、CO_2 等），尤其是酸性气体在热塑性塑料中渗透特性对 ESC 行为的影响未见公开文献报道。因此，油气混输介质条件下热塑性塑料管材的 ESC 机理如何目前尚不清楚。

4 油气田用热塑性塑料管材 ESC 研究展望

(1) 建立油气田用热塑性塑料管材 ESC 行为快速评价与分析方法。

评价与分析方法不仅对于一种材料的工程应用有着重要的把关作用，更能促进该种材料的性能优化以及新材料开发。目前热塑性塑料 ESC 行为评价方法主要以偏向于工程应用为导向、定性分析为目的，而没有过多的对 ESC 行为进行深入分析；另外，目前的评价方法基于一般环境，尚无酸性油气田用热塑性塑料 ESC 评价及控制体系。因此，需要探索适合于多场耦合下酸性油气田用热塑性塑料管材 ESC 行为的评价与分析方法，从而为该工况下热塑性塑料管材 ESC 机理探索提供更便捷的途径。

(2) 明确材料结构特性对其 ESC 行为的影响规律。

材料结构成分不同，将使裂纹形成及扩展路径发生变化，从而改变材料的 ESC 模式，这是不同热塑性塑料 ESCR 各异的根本所在。因此，探索材料结构特性对其 ESCR 性能的影响规律，建立包含材料结构在内的 ESC 模型，是理解不同类型热塑性塑料 ESCR 性能的关键。拟根据不同结构形态热塑性塑料 ESCR 的测试结果，确定不同材料 ESC 行为变化规律，进而阐明材料结构形态对 ESC 的影响机制。

(3) 揭示多场耦合下酸性油气田用热塑性塑料管材 ESC 机理。

多场耦合普遍存在于现实工程中，同样在油气集输领域，输送管道不仅承受着输送介质（如 H_2S/CO_2）的腐蚀作用，还承受着输送介质温度与压力等多个物理场相互叠加作用。通过对油田热塑性塑料管材失效分析可知，ESC 是导致管材失效的重要因素。然而，目前关于热塑性塑料 ESC 行为研究大多集中于单一场环境，并且尚无多场耦合下酸性油气混输用热塑性塑料管材的 ESC 机理研究。为此，只有认知该条件下 ESC 行为并揭示其规律，才能保障热塑性塑料管材安全应用于酸性油气田。

参 考 文 献

[1] Xuehua Cai, Dongtao Qi, Nan Ding, et al. Failure Analysis of RTP for Natural Gas Transportation in Changqing Oilfield [C]. International Pipelines and Trenchless Technology Conference Shanghai: American Society of Civil Engineers, 2009, pp. 2064-2068.

[2] Yong Bai, Fan Xu, Peng Cheng. Investigation on the Mechanical Properties of the Reinforced Thermoplastic Pipe (RTP) Under Internal Pressure [C]. The Twenty-second International Offshore and Polar Engineering Conference, June 17-22, Rhodes, Greece: International Society of Offshore and Polar Engineers, 2012.

[3] 韩方勇，丁建宇，孙铁民，等．油气田应用非金属管道技术研究 [J]．石油规划设计，2012，23 (6)：5-9.

[4] Ray L. Hauser. Environmental Stress Cracking of Commercial CPVC Pipes [J]. Journal of ASTM International, 2011, 8 (7): 1-7.

[5] 戚东涛，任建红，李厚补，等．H_2S 在多层聚合物复合管材中的渗透规律 [J]．天然气工业，2014，34 (5)：126-130.

[6] J. C. Arnold. Environmental effects on crack growth in polymers [J]. Comprehensive Structural Integrity Volume 6 : Environmentally Assisted Fatigue, 2003, 281-319.

[7] P. Sardashti, C. Tzoganakis, Polak, et al. Improvement of hardening stiffness test as an indicator of environmental stress cracking resistance of polyethylene [J]. J. Macromol. Sci., Part A: Pure Appl. Chem., 2012, 49 (9): 689-698.

[8] D. C. Wright. Environmental stress cracking of Plastic [M]. Rapra Technology, Shrewsbury. 1996.

[9] Alan Turnbull, Tony Maxwell. Test methods for Environment stress cracking of polymeric materials [R]. NPL Technical Review, 1999.

[10] H. T. Wang, B. R. Pan, Q. G. Du, et al. The strain in the test environmental stress cracking of plastics [J]. Polymer Testing, 2003, 22, 125 -128.

[11] Walter Michaeli, Jan Henseler. Environment stress creaking in Polycarbonate-Prediction of the long term behavior by the use of short time test [C]. SPE/ANTEC PROCEDINGS. 2007, 289-293.

[12] Joy J. Cheng, Maria A. Polak, and Alexander Penlidis. A Tensile Strain Hardening Test Indicator of Environmental Stress Cracking Resistance [J]. Journal of Macromolecular Science, 2008, 45 (8): 599-611.

[13] Paul Hinksman, David H. Isaac, Patrick Morrissey. Environmental stress cracking of poly (vinylidene fluoride) and welds in alkaline solutions [J]. Polymer Degradation and Stability, 2000, 68 (2): 299-305.

[14] PJ Deslauriers, DC Rohlfing. Estimating slow crack growth performance of polyethylene resins from primary structures such as molecular weight and short chain branching [J]. Macromol. Symp., 2009, 282 (1): 136-149.

[15] Lutfi F. Al-Saidi, Kell Mortensen, and Kristoffer Almdal. Environmental stress cracking resistance. Behavior of polycarbonate in different chemicals by determination of the time-dependence of stress at constant strains [J]. Polymer Degradation and Stability, 2003, 82: 451-461.

[16] A. Naceri. An analysis of moisture diffusion according to Fick's law and the tensile mechanical behavior of a glass-fabric-reinforced composite [J]. Mechanics of Composite Materials, 2009, 45 (3): 331-336.

[17] 韩冬雪．光氧老化和环境应力开裂对聚合物注塑制品协同作用研究 [D]．郑州：郑州大学，2010.

[18] J. H. Greewnood. Life Predietion in Polylners, ERA Teehnology Report No. 97-0782R, (1997).

[19] Joy J. Cheng, Maria A. Polak, and Alexander Penlidis. Polymer Network Mobility and Environmental Stress Cracking Resistance of High Density Polyethylene [J]. Polymer-Plastics Technology and Engineering, 2009, 48 (12): 1252-1261.

[20] A. B. Metzner. Rheology. Theory and Applications [J]. Journal of the American Chemical Society, 1959, 81 (6): 1520-1521.

[21] 李巧娟，谢大荣 . LDPE 环境应力开裂性研究 [J]. 高分子材料科学与工程，2003，19 (6)：153-155.

[22] P. Davis, S. Burn, S Gould. Fracture prediction in tough polyethylene pipes using measured craze strength [J]. Polym. Eng. Sci. , 2008, 48 (5): 843-852.

[23] V. Altstadt. The influence of molecular variables on fatigue resistance in stress cracking environments [J]. Advances in Polymer Science, 2005, 188: 105-152.

塑料合金在油田酸性盐水环境中的适用性研究

丁　楠　蔡　克　张　翔　李厚补

（中国石油集团石油管工程技术研究院，石油管材及装备材料服役行为与结构安全国家重点实验室）

摘　要：通过对新型耐高温塑料合金材料进行红外测试和DSC测试，确认其为尼龙/聚乙烯合金（PA/PE）。进而探讨了PA/PE材料在油田工况条件下的应用可行性，采用高温高压釜设备，研究了模拟油气田高温含硫盐水介质环境下的适用性。通过对比分析老化试验前后PA/PE的结构成分、质量、体积、力学性能和耐热性能的变化规律，发现PA/PE材料的耐高温性能优异，而且化学性能稳定，但是耐溶胀性能较差，介质的溶胀作用可以导致其拉伸性能的显著变化，而且H_2S和CO_2等气体组分的影响程度大于液体组分。

关键词：塑料合金；油田环境；高温；硫化氢；适用性

近年来国内油田的服役环境日益苛刻，逐步上升的含水率和温度，以及高含量的Cl^-、CO_2、H_2S等腐蚀性介质，给钢管的应用带来极大风险。目前，使用非金属与复合材料管材替代或修复钢制管道，已成为油田解决腐蚀问题的一种常用手段[1-4]。在各类非金属材料中，热塑性塑料既可以制成均质管道单独使用，也可以用作复合管的内衬层或基体材料，应用非常广泛。然而单一组分的热塑性塑料的性能难以同时满足耐腐蚀、耐高温，以及制造工艺的要求，而通过将两种或多种热塑性塑料利用物理共混或化学接枝的方法进行改性，制成塑料合金（也称为塑料混配物）后即可获得兼具各自优点的材料，从而改善或提高塑料性能。例如，目前在国内油田使用的塑料合金防腐蚀复合管，其内衬层是由氯化聚氯乙烯树脂、聚氯乙烯树脂、氯化聚乙烯树脂等材料制成的塑料合金，可用于输送70℃以下的油水介质。随着油田用户的认可和应用领用的拓宽，现有的塑料合金材料已经难以满足实际需求，在此背景下，国内某原材料制造商开发了一种新型塑料合金，以满足油田高温应用的需求。

本文参考ISO 23936-1：2009《与石油和天然气生产相关的介质接触的非金属材料——热塑性塑料》中的评定试验方法，利用高温高压釜设备研究此塑料合金材料在模拟油气田高温含硫盐水介质环境下老化前后结构成分、质量、体积、力学性能和耐热性能的变化规律，检验材料的适用性。

1　试样与试验方法

塑料合金试样是依据GB/T 1040.2—2006《塑料　拉伸性能的测定 第2部分：模塑和挤塑塑料的试验条件》模塑成型的1A型拉伸试样。适用性评价试验在美国Cortest公司生产的静态高温高压釜试验系统中进行，试验环境为气相总压 10MPa（H_2S 分压 1.0MPa，CO_2 分压 1.0MPa，其余 N_2），液相为 NaCl 水溶液（Cl^-浓度 140000mg/L），试验温度为 80℃，老化试验周期 168h，气相和液相中各放置 5 个试样，空白试样 5 个。

肉眼检查试样的外观；采用 BT224S 型电子天平测量试样的质量，其精度为 0.0001g；采用 ET-120SL 型电子密度计测量试样的体积变化；采用 CMT-4104 型万能试验机测试样的拉伸强度和断裂伸长率；采用 VERTEX70 型傅立叶红外光谱仪测试试样的成分；采用 XRD-300DL 型热变形及维卡软化温度测定仪测试试样的维卡软化温度；采用 TIME5410D 型邵氏硬度计仪测试试样的邵氏硬度；采用 AQ200 型差示扫描量热仪测试试样的热分析曲线。

2 结果与分析

2.1 成分分析

在进行各项性能测试前，首先对此塑料合金进行成分分析。空白试样的红外图谱（图 1）解析为：3301υ（NH）、2920υas（$-CH_2-$）、2852υs（$-CH_2-$）、1641υ（C=O）、1544［υ（CN）+ δ（NH）］、1467δ（CH_2）、1373δ（CH_2）、723［δ（NH）或 γ（$-CH_2-$）］，由此判定此试样的主要成分为聚酰胺和聚烯烃[5,6]。为了进一步确认，对空白试样进行了 DSC 扫描：第一轮扫描以 10℃/min 的速率从 25℃加热至 300℃，冷却后，以相同的条件进行第二轮扫描。DSC 扫描曲线如图 2 所示，第一轮和第二轮升温曲线的低温段均在 130℃左右出现一个吸热峰，这对应于聚乙烯的熔点；第一轮升温高温段在 220℃附近出现一个吸热峰，在第二轮升温扫描时此吸热峰分裂为两个相邻的吸热峰，这对应于尼龙的熔点和热行为特点[7,8]。综合红外分析和 DSC 测试结果，证实此塑料合金是尼龙（PA）和聚乙烯（PE）的共混物（以下简称PA/PE）。

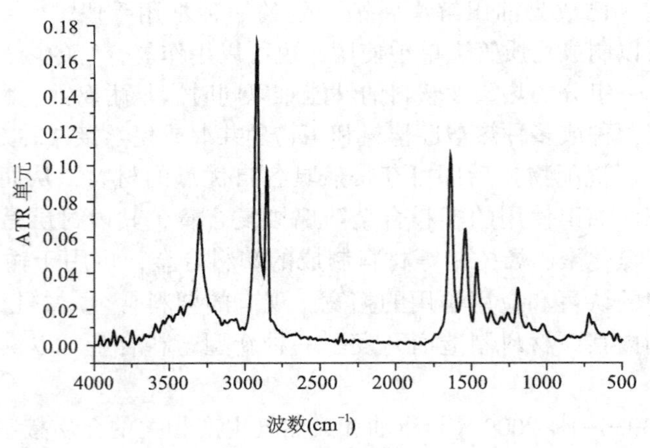

图 1 PA/PE 空白试样的红外谱图

2.2 油田介质相容性测试

2.2.1 外观变化

PA/PE 试样经高温高压釜介质相容性评价试验后，所有老化试样外形无变化，且试样表面无肉眼可见的缺陷；气相中的试样颜色由蓝色变为绿色，而液相中的试样颜色无变化。对比气液两相的组分差异，认为是气相中的 H_2S 气体导致了试样变色。

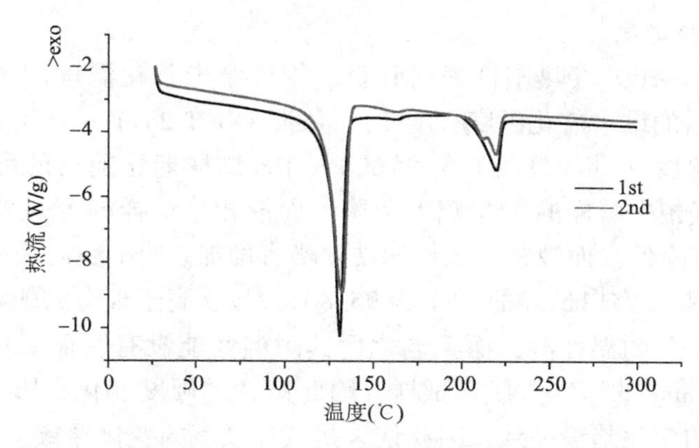

图 2　PA/PE 空白试样的 DCS 升温曲线

2.2.2　成分变化

使用同 2.1 节相同的设备和试验方法，分别对气相和液相环境中老化后的试样进行红外分析和 DSC 测试。试验结果显示，PA/PE 经气相和液相环境老化后，红外图谱中峰的数量和位置与图 1 中老化前试样的基本相同，这表明老化后试样的主要成分仍是 PA 与 PE，没有发现新生成的物质；而 DSC 扫描曲线中峰的数量、位置和面积也与图 2 中老化前试样的基本相同，这表明老化后试样中的 PA 与 PE 的含量和结晶程度也没有变化。

综合红外分析和 DSC 测试结果，可知老化试验没有改变试样的化学组成。因此，虽然不能排除在试验条件下水、Cl^-、H_2S 和 CO_2 等引起试样发生化学变化（如水解）的可能，但在试验周期内试样的主体成分没有明显改变。

2.2.3　质量和体积变化

由表 1 中的数据可知，老化后的试样出现了质量增加和体积增加的现象，而且气相中试样的变化程度大于液相中的试样。由于尼龙分子中具有亲水性的酰氨基，因而具有吸水性[9]；此外由于高分子材料分子间存在间隙，气体分子可以通过吸附和扩散作用进入高分子材料，研究早已证实了 H_2S、CO_2、CH_4、H_2O 等气体在 PE、PA、PVDF 等热塑性塑料中的渗透行为，而且高温对渗透行为有促进作用[10-13]。因此，PA/PE 在老化过程中吸附/吸收了 H_2S、CO_2 和水等气体/液体介质后产生了溶胀现象，导致了质量和体积增加；此外，试验结果表明相对于水和 Cl^- 而言，H_2S 和 CO_2 等气体组分对试样的溶胀效应更强。

2.2.4　力学性能变化

依据 GB/T 1040.2—2006 对 PA/PE 试样进行拉伸性能测试，拉伸速率为 50mm/min。如表 1 所示，相对与空白试样而言，老化后的试样的拉伸强度明显降低，同时断裂伸长率大幅增加，而且气相中试样的变化程度大于液相中试样。这表明，老化试验降低了 PA/PE 材料的拉伸强度，提高了材料的塑性。由于成分分析结果表明老化后试样没有发生化学变化，因此导致材料力学性能变化的主要因素应当是物理溶胀作用：试样吸收的水、Cl^-、H_2S 和 CO_2 等液体/气体小分子物质进入 PA/PE 材料内部后，可以在聚合物分子间的空隙位置聚集从而降低材料的致密性，其次可以对材料起到增塑剂的作用[14]，而且还可以通过氢键作用影响聚合物的分子链结构[15]，这些作用共同导致了材料力学性能的变化；而气相对 PA/PE 材料力学性能的影响大于液相，这同质量和体积变化的规律一致，由此也证明了上述基于溶胀作用的分析。

2.2.5 耐热性能变化

依据 GB/T 8802—2001《热塑性塑料管材、管件维卡软化温度的测定》B50 法，测试 PA/PE 试样老化前后的维卡软化温度（VST），依照 GB/T 2411—2008《塑料和硬橡胶 使用硬度计测定压痕硬度（邵氏硬度）》测试 PA/PE 试样老化前后的邵氏硬度。如表 1 所示，经过老化后，气相中试样的 VST 略有下降，而液相中试样的 VST 略有增加；老化后气相中的试样硬度略有降低，而液相中试样的硬度略有增加。VST 越高表明材料受热时的尺寸稳定性越好[16]，PA/PE 空白试样的 VST 为 95.8℃，显著高于现有的塑料合金复合管，因此 PA/PE 材料具有良好的耐温性能，而且老化后其耐温性能没有明显变化。VST 的测试原理是使用尖端面积为 $1mm^2$ 的压针头压入试样，因此试样的硬度变化将影响测试结果，而表 1 中试样 VST 同硬度的变化趋势一致，因此认为是试样硬度的变化导致了 VST 的变化。

表 1 PA/PE 试样老化前后性能变化

测试项目	老化环境		
	空白	气相	液相
Δm（%）	—	+2.83	+2.29
ΔV（%）	—	+2.02	+1.35
σ（MPa）	32.96	24.96	26.31
ε（%）	21.32	80.13	63.34
VST（℃）	95.8	92.9	97.4
邵氏硬度	67.14	67.02	67.74

3 结论

这种新型塑料合金的是 PA/PE 材料，其耐高温性能优于与现有的塑料合金复合管。模拟油田含硫盐水环境的老化试验表明，PA/PE 材料的化学性能稳定，但是介质的溶胀作用明显，而且 H_2S 和 CO_2 等气体组分的溶胀作用大于液体组分，在各项性能中拉伸强度和断裂伸长率对溶胀作用敏感。

参 考 文 献

[1] 韩方勇，丁建宇，孙铁民，等．油气田应用非金属管道技术研究［J］．石油规划设计，2012，23（6）：5-9.

[2] 赵小兵．非金属管材在油田集输系统的应用探讨［J］．油气田地面工程，2010，29（8）：55-56.

[3] 石鑫，羊东明，张岚．含硫天然气集输管网的腐蚀控制［J］．油气储运，2012，31（1）：27-31.

[4] 李忠，王宏军，卫杰．非金属管道在塔里木油田集油系统的应用［J］．油气储运，2003，22（1）：37-39.

[5] 王忠友，刘莹峰，周明辉，等．聚乙烯尼龙共混物的 PGC-MS 和 FT-IR 联用鉴别［J］．分析测试学报，2004，23（2）：92-94，97.

[6] 朱吴兰．红外光谱法鉴别不同种类的聚酰胺［J］．塑料，2009，38（3）：114-117.

[7] M. Z. Jandali，G. Widmann. 热塑性聚合物：第 1 版［M］．上海：东华大学出版社，2008：24，25，37，45，82，83.

[8] 王日辉，张建民，石晶．热历史对 UHMWPE 结晶度和熔点的影响［J］．合成树脂及塑料，2009，26

（5）：56-57，84.
[9] 宁冲冲，崔益华，吴银财，等．降低尼龙制品吸水率的研究进展［J］．塑料科技，2013，41（1）：105-109.
[10] Y. Makino，T. Okamoto，Y. Goto，et al. The Problem of Gas Permeation in Flexible Pipe［C］// paper 5745 presented at the 20th Annual OTC，2-5 May 1988，Houston，TX，United States. Houston：Offshore Technology Conference，1988.
[11] R. P. Campion，G. J. Morgan. The accurate measurement of elevated pressure gas permeation through polymers based on new specimen geometries［J］. Progress in rubber，plastics and recycling technology，2004，20（4）：351-374.
[12] T. R. Andersen，J. I. Skar，C. Hansteen. Permeability of methane，carbon dioxide & water in PA11 & PVDF used for flexible pipes［C］// paper 410 presented at the CORROSION 1999 conference，25 Apr 1999，San Antonio，TX，United States. Houston：NACE international，1999.
[13] Sijm Last，Steve Groves，Jean Rigaud，Carol Taravel-Condat，Jakob Wedel-Heinen，Richard Clements，et al. Comparison of models to predict the annulus conditions of flexible pipe［C］// paper 14065-MS presented at the OTC2002，6 - 9 May 2002，Huston，TX，United States. Houston：Offshore Technology Conference，2002.
[14] 何曼君，张红东，陈维孝，等．高分子物理：第3版［M］．上海：复旦大学出版社，2014：280.
[15] 朱本玮，朱华，吕桂英，等．环境湿度对尼龙66性能的影响及其时间效应［J］．现代塑料加工应用，2008，20（5）：5-7.
[16] Li H B，Li Q，Yan M L. Influence of Operation Procedures on Vicat Softening Temperature of Thermoplastic Materials［J］. Advanced Materials Research，2011，291-294：1820-1824.

Determination of Iron in Corrosion-resistant Alloy Tubing by Flow Injection Analysis

Shao Xiaodong

(CNPC Tubular Goods Research Institute, State Key Laboratory of Performance and Structural Safety for Petroleum Tubular Goods and Equipment Materials)

Abstract: Nickel corrosion-resistant alloy oil country tubular goods will be widely used in the exploitation of acid gas and oil wells. It is well known that iron is an important element in nickel corrosion-resistant alloy. In this paper, a simple chemiluminescence method, based on the enhancive effect of iron on the CL reaction between luminol and dissolved oxygen in a flow injection system, was proposed for determination of iron. The increment of chemiluminescence intensity produced is directly proportional to the iron concentration, and the linearity is obtained in the range of 6.0 ~ 700.0 ng/mL (R^2=0.9993), with the limit of detection (LOD) is 2.0 ng/mL ($3\times\sigma_{noise}$). The relative standard deviations (RSD) were less than 3.0% (n=5). At the flow rate of 2.0mL/min, a complete determination of iron, including sampling and washing, could be completed in 0.5 min, offering the sampling efficiency of 120 h^{-1} accordingly. The proposed procedure was satisfactory for the application to determine iron in G3 corrosion-resistant alloy tubing and coupling samples.

Keywords: Iron; Flow injection; Corrosion-resistant alloy

1 Introduction

With the increase desire for energy, the worldwide search for new sources of hydrocarbons is turning to the gas and oil wells containing high hydrogen sulfide. The maximum temperature in these wells can be up to 320℃, and the maximum pressure is up to 100MPa. Meanwhile, the partial pressure of hydrogen sulfide and carbon dioxide is very high. During exploitation, the elemental sulfur may be separated out. In order to meet the exploitation requirements of these wells with high temperature, high pressure and high partial pressure of hydrogen sulfide and carbon dioxide, the oil country tubular goods must adopt alloy materials with excellent resistance to high temperature and corrosion, good strength, plasticity, toughness, metallurgical stability, workability and solderability. The application scope in industry of nickel corrosion-resistant alloy is increasing, for example, in aerospace, nuclear power and ship products. In addition, many nickel corrosion-resistant alloys have excellent heat resistance, and become the ideal selection for resistance to corrosion and high temperature[1]. Therefore, the nickel corrosion- resistant alloy oil country tubular goods have excellent resistance to corrosion. This is because that the formed passive film on surface isolate the further contact between matrix and aggressive medium, protecting the matrix. It is not only has lower metal loss, but also can be subject to local corrosion, especially for resistance to hole corrosion or crevice corrosion, intergranular corrosion and stress corrosion. Thus, nickel corrosion-resistant alloy oil country tubular goods will be widely used in the exploitation of acid gas and oil wells. Continuous improvement of metallurgy and manufacturing technology promotes the development of nickel corrosion-resistant alloy. The content of chemical elements in nickel corrosion-resistant alloy directly influences the properties of alloy materials. In order to control the quality of nickel corrosion-resistant alloy products, the contents of trace impurities and alloy elements in nickel corrosion-resistant alloy should be determined. In alloy-making processes, the content of alloyed elements such as nickel, iron, manganese, chromium, molybdenum, vanadium, and so on, is required to be strictly controlled because it fundamentally determines the performance of nickel corrosion-resistant alloy. It is well known that iron is an important element in nickel corrosion-resistant alloy. Therefore, a rapid and precise analytical method for the determination of iron is essentially required for the production control of nickel corrosion-resistant alloy. In recent years, various chemical analysis techniques have distinct advantages in sensitivity, reproducibility, simplicity, cost effectiveness, flexibility and rapidity. A range of analytical techniques have been used for analysis of iron in various samples including UV-vis spectrophotometry, ion chromatography, atomic absorption spectrometry, inductively coupled plasma atomic emission spectrophotometry, and voltammetry[2-4]. Spectrophotometric methods occupy special position due to their simplicity, less expensive instrumentation and high sensitivity. A number of chromogenic reagents, such as *o*-phenylenediamine with hydrogen peroxide, *p*-anisidine with *N*, *N*-dimethylaniline, and ammonium thiocyanate have been reported for the determination of iron.

Chemiluminescence refers to the emission of light from a chemical reaction, which can occur in solid, liquid or gas systems. The fundamentals of chemiluminescence have been comprehensively reviewed in a number of textbooks and articles in recent years. Two main categories of chemiluminescence reaction have been described in the literature, direct and indirect[5]. Direct chemiluminescence can be represented by:

$$A+B \rightarrow [I]^* \rightarrow PRODUCTS+LIGHT$$

where A and B are reactants and $[I]^*$ is an excited state intermediate. The luminol reaction is an example of this form of chemiluminescence. In certain cases where the excited state is an inefficient emitter, its energy may be passed on to another species (a sensitizer, F) for light emission to be observed. This is called "indirect chemiluminescence" and is exemplified by the peroxyoxalate (light stick) reaction:

$$A+B \rightarrow [I]^* + F \rightarrow [F]^* \rightarrow F+LIGHT$$

Progress in flow injection chemiluminescence analysis has received much attention in chemical analysis due to the high sensitivity, rapidity and simplicity of this method[6]. The chemiluminescence method with different chemiluminescence systems has been exploited for the determination of iron. The emission of light observed when a solution containing luminol (5-amino-2, 3-dihydro-1, 4-phthalazine-dione or, more simply, 3-aminophthalhydrazide) and hydrogen peroxide. Luminol–hydrogen peroxide reaction as a classical chemiluminescence system is reported by different research groups, which is based on the oxidation of luminol by hydrogen peroxide. In this work, it is found that iron remarkably catalyzed the chemiluminescence reaction between luminol and hydrogen peroxide, and leading to fast chemiluminescence. On the basis of this a simple, sensitive and rapid procedure is developed for the determination of iron. The increment of chemiluminescence intensity produced is directly proportional to the iron concentration, and the linearity is obtained in the range of 6.0~700.0 ng/mL ($R^2=0.9993$), with the limit of detection (LOD) is 2.0 ng/mL ($3\times\sigma_{noise}$). The relative standard deviations (RSD) were less than 3.0% ($n=5$). At the flow rate of 2.0mL/min, a complete determination of iron, including sampling and washing, could be completed in 0.5 min, offering the sampling efficiency of 120 h^{-1} accordingly. The proposed procedure was satisfactory for the application to determine iron in G3 corrosion–resistant alloy tubing and coupling samples.

2 Experimental

2.1 Apparatus

A schematic diagram of the CL flow injection analysis system was shown in Fig. 1. A peristaltic pump was utilized to deliver all flow streams. PTFE tubing (1.0mm i. d.) was used as connection material in the flow system. A six–way valve with a loop of 100.0μL was employed for sampling. The flow cell was made by coiling 30.0 cm of colorless glass tube (2.0mm i. d.) into a spiral disk shape with a diameter of 2.0 cm and placed close to the photomultiplier tube (PMT) (Model IP28, Hamamatsu). The CL signal produced in the flow cell was detected without wavelength discrimination, and the PMT output was amplified and quantified by a luminosity meter (Model GD–1, Xi'an Remax Electronic Science – Tech. Co. Ltd.) connected to a recorder (Model XWT – 206, Shanghai Dahua Instrument and Meter Plant). The maximum of the fast CL emission at 425nm observed from the above reaction could be proved by a spectrofluorometer (Model F–4500, Hitachi Co. Ltd.). A UV–visible spectrophotometer (Model Lambda–40P, PE Co. Ltd.) was used and the optical path length was 1.0 cm in the wavelength range from 200nm to 400nm.

2.2 Reagents

All the reagents were of analytical grade, and the water used was purified in a Milli–Q system (Millipore, Bedford, MA, USA). Working standard solutions were prepared daily from the stock solution by appropriate dilution. Luminol (2.5×10^{-2} mol/L) was prepared by dissolving 4.4000 g luminol (Fluka, Switzerland) in 1.0 L of 0.1 mol/L NaOH solution. All chemicals including sulphuric acid, nitric acid (Xi'an Chemical Reagent Plant) and phosphoric acid (Sichuan Xilong Chemical Co. Ltd.) used in the experiments were of analytical reagent grade and were used as received. Solutions of ammonium persulfate, sodium nitrite (Xi'an Chemical Reagent Plant) and silver nitrate (Institute of Shanghai Fine Chemical Materials) were prepared daily by dissolving the reagents in deionised water that treated with a Milli – Q water purification system (Millipore, Bedford, MA, USA). Solutions of manganese were prepared from the steel certified reference materials (CRM) including GBW 01301, GBW 01302, GBW 01305, GBW 01307, GBW 01311, GBW 01312 (Institute of Anshan Iron and Steel Group Corporation), GBW 01354

(Institute of Fushun Iron and Steel Group Corporation), GBW 01358 (Shanghai Iron and Steel Institute), and GSBH 40084 (China Iron & Steel Research Institute Group). All working strength solutions in the experiment were prepared with deionised water.

2.3 Procedures

Weigh accurately 0.1000 g of drilled nickel corrosion-resistant alloy samples and dissolve the samples in 40mL mixture acid solution of 15mL hydrochloric acid, 15mL nitric acid, 5mL phosphoric acid and 5mL sulphuric acid in a 150-mL taper bottle. As shown in Fig. 1, flow lines were inserted into the sample, luminol, water carrier, and sodium hydroxide solutions, respectively. The pump was started at a constant speed of 2.0mL/min to wash the whole system until a stable baseline was recorded. Then 100.0 μL luminol solution was injected into the water carrier stream by injection valve, merged with the mixed solution stream of iron. The mixed solution in an alkaline medium was delivered into the CL cell, producing CL emission, detected by the PMT and luminometer. The concentration of the sample was quantified by the increment of CL intensity ($I=I_s-I_o$), where I_s and I_o were CL signals in the presence and in the absence of iron, respectively.

3 Results and discussion

3.1 Time profile of the CL reaction

The kinetic curve was examined by static method. The kinetic profile for CL intensity of luminol-dissolved oxygen reaction was tested using 5.0×10^{-5} mol/L luminol in 0.05 mol/L sodium hydroxide solution. As shown in Fig. 2, the CL signal of luminol-dissolved oxygen reached a maximum at 5 s after initiating the reaction, and tended to be vanishing in the following 20 s. Also, it can be seen that a scintillescent CL signal was detected in the presence of iron (10.0 ng/mL), and the CL intensity reached maximum at 3 s, then tended to be vanishing in following 16 s. It was also demonstrated that iron enhanced the CL reaction and increased the CL intensity greatly.

3.2 Effect of luminol concentration

The influence of luminol concentration on CL was examined. Through determining a series of standard solutions of iron (4.0×10^{-10} to 4.0×10^{-7} g/mL) by using different concentrations of luminol solution from 1.0×10^{-7} to 1.0×10^{-4} mol/L, it was found that luminol concentrations of 5.0×10^{-5} mol/L offered the higher sensitivity than any other concentrations of luminol. Therefore, 5.0×10^{-5} mol/L luminol was chosen for the subsequent experiment.

3.3 Effect of sodium hydroxide concentration

Owing to the nature of luminol CL reactions, which is more favored in alkaline medium, sodium hydroxide was introduced into the CL cell through a flow line to improve the sensitivity of the system. A series of sodium hydroxide solutions from 5×10^{-3} to 1.0×10^{-1} mol/L were tested in the presence of 1.0×10^{-8} g/mL iron and 5.0×10^{-5} mol/L luminol. The CL intensity (I) versus sodium hydroxide concentration plot reached a maximum at about 5.0×10^{-2} mol/L sodium hydroxide, thus this concentration was employed in subsequent experiments.

3.4 Effect of flow rate and the length of mixing tubing

The mixing tubes from 0.5 to 10.0 cm were employed respectively to test the effect of the length of mixing tube on CL intensity. By comparing the CL intensities by using different length of mixing tube in the presence of 10. 0 ng/mL iron, it could be observed that the CL intensity was much stronger using 5. 0 cm mixing tube than that using any other mixing tube. Accordingly, 5. 0 cm was then selected as the optimum length of mixing tube. The influence of flow rate on determination was examined by investigating the signal-to-noise ratio under different flow rates. The effect of flow rate on CL intensity was examined in the range from 0. 5 to 5. 0mL/min, and it was found that the flow rate of 2. 0mL/min offering the highest signal-to-noise ratio was then chosen as suitable condition considering analytical precision.

3. 5 Select dissolved acid for nickel corrosion-resistant alloy samples

The effect of the different acids for dissolve the nickel corrosion-resistant alloy samples was examined. It was found that a solution mixed with 15mL hydrochloric acid, 15mL nitric acid, 5mL phosphoric acid and 5mL sulphuric acid has a good effect for dissolve the nickel corrosion-resistant alloy. Thus, an acid solution mixed with 15mL hydrochloric acid, 15mL nitric acid, 5mL phosphoric acid and 5mL sulphuric acid was recommended for dissolve the nickel corrosion-resistant alloy samples in subsequent studies.

3. 6 Performance of proposed method for iron measurements

A series of standard solutions of iron were injected into the manifold depicted in Fig. 1. The increment of the CL intensity was found to be proportional with the concentration of iron, and the calibration graph was linear from 6. 0 ng/mL to 700 ng/mL with the detection limit of 2. 0 ng/mL ($3\sigma_{noise}$). And the regression equation is $\Delta I_{CL} = 1.032C + 19.03$ ($r^2 = 0.9993$). At a flow rate of 2. 0mL/min, a complete determination of analyte, including sampling and washing, could be accomplished in 0. 5 min with a RSD of less than 3. 0%.

3. 7 Interference studies

The extent of interference by diverse ions was determined by measuring the absorbance of solutions containing iron and various amounts of diverse ions. The criterion for interference was an absorbance value varying by more than +2% from the expected value of manganese alone. The results show that a large excess of cations and anions which are usually associated in the determination of iron do not interfere. The colorless metal ions do not interfere. The tolerable amounts of foreign species with respect to iron for interference at +2% level were over 200~300 for the colored metal ions. When the colored metal ions contents of not more than 200~300 times iron, it can be in some color solution adding a few drops of EDTA solution or sodium nitrite solution to fade the color, this solution as the measured reference solution to eliminate the interference. There was not obvious interference ions exist in G3 corrosion-resistant alloy samples.

4 Applications

4. 1 Determination of iron in G3 corrosion- resistant alloy samples

The recommended procedure has been applied satisfactorily to the determination of iron in G3 corrosion-resistant alloy samples. The samples were dissolved according to the following procedure: Weigh accurately 0. 1000 g of drilled nickel corrosion-resistant alloy samples and dissolve the samples in 40mL mixture acid solution of 15mL hydrochloric acid, 15mL nitric acid, 5mL phosphoric acid and 5mL sulphuric acid in a 150mL taper bottle. The results were summarized in Table 1 with the good accuracy and precision. To verify the results obtained by the proposed method, inductively coupled plasma atomic emission spectrophotometry method was applied to determine the samples according to the literature and the results obtained by the proposed method agreed well with that from the reference method.

Table 1 Results of determination of iron in G3 corrosion-resistant alloy samples*

Sample No.	Size (mm)	RSD (%)	Content (ω,%)	
			By the proposed method	By AES
1	ϕ88. 9mm×6. 45mm	1. 21	20. 15 ± 0. 40	20. 10 ± 0. 20
2	ϕ88. 9mm×6. 45mm	1. 99	20. 89 ± 0. 30	21. 01 ± 0. 10
3	ϕ88. 9mm×6. 45mm	1. 64	21. 72 ± 0. 30	21. 26 ± 0. 40
4	ϕ73. 02mm×5. 51mm	1. 55	26. 45 ± 0. 50	25. 93 ± 0. 50
5	ϕ73. 02mm×5. 51mm	1. 83	25. 37 ± 0. 40	25. 38 ± 0. 20
6	ϕ73. 02mm×5. 51mm	0. 57	26. 02 ± 0. 20	26. 29 ± 0. 30
7	ϕ88. 9mm×6. 45mm	0. 99	26. 71 ± 0. 30	27. 18 ± 0. 40
8	ϕ88. 9mm×6. 45mm	2. 02	27. 10 ± 0. 50	27. 23 ± 0. 20
9	ϕ88. 9mm×6. 45mm	1. 05	27. 53 ± 0. 30	27. 08 ± 0. 30

Notes: * The average of five determinations.

4. 2 Determination of iron in 825 corrosion- resistant alloy samples

The recommended procedure has been applied satisfactorily to the determination of iron in 825 corrosion-resistant alloy samples. The results were summarized in Table 2 with the good accuracy and precision. To verify the results obtained by the proposed method, inductively coupled plasma atomic emission spectrophotometry method was applied to determine the samples according to the literature and the results obtained by the proposed method agreed well with that from the reference method.

Table 2 Results of determination of iron in 825 corrosion-resistant alloy samples*

Sample No.	Size (mm)	RSD (%)	Content (ω,%)	
			By the proposed method	By AES
1	ϕ82. 5mm ×(4. 5+1) mm	2. 04	32. 65 ± 0. 80	33. 01 ± 0. 90
2	ϕ82. 5mm ×(4. 5+1) mm	1. 82	32. 25 ± 0. 70	32. 68 ± 0. 80
3	ϕ82. 5mm ×(4. 5+1) mm	1. 03	32. 97 ± 0. 90	32. 72 ± 0. 80
4	ϕ82. 5mm ×(4. 5+1) mm	2. 71	33. 12 ± 0. 50	33. 09 ± 0. 70
5	ϕ73. 02mm×5. 51mm	1. 21	35. 64 ± 0. 40	35. 11 ± 0. 60

续表

Sample No.	Size (mm)	RSD (%)	Content (ω,%)	
			By the proposed method	By AES
6	ϕ73. 02mm×5. 51mm	0. 89	36. 21 ± 0. 90	36. 58 ± 0. 90
7	ϕ73. 02mm×5. 51mm	2. 31	37. 05 ± 0. 80	36. 92 ± 0. 70
8	ϕ73. 02mm×5. 51mm	1. 88	37. 38 ± 0. 40	37. 92 ± 0. 50
9	ϕ73. 02mm×5. 51mm	1. 64	36. 23 ± 0. 50	36. 53 ± 0. 50

Notes: * The average of five determinations.

5 Conclusions

The presented CL method combined with FI technique offered prominent advantages including instrumental simplicity, high sampling efficiency, reducing reagents consumption, analytical sensitivity and selectivity compared with the existed methods. The satisfactory performance in the determination of iron in G3 corrosion-resistant alloy and 825 corrosion-resistant alloy demonstrated that the method was practical and suitable not only for quality control analysis but also for product analysis, confirming the promise for G3 corrosion-resistant alloy and 825 corrosion-resistant alloy research.

Acknowledgements

The author gratefully acknowledges the CNPC Tubular Goods Research Institute, and China National Quality Supervision, Testing and Inspection Center of Oil Tubular Goods.

References

[1] Guo X, Li J, Zhou B, Influence of shear temperature on chip formation of turning GH4169. J. Shanghai Jiaotong University, 2009, 43: 79-83.

[2] Senee K, Saisunee L, Napaporn Y, A simple and green analytical method for determination of iron based on micro flow analysis, Talanta, 2007, 73: 46-53.

[3] Divjak B, Franko M, Novic M, Determination of iron in complex matrices by ion chromatography with UV-Vis, thermal lens and amperometric detection using post-column, J. Chromatogr. A, 1998, 829: 167-174.

[4] Ugo P, Moretto L, Rudello D, Birrel E, Chevalet J, Trace iron determination by cyclic and multiple square-wave voltammetry at nafion coated electrodes. applicationto pore-water analysis, Electroanalysis, 2001, 13: 661-668.

[5] Barni F, Lewis S W, Berti A, Miskelly G M, Lagoa G, Forensic application of the luminol reaction as a presumptive test for latent blood detection, Talanta, 2007, 72: 896-913.

[6] Powe A M, Fletcher K A, Molecular fluorescence, phosphorescence, and chemiluminescence spectrometry, Anal. Chem., 2004, 76: 4614-4634.

Transport Behavior of Pure and Mixture Gas through Thermoplastic Lined Pipes Materials

Zhang Dongna Li Houbu Qi Dongtao Shao Xiaodong Cai Xuehua

(Tubular Goods Research Institute, State Key Laboratory of Performance and Structural Safety for Petroleum Tubular Goods and Equipment Materials)

Abstract: The transport behavior of pure and mixture gas (CO_2/CH_4) through the commonly used thermoplastic lined pipe materials polyethylene (HDPE), polyamide (PA) and polyvinylidene fluoride (PVDF) were studied by the differential pressure method. It was found that the permeability coefficient of the pure gas through HDPE, PA12 and PVDF all improved with the increasing temperature. The apparent activation energy of permeation (E_p) was calculated. The permeation of mixture gas with different volume fractions were also tested at different temperature, in which the deviation between test and ideal state was found. For HDPE and PVDF, the permeability coefficient of mixture gas is more than the pure gas because of the plastication of CO_2. But the permeability coefficient of mixture gas for PA12 is less than the pure gas, and this is owing to the tight interaction between chains and the competition between CO_2 and CH_4 for a limited number of active sites.

Keywords: Transport behavior; Gas; Thermoplastic; Pipes

1 Introduction

Thermoplastic lined pipes are used for the transporting of oil and natural gas because of its excellent corrosion resistance. With the growing usage amount of thermoplastic lined pipe, the permeability of gases in thermoplastic cannot be ignored. For the metal pipe lined by thermoplastic, the gas permeation phenomenon made corrosion of metal by the escaping acid gas. Another type of thermoplastic lined pipe is flexible pipe, which is manufactured by polymer lining and composite. For flexible pipe, the gas permeability leads to blister, which degrades the performance of the polymer. Besides, the permeability of gas through polymer may waste the natural gas resources.

The permeability of gases through polymer is essential for the selection and the design of the structures. Most of the gas transport properties were focus on the separation industry[1-4], such as the resource recovery of high-value organic vapors from industrial waste gas streams, higher hydrogen-containing off-gas streams from the petrochemical industry. The larger the difference between permeability of individual gas in the mix-gas, the more ideal the selectively is. There was some research concerned the permeation phenomenon of thermoplastic lining materials. B. Flaconnèche[5] studied the transport coefficients of five gases (He, Ar, N_2, CH_4 and CO_2) for three semicrystalline polymers, which were used as linear of flexible pipe. T. R. Andersen[6] measured the permeability of CH_4, CO_2 and water through plasticized polyvinylidene fluoride and plastized polyamide 11 for a number of temperatures and pressures, and shown that the permeation coefficients for CH_4 and CO_2 in plasticized PVDF are one decade higher than for deplasticized, but little work have been done on the mixture gas permeation behavior.

In this work, the gas transport coefficients of three thermoplastic lining materials were evaluated, and the materials are high density polyethylene (HDPE), polyamide (PA) and polyvinylidene fluoride (PVDF). The test gases were pure and mixed CO_2, CH_4.

2 Experimental

2.1 Materials

The materials tested were HDPE, PVDF and PA12, which were taken from extruded plane sheet by Evonik. The thicknesses of the sheets are 300 ~ 400μm. The three kinds of thermoplastic material were tested from 30℃ to 80℃, and the test gas were pure CO_2, CH_4 and the mixture of the two.

2.2 Permeability measurement

The determination of the transport coefficients of pure and mixture gas used the differential pressure method. The equipment was VAC-V2 gas permeability tester, which is followed by the standards of ISO 1505-1, ISO 2556 and ASTM D1434. Diffusion coefficient and solubility coefficient are obtained from the experimental curves, by using the time lag method[7]. Before the test, one side of the member is nearly vacuum (p_2), and the other side is about 0.1MPa (p_1). The pressure change of p_2 is measured, and the amount of gas (Q) were calculated by Equation (1), where V is the volume of lower pressure cavity, S is the test area of the membrane, T is the test temperature, T_0 and p_0 are 274.15K and 1.0133×10^5 Pa respectively.

$$Q=\frac{\Delta p}{\Delta t}\times\frac{V}{S}\times\frac{T_0}{p_0 T}\times\frac{24}{(p_1-p_2)} \tag{1}$$

The thickness of the membrane is d. The permeability coefficient (P) is calculated by Equation (2), and expressed in $cm^3\cdot cm/cm^2\cdot s\cdot Pa$.

$$P=\frac{\Delta p}{\Delta t}\times\frac{V}{S}\times\frac{T_0}{p_0 T}\times\frac{d}{(p_1-p_2)}=1.1574\times10^{-9}Q\times d \tag{2}$$

When a pressure is applied to one side of the membrane, before reaching the steady state, the flux and the concentration vary with time in every point inside the membrane. The curve represented gas amount versus time became straight during the steady state. The intercept between this line and the x axis is equal to:

$$\theta=\frac{d^2}{6D} \tag{3}$$

θ is the time lag, and from Equation (3), the diffusion coefficient (D) is obtained from the relation, which is expressed in cm^2/s. According to Fick's law[8], the solubility coefficient (S), expressed in $cm^3/cm^2\cdot s\cdot cm\cdot Hg$, is easily obtained from the relation:

$$S = \frac{P}{D} \tag{4}$$

3 Results and Discssion

3.1 Influence of the temperature for the pure gas transportation

Transport phenomena of CO_2 and CH_4 in HDPE, PA12 and PVDF were studied, in a temperature range from 30℃ to 80℃. The transport coefficients were shown in Table 1 and Table 2, which were intervening value among multiple tests. Figure 1 is the two forms of the pressure-time curve for p_2. The expression of a straight line means the steady state. For the first form, the steady state is reached after a time period, from which time lag and diffusion coefficient can be gotten. For the second form, diffusion coefficient and solubility coefficient can't be estimated because of the absence of time lag.

Table 1 Transport coefficient of CO_2 of HDPE, PA12 and PVDF

Polymer	T (℃)	P [10^{-14} cm^3 · cm/ (cm^2 · s · Pa)]	D (10^{-7} cm^2/s)	S [10^{-4} cm^3/ (cm^2 · s · cm · Hg)]
HDPE	30	8.918	/	/
	40	17.64	/	/
	60	26.83	0.925	38.70
	80	64.30	1.880	45.60
PA12	30	13.68	/	/
	40	24.41	1.340	24.27
	60	62.08	3.525	23.48
	80	125.7	/	/
PVDF	30	0.369	/	/
	40	1.699	0.20	11.4
	60	6.107	0.43	18.9
	80	18.42	1.29	19.1

Table 2 Transport coefficient of CH_4 of HDPE, PA12 and PVDF

Polymer	T (℃)	P [10^{-14} cm^3 · cm/ (cm^2 · s · Pa)]	D (10^{-7} cm^2/s)	S [10^{-4} cm^3/ (cm^2 · s · cm · Hg)]
HDPE	30	5.586	0.903	8.247
	40	10.29	1.578	8.770
	60	26.46	4.378	8.059
	80	67.71	12.04	7.498
PA12	30	2.876	0.485	7.911
	40	4.825	0.720	8.934
	60	13.83	1.923	9.597
	80	32.14	4.507	9.509

Continued

Polymer	T (℃)	P [10^{-14} cm^3 · cm/ (cm^2 · s · Pa)]	D (10^{-7} cm^2/s)	S [10^{-4} cm^3/ (cm^2 · s · cm · Hg)]
PVDF	30	0.070	/	/
	40	0.109	/	/
	60	0.848	0.167	6.767
	80	3.465	0.611	7.559

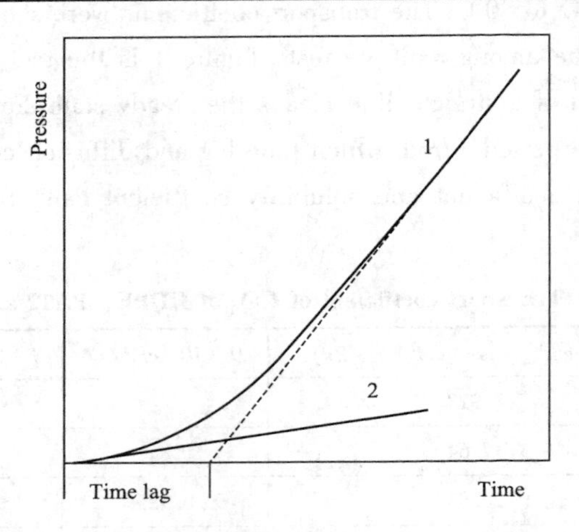

Fig. 1 Two forms of the pressure-time curve for p_2

The effect of test pressure could be ignored. Flaconnèche B.[5] found that no significant effect was noticed on the transport coefficients of pressure (4 to 10MPa), and Naito Y.[9] found that the permeability of CO_2 and CH_4 slight increase with the pressure from 0.1 to 13MPa in PE. In the test, the initial pressure of p_1 is 0.1MPa. The influence of temperature was discussed in this section. With improving the temperature, permeability coefficient for both CO_2 and CH_4 of HDPE, PVDF and PA12 are improved, and this phenomenon agrees well with the literature. The mobility of the polymer chain and the gas movement both increased at higher temperature, which accelerates the gas transport process. According to Arrhenius' laws[1]:

$$P=P_0\exp\left(\frac{-E_p}{RT}\right) \tag{5}$$

where P_0 is the front factor, E_p is the apparent activation energy of permeation, R is the gas constant, and T is the absolute temperature. Fig. 2 and Fig. 3 show the relationship between ln$P$ and $1/T$. E_p can be calculated by linear fitting, which was shown in Table 3.

Few studies concerning PA12 are available. The value of E_p for HDPE agrees well with the literature values[10]. It showed that E_p for PVDF in CO_2 and CH_4 are 30~36 and 60~62kJ/mol in Flaconneche B.'s research[5], and are 12.64 and 14.58 in Andersen T. R.'s research[6], which are different from the result in Table 3. In this test, PVDF is unplasticized, which is different from other researches. The adding of plasticizer improves the mobility of macromolecular chains and facilitates the diffusion of the gas molecules, and the effect of plasticizer for different gas is not the same. In this test, E_p of CO_2 for PVDF is higher than that of CH_4, and it means that the temperature dependency of P for CO_2 is higher in the unplasticized PVDF.

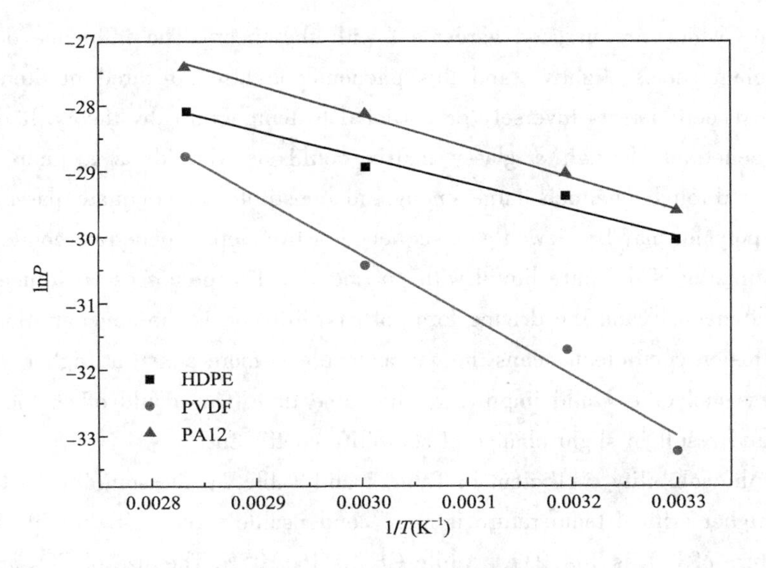

Fig. 2 Permeability of CO_2 versus reciprocal temperature

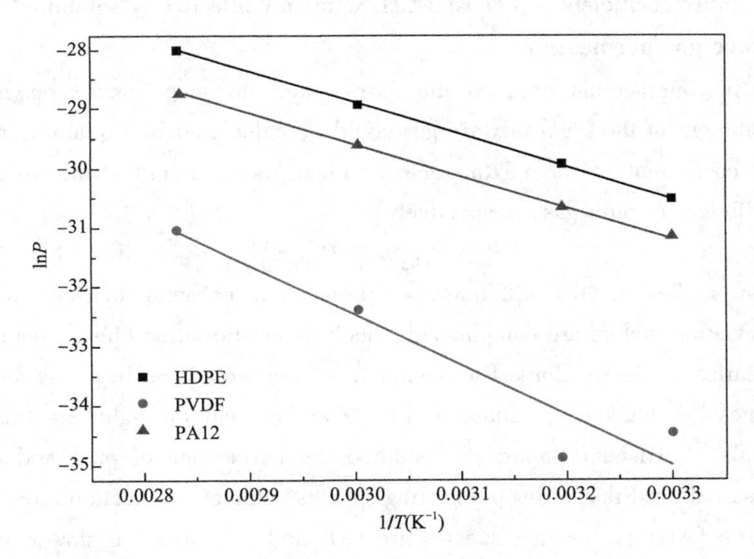

Fig. 3 Permeability of CH_4 versus reciprocal temperature

Table 3 The apparent activation energy of permeation

Gas	E_p (kJ/mol)		
	HDPE	PVDF	PA12
CO_2	32.61	75.46	39.38
CH_4	43.28	43.84	69.50

The trend of permeability coefficient at different temperature is the result of diffusion coefficient multiply by solubility coefficient. Diffusion coefficient is related to the size of the gas, while solubility coefficient is related to the critical temperature, T_c. The critical temperature is used as a scaling factor for gas condensability[1]. From Table 1 and 2, diffusion coefficient increased when the temperature rose, which are in good agreement with the theory. The influence of temperature on solubility coefficient seems slightly, and this phenomenon also appeared in literature[8]. But the value of solubility coefficient is inversely proportional to temperature by theory. It proposed that the diffusion of a penetrant through a glassy matrix could be viewed as a jump motion between neighboring holes through channels large enough to constitute an adequate passageway[11]. Vapor dissolution in a polymer may be viewed as a sequence of two steps: penetrate condensation to a pure liquid state and mixing of the pure liquid with polymer[12]. The processes of diffusion and solubility are existed simultaneously and the driving force of gas diffusion is the concentration difference. The improving of diffusion coefficient means the gas molecule is more active at higher temperature, and the number of gas molecule would improve at the same time. This could offset the difficulty of gas condensation, and result in slight change of solubility coefficient.

Comparing the solubility coefficient in Table 1 and Table 2, the solubility of CO_2 is more than CH_4. Gas with higher critical temperature is more condensable, and more soluble in polymers. The critical temperature of CO_2 is 304. 21K, while CH_4 is 191. 05K. The size of CO_2 and CH_4 is nearly the same. Permeability coefficient of CO_2 and CH_4 is mainly affected by solubility.

3. 2 Mixture gas permeation

If there is no interactions between the gases, and no special gas - polymer interactions, permeability coefficient of the ideal mixture gas could be calculated by Equation (6), where P_{id} is the permeability coefficient of the mixture gas in ideal state, x and P are volume fraction and permeability coefficient of pure gas, respectively[13].

$$P_{id} = P_{CO_2} x_{CO_2} + P_{CH_4} x_{CH_4} \qquad (6)$$

There have been studies on the difference of permeation behavior between pure and mixture gas[13-17]. The deviation factors are complicated, such as plasticization effect, coupling effect, and concentration polarization interactions. The deviation is more prominent in glassy membranes than in rubbery membranes[16,17]. Usually, Equation (6) can be used for light gas transport in rubber polymeric materials[14]. Hisham Ettouney[18] studied the permeation of pure and mixture gases in polysulfone. It was believed that faster permeating species enhance permeation rates of slower species for polysulfone. The faster permeating species are CO_2 and O_2, and the slower species are CH_4, N_2, air and so on. Like polysulfone, the permeation of the slower species of poly (methylmethacrylate), poly (ethylene terephthalate) all increase by faster species CO_2. But this behavior of polycarbonate is opposite, and it was assumed that CO_2 alters the segmental organization, which does not recover to its original state upon removal of CO_2[19]. It could be found that the transport behaviors of mixture gas through different materials are not the same. All the studies above were focused on the gas separation, and little research are available for the thermoplastic lined pipe material.

In this part, the volume fraction ratios of CO_2 and CH_4 in mixture gas are 1 : 9, 3 : 7, 5 : 5 and 7 : 3. The tested sheets are same as 2. 1. Systematic experiments were performed with the mixture gas of CO_2 and CH_4 for HDPE, PA12 and PVDF. The permeability coefficient of calculation by Equation (6) and tested are shown in Table 4~Table 6.

Table 4 Permeation coefficient of HDPE

ratio of CO_2 and CH_4	1 : 9		3 : 7		5 : 5		7 : 3	
P [$10^{-14}cm^3 \cdot cm/(cm^2 \cdot s \cdot Pa)$]	calculated	tested	calculated	tested	calculated	tested	calculated	tested
30	5. 92	5. 89	6. 59	7. 49	7. 25	8. 73	7. 92	13. 81
40	11. 02	9. 053	12. 50	12. 60	13. 96	14. 79	15. 43	24. 70
60	26. 50	29. 61	26. 57	35. 23	26. 65	40. 12	26. 72	47. 71
80	67. 37	73. 08	66. 69	76. 28	66. 00	94. 35	65. 32	122. 30

Table 5 Permeation coefficient of PA12

ratio of CO_2 and CH_4	1 : 9		3 : 7		5 : 5		7 : 3	
P [$10^{-14}cm^3 \cdot cm/(cm^2 \cdot s \cdot Pa)$]	calculated	tested	calculated	tested	calculated	tested	calculated	tested
30	3. 96	2. 85	6. 12	5. 40	8. 28	5. 42	10. 44	11. 70
40	6. 78	5. 05	10. 70	7. 46	14. 62	11. 64	18. 53	16. 30
60	18. 65	17. 22	28. 31	25. 12	38. 00	34. 37	47. 61	44. 89
80	41. 50	38. 47	60. 21	52. 97	78. 92	70. 79	97. 63	91. 35

Table 6 Permeation coefficient of PVDF

ratio of CO_2 and CH_4	1 : 9		3 : 7		5 : 5		7 : 3	
P [$10^{-14}cm^3 \cdot cm/(cm^2 \cdot s \cdot Pa)$]	calculated	tested	calculated	tested	calculated	tested	calculated	tested
30	0. 10	0. 48	0. 16	0. 72	0. 22	1. 15	0. 28	1. 98
40	0. 27	0. 69	0. 59	1. 61	0. 90	2. 70	1. 22	4. 29
60	1. 37	1. 88	2. 43	3. 34	3. 48	5. 33	4. 53	4. 63
80	4. 96	5. 29	7. 95	8. 99	10. 94	11. 89	13. 93	14. 03

There is a certain deviation between the tested and calculated value of permeability coefficient. Fig. 4 shows the permeation ratio θ, which defines as a ratio of the actual property P_{MIX} to the ideal property by Equation (7)[16]. Fig. 4 to Fig. 6 shows the permeation ratio θ of HDPE, PA12 and PVDF at different temperatures.

$$\theta=\frac{P_{MIX}}{P_{id}} \tag{7}$$

In Fig. 4, it could be found that the majority of the permeation ratio θ are above 1, which means the permeation is promoted by the mixed gas for HDPE. In the mixture gas, CO_2 acts as faster permeating species while CH_4 slower species. With increasing the CO_2 fraction, the permeation ratio is improved, and the permeation amount is enhanced. It could be found that the promote effect of CO_2 is obvious. The faster species CO_2 enhances permeation rate of slower species CH_4 because of the effect of plasticization. The easily condensability gas CO_2 could swell the plastic material, which increases the mobility of the molecule chain and the permeation amount. Besides, the relationship between temperature and permeation ratio is complex, and this will discuss in the following section.

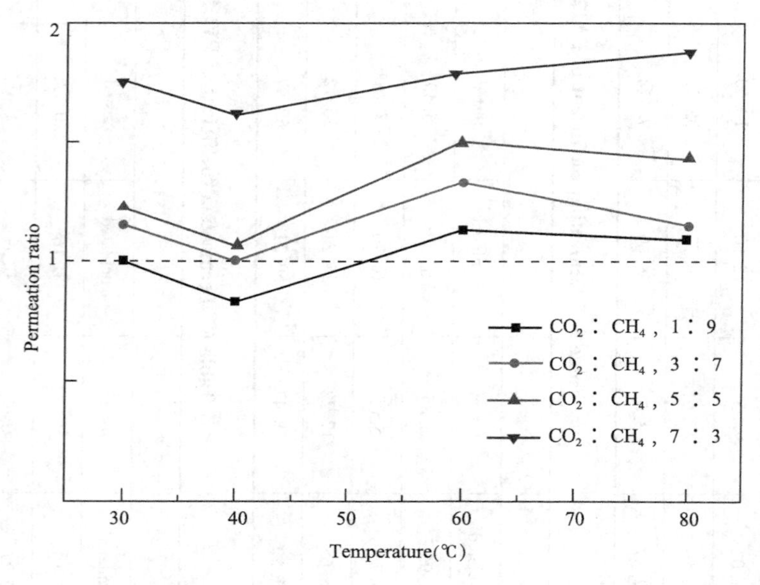

Fig. 4 The permeation ratio at different temperature for HDPE

Opposite phenomenon was found in Fig. 5. The permeation ratio is almost below 1 for PA12, which means the transport of mixture gas is hindered. As we know, amido bond (-NH-CO-) in PA12 can form hydrogen bonds, which bring tight interaction between molecule chains. The hydrogen bonds make the material difficult to be plasticized, so the effect of CO_2 should be ignored. The permeation amount of mixture gas is less than the pure gas, and this is related to the competition of two gases for the limited number of active sites in the polymer. Unlike the permeation ratio of HDPE and PVDF, the difference between calculation and tested is slight for PA12. It was found that the permeation ratio is nearly unity with increasing the temperature.

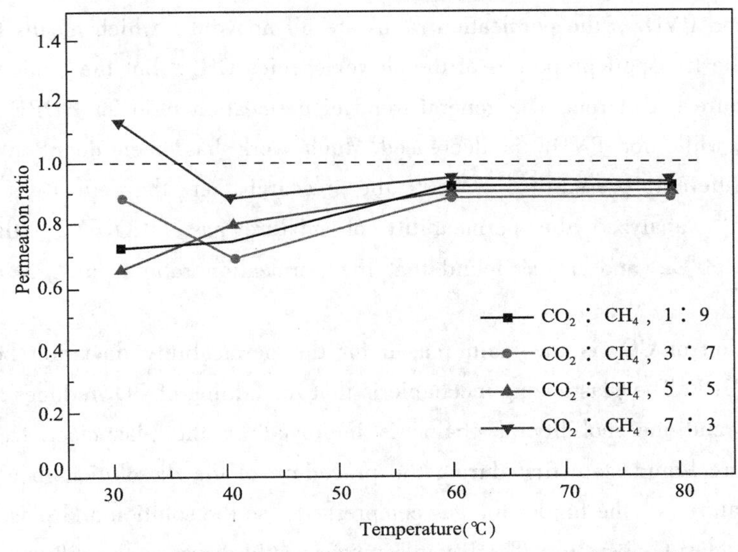

Fig. 5 The permeation ratio at different temperature for PA12

The situation of PVDF is similar as HDPE, that the permeation ratio is all above 1, but the maximum value of permeation ratio of PVDF is much more than HDPE. The barrier property of PVDF is excellent, and it is suitable to be used as flexible pipe lining. Although the permeation ratio is high, because of the less permeation coefficient for pure gas, the permeation amount is little. It could be found from Fig. 6 that the plastication of CO_2 in PVDF is obvious. The faster species CO_2 improve the permeation rate of the slower species CH_4 a lot, especially in the lower temperature. PVDF and HDPE are all nonpolar plastic, so there is little interaction between molecule chains, and the plastication of CO_2 in PVDF is greater than in HDPE. It could be found from Table 1 and Table 2 that, pure gas permeability coefficient in PVDF is much less than in HDPE, which make the plasticized effect in PVDF more prominent. It could also find from Fig. 6 that the permeation ratio decreased with the increasing of temperature.

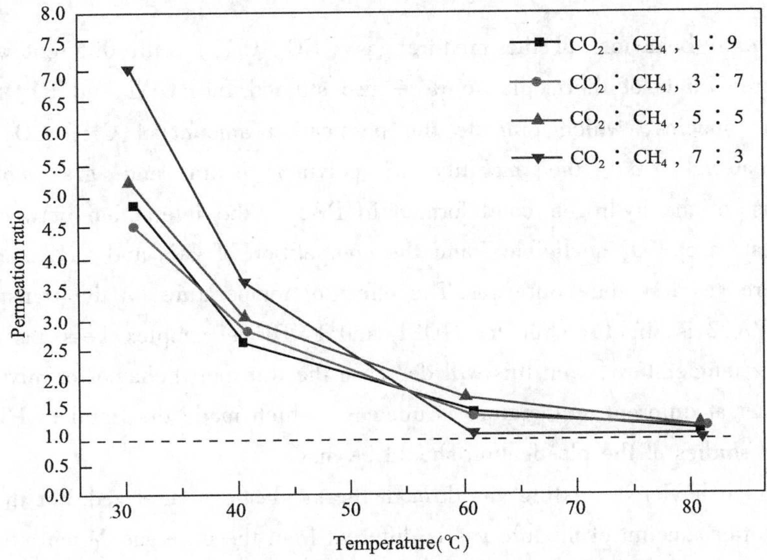

Fig. 6 The permeation ratio at different temperature for PVDF

For HDPE and PVDF, the permeation ratios are all above 1, which means the faster species CO_2 accelerated the transport properties of the slower species CH_4, but the tendency of permeation ratio and temperature is different. The general trend of permeation ratio for HDPE is increased with the temperature, while for PVDF is decreased. Much work has been done on the mixture gas permeation at different pressure[7,18], but the research on the effect of temperature is less. C. K. YEOM[16] analyzed the permeability of mixture gas (CO_2/N_2) through polymeric membrane at 30 ~ 60℃, and it was found that the permeation ratio is unity with increasing the temperature.

The plastication of CO_2 is the main reason for the permeability deviation between pure and mixture gas for HDPE. The plasticizing mechanic is that the adding of CO_2 reduces the force between molecule chains, and the mobility of chains is improved by the plasticizer. The gas should be condensed to a pure liquid state first during the procedure of the dissolution in a polymer[12]. The higher the temperature is, the harder the gas compressed, so the solution and plastication process is more difficult at higher temperature. The less plasticizer would decrease the diffusion process, which would decrease the permeability. But the effect of plasticizer and temperature is combined. The plastication at different temperature is unclear, which make the tendency not obvious. Unlike HDPE, the plastication of CO_2 for PVDF is more prominent. The solution of CO_2 in PVDF decreased at higher temperature is the main reason for the decrease of permeation ratio.

4 Conclusions

The permeation behaviors of the pure CO_2 and CH_4 through HDPE, PA12 and PVDF were studied by the differential pressure method. The transport parameters, such as diffusion coefficient and solubility coefficient were calculated by the time lag method. It was found that the permeability coefficient increased with increasing the temperature range from 30℃ to 80℃, so as the diffusion coefficient, and the solubility coefficient changed little. The temperature dependency of CO_2 for PVDF is most.

The permeation behaviors of the mixture gas (CO_2/CH_4) with different volume fractions through these three kinds of thermoplastic were also studied. In HDPE and PVDF, CO_2 acts as faster permeating species, which promote the permeation amount of CH_4. CO_2 plasticized the thermoplastic material, and the mobility of polymer chain and gas molecule are all improved. Because of the hydrogen bond formed in PA12, the interaction between PA12 chains made the plastication of CO_2 negligible, and the competition of CO_2 and CH_4 made the transport amount of mixture gas less than pure gas. The effect of temperature on the transport behavior of mixture gas for PA12 is slight, while for HDPE and PVDF is complex. Less gas dissolves in the polymer at higher temperature, and this will decrease the transport behavior of mixture gas. But the effect of plasticizer at different temperature is unclear, which made the trend in HDPE and PVDF different. Further studies of the plastication should be made.

The transport behavior of mixture gas through thermoplastic is analyzed. For the plastication of CO_2, the permeation amount of mixture gas is different from the pure gas. Much attention should be paid on the faster species, especially for the lining pipe made by HDPE and PVDF.

References

[1] Merkel T. C. , Gupta R. P. , Turk B. S. , et al. , J MEMBRANE SCI 191, 85 2001.

[2] Khanbabaei G. , Vasheghani-Farahani E. , Rahmatpour A. , CHEN ENG J 191, 369 2012.

[3] Sadrzadeh M. , Amirilargani M. , Shahidi, K. , Mohammadi T. , POLYM ADCAN TECHNOL 22, 586 2011.

[4] Pinnau I. , He Z. , J MEMBRANE SCI 244, 227 2004.

[5] Flaconneche B. , Martin J. , Klopffer M. H. , OGST 56, 261 2001.

[6] Andersen T. R. , Skar J. I. , Hansteen C. , CORROSION, Paper 410 1999.

[7] Flaconneche B. , Martin J. , Klopffer M. H. , OGST 56, 245 2001.

[8] Felder R. M. , Huvard G. S. , J Abnorm Social Psychol 65, 319 1980.

[9] Naito Y. , Mizoguchi K. , Terada K. , et al. , J POLYM SCI POL PHYS 29, 457 1991.

[10] Michaels A. S. , Bixler H. J. , J POLYM SCI 50, 413 2003.

[11] Tiemblo P. , Laguna M. F. , Garcia F. , et al. , MACROMOLECULES 50, 4156 1992.

[12] Alentiev A. Y. , Shantarovich V. P. , Merkel T. C. , et al. , MACROMOLECULES 35, 9513 2002.

[13] Wu F. , Li L. , Xu Z. , et al. , CHEM ENG J 117, 51 2006.

[14] Merkel T. C. , Bondar V. I. , Nagai K. , et al. , J POLYM SCI PHYS 38, 415 2000.

[15] Hughes R. , Jiang B. , GAS SEP PURIF 9, 27 1995.

[16] Yeom C. K. , Lee S. H. , Lee, J. M. , J APPL POLYM SCI 78, 179 2000.

[17] Pandey P. , Chauhan R. S. , PROG POLYM SCI 26, 853 2001.

[18] Ettouney H. , Majeed U. , J MEMBRANE SCI 135, 251 1997.

[19] Chiou J. S. , Paul D. R. , J MEMBRANE SCI 32, 195 1987.

复合材料增强管线钢管结构设计研究

张冬娜　戚东涛　邵晓东　丁　楠　李厚补　蔡雪华

（中国石油集团石油管工程技术研究院，石油管材及装备材料服役行为与结构安全国家重点实验室）

摘　要：针对复合材料增强管线钢管的结构设计，通过对样管水压试验的测试结果及有限元模型分析，得出了复合材料增强管线钢管的失效压力及设计压力计算方法。提出在预应力处理复合材料增强管线钢管时，增强层的应力值不能超过复合材料断裂强度的40%，钢层的应力值应小于钢材的最小拉伸强度。在设计压力下，复合材料层的厚度应保证应力低于其断裂强度的30%，钢层的厚度保证应力低于钢材的屈服强度。通过以上研究，得出了复合材料增强管线钢管的设计步骤及注意事项，为复合材料增强管线钢管的设计提供了依据。

关键词：复合材料增强；钢管；设计压力；设计方法；预应力

复合材料增强管线钢管是一种复合结构管道，通过在管线钢管外缠绕连续纤维增强复合材料，达到提高管道环向常压能力的目的，从而提高管道的输量。该管道除了承压能力高外，还具有成本低、止裂性好等优点。复合材料增强管线钢管主要包括三个结构层：内层钢管、过渡层和复合材料增强层，考虑到施工过程中管道的运输和安装对管道造成的影响，还可以在增强层外添加一个外保护层。复合材料增强管线钢管最初由 NCF 和 TransCanada 进行研究[1-3]，在我国的研究还处于起步阶段。

在之前对复合材料增强管线钢管的研究中，除了对管道进行水压试验外，由于钢层和复合材料层的模量差异，钢层承担了大部分的载荷，复合材料的增强作用发挥不充分，因此使用预应力法处理管道，即在首次加压时使管道产生一定的塑性变形，卸载后钢管存在一定的压应力而复合材料存在一定的拉应力，再次加压时复合材料的承载比例增加。通过对自制的复合材料增强管线钢管进行预应力（自紧）及水压爆破试验分析，得出了以下结论：（1）复合材料增强管线钢管的承压能力是钢层和复合材料层承载叠加的结果；（2）预应力处理后再次加压时管道的弹性区域增加，屈服强度提高；（3）预应力处理提高了复合材料层的承载比例，对最终的爆破压力没有影响；（4）管道失效时，复合材料层首先被破坏。

基于对自制管道的试验分析和有限元模型的计算，现对复合材料增强管线钢管的设计方法进行讨论与研究，为之后复合材料增强管线钢管的研究奠定基础。由于管道是复合结构，并且两个结构层的性能具有一定的差异，因此在进行管道设计的过程中，需要确定出复合材料增强管线钢管各结构层的厚度及自紧压力值。增强层的厚度保证工作压力、试验压力和操作压力下的管道的完整性，自紧压力起到重新分配结构层间载荷比例，增加复合材料利用率

基金项目：中国石油天然气集团公司科学研究与技术开发项目“复合材料增强管线钢管及非金属管道超前储备研究”，编号：2014B-3312。

的作用。在设计过程中，主要针对这两方面及两者之间的关系进行论述。

1 复合材料增强管线钢管的压力计算

复合材料增强管线钢管的主要承压层有两层，内层的管线钢管和复合材料增强层，由于管道在承压状态下，管道的环向应力（σ_s）为轴向应力（σ_a）的2倍，即$\sigma_s=2\sigma_a$，当环向应力满足承压要求时，轴向应力没达到管道的承载极限[4]。并且针对水平埋地管道，由于管道处于完全的约束装填，管道除了受到内压作用外，还受到土壤的压力和摩擦力，此时大口径管道的轴向应力很小，轴向应力远小于环向应力的1/2[5]，因此，提高管道环向承压能力非常必要，减少轴向应力的浪费。复合材料增强管线钢管通过复合材料的增强，提高管道环向应力，轴向应力完全由钢管承担。

在增强层加工的过程中，由于复合材料不承担轴向应力，因此缠绕角度近90°。缠绕层厚度由管道的压力要求及钢管的壁厚决定，管道的失效压力可由式1进行计算：

$$P_p=\frac{2}{D}\ (\sigma_b t+T_h W) \tag{1}$$

式中 P_p——复合管道的爆破压力，MPa；

D——钢管外径，mm；

σ_b——钢的拉伸强度，MPa；

t——钢管设计壁厚，mm；

T_h——复合材料环向拉伸强度，MPa；

W——纤维增强复合材料层的设计壁厚，mm。

使用式（1）可以计算出复合材料增强管线钢管的爆破压力，式中计算使用到的材料强度值均为材料的极限强度值。

以X65外径508mm，壁厚9.5mm的管线钢管为基础，制成了复合材料增强管线钢管样管，按照增强层厚度编号为P-1到P-5，其中在爆破压力的计算中，钢管的拉伸强度取535MPa，复合材料的环拉伸强度为1000MPa。

表1 不同增强层厚度复合材料增强管线钢管的爆破压力及最大操作压力

样管编号	增强层厚度（mm）	爆破压力（MPa）	
		计算值	实际值
P-1	2.60	30.2	30.4
P-2	3.58	34.1	34.5
P-3	4.15	36.3	37.1
P-4	7.01	47.6	43.8
P-5	7.30	48.8	44.7

在一系列复合材料增强管线钢管的爆破试验结果中，对比使用式（1）计算的爆破压力，当厚度比较薄时，如P-1，P-2及P-3，实际测得的爆破压力值比计算值偏大，这是由于在计算过程中，钢管的拉伸强度取的是X65管线钢的最低拉伸强度，复合材料的拉伸强度也是置信度为95%时的值，因此实际的强度可能比计算使用的强度值高，导致爆破压力实测值偏大。同时也发现对于P-4及P-5，计算值比试验结果大。复合材料环向拉伸强度的测试按照GB/T

1458—2008《纤维缠绕增强塑料环形试样力学性能试验方法》进行，标准中要求试样的厚度为1.5mm，在加工增强层的过程中，增强层较厚时会造成树脂含量分布不均匀，造成整体性能的下降，并且厚度越大，材料存在缺陷的可能性越大，因此拉伸强度值较按照标准测量值偏低，因此当增强层厚度偏大时，在计算时需要对复合材料强度值进行修正。

根据 CSA Z662 “Oil and gas pipeline systems” 中针对复合材料增强管道的部分，管道的设计压力如式（2）：

$$P=\frac{2}{D}\times (StT+T_{h}W)\times F\times L \tag{2}$$

式中 P——设计压力，MPa；

S——钢管最小屈服强度，MPa；

T——温度系数，取 1；

F——设计系数，取 0.5；

L——位置系数，取 1，其余参数同式（1）。

在之前的研究中，认为式（2）在计算复合管道设计压力时存在一定的问题：钢管的设计系数为 0.72，复合材料增强管线钢管的设计系数为 0.5，当增强层较薄时，按照式（2）计算的复合管道设计压力比未增强的钢管设计压力还要低。并且在计算过程中，管线钢的强度取值为最小屈服强度，复合材料的强度取值为环拉伸强度，为极限强度值，二者的取值存在一定的差异。当增强层厚度逐渐增加，计算过程中复合材料的部分所占的比例增大，按照极限强度计算的部分越来越大，并且在真实的管道运行过程中，复合材料承担应力不会超过其极限强度的 50%。因此认为应该对式（2）进行修改，可使用钢材和复合材料两个设计系数，如式（3）：

$$P=\frac{2}{D}(StTF_{s}+T_{h}WF_{c}) \tag{3}$$

式中 F_s——钢管部分的设计系数，取 0.72；

F_c——复合材料部分的设计系数，由于运行时复合材料的应力不超过其极限应力的 40%，考虑到管道的安全，此处取 0.3。

表 2 为使用两种公式计算的复合管道的设计压力，其中钢管的屈服强度为 450MPa，复合材料的拉伸强度为 1000MPa，并且在 Zimmerman T[1] 对复合材料增强管线钢管的研究中，将管道的爆破安全系数取为 2 或 2.5，即管道的爆破压力与最大操作应力的比值，按照两个安全系数计算的最大操作压力也在表 2 中。

表 2 管道设计压力

管道编号	设计压力		最大操作压力	
	式（2）计算结果	式（3）计算结果	爆破安全系数为 2	爆破安全系数为 2.5
P-1	13.5	15.2	15.2	12.2
P-2	15.5	16.3	17.3	13.8
P-3	16.6	17.0	18.6	14.8
P-4	22.2	20.4	21.9	17.5
P-5	22.8	20.7	22.4	17.9

当增强层较薄时，按照 CSA Z662 中的公式计算出的值小于式（3）的计算值，值得注意的是，未增强钢管的设计压力为 12.1MPa，按照式（2）增强层为 0 时计算的压力仅为 8.4MPa，当增强层厚度在 1.87mm 以下时，增强后复合管道的设计压力低于裸管的设计压力，这会造成管道成本的增加。当增强层较厚时，使用式（3）计算的值较低，按照对表 2 数据的分析，当增强层较厚时，更保守的计算更为安全，因此按照式（3）进行设计压力的计算比较合理。

CSA Z662 中要求操作压力不超过设计压力，对比表 2 中的数据，爆破安全系数取 2.5 时过于保守，对于 P-1，其最大操作压力接近于裸管，安全系数过大会造成一定浪费。当把按照式（3）计算的设计压力确定为最大操作压力时，按照爆破安全系数 2，此时的试验结果与计算值之间还有一定的余量，也说明式（3）的计算方法更为合理，与试验结果的趋势更为一致。尤其当增强层壁厚较厚时，按照式（3）计算的设计压力作为最大操作压力也较为合理，与此时爆破压力计算值偏大相比，此时的计算值更为安全。

2 自紧压力设计

通过管道的设计压力计算，对管道结构层参数有一个大致范围的估计。由于复合材料增强管线钢管结构层模量的差异，管道的应变下的应力分布不均匀，钢管的模量是复合材料的 3.5~5.5 倍，导致大部分的载荷还是由钢管承担，因此需要使用预应力法（自紧法）对复合结构管道进行处理。对复合材料增强的管道施加一定的内压力，管道发生屈服后卸载，此过程即为预应力处理过程。预应力处理完成后，此时内层的钢管由于屈服产生了塑性变形，不能完全回复至最初的尺寸，而复合材料层由于应变的变化会产生一定的拉伸应力，在拉伸应力的作用下钢层内有一定的压缩应力。当经过预应力处理的管道再次加压时，复合材料增强层承担更多的载荷，增强效果能更充分地发挥[6]。

图 1 为管道 P-3 预应力处理的压力/环向应变曲线，应变数据由应变计采集，预应力为 25MPa。从图中可以看出，管道在应变为 0.2%时出现屈服，25MPa 时的应变达到 0.590%，卸载后的残余应变为 0.345%。曲线中可以发现，屈服前曲线的斜率与卸载后的斜率非常近似，都是线弹性的表现。在 TransCanada 对复合材料增强管线钢管的研究中[1]，考虑到过大的塑性变形可能会对钢管造成的影响，因此对自紧压力大小进行了控制，保证预应力处理卸

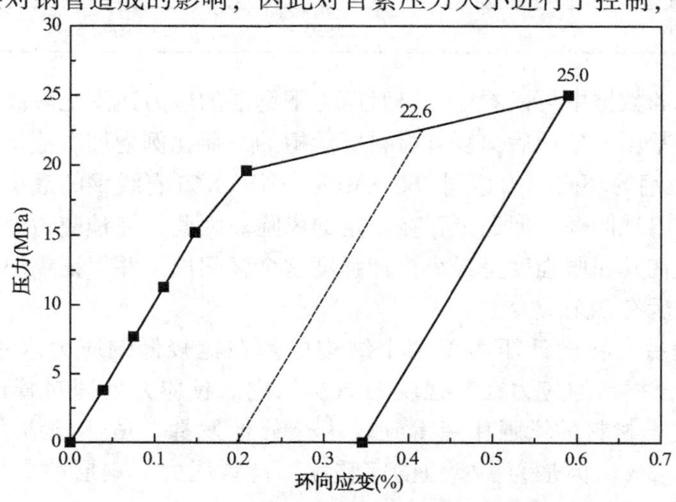

图 1　P-3 预应力阶段的压力/环向应变曲线

载后残余应变为0.2%，即发生塑性形变，并且形变量为塑性形变的最小值。图中虚线为按照发生0.2%残余形变时泄压后的曲线，此时的自紧压力值应为22.6MPa。

TransCanada在其研究中也指出，过大的屈服形变不会带来内层钢管的风险，因为复合材料的断裂延伸率低于2%，而钢管的断裂延伸率在20%左右，预应力处理阶段肯定要保证管道的完成性，在复合材料增强层完好的情况下，钢管可以保证其安全性。

但是值得注意的是，增强层的厚度、自紧压力、增强层应力的关系。出于经济角度考虑，增强层厚度尽可能薄，但管道应从安全角度考虑更多。由于复合材料增强管线钢管的结构与金属内胆纤维缠绕气瓶非常类似，因此也从此结构气瓶的研究中得到一些经验[7-9]。在GB/T 24160—2009《车用压缩天然气钢质内胆环向缠绕气瓶》中，有纤维应力比的概念，即缠绕气瓶最小设计爆破压力下纤维的应力与公称工作压力下纤维的应力比，针对玻璃纤维，要求纤维应力比大于等于2.75，也就是说公称工作压力下纤维的应力要小于最小设计爆破压力下的36.4%。在DOT-CFFC《铝内衬全缠绕碳纤维增强复合气瓶的基本要求》中要求纤维的应力分析应采用有限元分析的方法计算，在工作压力下，纤维（碳纤维或玻璃纤维）的最大拉伸应力不得超过最小爆破压力下的纤维轴向应力的30%。

由于预应力处理时管道的自紧压力超过了屈服压力，明显高于管道的工作压力，为了保证管道的完整性，缠绕层的厚度要保证管道在预应力处理阶段时，复合材料的应力值低于其最小拉伸强度的40%。表3为管道P-3在预应力分别为22.6和25MPa时，钢层和复合材料层在预压力及设计压力（17MPa）时的应力值。计算使用的软件为ANSYS有限元分析软件，计算单元选取Solid186，边界条件为在钢管一端施加一个轴向力，一端为Z方向的位移约束，模型内表面施加规定的内压力，模型两侧施加对称压力。

表3 P-3在预应力处理及设计压力下结构层的应力

自紧压力（MPa）	压力值（MPa）	钢层环向应力（MPa）	复合材料层环向应力（MPa）	复合材料层承担的应力比（%）
22.6	22.6	519.3	111.7	17.7
	17	390.1	89.7	18.7
25	25	560.8	221.5	28.3
	17	373.0	184.6	33.1

从有限元计算的数据中可以看出，自紧压力下钢层的应力达到比较高的应力值，高于屈服应力，更高的自紧压力处理后，复合材料层承担的载荷比例增加。值得注意的是，当自紧压力为25MPa时，钢的环向应力达到560.8MPa，高于X65管线钢的最小拉伸强度535MPa，虽然在试验中没有出现问题，但还是存在一定的风险。因此，在预应力处理阶段，自紧压力应使钢层的应力值在其屈服强度至最小拉伸强度这个区间内，并保证钢层发生完全的塑性变形，卸载自紧压力后有残余应变。

自紧压力处理后，在设计压力下两个结构层都在比较低的应力水平，钢层低于屈服压力450MPa，复合材料的应力在其强度的20%以内。预应力处理对管道应力重新分配的作用非常明显，复合材料的增强作用得到较为充分地发挥。虽然预应力处理后管道的残余形变可以超过0.2%，但是为了管道的完整性，自紧压力下钢层的应力应低于钢材的最小拉伸强度。

3 结构层厚度

复合材料增强管线钢管可以在承压能力较差管线钢管的基础上，提高管道的环向应力，从材料成本角度考虑，更适合使用玻璃纤维复合材料增强薄壁钢管，因此使用 ANSYS 有限元分析软件，计算了一系列结构层厚度不同，但设计压力同为 17MPa 的复合材料增强管线钢管，在 22.6MPa 自紧压力下，及卸载自紧压力的设计压力下，钢层和复合材料增强的应力情况，见表 4。

表 4 不同结构层厚度管道的应力分析

管道编号	结构层厚度（mm）		预应力 22.6MPa 时结构层应力值（MPa）		设计压力 17MPa 时结构层应力值（MPa）	
	钢层	复合材料层	钢层	复合材料层	钢层	复合材料层
P17-1	9.50	4.15	519.3	111.7	390.1	89.7
P17-2	7.10	6.73	545.8	260.1	377.1	231.7
P17-3	5.60	8.35	521.2	310.3	312.2	275.7
P17-4	4.80	9.21	500.0	327.3	263.6	288.0

管道 P17-1，即为在第二部分研究的管道 P-3，在 22.6MPa 的自紧压力下处于刚刚发生屈服的状态，对其余设计压力 17MPa 的三支管道，由于钢管厚度薄，在 22.6MPa 下时应变量更大，因此复合材料增强层的应力值更高。当自紧压力卸载后，在设计压力下，随着钢层的减薄和复合材料层的增加，两个结构层的应力更均匀，尤其对于 P17-4，钢层的应力低于复合材料层的应力。

在对钢管的水压爆破试验中，P17-1 的内层钢管的屈服压力为 19.5MPa，爆破压力为 21.6MPa。通过计算，钢管屈服和爆破时的应力分别为 521MPa 和 577MPa，使用实际的试验值可以计算出另外三支管道的屈服压力和失效压力。虽然钢层薄，增强层厚的复合材料增强管线钢管的应力分布为更期望达到的结果，但是自紧压力较钢管的爆破压力值高出很多，如 P17-4，壁厚 4.8mm 钢管的爆破压力为 10.9MPa，当自紧压力为 22.6MPa 时，是爆破压力的两倍多，虽然有限元计算的结果是钢管的应力低于钢的拉伸强度，但是还需进行试验，保证设计的接受度。

4 复合材料增强管线钢管设计步骤及注意事项

通过以上对复合材料增强管线钢管的分析，对于给定设计压力的复合管道的设计步骤如下：

（1）通过给定的设计压力，设计一组或几组结构层厚度。

（2）计算以上设计中残余应变为 0.2%时的自紧压力。

（3）分析自紧压力下钢层和复合材料层的应力值，其中钢层的应力不能超过钢材的最小拉伸强度，复合材料的应力值宜在其拉伸强度的 40%以内。

（4）分析设计压力下钢层和复合材料层的应力值，其中钢层的应力应低于其最小屈服强度，复合材料的应力值应在其拉伸强度的 30%以内。

（5）通过以上步骤筛选出结构层组合，并结合管道成本要求和具体实施情况，确定最后的设计。

针对复合材料增强管线钢管的结构特点，在其设计中应注意以下几个问题：

（1）预应力处理后的残余形变可以高于0.2%，但是要保证钢层在此时的应力低于最小拉伸强度。

（2）增强使用的钢管宜使用普通钢级的管线钢管，对于低钢级管线钢，其屈服强度$R_{p0.2}$和$R_{t0.5}$的值非常接近，而对于X80以上的管线钢，$R_{t0.5}$较$R_{p0.2}$高[10]，因此如果要达到预应力处理的效果就要使管道发生较大的变形，如要达到0.2%的残余形变，X80管线钢需拉伸至应变为0.5%，X100管线钢需拉伸至0.61%，大的形变要求自紧压力值高，增加了工艺的难度。从经济角度考虑，也更宜使用低钢级的管线钢。

（3）应注意钢管的轴向应力值，在预应力处理及设计压力下钢管的轴向应力应低于钢材的最小拉伸强度。

参 考 文 献

[1] Zimmerman T., Stephen G., Glover A.. Composite reinforced line pipe (CRLP) for onshore gas pipelines [C]. 2002 4th International Pipeline Conference, 2002, American Society of Mechanical Engineers: 467-473.

[2] Wolodko J.. Simplified methods for predicting the stress-strain response of hoop wound composite reinforced steel pipe [C]. 2006 International Pipeline Conference, 2006, American Society of Mechanical Engineers: 519-527.

[3] Salama M. M.. Qualification of fiber wrapped steel pipe for high pressure arctic pipeline [C]. OTC Arctic Technology Conference, 2011, Offshore Technology Conference: 1-11.

[4] 杨峰平，张良，罗金恒，等. 油气输送管典型应力状态下的屈服行为研究 [J]. 工程力学，2013 (11): 293-297.

[5] 林柏生. 缠绕玻璃钢管道轴向强度设计 [J]. 玻璃钢/复合材料，1999，(1): 8-11.

[6] 黄再满，蒋鞠慧，薛忠民，等. 复合材料天然气气瓶预紧压力的研究 [J]. 玻璃钢/复合材料，2001，5: 29-32.

[7] 赵立晨. 金属内衬纤维增强复合材料筒体设计 [J]. 宇航材料工艺，2007，37 (2): 45-47.

[8] 周海成，阮海东. 纤维缠绕复合材料气瓶的发展及其标准情况 [J]. 压力容器，2004，21 (9): 32-36.

[9] 沈军，谢怀勤，侯涤洋. 纤维缠绕聚合物基复合材料压力容器的可靠性设计 [J]. 复合材料学报，2006，23 (4): 124-128.

[10] 宫少涛，王爱民，吉玲康，等. X80与X100级管线钢屈服强度$R_{t0.5}$和$R_{p0.2}$的差异性研究 [J]. 焊管，2007，30 (5): 42-45.

复合材料增强管线钢管的预应力及水压爆破试验研究

张冬娜　戚东涛　丁　楠　邵晓东　魏斌　蔡雪华

（中国石油集团石油管工程技术研究院，石油管材及装备材料服役行为与结构安全国家重点实验室）

摘　要：天然气需求量的增加对长输管线的承压能力提出了更高的要求，而增加管道壁厚和提高钢级的方法都有一定的局限性，因此开发了复合材料增强管线钢管。此方法将连续玻璃纤维复合材料包覆在管线钢管外，达到提高管道环向承压能力的目的。针对此种结构的管道，使用ANSYS软件进行了有限元模型的计算，并以外径508mm的管线钢管为基础，制成了复合材料增强管线钢管样管，使用预应力处理管道后，对管道进行了水压爆破试验。结果表明：(1)复合材料增强管线钢管的承压能力是钢层和复合材料层承载叠加的结果；(2)预应力处理提高了复合材料层的承载比例；(3)管道失效时，复合材料层首先被破坏；(4)在试验的基础上，提出了复合材料增强管线钢管设计压力的计算方法，建议使用钢管和复合材料两个设计系数。

关键词：复合材料增强；钢管；水压试验；预应力；设计压力

随着我国工业的发展及对环境保护要求的提高，天然气作为一种清洁能源的用量也在逐年增加。而我国的天然气资源约60%集中在西部地区，由于需求量大，还有一部分天然气需要进口，包括中亚、缅甸和俄罗斯等地，这些都离不开天然气的输送工作。在天然气输送管道的研究中，为了提高管道的输量，提高管道的承压能力是研究的重点，目前常使用的方法包括增加管道壁厚，提高管道材料的钢级等。增加壁厚的同时也增加了管道的重量和成本，并且加大了焊接的难度，而对超高钢级的管线钢管而言，除了焊接工艺难度大，最大的问题就是其止裂能力差，一旦管道破坏产生裂纹，裂纹扩展可达数公里[1,2]。

因此，开发出复合材料增强管线钢管[3]，此种管道通过缠绕的方法将浸渍过树脂的连续纤维包覆在管线钢管表面，以提高钢管的环向承压能力。该技术最早由NCF公司开发，TransCanada针对此种类型的管道做了相应的研究，并将此种管道命名为CRLP（composite reinforced line pipe）[4-6]。

复合材料增强管线钢管在国内的研究刚起步，本文通过对自制的复合材料增强管线钢管进行有限元模型计算、预应力分析及水压爆破试验，分析各结构层在承压状态下的响应，研究复合结构管道的失效形式，并对管道设计压力的计算公式进行了讨论。

基金项目：中国石油天然气集团公司科学研究与技术开发项目“复合材料增强管线钢管及非金属管道超前储备研究”，编号：2014B-3312。

作者简介：张冬娜，女，工程师，博士；2013年毕业于西北工业大学并获博士学位；主要从事油气集输用非金属管的研发工作。

1 实验部分

1.1 试验管道及尺寸

试验用复合材料增强管线钢管为自制，图 1 为管道的示意图，其中内层为 X65 钢级，外径 508mm，壁厚 9.5mm，长 6m 的钢管，在钢管表面涂覆底漆后缠绕复合材料增强层，增强层使用的纤维为单向连续玻璃纤维，树脂为环氧树脂 E51，分别制造了两种增强层厚度的复合材料增强管线钢管样管。

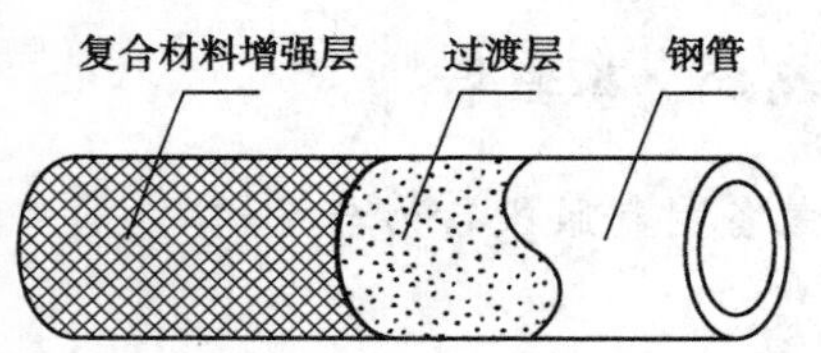

图 1 复合材料增强管线钢管示意图

1.2 试验准备

试验的加压设备为北京海德利森科技有限公司的水压爆破试验系统，在加压过程中使用东华测试 DH3817 应变仪记录复合管道应变变化。采用中航电测生产的箔式电阻应变计，使用 1/4 桥接法，电阻 R 为 350Ω，用 502 快速胶黏剂将应变计粘贴在打磨过的复合材料层外表面。应变计粘贴在管道中部，对管道的环向应变进行了测量。

1.3 试验设计

复合材料增强管线钢管在承压时，钢管和复合材料增强层在压力作用下产生一定的应变，过渡层将两个结构层粘接在一起，保证钢管和增强层的应变同步，两个结构层在承压状态下的应力只与其模量有关。钢管的模量为 206GPa，而复合材料的模量仅为 35～60GPa，二者相差很大，因此导致的结果是大部分载荷由钢管承担，复合材料的增强效果并不明显。

为了解决此问题，采用预应力处理法，即自紧法，对复合材料增强的管道施加一定的内压力，管道发生屈服后卸载，此过程即为预应力处理过程。预应力处理完成后，此时内层的钢管由于屈服产生了塑性变形，不能完全回复至最初的尺寸，而复合材料层由于应变的变化会产生一定的拉伸应力，在拉伸应力的作用下钢层内有一定的压缩应力。当经过预应力处理的管道再次加压时，复合材料增强层承担更多的载荷，增强效果能更充分地发挥[7]。

因此在复合材料增强管线钢管进行水压爆破试验前，先对管道进行预应力处理，通过模型计算出预应力大小。为了配合应变计采集数据，采用阶段保压的方式，每升高 4MPa，保压 1min。当压力加至预应力值后，卸载全部压力，再逐步加压至复合材料增强管线钢管失效。

1.4 有限元模型建模

针对复合材料增强管线钢管的结构，使用 ANSYS 软件建立有限元模型。在承压状态下，管道的环向应力（σ_s）为轴向应力（σ_a）的 2 倍，即 $\sigma_s = 2\sigma_a$，因此通过增强层的缠绕，提高管道的环向应力，轴向应力由钢管承担。过渡层的作用是将钢管和增强层粘接在一起，使两个结构层具有相同的应变，传递应力，避免预应力卸载后出现结构层分离的现象，并且起到钢管防腐的作用，但在管道承压过程中，没有增强的作用。

1.4.1 计算单元选择

管道结构属于旋转体，经筛选 ANSYS 单元库，选取了三维实体单元 Solid186[8]，见图 2，复合材料层和钢层之间采用粘接形式进行计算分析。Solid186 是一个高阶 3 维 20 点固体结构单元，单元通过 20 个节点来定义，每个节点有 3 个沿着 xyz 方向平移的自由度。Solid186 可以具有任意的空间各向异性，单元支持塑性、超弹性、蠕变、应力钢化、大变形和大应变能力。

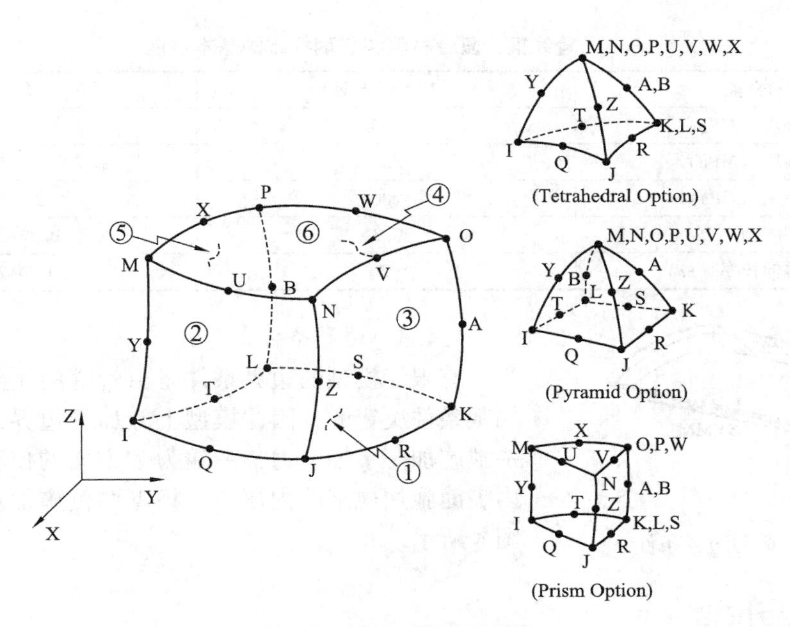

图 2　Solid186 单元示意图

1.4.2　几何模型与材料参数

复合材料增强管线钢管的结构为圆柱管，分析的是管道在受压情况的应力分布情况，因此整个结构还包括管道两头的密封板。由于两端的密封，整个结构为轴对称，管道受压也是均匀对称的。为了节约计算资源，截取管道中的一部分，即圆周方向 1/16，轴向长度 300mm，进行三维结构实体建模，如图 3 所示，由于结构简单，使用分析软件自动划分网格。

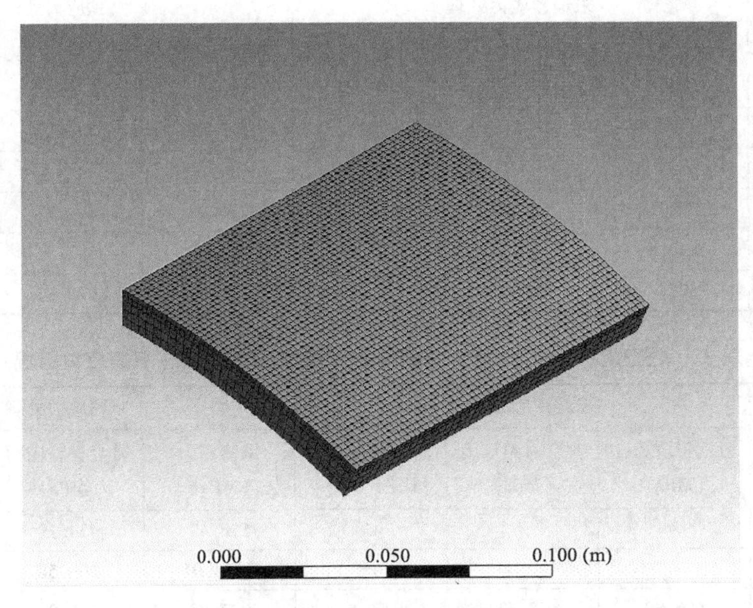

图 3　管道简化几何模型

由于复合材料只承担环向的载荷，因此纤维的角度为近 90°，即与管道轴向垂直。复合材料层和钢层的基本性能见表 1。

表 1　管线钢、复合材料和玻璃纤维的基本性能

材料性能	X65 管线钢	复合材料
屈服强度（MPa）	450	—
拉伸强度（MPa）	535	1045
弹性模量（GPa）	206	35~60
泊松比	0.3	0.22~2.7
拉伸断裂伸长率（%）	>20	1.5~2.0

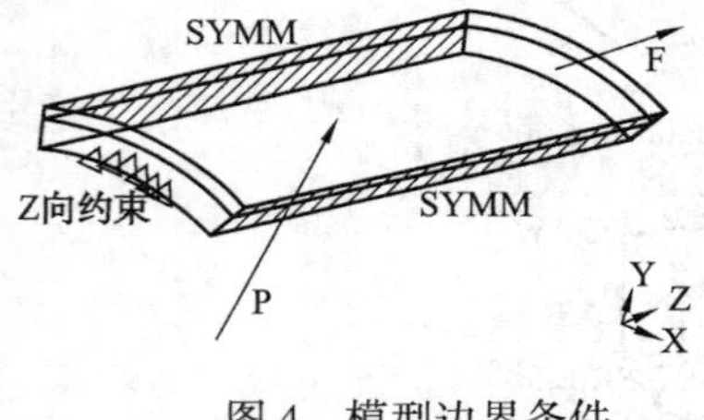

图 4　模型边界条件

1.4.3　边界条件

有限元模型的边界条件是由管道的实际约束条件和加载条件决定的，因此模型上施加的边界条件为在钢管一端施加一个轴向力，一端为 Z 方向的位移约束。模型内表面施加规定的内压力，模型两侧施加对称压力，如图 4 所示。

2　结果与讨论

2.1　有限元模型计算结果

计算了增强层厚度为 4mm 和 7mm 的复合材料增强管线钢管分别在使用和不使用预应力处理的情况下，钢层和增强层的应力大小，表 2 为增强层 4mm 的应力情况，表 3 为增强层 7mm 的情况。

表 2　增强层为 4mm 的复合材料增强管线钢管不同状态下的应力情况

管道内压力（MPa）	未经预应力处理			预应力处理		
	钢管应力（MPa）	复合材料应力（MPa）	复合材料载荷比例（%）	钢管应力（MPa）	复合材料应力（MPa）	复合材料载荷比例（%）
预应力 25	/	/	/	562.02	230.12	/
卸载预应力 0	/	/	/	-61.67	100.4	/
11.2	268.84	57.478	17.6	238.94	157.74	39.8
18.7	449.6	96.164	17.6	414.32	197.01	32.2
失效 35.5	572.0	927.0	61.8	571.13	928.15	61.9

表 3　增强层为 7mm 的复合材料增强管线钢管不同状态下的应力情况

管道内压力（MPa）	未经预应力处理			预应力处理		
	钢管应力（MPa）	复合材料应力（MPa）	复合材料载荷比例（%）	钢管应力（MPa）	复合材料应力（MPa）	复合材料载荷比例（%）
预应力 33	/	/	/	556.47	434.19	/
卸载预应力 0	/	/	/	-233.01	269.58	/
18.0	406.82	85.63	17.4	227.88	335.01	59.5
30.2	557.78	326.98	37.0	492.97	420.73	46.0
失效 45	540.04	928.02	63.2	541.87	925.83	63.1

注：表 2 及表 3 中的应力为各结构层所受的环向应力，以后不再对此进行说明。

表 2 和表 3 中选取的管道内压值除了预应力、卸载预应力及失效压力外，另外两个压力值都为复合管道二次加压时发生屈服前的压力，与管道服役状态吻合。从表 2 和表 3 的数据中可以看出，复合材料增强管线钢管在预应力处理后，钢管虽然发生了塑性变形，产生了正的应变，但是受到了压应力，其应力为负值。在相同的承压状态下，预应力处理后的管道复合材料承担的载荷比例明显增加，因此采用预应力法处理复合材料增强管线钢管的方法可以提高增强层的利用率，达到增强的目的。通过计算数据也说明，预应力处理不影响最终管道失效时的应力分布。

2.2 预应力阶段分析

自制了两种规格的复合材料增强管线钢管，采用 Minitest 4100 测厚仪对复合材料厚度进行了测量，每隔 1m 测一组，每组分别在相隔 90°的点测四个数值，至少测三组，取测量的最小值。其中按照设计厚度 4mm 制备的复合管的最终厚度为 4. 50mm，设计厚度 7mm 的管道最终厚度为 8. 10mm。由于在管道表面粘贴应变片需要对表面进行打磨，两个规格的复合材料增强管线钢管在打磨后的最小厚度分别为 4. 15mm 和 7. 06mm，两支复合管道分别编记为 P-1 及 P-2。

P-1 的预应力为 25MPa，P-2 的预应力为 33MPa，在预应力处理阶段，通过应变片采集的数据，绘制压力/应变曲线，分析此阶段管道的承压情况，如图 5 所示。

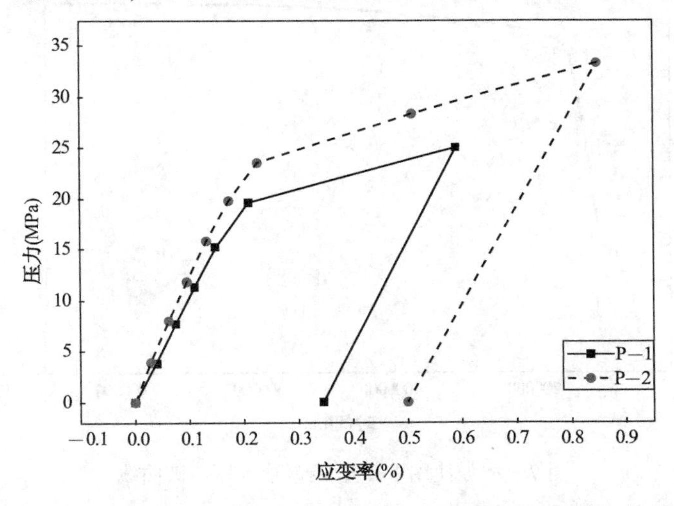

图 5 两种规格复合材料增强管线钢管预应力处理阶段的压力/应变曲线

由于应变片的数据在保压的时间段采集，为离散的数据点，两个曲线屈服时的应变都在 0. 2%左右，与管线钢的屈服应变一致，这是由于钢和复合材料在应力作用下有不同的响应。钢和复合材料的应力/应变曲线如图 6 所示，X65 管线钢的屈服强度为 450MPa，屈服后应力无明显的增加；复合材料的断裂伸长率在 1. 5%～2. 0%，在整个拉伸阶段应力与应变为线性关系，无屈服现象。

同时也对本试验中使用的钢管进行了水压爆破试验，压力/时间曲线如图 7 所示，从图中可以看出，钢管的屈服压力 19. 2MPa，爆破压力 21. 6MPa。在预应力处理的加压阶段，P-1与 P-2 的屈服压力分别为 19. 5 和 23. 6MPa。复合管道在屈服前主要是由钢管承载，曲线的形态也与钢管加载的形态一致。发生屈服后，钢管的曲线比较平直，爆破压力比屈服压力高 2. 4MPa，对于复合材料增强管线钢管，屈服后的压力增加明显。

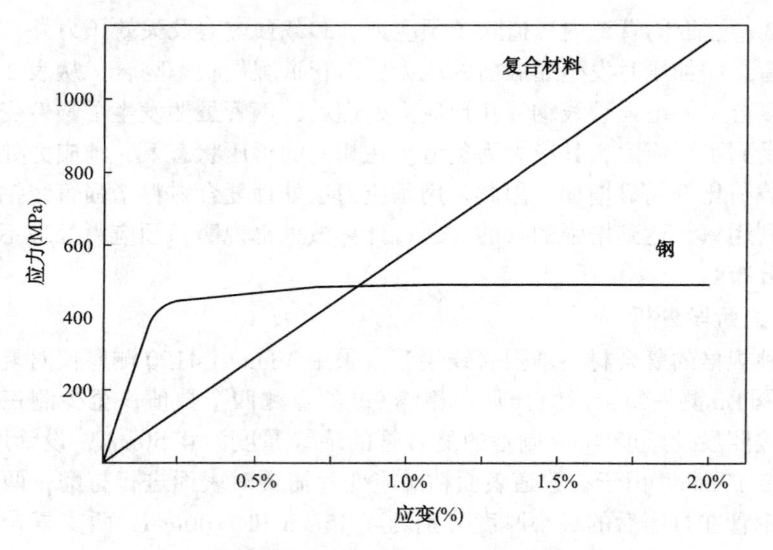

图 6　复合材料和钢的应力/应变曲线

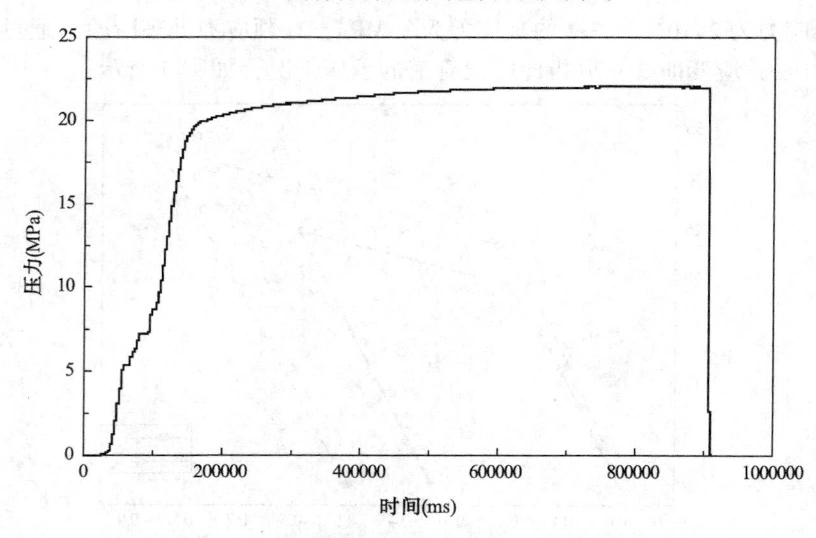

图 7　增强使用 X65 管线钢管水压爆破曲线

环向增强的钢管在一定内压力 P 下，满足

$$P=(\sigma_s T_s+\sigma_c T_c)/R \tag{1}$$

式中　σ_s，σ_c——分别为钢层和复合材料层的应力；

T_s，T_c——分别为钢层和复合材料层的厚度；

R——半径[9]。

结合图 6 中两种材料的应力/应变曲线，在复合材料增强管线钢管增压的过程中，钢层和复合材料增强层的增强作用是叠加的。在管道发生屈服前，钢层与复合材料都是线弹性的，两个结构层保持相同的形变，因此应力比即为两种材料的模量比。由于两种材料的模量相差很大，此阶段起主要承载作用的为钢管。管道的屈服压力即为屈服应变为 0.2%时管道的内压力，增强层厚度增强，屈服压力提高，但是由于其模量低，因此增加幅度有限。当管道屈服后，钢管的应力值趋于稳定，不会大幅度增加，但是复合材料还是线弹性的，承担载荷的比例越来越大。

2.3 水压试验过程

在预应力处理完成后，对复合材料增强管线钢管进行水压试验，加压至管道失效。表4为两种规格复合管道在水压试验过程的屈服压力与爆破压力值。由于钢具有20%的断裂伸长率，因此最终复合管道的破坏模式应是形变达到了复合材料的最大断裂伸长率，复合材料先被破坏，之后由于钢管不能独自承担现有的载荷，钢管随之失效。图8为P-1水压试验的失效照片，P-2的失效形式与P-1相似，都是复合材料增强层破坏，内层钢管在增强层破坏的位置膨胀，变形明显，由于X65管线钢韧性较好，最终钢管并未发生破裂。

表4　复合材料增强管线钢管水压试验结果

管道编号	增强层厚度（mm）	屈服压力（MPa）	失效压力（MPa）
P-1	4.15	25.5	37.1
P-2	7.06	33.0	43.8

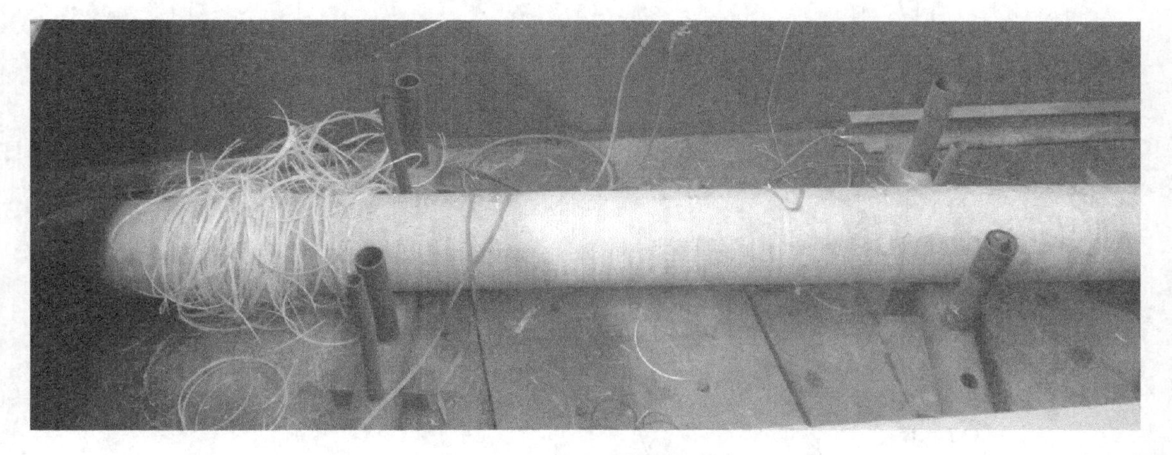

图8　P-1水压试验失效照片

有限元模型计算出的失效压力和最终水压试验的结果存在一定的偏差，一是由于在增强层缠绕的过程中，通过控制缠绕层的层数来控制厚度，最终厚度与计算值会存在一定的偏差，二是计算时所用到的材料性能参数与整体产品存在一定的差异，此外钢管实际的圆度也影响了计算的结果。虽然有限元模型的计算值与实际的试验值有一定的差异，但是误差在一定的范围内，都小于5%，模型的分析与计算对实际试验有指导的作用。

图9为P-2预应力处理及水压试验两个过程的压力/时间曲线，在预应力阶段，屈服压力为23.6MPa，预应力为33MPa，当第二次加压时，可以发现屈服点上移现象，此时的屈服压力达到33MPa。

图10为P-2的压力/应变曲线，应变数据使用应变计采集，与图8一致的是，从图10中可以明显看出预应力处理后的屈服点上移现象。在预应力卸载后，存在0.5%的残余应变，管道最终失效时的应变为1.75%，与复合材料的断裂伸长率一致。复合管道在发生第一次屈服时，管道的应力是钢管和复合材料发生0.2%形变时的应力组成的。当该管道发生第二次屈服时，由图10可得管道的累积形变为0.8%，在残余应变0.5%的基础上增加了0.3%，由于应力是材料模量和应变的乘积，此时的两个结构层中的应力大于管道发生第一次屈服时的应力。除了这部分应力外，钢层的应力应减去预应力结束后钢管残余的压应力，

复合材料的应力应加上残余的拉应力，由此可以看出，第二次屈服时复合材料承担应力的比例增加。结合表 2 和表 3 中残余应力的计算值，二次屈服时的两个结构层的应力之和大于第一次屈服时应力值，因此屈服压力提高。

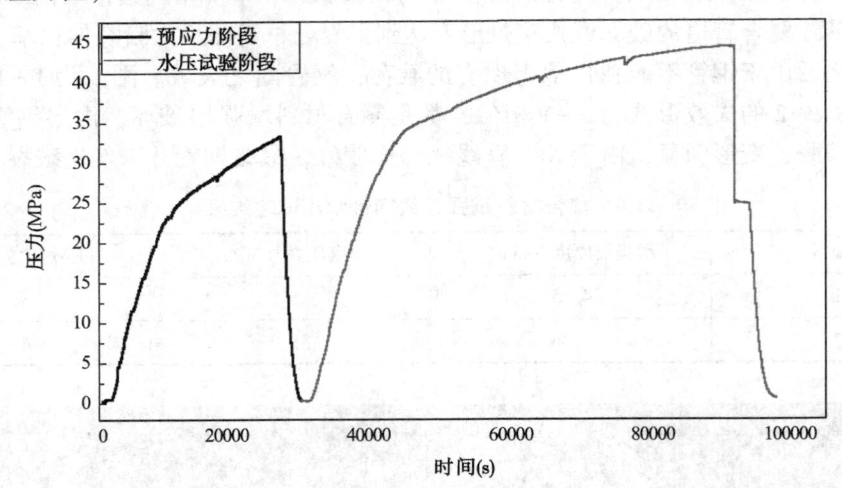

图 9　P-2 预应力及水压试验的压力/时间曲线

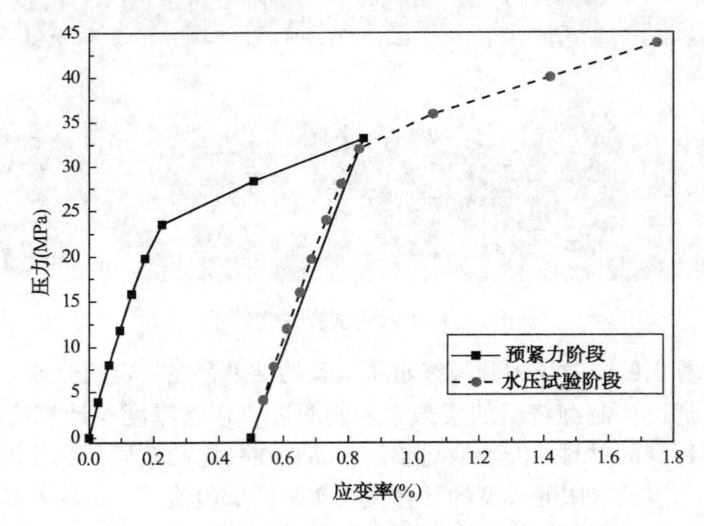

图 10　P-2 预应力及水压试验的压力/应变曲线

在两个加压阶段，屈服前曲线的斜率几乎一致，在屈服后曲线也几乎是沿着预应力处理阶段的斜率继续发展。结合表 2 及表 3 计算的内容，由于最终管道的破坏受到材料性能的限制，因此预应力处理对管道最终的破坏强度没有影响，只是影响了承载下应力在结构层之间的分配情况。

3　复合材料增强管线钢管的设计建议

复合材料增强管线钢管的结构与全钢钢管存在一定的差异，按照 CSA Z662 “Oil and gas pipeline systems” 中针对复合材料增强管道的部分，管道的设计压力如式（2）：

$$P=\frac{2}{D}\times\left(StT+T_{\mathrm{h}}W\right)\times F\times L \tag{2}$$

式中 P——设计压力，MPa；

D——钢管外径，mm；

S——钢管最小屈服强度，MPa；

t——钢管设计壁厚，mm；

T_h——复合材料环向拉伸强度，MPa；

W——纤维增强复合材料层的设计壁厚，mm；

T——温度系数，取 1；

F——设计系数，取 0.5；

L——位置系数，取 1。

由此式可以看出，复合材料增强管线钢管的设计压力是由钢管和复合材料两部分组成的。

根据式（2）对以 X65，外径 508mm，壁厚 9.5mm，增强层厚 4mm 和 7mm 的管道进行设计压力的计算，其中钢管的最小屈服强度为 450MPa，复合材料的环拉伸强度为按照 GB/T 1458—2008《纤维缠绕增强塑料环形试样力学性能试验方法》测试，为 1000MPa。

表 5 复合材料增强管线钢管的压力计算

项 目	式（2）计算的设计压力（MPa）	预应力（MPa）	设计压力下钢管的环向应力（MPa）	设计压力下复合材料的环向应力（MPa）
X65 钢管，4mm 增强层	16.3	25	357.34	184.07
X65 钢管，7mm 增强层	22.2	33	322.48	352.75

当管道的内压为设计压力时，通过有限元计算结果可以看出，复合材料和钢管的应力都在比较低的范围内，保证了管道的安全运行。

对比全钢的长输管道，其设计压力如式（3），

$$P=\frac{2}{D}\times S\times t\times F \tag{3}$$

式中 P——设计压力，MPa；

D——钢管外径，mm；

S——钢材的最小屈服强度，MPa；

t——钢管壁厚；

F——设计系数，取 0.72[10]。

对比式（2）和式（3），在计算中钢管的强度取值都为最小屈服强度，而复合材料取值为环向拉伸强度。通过表 5 的数据可以看出，管道在操作压力下，复合材料远远没有达到其拉伸极限强度。

按照 CSA Z662 中的计算方法，虽然加入了设计系数 0.5，但是在设计压力计算中，随着复合层厚度的增加，复合材料部分在计算中的比例增加，而复合材料的压力计算是按照环向拉伸强度，即极限强度计算的，钢管是按照最小屈服强度计算的，二者的取值方法存在一定的差异。尤其是当增强层比较薄时，由于设计系数 0.72 和 0.5 的差异，复合材料增强管线钢管的设计压力会低于未经增强的钢管。

在之前的分析中，复合材料增强管线钢管的承压能力是钢管和复合材料两部分的加和，预应力只影响应力分布而不影响最终的破坏强度，在此基础上，建议将复合材料增强管线钢管的设计压力分为两部分进行计算。其中一个部分即为管线钢管的承压部分，可以按照钢管的设计公式，设计系数取 0.72，另一部分为复合材料的承压部分。在对复合材料增强管线钢管的研究中，认为操作压力下复合材料的应力在其破坏强度的 40%以内时，可以不考虑蠕变破坏[4]，因此将设计系数定于 0.4 以内，并且从模型计算的结果也可以发现，设计压力下复合材料的应力值低于破坏强度的 40%。考虑管道的安全运行，将复合材料这部分的安全系数定为 0.3。此时设计压力如式（4）所示：

$$P=\frac{2}{D}\left(StTF_{s}+T_{h}WF_{c}\right) \tag{4}$$

式中 F_s——钢管部分的设计系数，取 0.72；

F_c——复合材料部分的设计系数，取 0.3。

其余参数同式（2），与表 5 中同规格的复合管道设计压力用此式计算分别为 16.8MPa 和 20.3MPa。以 X65，外径 508mm，9.5mm 壁厚的钢管为例，式（2）和式（4）计算的设计压力随增强层壁厚的变化如图 11 所示。

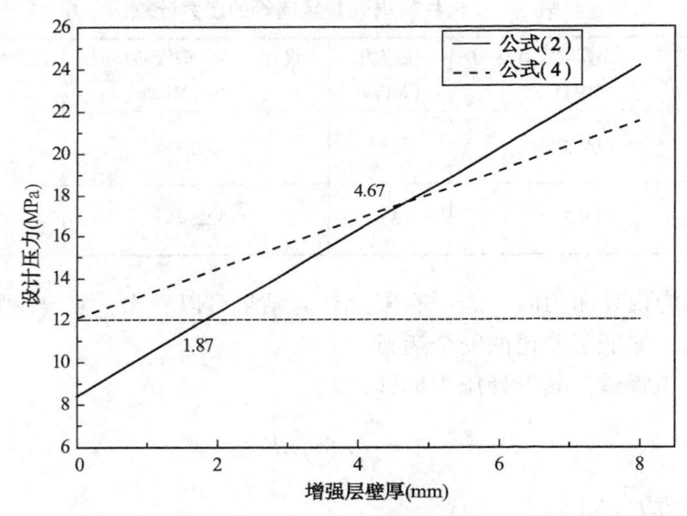

图 11 公式（2）、公式（4）计算不同增强层厚度的复合材料增强管线钢管的设计压力

图中虚线为裸管的设计压力，明显高于当增强层较薄时（低于 1.87mm）按照公式（2）计算的复合材料增强管线钢管的设计压力，也说明公式（2）存在一定的不合理性。当增强层厚度达到一定值时，图中即为 4.67mm，公式（4）的计算结果低于公式（2）。这是由于在公式（2）中，随着增强层厚度的增加，复合材料在计算中占的比例逐渐增大，而此部分是按照极限强度的 50%计算的，明显高于运行状态下复合材料部分所承担的应力值。因此，复合材料增强管线钢管的设计压力计算公式引入钢管和复合材料两个设计系数较为合理，但复合材料部分 0.3 的取值是否合适还需进一步进行研究。

4 结论

（1）使用 ANSYS 软件建立了复合材料增强管线钢管道的有限元模型，计算了压力下钢层与复合材料增强层的载荷，结果表明，预应力处理后复合材料承载比例增加。

（2）当对复合材料增强管线钢管首次加压时，屈服前主要由钢管承载载荷，屈服后复合材料的承载比例逐渐增加。

（3）预应力处理后，二次加压过程中复合管道的屈服压力升高，复合材料增强层的承载比例增强。

（4）复合材料增强管线钢管的失效形式是由于管道形变达到了复合材料破坏的伸长率，复合材料发生破坏，管道失效。

（5）复合材料增强管线钢管的设计压力计算公式应为钢管和复合材料承担压力的加和，设计系数应分别取值，其中钢管部分为 0.72，复合材料部分小于 0.4。

参 考 文 献

[1] Takeuchi I., Fujino J., Yamamoto A., et al. The prospects for high-grade steel pipes for gas pipelines [J]. Pipes and Pipelines International, 2003, 48 (1): 33-43.

[2] 李鹤林，吉玲康，田伟．高钢级钢管和高压输送：我国油气输送管道的重大技术进步［J］．中国工程科学，2010，12（5）：84-90.

[3] Laney P. Use of composite pipe materials in the transportation of natural gas [J]. Idaho International Engineering and Environmental Laboratory, Bechtel BWXT Idaho, 2002: 1-22.

[4] Zimmerman T., Stephen G., Glover A. Composite reinforced line pipe (CRLP) for onshore gas pipelines [C]. 2002 4th International Pipeline Conference, 2002, American Society of Mechanical Engineers: 467-473.

[5] Wolodko J. Simplified methods for predicting the stress-strain response of hoop wound composite reinforced steel pipe [C]. 2006 International Pipeline Conference, 2006, American Society of Mechanical Engineers: 519-527.

[6] Salama M. M. Qualification of fiber wrapped steel pipe for high pressure arctic pipeline [C]. OTC Arctic Technology Conference, 2011, Offshore Technology Conference: 1-11.

[7] 黄再满，蒋鞠慧，薛忠民，等．复合材料天然气气瓶预紧压力的研究［J］．玻璃钢/复合材料，2001，5：29-32.

[8] Altunisik A. Dynamic response of masonry minarets strengthened with fiber reinforced polymer (FRP) composites [J]. Natural Hazards and Earth System Science, 2011, 11 (7): 2011-2019.

[9] 赵立晨．金属内衬纤维增强复合材料筒体设计［J］．宇航材料工艺，2007，37（2）：45-47.

[10] 张宏，顾晓婷，赵丽恒．基于可靠性的油气管道设计系数研究［J］．焊管，2011，34（4）：58-62.

The improved mechanism performance of oil pump with micro-structured vanes

Li Ping[1] Xie Jin[2] Qi Dongtao[1] Li Houbu[1] Zhang XiaoJia[1]

(1. CNPC Tubular Goods Research Institute, CNPC Key Lab for Petroleum Tubular Goods Engineering; 2. South China University of Technology, School of Mechanical and Automotive engineering)

Abstract: Hard and brittle materials can be micro-machined by using the micro-tip shape of diamond wheel. In this paper, micro V-groove arrays with the depth ranging from 500 μm to 50 μm were micro-grinding on the tip of the graphite vane, the results showed that an accurate and smooth micro-structured array with a certain depth can enhance the vane pump efficiency. In order to explore the improved mechanism performance, the computational fluid dynamics (CFD) method was adopted to simulate the pump internal flow field, the micro-flow field between the internal wall of the stator and the vane end was underlined, the depth of the simulated micro-flow various from 0 to 500 μm. With the pressure difference between the inlet and the outlet for the micro-flow, a variational leakage was existed for the micro-groove array deeper than 50 μm because of the high velocity. When the depth was below 50μm, no leakage was found and the average flow velocity was greater than the normal vane and less than the end linear velocity. Furthermore, as the depth in the range of 10~50 μm, no leakage appeared but the average velocity is less than the depth of 50 μm. Thus, the fluid dynamic characteristics for the micro-groove array depth of 50 μm not just avoids leakage but also provides appropriate lubrication efficiency and enhances the oil pump efficiency.

Key words: Oil vane pump; Computational fluid dynamics; Numerical simulation

1 Introduction

Vane pump is a kind of volumetric pumps which widely used in machine tools and in hydraulic systems, such as transport gasoline, diesel oil, lubricating oil and other low viscosity of light fuel oil, because of the simple compact structure and the strong self-priming capacity. Vane pump consists mainly of a stator and a eccentric rotor, vanes, end covers and other parts. When the pump is rotating, the sliding vanes move in the rotor slots and ride against the inner wall of the stator by centrifugal force[1], these vanes, in conjunction with the inner wall of the stator, provide sealed chambers. The fluid enters into the pumping chambers from the suction port and is transported to the discharge port effectively. However, in some high pressure variable displacement vane pumps[2], evidence of wear tracks was found on the internal surface of the stator, as a consequence of the sliding contact between the stator and vanes, and the pump internal leakage frequently occur because of the pressure difference among the pumping chambers. In order to overcome the above problems, numerous authors studied the pump by conducting the numerical simulation or the experiment. Yong Lu's research [3] demonstrated that the improvement of volumetric efficiencies of the pump was the main factor to increase the prototype pump performance. SHAO Fei[4] used the computational fluid dynamics (CFD) numerical calculation method to analyze the flow characteristic in single-acting vane pump. WANG Xing-kun[5] also used the CFD studied the instantaneous flow and monitoring pressure of the double-acting vane pump. Xue liang[6] optimize the pump structure in view of the flow loss by loss Finite element simulation method.

Many studies have demonstrated that the vane pump structure characteristics and the contact between the stator and vanes is very essential, especially the micro-structure on the top of the vane have a great influence on the vane pump whole performance. In order to reduce wear, it is meaningful to use the surface technology to process microstructure on top of the blade and then to increase the volumetric efficiency and decrease maintenance costs of the pump. Now surface microstructure manufacture mainly depends on the UV (ultraviolet) embossed, laser, ion beam light, chemical etching processing technology and micro mechanical processing technology[7]. Diamond grinding wheel grinding method is an effective micro mechanical processing method for hard brittle materials[8]. The diamond grinding wheel V tip has micro-manufactured the micro-groove array onsingle crystal silicon, quartz glass, hard alloy steel[9,10].

In this paper, various depths of micro V-groove array were micro-ground on the top of the graphite vane to study the mechanism performance of improved oil pump. First, the CFD numerical calculation method was applied to study the internal flow field of the micro-structured pump. Second, the experiments on the pump performance were carried. Finally, the influence of the different dimensions of micro-groove arrays on the lubrication and the mechanical efficiency during pump working process were analyzed .

2 Experiment

2.1 Fabrication

Fig. 1 shows the finishing process of diamond grinding wheel micro-V-tip, the fabrication of V-tip diamond grinding wheel referred to the reference [10]. Where a is the diamond tool feed depth, v_f is the feed speed, n is the grinding wheel revolve speed. In this paper, a is of 1 ~ 20 μm, n *is* of 2000 ~ 3000r/min. The average abrasive particle size of the employed resin-bond diamond grinding wheel is 36 ~ 40 μm, the grinding wheel width is of 5 mm, 125 mm in diameter. Ceramic bond millstone is in the size of 100mm × 6mm × 50mm. The finished diamond grinding wheel is shown in Fig. 2.

Fig. 2 demonstrates the diamond grinding wheel V tips process on the vane top surface, walking along the equal interval paths in an alternative positive and negative direction. With several nanometers feed depth each time, the micron scale straight groove can be formed gradually. The grinding processing conditions is shown in Table 1.

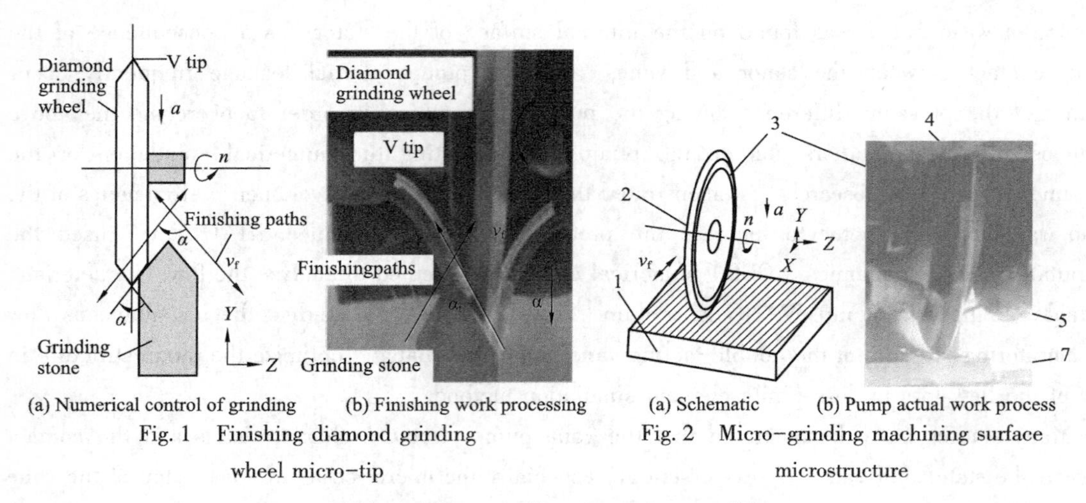

(a) Numerical control of grinding (b) Finishing work processing

Fig. 1 Finishing diamond grinding wheel micro-tip

(a) Schematic (b) Pump actual work process

Fig. 2 Micro-grinding machining surface microstructure

Table 1 The grinding processing conditions

CNC grinding machine	SMART B818
Diamond grinding wheel	The average abrasive particle size36 ~ 40 μm, resin-bond revolving speed n = 2000 ~ 3000r/min
Rough machining	The feed depth a = 5μm, the feed speed v_f = 500mm/min, the total feed depth $\sum a$ = 80 ~ 140μm
Finish machining	The feed depth a = 1μm, the feed speed v_f = 100mm/min, the total feed depth $\sum a$ = 10 ~ 20μm
Cooling liquid	water

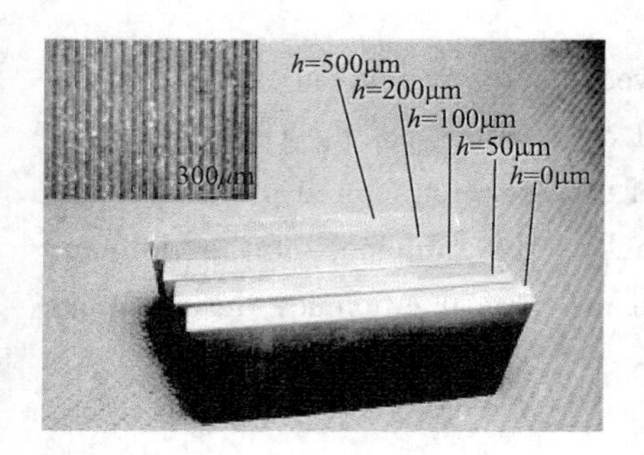

Fig. 3 The processed graphite vane microstructure surface samples

Five groups of vane samples was prepared with different micro-V-groove depth of 500μm, 200μm, 100μm, 50μm and 0 (no micro groove structure). And each group of sample consists six vanes with the same micro-V-groove structure. Fig. 3 shows the processed graphite vane microstructure surface samples.

2.2 Vane pump configuration

Fig. 4 shows the configuration and schematic diagram of the pump used in the experiment. It can be seen that the pump is consisted of a shaft, a rotor and six vanes. The locating ring provides radial force to the vanes to ensure that the blade tip is in contact with the inner

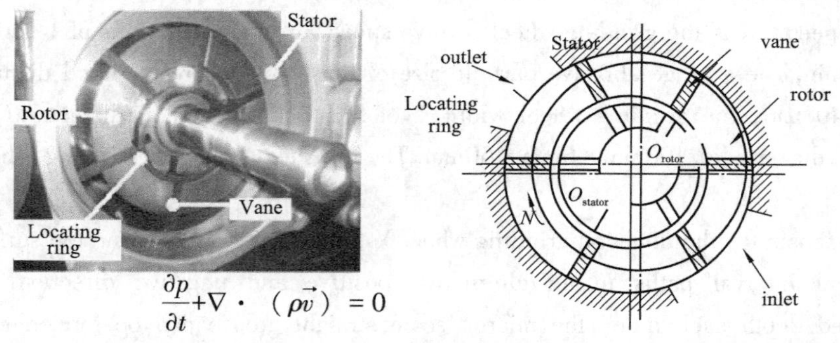

$$\frac{\partial p}{\partial t}+\nabla \cdot \left(\rho \vec{v}\right)=0$$

Fig. 4 Vane pump configuration

wall of the stator well. Thus, each hydraulic chamber is sealed by two adjacent vanes, two end covers between the rotor and the stator. As the rotor rotates in clockwise direction, the chamber volume starts increasing slowly from the inlet and decreasing gradually after passing the outlet. Thus, a pressure difference is produced and therefore induces oil enter into the pump from the inlet and flow out through the outlet effectively.

3 Results and discussion

Five pump performance test experiments were carried out after the preparation for vane samples. In each test experiment, one group of vane samples was installed in the same pump at one time. Then test the actual flow Q_o (L/min), the output power P_o (Watt, W) and the total efficiency ρ at different rotor speed. The actual flow Q_o means that the output flow of the working pump per unit time, the output power P_o is the product of the actual flow and the output pressure, the overall efficiency ρ can be calculated by the formula as follows:

$$\rho=\frac{P_o}{P_i}=\frac{\Delta p Q_o}{P_i} \tag{1}$$

Here Δp is the difference of import and export pressure.

Fig. 5 exhibits the pump actual flow changing with the rotor speed and the micro groove structure depth. It shows that the pump actual flow increases with the increasing rotor speed. The actual flow for the vanes with larger micro-V-groove depth (200 μm and 500 μm) are lower than that for the traditional vanes (no micro-groove array). This may be because the micro groove is too large to decrease the pump leakage then produce smaller volume efficiency. However, as the rotor speed is greater than 360 r/min, the actual flow for smaller micro groove depth (50 μm and 100 μm) vanes are larger than the traditional vanes, it may indicate that when the micro grooves decrease to a certain extent, it can both reduce the inner leak and reduce the interface friction between the vanes and the stator. This can be demonstrated in Fig. 6. The tendency of the output power is similar to the change of pump actual flow. It shows that the vanes with smaller micro-groove depth (50 μm and 100 μm) alternately have higher output power than the vanes with no micro-groove as the rotor speed exceed 360 r/min.

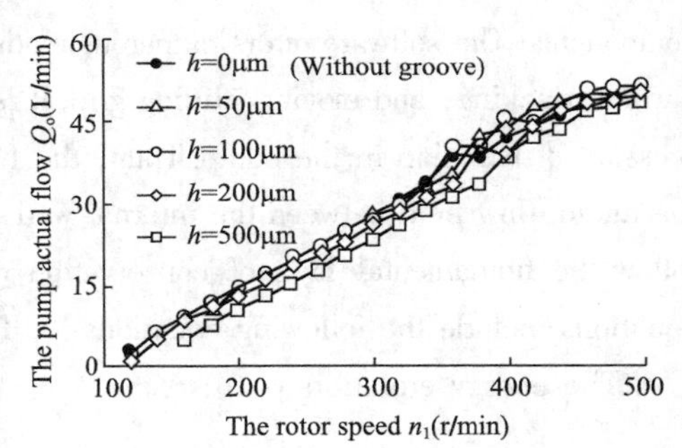

Fig. 5 The pump actual flow as a function of micro grooves structure depth

Fig. 7 shows the change law of the overall efficiency which is the reflection the pump power ratio. The case of micro-V-groove depth 500 μm vanes has the lowest overall efficiency. But as the micro-V-groove depth is above 200 μm, the overall efficiency is higher than the traditional vanes. Apparently, the vanes with the micro-V-groove of 100μm in depth has the highest overall efficiency. The reason can be explained by its highest actual flow and output power.

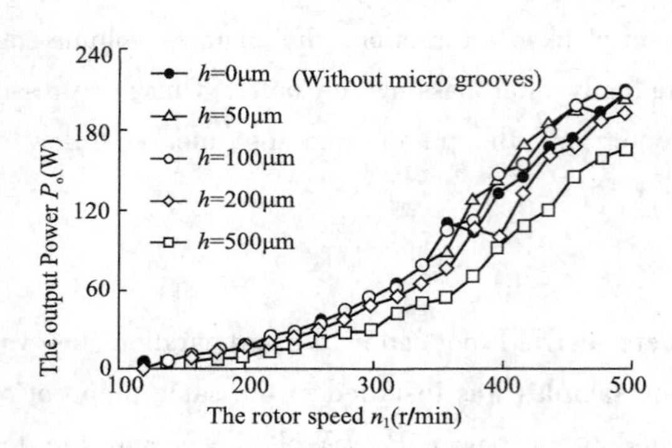

Fig. 6 The output power as a function of micro grooves structure depth

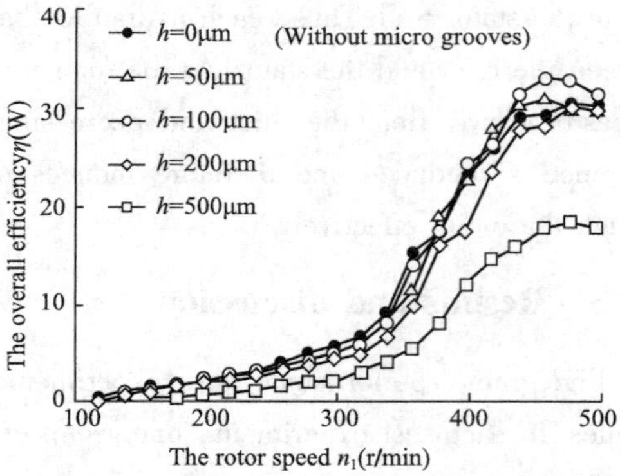

Fig. 7 The overall efficiency as a function of micro grooves structure depth

4 Analyse of the vane pump improved performance mechanism

Pumplinx is a CFD software that provides unique capabilities for the analysis and design of fluid pumps, motors, compressors, valves and other fluid flow devices and systems with rotating/sliding components. The software offers numerous modules including flow, heat, turbulence, species, Lagrangian particle tracking, and moving/sliding grid. It is credible to use the software to simulate the flow field and pressure distribution in the pump. Thus, the Fluent CFD software is utilized for numerical simulation of the micro-flow field between the internal wall of the stator and the vane end. The applied two software all follow the fundamental laws of conservation of mass, heat, and momentum. Therefore, the governing equations include the following equations[11]. The continuity equation:

The energy equation :

$$\frac{\partial}{\partial t}(\rho E_f+\nabla\cdot(\vec{v}(\rho E_f+P_f)))=\nabla\cdot(k_{eff}\nabla T_f) \tag{2}$$

where ρ is the fluid density and $\vec{v}$ is the velocity field in the whole domain. $E_f$is the enthalpy of the fluid, T_f is the temperature field, and k_{eff} is the conductivity coefficient of the fluid.

The momentum equation is:

$$\frac{\partial}{\partial t}(\rho\vec{v})+\nabla\cdot(\rho\vec{v}\vec{v})=-\nabla P_f+\nabla\cdot[\mu(\nabla\vec{v}+\nabla\vec{v}^T)]+\rho\vec{g}+\vec{F}_s \tag{3}$$

Where P_f is the pressure domain, g is the gravity force. F_s is the surface tension force with formula of , where κ is the curvature and the normal vector to the interface.

4.1 Calculation model

Schematic of the solution domains are shown in Fig. 8. Fig. 8a shows the pump interior whole flow of the Pumplinx simulation model. The computing size consistent with the real fluid domain size. The stator inner diameter is 98.506 mm and the rotor outer wall diameter is 88.938 mm, the eccentric distance is of 4.76 mm. Therefore, the nearest distance between the rotor and the stator, is only 0.024 mm, which can be deemed as a fixed calculation size between the vane top and the stator as the depth groove is various. The simulation rotating speed is set as 45 rad/s in clockwise direction, because the pump actual flow and output power are different at a higher speed. And the suction port is located at the lower left where the detected pressure is about 2 bar. The pressure of discharge port at the upper right is 0. 2 bar. Fig. 8b is the enlarged diagram of the interface between the vane top and the stator. In this simulation cases, the depth h changes from 0 to 500 μm, but the distance between the vane top and the stator maintains at 0. 024mm. Fig. 8c is the model for FLUENT simulation. Boundary conditions are defined as follows: The upper surface (stator) is considered as stationary wall, the jagged lower surface (vane) is regarded as moving wall, while the rotational speed is 45 rad/s, the left surface is the outlet, and the right surface inlet, the pressure value are set according to the Pumplinx simulation.

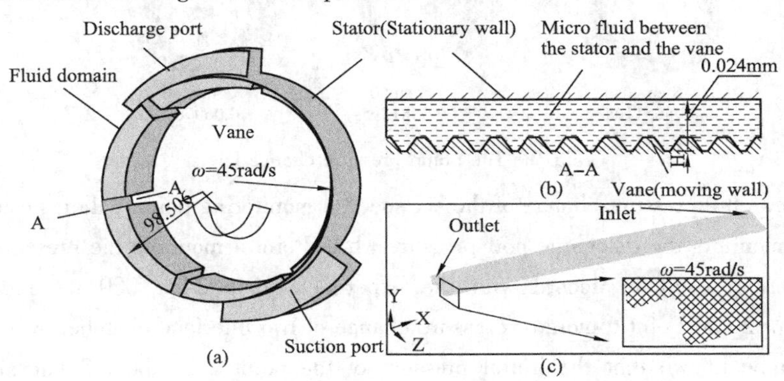

Fig. 8 The simulation initial parameters

Table 2 The initial parameters of the simulation

	Rotating speed ω (rad/s)	Density ρ (kg/m^3)	Dynamic viscosity η (kg · m/s)	suction port pressure P_s (pa)	discharge port pressure P_d (Pa)
Oil	45	725	0. 0007	240000	24000

4. 2 Simulation result

Fig. 9 is the change process of oil pressure. Because the pump is symmetrical along the B-B axis, the simulation of inner pressure change can be displayed for half. The pump operation period is 7. 2 circle per second. The pump chamber was partitioned into 10 chamber units to iterative compute, which means it will calculate ten times in every circle. Each chamber pressure periodically changes with the pump rotation and the inlet pressure is always lower than the outlet. As the fluid flow from the inlet to the outlet, the chamber volume increase firstly then decrease quickly which cause the pressure to change inversely. The pressure of the micro fluid between the blade tip and the stator can be roughly divided into the following three cases: Case (Ⅰ), shown in the top right of the figure, the inlet pressure of the fluid is about 2 bar greater than the outlet, because the top right chamber's volume is decreasing, this place pressure is the maximum pressure inside the pump, and the bottom left chamber's volume is increasing, the pressure is approximately equal to the suction port pressure. Case (Ⅱ), the inlet pressure is about 2 bar less than the outlet, as the outlet chamber fluid is incorporated into the discharge port. Case (Ⅲ), the pressure of the inlet is equal to that of the outlet, mainly because that they are approximately equal to the discharge port, or equal to the suction port pressure 2400 pa. in the work process of the pump, there are four vanes' fluid pressure basically belong to this situation.

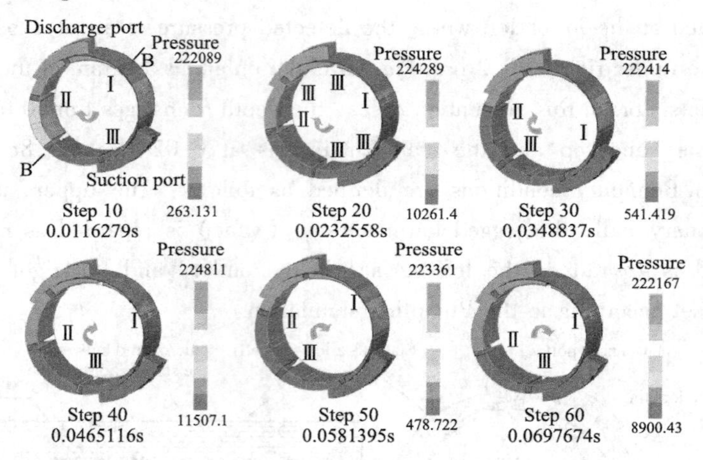

Fig. 9 The pump pressure change law

Fig. 10 shows the pressure change of the six specific monitoring points, their positions marked in Fig. 10. Point 1 monitors the discharge port pressure while Point 4 monitors the pressure of the suction port, the simulation result in accord with the experiment data 240, 000 pa and 24, 000 pa, respectively. Point 2 and Point 6 monitor pressure change of two interface chambers with the rotation of the pump. It can be known that the initial pressure of the point 2 is about 2 bar here, point 6 is approximate to 0, and after work for half a cycle, their pressure exchanged. This is because that the chamber is connected to the discharge port between the two places marked by point 3 and point 5, the pressure is affected heavily by the discharge port, once exceed the limit position, the pressure is closely to the suction port. What's more, the pressure for the two fluid regions marked by point 3 and point 5 also change as one falls another rises. However, it can be seen from figure 10, the time of the fluid under low pressure lasts longer than the high pressure. Thus, as one region pressure is reduced, another region pressure will be not improved rapidly.

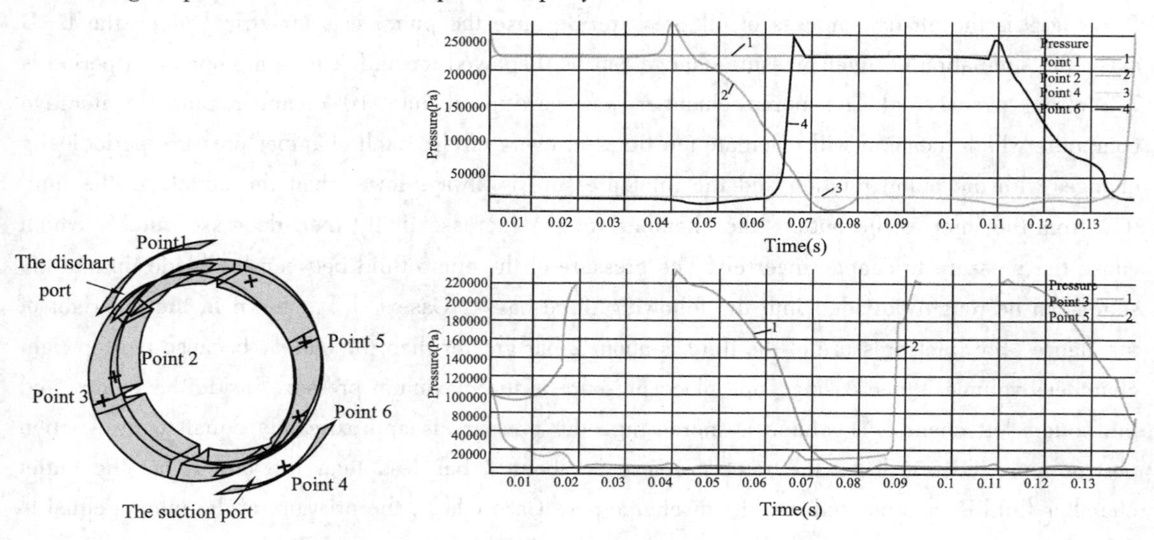

Fig. 10 The monitoring points pressure change law

4.3 Simulation

According to the Pumplinx simulation result, the pressure of the fluid between the blade tip and the stator can be divided into three cases according to the difference pressure for the inlet and the outlet, this article will continue to use FLUENT software to simulate and discuss the fluid dynamic characteristics for No. Ⅰ, No. Ⅱ and No. Ⅲ fluid condition. According to the actual working condition, the inlet pressure set 2 bar and outlet pressure set 0 bar.

4.3.1 Case Ⅰ

As the fluid pressure of the inlet is 2 bar and outlet pressure is 0, the micro fluid average velocity, the volumetric flow rate, and the total leak amount for 0 – 500 μm depth of the microstructures are summarized according to the simulation results showing in Fig. 11. The inlet and outlet average speed are equal in any case due to the fluid momentum conservation. The average speed is of 1.206 m/s for the traditional vanes, but only 0.803m/s for h = 10 μm. Increasing the microstructure depth, the liquid average velocity also increases, and the average speed are 1.60 m/s, 2.96 m/s for 50 μm and 100 μm, respectively. When the depth reaches up to 500 μm, the average speed is as high as 10.641 m/s. The average speed determines the volumetric flow rate, which has been demonstrated by Fig. 11 (b). In addition, the rotor external wall velocity, which is a standard velocity to estimate the liquid leakage, is of 2 m/s based on the rotor speed and the rotor radius. The total leak amounts for various microstructures displayed in Fig. 11 (b). The positive leak quantity means no leak. This happened for the microstructure depth that is less than 50 μm. When the depth is more than 50 μm, the leakage turned to be negative and the amount increases promptly.

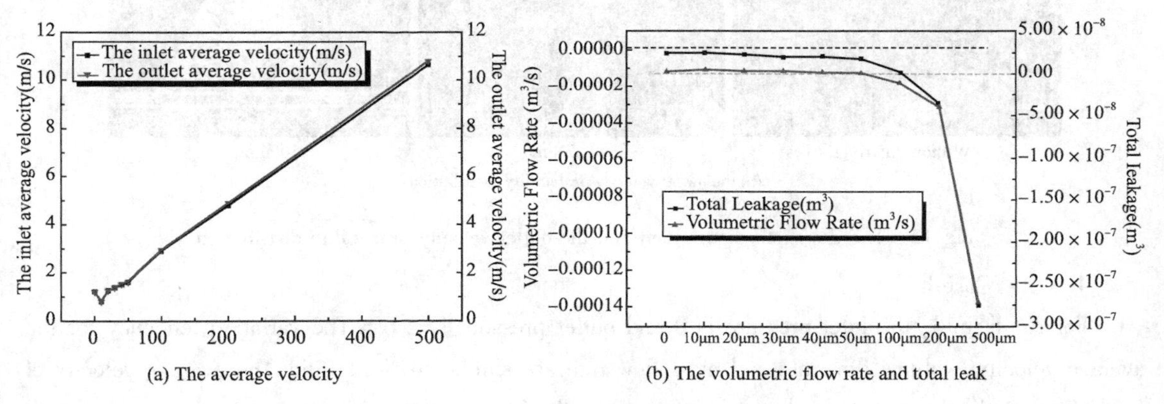

Fig. 11 The average velocity and the volumetric flow rate for inlet pressure 2 bar and outlet pressure 0

The flow resistance has a great influence on the fluid speed. The average speed for depth = 10μm is less than that for no microstructure case, mainly because the vane with this micro-grooves increase the flow resistance. In order to reduce the resistance but not sacrifice the capacity, it can be seen that the depth of 50 μm is just the watershed, because the average speed for 50 μm is close to the blade linear velocity 2m/s. As the depth greater than 50 μm, the leakage began and became seriously. Obviously, for the vanes with the micro-grooves of 500 μm, the inner speed is too high to lead to a serious leakage, this is the main reason for small actual flow, low output power and inefficiency.

Fig. 12 shows the distribution of the flow field and the outlet velocity. The fluid flow is in an obvious stratified flow state, the largest velocity of flow is near the vane tip, and then the velocity decrease linearly from the vane tip to the stator inner wall. But, when the vane tip has micro-grooves, the fluid flow characteristics is changed: the part of the liquid filling in grooves has a velocity higher than the vane tip. Similarly, the fluid against the stator inner wall also has the minimum velocity. This features can be acquired from the case of depth 50 μm, so the velocity is even larger near the wall of the fluid pressure, which is helpful for load bearing surface. Because the fluid pressure difference is helpful for vanes to bear the load. According to Reynolds equation, the lubrication effect applied by the oil in grooves is advantageous to reduce the interface friction. However, for the 500 μm case, the fluid velocity in the groove center is too high to avoid turbulence then causing inner leakage. Fig. 12 (b) revealed the velocity distribution for depth of 500 μm. It can be obviously seen that a velocity peak protrude clearly and the velocity for utter most of the liquid greater than 2 m/s, which is larger than the rotating speed. So the inner leakage cannot be avoid.

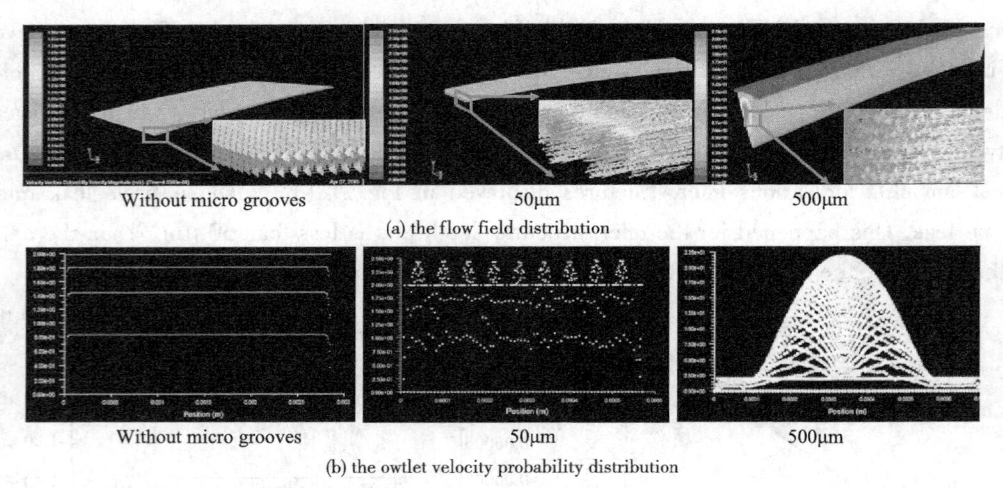

Fig. 12 The flow field distribution and the outlet velocity probability distribution

4.3.2 Case Ⅱ

For No. Ⅱ fluid, the inlet pressure is 0 and outlet pressure is 2 bar. The variation tendency for the average velocity, volume flow rate and mass flow rate are similar to No. 1 fluid. The average velocity of $h=10$ μm is still the minimum, but the difference is that the average velocity are relative smaller for all kinds of micro-groove depth. For instance, the average velocity is only 1.189 m/s for $h=100$ μm, smaller than the No. 1 fluid velocity 2.902 m/s. The Fig. 13 shows the flow rate for the surface without microstructure and with depth less than 50μm, which all of them are negative. Negative value means that the fluid flows out from the outlet. While the depth greater than 100 μm, the flow rate becomes positive. This may cause the oil reverse leakage.

Fig. 14 shows the outlet fluid dynamic characteristics for micro-groove of $h=100$ μm. The fluid near the rotor surface is consistent with the velocity of the rotor tip, marked by 1. However, the reverse flow appeared in the micro-groove center, and its velocity value is very larger, marked by 2. The reverse flow will cause heavy leakage.

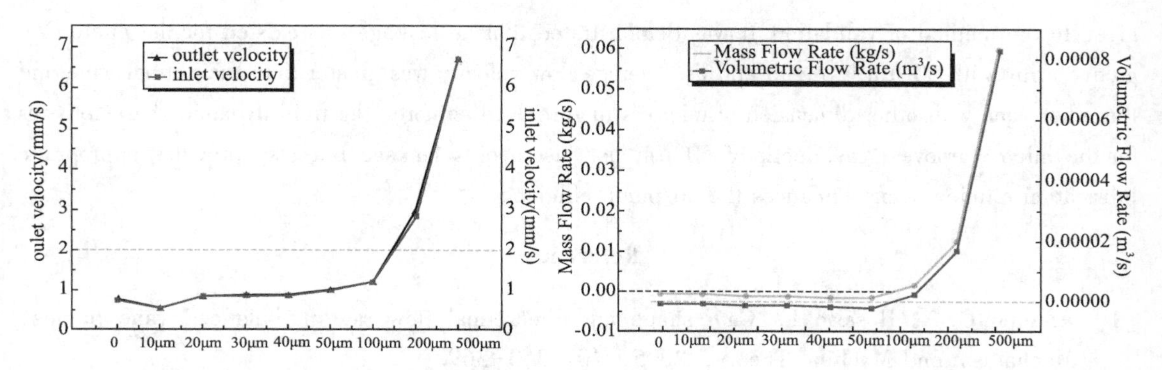

Fig. 13 The average velocity and flow rate as inlet pressure 0 and outlet pressure 2 bar

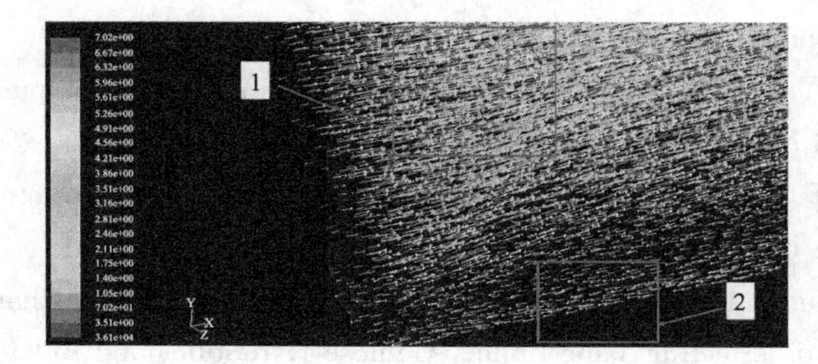

Fig. 14 Reverse leakage phenomenon

4.3.3 Case Ⅲ

The inlet pressure is equal to the outlet pressure for No. Ⅲ fluid condition. The average velocity of $h=10$ μm is smaller than the traditional surface. But the average velocity of that the depth h larger than 20 μm but smaller than 50 μm is higher than the traditional one and increases with the increasing groove depth, the average speed of $h=50$ μm reached to maximum. However, when the depth exceeds 100 μm, the average speed decrease quickly, the average velocity of $h=500$ μm decrease to minimum. The fact may be that micro-groove can hold the oil between the interface of vane tip and stator wall. When the micro-groove is too large, turbulence begins inner the interface as its micro space is open, even though the oil still exist in the micro-groove. When the micro-groove is too small, the oil film can't form to stand the centrifugal pull. So the micro-groove for depth of 50 μm is the optimal choose for this condition, because the oil film formed just enough to stand the centrifugal force in the process of the rotational, so there is no pressure difference between the inlet and the outlet, the oil inside the grooves is relative stasis.

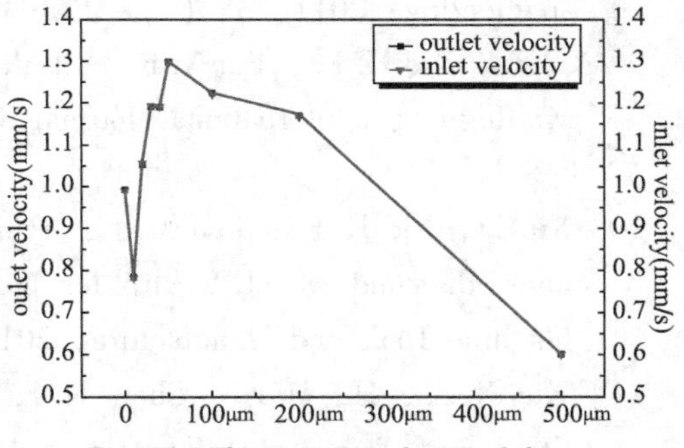

Fig. 15 The average velocity of the outlet and the inlet

5 Conclusion

The fluid dynamic characteristics for the oil vane pump with micro-V-groove arrays of 50-500 μm were simulated, and the corresponding actual working performance of the oil vane pump was given to the method of validation. It was demonstrated that no leakage was existed for the micro–V–groove array with 50 μm in depth and the average flow velocity was greater than the normal vane and the other vane with other dimension of micro–structure. Furthermore, the fluid dynamic characteristics for the micro–groove array depth of 50 μm not just avoids leakage but also provides appropriate lubrication efficiency and enhances the oil pump efficiency.

References

[1] Antonio G. *, Rosario L.. Cam shape and theoretical flow rate in balanced vane pumps, Mechanism and Machine Theory, 2005, 40: 353–369.

[2] Emiliano M., Alessandro A., Gianluca D., et al., On the wear and lubrication regime in variable displacement vane pumps, 2013, 306: 36–46.

[3] Lu Y., Zhao Y. Y., Bu G. X., et al., The integration of water vane pump and hydraulic vane motor for a small desalination system, Desalination, 2011, 276: 60–66.

[4] Shao F., Kong F. Y., Wang W. T., et al., Analysis of Flow Characteristic in Single–acting Vane Pump base on Dynamic Mesh Method, Fluid Machinery, 2011, 39 (8).

[5] Wang X. K., Hu D. F., Shang J. D.. Numerical Simulation and Characteristic Analysis of Flow Field in Double–acting Vane Pump, Chinese Hydraulics and Pneumatics, 2013, 5: 23–24.

[6] Xue L., Zhao Y. L., Zhang L. X., et al., Research on the Optimization of leak–proof Hydraulic Structure of Vane Pump, China petroleum machinery, 2014: 42–51.

[7] Xie J., Li P., Wu K. K., et al. Micro and Precision Grinding Technique and Functional Behavior Development of Micro–structured Surfaces, Journal of mechanical engineering, 49 (23).

[8] Xid J., Zhuo Y. W., Tan T. W.. Experimental study on fabrication and evaluation of micro pyramid–structured silicon surface using a V–tip of diamond grinding wheel [J]. Precision Engineering, 2011, 35 (1): 173–182.

[9] Xie J., Xie H. F., Liu X. R., et al. Dry micro–grooving on Si wafer using a coarse diamond grinding [J]. International Journal of Machine Tools and Manufacture, 2012, 61 (10): 1–8.

[10] Xie J., Xie H. F., Luo M. J., et al. Dry electro–contact discharge mutual–wear truing of micro diamond wheel V–tip for precision micro–grinding [J]. International Journal of Machine Tools and Manufacture, 2012, 61 (9): 44–51.

[11] Xie X. Z., Hu M. F., Chen W. F., et al. Cavitation Bubble Dynamics during Laser Wet Etching of Transparent Sapphire Substrates by 1064 nm Laser Irradiation. Journal of Laser Micro/Nanoengineering, 2013, 8 (3): 259–265.

五、其他

套管和油管标准的技术发展

方　伟　秦长毅　许晓锋　徐　婷　吕　华

（中国石油集团石油管工程技术研究院）

摘　要：概述了我国套管和油管技术和标准的发展历程，介绍了 API 套管和油管标准的特性与用途、标准溯源与技术内容的演变，通过对技术内容演变的分析，提出了标准实施中应注意的事项，建议有关各方应加强标准信息的跟踪和研究，及时掌握标准的变化，使国内生产企业和用户能及时应对。

关键词：套管和油管；标准；发展

套管和油管在油田使用十分广泛，在油气勘探开发中不可缺少。自 1929 年美国石油学会（API）制定出第一部管材标准 API 5[1] 以来，API 已制定了大量标准、规范、通报和推荐作法，由于注重产品的互换性、安全性，以及良好的工程实践和作业经验，长期以来一直为世界石油工业所普遍接受和采用，实际上已具有国际标准的性质，起到了国际标准的作用。世界上大多数国家的套管和油管生产和检验都采用 API 标准，我国从 1982 年起大规模正式直接或等同采用 API 套管和油管标准[2]。

套管和油管标准是 API 最早制定的标准之一，八十多年来不断修改、完善，换版达 70 余次之多，特别是 1988 年以 API Spec 5CT《套管和油管规范》第 1 版的形式出现后的二十几年时间内，在 API Spec Q1《质量纲要规范》的指导下，根据当前石油工业和相关工业技术的发展，已进行了八次较大幅度的修改。API Spec 5CT 是美国石油学会油田设备与材料标准体系中的核心标准，在石油管材标准项目中具有担纲的作用。

正是由于 API 标准的国际通用性，国际标准化组织（ISO）近年来将大量 API 标准转化为国际标准。ISO/TC67“石油天然气工业用材料、设备及海上结构”技术委员会 SC5“套管、油管和钻杆”分委员会根据 API Spec 5CT 并吸纳欧盟标准化委员会的意见于 1996 年发布了 ISO 11960 第 1 版[3]。这样就出现了针对同一标准化对象存在 API Spec 5CT 和 ISO 11960 两个国际通用标准。由于 API Spec 5CT 的修订速度明显快于 ISO 11960，因而出现了技术要求不同的两个国际通用标准并存的局面。这就有悖于标准化的目的在于维持正常经济秩序、促进贸易和技术交流这一基本原则，ISO/TC67 成员和 API/C1/SC5 成员迫切要求改变这一状况。在 API 和 ISO 缔结的先导性项目协议完成后，双方代表一致同意 API Spec 5CT 和 ISO 11960 联合行动，同期使用、同期修订、同期审议。2001 年 10 月 1 日，ISO 11960 第 2 版和 API Spec 5CT 第 7 版同时发布。

作者简介：方伟（1968—），女，高级工程师，硕士，主要从事石油管材的标准化工作。fangwei001@cnpc. com. cn。

质检公益性行业科研专项经费资助（编号：201510205-03）。

过去我国套管和油管一直依赖进口，在订货和检验中，主要以API标准为依据。由于早期的API标准主要注重于型式和参数的一致性，即对外观尺寸规定的比较严密，而对产品的技术要求、试验方法和验收规则规定的不严格、不严密，对材质要求很松，造成许多符合API标准的产品在使用中屡屡出现质量问题，不能适应我国油田现场的使用要求[4]。鉴于这种情况，管材研究所在参照API标准的基础上，结合其在套管的检验、失效分析和科研工作中所获得的数据、资料和科研成果，以及中国石油天然气总公司当时试行的“套管订货补充技术条件”，增补了一些API标准未作规定的重要内容，包括技术要求、试验方法和验收规则等，于1994年制定了行业标准SY/T 5989—1994《直焊缝套管国外订货技术条件》和SY/T 5990—1994《套管国外订货技术条件》。随着API标准对产品技术要求的不断提高及石油管材的国产化，管材研究所在对API标准跟踪研究的基础上，于1996年以等同采用API Spec 5CT第5版的形式制定了行业标准SY/T 6194—1996《套管和油管》。在该行业标准在油田、生产厂广泛应用的同时，ISO/API标准也在不断地修改完善，我们于2005年以等同采用ISO 11960：2001（第2版）/API Spec 5CT：2001（第7版）的形式将该标准上升为国家标准，制定了GB/T 19830—2005《石油天然气工业　油气井套管或油管用钢管》，后在ISO/API新版标准的基础上完成了该标准的两次修订。表1给出了API、ISO及我国套管和油管标准版次信息及相互关系。

表1　API、ISO、GB/SY套管和油管标准各版次及相互关系一览表

API标准		ISO标准		GB/SY标准
API 5CT：1988（第1版）				
API 5CT：1989（第2版）				
API 5CT：1990（第3版）				
API 5CT：1992（第4版）				SY/T 5989—1994、SY/T 5990—1994
API 5CT：1995（第5版）	IDT			SY/T 6194—1996
API 5CT：1998（第6版）		ISO 11960：1996（第1版）		
API 5CT：2001（第7版）	IDT	ISO 11960：2001（第2版）	IDT	SY/T 6194—2003、GB/T 19830—2005
API 5CT：2005（第8版）	IDT	ISO 11960：2004（第3版）	IDT	GB/T 19830—2011
API 5CT：2011（第9版）	IDT	ISO 11960：2011（第4版）		
API 5CT：2011（第9版）2012年9月勘误1		ISO 11960：2014（第5版）	IDT	GB/T 19830—××××

IDT——等同采用。

1　套管和油管产品标准的特性与用途

现行版API Spec 5CT的套管和油管分为4组、15个钢级和类型，针对不同钢级分别有三个产品规范等级的技术要求。按制造方法，又分为无缝管和电焊管两大类。除L80-9Cr、L80-13Cr、C90-1、T95-1、C110计5个钢级限定采用无缝工艺外，其他钢级除采用无缝工艺还可采用电焊工艺制造。其热处理工艺，除H、J、K、N80-1和M钢级外，其余10个钢级都必须进行全管体、全长淬火+回火处理，并对R、L、C、T等7个钢级规定了最低回火温度。对H、J、K、N、M和P共6个钢级只限定了S、P含量，而未规定其主要元素的化学成分。

该标准适用于油气井用钢质套管、油管、短节、接箍及附件等的生产、订货和检验，规定了钢管及附件的交货技术条件。

套管和油管不属于一般的冶金产品，而是在无缝管或棒材基础上经过深加工（压力加工、焊接、机加工、热处理等）的特殊冶金产品，实际上已经属于机加工产品的范畴。为满足使用要求，除对套管和油管用钢的化学成分、力学性能、冶金质量有严格的要求外，API 标准还对制成品的外径、内径、壁厚、圆度、直度及螺纹参数、接头的黏扣趋势、密封性能和结构完整性等都有很严格的要求。

2 API Spec 5CT 标准溯源与技术内容的演变

2.1 API Spec 5CT 标准溯源

API 套管和油管规范最初于 1929 年以规范 5 出版，后以 API 5A《三角螺纹套管和油管》再版发行，共发布 39 版[5]。本规范涉及的套管和油管产品有 H40、J55、K55、N80 钢级套管，J55 钢级平端尾管，H40、J55、N80 钢级油管，套管和油管短节，连接管等。

P 钢级套管和油管的暂行技术条件于 1955 年被采纳，最初列入 API 5A 中。以后，于 1960 年又将 P 钢级套管和油管从 5A 中抽出，新制定为 API 5AX《高强度套管和油管》，共发布 15 版[6]。规范涉及的套管和油管产品有 P110 钢级套管、P105 钢级油管、套管和油管短节、连接管等。

API 5AC《限定屈服强度套管和油管》最初于 1962 年 6 月被采纳。1963 年发布第 1 版，名称为《含硫环境用套管和油管》，共发布 16 版[7]。规范涉及 C75、L80、C90、C95 钢级套管，C75、L80、C90 钢级油管，油管短节，连接管等产品。

API 5AQ《Q125 钢级套管》第 1 版于 1984 年的标准化会议上通过，1985 年发布。后于 1987 年发布第 2 版[8]。规范涉及 Q125 钢级套管、套管接箍、套管短节、连接管等产品。

以上四项规范分别规定了不同钢级、规格套管、油管等产品的制造方法、热处理、材料的化学成分、机械性能、静水压试验等要求。其各次版本修订的主要内容有以下几方面：

（1）规格细化，在相临名义重量、壁厚等规格中增加中间规格套管和油管，使其产品规格更趋系统、完整。

（2）增加了材料理化性能检验批等规定，使试验更合理，操作性更强。

（3）增加了接箍类型如特殊间隙接箍、密封环接箍等，使产品类型更广泛。

（4）部分技术要求加严。

（5）增加部分技术要求及相应的试验方法。

（6）修正规范某些规格指标的数值。

2.2 API Spec 5CT 技术内容的演变

早期的 API 规范侧重于产品的互换性，对产品的质量要求重视不够。1982 年以来，一些用户向 API 反映，许多符合 API 标准的设备和材料在使用中发生了重大事故。很多石油公司和钻井承包商对 API 标准表示不满，开始在订货时提出补充技术条件。显然，用户提出的验收规范和技术条件必然是各种各样的，给制造商带来很大困难，反过来也导致用户成本的提高。因此，用户和制造商都给 API 施加压力，要求改革 API 标准化政策。鉴于上述原因，API 召开一系列会议，决定按 API Spec Q1 全面修订原有标准，其中包括在产品标准中加强和增补技术要求、质量指标和试验、验收规则。1988 年 3 月颁布的 API Spec 5CT 第 1 版，就是按 API Spec Q1 对已有的 API 5A、5AX、5AC 和 5AQ 进行整合、修订而成的，体现了API标准化政策变化的新精神和对质量的严格要求。除标准结构上的调整外，增补了部分钢级，加强了技术要求、试验方法和工艺控制等。例如，API 5A、5AC、5AX等规范对有害元素硫含量的一般规定为≤0.060%。改版后的API Spec 5CT对硫含量的一般限定改为≤0.030%，其中第2组、第4组的部分钢级则进一步限定硫含量≤0.010%。

过去，API标准对各种钢级套管和油管均无韧性要求。根据美国几家大石油公司和中国石油天然气总公司的提案，修订的API Spec 5CT正式增补了夏比冲击韧性的要求。1988年的API Spec 5CT第1版只对第4组Q125钢级规定了横向韧性要求，而第4版对除H40以外的所有钢级的接箍（含接箍毛坯、半成品）都规定了夏比冲击韧性要求。第5版对C90、T95等钢级的晶粒度、淬透性及硫化物应力腐蚀开裂试验等方面都比过去的版本有了更严格的要求。第7版则对管体冲击韧性、无损检验、水压试验等要求进一步提高。第8版建立了三个产品规范等级（PSL-1、PSL-2和PSL-3）的要求，增加了高抗泄漏接头的补充要求，调整了部分钢级的冲击韧性和硬度要求。第9版细化和提高了部分钢级的矫直、统计拉伸试验、硫化物应力开裂试验等要求[9]。

3 API Spec 5CT标准技术演变分析

API 5是美国石油学会最早制定的产品标准之一，经过80多年的发展、演变，目前已形成系统配套的系列标准，其中有关套管和油管的标准就有API Spec 5CT、API RP 5A3[10]、API RP 5A5[11]、API Spec 5B[12]、API RP 5B1[13]、API RP 5C1[14]、API TR 5C3[15]、API RP 5C5[16]、API Spec 5CRA[17]、API Std 5T1[18]等，涵盖了从制造、检验到使用的所有技术内容。

套管和油管在使用时，主要形式是组合成管柱。一个管柱少则几十根，多则几百根管子，所以互换性是套管和油管非常重要的性能要求。API从着手标准化工作至今，一直致力于保证石油工业材料、设备的互换性，并为此付出了极大的努力，也为石油工业的发展做出了十分突出的贡献。无论来自哪个国家、地区和厂家的套管和油管，只要是符合API标准要求的同规格产品，就可以互相代替而且使用性能相当。为了保证不同时期的管子也具有互换性，API在产品规格发生变化时都非常慎重，特别是影响互换性的螺纹尺寸、接箍尺寸、管子内外径等，几十年来从不改变。

从前面列举的API Spec 5CT各版内容的变化看，可见API标准技术发展规律有以下几个方面：

（1）钢级、规格方面，根据技术的发展水平和石油工业需求，钢级有增有减，规格增多减少。

（2）内容编排和编写格式方面，总的趋势是内容编排更合理，格式更接近ISO标准要求，并最终与ISO标准一致。

（3）技术要求项目只增不减，技术指标也有提高的趋势。

（4）检验要求更具可操作性。

（5）质量控制和过程控制方面的要求越来越多，可证实性要求越来越高。

保证产品的适用性是产品标准的目的之一。前面曾提到，互换性是API产品的重要性能之一。但是，随着石油工业的不断发展，套管和油管的服役条件越来越苛刻，仅满足互换性要求的套管和油管已不能适应油田生产需要。大多数石油公司在订购套管和油管时都根据需要在API标准的基础上提出补充技术条件。这就意味着API标准再不提高对产品使用性能方面的保证能力，难以适应石油工业发展的需要，也难以保持其在石油工业标准化方面的领先地位。另外，新技术、新工艺、新材料和新设备的涌现，为进一步提高套管和油管质量提供了条件。因而，API Spec 5CT 在材料的杂质含量、机械性能、尺寸精度和制造方法诸方面陆续提出了新的要求。

自 1924 年起，API 即开展质量认证。1985 年 1 月 API Spec Q1《质量纲要规范》颁布后，API 会标许可证审核就将其作为判定申请企业质量体系是否符合要求的准绳[19]。API Spec Q1 规定了促进广泛的安全和可互换产品的质量保证要求。凡是申请使用 API 会标的企业，必须首先通过质量体系审核，并在产品标准中增加了有关产品可追溯性、特殊过程等质量管理方面的要求，对测量仪器的校准、核查及其记录保存都做出了明确规定。通过认证的制造商可在按照 API 勘探开发标准规范生产的设备/产品上使用 API 会标。目前，全世界范围已有近 1000 家制造商取得了 API Spec 5CT 会标许可证（目前 API 会标认证的标准有 75 个，获得许可的有近 6800 家企业）。API 质量认证在促进石油工业材料和设备标准化发展、产品质量提高、安全性和互换性等方面起到了非常重要的作用。几十年的实践，铸就了 API 这块名牌。

质量认证是第三方依据程序对产品、过程或服务的符合性给予书面保证。产品质量认证的基础是产品标准，所以产品标准的使用者不仅包括产品的用户、生产者和经营者，而且还包括认证机构和检验机构。在预防为主、用户至上、以人为本和过程方法为主导思想的现代质量管理模式下，将用户需求融入产品标准的技术要求、逐渐增多为保证持续生产符合要求的产品而必不可少的质量控制手段是十分必要的。例如，对于特殊过程，不同的制造商、用户或审核员会有不同的理解，而它又是影响产品质量的关键过程。因此在产品标准中给予明确，就使各方意见得到统一。API 的产品标准中关于质量控制方面的要求越来越多，以弥补 API Spec Q1 的不足，或者说是 API Spec Q1 有关要求的具体化。这一点在近几年制修订的 API 标准中尤为突出。

4 标准实施中应注意的事项

从前面的分析中，我们可以了解 API 套管和油管标准发展的趋势。作为产品标准，API Spec 5CT 根据性能特性原则或者称为最大自由度原则，给生产企业留有充足的发展空间，以充分发挥生产企业的资源优势，获取最佳效益。但是，技术要求越来越高、质量控制越来越严、用户越来越挑剔，这些不争的事实确实是生产企业不得不面对的局面，这也是市场竞争的必然结果。我国已加入世贸组织，国内企业生产的石油管材也要参与世界竞争。但是，国内仍有企业将 API 标准的基本要求视为最高奋斗目标，千方百计地寻求在低水平上徘徊的出路；不是研究如何适应用户的要求，而是要用户削足适履。

API Spec 5CT 现有 4 组 15 个钢级，而我国取得 API 会标（5CT）使用权的 352 家企业生产平端或带螺纹和接箍套管和油管的有 38 家企业，其中生产第 1 组产品的 38 家，生产第 2 组产品的 3 家，生产第 3 组产品的 4 家，生产第 4 组产品的 2 家[20]。国内企业难以与国外企业竞争，其差距不仅仅存在于设备和技术，更主要的差距存在于市场观念、质量管理意识。生产企业在实施 API Spec 5CT 时应进一步深入研究标准，将标准中的技术要求作为设计输入，根据企业特点输出必需的过程控制文件。应强调的是，套管和油管有许多隐含的要求，它们出现在 API Spec 5B、API RP 5C3 等配套标准中，不可忽视。配备资源，满足标准规定。跟踪研究标准发展动态，应对新技术变化。

API Spec 5CT 在技术内容上的多次修改体现了用户期望，反映了用户至上的市场理念。标准不仅明确用户具有选择技术要求的自主权，而且越来越多地赋予用户以监督权。例如，索取有关试验、检验记录以验证产品质量，购方进厂监督过程控制等。API Spec Q1 中还规定购方可对关键过程实施停止点（Hold Point），即该过程只有购方代表在场时才可进行。我国套管和油管生产采用 API 标准已有 20 多年，但在实施标准方面还很不到位。订货时订单内容只有规格、钢级、数量、交货日期，几乎不加注任何技术要求的选择条款，更不提补充技术要求，基本上也不进行驻厂监造。到货验收时，只对管体进行外观检验，对两端螺纹尺寸进行抽检。近些年不少用户开始对新产品取样送质量监督检验机构进行试验，但周期性、批量的监督检验仍显不足。在使用过程中出现质量问题，不进行失效分析，提供不出足够证据说明管材有制造缺陷，多数情况下无果而终。

作为用户或制造商，均应充分利用法律法规赋予的权力保护自身的合法权益。技术标准经供需双方约定后就是保护用户权益的一把利器，通过贯彻实施产品标准从中受益。因此应做到以下几点：

（1）及时掌握标准动态。从 API Spec 5CT 版本更新情况可看出，API 标准 2~3 年（最多不超过 5 年）即修订一次。API 标准水平不断提高，得益于广开言路、集思广益、不断修改、持之以恒。无论是用户还是制造商，都应及时了解标准动态，一方面，用户在订货时和制造商组织生产时一定要采用标准的最新版本；另一方面，对于在使用和生产过程中发现的问题，可及时向国内相应的标准化归口管理部门反馈。

（2）深入研究标准，明确标准的规定要求，特别是修订后技术要求的变化。用户在订货时，需要注明的条款予以注明，需要与制造商协商的条款进行协商；制造商应针对标准技术要求的变化，及时调整和配备必要的设备，更新过程控制文件。

（3）充分利用标准中赋予用户的监督权。例如，让制造商提供检验、试验记录，提供试验合格证，派代表到制造商监督生产和检验等。对正常使用中发现有缺陷的产品可拒收。采用第三方驻厂监造，一方面可保证产品质量，另一方面可对生产过程中的不合格环节加以纠正和预防。

（4）配备资源，加强出厂和到货质量控制。国外资料表明，套管和油管经过无损检验发现不合格率约在 5%左右，如不进行检验，其经济损失可想而知。此外，管研院通过管材检验/试验，帮助油田挽回经济损失的案例很多，这说明检验/试验也是质量控制的重要手段。

（5）充分利用认证机构的监督职能。用户在采购套管和油管产品时，都首先选择通过 API 会标认证的制造商。API 在会标使用中明确规定，用户有责任将使用中发现的质量问题向 API 总部报告。API 要求用户向 API 报告会标产品所遇到的一切问题。API 对企业的年度监督审核或复审时，来自用户的投诉无疑是重中之重。2002 年 5 月，API 质量认证部还建立了用户投诉网站。但据了解我国油田用户即使遇到问题也很少向 API 报告，这无疑是对缺陷产品的一种放任。同时，对于制造商，为保证产品质量、扩大市场销售，可借助国内认证咨询机构的帮助，积极争取获得 API 会标许可证。

（6）参与 API 标准的制修订。通过数据收集和理论论证，将我国石油工业对套管和油管的某些特殊要求纳入 API 标准，使 API 标准能够体现我国的需求和利益。这已有成功的先例。

5 结论

（1）API Spec 5CT经过几十年的演变，其格式日渐规范，表征套管和油管使用功能适用性的技术要求有所增加，指标有所提高，质量控制方面的要求大量融入等诸多变化，不仅体现了用户期望，而且也是科学技术发展和现代质量管理理论的成果结晶。

（2）国内生产企业在实施 API Spec 5CT 时，应尽力使企业具有持续生产符合要求产品的能力，即在质量体系、资源配置、过程控制等更深层次上进一步下功夫，以不断提高市场竞争能力。

（3）国内套管和油管用户应把握好 API Spec 5CT 赋予的自主权和监督权，利用在市场经济中的主导地位，全面深入地做好套管和油管采购、进货的质量控制工作，维护自身权益。

（4）有关各方应进一步加强信息的收集、传递工作，及时掌握 API Spec 5CT 的变化，使国内生产企业和用户都能及时应对。

参 考 文 献

[1] Care and Use of Oil Country Tubular Goods：API 5［S］. 1st ed. 1929.

[2] 方伟，许晓锋，等. 油井管标准化及非 API 油井管标准体系［J］. 石油工业技术监督，2010，26（6）：20-23.

[3] Petroleum & Natural Gas Industries-steel Pipes for Use as Casing or Tubing for Wells：ISO 11960［S］. 1st ed. 1996.

[4] 樊治海，方伟，等. 实施标准化战略，促进“油井管工程”发展［J］. 石油工业技术监督，2004，20（2）：14-19.

[5] Sharp-Thread Casing and Tubing：API 5A［S］. 3rd ed. 1955 - 39th ed. 1987.

[6] High-Strength Casing and Tubing：API 5AX［S］. 1st ed. 1960 - 15th ed. 1987.

[7] Casing and Tubing for Sulfide Service：API 5AC［S］. 1st ed. 1963 - 16th ed. 1987.

[8] Q125 Casing：API 5AQ［S］. 1st ed. 1985，2nd ed. 1987.

[9] 方伟，秦长毅，等. API Spec 5CT 标准最新进展及主要技术内容变化［J］. 焊管，2012，35（8）：61-67.

[10] 套管、油管、管线管和钻柱构件用螺纹脂推荐作法：API RP 5A3［S］. 2009.

[11] 新套管、油管和平端钻杆现场检验推荐作法：API RP 5A5［S］. 2005.

[12] 套管、油管和管线管螺纹的加工、测量和检验规范：API Spec 5B［S］. 2008.

[13] 套管、油管和管线管螺纹的加工、测量和检验推荐作法：API RP 5B1［S］. 1999.

[14] 套管和油管的维护与使用推荐作法：API RP 5C1［S］. 1999.

[15] 套管、油管及用作套管或油管的管线管公式和计算及套管和油管使用性能表技术报告：API TR 5C3［S］. 2008.

[16] 套管和油管连接试验程序推荐作法：API RP 5C5［S］. 2003.

[17] 套管、油管和接箍毛坯用耐蚀合金无缝管规范：API Spec 5CRA［S］. 2010.

[18] 缺欠术语标准：API Std 5T1［S］. 2003.

[19] 方伟，秦长毅. API 会标在石油工业中的意义［J］. 石油工业技术监督，2000，16（12）：11.

[20] The API Composite List，2015，3.

1500t 复合加载试验系统技术改造及性能提升

李东风　王　蕊　韩　军　余　志　王亚龙

（中国石油集团石油管工程技术研究院）

摘　要：1500t 复合加载试验系统主要用于开展复杂工况油井管全尺寸复合加载试验，评价油井管性能。本文针对 1500t 复合加载试验系统轴向加载过程中存在的控制精度和加载速率不稳定等问题，分析问题产生的原因，针对性的提出对原系统的液压系统和控制系统进行升级改造。结果表明系统运行稳定，解决了原系统存在问题，增加了位移控制功能、各项报警和看门狗功能，提高了系统的安全可靠性、试验能力以及试验效率。

关键词：复合加载试验系统；轴向载荷；控制精度；加载速度；技术改造

1500t 复合加载试验系统的主要功能是模拟油田实际工况，对油井管全尺寸试样进行复合加载试验（包括轴向载荷、内压/外压、温度等载荷共同作用的复合载荷）[1]，研究油井管在实际工况下的力学行为与机理，评价各种油井管的结构完整性、密封完整性及工况适用性。随着服役时间的增加，该试验系统的轴向加载部分故障率增加，出现控制精度和加载速率不稳的现象，并且安全互锁功能不完善，无法满足高频次拉/压载荷循环试验的要求，严重影响了科研工作的开展，而试验系统的稳定性及可靠性是决定试验系统的能力、运行稳定性和安全性的重要因素[2-4]。因此，本文详细的分析系统问题产生的原因，对设备进行升级改造，提升该系统轴向载荷精度及稳定性的同时提高系统的安全可靠性，确保 1500t 复合加载试验系统性能满足要求的同时开发设备新功能，本文的研究对油田用油井管性能评价具有非常重要的意义。

1　试验系统改造原因分析及安全可靠性改造

1.1　改造原因分析

1500t 复合加载试验系统的加载速度及稳定性主要受液压系统及控制系统两部分影响，以下将通过这两方面进行分析。

（1）液压系统改造原因分析。

液压系统由加载回路和快速动作回路两个部分组成，其中加载回路是通过八个液压缸给试样施加拉伸或者压缩载荷，动作回路用于安装试样时现场快速动作。轴向载荷液压缸加载回路中，采用比例阀调节控制低油压，伺服阀调节控制高油压的控制方式，在施加载荷的过程中当油压由低压向高压转换时，控制阀会由比例阀切换到伺服阀，这易导致加载速率不恒定的问题，甚至会出现切换期憋压的现象。轴向载荷液压缸快速动作回路，动作速率过慢，致使安装试样的时间过长，严重影响试验效率。

作者简介：李东风（1979—），男，高级工程师，主要从事油管、套管的实验研究工作。

加载过程（特别是拉伸加载过程）中，试件突然断裂会造成油缸上腔压力突增的情况，这时，高压液会直接冲击主油箱，导致试验系统遭受巨大的冲击载荷，影响系统软硬件的使用寿命。

另外，由于长期服役造成液压缸磨损，8 个油缸之间存在串油现象，同时接头管路也存在渗漏现象，导致设备能力下降，最大拉伸载荷由原设计参数 1500t 降为 1020t，最大压缩载荷也由 1200t 降为 800t，此种情况造成试验系统轴向拉伸载荷保载不稳定。

（2）控制系统改造原因分析。

在苛刻工况条件下服役的油井管性能评价试验，需要试验系统快速响应的高精度闭环控制，准确和充足的数据采集记录，以保证能对试验结果提供有效的客观证明，因而控制和数据采集系统需要具有足够高的可靠性。

现存试验系统仅采用一台计算机进行控制和数据采集，不能够满足长周期实验的需求。经常出现计算机死机等情况而导致实验中止的情况，甚至是导致实验失败。同时在控制系统中由于当时进行设备设计时试验条件简单，系统未设计足够的位移控制功能，随着深井、热采井等负责工况油田的开采，对于位移控制的实验需求在逐渐增加，在控制、采集系统中也应相应增加与试样相匹配的控制功能。

1.2 安全可靠性改造

现存试验系统缺少安全护锁功能、各种报警指示以及系统信息反馈，尤其是关键故障报警信息，如液压站油位低报警，油温高报警，过滤器堵塞报警，压力超高报警，载荷超限报警等。另外，缺少安全防护屏障，该试验系统通常要进行高压高温实验，不设置安全防护屏障存在极大的安全隐患。因此，在设备升级改造过程中，提高设备的安全可靠性，对于系统的安全运行、试验操作者的人身安全都具有非常重要的意义。

2 1500t 复合加载试验系统升级改造方案

2.1 液压系统设计[5,6]

液压系统改造原理如图 1 所示，系统的加载回路采用带先导比例压力阀的遥控型变量泵和伺服阀组合控制的方案，远程控制变量泵的泵压恒定于某个值，然后通过伺服阀缓慢给油缸加载，并形成压力反馈的闭环控制。采用这种方式，由泵来衡定压力，伺服阀来保证供油量的精度，从两个方面控制油缸的工作压力稳定在允许误差范围之内，确保了控制精度，同时也解决了在两个阀之间切换所带来的问题，提升系统的控制精度，恒定了系统的加载速率。

为了避免加载过程（特别是拉伸加载过程）中试件突然断裂造成油缸上腔压力突增的情况，在每 4 个液压缸上腔加装了一个先导式溢流阀，能够在载荷突增时迅速打开，将高压液排至辅助油箱，避免高压液直接冲击主油箱。

为了实现快进退、慢加载的设计要求，快速动作回路大排量低压柱塞泵和带先导比例压力阀的遥控型变量泵提供动力，当需要试验系统快速动作油缸时，两台泵同时开启。为实现慢加载设计要求，低压泵停止工作。电控系统可以根据加载力计算出需要的泵压，由程序设定压力值给变量泵上的先导比例压力阀下达指令，使泵的输出压力稳定在需要值，同时泵自带的调节机构能自动调小柱塞泵的斜盘角度，使泵的输出流量维持在够用且能保证压力的状态下，既避免了无畏的能源消耗造成油箱升温过快和噪声过大，又避免了溢流流量过大带来的压力波动问题。节省能源，降低设备的噪声，在提升试验系统控制精度及加载速率问题的同时，改善了试验人员的工作环境。

在系统设计时采用双向锁，当保载时锁定液压缸的压力，避免接头或管路渗漏造成的压力下降情况，保持试验保载过程中载荷的稳定性。

在液压系统改造的同时维修液压缸，将八个液压缸按最小的杆径和最大的缸径进行统一的维修电镀处理，在解决串油问题的同时确保八个油缸产生的载荷一致。

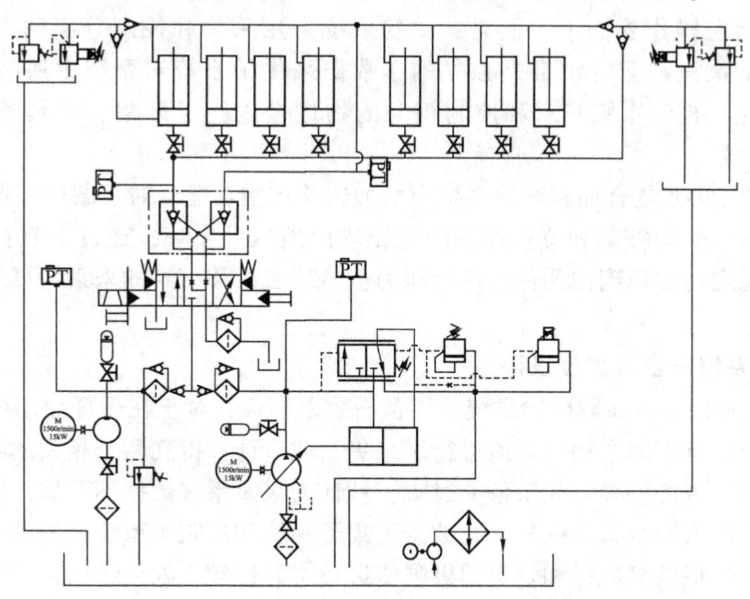

图 1　液压系统改造原理

2.2　控制系统设计

改造后的控制与采集系统原理图如图 2 所示。采用 PLC 与 NI 控制系统相结合的模式，NI 系统数据传输能力强，PLC 控制逻辑严密，有利于提高电控系统的可靠性，解决原有试

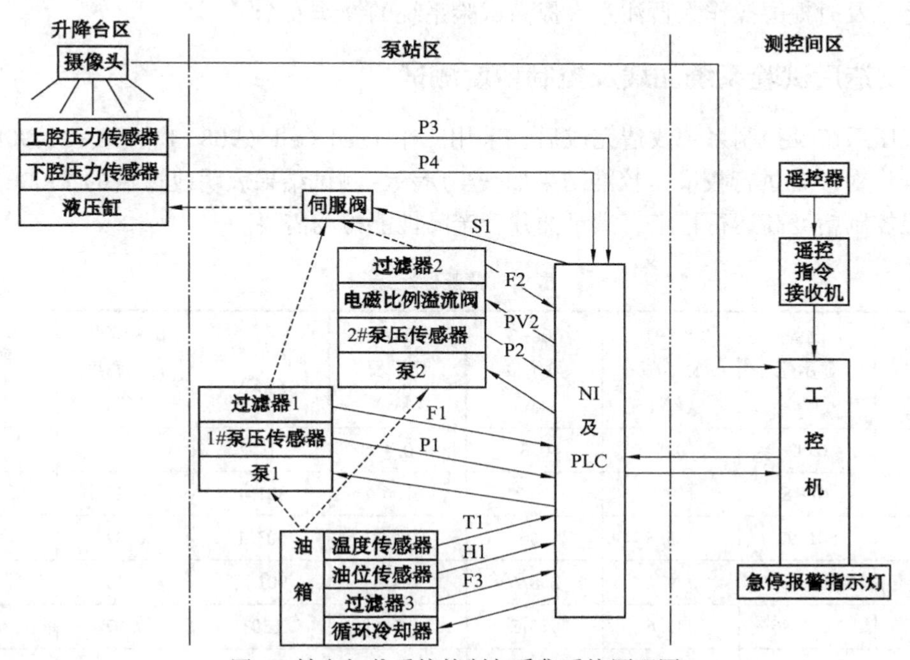

图 2　轴向加载系统控制与采集系统原理图

验系统由一台计算机进行控制及采集的弊端。

油压、油温、油位、过滤器污染程度等数据经由 NI 及 PLC 控制器上传到工控机，视频信号经由视频数据线直接传至工控机，遥控器指令通过接收机传至工控机再下达到各执行机构。工控机根据人工设定的参数能够自动执行加载流程。试验过程中，记录压力数据并实时显示在操作界面上。而且，试验系统采用采集和控制分离的方式，数据采集和控制系统在上位机和下位机都是独立的，数据采集子系统对系统中所有的物理量都进行了采集和存储，控制系统只对和控制相关的物理量进行了采集，并具备模拟和数字量输出功能。

在改造后的 1500t 复合加载试验系统中增加位移控制功能，对于载荷控制和位移控制采用 PID 控制方式。载荷控制和位移控制的过程均由液压泵实现，通过 PID 控制使得液压泵的输出的负载稳定地加载到被测件，通过压力信号或者位移信号的采集计算得到载荷大小或者位移数据。

2.3 试验系统安全可靠性设计

改造后的 1500t 复合加载试验系统采用备份数采模式，对于在闭环控制中需要用到的物理量，在同一个测点用两个物理通道进行了采集，或者计算得到两个相关的物理量，在控制过程中时刻判断二者的差异，如果偏差过大，就认为采集系统数据不可靠，系统故障，及时结束试验进程，进入安全状态恢复。在数据采集子系统和控制子系统之外设立一个独立运行的看门狗子系统，将检测数据和设定的极限值进行实时比较，如果超出，看门狗系统启动安全状态恢复进程，确保试验安全。

试验系统采用多点急停安全设计，除了控制子系统上位机的软件急停按钮，在监控中心和试验区域安装有 8 个硬件紧急停止按钮（ESD），只要有其中的一个被按下，都会停止试验，并启动安全状态恢复。同时可以在上位机软件中显示硬件急停的状态、部位，同时启动声光报警，及时提醒操作人员注意，提高试验系统的操作安全。

3 改造后试验系统加载及控制精度测试

在液压系统及控制系统改造完成后，采用三个 Load Cell（908-1、908-3、908-4 标准力传感器）对设备进行校准，校准结果如表 1 所示，测试结果表明改造后的 1500t 复合加载试验系统控制精度高运行稳定，满足油井管产品性能检测需要。

表 1 设备标定结果

程序设定值（kN）	设备显示值（kN）	908-1 显示值（kN）	908-3 显示值（kN）	908-4 显示值（kN）	示值和（kN）	试验系统控制精度（%）	系统误差（%）
0	0.92	0.5	0.3	0.4	1.2	/	/
300	300.8	119.9	75.2	106	301.1	0.27	0.10
600	601.92	225.8	166	211.3	603.1	0.32	0.20
900	901.39	329.4	258.6	313.4	901.4	0.15	0.00
1200	1200.86	433.6	353.8	415.6	1203	0.07	0.18
1500	1500.94	540	451.9	519.8	1511.7	0.06	0.72

续表

程序设定值（kN）	设备显示值（kN）	908-1显示值（kN）	908-3显示值（kN）	908-4显示值（kN）	示值和（kN）	试验系统控制精度（%）	系统误差（%）
1800	1800.21	639.3	545.6	617.5	1802.4	0.01	0.12
2100	2101.94	746.2	646.6	722.7	2115.5	0.09	0.65
2400	2399.97	843.9	740	819.1	2403	0.00	0.13
2700	2700.47	944.8	836.8	919.1	2700.7	0.02	0.01
2900	2900.32	1013.2	902.6	986.9	2902.7	0.01	0.08

4 结论

1500t复合加载试验系统在进行升级改造之后，解决原有试验系统存在的问题，改造后系统运行良好，试验系统安全性、可靠性及试验效率得到很大提高，同时进一步提升试验系统的试验能力。

（1）对原有1500t复合加载试验系统在试验过程中出现的问题进行分析，提出通过改造液压系统及控制系统，恢复试验系统出厂载荷设计能力，提升试验系统轴向载荷精度及稳定性，满足油田模拟工况需求，确保油井管质量评价数据准确可靠。

（2）对液压系统进行重新设计改造，提升系统的控制精度，恒定了系统的加载速率。进而提升系统的试验效率及硬件的使用寿命，减少后期维修成本。

（3）通过控制系统的设计改造，实现采集和控制分离的控制方式，提高系统的可靠性及运行速度，同时在系统中增加位移控制功能，便于试验系统开展非规试验项目，提升原有试验系统的试验能力。

（4）对试验系统进行安全可靠性设计，采用看门狗安全防护及多点急停的方式，提高试验系统及操作人员的安全。

（5）对改造后试验系统进行精度测试，试验结果表明，改造后试验系统精度及系统误差满足检测需求。

参 考 文 献

[1] 石油天然气工业　套管及油管螺纹连接试验程序：GB/T 21267［S］.2007.

[2] 王宏友.电气自动化控制设备的稳定性措施探究［J］.电子技术与软件工程.2014（1）：251.

[3] 邓帮飞，唐捷.电气自动化控制设备可靠性研究［J］.企业技术开发.2012（32）：102-103.

[4] 石建伟.浅析电气自动化控制设备的可靠性［J］.科技与企业.2013，08（263）：350.

[5] 液压系统通用技术条件：GB/T 3766［S］.1983.

[6] 液压系统元件技术条件：GB/T 7935［S］.2005.

大摆锤试验机能量测试结果影响因素研究

李　娜　陈宏远　张华佳　任继承　张庶鑫　蔺卫平　梁明华

（中国石油集团石油管工程技术研究院，石油管材及装备材料服役行为与结构安全国家重点实验室）

摘　要：为了提高大摆锤试验机测试结果的准确性，对大摆锤试验机能量测试中的影响因素进行了测试与分析。影响因素主要包括风阻及摩擦损失能量、不同初始能量及锤头速度、不同吸收能量获取方式以及试样的不同缺口形式等。研究结果表明，风阻及摩擦损失能量很小，操作时可忽略不计；不同初始能量及锤头速度对最终试验结果几乎无影响，试验时可根据试样厚度及钢级选择合适初始能量，减少提锤时间，提高工作效率；$E_{角度编码器}$ 与 $E_{表盘}$ 测量结果基本相同，$E_{角度编码器}$ 较 $E_{表盘}$ 更为准确，$E_{示波}$ 低于 $E_{角度编码器}$；压制缺口试样吸收能量显著高于人字形缺口试样。

关键词：大摆锤试验机；风阻及摩擦损失；初始能量；吸收能量；缺口形式

1　大摆锤试验机能量测试结果影响因素

自 1968 年落锤撕裂试验（DWTT）被 API 正式采纳以来，其试验结果被广泛用于对管线的断裂进行控制和预测，并用其作为衡量管线钢管抵抗脆性开裂能力的韧性指标之一。大摆锤试验机能量测试中的影响因素有很多，主要包括风阻及摩擦损失、不同初始能量及锤头速度的大小、不同吸收能量获取方式以及试样的不同缺口形式所得吸收能量的结果等。对这些影响因素进行测试与分析，对试验结果的准确性提供帮助。

2　影响因素的试验与分析

2.1　风阻及摩擦损失

在 20kJ、25kJ、30kJ、35kJ、40kJ 和 45kJ 不同初始能量下，对风阻及摩擦损失按照 ASTM E23 进行测量，每个初始能量下测试 3 次，试验结果见表 1。

表 1　风阻及摩擦损失测试结果

初始能量（kJ）	锤头速度（m/s）	风阻及摩擦损失（J）
20	5.28	40.65
25	5.90	48.84
30	6.47	55.16

作者简介：李娜（1981—），工程师，主要从事石油管材的质量检验工作。

续表

初始能量（kJ）	锤头速度（m/s）	风阻及摩擦损失（J）
35	6.99	63.40
40	7.47	71.91
45	8.01	80.19
50	8.35	88.51

由表 1 可看出，随着初始能量和锤头速度的增大，风阻及摩擦损失随之增加，但其差别相对于试样最终吸收能量微乎其微，可忽略不计。

2.2 初始能量及锤头速度

由表 1 可看出，初始能量与锤头速度是相对应的，初始能量越大，锤头速度就越快。落锤撕裂试验标准中规定锤头速度为 5~9m/s。在此范围内，对某实际应用的 X80 热轧钢板在室温下按照 SY/T 6476—2013，采用不同初始能量（20kJ、30kJ、40kJ 和 50kJ）进行吸收能量测试，每组包含 10 个试样，试验结果如图 1 所示。

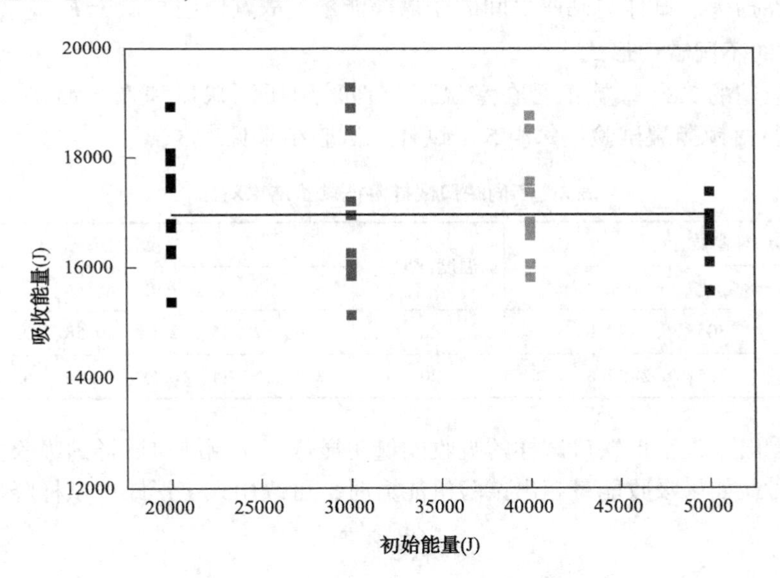

图 1 不同初始能量下的吸收能量

由图 1 可以得出，各初始能量下（即不同锤头速度）的吸收能量平均值线性回归方程为 $y=0.00001x+16952$，其斜率为 0.00001，说明不同初始能量及锤头速度对最终试验结果几乎无影响，试验时可根据试样厚度及钢级选择合适初始能量，减少提锤时间，提高工作效率。

2.3 不同吸收能量的获取方式

示波摆锤式落锤撕裂试验机共有 3 种吸收能量获取方式，即表盘度数 $E_{表盘}$、角度编码器计算所得能量 $E_{角度编码器}$ 及力—位移曲线积分所得能量 $E_{示波}$。对某实际应用的 X70 管线钢在不同温度下进行能量测试，同时读取 3 种不同方式所得能量，具体试验结果见表 2。

表 2　三种不同能量获取方式所得结果及对比　　（单位：kJ）

$E_{表盘}$	$E_{角度编码器}$	$E_{示波}$	$E_{角度编码器}-E_{表盘}$	$E_{示波}-E_{角度编码器}$
39.25	39.43	38.73	0.18	−0.70
28.25	28.60	28.34	0.35	−0.26
15.75	16.26	15.26	0.51	−1.00
13.20	13.64	12.87	0.44	−0.77
9.13	9.58	8.58	0.45	−1.00
7.50	8.04	8.04	0.54	0
2.38	2.77	2.21	0.39	−0.56
0.25	0.70	0.47	0.45	−0.23

由表 2 可看出，$E_{角度编码器}$与$E_{表盘}$均由势能差计算所得，其测量结果基本相同，角度编码器较表盘精度高，因此$E_{角度编码器}$较$E_{表盘}$更为准确。$E_{示波}$是由力—位移曲线积分所得，其中不包含风阻及摩擦损失、试样与砧座之间的摩擦等能量，故其值明显低于$E_{角度编码器}$。

2.4　试样的不同缺口形式

对某实际应用的 X80 螺旋钢管在室温下分别采用压制缺口和人字形缺口，按照 SY/T 6476—2013 进行落锤撕裂试验，每组 5 个试样，试验结果见表 3。

表 3　不同缺口试样落锤试验结果对比

试样参数		温度（℃）	吸收能量（kJ）	
缺口类型	长×宽×厚（mm×mm×mm）		单值	均值
压制缺口	305×76.2×22.0	20	21.06、28.15、22.64、20.58、19.09	22.30
人字形缺口	305×76.2×22.0	20	12.46、13.74、16.59、13.44、16.38	14.52

由表 3 可看出，人字形缺口试样的吸收能量明显低于压制缺口试样的吸收能量，当使用压制缺口试样时，如果吸收能量超出试验机能力时，可改用人字形缺口试样降低吸收能量后进行测试。

3　结论

（1）风阻及摩擦损失能量相对于试样最终吸收能量非常微小（≤1/500），操作时可忽略不计。

（2）不同初始能量及锤头速度对最终试验结果几乎无影响，试验时可根据试样厚度及钢级选择合适初始能量，减少提锤时间，提高工作效率。

（3）$E_{角度编码器}$与$E_{表盘}$均由势能差计算所得，其测量结果基本相同，角度编码器较表盘精度高，因此$E_{角度编码器}$较$E_{表盘}$更为准确。$E_{示波}$是由力—位移曲线积分计算所得，其中不包含风阻及摩擦损失能量、试样与砧座之间的摩擦能量等，故其值明显低于$E_{角度编码器}$。

（4）对比人字形缺口和压制缺口试样的吸收能量试验结果，压制缺口试样吸收能量显著高于人字形缺口试样。

参 考 文 献

[1] EIBER R J. Correlation of Full-scale Tests with Laboratory Test [C] //Proceeding of 2nd Symposium on Line Pipe Re-search. Texas, Dallas: Pipeline Research Committee of AGA L30000, 1965: 83-118.

[2] 陈宏达，霍春勇，冯耀荣，等．管线钢落锤撕裂试验方法的建立、应用及发展［J］．钢铁研究学报，2005，17（6）：1-5.

[3] Standard Test Methods for Notched Bar Impact Testing of Metallic Materials：ASTM E23—12c［S］.

[4] 国家能源局．管线钢管落锤撕裂试验方法：SY/T 6476—2013［S］．北京：石油工业出版社，2013.

[5] 郭鸿飞．对摆锤式冲击试验机检定过程的分析［J］．机械管理开发，2014（4）：65-67.

[6] 刘慕双，孙占刚．冲击试验机摆锤特征参数的优化及评价［J］．机械科学与技术，2006（11）：75-77，127.

[7] 洪刚，张庄，任立志．摆锤试验机冲击吸收功的计算机测量方法［J］．物理测试，2005（1）：33-35.

[8] 周海．RPSW/A 摆锤示波冲击试验机数据采集及处理［D］．昆明：昆明理工大学，2006.

[9] 类成华，杨雷岗，孙继松，等．仪器化冲击试验机进行 V 型缺口欧标冲击试验［J］．物理测试，2015（5）：12-15.

[10] 邱自学，袁江，姚兴田．可实现多参数测试的落锤式冲击试验机［J］．仪表技术与传感器，2006（8）：59-61.

水压疲劳试验应变采集系统的不确定度分析

王　蕊　李东风　杨　鹏　王亚龙　韩　军

（中国石油集团油管工程技术研究院）

摘　要：为了评估管道疲劳寿命及疲劳裂纹扩展规律，通常采用实物水压疲劳试验方法进行试验研究。本文介绍了油气输送管道水压疲劳试验方法，并对水压疲劳试验过程中，应变测量数据采集系统测量不确定度进行了详细分析。重点讨论了不确定度分析的方法，并根据水压疲劳试验得出应变测量结果的不确定度。本文的研究结果为油气输送管道水压疲劳试验结果的准确分析判断提供依据。

关键词：水压疲劳；应变采集；测量；不确定度

油气输送管道的安全输送，取决于系统的每一个环节，一些承压元件很难检测内部结构的缺陷，尤其是投产以后的管线，现场条件更难测量其内部缺陷，因此，最有效的方式是通过水压疲劳试验的方法确定油气管道的疲劳寿命。水压疲劳试验方法是对密闭的钢管试样内部循环加压及降压，模拟实际管线的输送压力波。试验过程中，测定钢管的环向变形情况、裂纹扩展、裂纹尖端张开位移等特征值，以此来评估钢管的疲劳寿命及裂纹扩展规律[1-3]。应变采集系统是水压疲劳测试中不可缺少的设备，随着数据采集系统软、硬件的发展，极大的促进了采集系统在静态及动态测量系统中的广泛应用，同时也使得数据采集的误差逐渐减小[4-6]。虽然数据采集系统测量结果的精度和准确性不断提高，然而试验过程中测量误差客观存在，任何测量结果都具有不确定度[7-10]，而应变采集系统是水压疲劳测试中必不可少的测量设备，对应变采集系统测量不确定度进行分析，有助于对钢管水压疲劳试验结果及管道在疲劳载荷情况下使用寿命进行更准确、合理的分析。

本文针对钢管的水压疲劳试验，运用不确定度分析方法，分析了环向应变测量数据采集系统测量结果的不确定度来源，对试验系统进行了测量不确定度评定，并给出了钢管的水压疲劳试验测量的不确定度。

1　不确定度分析与评定的方法

测量不确定度评定通常包括以下步骤：

（1）建立测量模型。

输入量 x_1、x_2、$\cdots x_N$ 与输出量 y 间的函数关系：$y=f$（x_1、x_2、$\cdots x_N$）。

作者简介：王蕊，女，工程师，1979年10月出生，工学博士学位，目前从事石油管全尺寸质量检测与评价工作。

（2）测量不确定度的来源分析。

测量模型中应包含所有应该考虑的影响量，而每一个影响量对测量结果来说都是一个测量不确定度分量，也是进行不确定度分析的来源所在。不确定度的来源包括测量方法、测量设备、环境条件以及操作人员等方面。

（3）标准不确定度分量 u（x）的评定。

有两种评定方法：

A 类评定：当测量次数足够多时，用贝塞尔公式评定标准不确定度。

$$s_n = \sqrt{\frac{\sum_{i=1}^{n}(x_i - \bar{x})^2}{n-1}}$$

B 类评定：用不同于观测列的统计分析来评定标准不确定度。

u（x）$=a/k$（a 为区间半宽，k 为包含因子）；

k 由概率分布决定，常用的均匀分布 $k=\sqrt{3}$。

（4）计算合成标准不确定度。

当测量结果是由若干个输入量 x_i 求得，且输入量间彼此独立或不相关时，按不确定度传播律即各输入量的方差和计算其合成标准不确定度。

（5）扩展不确定度 U_p 的评定。

扩展不确定度是确定了被测量估计值区间半宽度的量，$U_p=k_p\times u_c$（x），其中 k_p 为给定概率 p 的包含因子。

2　水压疲劳试验工作原理

水压疲劳试验系统由水压增压系统，载荷控制系统，数据采集系统以及试验管组成，系统示意图如图 1 所示。水压增压系统，主要采用带溢流阀的高低压水泵，通过控制系统控制电液伺服阀及压力传感器，实现对试验载荷进行闭环控制，从而对试验样品进行增压、保压、降压等压力载荷控制及压力数据采集。应用该系统可完成静水压试验、常幅疲劳以及随机疲劳谱等实物水压疲劳试验。该系统测量部分的主要设备是多通道应变仪，实时采集、记录试验过程中的应变变化。试验过程中通过压力、动态应变曲线等实时曲线监测试验过程。

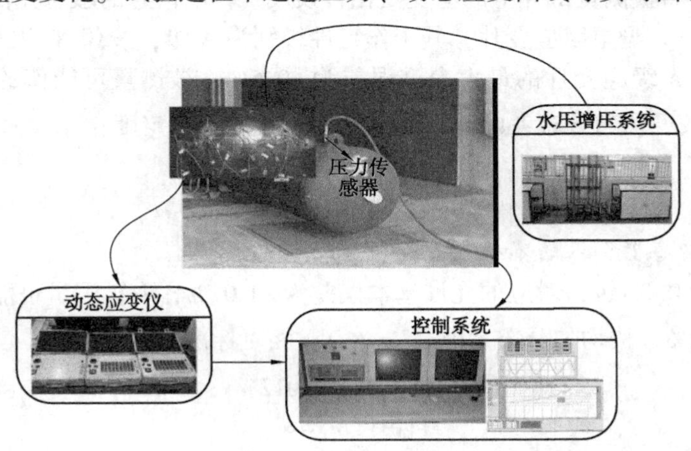

图 1　水压疲劳试验系统组成示意图

3　水压疲劳试验应变采集系统的不确定度评定

3.1　测量模型

水压疲劳试验系统测量方法为直读法，其测量模型为：

$$y=x$$

式中　y——应变测量结果；

x——动态应变仪的示值。

3.2　不确定度来源

影响水压疲劳试验过程中动态应变的因素包括动态应变仪、应变片、压力传感器、接触电阻、连接线电阻及工作环境、系统的随机因素等。水压疲劳试验系统软硬件及应变片本身所固有的系统因素会引入不确定度，另外，系统的随机因素（人为操作等）也会引入的不确定度。以粘贴应变片为例，由人为操作所导致的随机影响因素包括：应变片的粘贴方位、应变片的粘贴角度、粘贴质量、有无气泡、试样的表面质量是否达到测量要求等，这些都会引入不确定度。本文根据各影响因素的特性，对动态应变仪、应变片、压力传感器、接触电阻、连接线电阻及工作环境因素等因素采用 B 类评定，对系统的随机因素引入的不确定度（人为操作等）采用 A 类评定，进行水压疲劳试验不确定度分析。

3.3　标准不确定度分量的评定

3.3.1　动态应变仪的量化不确定度

DH3817 动静态应变测试系统的最大量程为 ± 30000με，对应的电压测量范围为（-30~30mV），A/D 分辨力为 14 位，该动静态应变测试系统量化误差：

$$\mathrm{LSB}=(60/2^{14})\%=0.0037\%$$

因此分辨力造成的误差在（-0.5LSB，0.5LSB）范围内服从均匀分布（置信因子为 $k=\sqrt{3}$），最大误差区间半宽度 $a_1=0.5\mathrm{LSB}=0.0018\%$，则量化不确定度：

$$u_1=a_1/\sqrt{3}=0.001\%$$

$$\text{自由度 } v_1=\infty$$

3.3.2　应变片引入的不确定度

应变片是一种将被测量试样上的应变变化转换成一种电信号的敏感元件，试验中使用的是 BE120-3BB（11）型电阻应变片，其工作特性级别为 A 级，按照 BT 13992—1992《电阻应变计》的标准，A 级应变片的最大允许误差为±0.5%，被测量可能值的区间半宽的相对值为 $a_2=0.5\%$，按照均匀分布考虑（$k=\sqrt{3}$），相对标准不确定度：

$$u_2=a_2/\sqrt{3}=0.29\%$$

$$\text{自由度 } v_2=\infty$$

3.3.3　压力传感器引入的不确定度

压力传感器为 0.02 级，对应的允许基本误差为±0.02%，取被测量可能值的区间半宽的相对值为 $a_3=0.02\%$，按均匀分布考虑（$k=\sqrt{3}$），相对标准不确定度：

$$u_3=a_3/\sqrt{3}=0.012\%$$

$$\text{自由度 } v_3=\infty$$

3.3.4　接触电阻引入的不确定度

接触电阻 ΔR 一般在 $10^{-4}\,\Omega$ 量级，根据公式 $\Delta R/R=K\varepsilon$，可计算出相应的应变量大小，

$R=120\Omega$、$K=2.14$、$\Delta R=10^{-4}\Omega$ 时，其应变为 0.3894με，相对于满量程 30000με 的误差为±0.0013%，按照均匀分布考虑（$k=\sqrt{3}$），取被测量可能值的区间半宽的相对值为 $a_4=0.0013\%$，因此接触电阻引入的不确定度：

$$u_4=a_4/\sqrt{3}=0.00075\%$$

$$\text{自由度 } v_4=\infty$$

3.3.5 连接线电阻及环境温度引入的不确定度

应变采集系统就是通过应变片阻值的变化实现应变测量，该系统测量桥的桥臂为 120Ω 的精密电阻，而应变采集系统一般距离被测试样比较远，在本试验系统中单根导线长度 50m，经过实际测量输入导线的总电阻约 6Ω，导致作用臂的电阻在零点时不为 120Ω，这就由连接线电阻引入一个不确定度。在该应变采集系统中，通过长导线电阻修正将连接线电阻减去，因此，由连接线电阻引入的不确定度可以忽略。

在对被测试样进行应变采集过程中，试验现场的工作环境温、湿度对测量结果引入一个不确定度，在该应变采集系统中通过温度补偿模块实现温湿度补偿，因此由环境温度引入的不确定度可以忽略。

3.3.6 由随机影响引入的不确定度

采用 A 类评定方法，确定应变采集系统由随机影响因素引入的不确定度。选取外径 1219mm，壁厚 26.4mm 的 X80 直缝焊管实物水压疲劳试验一组实测动态应变（20 个测量数据）。试验采用恒幅加载，加载应力比为 0.1，最大载荷选取 26.1MPa，最小载荷 2.61MPa，实测数据如表 1 所示。

表 1 动态应变实测数据 单位：με

1	2	3	4	5	6	7	8	9	10
570.62	570.58	574.12	576.68	579.37	577.67	578.23	580.75	588.42	578.37
11	12	13	14	15	16	17	18	19	20
590.93	580.43	587.42	581.21	584.75	584.76	588.30	590.84	590.39	581.23

算数平均值为：

$$\bar{x}=\frac{\sum_{i=1}^{n}x_i}{n}=581.75$$

根据贝塞尔公式：

$$s=\sqrt{\frac{\sum_{i=1}^{n}(x_i-\bar{x})^2}{n-1}}=0.94$$

因此，得到标准不确定度 $u_5=0.94\%$。

自由度 $v_5=18$。

3.4 合成标准不确定度

由于上述不确定度分量各不相关，因此，合成标准不确定度：

$$u_c=\sqrt{u_1^2+u_2^2+u_3^2+u_4^2+u_5^2}=0.96\%$$

根据韦—萨公式，其有效自由度为：

$$\nu_{\mathrm{eff}}=\frac{u_{\mathrm{c}}^{4}}{\sum_{i=1}^{n}\left(\frac{u_{i}^{4}}{\nu_{i}}\right)}=19.5\approx 20$$

3.5 扩展不确定度

扩展不确定度公式为 $U_{\mathrm{p}}=k_{\mathrm{p}}u_{\mathrm{c}}$，当 x 接近正态分布时，k_{p} 可用 t 分布临界值表查到，$k_{\mathrm{p}}=t_{\mathrm{p}}$（$\nu_{\mathrm{eff}}$），取包含概率 $p=95\%$，自由度 $\nu_{\mathrm{eff}}=20$，查 t 表得：

$$k_{\mathrm{p}}=t_{95}(20)=2.09$$

所以 $U_{\mathrm{p}}=k_{\mathrm{p}}u_{\mathrm{c}}=2.09\times0.96=2.0\mu e$

经过上述评定，在开展水压疲劳试验应用应变采集系统进行应变测试时，包含概率为95%的扩展不确定度是 $U_{\mathrm{p}}=2.0\mu e$，$\nu_{\mathrm{eff}}=20$。

4 试验结果分析

结合上述不确定度分析，可获得试验结果如表 2 所示。结果表明，由随机影响因素引入的不确定度、应变片引入的不确定度是影响测量结果准确性的重要因素。因此，在试验过程中应严格按照标准要求及操作规程完成试验准备工作，避免由人为的随机因素对测量结果引入的不确定度。同时对使用的应变片必须确保应变片的质量，注意其敏感栅有无锈斑，缺陷，是否排列整齐，基底和覆盖层有无损坏，引线是否完好等方面，同一次测量的应变计，灵敏系数必须相同，阻值相差不得超过±0.5，减少由应变片引入的不确定度，确保试验结果的准确性。

表 2 应变采集系统标准不确定度汇总

标准不确定度分量	不确定度来源	评定方法	标准不确定度
u_1	动态应变仪的量化不确定度	B 类	0.001%
u_2	应变片引入的不确定度	B 类	0.29%
u_3	压力传感器引入的不确定度	B 类	0.012%
u_4	接触电阻引入的不确定度	B 类	0.00075%
u_5	由随机影响因素引入的不确定度	A 类	0.96%

5 结论

本文通过对水压疲劳试验应变采集系统中的不确定度分析，阐述了试验不确定度来源，并对其不确定度进行了评定。结果表明，人为因素和应变片的质量直接影响试验结果的准确性。通过提高人员粘贴应变片水平并控制应变片质量，可以有效减少试验结果误差，为准确评估油气输送管道的安全性提供依据。

参考文献

[1] 刁顺，冯耀荣，庄传晶，等. 几种油气输送管材料的疲劳特性与管道寿命预测［J］. 中国安全科学学报，2008，18（1）：123-130.

[2] 潘家华. 油气管道的风险分析续（一）［J］. 油气储运，1995，14（4）：1-7.

[3] 管线钢管水压疲劳试验方法：Q/SY TGRC 61［S］.
[4] 陈瑾. 数据采集硬件如何避免缺陷与误差［J］. 电子技术. 2002,（9）：63-64.
[5] 梁志国. 数据采集系统误差限的测量不确定度［J］. 计量技术，2002,（9）：45-48.
[6] 齐征. 测量不确定度与测量误差之比较及应用优势［J］. 研究与探讨. 2010,（10）：92-93.
[7] 丁富连，王承忠，陈卓人，等. 拉伸试验中测量的不确定度评定［J］. 宝钢技术. 2003,（3）：47-54.
[8] 邓俊，邓莉. 液压导管振动疲劳试验不确定度分析［J］. 重庆理工大学学报（自然科学）. 2013，27（9）：76-79.
[9] 丁颖，李浚圣. 数据采集系统中的不确定度分析［J］. 沈阳大学学报，2006，18（2）：22-25.
[10] 全国法制计量管理计量技术委员会. 测量不确定度评定与表示：JJF 1059. 1—2012［S］. 北京：中国标准出版社，2013.

非牛顿流体石油管流动研究进展

李孝军　冯耀荣　刘永刚　林　凯　刘文红

（中国石油集团石油管工程技术研究院，石油管材及装备材料服役行为与结构安全国家重点实验室）

摘　要：从研究非牛顿流体流动数学模型及其求解方法、原油圆管分层流与非定常流、环空管流流动规律、非牛顿流体流动流态判别及非牛顿流体流动实验等5个方面，综述了21世纪始10余年国内外非牛顿流体石油管流动现状。非牛顿流体流动本构方程的建立和流变参数的确定，是研究非牛顿流体流动和流变特性的基础，对发展非牛顿流体力学理论和解决生产技术问题都至关重要。总结前人研究成果的基础上，根据非牛顿流体石油管流动的发展状况，提出将高分子材料学、电磁场理论、弹塑性力学理论、混沌学、现代计算机有限元数值模拟等学科与高等流体力学相结合的方式，寻求非牛顿流体管流流动中分层流混输技术、非定常流动稳定性与流动规律、环空紊流流动规律和流态判据的相关性与合理性等亟待解决问题的思路，以实现节能降耗、增效增产。

关键词：非牛顿流体；数学模型；管流；环空流；流态；仿真试验

石油工业中的钻井液、水泥浆、含蜡原油和聚合物水溶剂等都是非牛顿流体，其各种流动现象的理论和实验研究一直受到人们的重视，但目前还未找到适用于所有非牛顿流体的本构方程[1-4]。现在的研究方法是把非牛顿流体分成若干类，如非时变性流体、黏弹性流体、触变性流体和震凝性流体等，通过理论和实验研究找到各自的本构方程，分别研究它们的流动现象，得出各种流动问题的解[5-8]。目前，非牛顿流体石油管流动研究处于世界领先地位的挪威，也只研究了在某些特殊情况下一些简单流动现象，如管内流动的解析解[9-12]。然而，这些初步的进展为这些流体的其他流动现象研究提供了方法和理论基础[13-16]。非牛顿流体在水平井和稠油开采井中的环空流动，从层流到过渡流再到紊流的变化规律，受流体流变性、井眼与管柱内外径比和管柱偏心度的影响。采出到地面管线的较大流性指数非牛顿流体与低黏牛顿流体的混输，可实现节能降耗，提高管输效率。非牛顿非定常流处于混沌状态，可着眼于海洋管非牛顿水合物段塞流的研究。而且，随着现代计算机技术的飞速发展，精确的数值模拟对非牛顿流体流动研究也起着至关重要的作用，为21世纪非牛顿流体石油管流动机理研究带来了新的契机。

基金项目：中国石油天然气集团公司应用基础研究项目“复杂工况气井油套管柱失效控制与完整性技术研究”（编号：2014A—4214）。

作者简介：李孝军（1982—），男，工程师，工学博士（后），2012年博士毕业于西南石油大学油气井工程专业，现在中国石油集团石油管工程技术研究院主要从事油井管与管柱工程技术方面的研究。地址：陕西西安锦业二路89号石油管工程技术研究院综合楼401室；电话：029-81887840；E-mail：lixiaojun003@cnpc.com.cn。

1 非牛顿流体数学模型及其求解方法

长期以来，非牛顿流体数学模型的研究和求解是高校和科研院所的研究领域之一，能量耗散模型也是其研究热点。2002 年，张劲军和严大凡[1]根据剪切率和能量耗散率定义，建立幂律流体湍流剪切率与能量耗散率的关系式，提出了一种幂律流体管内湍流能量耗散率的近似算法。2004 年，殷谷良和董柏青[2]讨论一类带 p 幂增长耗散位势的非牛顿模型解的渐近性态，利用改进的 Fourier 分解方法证明了其解在 L^2 范数下衰减率为 $(1+t)^{-n/4}$。2010 年，张其亮和张兴伟等[3]利用 Stokes 算子的谱分解法和 L^p-L^q 估计研究一类三维不可压缩非牛顿流体弱解的最优代数衰减速率，证明了初速度在 L^2 范数下其弱解的衰减率为 $t^{-5/4}$。由此，得到了非牛顿流体在剪切作用下流速随时间衰减的最优代数衰减率。

在非牛顿流体流动方程解的存在性证明基础上[4-7]，清华大学朱克勤和东北石油大学崔海清各自带领的科研团队在非牛顿流体石油管流动建模求解方面做了大量的深入研究。朱克勤等[8-10]提出了一种用弹簧和油壶连接的分形网络结构来比拟分数元模型的应力-应变特性，利用 Heaviside 运算微积，证明了该结构对应的黏弹性流体为 1/2 阶导数的分数元，还导出分数元模型的圆管起动流的解析解；研究了同心环空宾汉流体的 C-P 流，得到了不同流动类型的解析解，发现了 2 种新的流型，依据宾汉数、轴向库艾特数和半径比 3 个无量纲量，从数理视角划分出了 8 种不同的速度分布形式；推导了圆管内幂律流体分层 Poiseuille 流动的控制方程和基本流动解析解，进而得到该流动的小扰动线性控制方程，采用切比雪夫配点法引入虚拟网格求得了扰动方程的数值解，解决了柱坐标系的奇点影响剪切稀化流体分层 Poiseuille 流动数值求解的问题。东北石油大学崔海清带领的科研团队[11,12]首次利用待定系数法变换幂律流体在管内做行星运动的环空流方程，建立了给定流量情况下幂律流体在内管做行星运动的环空流压力梯度和内管壁所受流体作用力的数学模型，导出了内管壁上的法向应力差、切向应力和扭矩的计算公式，并利用有限差分法对其数值计算。

21 世纪初国外专家克服边界服役条件复杂和非牛顿流体非连续性的难点，寻求数值算法解决其数模难题。2003 年，Filip P. 和 David J.[13]分析了同心环空中符合 R-S 模型的非牛顿流体轴向层流，用几何、动力和流变参数获得的半解析准则多元化表示段塞流区相关位置的所有可能情况。2012 年，Wilson H. J.[14]提出非牛顿流体的非连续性和不稳定性特征虽已得到很好的实验验证，至今却难以完全用数值模拟和解析法来描述。近年来，挪威专家 Gjerstad K. 带领科研小组[15,16]建立了 H-B 非牛顿流体圆管层流近似模型，其摩擦压力梯度满足显性连续可微，可用于控制、实时优化和预测钻井操作，与其他隐性算法相比较精度更高。

10 余年来，从非牛顿流体油管流动数学物理模型的建立和初边值条件的确定，到解的存在性和唯一性证明及模型求解，都有深入研究[1-16]。非牛顿流体的非连续性和不稳定性特征，亟须结合实际工况实验结果开展数模仿真来进一步揭示，得到半经验半解析式来进一步预测这些特征。笔者曾从非牛顿油管流动的物理现象出发，采用不需判断原油流变性质的赫巴流模式本构方程，确定其屈服值、稠度系数和流变指数。结合细管式流变仪测得的原油流量与压降数据，回归得到该原油的赫巴模式半解析式，实现原油的精细、经济输送[17]。采用分层雷诺数建立非牛顿流体赫巴流在管流和环空流的流态稳定性计算模型，找到非牛顿流油管流动过渡流区的最易失稳位置，得到了可供现场钻井液合理使用的流态判据[18]。

2 非牛顿流体油井管流动

油气勘探开发过程中，涉及了大量的非牛顿流体环空管流，诸如钻井液携岩携水循环机理、抽油杆泵和螺杆钻具的往复运动及工具润滑剂选用等。流道空间的几何非对称性，导致了非牛顿环空管流流态及流动规律的复杂性。

2.1 环空分层流

非牛顿流体偏心环空层流中，极值流速所在坐标位置偏向内管，而且随环空间隙的增大而越偏向内管，宽间隙处的极值流速大于窄间隙处的极值流速。影响宽边极值流速与窄边极值流速比值的主要因素是偏心度 ε 和流性指数 n，偏心度 ε 越大、流体非牛顿性越强，宽窄边的极值流速相差也越大。同种流体在不同环行空间中流动时，内外径之比越小，速度分布偏离几何中心越远。一般情况下，内外径之比大于 0.8 时，可认为流动对称于环行空间的几何中心柱面而不会引起太大的误差。

2.2 环空过渡流

在偏心环空轴向流动中，环空断面上不同位置的流质发生失稳的临界水力条件，反映了该点的流动强度与紊动条件。流质在环空中处于局部紊流时，断面上分布有层流、过渡流及紊流三种状态。国外学者主要研究了水平井钻柱偏心环空旋进时非牛顿流体钻井液的局部紊流特征。2006 年，Kim Y. J. 等[19]模拟了钻柱旋转的直径比为 0.52 的同心环空非牛顿流体过渡流特征，发现过渡流可以揭示雷诺数与罗斯比数同表面摩擦系数的关系，进而理解过渡流不稳定机理。国内学者主要研究了稠油开采时非牛顿流体环空油流的局部紊流特征。2007 年，李阳等[20]对稠油井抽油杆环空流模型的研究结果表明，环空偏心度是引起流场分布不均匀的主要参数，调整环空几何结构或流体流变性均可获得设计的局部紊流流场。

2.3 环空紊流

工程中普遍存在的环空紊流现象国内研究较少，较深入的工作也仅为紊流压降的量纲和实验分析[21]，无法揭示流场分布的规律和特性。在全断面充分紊流条件下，剪切应力主要由雷诺应力 τ'构成，利用普朗特动量交换理论，环空不同切向处的紊动强度可由边界层厚度及摩阻流速给予描述[22]。以非牛顿流体流变学理论和人工举升理论为基础，按照不同环空流动规律，李阳等[20]给出稠油井从地层到井口垂直井筒流动的运动方程和边界条件并求解，得到稠油井幂律流流动视黏度模型，为稠油井非牛顿流体流动优化设计提供理论参考。2010 年，贾泽琪和韩洪升[23]将数值模拟和 PIV 技术结合用于非牛顿流体在偏心环空非定常流动研究，为掌握聚驱井产出液在抽油杆轴向往复运动的井筒中的流动规律提供有价值的理论参考。

国外研究环空紊流现象主要涉及水平井钻柱偏心环空旋进时非牛顿流体钻井液的携岩携液能力和螺旋流特征。2004 年，Savins J. G. 和 Wallick G. C. [24]提出定量预测剪切黏滞性流体在螺旋流场中轴向泵排、压力梯度、角速度和扭矩耦合作用下的流动规律，其中最有趣的结果是耦合作用下的轴向流动阻力小于螺旋流的，即在指定轴向压力梯度下，螺旋流的轴向泵排比纯环空流场的要高。2010 年，Ozbayoglu E. M. 和 Sorgun M. [25]对钻柱旋转的偏心水平环空中非牛顿流体摩擦压力损耗修正因子进行精确研究，通过大量不同非牛顿钻井液在不同钻柱转速下的试验数据，发现钻柱转速对环空压耗的影响是独立的。

3 非牛顿流体输送管流动

3.1 圆管分层流

圆管分层流是圆管两相流的一种基本流型，是下倾管中出现的主要流型，也是其他流型产生的基础。2000年，郑永刚等[26]提出圆管分层流新模型，利用该模型研究了非牛顿流体在圆管中层流-紊流分层流动，得出其速度场分布的解析式，进而研究了分层流的阻力规律，为分层掺气减阻提供了理论依据。在输送流性指数 n 较大的高黏幂律流体时，适当添加天然气或水等黏度较小的牛顿流体，形成幂律—牛顿流体圆管稳定的分层层流，比单一输送高黏幂律流体高效节能。添加适当的低黏牛顿流体可增加高黏幂律流流量最高达40%，而且添加的牛顿流体黏度越小，则比单一输送高黏幂律流体的节能效率越高。2003年，贺成才[27]通过精确地数值计算确定一个最佳的低黏牛顿流体的添加量，比单一输送流性指数 n 较大的高黏幂律流体增加流量达15.2%~38.25%。然而，在输送流性指数 n 较小的高黏幂律流体时，无论添加什么样的低黏流体都不可能实现高黏幂律流体的流量增加。

国外学者在非牛顿流体圆管分层流方面的理论研究远超国内水平，21世纪以来挪威学者在此领域研究尤为突出。2007年，Holmås K. 和 Biberg D.[28,29]发现通过水力相似转换得到的 Biberg 分层流管流等效直径与良好的压力梯度的 CFD 仿真结果较好吻合，表明明渠流对圆管流的转换是有效的建模工具，为大型管道系统的油气分层流模拟精度提供了潜力。2013年，Lawrence C. 等[30]建立了凝析气和轻质油的油气水多相流新力学模型，修正了因管径和流体性质的比例放大行为，降低了与流体性质和管线尺寸相关的显著影响试验数据合理性的不确定性。凝析气重点研究了三维预集成三相分层流模型，改进湍流闭包和液滴与气泡传输的分散模型，作为预测凝析气管线中液体积聚发生的基础。

如何实现高含蜡高凝高黏原油圆管分层流的高效流动，已成为减小输送能耗提高效益的关键，研究人员大都通过使用添加剂或改变输送条件以破坏原油自身分子结构改变其物理属性，来达到对原油除蜡降凝降黏，从而影响圆管分层流流型流态的转变。挪威学者对圆管三维多相分层流相间交互作用机理的研究，或为进一步提高非牛顿流体管输效率搭建平台。

3.2 非定常流

20世纪国内学者对非牛顿流体圆管非定常流动的研究[31,32]可知，非牛顿黏弹性流体在管内的不定常流动，弹性效应对其流动稳定性的影响，取决于流动类型和模型的选择。从扰动中获取弹性能时，流动更稳定，反之更不稳定。研究时有必要区分两种不同的流动：在材料和空间意义上都是定常的流动；只在空间意义上定常，而在材料意义上非定常的流动。前者可只限于注意现在时刻扰动是在增长还是在衰减，后者需考察材料元的变形历史、提取或释放弹性能的能力。这种能力取决于它的储能能力和已储能的大小。

近年，国内外在此方面的研究成果很少。2008年，王廖沙等[33]给定突扩管前后管直径比观察分歧流动现象，研究了恒剪切速率下不可压缩非牛顿流体在突扩管中的流动行为。2010年，崔海清科研团队[11,12,34]建立了变系数二阶流体在内管做轴向往复运动的偏心环空中非定常流的瞬时压力梯度方程和时均压力梯度公式，并给出相应的数值计算方法。2011年，Gainville M. 和 Sinquin A.[35]在第7届天然气国际会议上公布了输送管层流与紊流的水合物段塞流特征。

非牛顿非定常流动规律常成混沌状态，很多流动参数理论解的存在性和唯一性尚待证明，需要借鉴国外有限成果进一步对非牛顿流体微观结构和流变特性的相互关系与数学模型的建立上深下功夫，海洋管非牛顿水合物段塞流便是其中一个方向。

4 非牛顿流体管流流态判别

自雷诺 1883 年提出临界雷诺数判别层流与紊流临界状态以来，据层流稳定性现象提出的判别流态转变的准则主要有：雷诺提出的雷诺数 Re、瑞安和约翰逊提出的稳定性参数 Z、汉克斯提出稳定性参数 Ha、Mishra 和 Tripthi 提出的稳定性参数 X、岳湘安以涡流模型为基础提出的稳定性参数 Y 等。牛顿流体的临界雷诺数有明确的临界值，而非牛顿流体的临界雷诺数均为其流变参数的函数。事实上，只有牛顿流体与幂律流体圆管流的临界雷诺数值有实验依据，而其环空管流的临界雷诺数则属于外推。

目前，非牛顿流体管流流态稳定性判别理论按其基本模型可分为局部稳定性理论和整体稳定性理论。局部稳定性理论可从微观上精确分析非牛顿流体管流中紊动初始点，整体稳定性理论则从宏观上表征流体稳定性。从整体稳定性理论出发，在前人成果基础上[28,36-39]，归纳出幂律流体和宾汉流体分别在圆管和环空中的流态判别 M—T 稳定性参数 X 表达式（1）～（4），涡流模型稳定性参数 Y 表达式（5）～（8），以及幂律流体流态判别平均视黏度雷诺数 Re_u 表达式（9）～（10）。M-T 稳定性参数 X 一般用来分析判断实验时流体的流态，涡流模型稳定性参数 Y 考虑了涡旋流体宏观性质上的随机性与脉动，平均视黏度雷诺数 Re_u 则考虑了断面流速变化对宏观流态的影响。关于局部稳定性参数 Z 的表达式可参见文献[18]详述。针对某种流动条件下的圆管非牛顿流体流动，可以先通过试验测定其相应的稳定性参数临界值，然后依据上述幂律或宾汉流体的流态判别式，外推判别不同流变性、过流几何尺寸和初边值下相应非牛顿流体的流态。

4.1 M-T 稳定性参数 X

4.1.1 幂律流体稳定性参数 X

圆管中

$$X_p=\frac{3}{16}\frac{(3n+1)^2}{(2n+1)(5n+3)}\frac{8^{1-n}\rho u^{2-n}D^n}{K\left(\frac{3n+1}{4n}\right)^n} \tag{1}$$

式中 n——流性指数；

K——稠度系数；

ρ——液体密度；

u——速度分布；

D——圆管内径；

p——下标 p 为圆管。

环空中

$$X_a=\frac{35}{72}\frac{(2n+1)^2}{(3n+2)(4n+3)}\frac{12^{1-n}\rho u^{2-n}(D_a-d_a)^n}{K\left(\frac{2n+1}{3n}\right)^n} \tag{2}$$

式中 D_a——环空管外径；

d_a——环空管内径；

a——下标 a 为同心环空。

4.1.2 宾汉流体稳定性参数 X

圆管中

$$X_p=\frac{\rho uD}{\eta}\frac{8y_p}{\left(1-\frac{3}{4}\alpha+\frac{1}{4}\alpha^4\right)^2} \tag{3}$$

式中　η——塑性流体的塑性黏度；

α——核隙比；

y——与流速垂直方向的坐标；

$$y_{\mathrm{p}}=\frac{1}{16}(1-\alpha)^{6}\alpha^{2}+\frac{1}{64}-\frac{19}{280}\alpha+\frac{3}{80}\alpha^{2}+\frac{13}{40}\alpha^{3}-\frac{29}{32}\alpha^{4}+\frac{9}{8}\alpha^{5}-\frac{61}{80}\alpha^{6}+\frac{11}{40}\alpha^{7}-\frac{93}{2240}\alpha^{8}$$

环空中
$$X_{\mathrm{a}}=\frac{\rho u(D_{\mathrm{a}}-d_{\mathrm{a}})}{\eta}\frac{35y_{\mathrm{p}}}{16\left(1-\frac{3}{2}\alpha+\frac{1}{2}\alpha^{3}\right)^{2}} \tag{4}$$

式中
$$y_{\mathrm{a}}=\frac{1}{8}(1-\alpha)^{6}\alpha+\frac{2}{35}-\frac{2}{5}\alpha+\frac{6}{5}\alpha^{2}-2\alpha^{3}-2\alpha^{4}+\frac{6}{5}\alpha^{5}+\frac{2}{5}\alpha^{6}-\frac{2}{35}\alpha^{7}$$

4.2 涡流模型稳定性参数 Y

4.2.1 幂律流体稳定性参数 Y

圆管中
$$Y_{\mathrm{p}}=\frac{(3n+1)^{3}}{2(2n+1)(3n+2)(4n+3)}\frac{8^{1-n}\rho u_{\mathrm{m}}^{2-n}D^{n}}{K\left(\frac{3n+1}{4n}\right)^{n}} \tag{5}$$

式中　u_{m}——截面平均流速。

环空中
$$Y_{\mathrm{a}}=\frac{256}{2835}\left(\frac{2n+1}{n+1}\right)^{3}\frac{12^{1-n}\rho u^{2-n}(D_{\mathrm{a}}-d_{\mathrm{a}})^{n}}{K\left(\frac{2n+1}{3n}\right)^{n}} \tag{6}$$

4.2.2 宾汉流体稳定性参数 Y

圆管中
$$Y_{\mathrm{P}}=\frac{\rho uD}{\eta}\frac{16y_{\mathrm{p}}}{\left(1-\frac{3}{4}\alpha+\frac{1}{4}\alpha^{4}\right)^{2}} \tag{7}$$

式中
$$y_{\mathrm{p}}=\frac{2}{105}-\frac{11}{120}\alpha+\frac{3}{20}\alpha^{2}-\frac{1}{24}\alpha^{3}-\frac{1}{6}\alpha^{4}+\frac{9}{40}\alpha^{5}-\frac{7}{60}\alpha^{6}+\frac{19}{840}\alpha^{7}$$

环空中
$$Y_{\mathrm{a}}=\frac{\rho u(D_{\mathrm{a}}-d_{\mathrm{a}})}{\eta}\frac{79y_{\mathrm{a}}}{105\left(1-\frac{3}{2}\alpha+\frac{1}{2}\alpha^{3}\right)^{2}} \tag{8}$$

式中
$$y_{\mathrm{a}}=\frac{1}{24}-\frac{1}{4}\alpha+\frac{5}{8}\alpha^{2}-\frac{5}{6}\alpha^{3}+\frac{5}{8}\alpha^{4}-\frac{1}{4}\alpha^{5}+\frac{1}{24}\alpha^{6}$$

4.3 幂律流体平均视黏度雷诺数 Re_{u}

圆管中
$$Re_{\mathrm{up}}=\frac{8^{1-n}\rho u^{2-n}D^{n}}{2K\left(\frac{1+n}{1+3n}\right)\left(\frac{3n+1}{4n}\right)^{n}} \tag{9}$$

环空中
$$Re_{\mathrm{ua}}=\frac{12^{1-n}\rho u^{2-n}(D_{\mathrm{a}}-d_{\mathrm{a}})^{n}}{1.5K\left(\frac{1+n}{1+2n}\right)\left(\frac{2n+1}{3n}\right)^{n}} \tag{10}$$

从圆管和环空幂律流体稳定性参数式可知，幂律流稳定性与流变指数、稠度系数、密度、流速和过流断面当量直径有关。同种非牛顿流体流动，在相同流速下其稳定性是一致的；但流速不同时，会引起流变指数和稠度系数的差异，导致其稳定性也有显著差异，这也是非牛顿流体难以统一进行流态区域划分的原因。在相同条件下，容易证明判别非牛顿流体流态的岳湘安涡流稳定参数 Y 比 M—T 稳定性参数要大，说明考虑涡流扰动时非牛顿流体更容易成为混沌紊流。宾汉流体受启动压力环空宽窄间隙比的影响，会在环空流宽窄间隙处的稳定性差异会较大。从整体稳定性和工程应用来说，也用幂律流体视黏度雷诺数对非牛顿流体流态进行粗略判别。不过，由于非牛顿流体流动的流变性差异，在研究流态变化机理时还需结合上述 3 种流态判别法，在试验和实践比对基础上具体情况具体分析。

5 非牛顿流体流动实验

非牛顿流体流变测量，一般是在一定条件下，对流体施加切应力，跟踪其受力响应而得，具有非单项性、非单值性和非可逆性的特点。2006 年，张金亮等[40]研究了辽河油田超稠油的流变特性，通过试验确定该超稠油的流体类型，找出超稠油牛顿流与非牛顿流的转变点，归纳出其流变方程；王志华[41]通过室内建立的小型管流结蜡环道实验装置，研究了大庆肇源油田高凝原油在输油管道中的蜡沉积速度，考察了蜡沉积规律受温度的影响，并利用差热扫描量热仪从微观角度分析测试了蜡沉积物特性，探讨了结蜡机理和影响管壁结蜡的因素。2007 年，张劲军等[42]用旋转黏度计对原油在温度扫描条件下黏温关系进行实验研究，建立了特征温度与剪切过程黏性流动熵产的经验关系式。2010 年，贾泽琪和韩洪升[23]开展了对幂律流体在内管做轴向往复运动的偏心环空非定常流的数值模拟，建立了一套可调偏心度、冲程和冲次的垂直环空管道装置对其模拟结果进行 PIV 测试验证。

国外开展非牛顿流体石油管流的全尺寸试验比国内要早，更早的考虑了服役环境对非牛顿流体流动的影响。2003 年，Ozbayoglu E. M. 等[25]用 Tulsa 大学钻井研究计划资助的井斜角从 70°~90°的模拟井眼 8″-4 1/2″的全尺寸封闭流动装置开展假塑性泡沫携岩试验，在指定流量和机械钻速下，岩屑床厚度随泡沫质量增加而增加，与该范围内的井斜角没有关系。2008 年，Tulsa 大学 Ahmed R. 和 Miska S.[43]通过室内试验和现场测量证明钻柱旋转会影响非牛顿流体在环空的摩擦压力损失，采用新开发的试验设备对 5 种不同测试流体在 4 个不同环空结构中的流变特征进行试验，并对同心环空螺旋流进行了理论分析。2010 年，Japper-Jaafar A. 等[44]用激光多普勒测速仪测得非牛顿流体与牛顿流体在同心环空流动的平均和均方根轴向流速分布与压降数据，更多剪切稀化流体的过渡流在更大范围的雷诺数区域观测到。2014 年，Erge O. 等[45]考虑 H-B 流体在钻柱旋转下的摩擦压力损失，开展受压钻柱在水平井中的流动试验，提出了 H-B 流体在同心和偏心环空中从层流到紊流区域流态的新相关性。

21 世纪计算机技术与数值分析等学科的发展，计算机编程和数值模拟软件计算非牛顿流体流动问题成为热点，在工程允许误差范围内使用，解决了大量工程实际问题，成为了非牛顿流体实验手段的重要补充，从而推动了非牛顿流体的数值理论发展和工程应用。

6 建议

(1) 非牛顿流体偏心环空轴向流中，内外径之比大于 0.8 时，可假设流动对称于环形空间的几何中心柱面而不会引起太大误差。偏心度 ε 是引起流场分布不均匀的主要参数，可调整环空几何结构或流体流变性获得设计流场。目前，对几何轴对称的非牛顿流体非线性流动，还可以用弹性数和无量纲相似数描述其流动状况，难点在于像内外径之比小于 0.8 的偏心环空非牛顿流体流动机理就很难描述，需要运用混沌理论结合数模仿真技术进行评估。

（2）非牛顿流体圆管分层流混输技术中，可结合高分子纳米材料特性和电磁场理论，研制一种智能流体。纳米材料混入非牛顿流体中改变其流变性和磁性，电磁场可控制含纳米材料的流动屈服，这种智能流体可实现增大流性指数 n 较小的高黏幂律流体在分层流时的流量，以实现原油的低温输送。

（3）非牛顿流体圆管非定常流中，国外在黏弹性流体纯弹性不稳定现象方面做了大量深入研究，国内在借鉴其成果的基础上，可从高等流体力学、弹塑性力学、时均流、能量扰动及混沌学等基础理论出发，逐步建立非牛顿流体在空间意义上定常、材料意义上非定常流动稳定性的基础理论，探索非牛顿流体非定常流动规律。

（4）非牛顿流体管流稳定性研究中，非牛顿流体流动对附加扰动非常敏感，非牛顿流体流动的复杂性导致了难以准确描述其流态变化，还需通过大量可视化试验来深入研究其流型流态变化机理。

参 考 文 献

[1] 张劲军，严大凡．幂律流体在圆管内湍流流动剪切率的近似计算［J］．水动力学研究与进展，2003，18（2）：248-252.

[2] Yin G L, Dong B Q. Asymptotic behavior of solution to equations modelling non-Newtonian flows［J］. J. Math. Res. Exposition, 2006, 26（4）：699-706.

[3] 张其亮，张兴伟，宋娟，等．三维非牛顿流体动力学方程衰减性的一个注记［J］．高校应用数学学报A辑，2010，25（1）：122-126.

[4] 谭启建，冷忠建．一类退缩非线性自由边值问题弱解的存在唯一性［J］．数学杂志，2006，26（6）：657-664.

[5] 赵辉．有界区域上 p（x）-Laplacian 问题解的存在性［D］．哈尔滨：哈尔滨工业大学，2006.

[6] 赵永杰．一个非自治不可压缩非牛顿流体在局部一致空间中的研究［D］．兰州：兰州大学，2008.

[7] 郭春晓．随机非牛顿流解的适定性及其动力系统的研究［D］．绵阳：中国工程物理研究院，2010.

[8] 朱克勤，杨迪，胡开鑫．黏弹性流体的分数元模型及圆管起动流［J］．力学季刊，2007，28（4）：521-527.

[9] Liu Y Q, Zhu K Q. Axial Couette-Poiseuille flow of Bingham fluids through concentric annuli［J］. J. Non-Newtonian Fluid Mech., 2010, 165（21-22）：1494-1504.

[10] 孙学卫．剪切稀化流体分层流动的线性稳定性研究和数值模拟［D］．北京：清华大学，2011.

[11] 修德艳．幂律流体在内管做行星运动的环空中流动的压力梯度［D］．大庆：大庆石油学院，2007.

[12] 裴晓含．幂律流体在内管做行星运动的环空中流动时内管壁的受力分析［D］．大庆：大庆石油学院，2007.

[13] Filip P, David J. Axial Couette-Poiseuille flow of Power-Law viscoplastic fluids in concentric annuli［J］. J. Pet. Sci. Eng., 2003, 40（3-4）：111-119.

[14] Wilson H J. Open mathematical problems regarding non-Newtonian fluids［J］. Nonlinearity, 2012, 25（3）：45-51.

[15] Gjerstad K, Time R W, Bjørkevoll K S. Simplified explicit flow equations for Bingham plastics in Couette-Poiseuille flow-for dynamic surge and swab modeling［J］. J. Non-Newtonian Fluid Mech., 2012, 175-176：55-63.

[16] Gjerstad K, Ydstie B E, Time R W, et al. An explicit and continuously differentiable flow equation for Non-Newtonian fluids in pipes［J］. SPE J. 165930, 2013, 19（1）：78-87.

[17] 李孝军，练章华，赖天华，等．赫—巴模式在原油管流测量中的应用［J］．油气田地面工程，2008，27（5）：18-19，21.

[18] 李孝军，练章华，陈小榆，等．赫—巴流体稳定性参数 Z 的解析解分析［J］．西安石油大学学报（自然科学版），2010，25（1）：57-60.
[19] Kim Y J，Yoon C H，Park Y C，et al. Vortex flow study on Non-Newtonian fluids in concentric annulus with inner cylinder rotating［C］. Proc. 16th Int. Offshore Polar Eng. Conf.，San Francisco，California，USA，2006：835-839.
[20] 李阳，张凯，王亚洲，等．稠油油井幂律流体流动视黏度模型［J］．石油勘探与开发，2007，34（5）：616-621.
[21] 岳湘安，陈家琅，刘宏，等．泥浆模拟液在偏心环形空间中的紊流［J］．大庆石油学院学报，1991，15（1）：33-39.
[22] 王艳辉．偏心环空非牛顿流体紊动场特性的研究［J］．水动力学研究与进展，1997，12（2）：150-156.
[23] 贾泽琪．偏心环空中非定常流场数值模拟及 PIV 实验研究［D］．大庆：大庆石油学院，2010.
[24] Savins JG，Wallick GC. Viscosity profiles，discharge rates，pressures，and torques for a rheologically complex fluid in a helical flow［J］. AIChE J.，2004，12（2）：357-363.
[25] Ozbayoglu E. M.，Sorgun M. Frictional pressure loss estimation of Non-Newtonian fluids in realistic annulus with pipe rotation［J］. J. Canadian Pet. Technol.，2010，49（12）：57-64.
[26] 郑永刚，谢翠丽，等．非牛顿流体在圆管中层流-紊流分层流动规律［J］．四川大学学报（工程科学版），2000，32（3）：1-4.
[27] 贺成才．幂律—牛顿流体圆管分层层流的数值模拟［J］．天然气与石油，2003，21（1）：18-21.
[28] Biberg D. An engineering model for two-phase stratified turbulent duct flow［J］. Multiphase Sci. and Tech.，2004.
[29] Holmås K，Biberg D. Comparison of the Biberg analytical depth-integrated two-phase stratified flow model with CFD simulations［C］. 13th Int. Conf. on Multiphase Prod. Technol.，Cannes，France，BHR Group June 2007：121-137.
[30] Lawrence C，Nossen J，Skartlien R.，et al. Mechanistic models for three-phase stratified and slug flows with dispersions［C］. 16th Int. Conf. on Multiphase Prod. Technol.，Cannes，France，BHR Group June 2013：283-296.
[31] 刘慈群，黄军旗．非牛顿流体管内不定常流的解析解［J］．应用数学和力学，1989，10（11）：939-946.
[32] 陈文芳，范椿．非牛顿流体流动的不稳定性［J］．力学进展，1985，15（1）：49-53.
[33] Wang L S，Jiang J F，Hong R Y，et al. Transition to asymmetry of Non-Newtonian fluids：in an abrupt expansion［J］. Comput. Appl. Chem.，2008，25（12）：1473-1476.
[34] 李楠．黏弹性流体在内管做轴向往复运动的偏心环空中非定常流的压力梯度［D］．大庆：大庆石油学院，2010.
[35] Gainville M and Sinquin A. Hydrate slurry characterization for laminar and turbulent flows in pipelines［C］. Proc. of the 7th Int. Conf. on Gas Hydrates，Edinburgh，Scotland，United Kingdom，17-21 July，2011.
[36] 李兆敏，蔡国琰．非牛顿流体力学［M］．东营：石油大学出版社，2001.
[37] 刘乃震，王廷瑞，刘孝良，等．非牛顿流体的稳定性及其流态判别［J］．天然气工业，2003，23（1）：53-57.
[38] 何世明，罗德明，虞海生，等．判别液体流态的层流稳定性理论［J］．天然气工业，2000，20（5）：67-69.
[39] 刘崇建，刘孝良，柳世杰．非牛顿流体流态判别方法的研究［J］．天然气工业，2001，21（4）：49-52.
[40] 张金亮，王为民，申龙涉，等．辽河油田超稠油流变特性的试验研究［J］．油气田地面工程，2006，

25（7）：11.

［41］王志华．高凝原油管道输送蜡沉积规律实验研究［J］．特种油气藏，2006，13（5）：91-93.

［42］张劲军，朱英如，李鸿英，等．含蜡原油特征温度实验研究［J］．石油学报，2007，28（4）：112-114.

［43］Ahmed R，Miska S. Experimental study and modeling of yield Power-Law fluid flow in annuli with drillpipe rotation［C］．SPE Drill. Conf. 112604，2008.

［44］Japper-Jaafar A，Escudier MP，Poole RJ. Laminar，transitional and turbulent annular flow of drag-reducing polymer solutions［J］．J. Non-Newtonian Fluid Mech.，2010，165：1357-1372.

［45］Erge O，Ozbayoglu E M，Miska S Z，et al. The effects of drillstring eccentricity，rotation and buckling configurations on annular frictional pressure losses while circulating yield Power Law fluids［C］．SPE Drill. Conf. 167950，2014.

实体膨胀管润滑减阻工艺优选研究

刘 强 冯耀荣 吕 能 白 强 武 刚

（中国石油集团石油管工程技术研究院，石油管材及装备材料服役行为与结构安全国家重点实验室）

摘 要：实体膨胀管技术是节省钻完井成本的一种有效手段正逐步成为钻井新技术的发展方向。膨胀管膨胀过程中的摩擦是一个非常重要的问题，它影响到膨胀的难易程度和胀后套管的性能，因此选择一种合适的润滑介质并且降低膨胀过程中的摩擦系数对膨胀管技术来说非常重要。本文使用摩擦模拟和实物试验相结合的方法，研究了在铅基复合润滑涂层、二硫化钼复合润滑涂层和没有润滑脂的三种条件下普通 N80 膨胀管材料的摩擦性能，使用 S—4800 扫描电镜对不同载荷不同实验时间下的磨损情况进行了分析，用实物试验进行了验证，并讨论了两种不同润滑脂润滑条件下的磨损机理。通过实验可以得出两种润滑脂的加入可以显著降低膨胀过程中膨胀管材料的摩擦系数，磨损机制也从原来的磨粒磨损变成了黏着磨损，铅基复合润滑涂层由于摩擦系数更低具有最好的润滑效果，可以显著降低膨胀阻力，提高膨胀管材料在膨胀过程中的耐磨性能。

关键词：膨胀管技术；磨损性能；润滑脂；摩擦系数；实物试验；磨损机制

1 研究背景

膨胀管技术是近二十年发展起来的一项具有划时代意义的石油钻井技术。所谓的膨胀管技术就是将可膨胀的套管下到井下，通过冷变形方式使得膨胀管的直径扩大到要求的尺寸，从而达到节省井眼直径的目的。该技术可以广泛地应用于完井、固井作业，套管内补贴、用作应急尾管、尾管悬挂器[1-4]。该技术的出现从根本上改变了未来深井和复杂地质条件的钻井采油工艺。实际工程应用表明，可膨胀管技术能够大幅度降低钻采成本和缩短施工时间，被称为石油钻采行业的一次技术革命[5-7]。

近年来，国内外的学者们在膨胀管技术的材料、工艺和评价方面做了大量的研究工作。天津大学的 Xu Ruiping，Liu Wenxi 等研究了一种新型的膨胀管材料，并且提出了一个高性能膨胀管材料的设计准则强塑性[8]；阿联酋石油研究所的 A. C. Seibi 等人用模拟计算和试验的方法研究了用铝管和钢管做实体膨胀管的区别[9]，西南石油大学的尹虎等人建立了实体膨胀管力学模型及膨胀力计算模型，研究得到了膨胀锥角和摩擦系数的关系，但在试验中没有考虑润滑条件[10]，Weatherford 公司的 Richard Delange 等人建立了一种新型的弯曲膨胀实验装置和试验方法来模拟井下工况对膨胀管的连接螺纹进行评价试验，但是在试验中也没有考虑到膨胀锥和膨胀管之间的润滑减阻问题[11]。

膨胀管膨胀过程中的摩擦是一个十分重要的因素，它关系到膨胀过程的难易程度和膨胀后管材的力学性能，并且，膨胀锥和管材之间的润滑状态直接决定了膨胀过程摩擦系数的大小，因此改善膨胀过程的润滑状态，选用合适的润滑减阻工艺，降低摩擦因素，对膨胀过程非常重要，但膨胀管膨胀过程中的润滑减阻工艺的研究未见研究报道。本文以一种常用的N80钢级膨胀管为研究对象，采用有/无润滑涂层摩擦磨损学实验室研究和实物膨胀试验相结合的方法，对无润滑和不同润滑工艺下膨胀管的摩擦磨损性能、摩擦系数以及实物润滑效果进行试验研究，对比不同润滑涂层下的膨胀力、外径回弹率和长度变化等重要参数，优选出适合的润滑减阻工艺。

2 试验材料及方法

本文所用的膨胀管材料来自西安三环科技开发总公司研制的N80钢级膨胀套管，化学成分见表1，显微组织为铁素体+珠光体+少量粒状贝氏体，如图1所示，晶粒度为11级。在管体上切割10mm×10mm×6mm的立方体试样，并将试样所有表面在1000#砂纸上磨平，将试样在丙酮溶液中清洗后测重。试验所用摩擦环材料为GCr15，经淬火及低温回火处理，硬度为60~62HRC，与膨胀锥的材质相同，摩擦环的尺寸如图2所示。

图1 膨胀管材料的显微组织

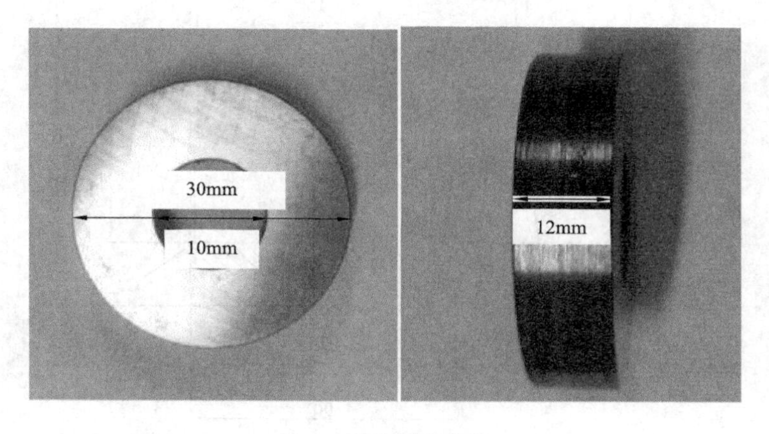

图2 摩擦环示意图

本文所选的润滑减阻工艺为膨胀管内壁涂刷铅基复合润滑涂层和二硫化钼复合润滑涂层，这两种涂层均为西安三环科技开发总公司所研制。其中铅基复合润滑涂层的成分为：含铅粉体积分数 15%~35%，锌粉 10%~25%，石墨 20%~45%，含铜粉 10%~20%，其余为普通润滑脂；二硫化钼复合润滑涂层中含二硫化钼粉含量为 10%~25%，钛粉 10%~15%，其余为普通锂基润滑脂。

将试验膨胀管试样分为 3 组，A 组不加任何润滑剂，B 组试样表面涂抹二硫化钼复合润滑涂层，C 组表面涂抹铅基复合润滑涂层，使用磨损试验机并装上 GCr15 材质的摩擦环分别在 10N、30N、50N、80N 的载荷下对 3 组试样进行对磨实验，摩擦试验时间分别为 25min，50min，75min 和 100min，在实验的过程中 B 组和 C 组不停的在试样表面添加相应的润滑脂，以保证试验的准确性。试验结束后将试样取出使用 KQ—250DB 型超声波清洗器清洗 5min 并吹干，然后称重以计算磨损量。使用 S—4800 扫描电子显微镜对试验后的试样磨痕形貌进行分析。

表 1　膨胀管材料的化学成分（wt%）

试样	C	Si	Mn	P	S	Nb	Cr	Mo	Ti	Al
N80 膨胀管	0. 16	0. 31	1. 07	≤0. 01	≤0. 008	0. 06	0. 21	0. 22	0. 02	0. 11

实物膨胀试验选用的膨胀管材料为与摩擦磨损试验一样的 φ140mm×8mm×1000mm 膨胀管 3 根，编号编为 RH—0、RH—1、RH—2，分别对应无润滑涂层、铅基复合润滑涂层和二硫化钼复合润滑涂层。

实物膨胀试验选用的膨胀管见图 3，膨胀锥大端外径为 137. 8mm，实物试验评价方法依据 SY/T 6951—2013《实体膨胀管》，使用管研院的膨胀管实物评价系统采用机械式膨胀的方法对膨胀管进行实物试验评价，实物评价系统见图 4，实验项目及实验方法见图 5。

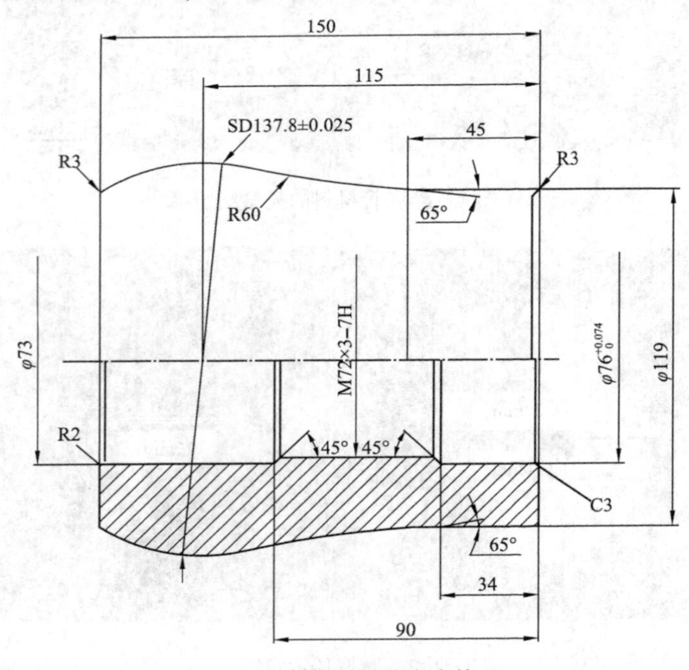

图 3　实验所用的膨胀管

图 4　试验所用的全尺寸膨胀管评价装置

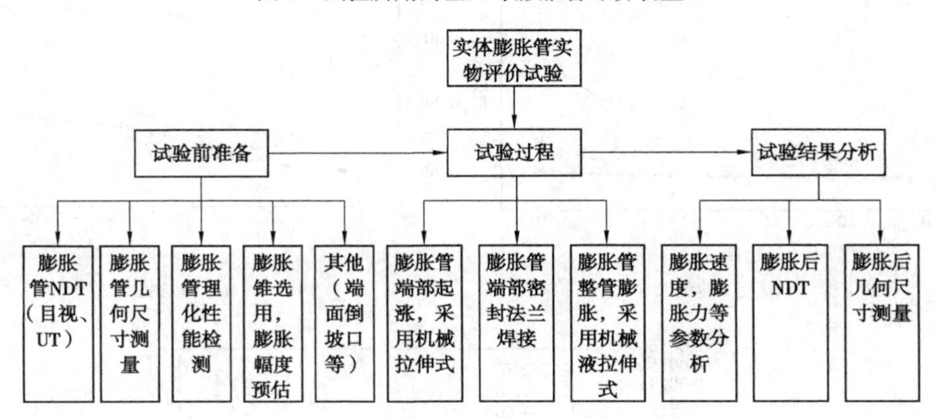

图 5　实体膨胀管实物性能评价方法

由于本实物试验重点对不同润滑减阻工艺下的膨胀力、外径和长度进行对比分析，因此对 3 根膨胀管的外形尺寸测量分析非常关键，外形尺寸检测方法是距实体膨胀管一端的端面开始取第一个测量截面，沿轴向间隔 200mm 依次标记，共计 5 个截面（0~5 截面）；每个截面上测量 8 个点（A—H）的壁厚值，4 个位置（A—E、B—F、C—G、D—H）的外径值，如膨胀管测量示意图见图 6，记录每个点的距离、壁厚和外径，在膨胀前后进行对比。

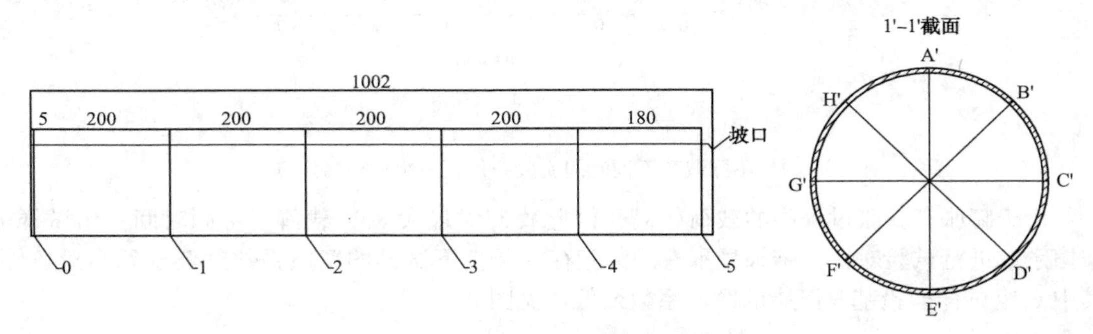

图 6　膨胀管几何尺寸测量示意图

3 实验结果

3.1 摩擦磨损性能

将A，B，C三组试样分别在10N、30N、50N、80N的载荷下进行对磨实验，试验时间分别为25min，50min，75min和100min，磨损重量损失量结果见图7，结果显示每组试样的磨损重量损失量均随着载荷的增大而增大，在同一载荷下，磨损时间越长磨损量越大。对最高80N载荷条件下的三组试样进行对比可以看出，在没有润滑剂存在的条件下，膨胀管材料的磨损量远远大于有润滑脂润滑的材料磨损量，A组试样的磨损量大约是C组的近140倍，而不同润滑脂润滑的材料磨损量相差不大，见图8。

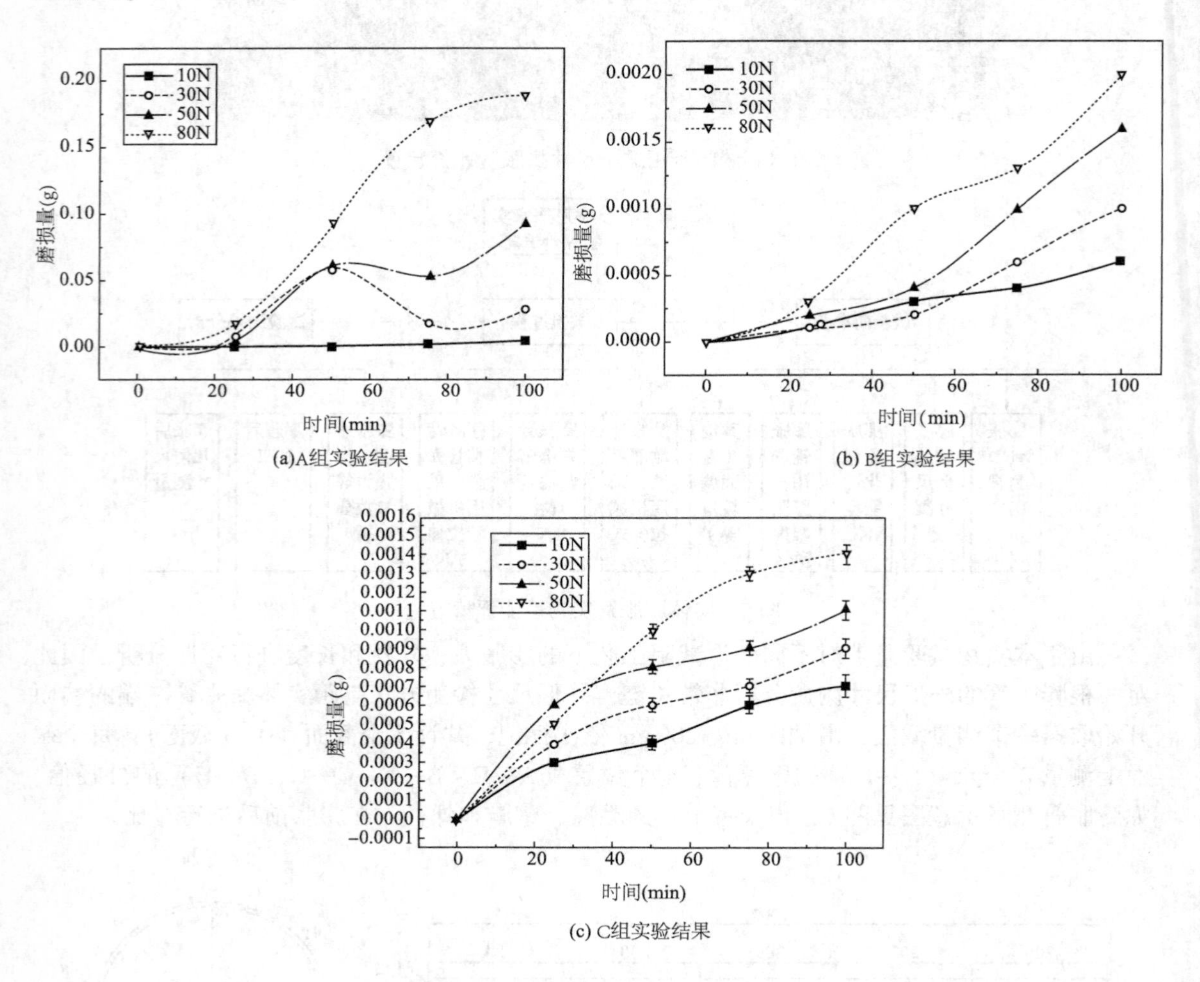

图7 不同载荷不同时间实验条件下三组试样磨损量

由于膨胀管膨胀过程中的载荷较高，因此我们对最大80N载荷下不同时间三组试样的摩擦系数进行试验对比，结果显示有润滑脂存在条件下试样的摩擦系数远低于无润滑条件，其中C组试样即铅基润滑脂的摩擦系数最低，见图9。

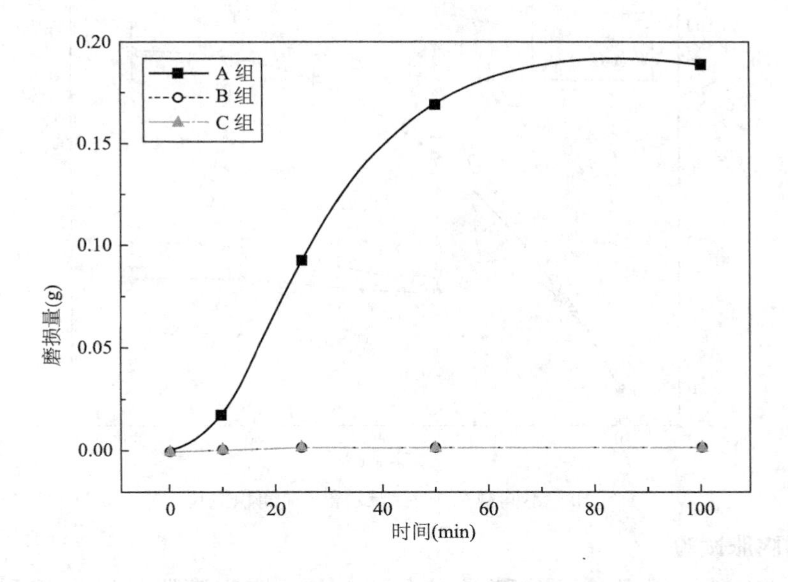

图 8　在 80N 载荷下三组试样磨损量对比

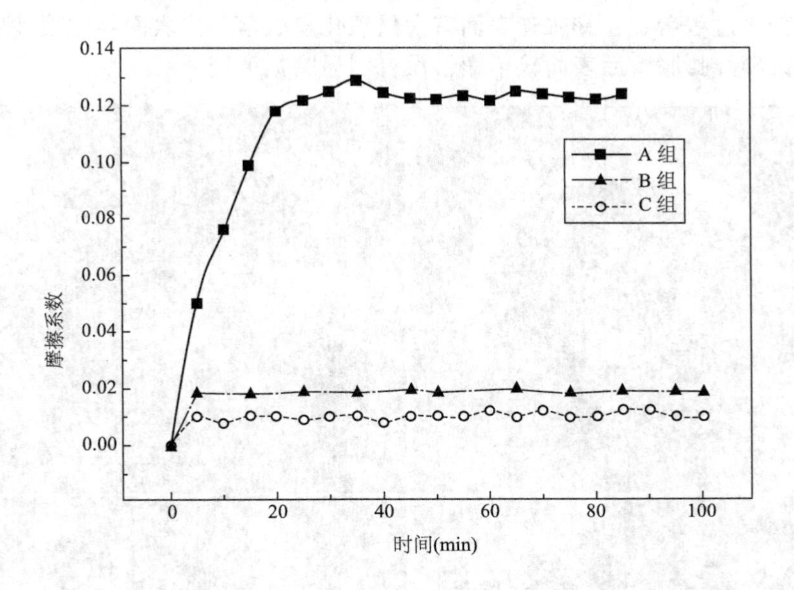

图 9　在 80N 载荷下三组试样摩擦系数对比

3.2　摩擦形貌观察

对三组试样在不同载荷下的磨损表面进行观察，并计算试样磨损面宏观面积百分比，结果见图 10，可以看出当加载载荷较低时，三组试样的磨损面积基本相同，为表面积的 28%附近，随着载荷的升高，没有润滑介质的表面开始发生大量的磨损，当载荷升至 80N 时磨损面积达到 90%，相反，使用了润滑脂的 B 组和 C 组在润滑脂的作用下显著降低了磨损面积，其中使用铅基润滑脂的 C 组试样在载荷为 80N 时磨损面积最小只有 28%。

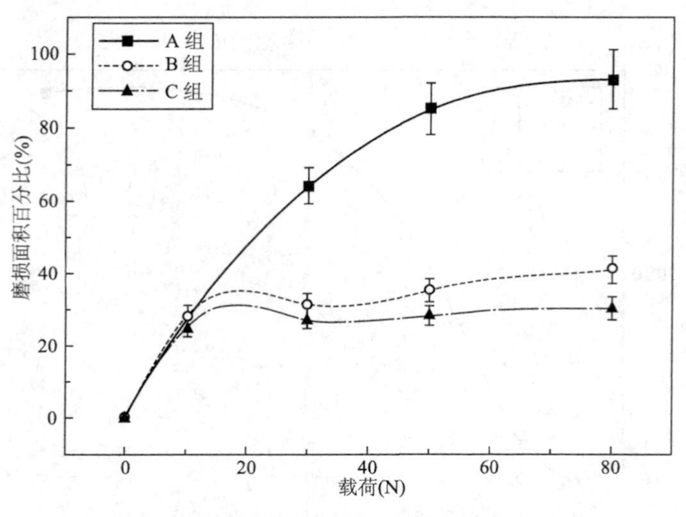

图10　不同载荷下三组试样磨损面积对比

3.3　实物膨胀试验

考虑到要对三种润滑条件下的膨胀力大小，因此实物膨胀试验采用同一膨胀速度200mm/s进行膨胀试验，实验过程见图11，从实验过程来看，膨胀锥进入膨胀管及膨胀过程比较顺利，膨胀过程稳定，膨胀管表面有大量氧化皮脱落，内表面经过喷砂后脱落的氧化皮较少，膨胀锥离开膨胀管后表面较干净，没有明显磨损痕迹。

图11　膨胀管实验过程

对三种不同润滑减阻条件的膨胀管起胀力和膨胀过程的膨胀力进行采集和整理，结果见图 12 和图 13。

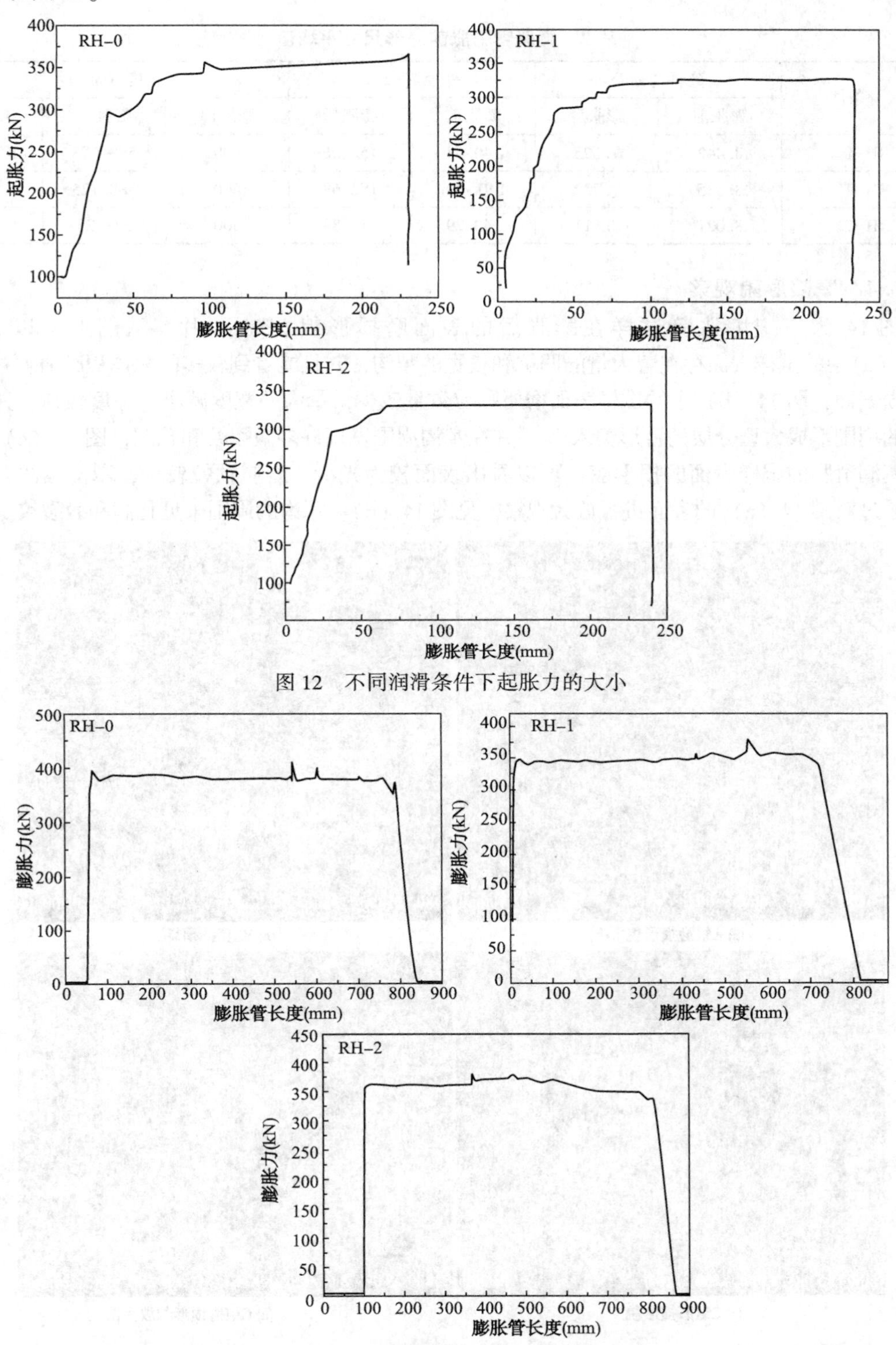

图 12 不同润滑条件下起胀力的大小

图 13 不同润滑条件下膨胀力的大小

实物膨胀试验后，对膨胀管的膨胀后的外径、壁厚进行测量，并进行统计分析，结果见表2。

表2　膨胀管膨胀前外形尺寸平均值

润滑条件	壁厚（mm）		外径（mm）		长度（mm）		
	膨胀前	膨胀后	膨胀前	膨胀后	膨胀前	膨胀后	变化率
RH-0	8.442	7.625	140.53	152.06	600	588.875	1.85%
RH-1	8.535	7.722	140.486	152.68	600	588.125	1.98%
RH-2	8.021	7.511	140.29	152.88	600	588.25	1.96%

3.4　摩擦形貌观察

图14为三组膨胀管试样在膨胀后的表面磨损形貌SEM照片，从图上可以看出，图14（a）中的试样表面存在着大量的凹坑和较宽的犁沟，并且能看到较多的大体积磨屑颗粒残存在磨损表面，图14（b）中的试样表面磨屑颗粒数量较少，犁沟的宽度减小，深度较深，并且在犁沟的周围形成大型分层的舌形契入[12]，并在犁沟周围发现许多微裂纹和孔洞；图14（c）为使用铅基润滑脂的试样表面磨损形貌，可以看出表面较为光滑，磨屑颗粒较少，犁沟较浅并且狭窄，通过对图14（c）的表面进行放大观察，见图14（d），犁沟的周围未见孔洞和微裂纹。

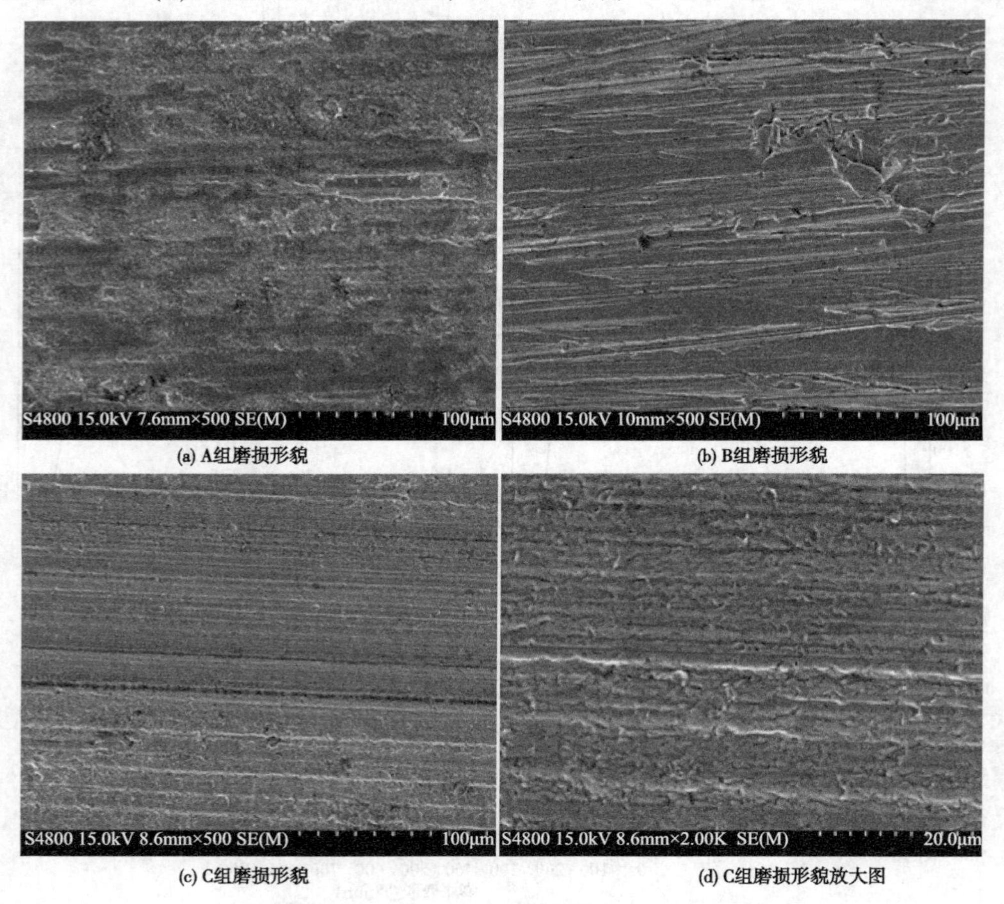

(a) A组磨损形貌　(b) B组磨损形貌

(c) C组磨损形貌　(d) C组磨损形貌放大图

图14　在80N载荷下三组试样表面磨损SEM显微照片

4 讨论分析

从以上实验结果可以看出，有润滑脂存在条件下膨胀管材料在膨胀过程中的摩擦系数远低于无润滑介质，磨损量也小很多。

对不同润滑减阻工艺实物试验的膨胀管起胀力和膨胀力大小进行对比，结果见图 15，可以看出，在同样的实验条件下，铅基复合润滑涂层具有最低的起胀力和膨胀力，二硫化钼复合润滑涂层次之，无润滑涂层的膨胀力最大，比铅基复合润滑涂层平均大 40~50kN。

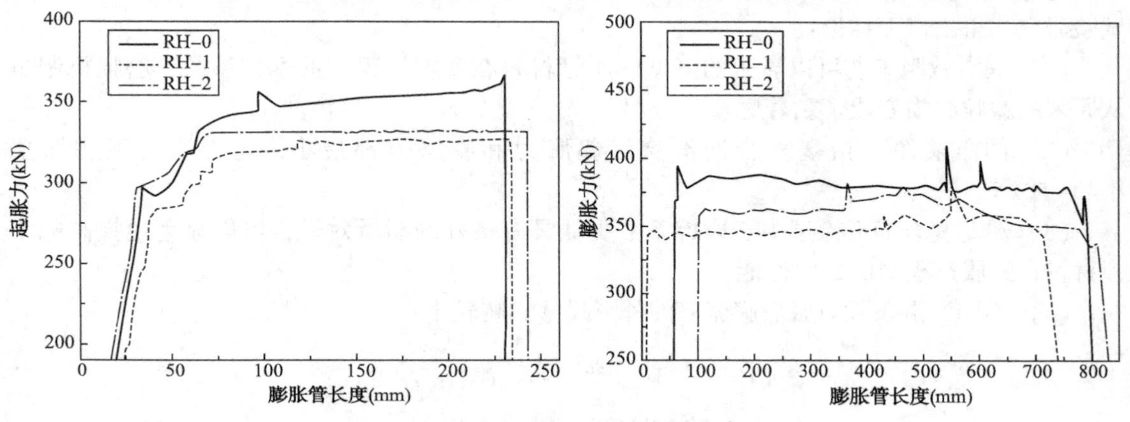

图 15 不同润滑条件膨胀管膨胀力对比

对不同润滑条件下的膨胀管壁厚和外径变化进行分析，可以发现不同的润滑涂层条件对膨胀后的壁厚和外径影响较小，没有明显的规律。对不同润滑条件下的膨胀管胀后长度缩短率进行对比，可以看出，随着润滑条件的不断改善，膨胀管膨胀后的长度收缩率逐渐增大，但是差距不大，如图 16 所示。

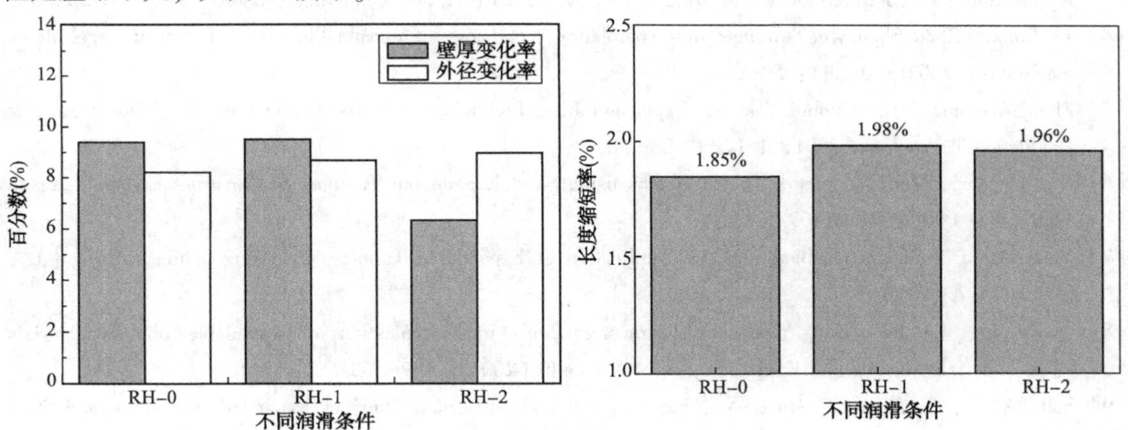

图 16 不同润滑条件下膨胀管外形尺寸变化对比

从 3.4 的表面磨损形貌 SEM 照片可以看出，试样表面存在着大量的凹坑和较宽的犁沟，并且能看到较多的大体积磨屑颗粒残存在磨损表面，这是由于试样在摩擦过程中表面的部分合金黏附到对磨环上，经过反复转移和挤压等这些颗粒发生加工硬化、疲劳及氧化等脱落并形成游离磨屑，磨屑会以磨粒的方式对合金表面产生显微切削作用，由于磨屑体积较大使得磨损表面产生较宽的磨损犁沟，并且在犁沟的周围产生微裂纹，所以当载荷增大时磨损量增加很快，并且摩擦系数也相当高，此时的磨损机制为黏着磨损和磨粒磨损的混合磨损机制，磨粒磨损占主导地位。当加入不同的润滑脂时，在润滑脂的作用下磨损表面的磨屑颗粒数量大量减少，同时磨屑颗粒也以小体积为主，使得犁沟的宽度大为减少，同时润滑脂里面的铅粒子、石墨粉和铜粒子等对材料表面进行抛光，膨胀过程的摩擦力减小并且摩擦系数降低，此时的磨损机制由磨粒磨损转变为黏着磨损[13]。

5 结论

通过对膨胀管不同润滑条件下膨胀管的摩擦磨损性能、摩擦系数以及实物润滑效果进行试验研究，得出以下结论：

（1）润滑减阻工艺可以显著的降低膨胀管材料在摩擦过程中的摩擦系数，磨损机制也从原来的磨粒磨损变成了黏着磨损。

（2）同样条件下铅基复合润滑涂层起胀力和膨胀力都最小，比无涂层下膨胀要小40~50kN。

（3）铅基复合润滑涂层由于摩擦系数更低具有最好的润滑效果，可以显著地提高膨胀管材料在膨胀过程中的耐磨性能。

（4）不同润滑条件对胀后膨胀管的外形尺寸影响较小。

参 考 文 献

[1] Oladele O. Owoeye, Leste. O. Aihevba, R. A. Hartmann, et al. Optimisation of Well Economics By Application of Expandable Tubular Technology. SPE. IADC/SPE 59142.

[2] Chan L. Daigle, Donald B. Campo, Carey J. Naquin, et al. Expandable Tubulars: Field Examples of Application in Well construction and remediation, SPE 62958.

[3] Chen Jing, Xiong Qingshan, Peng Mingwang. Technology for Well Cementing with Expandable Tube and Its Application [J]. Exploration Engineering (Rock & Soil Drilling and Tunneling), 2008, 8: 19-21.

[4] Li Junbo, Zhao Shengyin. Two new well completion technology of expandable tube [J]. China petroleum machinery, 2002, 30 (9): 59-60.

[5] Zhang Wenhua, Liu Guohui, et al. Expansion Pipe Technology And Its Application [J], Oil Drilling & Production Technology, 2001, 1 (23): 28-31.

[6] Karl Demong, Mark Rivenbark. Breakthroughs using Solid Expandable Tubulars to Construct Extended Reach Well [J], IADC/SPE 87209.

[7] Yash Gupta, Sudeepto N. Banerjee. The Application of Expandable Tubulars in casing While Drilling [J], SPE 105517.

[8] Xu Ruiping, Liu Jie, Zhang Yuxin, et al. Transformation- induced plasticity of expandable tubulars materials [J], Materials Science and Engineering A, 438 - 440 (2006): 459-463.

[9] Seibi A. C., Barsoum I., Molki A., Experimental and Numerical Study of Expanded Aluminum and Steel Tubes [J], Procedia Engineering, 10 (2011): 3049-3055.

[10] Yin Hu, Li Qian, Li Lintao. Research on mechanical model of solid bulged tube [J], Drilling & Production Technology, 2011, 34 (4): 59-62.

[11] Richard Delange, Shaohua Zhou, Scott Osburm. An Apparatus and Test Procedure for Qualifying Expandable Connections for Challenging Applications [J], SPE/IADC 125667.

[12] Panin V., Kolubeav A., Tarasov S., et al. Subsurface layer formation during sliding friction [J]. Wear, 2002, 249 (1): 860-867.

[13] Mishina H., Atmospheric characteristics in friction and wear of metals [J], Wear, 1992, 152 (1): 99-110.

壁薄焊缝 TOFD 检测可行性讨论

姚　欢[1]　刘　琰[1]　罗全恒[1]　胡江峰[2]　杨　涛[2]　徐生东[1]

（1. 中国石油集团石油管工程技术研究院；2. 中石油管道联合有限公司西部分公司）

摘　要：超声波 TOFD 成像检测多用于大壁厚焊缝检测中，但油气输送管道存在较多壁厚小于 12mm 的薄壁焊缝，本文分析了 TOFD 检测盲区，并针对薄壁焊缝进行了盲区计算分析，根据结果讨论了 TOFD 检测技术对薄壁焊缝检测的可行性。

关键词：油气管道；超声波检测；薄壁焊缝；TOFD 检测

超声波衍射时差法 Time of Flight Diffraction （TOFD）检测技术在无损检测中的应用越来越成熟和广泛。在焊缝的检测中，超声 TOFD 检测具有检测效率高、缺陷检出率高、定位定量准确等优点，并是目前唯一能够准确给出缺陷自身高度的无损检测方法，而缺陷自身高度是管道安全评价计算中所需要的关键参数之一。但是，TOFD 检测中在工件扫查面附近和工件底面附近都存在盲区，即上表面盲区和下表面盲区，一直是 TOFD 检测技术的一个弊端。

与射线检测不同，TOFD 检测更擅长于厚壁焊缝检测，而目前国内石油工业油气集输管道壁厚基本分布在 6~30mm，同时国内最新发布的 TOFD 检测标准 NB/T 47013.10—2015 中规定，TOFD 检测范围为：12mm≤工件公称厚度≥400mm[1]，这就将很多薄壁集输管道焊缝检测划定在了范围之外（这里暂将壁厚小于 12mm 的焊缝定义为薄壁焊缝）。国外 TOFD 相关检测标准中检测厚度范围也不尽相同，具体如表 1 所示。

表 1　不同标准所规定 TOFD 检测范围

序号	标准名称	壁厚范围	备注
1	BS7706（英国标准）	/	未明确要求
2	ENV583-6（欧洲标准）	6~300mm	
3	CEN/TS14751（欧洲标准）	6~300mm	超过范围需试验证明
4	ASTM2373（美国标准）	9~300mm	超过范围需试验证明
5	ASTM2235-9（美国标准）	≥12.7mm	
6	NB/T 47013.10（中国标准）	12~400mm	

1　上表面盲区计算分析

根据 TOFD 检测原理，靠近扫查面的近表面缺陷信号可能隐藏在直通波信号之中，导致缺陷无法有效识别，因此将直通波所覆盖的深度范围定义为上表面盲区。声速定义为 c，探

作者简介：姚欢，男，工程师，生于 1982 年，中国石油天然气集团公司管材研究所无损检测中心，主要从事现场无损检测和技术服务工作。联系方式：029-81887945　13991348096。传真：029-81887953。邮箱：yaohuan@ cnpc. com. cn。

头中心间距（PCS）为 $2s$，直通波传输时间为 t_L（等于 $2s/c$），直通波脉冲宽度时间 t_P 可从振幅的10%处截取得到。这里取直通波1.5倍周期，则盲区深度可按下式[2]算出：

$$d=\left[\left(\frac{ct_P}{2}\right)^2+cst_P\right]^{\frac{1}{2}} \tag{1}$$

根据计算公式（1），若使用10MHzϕ3mm折射角为60°的探头，这里 t_P 取直通波1.5倍周期，为1.5μs，纵波声速 $c=5.9\text{mm}/\mu\text{s}$。则计算以下薄板焊缝上表面检测盲区如表2所示。

表2　探头10MHz折射角60°时上表面盲区计算

板厚（mm）	折射角（°）	PCS（mm）	上表面盲区（mm）
12	60	28	3.5
10	60	23	3.2
8	60	19	2.8
6	60	14	2.5

根据计算公式（1），若使用10MHzϕ3mm折射角为45°的探头，则以下厚度薄板焊缝上表面检测盲区如表3所示。

表3　探头15MHz折射角45°时上表面盲区计算

板厚（mm）	折射角（°）	PCS（mm）	上表面盲区（mm）
12	45	16	2.2
10	45	14	2.1
8	45	11	1.8
6	45	8	1.6

通过以上的计算分析可以看出，为减小TOFD上表面检测盲区，可以采取的措施包括：增大探头频率，使用短脉冲探头，减小探头折射角度，减小PCS等。但实际上TOFD检测信号中的直通波、底面回波和不同缺陷的衍射波沿着不同路径传播，它们的频率成分存在较大的差异，其各自信号的分辨率也存在差异，造成声波传播时间的测量精度不同。因此检测盲区影响因素较多，纯计算方法得出的盲区有一定误差，实际的盲区深度最好用对比试块确定。

2　下表面盲区计算分析

同样，TOFD检测中焊缝下表面也存在一个盲区，靠近焊缝中心下表面的缺陷信号可能会隐藏在底面回波信号中，从而无法有效区分。定义底面回波深度为 D，传输时间为 t_D，底面回波时间宽度从振幅的10%处截取获得，其值为 t_P。这里取直通波的一倍周期[3]，则盲区高度 h 可按照下式计算：

$$d=\left[\left(\frac{c}{2}\right)^2\ (t_D+t_P)^2-s^2\right]^{\frac{1}{2}}-D \tag{2}$$

分析得出底部缺陷的上尖端信号应领先于底面回波，不应被底面回波信号所淹没。所以

TOFD 检测底面盲区是由缺陷信号与底面回波不重叠度的分辨能力造成，这种分辨力取决于 TOFD 扫描图像的分辨率和肉眼观察力，同时与缺陷信号的强弱、底面回波强度等因素有关。因此，上面所列焊缝中心的底面盲区计算公式不够准确，甚至可以不计算，因为该盲区很小，一般不超过 1mm，甚至小于 0. 5mm。

实际上底面缺陷还有另外一种情况，即偏离于焊缝中心位置的缺陷，这种情况需要依据椭圆方程求得检测盲区，检测过程中通常使用对比试块，采用试验的方法测定盲区高度，这里不做讨论。

3　薄壁焊缝 TOFD 检测解决方法讨论

3. 1　横波超声 TOFD 检测技术

通常 TOFD 检测选用衍射系数较高的纵波，在直通波和底面反射波之间会出现缺陷的衍射波，而在这个区域之后会出现复杂的波型转换，转换后的得到的横波因为波幅较弱往往被忽略。印度研究人员尝试将衍射横波应用到 TOFD 超声检测中，称为横波衍射时差技术 (S-TOFD)。在普通碳钢材料中，横波传播速度约为纵波声速的一半，这样在横波 TOFD 技术中，参考波的底面反射纵波和底面反射横波的时间差值，比纵波 TOFD 技术中的直通波和底面反射纵波的时间差值大 2 倍左右。通过计算分析和实验研究，当缺陷尖端的埋藏深度不超过工件厚度的 2/3 时，缺陷尖端衍射横波在底面反射纵波之后被探头接收，这样横波 TOFD 技术就能够检测薄壁材料中的缺陷以及近表面缺陷。但因为横波的衍射系数较小，缺陷的衍射横波很可能被噪声淹没，无法识别，分辨缺陷的衍射横波信号对检测设备有更高的要求，所以横波衍射时差技术（S-TOFD）距离实际检测应用，还需要更多的研究和试验。

3. 2　超声 TOFDW 检测技术

哈尔滨工业大学现代焊接生产技术国家重点实验室为了解决超声 TOFD 检测中表面盲区问题，提出一种基于纵波三次反射的超声 TOFDW 检测模式[4]，原理如图 1 所示。对比常规超声 TOFD 检测方法，TOFDW 技术以纵波一次底面反射波和三次反射波作为参考，二者之间的时域为缺陷的有效识别范围。TOFD 检测近表面缺陷识别能力的高低由缺陷回波与其相邻波声程差决定的，纵波一次底面反射波与三次反射波的声程差比常规超声 TOFD 的直通波与一次底面反射波的声程差要大。实验结果表明，TOFDW 方法能够检测到埋藏深度为 1. 0mm 的人工缺陷，同时缺陷近表面埋藏深度测量的平均绝对值误差不超过 0. 3mm。

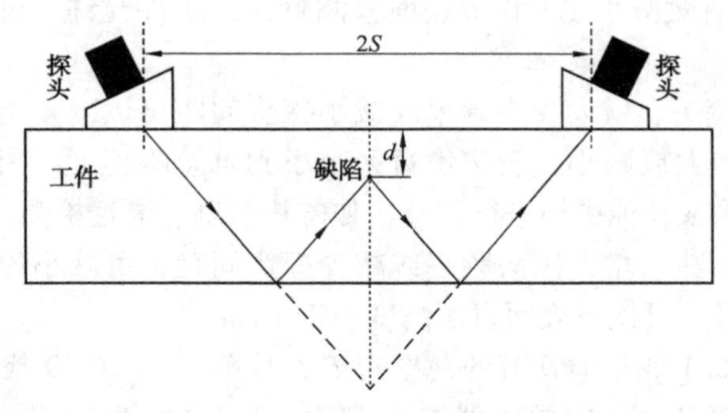

图 1　纵波三次反射技术（TOFDW）原理示意图

3.3 超声 TOFDR 检测技术

在常规 TOFD 检测基础上，南昌航空大学无损检测技术教育部重点实验室提出了基于二次波的综合超声衍射反射回波方法（TOFDR），检测原理如图 2 所示，以及 TOFDR+TOFDW 检测方法。将超声 TOFDR 方法和 TOFDW 方法结合使用能更好地发现试件近表面缺陷，该方法可以识别 1mm 深度的缺陷，而且 TOFDR+TOFDW 方法与传统 TOFD 检测相比，可以选择频率相对较低的检测探头。

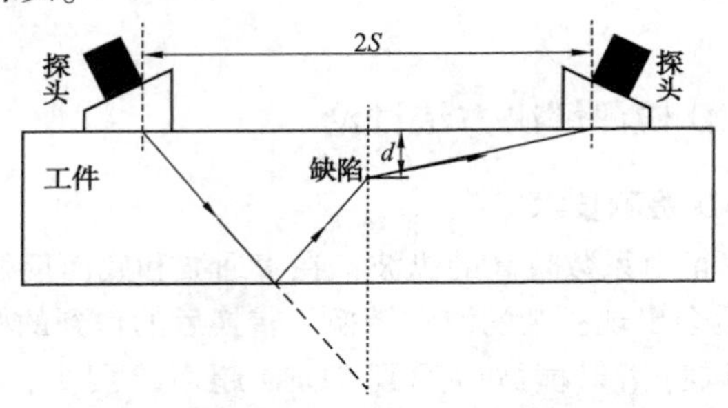

图 2　二次波的综合超声衍射反射回波方法（TOFDR）原理示意图

3.4 超声爬波检测技术

超声爬波是在自由表面的位移有垂直分量的纵波。当纵波以第一临界角附近的角度入射到界面时，就会在第二介质中产生表面下的纵波，即为爬波。爬波产生原理如图 3 所示。由于爬波在传播过程中，能量主要集中在工件表面以下某个范围内，在其往下传播的过程中，能量衰减急剧，因此爬波只能用于表面及近表面的缺陷检测。虽然超声爬波检测有着无法测量缺陷深度、对水平位置的测定也存在一定误差的缺点，但在目前对工件表面盲区无有效解决方案的情况下，超声爬波检测也是一种解决 TOFD 表面盲区检测的有效辅助方法[5]。

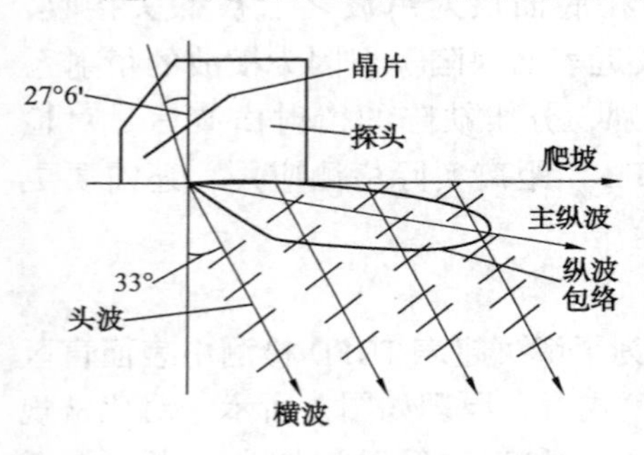

图 3　爬波探头声场示意图

4 结论

（1）针对工件近表面缺陷的检测，为了使探头的主声束反射到近表面缺陷上，保证缺陷波具有足够的检测能量，应选择探头折射角较小的探头（如 45°探头），可有效减小 TOFD 近表面检测盲区，条件允许，可选择折射角度更小的探头。

（2）使用短脉冲探头、增大探头频率或减小探头间距可以减小上表面盲区深度，但上述参数不能无限制的增大或减小，无法使盲区减小到可忽略程度。因为上表面盲区无法忽略，实际检测时需要采取其他措施进行弥补，如磁粉检测、渗透检测。

（3）使用短脉冲探头、增大探头频率或减少探头间距，可减小焊缝中心部位的下表面盲区，如参数选择恰当，可使下表面盲区高度小于 1mm。

（4）针对我国石油工业集输管道的薄壁焊缝，目前国内 TOFD 检测设备和检测工艺无法完全克服上下表面盲区，文中所提到的 TOFDR、TOFDW 等检测方法仍处于试验摸索阶段，对于薄壁焊缝检测仍需要辅助其他检测手段。

参 考 文 献

[1] 超声波衍射时差法检测规程：NB/T 47013.10—2015 [s].

[2] 采用衍射时差法超声波检测的标准实施规程：ASTM E2373—2004 [s].

[3] 强天鹏，肖雄，等.TOFD 技术的检测盲区计算和分析 [J].无损检测，2008，30（10）：738-740.

[4] 卢超，王鑫，陈振华.近表面缺陷的超声 TOFDR 和 TOFDW 检测 [J].失效分析与预防，2012，7（3）：153-157.

[5] 刘礼泉，郑晖，邬冠华.超声衍射时差法检测表面盲区分析及盲区内缺陷的超声爬波检测工艺和应用 [J].无损检测，2013，35（7）：42-46.

第二篇　成果篇

一、省部级以上科技成果简介

1. 中缅天然气管道设计施工及重大安全关键技术研究与应用

中国石油天然气集团公司科学技术进步奖
CNPC Science and Technology Progress Award Certificate

证 书

为表彰中国石油天然气集团公司
科学技术进步奖获得者，特颁发此证书。

获奖项目：中缅天然气管道设计施工及重大安全关键技术研究与应用

获奖单位：中国石油集团石油管工程技术研究院

奖励等级：特等奖

证 书 号：2016-J-0-02-D03

中国石油天然气集团公司

2016年12月8日

本项目来源于实验室承担的中国石油天然气集团公司重大科技专项“西气东输二线管道断裂与变形控制关键技术研究”，中国石油天然气集团公司“十二五”技术开发项目“特殊地区管道建设关键技术研究”和应用基础项目“高强度管道失效控制的应用基础研究”等课题研究。

目前建设的一些管道项目沿线地质条件十分复杂，如西气东输二线穿越22条活动断层，中亚D线途经大量的9度强震地区，中缅管道将通过9度强震区和5条活动断裂带，并途经多个采矿沉降地区，施工难度极大，被称为世界管道建设史上难度最大的工程之一。这些地质条件对管道的设计、材料、施工都提出了巨大的挑战。因此，为保证可能发生较大变形地段的X70钢级管道安全，在局部可能发生较大变形的地段采用了大变形钢管。针对西二线、西三线、中亚D线用X80大变形钢管，以及中缅用X70大变形钢管，通过全尺寸弯曲试验和变形行为研究、批量试制程序和质量控制及评估方法研究、大变形钢管冷弯管技术研究等相关的技术攻关，攻克了国产大变形管线钢的关键技术，形成国产大变形钢管变形行为预测、生产制造、性能控制等系列应用技术，实现大变形钢管的技术突破，推动国产X70、X80大变形钢管的生产应用水平，并最终保证了上述油气管道工程的顺利实施。主要创新点：（1）研究提出了抗大变形钢管的关键技术指标和检测评价方法；（2）通过单炉和小批量试制/千吨试制，确定了生产工艺的可行性及稳定性，提出了大变形钢管的小批量试制流程、质量控制方案；（3）通过钢管的屈曲变形仿真计算和全尺寸压缩弯曲试验，确定了基于应变地区用钢管屈曲应变指标；（4）在国内外首次提出了时效试验方法细则；（5）通过课题研究，撰写完成了X70、X80抗大变形钢管的工程技术标准，为高应变直缝埋弧焊管在基于应变设计地区许用应变确定提供了依据。

本项目获2016年中国石油天然气集团公司科技进步特等奖。

2. 基于应变的稠油蒸汽热采井套管柱设计方法及工程应用技术

项目围绕稠油热采井套损治理和失效预防，依托2011年中国石油天然气集团公司“油套管柱完整性技术”（2011A-4208）”项目，深化2005年中国石油天然气集团公司国际合作项目“油气管道/柱风险评价”（05D40101）项目内热采井套损机理研究，与新疆油田、辽河油田及中海油热采井套损治理相结合，联合渤海装备、衡阳钢管及新冶钢集团等制造单位联合攻关，形成本项技术成果，实现了套损机理、管柱设计、管材选用评价、工程配套技术及标准化一体化成果，为以蒸汽热采为主要方式的稠油工业高效、安全、经济开发提供了成套技术支撑。

通过套损普查与分析，掌握了套损主要模式及失效机理，指出套损主要来源于长期服役产生的热-弹塑性变形。井下超过180 ℃的大温差将造成管材热屈服，过量塑性变形将导致管材缩径及断裂。塑性变形伴随着持久性包申格效应，产生循环硬化，由于螺纹连接部位应力/应变集中持续增加，螺纹接头将发生脱扣或者断裂。油田大量使用的API偏梯形螺纹和螺纹脂，在200 ℃以上环境将失去密封能力，导致注入蒸汽泄漏，被井下泥岩层吸收，引发地层间大的横向载荷，导致管柱发生剪切变形及错断。

针对循环变温服役环境，明确了套管柱累积应变构成，建立了套管柱需求应变计算方法和试验测试中的管柱应变判断系统及方法。依据套管热弹塑性变形及规律，依据材料均匀变形能力，建立了许用应变计算方法。针对注汽、采油及动态热循环不同状态，建立了套管柱设计及安全评价三准则模型。针对管柱螺纹连接部位的应力/应变集中效应，突破API Spec 5CT套管标准约束，提出“管体-管端-螺纹连接”强度错配新方法，在确保管体发生塑性变形范围内，管端及螺纹连接应力不超过其屈服强度，预防脱扣及断裂失效。据此，提出管端镦粗、二次热处理工艺，及管材局部加速冷却工艺，实现了管柱强度错配设计。依据上述理论指导和专利技术联合制造厂家开发了80SH、90SH及110SH三个钢级热采套管新产品，针对热采井仅需维持注汽阶段的气密封，考虑现场测井接箍定位需要，开发了台肩型和气密封特殊螺纹两种接头新产品。

中国石油天然气集团公司技术发明奖
CNPC Technology Invention Award Certificate

证 书

为表彰中国石油天然气集团公司技术发明奖获得者，特颁发此证书。

获奖项目：基于应变的稠油蒸汽热采井套管柱设计方法及工程应用技术

获 奖 者：韩礼红

奖励等级：二等奖

证 书 号：2016-F-2-05-R01

2016 年 12 月 8 日

在工程配套技术方面，从螺纹密封面检测方法、装备及实物模拟试验工装设计等方面入手，建立了蒸汽热采井套管柱工况模拟试验评价方法，完成了 80SH、90SH、110SH 及不同结构螺纹接头产品的试验评价。通过高温应力松弛效应分析，取消了传统的预应力固井，简化了工艺。针对技术套管与套管头焊接需求，建立了焊接接头适用性评价方法。针对注汽间隙管柱测井及频繁清洗需求，研发了同心管射流负压冲砂洗井方法。

本项技术成果获得国家授权发明专利 14 项，授权实用新型专利 16 项，开发套管新产品 3 项、螺纹接头新产品 2 项、形成管柱强度错配新工艺 2 项，制定并发布石油行业新标准 4 项，发表科技论文 17 篇。相关科技成果获得 2015 年《中国石油报》报道，并受邀在 2013

年世界石油大会和2016年SPE加拿大稠油国际会议上进行了学术报告，获得国内外专家高度评价。

研究成果在新疆油田完成了8口井现场试验，与同期常规井相比变形降低42%，实现了五年零套损效果。在辽河油田蒸汽吞吐井组转火驱井方面开展了10余口井套损机理及承载能力分析，为火驱井点火工艺优化提供了技术支撑。在渤海湾热采井套损治理方面，完成了多井次变形及失效分析。在渤海装备、衡阳华菱及新冶钢等制造企业获得推广应用，在热采井套管新产品开发、现有管材工艺改进及常规套管质量提升等方面发挥了重要作用。油田工程应用配套技术在油田稠油开发方面获得规模应用，为油田高效、安全开发节约了大量资金。自2005年以来，技术成果综合应用经济效益达4亿元，显著支撑了油田经济型开发。

本项目获2016年中国石油天然气集团公司技术发明二等奖。

3. 地下储气库运行安全保障技术研究

本项目为2012年中国石油天然气集团公司重点科技项目“地下储气库运行与安全保障技术研究”（2012B-3407），主要针对储气库管柱完整性评价、在役溶腔稳定性评价、压缩机故障诊断、选材及安全评价标准体系等开展研究攻关，属于油气储运安全技术领域。

本项目以我国目前已建的枯竭油气藏型和盐穴型两类储气库为工程依托，将消化吸收和自主创新相结合，针对当前储气库面临的管柱评价依赖国外、在役溶腔稳定性评价缺乏工具手段、压缩机故障诊断预警难、选材及安全评价标准空白等运行安全保障技术需求，针对储气库运行安全保障技术开展了理论探索、技术开发和现场应用研究。通过项目研究攻关，在国内首次建立了地下储气库管柱完整性评价和储气库压缩机故障自适应诊断技术，建立了在役老腔稳定性评价方法，首次开发了专用评估软件，编制了管柱选材、安全评价、风险评估、声纳检测等标准规范，填补国内空白，支撑了储气库运行安全保障技术体系。

项目取得了系列创新性研究成果，包括基于三轴强度的含缺陷储气库管柱剩余强度评价方法、基于数值模拟和实物试验的储气库管柱螺纹密封性完整性评价方法、基于层次分析法的地下溶腔稳定性模糊综合评价方法、单腔和双腔稳定性数值分析方法、适用于变工况条件下的储气库注采压缩机自适应故障诊断技术、盐穴型储气库完整性评价软件、基于全井段测井数据的地下储气库老井套管柱强度分析软件、储气库注采压缩机自适应诊断系统、以及管柱选材及安全评价标准等。

本项目制定并发布企业标准5项，开发专用软件3套，申请国家专利8项，其中发明专利7项，授权发明专利5项，发表学术论文16篇，其中SCI/EI检索3篇，项目成果进一步完善和丰富了储气库的运行安全保障技术体系，对安全工程学科的发展具有重要推动作用。

本项目采用边开发、边应用的模式，不仅为地下储气库安全运行提供了技术保障，而且促进了我国地下储气库安全管理的技术进步。研究成果已成功应用于西气东输、北京管道、长庆油田、辽河油田、新疆油田、大港油田等储气库的管材选用和管柱评价，应用于大张坨储气库德莱塞蓝压缩机整机及关键部件、京58和永22储气库群的4台往复式压缩机故障诊断，以及金坛西1和西2井的老腔稳定性分析评价，为储气库现场管柱密封性、耐蚀性以及管材成本合理控制、管柱适用工况条件和运行检测周期确定、压缩机故障诊断、老腔稳定性提供了合理化建议和决策依据，为储气库安全运行提供了技术保障，产生直接经济效益3000多万元，节约投资约3亿元，经济与社会效益显著。同时，随着我国在建储气库大量投入注采运行和未来储气库建设的重大需求，项目成果将具有广阔的应用前景。

本项目获2016年中国石油天然气集团公司科技进步二等奖。

中国石油天然气集团公司科学技术进步奖

CNPC Science and Technology Progress Award Certificate

证 书

为表彰中国石油天然气集团公司
科学技术进步奖获得者，特颁发此证书。

获奖项目：地下储气库运行安全保障技术研究

获奖单位：中国石油集团石油管工程技术研究院

奖励等级：二等奖

证 书 号：2016-J-2-40-D01

2016年12月8日

4. 海底管道腐蚀控制技术研发及应用

本项目属于油气田开发领域应用基础研究类项目，主要为海底管道安全运行提供理论依据和解决措施。

本项目以我国东南沿海海洋油气开发为工程依托，紧密结合海底管道防腐设计、材质性能、施工工艺及安全运行的实际需求，开展了系统性的理论分析、技术开发和现场应用研究，突破了系列影响海底管道本质安全的技术难题，项目取得了五项创新成果，总体技术达到国际先进水平。

中国石油天然气集团公司科学技术进步奖

CNPC Science and Technology Progress Award Certificate

证　书

为表彰中国石油天然气集团公司科学技术进步奖获得者，特颁发此证书。

获奖项目：海底管道腐蚀控制技术研发及应用

获奖单位：中国石油集团石油管工程技术研究院

奖励等级：二等奖

证 书 号：2016-J-2-39-D01

2016年12月8日

（1）创新了基于全生命周期的海底管道防腐设计新方法。即以案例分析为基础，全生命周期视角选取腐蚀环境、模拟实验和模型预测相结合定性管材耐蚀区间，辅以材质、工艺、药剂等多种防腐措施优化和完整性管理配套统筹考量进行防腐设计，打破了传统采用极端工况设计后技术富余经济不足的局限。研究成果为东海西湖凹陷区块八条海管设计了经济适用的防腐方案，确保了具有战略意义管道按时建成并多年运行未出事故。

（2）拓展了海底管道内防腐产品及腐蚀评价装置体系。建立了海洋环境用双金属复合管性能评价方法和技术指标，编制了产品订货技术条件；开发了 TG520 高温 CO_2 缓蚀剂，解决了带压套管防腐问题；国内首次设计了多相流腐蚀模拟实物装置和顶部腐蚀评价系统，摆脱了复杂流态腐蚀评价难的困扰。研究成果丰富了海底管道防腐选材评价手段，保障了复合管订货、验收和安全使用，确保了崖城作业区套管诱喷作业安全施工。

（3）开发了海底管道外防腐补口技术及专用牺牲阳极产品。设计了干湿两用热收缩带的补口底漆和聚酰胺酯泡沫填充工艺，制定了补口工艺规程，形成了海管快速补口技术，解决了高温高湿高腐蚀环境补口施工质量控制问题，开发了海泥环境用牺牲阳极材料，提升了管道防腐质量。研究成果保障了西气东输二线香港支线补口施工安全，累计完成补口1613道，一次性报检合格率100%。

（4）完善了海洋用双金属复合管施工工艺。搞清了S形铺管法对海洋环境用复合管材性能影响规律，给出安全铺管弯曲半径，开发了端部堆焊结构及现场焊接工艺，形成了焊接工艺评定技术，解决了海洋环境用复合管焊接问题，确保了铺管施工安全。研究成果促成了双金属复合管在崖城、番禺等海洋领域的安全应用，为海洋环境双金属复合管材防腐应用打开了新局面。

（5）创新了海底管道运行风险检测及快速维修技术。开发的海管防腐层检测装置填补了国内空白，首次实现对外防腐层非接触式连续检测评价，编制的维抢修施工作业规范，进一步提升了管道维护水平。研究成果保证了冀东油田南堡作业区两条海底输油管道近10km外防腐层状况检测，解决了管道防护状况无法全面掌握的难题。

本项目申请国家专利17项，其中发明专利11项（授权7项）、实用新型专利6项，制定行业及中国石油天然气集团公司企业标准4项，在国内外核心期刊共发表文章20余篇（SCI/EI收录4篇），形成专著2部。

项目研究成果先后在西气东输二线香港支线、冀东油田南堡作业区及中海油平湖、崖城和番禺得到应用，建立了从管道防腐设计、性能评价、施工工艺到后期运行保障技术体系、开发了系列管道防腐产品及检测评价装置，丰富和完善了海底管道安全运行管理机制，提升了海底管道本质安全保障水平，保障了我国油气战略开发进程，产生了显著的社会和经济效益，项目实施以来产生直接经济效益超过3.2亿元。

本项目获2016年中国石油天然气集团公司科技进步二等奖。

5. 特殊用途管线钢管应用技术研究

本项目为中国石油天然气集团公司重大科技专项研究课题“西气东输二线管道断裂与变形控制关键技术研究”的关键研究内容，是材料科学与工程和石油管工程领域的交叉学科，属于材料科学技术领域，适用于高强度大口径天然气特殊用途钢管（抗大变形钢管、低温环境用钢管和管件、高屈强比X80钢管、冷弯管）的生产开发、质量改进和质量控制，确保管道运行安全。

本项目针对基于应变设计地区用抗大变形钢管、低温环境用钢管和管件、高屈强比X80钢管、冷弯管等特殊用途管线钢管开展应用技术研究，通过系统性的理论探索、技术开发和工程应用，结合我国天然气管道服役环境及运行条件的特点，形成了特殊用途管线钢管的性能控制、生产制造方面的系列应用技术。解决了抗大变形钢管，低温用钢管、弯管及管件，冷弯管的多项关键技术问题，包括X70抗大变形钢管全尺寸弯曲试验技术、管道应变控制时屈曲应变预测方法、高寒地区站场裸露钢管及管件低温脆性断裂控制技术、X80钢管屈强比安全性影响、X80冷弯管应变时效规律、X80冷弯管用母管关键技术指标、X80冷弯管质量控制等核心技术和方法。

本项目发表论文17篇，发布标准3项，获得授权专利10项。利用课题研究成果，为管道建设公司提供了及时有效的技术支持，同时为厂家提供了全面的技术服务，促进了管线建设工程的进展。通过课题研究，取得了一系列创新性的研究成果，主要创新成果如下：(1)揭示了承压管道应变容量与钢管规格、工作压力和管材基本性能的相关性，创立了管道应变控制时屈曲应变预测公式，并确定了工程应用的控制指标。建立的钢管屈曲应变容量预测公式包含了应力比和屈服强度，经全尺寸内压弯曲试验证实，其准确度优于国外现行预测公式。(2) 研究建立了高寒地区站场裸露钢管和管件的最低壁温计算方法，制定了低温脆断控制方案。(3) 攻克了屈强比影响管道安全的技术难题，得出了屈强比和管道壁厚的简化公式。利用该模型掌握了屈强比对管道承载能力和管道安全性的影响规律，解决了屈强比影响管道安全的问题。(4) 研究获得冷弯工艺和应变时效对X80钢管性能的影响规律，提出了X80钢级冷弯管母管技术要求和X80冷弯管质量控制指标。(5) 通过课题研究，撰写完成了多个抗大变形钢管相关的工程技术标准，应用于西气东输二线等工程中。

本项目取得的研究成果已被纳入西气东输二线、三线，中亚C、D线，中缅管线等系列技术标准中，并获得了工程应用。本项目成果解决了长期困扰高压天然气管道工程建设及运行安全的技术难题，在西二线、三线及中缅管线工程中的应用避免了管道敷设长距离绕行的问题，节约了大量的管道建设成本。本成果可在油气管道建设和运行过程中全面应用和推广，不仅可用于新建油气管道的可行性研究、设计、管材采购、质量控制等多个环节，也可用于在役管线的安全评价和完整性管理。我国正处于油气管线建设的高峰期，因此本成果具有广阔的应用前景。

本项目获2016年陕西省科学技术二等奖。

6. 输气管道提高强度设计系数工程应用研究

该成果为2012年中国石油天然气集团公司“第三代大输量天然气管道工程关键技术研究”重大专项课题八“2012E—2801—08”的主要研究成果，属于油气储运工程学科和能源安全技术领域。

课题以西气东输三线为依托工程，通过国内外设计标准及工程应用现状调研、现场检测、统计分析、试验研究、理论计算等研究手段，对输气管道提高设计系数的断裂控制、设计及施工技术、安全可靠性、风险水平、完整性管理技术等开展配套技术研究，编制0.8设计系数管材关键技术指标体系、现场焊接及施工技术规范、完整性管理措施。

欧美国家也已开展了相关研究及探索，美国和加拿大已有建设0.72以上设计系数管道工程，设计系数主要为0.72~0.78，但0.8设计系数的应用却很少，仅Alliance在0.72设计系数通过后期提压至0.8设计系数。而我国自1994年首次GB 50251《输气管道工程设计规范》颁布实施以来，一直沿用了一级地区0.72设计系数，在较高设计系数技术方面则未开展相关研究，更没有相关工程经验。

课题调研了0.8设计系数在国外的标准、研究和应用情况，对比了国内外X80钢级钢管实物质量，评估了提高我国输气管道强度设计系数的安全可行性，开展了0.8设计系数X80管道断裂控制技术、管材质量评价技术、现场焊接技术、100%最小屈服强度试压技术、施工质量控制、完整性管理措施等关键技术攻关，指导开发出0.8设计系数管道用X80螺旋缝埋弧焊管，并为西三线0.8设计系数示范工程提供12.5×10^4t钢管，制定了《西气东输三线0.8设计系数管道用X80螺旋缝埋弧焊管技术条件》《西气东输三线西段一级地区荒漠无人区提高强度设计系数技术规范》等企业标准及0.8设计系数管道现场焊接工艺规程，并在此研究基础上修订了 GB 50251《输气管道工程设计规范》。共申报专利 5 项，制修订标准 3 项，焊接工艺规程 8 项，发表论文 16 篇。

课题采用边研究、边应用的模式，为 X80 钢级 0.8 设计系数示范工程建设提供技术支撑和保障。X80 钢级 260km 示范工程强度设计系数从 0.72 提高到 0.8，钢管壁厚由 18.4mm 减小到 16.5mm，壁厚减薄 10.3%，直接节约管材 1.25×10^4t，节约投资约 1 亿元，另外间接降低了运输、焊接等费用。0.8 设计系数技术在西四线、西五线、中俄输气管线一级地区都要进行全面推广，直接经济效益可达数十亿，应用前景广阔。

本项目获 2016 年陕西省科学技术二等奖。

7. 含缺陷钢质管道复合修复技术及工程应用研究

本项目为中国石油集团石油管工程技术研究院 2012 年院级技术开发项目。

管道修复技术在国外一般被称为“3R 技术”，即 Repair，Rehabilitation，Replace（修补、修复及更换管段）。纤维复合材料修复补强技术是 20 世纪 90 年代发展起来的一种结构补强技术。近年来，对多种修复产品开挖验证其修复效果的检测研究发现，修复点在补强修复并回填一段时间后，出现修复材料与管体脱黏、分层、空鼓、压边搭边和边界无封口或封口不完整等问题，说明目前市场上的国内外补强修复产品存在修复材料本身和施工工艺问题。

本项目针对近年来复合材料修复补强技术现场施工和开挖验证发现的问题，将消化吸收和自主创新相结合，进行了系统性的理论探索、技术开发和现场应用研究，建立了从修复材料体系改进提高，施工工艺整体优化和现场应用后修复效果和施工质量检测评价指标体系建立和操作实施成套的管道修复管理和维护技术。结合最新固化剂、增韧剂等化工产品，分别完成了对缺陷填充材料、层间胶黏剂和绝缘底胶的配方改进研究，并进行了现场修复验证和全尺寸水压爆破验证，新体系与原专利产品相比，其主要技术指标大幅度提高。对现场修复层和管体脱黏问题进行原因分析，针对现场管体表面处理不合格、打磨除锈不彻底等问题，提出了带锈转化表面处理技术，带锈转化液的研制大大减少现场表面处理工序，提高施工效率，降低人工打磨成本，有效增加修复层与管体的黏结力和防腐性能。对现场配胶、混胶和涂胶三个阶段问题进行了分析，提出并研制了预浸料复合材料修复技术，确保固化后复合材料具有稳定的抗拉强度和弹性模量，能够有效分担管道承受的载荷，提高修复补强技术的有效性及修复工程质量。在国内提出了采用复合修复层均匀加压固化技术解决现场发现的分层、空鼓和脱黏等问题，分析了均匀加压优点和确定了加压气囊加压方式，根据现场施工要求，提出了加压气囊相关技术参数和指标要求。考虑现场检测的准确性和可操作性，在国内首次建立了系统的管道复合材料修复补强技术检验评价指标体系，包含修复设计参数、材料本身及施工质量三部分检测评价内容，共 4 个一级指标和 16 个二级指标，并形成了标准。

本项目申请国家发明专利 8 项，授权 8 项，制定标准 2 项，研发装置 2 套，发表论文 6 篇，其中 EI 收录 2 篇，进一步完善和丰富了油气管道完整性技术体系，对安全工程学科的发展具有重要推动作用。

项目研究成果已在西南油气田、青海油田、西部管道、陕西省天然气等 10 余个管线和站场中得到成功应用，为油气管线和场站管道的安全管理提供了有力的技术支持和科学依据。依据本项目研究成果现场修复补强管体缺陷 30 多处，现场检测评价修复点 42 处，横向技术服务合同额近 200 万元，3 年来预防油气泄漏等安全事故 7 起，节约停输维抢修直接和间接经济损失约 7000 万元，保障了油气管道及站场安全平稳运行。本项目背景突出，目标明确，与工程结合紧密，具有广阔的市场推广价值。

本项目获2016年中国石油天然气集团公司技术发明三等奖。

8. 炼化典型在役换热器管束腐蚀防护和完整性评价技术研究及工业应用

本项目为2011年中国石油天然气集团公司"十二五"重大科学研究与技术开发项目"炼化典型在役换热器管束腐蚀防护和完整性评价技术研究及工业应用"的研究成果，涉及炼油化工、安全工程、石油管工程领域等交叉学科。主要针对炼油装置常减压换热器的频繁腐蚀失效及停工停产甚至人员伤亡等问题，研究解决在役换热器管束缺陷的检测、腐蚀的控制以及完整性评价与预测等方面的技术难题，提高装置运行的安全可靠性和经济效益。

项目以兰州石化、长庆石化为工程依托，紧密结合炼油厂常减压装置安全运行的实际需求，在换热器管束腐蚀防护及完整性评价方面开展了系统性的理论探索、技术开发和现场应用研究，解决了常减压换热器管束腐蚀失效、腐蚀预测、完整性评价及在役管束无损检测等关键问题。通过项目攻关，识别了典型换热器主要失效模式，实现了炼化换热器管束的腐蚀环境分级，揭示了换热管束典型环境下的腐蚀机理及规律，实现防腐蚀选材及设计的优化，编制了炼化工况环境下的选材指南，同时开发了针对性强的有效防腐措施。建立了炼化换热管束选管堵管方法及剩余强度、剩余寿命的完整性评价方法，制订了无损检测技术导则及完整性评价技术标准规范及软件，并开发了腐蚀数据管理系统数据库，填补了国内该技术领域的空白。

本项目采用边开发边应用的模式，不仅为炼化换热器的安全运行提供了技术保障，而且促进了我国炼化企业安全管理的技术进步。项目取得的研究成果在炼化企业尤其在兰州石化得到成功应用。通过无损检测技术的优化组合对检修换热管束进行检测，准确率从原来的不足60%提高到90%以上。现场制定合理有效的选管及堵管方案，结合腐蚀预测模型计算结果及完整性评价技术，准确找到存在腐蚀风险及安全隐患的管束，提高了炼厂检修的工作效率，节约了费用。同时有效解决了炼厂检修存在的维修盲点，大大降低了开工运行后发生泄漏的风险，保障了炼厂换热器的长周期安全平稳运行，经济效益和社会效益显著。

本项目共申报国家专利10项（其中发明专利5项），授权6项；发表技术论文15篇（其中SCI、EI 4篇），制定标准及规范4项，开发软件2套，开发装置1套。进一步完善和丰富了炼化换热器腐蚀防护与完整性技术体系，对炼化装置安全保障技术的发展具有重要推动作用。

本项目获2016年陕西省科学技术三等奖。

9. 油气集输用非金属管标准体系研究及标准制定

近年来，油田介质环境日益苛刻，含水率、温度逐步上升，Cl^-、CO_2、H_2S等腐蚀性介质含量升高，给钢管的应用带来极大风险。非金属及复合材料管材（以下简称非金属管）成为油田地面集输管网腐蚀问题的重要解决方案之一。据统计，当前国内各油田非金属管的用量已超过20000km，占所有地面集输管线的10%以上，且以每年10%的速度快速增加，大大促进了非金属管产品种类的拓展和应用技术的进步。

伴随着非金属管用量的大幅增加，种类日益增多，非金属管在产品、测试方法、质量控制、施工验收等领域标准的严重缺失或滞后逐渐成为制约我国非金属管产品进一步推广应用的巨大瓶颈，也逐渐成为油气输送领域技术进步和非金属管行业进步的桎梏。为了从根本上解决阻碍我国油田用非金属管发展的核心问题，本项目充分调研国内外非金属管标准规范及其应用状况，构建了油气集输用非金属管的标准体系。通过开展系统的试验研究及验证，确定了油气集输用非金属管的强度、理化性能、服役性能、相容性能等关键性能判定指标，并开发设计出介质相容性、耐磨、结垢结蜡、气体渗透、弯曲等实物管材评价方法。提出了油田现场非金属管设计、施工、验收等关键环节控制点，制定出20项非金属管系列标准，有效解决了油田非金属管无标可依或标准滞后问题，显著促进了非金属管在国内油田的规模化应用，推动了我国石油工业用非金属管的生产与应用技术进步。

本项目取得4项创新性成果：（1）首次从产品类别和应用过程两个维度建立油气集输用非金属管标准体系，为我国油田用非金属管标准的顶层设计提供了依据，对标准制修订的规划和发展提供明确指导。（2）首次设计开发了模拟油田现场应用环境的非金属管介质相容性、长期服役、耐磨、结垢结蜡、气体渗透、弯曲等关键性能的测试评价技术和相关设备，为产品完整性和可靠性评价建立坚实的平台。（3）首次明确非金属管强度、理化性能、耐温性能、连接性能、长期服役性能、油田介质相容性能等关键参数指标，为非金属管产品的量化评估提供指南。（4）制定20项油气集输用非金属管系列标准，填补了我国石油工业用非金属管的多项标准空白。

本项目建立的各类非金属管标准广泛应用于产品的检测评价、设计采购、施工验收等领域，给产品在油气田中的应用、推广及质量优化提供了基本保障，为非金属管制造商创造了巨大的经济效益；形成的检测评价技术广泛应用于油气田特定环境下非金属管的设计、选材评价，并推广应用至新型非金属管产品（如海洋管）的制造和评价，有效解决了对拟用非金属管评估的重大技术难题；制定的各类非金属管施工验收标准广泛应用于延长油田、长庆油田、塔里木油田、塔河油田等非金属管工程建设，有效降低了施工不规范带来的应用风险，确保油田获得良好的经济效益和社会效益；项目构建的非金属管标准体系填补了国内空白，规范了我国油气集输用非金属管行业发展，有效推动了我国石油天然气工业的健康、安全、可持续发展，创造了良好的社会效益。

项目成果实施以来，取得新增产值总额逾2.0亿元，新增利润3486万元，节支总额8350万元，产生的总经济效益约为3.25亿元。项目共制定标准20项：其中国家标准4项，行业标准12项，企业标准4项。发表论文16篇：其中SCI收录3篇，EI收录7篇。申报专利13项：其中发明专利10项，实用新型专利3项，获授权7项。

本项目研究成果对于推动我国石油工业非金属管的快速发展，制订非金属油气管道领域标准发展规划，最终建立我国石油工程完善的非金属管标准体系具有重要意义；同时对规范我国油田用非金属管市场，有效解决地面集输管网中的腐蚀问题具有长久的指导意义。

本项目获2016年陕西省科学技术三等奖。

二、授权专利目录

序号	专利名称	专利号	专利类别
1	一种油套管用缺陷定量无损检测设备	2011102168319	发明专利
2	一种膨胀管膨胀装置	2012101241966	发明专利
3	一种管线钢中M/A岛组织的显示方法	2012101394514	发明专利
4	一种管线钢中M/A岛组织的面积含量评定方法	2012101501041	发明专利
5	便携式石油管外螺纹锥度测量仪	2012102424644	发明专利
6	便携式石油管内螺纹锥度测量仪	201210242470X	发明专利
7	一种油田管材用高性能防腐涂料及涂覆方法	2012103943130	发明专利

续表

序号	专利名称	专利号	专利类别
8	管道环焊缝的全位置自动焊接方法	2012104400451	发明专利
9	地下储气库完井管柱内压疲劳试验装置及其试验方法	2012105005604	发明专利
10	油气输送管道及制备方法	2012105181165	发明专利
11	一种含稀土的高强韧钻杆及其制备工艺	2012105600845	发明专利
12	一种油气田用缓蚀剂及其制备方法	2012105898869	发明专利
13	一种油气田用缓蚀剂及其制备方法	2012105900163	发明专利
14	管线钢平面应变断裂韧性和安全临界壁厚的确定方法	2013100084604	发明专利
15	一种接箍	2013100177929	发明专利
16	一种获取一级地区油气输送管试压压力的方法及装置	2013100280008	发明专利
17	一种抗硫化物应力开裂的油套管	2013100283595	发明专利
18	一种地下储气库地震危害的预测系统和方法	2013100320908	发明专利
19	一种带环形裂纹的圆棒拉伸试样及其制备方法	2013100816157	发明专利
20	V150 钢级油井管全尺寸实物试验试样制备方法	2013101099594	发明专利
21	稠油蒸汽吞吐热采井用套管的包申格效应评测方法	2013101343292	发明专利
22	一种双金属复合管环焊缝焊接方法	2013102013035	发明专利
23	评价溶解态二氧化碳在防腐涂层中扩散及渗透性能的方法	2013102323737	发明专利
24	外螺纹接头中径测量仪、杆、定位样板及中径测量方法	201310253536X	发明专利
25	内螺纹接头中径定位样板及中径测量方法	2013102536362	发明专利
26	利用单边缺口拉伸试验测量管线钢断裂韧性的方法	2013104793277	发明专利
27	用于管道止裂韧性测量的装置及其测量方法	2013104793582	发明专利
28	一种热采井套管材料选用方法	2013105447498	发明专利
29	一种确定管道近中性 pH 值应力腐蚀开裂敏感区段的方法	2013105906588	发明专利
30	全尺寸非金属管材气体渗透性能的测试装置及其测试方法	2013106940980	发明专利
31	确定热采井套管柱总应变的方法	2013107142460	发明专利
32	一种热采井套管柱应变判断系统及其方法	2013107145153	发明专利
33	用于提高钻杆内螺纹接头耐磨性的方法	2013107460425	发明专利
34	一种定位取样尺	2013107467693	发明专利
35	一种换热器管束的检测方法	2014100199250	发明专利
36	一种研究大口径热挤压成型三通壁厚的测试方法	2014101122129	发明专利
37	低毒曼尼希碱化合物、由其制备的酸化缓蚀剂及其制备方法	2014101173065	发明专利
38	一种生产油气管道用弯管的两步感应加热装置及方法	2014106410705	发明专利
39	一种双金属复合管环焊缝对焊焊接方法	2014106455710	发明专利
40	一种铝合金钻杆管体与钢接头的连接结构	2014106458367	发明专利
41	一种 $H_2S-HCl-H_2O$ 体系用高温缓蚀剂及其制备方法	2014106710288	发明专利

续表

序号	专利名称	专利号	专利类别
42	一种超深井用超高强度铝合金钻杆管体及其制造方法	2014106934999	发明专利
43	一种涂层湿膜厚度测试装置	2015208458982	实用新型
44	一种油井管屈曲管柱的高温腐蚀和冲蚀试验装置	2015208931853	实用新型
45	一种提高 ERW 焊管沟槽腐蚀准确性的实验室测试装置	2015209324487	实用新型
46	钢套筒柔性止裂器	2015209817747	实用新型
47	一种复合加载应力腐蚀试验装置	2015209872655	实用新型
48	输气钢管全尺寸气体爆破试验用聚能切割装置	2015210476211	实用新型
49	一种碳纤维复合材料止裂器	2015210896241	实用新型
50	一种输气钢管全尺寸气体爆破试验断裂速度测量装置	2015211302734	实用新型
51	一种弹簧引伸计悬臂杆定位装置及弹簧引伸计	2015211303281	实用新型
52	一种大直径、高钢级管线钢套筒柔性止裂器	2015211409445	实用新型
53	钢管连接器	2016202501790	实用新型
54	用于评价油气管道在高流速下冲刷腐蚀的试验装置	2016203314151	实用新型
55	一种便携式钢管耐腐蚀性能测试装置	2016203564797	实用新型